Biomolecular and Bioanalytical Techniques

Biomolecular and Bioanalytical Techniques

Theory, Methodology and Applications

Edited by

Vasudevan Ramesh
School of Chemistry
University of Manchester
UK

This edition first published 2019

Registered Offices
John Wiley & Sons, Inc., 111 River Street, Hoboken, NJ 07030, USA
John Wiley & Sons Ltd, The Atrium, Southern Gate, Chichester, West Sussex PO19 8SQ, UK

Editorial Office
John Wiley & Sons Ltd, The Atrium, Southern Gate, Chichester, West Sussex PO19 8SQ, UK

For details of our global editorial offices, customer services, and more information about Wiley products visit us at www.wiley.com.

Wiley also publishes its books in a variety of electronic formats and by print-on-demand. Some content that appears in standard print versions of this book may not be available in other formats.

Library of Congress Cataloging-in-Publication Data

Names: Ramesh, Vasudevan, editor.
Title: Biomolecular and bioanalytical techniques : theory, methodology and applications / edited by Vasudevan Ramesh, School of Chemistry, University of Manchester, Manchester,U.K.
Description: Hoboken, NJ : Wiley, 2019. | Includes bibliographical references and index. |
Identifiers: LCCN 2018059850 (print) | LCCN 2019001362 (ebook) | ISBN 9781119483984 (Adobe PDF) | ISBN 9781119484011 (ePub) | ISBN 9781119483960 (hardcover)
Subjects: LCSH: Molecular biology–Technique.
Classification: LCC QH506 (ebook) | LCC QH506 .B5534 2019 (print) | DDC 572.8/38–dc23
LC record available at https://lccn.loc.gov/2018059850

Cover Design: Wiley
Cover Images: Top: Courtesy of Chad Brautigam;
Bottom: Courtesy of Carolyn Moores;
Background: © Nattzkamol/Shutterstock

Set in 10/12pt WarnockPro by SPi Global, Chennai, India

Printed and bound in Spain by Graphycems

10 9 8 7 6 5 4 3 2 1

Contents

List of Contributors

Nathan N. Alder
Department of Molecular and Cell Biology
University of Connecticut
Storrs, CT
USA

Elaine Armstrong
Health & Safety Services, Compliance & Risk
University of Manchester
UK

Daniela Barillà
Department of Biology
University of York
York YO10 5DD
UK

Arnaud Baslé
Institute for Cell and Molecular Biosciences
University of Newcastle
UK

Chad A. Brautigam
Department of Biophysics
UT Southwestern Medical Center
Dallas, TX
USA

Richard C. Brewster
Institute of Quantitative Biology, Biochemistry and Biotechnology
School of Biological Sciences
University of Edinburgh
UK

Mark Carlile
School of Pharmacy and Pharmaceutical Sciences
University of Sunderland
UK

Ka Lung Andrew Chan
School of Cancers and Pharmaceutical Science
Institute of Pharmaceutical Science
King's College London
UK

Tony Cheung
School of Chemistry
University of Manchester
UK

Graeme L. Conn
Department of Biochemistry
Emory University School of Medicine
Atlanta, GA
USA

Valerie J. Gillet
Information School
University of Sheffield
UK

Sophia C. Goodchild
Department of Molecular Sciences
Macquarie University
Sydney, NSW
Australia

Nicholas J. Harmer
Living Systems Institute
University of Exeter
UK

Finbarr Hayes
Faculty of Biology, Medicine and Health
The University of Manchester
UK

Krishanthi Jayasundera
Department of Molecular Sciences
Macquarie University
Sydney, NSW
Australia

Blagojce Jovcevski
School of Physical Sciences
University of Adelaide
SA
Australia

Hsueh-Fen Juan
Department of Life Science
Graduate Institute of Biomedical Electronics and Bioinformatics
National Taiwan University
Taipei
Taiwan

Richard J. Lewis
Institute for Cell and Molecular Biosciences
University of Newcastle
UK

W. John Lough
School of Pharmacy and Pharmaceutical Sciences
University of Sunderland
Chester Road
UK

Szymon W. Manka
Institute of Structural and Molecular Biology
Birkbeck College
London
UK

Carolyn A. Moores
Institute of Structural and Molecular Biology
Birkbeck College
London
UK

Raymond T. O'Keefe
Faculty of Biology
Medicine and Health
University of Manchester
UK

David J. Parry-Smith
Wellcome Sanger Institute
Hinxton, Cambridgeshire
UK

Tara L. Pukala
School of Physical Sciences
University of Adelaide
Adelaide, SA
Australia

Vasudevan Ramesh
School of Chemistry
University of Manchester
UK

Maria Reif
Physics Department T38
Technical University of Munich
Garching
Germany

Alison Rodger
Department of Molecular Sciences
Macquarie University
Sydney, NSW 2109
Australia

Huw B. Thomas
Faculty of Biology, Medicine and Health
University of Manchester
UK

Mirella Vivoli Vega
Department of Biomedical Experimental and Clinical Sciences
University of Florence
50134 Florence
Italy

Stephen Wallace
Institute of Quantitative Biology, Biochemistry and Biotechnology
School of Biological Sciences
University of Edinburgh
UK

Martin Zacharias
Physics Department T38
Technical University of Munich
Garching
Germany

Preface

Fundamental research in chemical and biological sciences has witnessed significant advances in recent years with the boundary becoming blurred between the two rival, traditional disciplines. Concomitantly, rapid development of interdisciplinary research techniques in support of such advances is taking place at regular intervals. In addition, taught courses delivered at universities are being designed to equip the younger generation with knowledge and skills needed to pursue interdisciplinary research.

A sound knowledge of the biomolecular structure aided by appropriate bioanalytical techniques is indispensable to elucidate the mechanism of action of biological macromolecules and their function at large. The present book intends to provide a good introduction and understanding on a range of biomolecular and bioanalytical cum biophysical techniques and is primarily aimed at the advanced undergraduate (years 3 and 4) and MSc students who are fresh to research. Nowadays, short-term research projects (team and/or individual based) form an essential component of the taught curriculum and make a sizeable contribution towards assessment in modern Chemistry and Biology courses. Hence, it is important that prospective research students acquire sufficient background and skills in various research techniques before graduating. The book should also be suitable for those contemplating an advanced postgraduate (PhD) and postdoctoral research career in Chemistry, Biochemistry, Biophysics and Pharmaceutical Chemistry in academia and industry.

The book is research focused, interdisciplinary and comprehensive with a broad selection of topics drawn from contemporary Chemistry and Biology research, in a single volume. The order of chapters and the depth of coverage make the book well balanced and well connected with a gradual, smooth flow of information and increasing knowledge. It attempts to bridge the gap between introductory core material taught in years 1 and 2 undergraduate levels and advanced research texts; thus prior knowledge of the former is expected to derive a complete understanding of the present book.

In many ways, the present book is the first of its kind with an international team of authors of considerable teaching and research experience, each contributing a single chapter on a specific technique of her/his expertise. The format of each chapter is broadly similar with greater emphasis on experimental procedures accompanied by case studies rather than theory. Obviously, it is not possible to elaborate on every technique in critical detail but they are sufficient to provide the necessary background and detailed experimental procedures to afford the reader a good understanding of the various techniques before applying them. Further, each chapter is accompanied by a selection of books recommended for further reading and a comprehensive list of references pertinent to the technique discussed. The suitability of the book as lecture material will depend on the intended learning objectives and outcomes of the taught course as determined by the lecturer.

The book is broadly organised into five parts after a short, important chapter on Health and Safety (H&S). In addition to common laboratory safety, safety in a research laboratory requires additional knowledge of chemical and biohazards and risks associated with special techniques and high end equipment.

The first part covers Chapters 2 to 5 on Chemoinformatics, Bioinformatics, Gene expression analysis and Proteomics. These chapters describe the computational tools available to search relevant databases for molecules with the likelihood of biological activity or placing sequence data (DNA, RNA, proteins) in a biological context or design experiments for the cloning or analysis of genes expressed in eukaryotic cells or compare the proteomes of different biological samples.

The next set of chapters (6 to 8) covers Protein expression and purification, Chromatography and separation techniques in Biology and Synthetic methodology in chemical biology describe in detail the recombinant genetic and synthetic methods used for the preparation of proteins and small peptide molecules, respectively, and the various chromatographic methods used for the separation and purification of such molecules.

The subsequent set of chapters (9 to 12) covers Mass spectrometry, Analytical ultracentrifugation, Light scattering, Reaction kinetics in Biology and Isothermal titration calorimetry form the core analytical techniques for characterisation of biomacro molecules through the measurement of mass-to-charge ratios of gas phase ions, the quantitative determination of molecular size and shape and determination of the steady-state kinetic parameters of enzymes and the investigation of biomolecular interactions to define the thermodynamics of binding in order to understand basic biological processes. Prior knowledge about these properties of a biomolecule should provide the basis for its three-dimensional structure determination.

Chapters 13 to 15 on Molecular Spectroscopy (Infra-red, Raman, Fluorescence, Circular dichroism spectroscopies) discuss how molecular vibrations manifested by distinct IR and Raman spectral bands can be correlated with functional groups and their role in biomedical research, application of a number of specific steady-state fluorescence-based techniques to analyse the structures, associations and conformations of biological macromolecules and determination of protein secondary structure using circular dichroism.

Finally, the last four comprehensive chapters (16 to 19) on Biomolecular structure determination (X-ray, NMR and Cryo-TEM, Molecular Simulation) describe the (i) determination of the high resolution structures of proteins and their complexes in the crystalline state (X-ray), (ii) determination of the three-dimensional structure and conformational analysis of DNA in the solution state (NMR), (iii) visualisation at atomic resolution of a wide range of macromolecular complexes in frozen hydrated form (Cryo-TEM) and (iv) the physical basis of computer modelling and principles of molecular simulations to extract structural and thermodynamic quantities.

I hope the book will prove a useful text for students, researchers and lecturers alike.

A book of this type involving a number of different techniques would not have been possible without the valuable contribution and continued support of all the authors and I remain very grateful to them. I also thank the numerous reviewers with their helpful comments and corrections on different chapters. During the course of my academic career, I had the privilege of teaching Biophysical Chemistry to undergraduate and postgraduate students, of varying backgrounds, mostly at the University of Manchester, UK, and benefited from their comments and criticisms. I am also grateful to the former postgraduate (PhD and MPhil) members of my research group whose enthusiastic support gave me the impetus to develop ideas for this book project. Finally, I thank Wiley (UK) for their interest and support throughout.

Vasudevan Ramesh
School of Chemistry
University of Manchester, UK

1

Principles of Health and Safety and Good Laboratory Practice

Elaine Armstrong

Health & Safety Services, Compliance & Risk, University of Manchester, Oxford Road, Manchester, M13 9PL, UK

1.1 Introduction

Scientific research, by definition, involves carrying out novel work to further scientific knowledge and, in pursuit of this activity, new techniques are developed and applied. The object of this chapter is to discuss a set of principles and guidelines that, when followed, will provide a safe and healthy environment for researchers, which will, in turn, facilitate and promote good science.

Chemical and biological laboratories are potentially very hazardous places in which to work. In recent years there have been a number of very serious accidents in academic laboratories with tragic and sometimes fatal consequences for those involved. These include fatalities in 2009 when a researcher died of extensive burns due to contact with a pyrophoric chemical [1] and in 2011 when a researcher was asphyxiated in an oxygen depleted atmosphere caused by evaporation of liquid nitrogen into a non-ventilated space [2]. Serious injuries were caused to a graduate student in 2010 who was grinding energetic material that exploded [3] and a researcher lost an arm when a pressure vessel exploded in 2016 [4].

The risk of accidents and injury can be significantly reduced by researchers being aware of any potential hazards and working with care and attention to detail. Prior to commencing any work activity, it is very important, and time well spent, for researchers to familiarise themselves with all available information about the materials, equipment and processes that they will be using during the course of their work. The safety of everyone in the laboratory is largely determined by each individual's work practices.

1.2 Good Laboratory Practice

Good Laboratory Practice, or GLP, is a series of behaviours that is designed to prevent accidents, many of which will be described in the specific procedures developed by administrators or principal investigators for use in their laboratories. However, some general guidelines are given below:

- Do not eat, drink, smoke or apply cosmetics in the laboratory.
- Wash and dry hands before leaving the laboratory.
- Wear shoes with a closed toe – no sandals or flip flops.
- Wear personal protective equipment (PPE) that is required by the relevant risk assessment, properly (safety spectacles worn on top of the head do a poor job protecting eyes from chemical splashes).
- Cover any broken skin with suitable dressings.
- Keep benches and fume cupboards clear of unnecessary equipment, which leaves room for carrying out the work and will minimise the effect of any accidents.

Biomolecular and Bioanalytical Techniques: Theory, Methodology and Applications, First Edition. Edited by Vasudevan Ramesh.

- Ensure that all chemicals are properly labelled with the name of the chemical and any hazard information and, for samples, the owner's name, date of preparation and quantity.
- Replace lids and stoppers.
- Return chemicals to their dedicated storage areas after use.
- Check chemical stock and equipment that is not in regular use periodically to ensure it is in good condition and specific storage conditions are being met (e.g. certain chemicals should not be allowed to 'dry out').
- Store chemicals safely in appropriate storage spaces.
- When carrying large bottles of solvent, always use suitable carriers and do not lift large bottles solely by the neck.
- Keep substances that are incompatible with each other apart and in separate storage spaces, and label them clearly.
- Comply with local restrictions on the amount of highly flammable and flammable materials (which includes waste).
- If equipment becomes faulty, take it out of service, label it and report it to someone who will arrange for its repair.
- Use all equipment in accordance with the manufacturer's instructions.
- Dispose of all out of date and/or unwanted chemicals and equipment safely, on a timely basis and according to local procedures.
- Inspect any glassware before use and do not use any that is broken, chipped or cracked, as this might either directly cause injury to the researcher or fail catastrophically in use.
- Follow any local rules and guidance about working alone.
- Follow any local rules and guidance about working out of hours.

In addition to using GLP, there is a lot of other information available to assist researchers in how to work safely. Much of this will be detailed in the local arrangements for the facility (including standard operating procedures, existing risk assessments, laboratory scripts), safety data sheets (SDSs) for chemicals, instructions for the use of kits in microbiology, user manuals for equipment, etc., and other texts [5, 6].

1.3 Risk Assessment

Risk assessment is a tool used to develop ways of working to minimise the risk of causing harm to people and damaging facilities. Carrying out a risk assessment is a fundamental requirement in most health and safety regulations [7–12]. However, this requirement can result in a number of separate assessments being carried out for different parts of the same process, when actually all the requirements could be captured in a single 'holistic' risk assessment. Risk assessments must be carried out by 'competent' people. (Competent people are those who have sufficient knowledge, ability, training and experience in their field to be able to advise on the safest way to carry out the task that is being assessed.) Principal investigators, laboratory supervisors as well as safety advisors and officers should be able to assist with the process.

It is pertinent here to differentiate between hazard and risk.
A hazard is something that has the potential to cause harm.
A risk is the probability or likelihood of a hazard causing harm.

Before starting work, it is necessary for the people involved to be able to:

- Recognise and identify any hazards associated with the work – these hazards can be associated with materials, equipment, the environment in which it is being done and the people carrying it out – see Table 1.1 for examples of common hazards in laboratories.

Table 1.1 Common laboratory hazards.

Source	Hazard
Chemicals	Can cause different types of health effects (e.g. carcinogenic, toxic, corrosive, etc.) depending on the chemical and what form it is in (e.g. solid, dusty solid, liquid, gas, etc.), when coming into contact with a person by inhalation, direct contact and, less commonly, injection and ingestion.
Biological materials and genetically modified organisms	Contamination and possible infection, release to the environment.
Glassware	Cuts when handling broken glassware and following instances of catastrophic failure due to being put under vacuum or pressure. Leakage of harmful compounds if glassware breaks in use.
Vacuum apparatus and glassware attached to it	May implode violently. Risk of formation of liquid oxygen if the vacuum system is left open to air with the cold trap in position, which promotes combustion if exposed to flames, sparks and organic materials, including grease.
Pressure apparatus and glassware attached to it	May explode violently.
Cryogenic liquefied gases	Asphyxiation due to oxygen depletion, cold burns, explosion and ejection if heated up quickly.
Compressed gas cylinders	May leak harmful gases or discharge violently (cylinder may become a projectile if not suitably restrained).
Electrical equipment	Danger of fires caused by sparking in areas where there are highly flammable vapours (many solvents), electric shock, electrocution and burns if equipment is poorly maintained or there are exposed connections.
Ultraviolet light	Causes burns and can damage eyesight if suitable protective eyewear is not worn.
X-rays and radioactive isotopes	Cause burns and cell mutation.
Lasers	Can cause burns and blindness if laser beam enters the eye either directly or by reflection.
High magnetic fields	Can be dangerous for people with pacemakers (overrides the working settings to test mode) or ferromagnetic implants, which can be drawn to the magnet. High magnetic fields can also wipe mobile electronic devices including key fobs, mobile devices.
Robotic equipment (e.g. autosamplers)	Contact with moving parts can cause crush injuries, puncture wounds and amputation.
Sharps (needles, scalpels, microtomes, cryostats, pipettes)	Puncture wounds and cuts that may also introduce chemical and/or biological material directly into the body.

- Assess the risks to people posed by the hazards. This includes identifying who could be harmed, how they may be harmed and how severe the harm could be. The hazards that could cause the most severe harm and those that could cause harm to the highest number of people are the ones that must be prioritised when thinking about ways to prevent the harm occurring.
- Reduce and mitigate the risks by adopting ways of working that prevent the hazards coming into contact with people. There is a standard hierarchy of ways to reduce and control hazards, which is shown in Figure 1.1. The most effective way of controlling a hazard is to eliminate it altogether, which is often quite difficult, but must be considered first.
  i) *Substitution* could involve replacing a substance in one form with the same substance in a less hazardous form (e.g. replacing a very dusty powder used to make a solution, to obtaining the solution already made) or substitution of one chemical with another. The overall level of hazard does need to be assessed

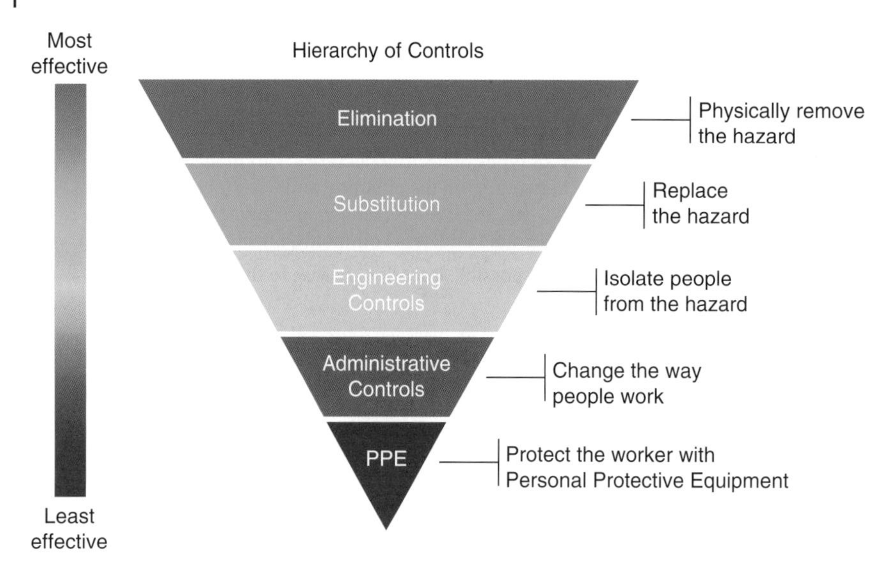

Figure 1.1 Hierarchy of control. Source: Image taken from the Institute of Occupational Health and Safety.

carefully as a very toxic chemical could be substituted by a less toxic one that presents a higher level of physical risk – e.g. is more highly flammable.

ii) *Engineering controls* are commonly used to isolate people from the hazard. This type of control includes totally enclosing a process, e.g. in a glove box or Class 3 microbiological safety cabinet, use of interlocks that automatically switch off lasers or X-rays when a portal is open or providing local exhaust ventilation (fume cupboards, capture hoods situated over equipment, etc.).

iii) *Administrative controls* depend on people acting in a certain way and working to standard operating procedures and risk assessments. Personnel should also have had sufficient instruction, information, training and supervision to carry out the work safely.

iv) *Personal protective equipment*, or PPE, is the lowest level of control. PPE has a vital role in protecting researchers from hazards, but it does have some drawbacks, which include failing in a dangerous situation and only protecting the person wearing it. It is only effective if it is worn properly, it fits and is correctly specified. A management system for PPE is required to ensure that the correct equipment is specified and procured and people are trained in its use. If there are several items needed in combination, they must be compatible with each other and not increase the overall danger. It must fit the person for whom it was procured. Workers need to be taught how to use the equipment and what limitations it has (e.g. some laboratory safety spectacles are not manufactured to provide protection against ultraviolet light), how to put the equipment on and, importantly, how to take off contaminated equipment safely. Arrangements are needed for cleaning, storage, inspection, repair, replacement and the eventual safe disposal of the PPE. If respiratory equipment is needed, then face fit testing must be carried out.

- Prepare for emergencies. This is to provide guidance, before it might be needed, about how to deal with the consequences of something going wrong. This could include the procedure for evacuating a space and having a trained spill team with suitable equipment for clearing up after an accident, as well as the provision of people who have been trained in first aid and have the necessary equipment to deal with various injuries, etc.

As part of the risk assessment process, it is necessary to record the significant findings of the risk assessment. Many institutions will have their own templates for recording the risk assessment and guidelines for how long risk assessments must be kept.

There are general scenarios when a risk assessment should be reviewed, which include:

i) After a set period of time, which should be made clear in the local arrangements
ii) After an adverse event (accident or incident)
iii) If there is a change in the law (regulations)
iv) If there is a change in the equipment, methods, chemicals, etc. being used
v) If a vulnerable person starts doing the work (this could be someone whose health status changes, e.g. pregnancy, immunocompromisation, medication, etc.), or a young person (under age 18 in the UK), or perhaps an older person returning to the laboratory.

What is important to remember, however, is that the risk assessment should be a working document that provides information to help keep people safe and prevent harm. New workers and visitors need to be given the relevant information by their supervisors or laboratory managers.

1.4 Chemical Risk Assessment

There is extensive information available about the hazards posed by commercially available chemicals (and mixtures of chemicals provided as kits for applications in life sciences and molecular biology). Prior to using any chemicals, these data should be examined. There is a standard set of data that is provided in a SDS in 16 clearly labelled sections. This information includes: hazard identification (Section 3), first aid measures (Section 4), fire and explosion data (Section 5), how to clear up after an accidental release (Section 6), handling and storage requirements (Section 7), exposure controls and personal protection (Section 8), physical and chemical properties (Section 9), stability and reactivity data (Section 10), toxicological information (Section 11) and disposal considerations (Section 13). This is all valuable information when completing a chemical risk assessment that combines the assessment of health risks posed by chemicals with the physical risks of fire and explosion, and will inform how and where the substance should be handled, used, stored and disposed of.

For new or novel compounds produced during the course of the work, detailed hazard information will not be known, but by consideration of information relating to analogous compounds some indications of the hazards can be predicted, but the aim must be to avoid contact by use of controls including local exhaust ventilation and suitable PPE. All products and by-products of reactions should be considered in the risk assessment.

In an ideal world, the chemical risk assessment should form one part of a risk assessment that would cover a procedure from beginning to end, including the activities and apparatus used in the middle, but many laboratories choose to conduct a chemical risk assessment separately and refer to other assessments or standard operating procedures for the equipment, which is also perfectly acceptable.

There are five steps in the risk assessment process.

1.4.1 Step 1: Identify the Hazards

The majority of the hazard information required for the chemicals in a chemical risk assessment is given in the SDS. SDSs are available on line as well as being provided in paper copy the first delivery of the chemical from a new supplier.

1.4.2 Step 2: Identify Who Could Be Harmed and How

In most instances, the people who could be harmed may include, in addition to the worker carrying out the procedure, colleagues sharing workspaces and maintenance and cleaning staff. How they may be harmed depends on the hazardous properties of the chemical, how it gets on or into the body and whether it causes local, short lived effects (acute) at the site of contact or whether it is stored in the body and concentrations build up over time and effects are long lasting (chronic) or people become sensitised to it. These types of effects are described in the SDS.

Chemicals can enter the body by one of four ways, the most common being inhalation where powders, vapours, fumes, etc., are breathed in. The second most common method is by direct contact, i.e. splashes to skin or mucosal membranes. Injection via sharps injuries or into uncovered open wounds (including uncovered cuts, grazes or patches of broken skin due to some medical conditions) is less common, but occurs, as is ingestion of the chemical by mouth.

1.4.3 Step 3: Decide What Controls Are Needed and Whether More Could Be Done

Understanding the route of exposure is critical in determining the controls to prevent entry to the body. Working through the hierarchy of controls mentioned earlier, the question to be considered is whether the substance can be eliminated entirely or whether it can be substituted with something less harmful or a less harmful variant. A useful illustration is to consider the various forms of sugar. This occurs in large individual crystals, small crystals (granulated and caster sugar), fine powder (icing sugar) and liquid (glycerol). Of these forms, the ones that would be selected to prevent entry to the body by inhalation would be either glycerol or large individual crystals, with icing sugar the most likely to be inhaled.

Special consideration should also be given to work with manufactured nanomaterials, about which separate guidance is available [13].

The amount of substance to be used will also affect the suitability of the controls. Smaller amounts of substance or more dilute substances present less of a hazard than larger amounts or more concentrated substances. This is true whether considering the effects on health or considering runaway thermal reactions. The time of contact also has an effect on how much harm the substance could cause.

Engineering controls used to prevent inhalation include the use of glove boxes or Class 3 microbiological safety cabinets (total containment), fume cupboards (for operator protection from chemicals), microbiological safety cabinets (to protect operator and environment from aerosols containing biological materials) or other local exhaust ventilation. Local exhaust ventilation also has an important role in extracting and diluting any fumes and vapours present, which on a laboratory scale (less than 500 ml) should prevent a build-up of explosive atmospheres. Where larger amounts of flammable materials or explosive materials are in use, any electrical equipment should be intrinsically safe.

Administrative controls include the use of standard operating procedures, training, information and supervision.

Use of PPE is the last control measure to be considered. One of the main causes of injury in chemical laboratories is where PPE is either not worn as it was designed to be or is not correctly specified. In most synthetic chemistry laboratories, wearing of laboratory coats, gloves and safety spectacles is standard practice. However, there are always a number of instances of chemicals getting into researchers' eyes underneath or around standard safety glasses, which may mean that a better type of eye protection, e.g. fully enclosing goggles or visor, may be more appropriate. Different types of gloves have different chemical resistance to commonly used reagents. The information about the chemical resistance of gloves is available on glove manufacturers' websites and is included in more recent versions of SDSs. If splash protection is all that is required, disposable gloves of the most appropriate type should be used and be changed regularly.

An important part of the risk assessment is the planning for dealing with an adverse event, for example having access to suitable first aid assistance, spill kits (and, on occasion, fire extinguishers, provided that users have been trained to use them and they can do so without putting themselves and others at risk).

A number of commonly used chemical reagents can cause fire and explosion hazards. This information will be detailed in Sections 5, 9 and 10 of the SDS. Some chemicals are pyrophoric, which means they spontaneously ignite in air and/or when exposed to moisture in the atmosphere. Examples of these are noted in Table 1.2, although this is not an exhaustive list.

Other common reagents that contain very reactive (explosive) functional groups are summarised in Table 1.3. A common cause of explosive accidents in the laboratory is the mixing of nitric acid with organic solvents. Other mixtures that have caused severe accidents are liquid oxygen and liquid air in the presence

Table 1.2 Common pyrophoric hazards.

Alkali metals, e.g. sodium, potassium (and sodium–potassium alloy), lithium, caesium
Metal hydrides and alkylated derivatives, e.g. potassium and sodium hydrides, lithium aluminium hydride, alkyl aluminium hydrides, silanes
Finely divided metals, e.g. magnesium, titanium, bismuth, titanium, zirconium
Hydrogenation catalysts (especially those that have been used), e.g. raney nickel, palladium carbon, platinum carbon
Alkylated metal halides, e.g. dimethyl ammonium chloride
Solutions of alkyl metals, e.g. tertiary butyl lithium, concentrated normal butyl lithium, trimethyl aluminium

Table 1.3 Functional groups with explosive hazards.

Inorganic and organic peroxides and peracids
Inorganic and organic chlorates and perchlorates
Organic and inorganic nitro $(NO_2)^-$ and nitrates $(NO_3)^-$
Organic and inorganic azides
Fulminates
Acetylene and metal acetylides
Metals bonded with carbon in organometallic compounds
Azo and diazo compounds

of organic materials, alkali metals coming into contact with chlorinated solvents and mixing strong oxidising agents with reducing agents. Further information can be found in *Bretherick's Handbook of Reactive Chemicals* [14].

The chemical risk assessment should cover all aspects of the procedure, from starting with the raw materials, including their storage requirements, to the final disposal from the laboratory. Sufficient controls are needed to reduce the risks of exposure to the lowest reasonably practicable level.

1.4.4 Step 4: Record and Implement the Significant Findings

The significant findings should be recorded, either on a local template or written down elsewhere. The controls that are stated in the record of significant findings should be implemented while the procedures are being carried out. Everyone who could be affected by the work that is covered in the risk assessment should be made aware of the content of the risk assessment.

1.4.5 Step 5: Decide When an Assessment Needs to Be Reviewed

Note that review does not necessarily mean re-do. The review could just be a check for reassurance that the assessment is still valid. However, if there has been an accident or change of personnel, etc., as described in Section 1.3, the assessment may need to be updated or re-done.

1.5 Biological Materials and Genetically Modified Organisms

Where work with biological materials and genetically modified organisms (GMOs) is being carried out, there will be a 'competent person' appointed to assist with compliance with statutory duties, including the provision of advice and assistance with developing risk assessments. This person is often referred to as the Biological Safety Officer (BSO) or Biological Safety Advisor (BSA).

Any measure used to control risk must be underpinned by the principles of good microbiological practice, including the use of good aseptic technique.

1.5.1 Biological Hazards

Biological materials that may present a hazard to human health are also known as biological agents. They are described as 'microorganisms, cell cultures or human endoparasites, whether or not genetically modified, which may cause infection, allergy, toxicity or otherwise create a hazard to human health'.

When dealing with biological agents, the main principles of risk assessment described in Section 1.3 are still applicable and risk assessments must be carried out before any work starts. Biological agents are categorised into four hazard groups (HGs) based on their ability to infect healthy humans and the level of harm that they can cause [7b]. Classification of the biological agent to be used is an early consideration when developing the risk assessment. An Approved List of biological agents and their classification is available on the HSE website [15]. If the agent is not listed, consult your local BSO who will be able to assist.

- *Hazard group 1*: Unlikely to cause human disease.
- *Hazard group 2*: Can cause human disease and may be hazardous to employees. It is unlikely to spread to the community and there is usually effective prophylactic treatment available.
- *Hazard group 3*: Can cause severe human disease and may be a serious hazard to employees; it may spread to the community, but there is usually effective prophylactic treatment available.
- *Hazard group 4*: Causes severe human disease and is a serious hazard to employees; it is likely to spread to the community and there is usually no effective prophylaxis or treatment available.

An advantage of this classification is that there are specific containment measures detailed for work with each HG that are appropriate to the level of harm that they can cause. These measures must be taken into account when developing risk assessments, along with the type of work being carried out, the amount or titre of the biological agent being used, the procedures undertaken, the equipment used (e.g. the use of sharps could provide a route of direct entry of the biological agent via injection) and methods for disinfection and decontamination. Waste that has been generated as a result of work with biological agents must be suitably treated before leaving the premises (either by treating with disinfectant or autoclaving, or both) so that it no longer presents a biological hazard. This will usually be a separate waste stream that is collected in receptacles that are suitably labelled and will be removed for incineration by a licensed contractor.

Work with HG 3 and HG 4 agents requires the use of specialist facilities.

Further guidance about managing the risks from biological agents is available from the HSE [8].

1.5.2 Genetically Modified Organisms (GMOs)

Genetic modification refers to the introduction and incorporation of new combinations of genetic material (which can be derived from existing organisms or synthetically made) into a recipient organism in which they do not naturally occur. The introduced genetic material is capable of stable incorporation and/or continued propagation in the recipient organism.

The control of genetic modification work with microorganisms is intended to reduce any risks to humans *and the environment* to the lowest level reasonably practicable from any product (e.g. over expressed protein, the vectors used, the genetically modified microorganism [GMM], etc.) arising from the work.

The use of genetic modification is subject to regulation in the UK by the Health and Safety Executive (HSE); further details about safe working and how to carry out risk assessments is provided in the guidance to the Genetically Modified Organisms (Contained Use) Regulations 2014 [9] and the Control of Substances Hazardous to Health Regulations [7b]. Any premises where contained use (CU) is taking place must be notified to the HSE prior to any work taking place.

In the UK, CU refers to any activity involving GMOs where barriers are used to limit contact with and protect humans and the environment. These barriers can be physical, chemical or biological and are frequently combinations of these. If there are no barriers in place the activity should be carried out under the provision of the Genetically Modified Organisms (Deliberate Release) Regulations 2002 [16], which at the time of writing are overseen by the Department of the Environment, Farming and Rural Affairs (DEFRA).

Physical barriers are anything used to prevent escape or exposure to the GMO and can include buildings, rooms, containers, equipment or physical processes such as ventilation or ultraviolet irradiation.

Chemical barriers are chemicals used to inactivate or destroy a GMO before waste disposal or the use of chemicals to prevent escape.

Biological barriers are where a GMO has inherent or engineered characteristics that mean it is attenuated, disabled or rendered unable to survive outside a specialised environment. To be included as control measures in the risk assessment, these characteristics should be robust and stable.

Competent advice must be sought on the development of the risk assessment before the CU work can commence. Working through the structured risk assessment process is very important as it is used to classify the CU into one of four risk classes based on the highest containment level (described in Schedule 8 of the Genetically Modified Organisms (Contained Use) Regulations 2014) selected in which to carry out the work.

- *Class 1*: The GMM presents no or negligible risk.
- *Class 2*: The GMM presents low risk.
- *Class 3*: The GMM presents moderate risk.
- *Class 4*: The GMM presents high risk.

If the CU assessment classifies the work as Class 1, the risk assessment can be signed off by the local BSO, but at Classes 2 or higher, in the United Kingdom the risk assessment must be scrutinised and accepted by a GM committee before work commences. Work at Classes 2 or higher also requires a notification to be submitted to the HSE.

The risk assessment must consider all aspects of the planned work, including handling, transport, work area decontamination, inactivation of the GM microorganisms, disposal and waste management. Other independent advice on the risk assessment process for CU is available in the Scientific Advisory Committee on Genetically Modified Organism (SACGM) Compendium of Guidance [17].

1.6 Vacuum Apparatus, Pressure Systems and Associated Glassware

Glass apparatus used under vacuum or under pressure is prone to sudden energetic failures, which is why it is important to regularly check glassware for cracks, chips, etc., and not use it if it is damaged. One way of mitigating the effects of implosion or explosion is to either surround the glassware with polynet or tape. As mentioned in Table 1.1, care must be taken to ensure that the cold trap is removed before vacuum equipment is opened to air after use.

Pressure systems, e.g. autoclaves, may need to have a written scheme of examination and a statutory inspection if they fall under the Pressure Systems Safety Regulations [10] in the UK. They must not be used if they do not have a current valid certificate.

1.7 Cryogenic Liquefied Gases

Cryogenic gases have been cooled until they either solidify (carbon dioxide) or liquefy (nitrogen, helium). They can cause cold burns on contact, so the risk assessments for using these materials involve controls to ensure that they do not come into contact with people. The liquefied gases in particular displace large amounts of air as they warm to room temperature and can be an asphyxiation risk. Oxygen depletion monitoring may be in use and good ventilation is required where cryogens are used and stored. As carbon dioxide 'boils off' it would reach a fatal concentration before oxygen depletion became the primary health hazard.

1.8 Compressed Gas Cylinders

Compressed gas cylinders present several risks – they are usually quite heavy and can be awkward to move, even with a cylinder trolley. They may leak harmful gases and should never be accompanied by a person if they are transported in a lift. They may also discharge violently, presenting a kinetic hazard, so should be suitably restrained in use, transport and storage.

1.9 Electromagnetic Radiation

Radiation from the electromagnetic spectrum has been utilised in a number of techniques used for the characterisation of molecules. Electromagnetic radiation can be categorised into non-ionising and ionising radiation.

1.9.1 Low Energy or Non-ionising Radiation That Does Not Cause Ionisation of Atoms and Molecules

This type of radiation includes microwaves, infra-red (IR) and ultraviolet (UV) light. Microwave radiation and IR are generally very safe and are used to either heat things up in a laboratory or to ascertain information about the rotational states of molecules in the gas phase. UV radiation is used in molecular spectroscopic techniques, which is generally very safe.

High energy UV, which can ionise valence electrons (though still classed as non-ionising at wavelengths of 125 nm or less), is used in other laboratory applications where the light source can be quite intense, e.g. use of high powered UV lights in chemical synthesis and in light boxes or transilluminators, which are used to visualise substances that fluoresce at these wavelengths.

The use of high energy UV sources requires strict controls, which can include blacking out fume cupboards where a UV lamp is being used in photochemical reactions and standard operating procedures. As UV is also known as black light, which may not be visible to the naked eye, it is very important that there are clear indications, supported by signage if necessary, if a source is on.

Any safety eyewear has to be rated for protection against the specific wavelength of UV being used. It may also be necessary to work behind protective UV filtering shields to protect exposed skin on the face and neck, as well as the eyes. UV has been the cause of a number of eye injuries in laboratories due to its effects on corneal tissue, which have been quite serious.

1.9.2 Ionising Radiation

Ionising radiation causes the ionisation of atoms and molecules and has important uses in structure determination, of which X-ray diffraction is a common technique. Work with ionising radiation falls under specific legislation in the United Kingdom [11]. Local rules are required for all work in these areas and standard operating procedures must be followed. Most modern equipment generates X-rays as required and there are interlocks on the equipment that shut down the generator when doors and hatches in the equipment are opened for sample placement, etc. Specialist training for use of this type of equipment is required for authorised users.

Radioactive isotopes are often used as labels and tracers and sources of ionising radiation. Their use in the United Kingdom also falls under specific legislation that requires a risk assessment to be conducted before any work is carried out. All isotopes must be accounted for at every stage in their receipt, storage, use and disposal with records readily available for inspection by the competent authority (Environment Agency in the UK). Local rules, strict safety controls and written protocols are required for work with these sources, which not only deal with methods of work but also detail remedial actions in case of emergency and waste treatment/disposal. Regular monitoring of areas where open sources are used to check for spills is required.

1.10 Lasers

Lasers produce monochromatic beams of electromagnetic radiation, which are all in phase and moving in the same direction. These beams are often very narrow and can be focused, which produces small areas of high irradiance. Thus they can provide significant hazards for those working with them. They can cause thermal burns and photochemical injuries. If the products of the laser interaction with materials are toxic, local exhaust ventilation may be required.

Damage caused by lasers is dependent on a number of factors, including the energy of the laser, the wavelength of the radiation produced, the length of time the target is irradiated, etc. There should be access to a specially trained laser safety advisor where high powered lasers are used.

There are a number of standard controls used to prevent harm to the operator. The laser should have an enclosed beam path. Key switches and interlocks should be used to prevent access to the beam path when the lasers are in use. All surfaces and tools associated with the laser system should have matt surfaces to prevent stray reflections. Operators must ensure any jewellery or reflective clothing is covered. When the experiment is being set up or adjusted, the use of a low power visible laser beam to make the adjustments should be considered. If this is not possible, use the laser on its lowest power setting and ensure that appropriate goggles are used to protect eyes. Local rules will generally apply.

Electrical safety is very important; high voltage power supplies are commonly required to run lasers and should be earthed. Many lasers also use water for cooling and possible leaks combined with the proximity to high voltage power supplies can create hazardous conditions.

1.11 High Magnetic Fields

High magnetic fields are commonly encountered in laboratories where there are nuclear magnetic resonance (NMR) and electron paramagnetic resonance (EPR) spectrometers. These techniques use superconducting magnets to create very high magnetic fields used to determine chemical structure.

High magnetic fields in general have not been associated with adverse biological effects in humans, but may cause problems for those with heart pacemakers and other medical implants. It is not recommended that anyone with a medical implant gets closer to a magnet than the 5 Gauss (5 G) line, which should be clearly marked with a barrier if it lies outside the magnet. Many newer magnets are very well shielded and the 5 G line is within the casing of the magnet, but there are still many older magnets in use. Any ferrous tools will also be attracted to the magnet and may be drawn in; the closer they are to the magnet, the faster they will be travelling. If they make contact with the magnet case, the magnet may become unstable (alternatively this could occur when someone is trying to remove something that is 'stuck' to the magnet).

If the magnet becomes unstable (they often contain liquid helium in a core surrounded by liquid nitrogen), then a 'quench' may occur, which results in a large amount of cryogenic liquid being released very quickly into the room and can cause oxygen displacement and asphyxiation. Due to this possibility, facilities with superconducting magnets should be fitted with oxygen depletion monitors that have audio and visual alarms. Preferably, the monitors should have a continuous recording facility and be hard wired rather than battery operated. These should reset automatically when the oxygen concentration is restored.

While not safety related, it is useful to note that high magnetic fields can wipe magnetically stored information, e.g. car key fobs, bank cards, mobile phones, tablets, etc.

1.12 Sharps

Contact with needles, scalpels, microtome blades and cryostats make up a significant proportion of laboratory accidents. They will transfer any contaminants (which could be chemicals or biological agents) directly into the body of the person who is injured by them. It is therefore important to keep clean needles and blades shielded until the point of use and to be shown the techniques to use these items safely. Used needles and blades should be discarded immediately into a sharps container. For certain biological applications it may be stipulated in the risk assessment for the work that sharps must not be used.

1.13 Ergonomic Issues

Many of the techniques used in chemical and biological research involve workers interfacing with computers and other visual display equipment. While these activities seem very low risk when compared with other potential hazards that are associated with laboratories, development of work related musculoskeletal disorders, more specifically work related upper limb disorders (WRULDs), while not life threatening, can indeed be life changing. The most common problems are seen as a result of the human interface with display screen equipment; however, there has been a notable increase of workers presenting upper limb disorders from other common repetitive actions such as pipetting.

Early warning signs of WRULD are pain, soreness, numbness, tingling in hands, wrists or forearms and/or unexplained clumsiness. If any physical discomfort is noted, the line manager should be contacted, who should then make a referral to the Occupational Health provider. A prompt referral is necessary in order to prevent irreversible damage. Meanwhile, adaptations should be made to the type of work being undertaken.

1.13.1 Display Screen Equipment

There is very comprehensive guidance provided by the HSE on setting up computer workstations [12]. While this guidance is mainly used for setting up computers in office environments, the principles can be adapted to setting up display screens that are used to control instrumentation and for results analysis. Workstation assessments must be carried out. Workers must be trained in how to adjust chairs and the computer hardware for optimal comfort. Very small alterations in the set-up of a workstation can make a very big difference to the comfort of the worker.

Workstation Setup

i) The chair should be adjusted so that the user's back is supported by the backrest.
ii) The worker's forearms are approximately horizontal.
iii) The worker's knees are level with their hips.
iv) The worker's feet rest flat on the floor or on a footrest.
v) The worker's eyes are approximately the same height as the top of the screen, shown in Figure 1.2.
vi) The mouse should be positioned close to the worker and the keyboard and there should be no need to over-reach to use it, shown in Figure 1.3.
vii) When using the mouse the worker's wrist is straight, with the arm relaxed and supported.
viii) The keyboard should be positioned so that there is sufficient room in front of it to support the hand and wrist, shown in Figure 1.4.
ix) The wrists should be kept as straight as possible.
x) The worker's elbows should be at approximately 90°.
xi) Workers should take regular rest breaks when using display screen equipment and if possible intersperse use of the display screen equipment with other tasks.

Figure 1.2 Correct seating position.

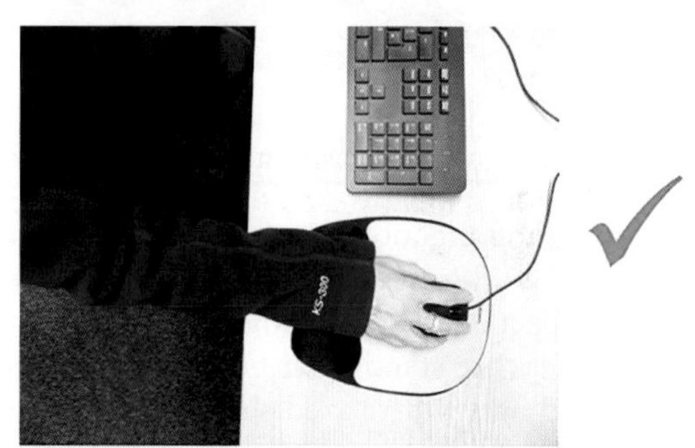

Figure 1.3 Position of the mouse.

Figure 1.4 Position of the hands in relation to the keyboard.

1.13.2 Laboratory Ergonomics

Incorporating ergonomic principles into laboratory work could help prevent some of the symptoms of WRULD. Work/rest schedules, task rotation and the type of equipment or tools being used can all have a direct impact on the risk of injury.

If workers are standing at benches, the work should be positioned close to elbow height and be able to be carried out with relaxed shoulders. There should be sufficient knee and foot clearance so workers can stand naturally. If workers are seated, there should be sufficient room for legs and knees under the bench or microbiological safety cabinet. Chairs should be adjustable to accommodate workers of different stature and footrests, or integral footrings may be required. Work equipment should be positioned within arms' reach from the edge of the bench. When using microscopes, it should not be necessary to stretch or stoop and the worker's neck, shoulder and back should be in a neutral position, with the back being supported by a chair. The microscope should be positioned so that it is within easy reach of the worker. It is advisable to ensure that workers' arms are supported by the worksurface, chair armrests, etc., for periods of prolonged work. Regular breaks should be taken when work requires intense concentration and being in one position for long periods of time.

There has been a noticeable increase of WRULD related to prolonged periods of pipetting. This might be expected as pipetting can involve prolonged periods of time making very small but repetitive movements. Key to avoiding this is being able to plan work and vary tasks where possible. Regular breaks should be taken and the opportunity to stretch fingers, hands and arms should not be missed.

The ideal pipetting position is that the pipette is held close to the body and all peripheral equipment required is within easy reach. Workers should receive training on pipetting good practice from more experienced colleagues or pipette academies, which can often be arranged via pipette suppliers.

If a lot of pipetting is to be carried out, the use of more advanced pipettes should be considered, e.g. electronic or latch-mode or multichannel pipettes. All pipettes should be regularly serviced and calibrated. Clamps and holders should be made available to support test tubes and vials so they do not need to be hand held for long periods.

References

1 Jylian N Kemsley; Chemical and Engineering News, 2009, 87(31), 20–34. American Chemical Society.
2 http://press.hse.gov.uk/2017/nhs-trust-and-imperial-college-london-fined-after-death-of-worker
3 Jylian N Kemsley; Chemical and Engineering News, 2010, 88(34), 34–37. American Chemical Society.
4 Jylian N Kemsley; Chemical and Engineering News, 2016, 94(28), 5. American Chemical Society.
5 Hill, R.H. Jr., and Finister, D.C. (2010). *Laboratory Safety for Chemistry Students*. Wiley. ISBN: 9780470344286.
6 Leonard, J., Lygo, B., and Procter, G. (2013). *Advanced Practical Organic Chemistry*. 3e. CRC Press. ISBN: 9781439860977.
7 (a) Management of Health and Safety at Work Regulations 1999, www.hse.gov.uk/pubns/hsc13.pdf. (b) Control of Substances Hazardous to Health Regulations 2002 (as amended), L5, 6e. HSE Books. ISBN: 9780717665822. (c) Dangerous Substances and Explosives Atmospheres Regulations 2002, L138, 2e. HSE Books. ISBN: 9780717666164. (d) Personal Protective Equipment Regulations 1992, L25, 3e. HSE Books. ISBN: 9780717665976.
8 Safe Working and the Prevention of Infection in Clinical Laboratories and Similar Facilities. HSE Books. ISBN: 9780717625130.
9 The Genetically Modified Organisms (Contained Use) Regulations 2014, Guidance on Regulations, L29 HSE Books. www.hse.gov.uk/pubns/books/l29.htm
10 *Pressure Systems Safety Regulations 2000, ACOP*, L122, 2e. HSE Books. ISBN: 9780717666447. www.hse.gov.uk/pubns/books/l122.htm.
11 Work with Ionising Radiation, Ionising Radiations Regulations 2017, ACOP, L121. 2e. HSE Books. ISBN: 9780717666621., www.hse.gov.uk/radiation/ionising/legalbase.htm
12 Work with Display Screen Equipment. Health and Safety (Display Screen Equipment) Regulations 1992 as amended by the health and safety (Miscellaneous Amendment) Regulations 2002. Guidance, L26. ISBN 9780717625826. www.hse.gov.uk/pubns/books/l26.htm
13 UK Nanosafety Group (2016). *Working Safely with Nanomaterials in Research and Development*, 2e. http://www.safenano.org/uk-nanosafety-group.
14 Urben, P. and Bretherick, L. (2006). *Bretherick's Handbook of Reactive Chemical Hazards*, 7e. Elsevier. ISBN: 9870123725639.
15 The Approved List of biological agents www.hse.gov.uk/pubns/misc208.pdf
16 The Genetically Modified Organisms (Deliberate Release) Regulations 2002 www.legislation.gov.uk/uksi/2002/2443/contents/made
17 The Scientific Advisory Committee on Genetically Modified Organisms (SACGM) Compendium of Guidance www.hse.gov.uk/biosafety/gmo/acgm/acgmcomp

2

Applications of Chemoinformatics in Drug Discovery

Valerie J. Gillet

Information School, University of Sheffield, Regent Court, 211 Portobello, Sheffield, S1 4DP, UK

2.1 Significance and Background

The term chemoinformatics first appeared in the literature 20 years ago following the introduction of automation techniques within the pharmaceutical industry for the synthesis and testing of compounds in drug discovery. The use of combinatorial chemistry and high throughput screening techniques resulted in a vast increase in the volumes of structural and biological data available to guide decision making in drug discovery and led to the following definition of chemoinformatics:

> The mixing of information resources to transform data into information, and information into knowledge, for the intended purpose of making better decisions faster in the arena of drug lead identification and optimisation [1].

Chemoinformatics is now recognised as a discipline, albeit one that falls at the intersection of other disciplines such as chemistry and computer science. It is also a discipline with fuzzy boundaries; for example, the distinction between computational chemistry and chemoinformatics is not always clear. A feature of chemoinformatics today is that it typically involves the analysis of large datasets of compounds. That said, many of the techniques embodied within chemoinformatics have much earlier origins [2], starting with the representation of chemical structures in databases more than 50 years ago. Much of the early activity in chemoinformatics was based on proprietary data and commercial or proprietary software. However, more recently the availability of very large public data sources such as ChEMBL and PubChem, together with the increasing number of open source software tools [3], means that chemoinformatics techniques are now more mainstream and accessible within academia than was the case previously. Furthermore, the techniques have now extended beyond drug and agrochemicals discovery to a much wider range of chemistry domains including the food industry and materials design.

This chapter provides an overview of chemoinformatics techniques that are commonly applied in early stage drug discovery. Following a discussion of some basic foundations relating to structure representation and search, the main focus will be on virtual screening. Virtual screening is the computational equivalent of biological screening and is used to prioritise compounds for experimental testing.

2.2 Computer Representation of Chemical Structures

The common language of organic chemistry is the two-dimensional (2D) structure diagram as shown in Figure 2.1. Most chemical information systems, including web-based systems, include user-friendly

Biomolecular and Bioanalytical Techniques: Theory, Methodology and Applications, First Edition. Edited by Vasudevan Ramesh.

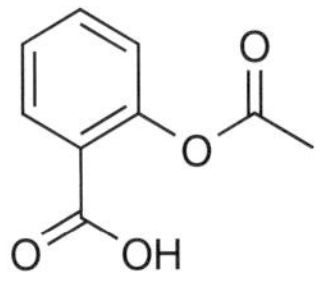

Figure 2.1 Structure diagram of aspirin.

structure drawing programs that enable graphical input of query structures and allow high quality images to be produced. For example, the standalone ChemDraw package was used to produce the image of aspirin shown in Figure 2.1 and the JMSE molecular editor allows molecular editing within web browsers [4]. While these programs enable structures to be drawn on a computer, the images themselves have little value for chemoinformatics applications since they do not contain chemical meaning; they have to be converted to other forms for computer processing.

A widely used method for the representation of chemical structures is the line notation, of which the SMILES (Simplified Molecular Input Line Entry System) notation [5] is most common. In SMILES, 2D structures are represented by linear strings of alphanumeric (that is, textual) characters. SMILES strings are based on a small number of simple rules, they are relatively easy to understand, and they are compact in terms of data storage. For these reasons, they are widely used for transferring compounds between systems and users, and for entering structures into chemoinformatics applications. In SMILES, atoms are represented by their atomic symbols. Upper case characters are used to represent aliphatic atoms (C, N, S, O, etc.) and lower case are used for aromatic atoms. Hydrogen atoms are implicit. Single bonds are inferred between adjacent atoms, as are aromatic bonds; double bonds are represented as '='; and triple bonds as '#'. Additional rules enable stereochemistry to be encoded. A SMILES string can be constructed by 'walking through' the bonds in a structure diagram visiting each atom once. Rings are encoded by 'breaking' one of the ring bonds and attaching a pair of integer values, one to each of the two atoms of the broken bond. Branch points, where more than one path can be followed, are encoded using parentheses. A SMILES representation of aspirin is shown in Table 2.1, along with InChI and InChIKey notations, which are described below.

2.3 Database Searching

One of the fundamental concepts in chemoinformatics is the representation of chemical structures as graphs. While molecular editors and line notations allow easy input and sharing of molecules, the storage and retrieval of structures from databases is based on the mapping of a chemical structure to a mathematical graph. A graph consists of nodes that are connected by edges. Graph theory is a branch of mathematics in which many well-established algorithms exist that can be used to facilitate analysis and searches of databases of compounds. In a molecular graph, the nodes correspond to atoms and the edges to bonds. The nodes and edges can have properties associated with them, for example, atom type and bond type. Molecular graphs are represented as connection tables, which essentially list the atoms and bonds contained in the structure. A number of different connection table formats exist that vary in the way in which the information is stored. A common format is the MDL Molfile, which consists of a header section, an atom block, with one row for each atom, and a bond block, in which each bond in the structure is listed once only [6]. The Molfile for aspirin is shown in Figure 2.2 with the different blocks of information highlighted. Connection tables can also be aggregated into a single file to allow sets of molecules to be exchanged or input to applications.

Chemical databases can be searched in different ways. An exact structure search is used to retrieve information about a specific compound from a database, for example, to look for a synthetic route or to see if a compound is available to purchase from a chemical supplier. An exact structure search is also required

Table 2.1 Line notation representations for aspirin.

SMILES	OC(=O)c1ccccc1)OC(=O)C
InChI	InChI = 1S/C9H8O4/c1-6(10)13-8-5-3-2-4-7(8)9(11)12/h2-5H,1H3,(H,11,12)
InChIKey	BSYNRYMUTXBXSQ-UHFFFAOYSA-N

```
Aspirin.mol
  ChemDraw04041813452D

 13 13  0  0  0  0  0  0  0  0999 V2000
   -1.7862    0.2063    0.0000 C   0  0  0  0  0  0  0  0  0  0  0  0
   -1.7862    1.0313    0.0000 C   0  0  0  0  0  0  0  0  0  0  0  0
   -1.0717    1.4437    0.0000 C   0  0  0  0  0  0  0  0  0  0  0  0
   -0.3572    1.0313    0.0000 C   0  0  0  0  0  0  0  0  0  0  0  0
   -0.3572    0.2063    0.0000 C   0  0  0  0  0  0  0  0  0  0  0  0
   -1.0717   -0.2062    0.0000 C   0  0  0              0  0  0  0  0
   -1.0717   -1.0312    0.0000 C   0  0  0              0  0  0  0  0
   -1.7862   -1.4437    0.0000 O   0  0  0  0  0  0  0  0  0  0  0  0
   -0.3572   -1.4437    0.0000 O   0  0  0  0  0  0  0  0  0  0  0  0
    0.3572   -0.2062    0.0000 O   0  0  0  0  0  0  0  0  0  0  0  0
    1.0717    0.2063    0.0000 C   0  0  0  0  0  0  0  0  0  0  0  0
    1.7862   -0.2062    0.0000 C   0  0  0  0  0  0  0  0  0  0  0  0
    1.0717    1.0313    0.0000 O   0  0  0  0  0  0  0  0  0  0  0  0
  1  2  2  0
  2  3  1  0
  3  4  2  0
  4  5  1  0
  5  6  2  0
  6  1  1  0
  6  7  1  0
  7  8  2  0
  7  9  1  0
  5 10  1  0
 10 11  1  0
 11 12  1  0
 11 13  2  0
M  END
```

Header

Atom block

Bond block

Figure 2.2 The MDL Molfile for aspirin. The third line of the header indicates that there are 13 atoms and 13 bonds, respectively. The atom block lists each atom showing the *x*, *y* and *z* coordinates (in this case the *z* coordinate for all atoms is zero, indicating that the connection table has been derived from a 2D representation) and the atom type. The remaining fields for the atom are all set at zero here; however, these can be used to store additional information such as stereochemistry and charge, etc. The bond block lists each bond once; the first row indicates that atom 1 is bonded to atom 2 by a double bond (shown by the label 2).

during the registration process for new compounds; before adding a new compound to a database it is necessary to ensure that it is not already present. In graph theory terms, an exact structure search can be solved using a graph isomorphism algorithm, that is, an algorithm for determining if two graphs are the same. While this may appear simple at first glance, it is actually a difficult problem to solve since the atoms in a connection table can be presented in any order. Similarly, for a SMILES string, the start point for the walk through a structure can begin with any atom, so that typically there are many valid SMILES representations for a given molecule. For example, alternative SMILES representations for aspirin to that shown in Table 2.1 are c1cccc(OC(=O)C)c1C(=O)O) and CC(=O)Oc1ccccc1C(=O)O).

In principle, exact matching could be achieved by renumbering the connection table of each database structure in all possible ways and testing each for identity with the query molecule. However, this approach is computationally infeasible since there are $N!$ different ways of numbering a connection table consisting of N atoms. By way of example, there are more than 6×10^9 different numberings for aspirin which consists of 13 heavy atoms. Instead, a canonical representation is generated, which is a unique ordering of the atoms in a molecular graph. This can be achieved using the Morgan algorithm [7] or a variant thereof. The Morgan algorithm is an iterative process that involves assigning connectivity values to the atoms in order to differentiate them, as illustrated in Figure 2.3. In the first iteration, each atom is assigned a value according to the number of non-hydrogen atoms it is connected to, that is, to its number of neighbouring atoms. In subsequent iterations, each atom is assigned the sum of the connectivity values of its neighbours. The iterations continue until

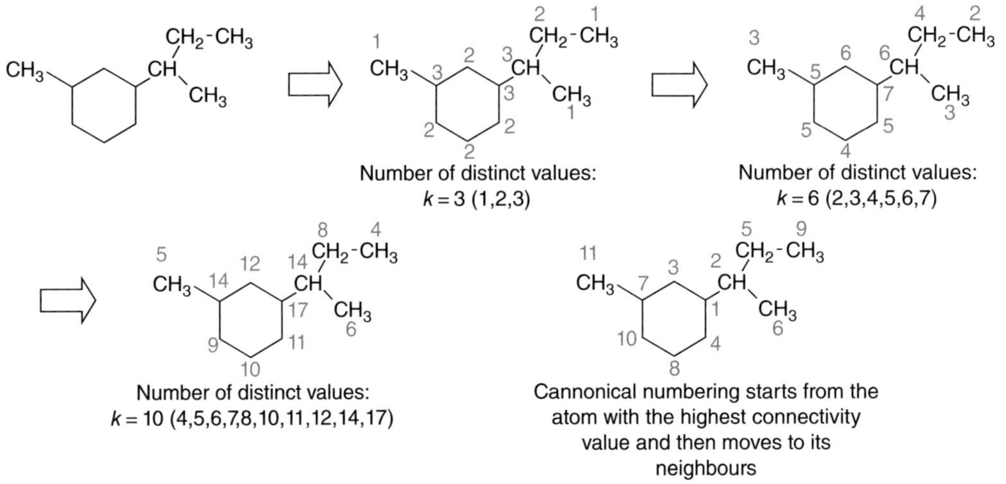

Number of distinct values: $k = 3$ (1,2,3)

Number of distinct values: $k = 6$ (2,3,4,5,6,7)

Number of distinct values: $k = 10$ (4,5,6,7,8,10,11,12,14,17)

Cannonical numbering starts from the atom with the highest connectivity value and then moves to its neighbours

Figure 2.3 Illustration of the use of the Morgan algorithm to produce a canonical representation.

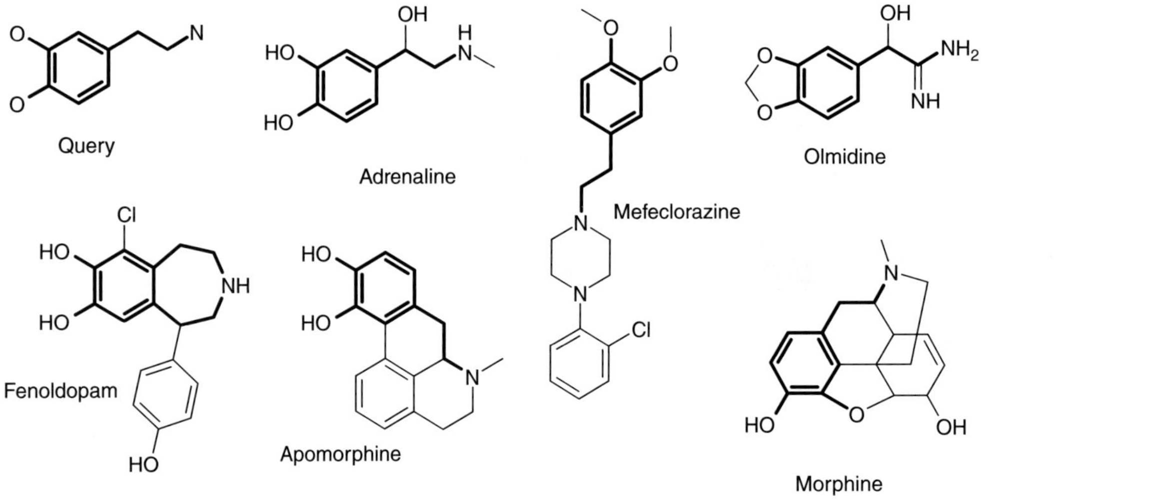

Query Adrenaline Mefeclorazine Olmidine Fenoldopam Apomorphine Morphine

Figure 2.4 Example of dopamine as a substructural query along with some compounds that contain the substructure, with the substructure highlighted in bold.

the number of different connectivity values is at a maximum. The atoms are then numbered as follows: the atom with the highest connectivity value is assigned as the first atom, its neighbours are then numbered in decreasing order of connectivity values and so on. If a tie occurs, then additional properties of the atoms are taken into account.

A substructure search is the process of identifying all molecules that contain a particular substructural fragment, for example, the dopamine substructure as shown in Figure 2.4. In graph theory, this is the problem of subgraph isomorphism, that is, determining if one graph is contained within another. Efficient algorithms for subgraph isomorphism exist and have been adapted to substructure search [2], but for large databases they are too slow to be used in isolation. Therefore, most substructure search procedures consist of two stages. The first is a screening step, in which fast bit string operations are used to eliminate compounds that cannot possibly match the query structure. Those compounds that remain then undergo the more time consuming subgraph isomorphism search. Compound screening is based on what have become known as molecular fingerprints. These are binary vectors where each position in the vector is associated with a molecular fragment and is set to '1' if the fragment is present within a structure and to '0' otherwise. There are two types of molecular fingerprints: dictionary-based and hash-based. In dictionary-based fingerprints, a dictionary of fragments is pre-compiled with each fragment mapped to a particular bit in the bit string. There is, therefore, a one-to-one mapping between the fragment and bit position, as shown in Figure 2.5a. An example of dictionary-based fingerprints is MACCS [8]. In hashed fingerprints, the fragments are generated algorithmically, for example, to include all paths of atoms up to a given length. A series of hashing algorithms is then applied, each of which generates a number corresponding to a bit in the bit string. Thus, there is a many-to-many mapping between fragments and bit positions, as shown in Figure 2.5b.

For a database search, fingerprints are pre-generated for all the compounds in the database. A fingerprint is generated for the query at run time and only those compounds whose fingerprints have bits set for all of the

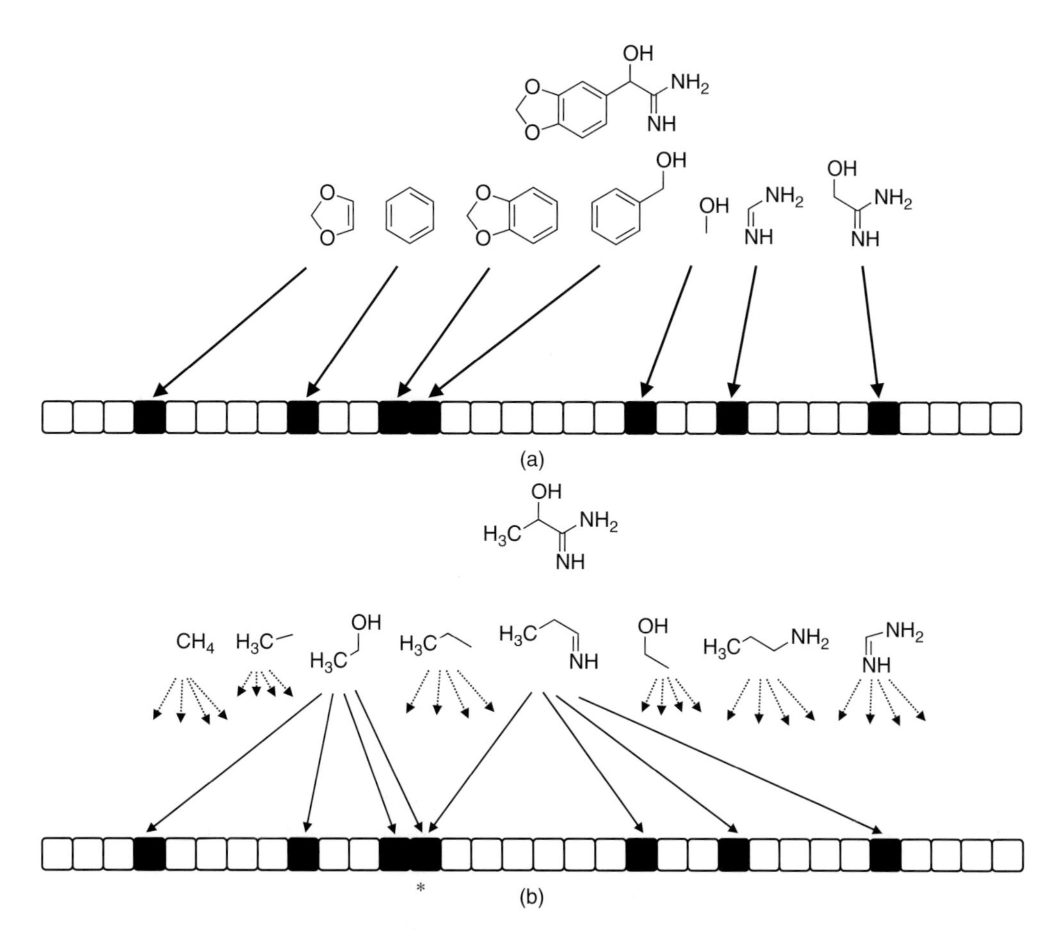

Figure 2.5 Illustration of 2D fingerprints. (a) A dictionary-based fingerprint where each bit in the fingerprint represents a particular substructural fragment. A bit is set on '1' (shown as solid) if the fragment is present in the structure and to '0' (shown as unfilled) if it is absent. (b) A hashed fingerprint where all linear paths up to a specified length are extracted from the structure. Each fragment is then processed by a number of different hashing algorithms and the resulting bits are set. Each fragment gives rise to a number of bits in the fingerprint and bit collisions can occur (shown by the * symbol), where a given bit is set by more than one fragment.

'1' bits in the query are passed forward to the subgraph isomorphism search. Both types of fingerprints have been shown to be effective for screening out the vast majority of compounds that do not match a query.

The final search technique in common use in chemoinformatics is similarity searching. This forms one of the family of methods used for virtual screening and so is discussed below.

2.4 Practical Issues on Representation

The analogy between chemical structures and mathematical graphs is extremely useful and forms the basis of many chemoinformatics applications; however, it is not perfect. This is due to the complexities associated with chemical representation. Particular issues relate to tautomerism, aromaticity, ionisation state and mesomerism, whereby a given structure may be represented (and exist) in different forms [9], as illustrated in Figure 2.6. In some instances, it may be beneficial to recognise different forms as the same structure, for example, for registration in databases. The usual way this is handled is to apply somewhat arbitrary 'business rules' whereby a structure is converted to a standard form for storage, with the same rules being applied to a query structure [10]. The particular set of rules that is applied may differ from one system to another [9b], which can cause problems when the output of one system or program is used as the input to another. In other cases, it may be desirable to treat different forms as different structures due to their different properties. For example, the shift of a hydrogen from one atom to another can change an atom from a hydrogen bond donor to an acceptor, which can affect how a molecule may interact with a receptor. Hence for some applications it is more appropriate to enumerate the different protonation states and tautomeric forms of each compound. These issues remain the topic of much debate.

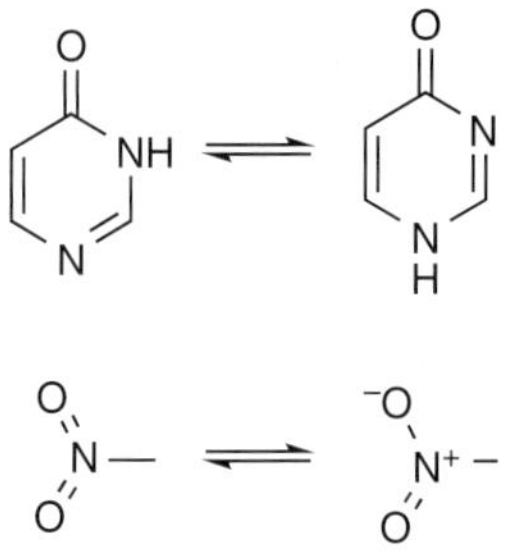

Figure 2.6 Different tautomers of four-pyrimidone (top) and alternate forms of the nitro group.

The complexities associated with the representation of chemical compounds are such that careful handling of structures is required to ensure the effective operation of chemoinformatics applications. As a consequence, structure cleaning is a common step in any chemoinformatics workflow and a typical data curation workflow is described by Fourches et al. [11].

SMILES is a proprietary system and the published algorithm for generating canonical SMILES is incomplete. The result is that several different canonicalisation algorithms have been developed that differ slightly from one another so that there is no single standard [12]. To overcome this issue, a more recent line notation called InChI has been developed as a non-proprietary, unique standard chemical identifier. A hierarchical approach is followed that enables chemical structures to be encoded at different levels of granularity. The highest layer encodes connectivity, with subsequent layers encoding more detailed information such as charge, stereochemistry, isotopic information, hydrogens, etc. The hierarchical approach enables different mesomeric forms of a structure to be given a single unique identifier. The InChIKey is a condensed digital representation of an InChi developed for web-based searching [13].

2.5 Virtual Screening

Virtual screening refers to the application of computational (in-silico) tools to predict the properties of compounds in order to prioritise them for experimental testing. Virtual screening is usually applied in the lead generation phase of drug discovery with the aim of identifying compounds with a desired biological activity. These compounds would then be taken forward to lead optimisation with the aim of improving their properties, for example, their potency and ADME (absorption, distribution, metabolism and excretion) properties.

A wide range of different virtual screening techniques is available with the method of choice depending on the information available about the biological target of interest [2, 14]. Ligand-based virtual screening refers to techniques that are used when the three-dimensional (3D) structure of the protein target is unknown. When one or more active compound is available, for example, a competitor compound or a compound in the literature, then similarity searching can be used to identify compounds that are structurally similar to the known active. When multiple active compounds are known then it may be possible to build a pharmacophore that represents the spatial arrangement of functional groups required for a molecule to bind to a receptor. The pharmacophore can then be used to query a database to find other molecules that may exhibit the pharmacophore. If both active and inactive compounds are known, for example, following a round of experimental testing, then machine learning techniques can be used to develop a structure–activity relationship model for making predictions about unknown compounds. Structure-based virtual screening techniques are used when the 3D structure of the target is known, with the most common technique being protein–ligand docking. Each of these techniques is described in more detail below.

2.6 Ligand-Based Virtual Screening

2.6.1 Similarity Searching

Perhaps the simplest virtual screening method is that of similarity searching, first developed in the 1980s [15]. Given a molecule of interest, for example, one that is known to exhibit a desired biological activity, it can be used as a query molecule to rank order, or prioritise, compounds in a database that are most similar to it. The premise on which similarity searching is based is the similar property principle [16], which states that molecules that are structurally similar are likely to exhibit similar activities. Thus, the top ranking compounds should have an increased likelihood of exhibiting similar activity to the query compared to a set of compounds selected at random. While there are exceptions to this rule, where a small change in structure leads to a large change in potency (a phenomenon that has become known as an activity cliff [17]), the similar property principle is the cornerstone of all medicinal chemistry.

Similarity searching requires a way of quantifying the similarity between the query and each compound in the database. A similarity measure consists of three components: a set of molecular descriptors, an optional weighting scheme, whereby some of the descriptors can be given more emphasis than others, and a similarity coefficient, which is used to quantify the similarity based on the molecular descriptors and their weights (if applicable).

Many different molecular descriptors have been developed for similarity searching. They can be divided into whole molecule descriptors, 2D fingerprints and descriptors derived from a 3D representation. Whole molecule properties are typically represented by real valued numbers or integers and represent physicochemical properties, such as log P and molecular weight, or counts of features such as the numbers of hydrogen bond donors and acceptors. Whole molecule descriptors can also include topological descriptors, which are derived from the 2D graph representation of a molecule. A topological index typically captures some element of the size and shape of a molecule into a single number. For example, the Wiener index is the average pairwise topological (through bond) distance between all pairs of atoms within a structure and gives a measure of the amount of branching in a molecule.

A single property is unlikely to be sufficiently discriminating to be useful in similarity searching and so whole molecule descriptors are usually combined into a vector of values. Some pre-processing of the descriptors is usually required. Firstly, standardisation should be used to place all the descriptors on to the same scale and ensure that subsequent calculations are not dominated by descriptors that operate over a greater range of values. For example, the typical molecular weight for a drug-like molecule is in the order of a few hundred Daltons, whereas, according to Lipinski rules, the log P of a compound intended for oral absorption should be <5 [18]. Standardisation can be achieved using the Z score, in which a value is transformed such that the mean value of the dataset is at zero and ± 1 represents one standard deviation from the mean. Another common approach is to scale the descriptors to fall in the range zero to one. It is also usual to remove highly correlated variables and variables that show limited variance across the dataset. More sophisticated data reduction techniques can also be used, such as a principal components analysis, which transforms the original descriptors into a smaller number of orthogonal descriptors that are linear combinations of the originals.

Despite being developed for substructure search, 2D fingerprints have proved to be very effective at identifying molecules that are structurally similar and are the most commonly used descriptors for similarity searching. In addition to the dictionary-based and hashed-based fingerprints described above, other variants include atom pair descriptors and circular fingerprints. An atom pair encodes each atom according to its properties such as element type and number of pi electrons together with the shortest bond distance between the two atoms [19]. A molecule is represented by all of its constituent atom pairs, which can be mapped to a binary vector. In circular fingerprints, each atom is represented by its element type together with its neighbouring atoms, see Figure 2.7. The radius/diameter that is used to capture the neighbours can be varied, for example, ECFP2 descriptors use a diameter of two and include the nearest neighbours only, whereas ECFP4 descriptors also encode the neighbours of the nearest neighbours at diameter 4 [20]. As for atom pairs, circular fingerprints can be mapped to a binary vector. Recent comparisons of different 2D fingerprints for similarity searching are provided by Riniker and Landrum [21] and O'Boyle and Sayle [22].

The role of weighting schemes in similarity search has been much less studied than either the types of descriptors used or the similarity coefficient. Perhaps the most studied approach has been to count the number of occurrence of the fragments when constructing a fingerprint, rather than simply recording the presence or absence of fragments. Although some evidence exists that the use of weighting schemes can be effective in some situations [23], the most common approach to similarity searching is to ignore the use of weighting schemes.

Similarity coefficients provide a way of quantifying the similarity between a pair of molecules based on their molecular descriptors. For binary fingerprints, the most common similarity coefficient is the Tanimoto coefficient, which is defined as

$$S_{A,B} = \frac{c}{a + b - c}, \tag{2.1}$$

where c is the number of bits set to '1' in common, a is the number of bits set to '1' in molecule A and b is the number of bits set to '1' in molecule B. For binary fingerprints, the Tanimoto coefficient ranges from 1 (when the fingerprints are identical) to 0 (when there are no bits in common). Thus, the molecules in a database are ranked on decreasing similarity value to the query. Note that whereas identical structures will give rise to identical fingerprints the opposite is not true; identical fingerprints does not necessarily imply identical structures.

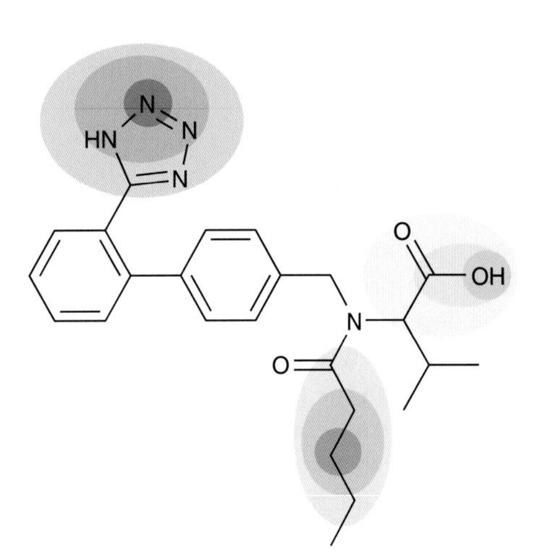

Figure 2.7 An illustration of circular fingerprints at different radii. The overlapping shaded areas of decreasing colour intensity correspond to fingerprints of radius 0, 1 and 2 (or diameter 0, 2 and 4).

For descriptors that are based on continuous values such as physicochemical properties, it is more usual to calculate the distance between molecules using, for example, Euclidean distance, which is defined as

$$D_{A,B} = \sqrt{\sum_{i=1}^{N} (x_{i,A} - x_{i,B})^2}, \tag{2.2}$$

where $x_{i,A}$ and $x_{i,B}$ are the values of descriptor i in molecules A and B, respectively, and there are N different descriptors. In virtual screening, the molecules in a database will then be ranked on increasing distance to the query molecule.

2.6.2 Scaffold Hopping

2D fingerprints have proved to be surprisingly effective in similarity searching applications, but evidence suggests that they are good at finding close analogues and less effective at finding compounds that belong to different chemical series. Moving from one chemical series to another is often referred to as scaffold hopping [24]. There are a number of reasons why scaffold hopping can be beneficial. One is to enable new intellectual property (IP) to be established, by moving away from the patent coverage associated with the query compound. Another reason might be to replace some parts of the query structure that give rise to undesirable properties, for example, unwanted side effects. A third reason might be related to compound synthesis; for example, the query compound may have some characteristics that make it unsuitable for scale-up as a drug compound.

A number of descriptors have been developed that aim to focus on features of the molecules that could be responsible for receptor binding with less emphasis given to the exact connectivity or skeleton of a structure. For example, Similog keys [25] use an atom typing scheme in which key atoms are described according to the presence or absence of four properties: hydrogen bond donor, hydrogen bond acceptor, bulkiness and electropositivity. Atom triplets (sets of three atoms) consisting of atom types combined with the through bond distances between the atoms form DABE keys, see Figure 2.8. A molecule descriptor is then constructed by counting all possible DABE keys contained within a structure. CATS descriptors are related but are based on atom pairs with the individual atoms represented according to property rather than element type [26].

Another approach is that of reduced graphs. A reduced graph is an abstract representation of a molecule in which groups of atoms are replaced by nodes and edges are formed between the nodes in order to retain the topology of the original structure. There are many ways in which chemical graphs can be reduced. For drug discovery applications the aim is usually to capture groups of atoms according to their functionality so that compounds with similar bioactivity are identified as similar, irrespective of their exact chemical scaffolds. Thus, typical node definitions include hydrogen bond donor groups, hydrogen bond acceptors, aromatic rings, etc. See Figure 2.9 for an example. Reduced graphs have been used in a number of applications including similarity searching, identifying structure–activity relationships and as a way of browsing the content of clusters of compounds [27].

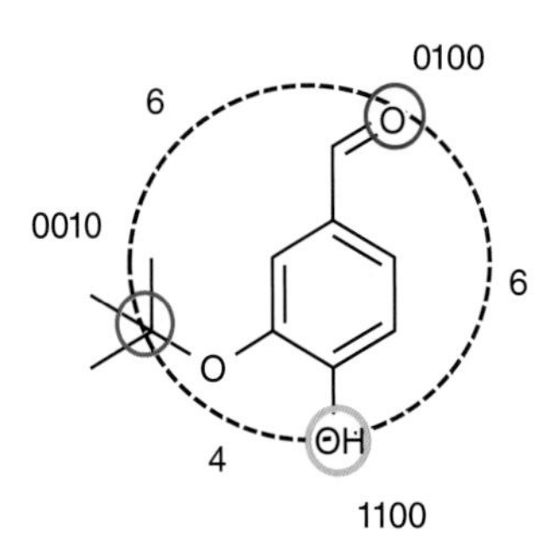

Figure 2.8 Similog keys. The red atom is assigned as a hydrogen bond acceptor; the green atom is both donor and acceptor and the blue atom is assigned as bulky.

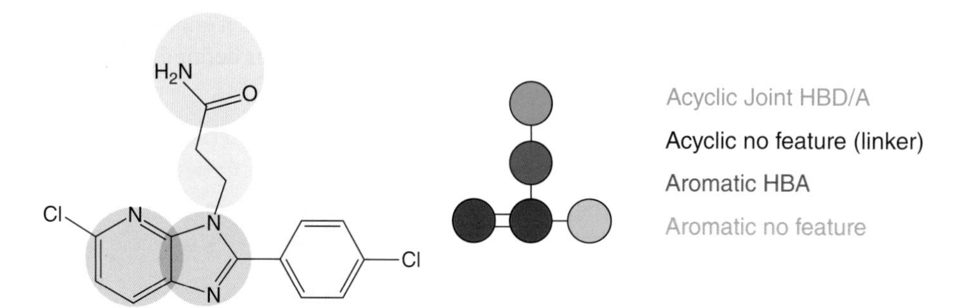

Figure 2.9 A chemical structure and its reduced graph representation. Nodes have been identified as two fused aromatic rings with the hydrogen bond donor character in red (the fusion is shown by a '=' in the reduced graph), an aromatic ring with no hydrogen bonding feature (orange), an acyclic linker node (blue) and an acyclic node with both hydrogen bond donor and hydrogen bond acceptor characteristics (green).

2.6.3 3D Similarity Search

Given that drug–receptor binding is a 3D event, there has been considerable interest in developing similarity methods that are based on 3D representations of molecules [28]. 3D similarity searching can be divided into alignment-free methods and those that require the molecules to be superimposed prior to calculating their similarity. Alignment-free methods typically involve abstracting the 3D features of a molecule, such as interatomic distances and/or angles, into a vector representation that is independent of the orientation of the molecule in 3D space. This then allows descriptors generated from different molecules to be compared directly. An example of this approach is ultrafast shape recognition (USR) in which a molecule is represented by 12 descriptors, which are statistical moments derived from the atomic distances with a molecule [29]. Four reference locations are generated from the atomic coordinates and include the geometric centre of the molecule together with atoms at the extremes. The distances of all atoms to each reference point are collected as a histogram from which the average, the standard deviation and the skewness are extracted to give the 12 descriptors.

Perhaps the most well-known of the alignment-based methods is the ROCS (rapid overlay of chemical structures) program [30]. The ROCS algorithm compares molecules by measuring the common volume they occupy and is based on earlier work by Grant, Gallardo and Pickup in which atomic volumes are represented by Gaussian functions [31]. The use of Gaussians enables the overlap volume to be calculated rapidly using analytical methods. A Tanimoto-like function is used to convert the volume overlap to a similarity score as follows:

$$S_{A,B} = \frac{V_c}{V_a + V_b - V_c}, \tag{2.3}$$

where V_c is the common overlap volume, V_a is the volume of molecule A and V_b is the volume of molecule B. Although the calculation for a given superposition is relatively fast, the method requires that the superposition of the pair of molecules that maximises the overlap is found prior to calculating the similarity; hence ROCS is typically slower in operation than alignment-free methods. In addition to shape-based similarity, ROCS also has a chemistry-based alignment method known as 'colour', which takes account of the properties of atoms; for example, atoms that can act as hydrogen bond donors or acceptors. The combination of shape and chemistry matching is known as ComboROCS.

Although 3D similarity methods are appealing conceptually, the issue of conformational flexibility makes them considerably more complex than 2D methods. Most methods handle conformational flexibility by pre-computing an ensemble of conformers for each molecule. The aim is to sample conformational space at a resolution that is sufficient to include all low energy conformations but not so exhaustive that excessive numbers of conformers are produced, thereby greatly increasing the computational time required to process them [24c]. Typical sampling strategies are based on a threshold strain energy, root mean square deviation of atom positions or simply on the maximum number of conformers permitted. Ideally the sampling method will result in something similar to the bioactive conformation of a molecule being represented; however, it may be that the bound conformation is not close enough to an energy minimum for this to occur. Effective conformational sampling remains a challenging area for 3D methods [32].

2.6.4 Pharmacophore-Based Virtual Screening

A pharmacophore is the spatial arrangement of functional groups, or features, required for a small molecule to bind to a receptor [33]. A pharmacophore is usually expressed according to features that describe the types of interaction made, rather than specific functional groups. For example, typical features are hydrogen bond donors, hydrogen bond acceptors and hydrophobic and aromatic groups. Figure 2.10 shows a set of pharmacophore features for a series of CDK2 ligands. In this example, the alignment was generated from a series of ligands for which 3D structures of the protein–ligand complexes are available by superimposing the complexes according to the binding site atoms of the proteins and extracting the ligands. The pharmacophoric features were then determined as those features that occur in all of the ligands at the same location within the binding site. These include hydrophobic features that are centred on the rings and donor and acceptor features shown as projections from the relevant heavy atoms.

Pharmacophore methods are usually used when the 3D structure of the target is unknown. The aim is to explore the conformational space of a set of active compounds while simultaneously aligning them such that similar features in the different compounds are overlaid. This process is carried out in the absence of the 3D structure of the receptor. The underlying assumption is often made that all of the active compounds bind to the target in the same way. It is important the actives that are used to build a pharmacophore are chosen carefully. For example, they should be sufficiently diverse that spurious features are not identified, that is, features that occur in all of the molecules but that are not involved in binding. Also, ideally at least one of the actives should be reasonably rigid to avoid finding promising looking overlays that do not correspond to the binding conformations. A wide number of pharmacophore identification methods have been developed [33a and d, 34], with their performance being assessed on the ability to reproduce known binding poses [35].

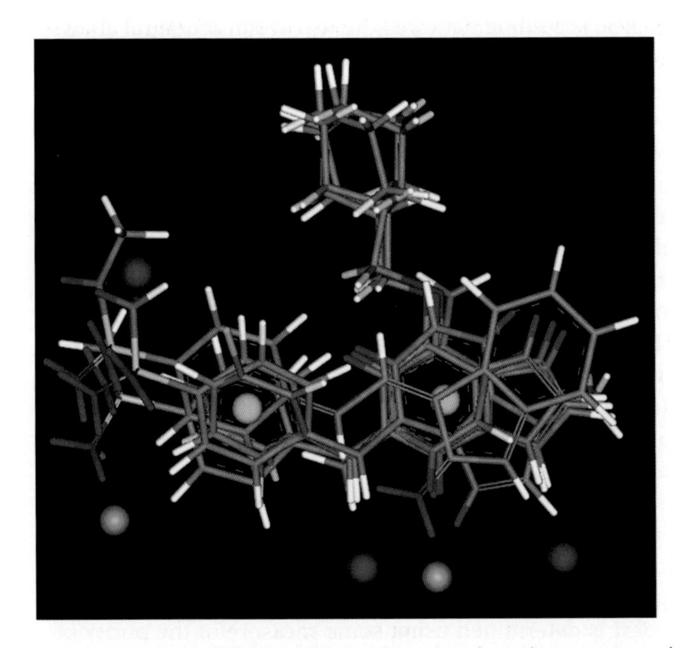

Figure 2.10 A set of CDK2 inhibitors aligned to show their common pharmacophoric features.

The difficulties associated with pharmacophore identification are such that it is rarely possible to determine the true pharmacophore directly and, therefore, most pharmacophore generation programs will generate a set of plausible hypotheses that should then be evaluated. This can be done by using a holdout set, that is, omitting some of the active compounds when generating the hypotheses and choosing a hypothesis that is consistent with those left out. Once a pharmacophore has been identified and validated, it can then be used to search a database to identify those compounds that match the pharmacophore. As this is a 3D technique, it is important that conformational flexibility is taken into account when searching the database. It is also important that the same feature definitions are used when searching as were used to identify the pharmacophore. These may sound like obvious conditions, but they are sometimes overlooked. As for other 3D methods, it is important that protonation and tautomer states are considered appropriately since these can determine whether a given atom can act as a hydrogen bond donor or a hydrogen bond acceptor.

2.6.5 Structure–Activity Relationship Modelling

When both active and inactive compounds are available, machine learning techniques can be used to learn a model of activity that can then be applied to make predictions about previously unseen compounds. Quantitative structure–activity relationship (QSAR) modelling dates back to the work of Hansch, who correlated biological activity with physicochemical properties using multiple linear regression. This approach was extended later by Free and Wilson to the use of structural properties as descriptors rather than physicochemical properties. While these approaches are still in use they are generally restricted to building local models relevant to a small structurally homogeneous dataset, as seen in lead optimisation [36].

For virtual screening applications, the aim is to use machine learning techniques to build structure-activity models that can be applied to a wide variety of drug-like molecules that are structurally heterogeneous [37]. The models are developed using training data that is typically labelled as active and inactive, with the focus being on classifying test compounds as active or inactive, or rank-ordering compounds on their probability of being active. The popularity of this approach has grown in recent years due to a number of factors including the rapid growth in publicly available data about compounds and their properties and the development of a wide variety of sophisticated nonlinear machine learning algorithms.

The first application of machine learning to biological screening data was the use of substructural analysis (SSA) [38]. In SSA, each molecule in a training set of active and inactive compounds is characterised by a set of binary descriptors such as 2D fingerprints. A set of weights is then calculated, one for each fragment (or bit position) in the fingerprint. A fragment's weight reflects the probability that a molecule containing that fragment will be active; for example, the weight may be the fraction of the actives in the training set that contain the fragment. A previously unseen compound is scored by combining the weights of the fragments it contains. The resulting score represents the probability that the test compound will be active and SSA scores can be used to rank-order a database of compounds. SSA is closely related to the naïve Bayesian classifier that has become popular in chemoinformatics recently. Other commonly used machine learning methods include k-nearest neighbours (kNN), decision trees, random forests and support vector machines.

kNN is conceptually very easy to understand. Each test set compound is compared with all training set compounds to find its k nearest neighbours. The class membership of a test compound is then predicted to be the majority class amongst the k nearest neighbours, as shown in Figure 2.11. Implementing kNN requires a definition of similarity (that is, a set of descriptors and a similarity coefficient) and the value of k to be specified; k is usually chosen as an odd number in order to avoid the use of ties.

A decision tree consists of a set of rules that is used to associate specific features or descriptor values with a classification label. For example, each rule may correspond to the presence or absence of a particular feature or a particular range of descriptor values. A decision tree is constructed using training data, which is progressively split into subsets with the aim of separating the two classes. At each decision point, the descriptor or variable that gives the best split is chosen, where best is determined using some measure of the purity of a split. The same procedure is then applied to each of the subsets that are produced. One way of determining

Figure 2.11 Example of kNN. The training examples are shown in red (representing active compounds) and green (representing inactive compounds). The test instance is shown in blue with its three nearest neighbours identified. The test compound is assigned as inactive since this is the majority class membership of its three nearest neighbours.

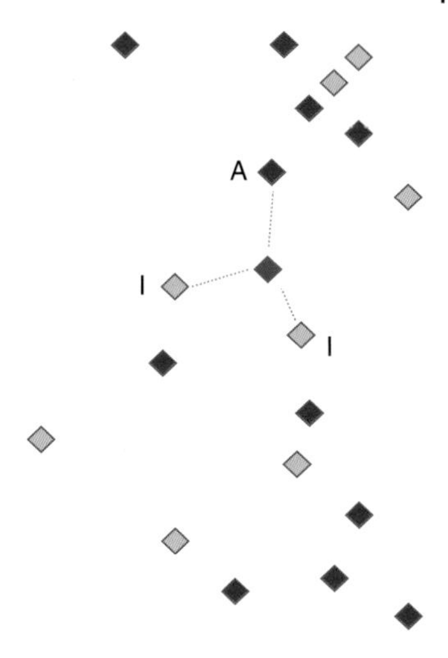

the best split is to use entropy, which measures the extent of disorder in a set. If there are N molecules in i classes with each class containing n_i molecules, the entropy is

$$I = -\sum p_i \ln p_i \quad \text{where } p_i = \frac{n_i}{N}. \tag{2.4}$$

The aim is to choose the split that minimises entropy. Once a tree has been constructed a test compound is classified by 'dropping' it down the tree until a terminal, or leaf, node is reached. The test compound is then assigned to the class according to the distribution of training examples in the node; for example, it is assigned to the majority class represented in the leaf node.

Decision trees generate models that are easy to interpret, but they are prone to overfitting and they are sensitive to small changes in the training data. These factors can limit their performance as predictors. One way of improving performance is to use an ensemble of decision trees. Random Forest is one such approach where many classification trees are grown using different random samples of the training data. The randomisation occurs at two levels. One is that a random subset of the features is available at each node and, second, each tree is based on a bootstrap sample of the data whereby for a set of N molecules, N are chosen at random from the data with replacement. Bootstrapping means that each molecule may be chosen zero, one or more

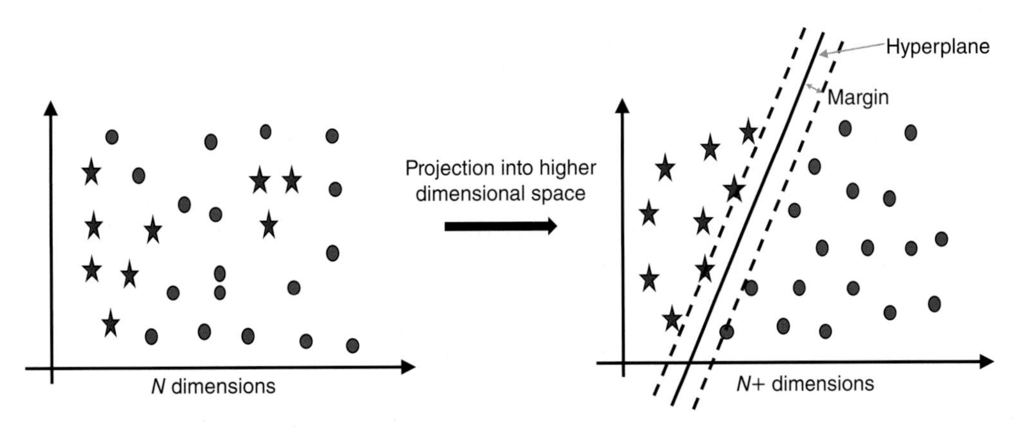

Figure 2.12 Support vector machine.

than one times. The trained ensemble of trees, the forest, is then used to make predictions on test data. A test compound is passed through each tree with its classification being determined by the majority vote. The proportion of votes cast for a particular class can be used as a measure of the confidence of the prediction.

Support vector machines (SVMs) have been widely used in chemoinformatics, especially for classification problems. The aim is to find the best separation between two classes of compounds such that each class lies on the opposite side of a hyperplane, as shown in Figure 2.12. There is usually an infinite number of hyperplanes that could be constructed and in SVMs the hyperplane that maximises the margin between the two classes is chosen, the aim being to maximise the margin while minimising the number of misclassified samples. It is not usually possible to find a linear separation of the compound classes using the original descriptor space and so what is known as the 'kernel trick' is used to transform the original descriptors into a higher dimensional space where they are linearly separable. The radial basis function has been shown to perform well and is therefore widely used [39].

2.7 Protein–Ligand Docking

When the 3D structure of the protein target is known, the most widely used virtual screening technique is protein–ligand docking. The Protein Databank (PDB) is a public repository of 3D structures of macromolecules including proteins, DNA and RNA, the majority of which have been determined using experimental techniques such as X-ray crystallography, cryoelectron microscopy or NMR techniques [40]. In the context of virtual screening, the aim of docking is to evaluate small molecules, or ligands, within a database and prioritise those that are most likely to bind to the protein [41]. In order to bind, a small molecule should have complementary steric and electrostatic properties to a binding site on the protein. For each ligand, docking involves identifying the bound conformation of the ligand, determining the relative orientation of the ligand within the protein binding site and then estimating the binding energy of the ligand, so that the database of ligands can be rank-ordered [14a]. Protein–ligand docking is a complex problem, especially accurate scoring, and, as for other virtual screening methods, the aim is to reduce the number of false negatives in the prioritised compounds rather than selecting high affinity compounds directly.

Many docking programs have been developed that adopt different approaches to the steps outlined above. The conformational space of the ligands can be explored by pre-computing ensembles of conformers, as described above in 3D similarity methods, with each conformer docked to the protein as a rigid body, as in DOCK [42]. The alternative approach is to explore the conformational space of the ligand simultaneously with the docking procedure, that is, at the same time as exploring different orientations of the ligand within the binding site. This can be achieved using stochastic methods such as simulated annealing or genetic algorithms as in GOLD [43] and AutoDock [44] or by using an incremental construction method whereby a ligand is broken into fragments and reconstructed within the binding site with conformational space explored in a stepwise manner [45]. Identifying the binding pose also requires some degree of protein flexibility to be taken into account and most docking programs now allow movement of the protein sidechains within the binding pocket. However, handling full protein flexibility is limited to molecular dynamics simulations, which are far too computationally demanding to be used in virtual screening.

A scoring function can be divided into force field, empirical and knowledge-based scoring functions [41b]. Force-field methods use classical molecular mechanics methods to estimate binding energies directly by summing the contributions of non-bonded terms such as electrostatic and van der Waals energy terms between all atoms in the two molecules in the complex. Empirical scoring functions adopt a QSAR-type approach in which the weights associated with various energy terms are fitted to experimental data using, for example, multiple linear regression. The typical terms include hydrogen bonding interactions, hydrophobic contact terms, desolvation and entropy terms with the experimental data consisting of measured binding affinities for protein ligand complexes. Empirical scoring functions are straightforward to calculate but their accuracy is dependent on the relationship between the training data used to derive the function and the data used for virtual screening. Knowledge-based scoring functions are based on statistical data extracted from protein–ligand complexes with the underlying assumption being that frequently occurring interatomic distances represent favourable contacts. Pairwise energy potentials are derived from known complexes and a given binding pose is scored by summing the contributions for individual atom pairs.

There are many examples of the successful applications of protein–ligand docking in the literature; see, for example, [41c]. However, considerable care needs to be exercised when setting up a docking-based virtual screening protocol. This includes careful selection and preparation of both the protein target and the database of ligands [41a and c]. First it is necessary to select the most suitable protein structure. Ideally this should be a structure determined using X-ray crystallography and be of high resolution. Consideration should also be given to which conformation(s) of the protein to use; for example, there may be multiple 3D structures of the protein in different conformations with and without ligands being bound. When a protein–ligand complex is available, the binding site is already known; in the absence of a complex, various programs are available to predict the binding site [41b]. It has been shown, however, that dockings based on structures for which protein–ligand complexes exist generally perform better than those based on apo (unbound) protein structures. A key factor in this is the ability of a binding site to adapt its shape on ligand binding, a process known as induced fit [46].

Preparation of the protein includes adding hydrogen atoms (since these are not present in structures extracted from the PDB) and specifying the correct protonation and tautomeric states of the binding site residues. This is particularly important since they determine the interactions that can be made with a ligand. Consideration also needs to be given to water molecules; that is, whether or not these should be eliminated from the binding site or if key water molecules should be conserved. The ligands also require careful handling. If they originate as 2D structures then they must first be converted to 3D structures. For chiral molecules, the stereochemistry should be handled appropriately and, if not specified, then all enantiomers should be enumerated. The most effective way of handling the different protonation and tautomeric states of ligands is to enumerate the different possibilities. There is a growing number of publicly available datasets that are available for docking; for example, the ZINC database now consists of over 100 million commercially available compounds that are already prepared for docking [47]. Finally, if a rigid docking algorithm is to be used then low energy conformers should be pre-generated [48].

Given the complexities of setting up a docking run, it can be beneficial to run retrospective tests using known ligands and a carefully selected set of decoys. Such a test should be able to reproduce the known binding modes for protein–ligand complexes, as shown in Figure 2.13, and should also result in the known

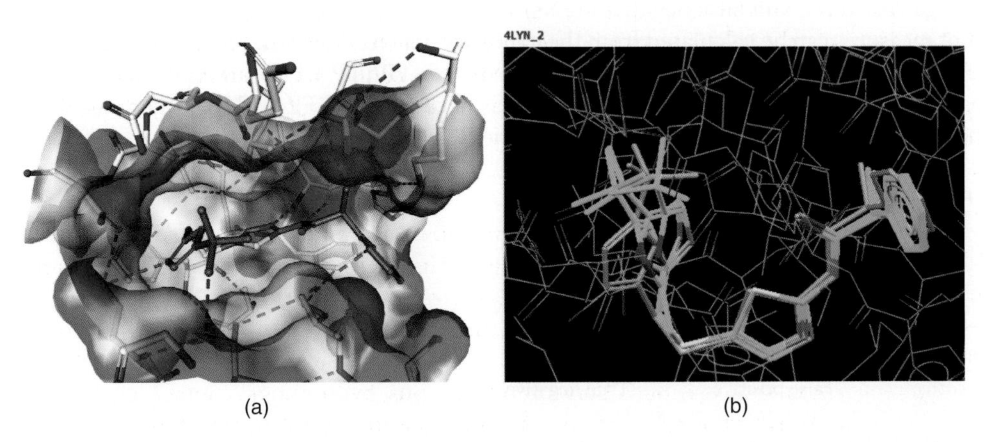

Figure 2.13 (a) Protein-ligand complex 4LYN, a cyclin-dependent kinase 2 complex. Image generated using the NGL Viewer [49]. (b) Poses generated by the docking program GOLD are superimposed on the X-ray crystal pose, which is shown in pink.

actives being ranked above the decoy compounds. If this cannot be achieved for retrospective data then it is unlikely that the docking setup will be effective in a prospective virtual screen.

As indicated above, the accurate prediction of binding affinity remains a challenge. Scoring functions are generally considered to be successful in predicting the correct binding pose for a small molecule but they are less effective at predicting relative binding affinities. Therefore, post-docking analysis procedures are often used to select the final compounds for experimental screening rather than relying on the docking score alone. This can involve visual inspection as well as the use of automated methods, for example, to prioritise preferred interaction patterns [14a]. There continues to be a steady stream of incremental improvements to scoring and other aspects of docking, such as the ability to handle new interactions terms, metal coordination and improved handling of water [41c].

2.8 Evaluating Virtual Screening Methods

Given the wide variety of different descriptors and methods available for virtual screening along with the parameterisation required by many methods, it is important to evaluate virtual screening performance in order to establish the best protocol. Performance can be evaluated using retrospective data, that is, data for which the correct answers are already known. For methods that generate a ranking of molecules, such as similarity searching and protein–ligand docking, the usual measures include the enrichment factor and area under the curve (AUC). The enrichment factor determines the increase in actives in the top few positions of the ranked list relative to their being distributed evenly throughout the ranked list. The AUC takes account of the distribution of actives over the entire list, with the value 1 indicating that all of the actives are ranked at the top and a value of 0.5 indicating an even distribution throughout the list. A number of variants of the AUC have been developed that give greater weight to the earlier part of the ranked list since this is most important for virtual screening, where the aim is to select a small subset of the available compounds.

For machine learning methods, the usual approach is to divide the available data into two sets: a training set that is used to build the model and an external test set that is used to assess its performance. The training data can be further split into training and validation data, with the validation set used to tune the parameters of the model. The performance of the model is calculated on the external test set. A key consideration for machine learning methods is that predictions made by the resulting models should only be considered reliable for compounds that have similar characteristics to the training data, a concept that has become known as the applicability domain of a model [50]. For classification problems, performance is typically assessed using elements of the confusion matrix (Table 2.2), where predictions are assigned as true positives (TPs), false positives (FPs), true negatives (TNs) and false negatives (FNs).

A variety of different measures can be calculated from the confusion matrix. For example, sensitivity, or the TP rate, is the proportion of the positive class that is correctly predicted (TP/(TP + FN)) and specificity, or the TN rate, is the proportion of the negative class that is predicted as negative (TN/(FP + TN)). Other measures include accuracy, the F1 score and Matthew's correlation coefficient (MCC).

Table 2.2 A confusion matrix.

		Predicted class	
		Positive	**Negative**
Actual class	*Positive*	True positive	False negative
	Negative	False positive	True negative

2.9 Case Studies of Virtual Screening

Many successful applications of virtual screening have now been reported in the literature and have been summarised in various reviews; see, for example, [14a, 41b and c, 51]. Two examples are provided here.

2.9.1 Ligand-Based Virtual Screening: Discovery of LXR Agonists

Temml et al. [52] used a combination of pharmacophore modelling and 3D shape matching to identify selective agonists of liver X receptor (LXR) as potential regulators of cholesterol metabolism. The study involved a set of six pharmacophore hypotheses that had been generated in an earlier study using Discovery Studio and had been experimentally validated as LXR agonists. The active compounds found using these pharmacophores had shown general LXR activity but had not been tested for selectivity. Therefore, a test set of compounds was assembled and sorted on their LXR subtype selectivity. Conformers were generated for these compounds and the resulting library was screened against the six pharmacophore models. Three of the models successfully identified a significant number of highly selective compounds. The most selective of the test compounds was also used as a query in a shape-based search using the ROCS program with the ComboScore, which combines shape and chemistry features. The ROCS search was conducted against the Specs library consisting of around 200 000 compounds. A maximum of 400 conformers was generated for each compound in the database using the OMEGA program. The top 500 molecules found using ROCS were then filtered using the three selective pharmacophores. For the pharmacophore, matching a new conformer database was generated using the same software suite that had been used to generate the original hypotheses. The pharmacophore filtering reduced the compound set to 56, of which 10 were selected for biological testing. The combined pharmacophore and shape-based virtual screening procedure identified some selective LXR agonists and was therefore deemed to be a success.

2.9.2 Combined Ligand- and Structure-Based Approaches: Pim-1 Kinase Inhibitors

Pim-1 has been found to be involved in a number of signalling pathways and is implicated in multiple human cancers. Ren et al. [53] used a combination of ligand- and structure-based virtual screening methods in a hierarchical virtual screening procedure and were successful in identifying novel inhibitors of Pim-1. The virtual screening methods they used were SVM modelling, pharmacophore modelling and protein–ligand docking.

An SVM model was built using a set of 500 known Pim-1 inhibitors and a set of 37 000 presumed inactive compounds. The model was based on a set of 50 molecular descriptors consisting of topological descriptors and physicochemical properties. The SVM model was trained using cross-validation and evaluated on an independent test set. The resulting model showed high accuracy on both the training set and the test set and was then used as the first virtual screen in the hierarchy to screen a database of 20 million compounds; 56 500 compounds were retained.

The pharmacophore model was built using eight Pim-1 inhibitors. Four hypotheses were generated and the top scoring one, consisting of one hydrogen bond acceptor, one hydrogen bond donor and one hydrophobic aromatic feature, was validated on a test set consisting of known actives and inactives. A high percentage (83%) of the actives mapped to the hypothesis. The molecule outputs from the SVM classifier were mapped to the pharmacophore hypothesis and a subset of 10 700 compounds were retained. These were then docked to the protein binding site using the GOLD docking program. The docking protocol was optimised prior to running the virtual screening. A high resolution protein–ligand complex was chosen. Seven compounds that were known to bind to the protein were docked to it with the docking parameters adjusted, including the choice of scoring function, until the known binding poses were reproduced. The 935 top scoring compounds found by

GOLD were inspected visually and 47 were chosen for experimental testing in an in vitro biochemical assay. Five compounds were identified as novel Pim-1 inhibitors.

2.10 Conclusions

This chapter has provided an introduction to the fundamental concepts that underpin many chemoinformatics applications. The starting point was to consider representation issues as a foundation before describing basic search techniques. The main focus was then on the variety of virtual screening techniques that are routinely used in the process of identifying lead compounds. Although these were presented as distinct approaches, the brief application examples highlighted above show that many drug discovery programs follow a cascaded approach with multiple in-silico approaches used. This is especially the case when very large libraries of compounds are explored, with the less computationally demanding approaches used first to reduce the number of compounds that undergo more complex analysis such as docking.

The focus here has been on lead generation, but it should be acknowledged that chemoinformatics techniques impact on all stages of the drug discovery pipeline and there is a large number of topics that have not been discussed, for example, computational filtering and library design, chemical space exploration including diversity analysis and clustering, toxicity prediction, reaction searching, de novo design and assessment of synthetic accessibility, to name a few. Another limitation is the focus on the traditional single target compound approach whereas drug discovery is now shifting towards multitarget approaches. For example, machine learning methods are being developed for multitask prediction where the aim is to predict the response to a number of different endpoints using deep learning methods [54]. What is clear is that with the growing volumes of data, chemoinformatics techniques are required more than ever to guide decision making in drug discovery.

References

1 Brown, F.K. (1998). Chemoinformatics: what is it and how does it impact drug discovery? In: *Annual Reports in Medicinal Chemistry*, vol. 33 (ed. J.A. Bristol), 375–384. Academic Press.

2 Willett, P. (2011). *Wiley Interdiscip. Rev.-Comput. Mol. Sci.* 1: 46–56.

3 (a) Gonzalez-Medina, M., Naveja, J.J., Sanchez-Cruz, N., and Medina-Franco, J.L. (2017). *RSC Adv.* 7: 54153–54163. (b) Kooistra, A.J., Vass, M., McGuire, R. et al. (2018). *ChemMedChem* 13: 614–626. (c) Humbeck, L. and Koch, O. (2017). *ACS Chem. Biol.* 12: 23–35.

4 Bienfait, B. and Ertl, P. (2013). *J. Cheminf.* 5: 24.

5 (a) Weininger, D. (1988). *J. Chem. Inf. Comput. Sci.* 28: 31–36. (b) Weininger, D., Weininger, A., and Weininger, J.L. (1989). *J. Chem. Inf. Comput. Sci.* 29: 97–101.

6 Dalby, A., Nourse, J.G., Hounshell, W.D. et al. (1992). *J. Chem. Inf. Comput. Sci.* 32: 244–255.

7 Morgan, H.L. (1965). *J. Chem. Doc.* 5: 107–113.

8 Durant, J.L., Leland, B.A., Henry, D.R., and Nourse, J.G. (2002). *J. Chem. Inf. Comput. Sci.* 42: 1273–1280.

9 (a) Sayle, R.A. (2010). *J. Comput. Aided Mol. Des.* 24: 485–496. (b) Warr, W.A. (2010). *J. Comput. Aided Mol. Des.* 24: 497–520.

10 Hersey, A., Chambers, J., Bellis, L. et al. (2015). *Drug Discovery Today: Technol.* 14: 17–24.

11 (a) Fourches, D., Muratov, E., and Tropsha, A. (2010). *J. Chem. Inf. Model.* 50: 1189–1204. (b) Fourches, D., Muratov, E., and Tropsha, A. (2016). *J. Chem. Inf. Model.* 56: 1243–1252.

12 O'Boyle, N. (2012). *J. Cheminf.* 4.

13 Warr, W.A. (2015). *J. Comput. Aided Mol. Des.* 29: 681–694.

14 (a) Rognan, D. (2017). *Pharmacol. Ther.* 175: 47–66. (b) Vogt, M. and Bajorath, J. (2012). *Bioorg. Med. Chem.* 20: 5317–5323.

15 (a) Willett, P. (2014). *Mol. Inf.* 33: 403–413. (b) Maggiora, G., Vogt, M., Stumpfe, D., and Bajorath, J. (2014). *J. Med. Chem.* 57: 3186–3204.
16 Johnson, M.A. and Maggiora, G.M. (1990). *Concepts and Applications of Molecular Similarity*. New York: Wiley.
17 Maggiora, G.M. (2006). *J. Chem. Inf. Model.* 46: 1535–1535.
18 Lipinski, C.A., Lombardo, F., Dominy, B.W., and Feeney, P.J. (1997). *Adv. Drug Delivery Rev.* 23: 3–25.
19 Carhart, R.E., Smith, D.H., and Venkataraghavan, R. (1985). *J. Chem. Inf. Comput. Sci.* 25: 64–73.
20 (a) Rogers, D., Brown, R.D., and Hahn, M. (2005). *J. Biomol. Screen.* 10: 682–686. (b) Rogers, D. and Hahn, M. (2010). *J. Chem. Inf. Model.* 50: 742–754.
21 Riniker, S. and Landrum, G.A. (2013). *J. Chem.* 5.
22 O'Boyle, N.M. and Sayle, R.A. (2016). *J. Chem.* 8: 14.
23 Arif, S.M., Holliday, J.D., and Willett, P. (2009). *J. Comput. Aided Mol. Des.* 23: 655.
24 (a) Brown, N. (2014). *Mol. Inf.* 33: 458–462. (b) Hu, Y., Stumpfe, D., and Bajorath, J. (2017). *J. Med. Chem.* 60: 1238–1246. (c) Schuffenhauer, A. (2012). *Wiley Interdiscip. Rev.: Comput. Mol. Sci.* 2: 842–867.
25 Schuffenhauer, A., Floersheim, P., Acklin, P., and Jacoby, E. (2003). *J. Chem. Inf. Comput. Sci.* 43: 391–405.
26 Reutlinger, M., Koch, C.P., Reker, D. et al. (2013). *Mol. Inf.* 32: 133–138.
27 Birchall, K. and Gillet, V.J. (2011). Reduced graphs and their applications in chemoinformatics. In: *Chemoinformatics and Computational Chemical Biology*, vol. 672 (ed. J. Bajorath), 197–212.
28 (a) Finn, P.W. and Morris, G.M. (2013). *Wiley Interdiscip. Rev.-Comput. Mol. Sci.* 3: 226–241. (b) Nicholls, A., McGaughey, G.B., Sheridan, R.P. et al. (2010). *J. Med. Chem.* 53: 3862–3886. (c) Shin, W.-H., Zhu, X., Bures, M.G., and Kihara, D. (2015). *Molecules* 20: 12841–12862.
29 Ballester, P.J. (2011). *Future Med. Chem.* 3: 65–78.
30 Rush, T.S., Grant, J.A., Mosyak, L., and Nicholls, A. (2005). *J. Med. Chem.* 48: 1489–1495.
31 Grant, J.A., Gallardo, M.A., and Pickup, B.T. (1996). *J. Comput. Chem.* 17: 1653–1666.
32 (a) Scior, T., Bender, A., Tresadern, G. et al. (2012). *J. Chem. Inf. Model.* 52: 867–881. (b) Hawkins, P.C.D. (2017). *J. Chem. Inf. Model.* 57 (8): 1747–1756.
33 (a) Leach, A.R., Gillet, V.J., Lewis, R.A., and Taylor, R. (2010). *J. Med. Chem.* 53: 539–558. (b) Guner, O.F. and Bowen, J.P. (2014). *J. Chem. Inf. Model.* 54: 1269–1283. (c) Caporuscio, F. and Tafi, A. (2011). *Curr. Med. Chem.* 18: 2543–2553. (d) Langer, T. (2010). *Mol. Inf.* 29: 470–475.
34 Vuorinen, A. and Schuster, D. (2015). *Methods* 71: 113–134.
35 (a) Braga, R.C. and Andrade, C.H. (2013). *Curr. Top. Med. Chem.* 13: 1127–1138. (b) Sanders, M.P.A., Barbosa, A.J.M., Zarzycka, B. et al. (2012). *J. Chem. Inf. Model.* 52: 1607–1620. (c) Giangreco, I., Cosgrove, D.A., and Packer, M.J. (2013). *J. Chem. Inf. Model.* 53: 852–866.
36 (a) Lewis, R.A. and Wood, D. (2014). *Wiley Interdiscip. Rev.-Comput. Mol. Sci.* 4: 505–522. (b) Cherkasov, A., Muratov, E.N., Fourches, D. et al. (2014). *J. Med. Chem.* 57: 4977–5010.
37 (a) Lavecchia, A. (2015). *Drug Discovery Today* 20: 318–331. (b) Mitchell, J.B.O. (2014). *Wiley Interdiscip. Rev.-Comput. Mol. Sci.* 4: 468–481.
38 Cramer, R.D., Redl, G., and Berkoff, C.E. (1974). *J. Med. Chem.* 17: 533–535.
39 Yao, X.J., Panaye, A., Doucet, J.P. et al. (2004). *J. Chem. Inf. Comput. Sci.* 44: 1257–1266.
40 Berman, H.M., Westbrook, J., Feng, Z. et al. (2000). *Nucleic Acids Res.* 28: 235–242.
41 (a) Forli, S. (2015). *Molecules* 20: 18732–18758. (b) Ferreira, L.G., dos Santos, R.N., Oliva, G., and Andricopulo, A.D. (2015). *Molecules* 20: 13384–13421. (c) Irwin, J.J. and Shoichet, B.K. (2016). *J. Med. Chem.* 59: 4103–4120.
42 Ewing, T.J.A., Makino, S., Skillman, A.G., and Kuntz, I.D. (2001). *J. Comput. Aided Mol. Des.* 15: 411–428.
43 Jones, G., Willett, P., Glen, R.C. et al. (1997). *J. Mol. Biol.* 267: 727–748.
44 Morris, G.M., Huey, R., Lindstrom, W. et al. (2009). *J. Comput. Chem.* 30: 2785–2791.
45 Kramer, B., Metz, G., Rarey, M., and Lengauer, T. (1999). *Med. Chem. Res.* 9: 463–478.
46 McGovern, S.L. and Shoichet, B.K. (2003). *J. Med. Chem.* 46: 2895–2907.
47 Sterling, T. and Irwin, J.J. (2015). *J. Chem. Inf. Model.* 55: 2324–2337.

48 Ebejer, J.P., Morris, G.M., and Deane, C.M. (2012). *J. Chem. Inf. Model.* 52: 1146–1158.
49 AS Rose, AR Bradley, Y Valasatava, JM Duarte, A Prlić and PW Rose. Web-based molecular graphics for large complexes. ACM Proceedings of the 21st International Conference on Web3D Technology (Web3D '16): 185–186, 2016. https://doi.org/10.1145/2945292.2945324.
50 Hanser, T., Barber, C., Marchaland, J.F., and Werner, S. (2016). *SAR QSAR Environ. Res.* 27: 893–909.
51 (a) Lavecchia, A. and Di Giovanni, C. (2013). *Curr. Med. Chem.* 20: 2839–2860. (b) Danishuddin, M. and Khan, A.U. (2015). *Methods* 71: 135–145. (c) Kumar, A. and Zhang, K.Y.J. (2015). *Methods* 71: 26–37.
52 Temml, V., Voss, C.V., Dirsch, V.M., and Schuster, D. (2014). *J. Chem. Inf. Model.* 54: 367–371.
53 Ren, J.-X., Li, L.-L., Zheng, R.-L. et al. (2011). *J. Chem. Inf. Model.* 51: 1364–1375.
54 (a) Ekins, S. (2016). *Pharm. Res.* 33: 2594–2603. (b) Gawehn, E., Hiss, J.A., and Schneider, G. (2016). *Mol. Inf.* 35: 3–14. (c) Ramsundar, B., Liu, B., Wu, Z. et al. (2017). *J. Chem. Inf. Model.* 57: 2068–2076. (d) Zhang, L., Tan, J., Han, D., and Zhu, H. (2017). *Drug Discovery Today* 22: 1680–1685.

Further Reading

Bajorath, J. (2014). *Chemoinformatics for Drug Discovery*. Wiley.
Brown, N. (2016). *In Silico Medicinal Chemistry: Computational Methods to Support Drug Design*. The Royal Society of Chemistry.
Engel, T. and Gasteiger, J. (eds.) (2018). *Chemoinformatics. Basic Concepts and Methods*. Weinheim: Wiley-VCH.
Leach, A.R. and Gillet, V.J. (2007). *An Introduction to Chemoinformatics*. Dordrecht: Springer.
Sotriffer, C. (ed.) (2011). *Virtual Screening: Principle, Challenges, and Practical Guidelines*. Weinheim: Wiley-VCH.

3

Bioinformatics and Its Applications in Genomics

David J. Parry-Smith

Wellcome Sanger Institute, Hinxton, UK

3.1 Significance and Short Background

This chapter introduces the field of bioinformatics, which is a scientific discipline dealing with the analysis of biological data. More specifically, we deal with bioinformatics as it is applied to the field of genomics.

Biological data can take many forms, including DNA, RNA and protein sequence information. It can encompass higher level collections of data and analyses of data. These include databases of structurally and functionally relevant sequence patterns and databases of small molecule ligand binding sites. It can also encompass imaging of a wide variety of processes, including X-ray diffraction data and images of the three-dimensional structures of DNA, RNA and protein complexes. New forms of biological data are being generated all the time as new experimental approaches are developed. Analysis of the data derived using these techniques is underpinned by a sound understanding of how bioinformatics relates to the functioning of the cellular machinery, whole organisms (e.g. genetics) and even the evolution of species (e.g. phylogenetics).

Data analysis has always been fundamental to scientific understanding. Observation leads to classification and generalisation. Rules emerge that enable us to explain the way systems behave now and to predict how they may behave in the future. Such systems range from tracking the course of the planets and stars across the sky to the quantum behaviour of fundamental particles in an atom of helium, say. The development of technology has a major impact on the amount of data available to review and analyse.

In the field of bioinformatics, data gathered on the primary sequence of proteins and the order of the bases comprising the sequences of the DNA of genes have resulted in substantial repositories that are freely available to the public for exploration. The teams of scientists that isolated and cloned the individual genes whose sequences were deposited in these data resources published their work in the scientific literature, at the same time depositing their data in a *sequence database*. Scientists interested in determining the three-dimensional structure of the protein that the gene expresses also deposited their data upon publication, but in databases more appropriately designed to hold such data. Bioinformatics is fundamentally involved in providing the means for collating (gathering), analysing and curating (maintaining or looking after) the databases of structural coordinates and DNA or protein sequence. These resources are referred to as *primary resources* because they contain the actual data determined by the experimental science. *Secondary resources* are databases that corral information gleaned from analysis of the primary resources – such as a database of gene families or a database of conserved patterns of residues in protein families (sometimes referred to as sequence motifs defining functional or structural domains).

In the late twentieth century, a more holistic approach was taken to the production of data that contributes to the primary resources. Instead of focusing on individual genes, whole genomes of organisms important to medical research were sequenced. This resulted in a very rapid expansion of the sequence

Biomolecular and Bioanalytical Techniques: Theory, Methodology and Applications, First Edition. Edited by Vasudevan Ramesh.

data available in sequence databases of all types. Annotation of the databases became a key function of bioinformatics. The major centres for warehousing sequence data (EMBL-EBI in Europe, Genbank in the USA) have substantial ongoing programmes of work in annotation. Whereas, formerly, only the sequences of the cloned genes tended to be available, now sequence information related to those parts of the genome involved in the control of gene expression and functional but non-coding regions is readily accessible. This is because whole genomes are being sequenced. In addition, other parts of the genome (previously referred to as 'junk DNA', but in fact whose function has simply yet to be determined) are now available for analysis.

It took a long time to sequence the first whole genomes compared to the current rate of sequencing. The human genome sequence (itself a mosaic of multiple individuals and not representative of any single person) took many years to complete, being officially begun in 1990 and declared complete in 2003, with initial publication in 2001 [1]. Now, in the second decade of the twenty-first century a whole genome sequence can be obtained for a specific cell line in a few days. The bioinformatics required to deal with the data from such experiments has had to be developed to cope with the amount of data involved. It is interesting to note that although computing capacity has increased to a staggering extent over the last few years, sequencing capacity has far exceeded this increase. In fact, today we do not normally think of library construction and sequencing for *whole genome sequencing* (WGS) to be experimental, as systems are available that enable a factory approach to WGS. Projects to generate WGS for the UK BioBank of 50 000 genomes (www.ukbiobank.ac.uk), 100 000 genomes (www.genomicsengland.co.uk) and 2 million genomes (AstraZeneca in partnership with other providers) are in progress [2]. Other commercial approaches (e.g. Human Longevity, Inc.) are amassing substantial proprietary databases of genome sequence data.

3.1.1 Big Data

The term *big data* is applied to large volumes of biological data, as it is to data derived from financial or astronomical domains. Each area generates petabytes of data on a regular basis. As in all other areas of big data endeavour, the computational techniques appropriate to the size of these data streams are appropriate to apply. These include artificial intelligence techniques of linear and non-linear regression, classification and machine learning.

3.1.2 Computational Challenges

There are significant challenges in analysing sequence data for genomics. These include:

- Computer hardware utilisation, for example, exploitation of graphics chips (GPUs) designed for rapid matrix calculations (as demanded by machine learning)
- Data storage, covering parallelised file systems, object stores and hierarchical storage
- Software development, necessitating the expertise in writing code for making effective and efficient use of hardware and ever increasing volumes of data.

Cloud-based approaches are coming to the fore, with commercial organisations offering viable solutions and research institutions implementing internal cloud-based flexible compute environments. This approach enables a timely response to the rapidly changing flow of data and the information and knowledge derived from it. In terms of software advances, programming for machine learning and artificial intelligence has much to contribute in identifying patterns and trends when analysing large quantities of genomic data. Comparison of whole genome sequences and variant call files (VCFs) resulting from these large sequencing projects are areas of active research [3].

3.1.3 Bioinformatics Roles

Bioinformatics scientists who assist in the design of experiments for research programmes both large and small rely on the work of the primary resource annotators. These annotators are proficient at reviewing the literature related to the part of the genome they are annotating, and running and interpreting the results of sequence alignment software. A second group of bioinformaticians, who are skilled at using the many tools of bioinformatics, assist in the selection and generation of oligonucleotides (e.g. polymerase chain reaction [PCR] or sequencing primers) for generating genetic engineering designs. These designs are key to the execution of experimental work in the laboratory. They use the annotation of genomic sequences by focusing on individual genes to support experimental design.

As the scope of the research programmes becomes more ambitious, higher throughput systems are put in place, including laboratory automation and high throughput techniques based on multiwell plates (e.g. 96-well and 384-well plates). The amount of material generated by these high throughput processes demands ever higher capacity in sequencing capability to confirm what has happened in these experiments (known as *genotyping*). Thus there are additional bioinformatics roles involved in putting together pipelines of bioinformatics tools (aligners, quantifiers, visualisations, further database storage of results, laboratory information management systems [LIMSs], reporting systems and so on).

Additional roles encompass the analysis involved in the related areas of *transcriptomics* (the study of the complete transcript set from cells) and *epigenomics* (the study of all the *epigenetic* modifications to the cell's genetic material, including methylation of DNA and histone modification). These roles may involve assessing which genes in a whole genome are affected when a cell is subjected to some stress, like a drug or heat or light. Again, annotation is key to accurately map the transcripts back to the correct gene or non-coding RNA. Epigenomics and transcriptomics data also drive a more accurate annotation of the genome.

Therefore, some bioinformatics scientists will be intimately involved in experimental design and, indeed, many laboratory-based scientists will have significant bioinformatics skills in their area of expertise. Even now, most experiments contain at least some bioinformatics analysis. The sheer bulk of data means that it is often computational analysis that leads the experimental hypotheses, or at least allows for more rapid removal of hypotheses that are wrong. Other bioinformatics scientists will be much farther removed from the molecular biology in the laboratory. Some will need advanced computational and software development skills; others may be able to rely on publicly available or commercial bioinformatics tools to support them in their experimental design or analytical work.

Different business needs will also define the role of the bioinformatics scientist. In a pharmaceutical preclinical research team, bioinformatics may be used to understand the evolutionary relationships between members of a family of genes. This helps determine whether a specific gene is being targeted by a chemical compound (e.g. a drug) or whether the compound may affect an unintended but closely related member of the protein family. Bioinformatics may be used in the following ways (among others):

- Curate in-house databases of drug targets (generally proteins, which are normally enzymes catalysing key reactions).
- Select antigenic sites for potential antibody-based therapies.
- Design gene editing protocols in a precision genome engineering context to support a drug discovery programme.

Some research contexts may require extensive knowledge of structural biology and protein DNA complexes as well as complexes involving RNA, other proteins or small molecules. This area pushes more into the disciplines of biophysics and computational chemistry, which are both close scientific neighbours of bioinformatics. Images and the metadata that are assigned to them by technicians and clinicians are important in the context of clinical bioinformatics.

Whatever the context of the bioinformatics, there are some fundamental theoretical concepts that should be understood by all scientists as being characteristic of bioinformatics as a science.

3.2 Theory/Principles

3.2.1 Molecular Biology

Whatever the type of bioinformatics role and whatever the scale of data to be analysed, all must understand the fundamental molecular biology that underpins the biology and the generation of biological data. Core to this understanding is the *central dogma of molecular biology*: 'DNA makes RNA makes protein' (a useful, if oversimplistic, memory phrase), but more formally 'DNA is *transcribed* to produce RNA which is *translated* to produce protein'. Fundamentally, the central dogma is about the flow of information from DNA, via RNA to the protein product (as stated by Francis Crick in 1956).

From the point of view of bioinformatics, there is a one-way flow of information from DNA to DNA, from DNA to RNA and from RNA to protein. There is no flow of information from protein back to RNA or DNA. Proteins can and do interact with DNA and RNA but there is no information transfer. The information we are describing here is the sequence of DNA, segments of which are transcribed to produce an RNA template. This is then translated three bases (a *codon*) at a time into the *amino acids* that form the single polypeptide chain of a protein. Knowing the sequence of bases in a stretch of DNA (or RNA) allows us to predict the resulting protein sequence. Knowing a protein sequence does not allow us to predict with any certainty the sequence of bases that generated that protein. This is because the *genetic code* is a redundant code with multiple codons for the same amino acid residue. There are only two exceptions: there is only one codon for methionine (M), which has a special role as the first residue in a nascent polypeptide sequence in RNA translation, and only one codon for tryptophan (W). These two amino acids are generally the least commonly observed residues, although, in individual protein sequences, biophysical requirements may mean that other residues have a lower relative abundance. In fact, there is a large degree of variability in the total abundance of amino acids in a protein. Cysteine is a specific example of an amino acid that is more commonly observed in the shorter sequences of the polypeptide hormones. Here, the strong covalent bonding of the disulphide bridge is necessary to stabilise the structure of the smaller molecule in an extracellular environment. This is one of the major reasons for the existence of disulphides. In genomic bioinformatics, we are always mindful that the workhorses of the cell are proteins.

You should use reliable resources from the reference list at the end of this chapter to review the following topics:

- The structure of DNA
- The four DNA bases (A, T, C, G)
- Complementary base-pairing in DNA
- The four bases of RNA (A, U, C, G) and their base-pairing
- The production of messenger RNA (mRNA) from the DNA template in the genome
- The production of the primary protein sequence (consisting of the 20 naturally occurring amino acids: A, C, D, E, F, G, H, I, K, L, M, N, P, Q, R, S, T, V, W, Y)
- The physical characteristics of the amino acid residues (see Table 3.1).

Note that generally in bioinformatics the one-letter codes for the amino acid residues are used. See Table 3.1 for the one letter, three letter and full names grouped according to physical characteristics along with their codons.

Fundamental to the skill of bioinformatics is the ability to handle sequence data either for an individual gene or part of a gene and its protein product. Therefore, we need to be able to convert information at DNA or RNA levels to protein sequence efficiently. We can then work on multiple genes, or indeed whole genomes, effectively. This process is known as three-frame translation (or six-frame if both strand directions are taken

Table 3.1 The one-letter, three-letter and full names of the amino acid residues.

One-letter code	Three-letter code	Name	No. of codons	1st base	2nd base	3rd base	Properties
F	Phe	Phenylalanine	2	T	**T**	T, C	Aromatic, hydrophobic
L	Leu	Leucine	6	T C	**T** **T**	A, G T, C, A, G	Aliphatic, hydrophobic
I	Ile	Isoleucine	3	A	**T**	T, C, A	Aliphatic, hydrophobic
M	Met	Methionine	1	A	**T**	G	Hydrophobic
V	Val	Valine	4	G	**T**	T, C, A, G	Aliphatic, hydrophobic
S	Ser	Serine	4	T	**C**	T, C, A, G	Ambivalent
P	Pro	Proline	4	C	**C**	T, C, A, G	Hydrophobic, small
T	Thr	Threonine	4	A	**C**	T, C, A, G	Ambivalent, tiny
A	Ala	Alanine	4	G	**C**	T, C, A, G	Ambivalent, tiny
Y	Tyr	Tyrosine	2	T	**A**	T, C	Ambivalent, polar
H	His	Histidine	2	C	**A**	T, C	Basic, hydrophilic
Q	Gln	Glutamine	2	C	**A**	A, G	Ambivalent, polar
N	Asn	Asparagine	2	A	**A**	T, C	Ambivalent, polar
K	Lys	Lysine	2	A	**A**	A, G	Basic, hydrophilic
D	Asp	Aspartate	2	G	**A**	T, C	Acidic, hydrophilic
E	Glu	Glutamate	2	G	**A**	A, G	Acidic, hydrophilic
C	Cys	Cysteine	2	T	**G**	T, C	Ambivalent, small
W	Trp	Tryptophan	1	T	**G**	G	Hydrophobic, Large
R	Arg	Arginine	6	C A	**G** **G**	T, C,A, G A, G	Basic, hydrophilic
S	Ser	Serine	2	A	**G**	T, C	Ambivalent, tiny
G	Gly	Glycine	4	G	**G**	T, C, A, G	Ambivalent, smallest
*	—	Stop	1	T	**G**	A	(See legend)

The codons for the residues can be deduced by scanning across the row for the residue and writing down the first and second bases shown and then adding the base(s) from the third base column to each pair. Therefore, for Arginine, if the first base is A, the second is G followed by either A or G (AGA, AGG). Note that it is the second base that is most critical in determining the properties of the amino acid. The third base can vary considerably, while often still resulting in the same residue. TGA was found in 2007 to code for Sec (selenocysteine) but requires a cofactor to be present. TAG can likewise code for pyrrolysine. Premature stop codons (TGA) cause shortened forms of polypeptides to be synthesised and are hence known as nonsense codons. The properties column is derived from the classification in [4]. * is the one letter amino acid code for the TGA codon.

into account). The longest *open reading frame* (ORF) – that is, the longest sequence of residues translated without encountering a *stop codon* – is generally taken to be the standard protein product of the nucleic acid sequence. See Figure 3.1 for an illustration of the translation tracks turned on in the Ensembl genome browser and Figure 3.2 for a closeup zoomed in detail in which the translations in all six reading frames are made clear.

3.2.2 Gene Structure

Different genomes have different levels of complexity. The human genome is perhaps surprisingly not the most complex genome we know. It has a relatively low complement of genes (around 23 000) compared to some other organisms. It is the complex structure of human genes that makes for a combinatorial explosion

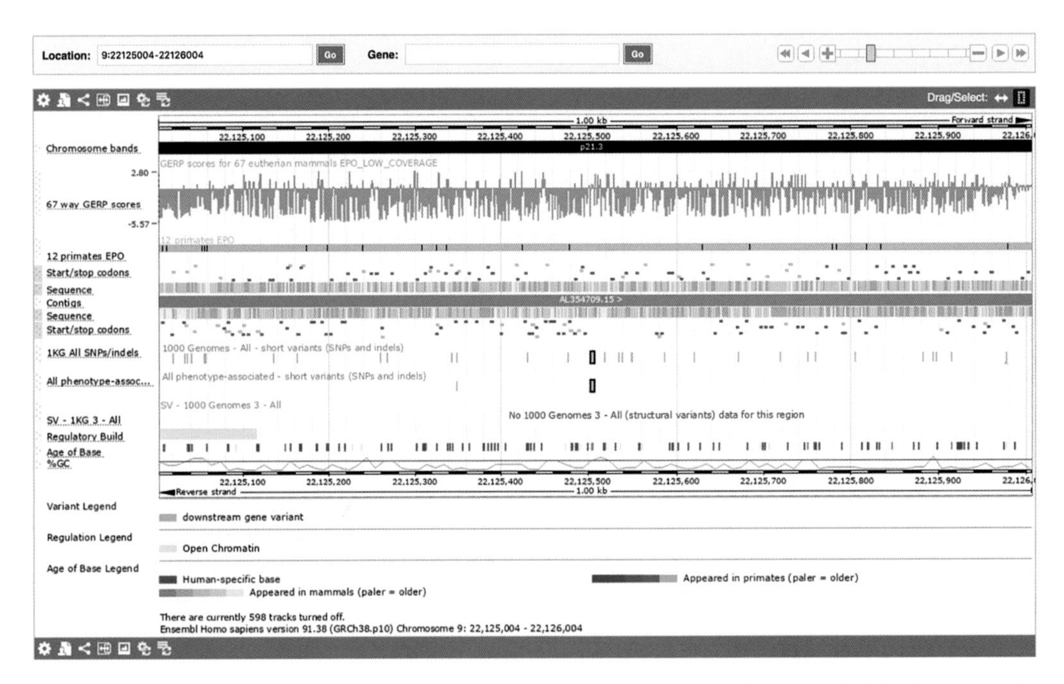

Figure 3.1 An example of a genomic region visualised in the Ensembl genome browser. The blue bar across the middle of the view gives the contig ID and location. A contig is a fundamental unit of assembled sequence derived by aligning individual sequence reads. The specific assembly technique used depends on the underlying sequencing technology: typically, Sanger sequencing, or next generation sequencing (NGS). The coloured display immediately above and below the blue line indicates the colour coded sequence of the forward and reverse strands of the DNA.

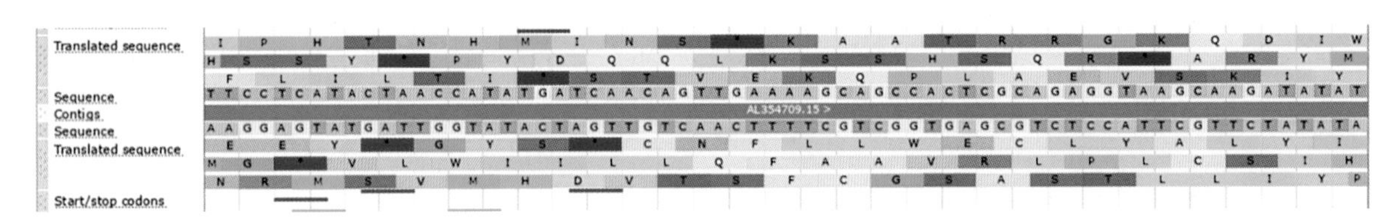

Figure 3.2 A zoomed-in section from Figure 3.1. When the zoom level changes, additional tracks become active that were not apparent in Figure 3.1, including the three possible translations on the forward strand shown above the contig bar and the three translations on the reverse strand below it. The colour coding of the DNA bases and the amino acids in the diagram can be a useful visual aid, or it can be turned off. Many different colouring schemes have been proposed and the selection of specific schemes can be helpful in certain analyses.

of gene and protein products. The processes involved result from regulation of gene expression and splicing of exons and introns. The memory phrases 'exons are expressed' and 'introns interrupt' can be helpful. Introns are eventually spliced out by the editing machinery of the cell. Many human genes, for example, have multiple splice variants that are made up by editing together specific exons of one particular gene. However, the most complex example of splice variation known is the gene DSCAM in *Drosophila melanogaster*. With 38 000 splice forms, this single gene has more splice variants than there are genes in the entire *D. melanogaster* genome. This can make analysis a complex process. Not all splice variants are known and many are tissue, or even cell specific – that is, they are only expressed in certain cells and at certain stages of the *cell cycle* (or upon cell–cell signalling or stresses). When conducting an analysis, we may refer to the *canonical sequence*. This is a sequence that reflects the most commonly observed base at a position.

3.2.3 Software Development Note

Bioinformatics scientists involved in programming applications for analysis should note that application programming interfaces (APIs) are available to assist in gathering the data required for conducting a bioinformatics analysis. These can be used in scripts and programs written in scripting languages like Perl and Python. For example, in the ENSEMBL core API, the canonical sequence refers to a commonly observed arrangement of exons in a particular transcript. Other transcripts may be observed, or might be predicted to occur based on evidence of splicing observed in sequencing projects. The API calls can be used to determine which transcript is most appropriate to use. Discussion with domain specialist scientists may also be necessary at this point in a project.

3.3 Databases

In Section 3.1 we talked about databases without giving a formal definition. *A database is an organised or structured collection of data held in a computer*. The structured nature of the database means that it will be made up of entries consisting of the same overall form but where the detail differs between entries. A simple database format might consist of records with 'internal_id', 'gene_symbol', 'description' and 'sequence' attributes. Attributes are sometimes also known as 'fields' or 'keywords'. There may be hundreds of entries using these keywords as *field tags* with a separator (perhaps a colon) followed by the data. Alternatively, the format can be specified separately and the data would then be expected to follow this format for each record. The format for a gene in the European Nucleotide Archive (ENA) is more complicated because it deals with complex data relationships and cross-references and has evolved over time. The opening of an entry is shown in Figure 3.3 and the conclusion of the entry – including the cDNA sequence – in Figure 3.4. Note that, in molecular biology, cDNA is DNA synthesised from an mRNA transcript using an enzyme termed 'reverse transcriptase' (RT). In bioinformatics, cDNA is used to refer to an mRNA transcript's sequence, expressed as DNA bases (GCAT) rather than RNA bases (GCAU). In this database entry, observe that the sequence is in lower case letters and, even though the entry is mRNA, it uses the alphabet of DNA ('agct') rather than RNA ('agcu').

Sequence databases are held in a compressed flat file format (meaning a human readable text-based format) with keywords indicating the different fields within an entry. The compression saves space on file servers and reduces transfer time using file transfer systems, such as 'ftp', which is a program that can be used to transfer files between computers using the file transfer protocol (FTP). This type of flat file can be used to populate a database management system (DBMS). The role of the DBMS is to ensure integrity of the data (the data only changes when we want it to) and efficiency of access to the data in the database from programs (such as are implemented by web-based query systems and genome browsers). Logical components of DBMSs are

```
ID   AF151109; SV 2; linear; mRNA; STD; HUM; 642 BP.
XX
AC   AF151109;
XX
DT   02-NOV-1999 (Rel. 61, Created)
DT   08-FEB-2018 (Rel. 135, Last updated, Version 4)
XX
DE   Homo sapiens putative BRCA1-interacting protein (BRIP1) mRNA, partial cds.
XX
KW   .
XX
OS   Homo sapiens (human)
OC   Eukaryota; Metazoa; Chordata; Craniata; Vertebrata; Euteleostomi; Mammalia;
OC   Eutheria; Euarchontoglires; Primates; Haplorrhini; Catarrhini; Hominidae;
OC   Homo.
XX
RN   [1]
RP   1-642
RA   Wang Q., Zhang H., Greene M.I.;
RT   "BRIP1, a candidate for BRCA1-interacting protein";
RL   Unpublished.
XX
RN   [2]
RP   1-642
RA   Wang Q., Zhang H., Greene M.I.;
RT   ;
RL   Submitted (13-MAY-1999) to the INSDC.
RL   Pathology, University of Pennsylvania, 252 John Morgan Building, 3600
RL   Hamilton Walk, Philadelphia, PA 19104, USA
XX
RN   [3]
RC   Sequence update by database staff to remove vector contamination
RP   1-642
RA   Wang Q., Zhang H., Greene M.I.;
RT   ;
RL   Submitted (06-FEB-2018) to the INSDC.
RL   Pathology, University of Pennsylvania, 252 John Morgan Building, 3600
RL   Hamilton Walk, Philadelphia, PA 19104, USA
XX
```

Figure 3.3 Extract of an entry from the European Nucleotide Archive (ENA) illustrating its structure. Here the individual records of the entry are tagged with two-letter codes: 'XX' indicates a blank line and 'ID' is the first line and follows a specific format of its own with fields separated by semicolons.

indicated in Figure 3.5. See Figure 3.6 for an example of part of a relational database schema visualisation that indicates relationships between tables and attributes in a LIMS.

Database formats are specific to the type of data being stored. Use online resources to explore a number of different types including a protein sequence, DNA sequence, protein structure and protein families. This will help to gain an understanding of the breadth of data that is being managed in these databases.

3.3.1 Accessing and Using Data

Data are made available using the Internet and are typically accessed using various web browsers from many providers. Much of the data used in bioinformatics is available publicly and freely via file download sites. These are typically FTP servers, but as file sizes get ever larger alternative methods of distribution are sometimes preferred. Cloud-based services from a number of vendors are available that provide a flexible compute capability alongside databases. Data can thus be analysed in one place without having to replicate it across multiple sites. There are, of course, costs associated with computing in the cloud as there are in bringing data in-house and providing computational facilities to analyse it.

```
CC   On Feb 6, 2018 this sequence version replaced AF151109.1.
XX
FH   Key             Location/Qualifiers
FH
FT   source          1..642
FT                   /organism="Homo sapiens"
FT                   /mol_type="mRNA"
FT                   /cell_line="HeLa"
FT                   /db_xref="taxon:9606"
FT   gene            <1..642
FT                   /gene="BRIP1"
FT   CDS             <1..346
FT                   /codon_start=2
FT                   /gene="BRIP1"
FT                   /product="putative BRCA1-interacting protein"
FT                   /db_xref="GOA:Q9P0J6"
FT                   /db_xref="H-InvDB:HIT000071591.13"
FT                   /db_xref="HGNC:HGNC:14490"
FT                   /db_xref="InterPro:IPR000473"
FT                   /db_xref="InterPro:IPR035977"
FT                   /db_xref="PDB:3J7Y"
FT                   /db_xref="PDB:3J9M"
FT                   /db_xref="PDB:5OOL"
FT                   /db_xref="UniProtKB/Swiss-Prot:Q9P0J6"
FT                   /protein_id="AAF04788.2"
FT                   /translation="RVRACGRIHHNMANLFIRKMVNPLLYLSRHTVKPRALSTFLFGSI
FT                   RSAAPVAVEPGAAVRSLLSPGLLPHLLPALGFKNKTVLKKRCKDCYLVKRRGRWYVYCK
FT                   THPRHKQRQM"
XX
SQ   Sequence 642 BP; 216 A; 153 C; 135 G; 138 T; 0 other;
     ccgggtgaga gcgtgcggcc ggattcacca caacatggca aatcttttta taaggaaaat        60
     ggtgaaccct ctgctctatc tcagtcgtca cacggtgaag cctcgagccc tctccacatt       120
     tctatttgga tccattcgaa gtgcagcccc cgtggctgtg gaacccgggg cagcagtgcg       180
     ctcacttctc tcacccggcc tcctgcccca tctgctgcct gcgctggggt tcaaaaacaa       240
     gactgtcctt aagaagcgct gcaaggactg ttacctggtg aagaggcggg gtcggtggta       300
     cgtctactgt aaaacccatc cgaggcacaa gcagagacag atgtagaccc tttccctcca       360
     gagtcacgca catactcgtc atcgcatcac ttgggagaat ggttgtatct tatggaagga       420
     attatcacat caaggagtca ggggaaagtg actggaagca aacgccctaa aagttaccca       480
     tcacgtttca gtgtaaatga gtaactatag aagacattgc gttatcttat ttccaaaacg       540
     ttccaactaa aaaacatttt cctattaaaa tagaccttcc gaaaaaaaaa aaaaaaaaaa       600
     aaaaaaaaaa aaaaaaaaaa aaaaaaaaaa aaaaaaaaaa aa                          642
//
```

Figure 3.4 The ending of the ENA entry from Figure 3.3. There are many feature 'FT' records that are formatted to give detailed information on the origin of the sequence, organism, at which stage in the DNA→RNA→protein information flow we are. There are also several cross-references to other databases ('db_xref' fields). The sequence itself is presented in a singly tagged 'SQ' record. The entry ends with the '//' separator. The ENA database files can be downloaded and all the entries are separated in this way. This format is called a flat file, as opposed to the format it takes when loaded into a relational database management system (RDBMS) such as that shown in Figure 3.5.

You should be familiar with at least one web-based genome browser. The browsers at www.ensembl.org and genome.ucsc.edu are good places to start exploring and we use the Ensembl genome browser to illustrate several techniques in the following section.

3.4 Techniques

The first technique we will look at is the use of a *genome browser* and a gene information database to understand something about the human BRCA1 gene.

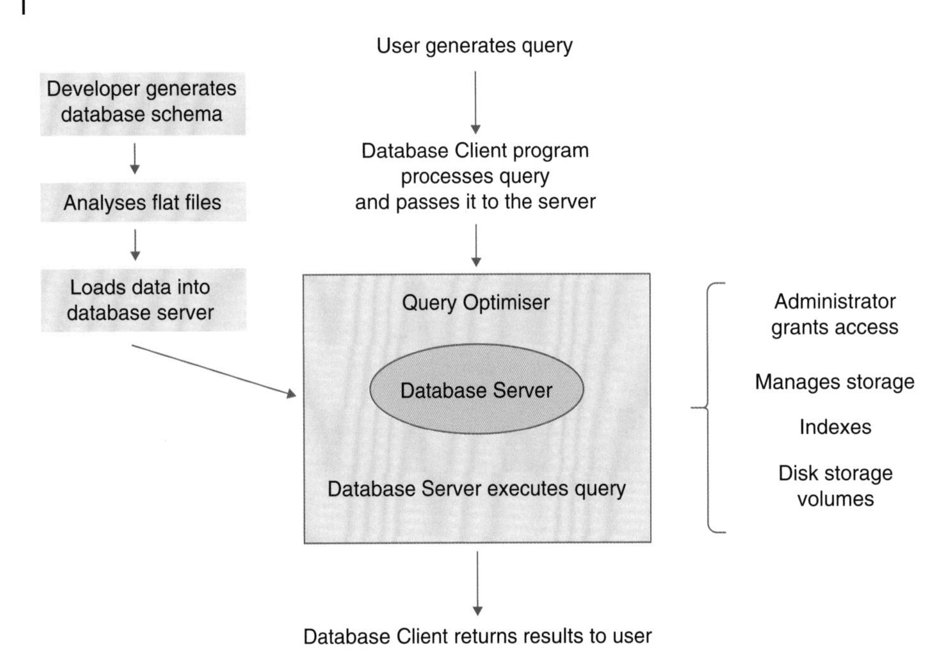

Figure 3.5 The main features of a database management system (DBMS) comprise the database server and associated administrative features that permit multiuser access to the system. Users access the system through a command interface or web-based interface. Developers have a programmatic interface that can be used to develop scripts that can access the database server to store and retrieve data. Relational database management systems (RDBMSs) use SQL (structured query language) to form queries that are processed by the database client software and passed to the database server for optimisation and execution. Flat files, such as those illustrated in Figures 3.3 and 3.4, can be parsed by specially developed scripts and loaded into the database using a schema specifically designed to hold the data and express the relationships between attributes. SQL provides a flexible tool for querying such data.

3.4.1 Genome Browsers

Here we will use the www.ensembl.org site, which provides a fully capable genome browser enabling users to explore the publicly available data for specific genomes. Often bioinformaticians will work on a single gene or family of genes. Scientific colleagues working on a project will already have the specific code to look up for a gene. As a bioinformatician, you will want to become familiar with the background to the project, which suggests a wider search. Assuming that the project is researching breast cancer genes, we would probably first use a trusted web search engine to get a general impression of the work going on in the general area of 'breast cancer genes'. This may seem obvious and unnecessary but the effort that large companies put into development of search engines can shortcut a lot of hard bioinformatics work as new resources are coming online all the time.

Next, use a tool like the Open Targets Platform (which is freely accessible at www.targetvalidation.org) to perform a more targeted search resulting in the output shown in Figure 3.7.

The first thing to be aware of is that gene symbols change over time. What we currently call BRCA1 has synonyms in the scientific literature. These synonyms are listed in the Ensembl page for the human BRCA1 gene. The Ensembl page (a portion of which is shown in Figure 3.8 and a part of the scrollable interactive view in Figure 3.9) is a very functional and beautiful representation of a great many pieces of data brought together to provide a rich environment for genomic exploration. Note the number of transcripts and the link to splice variants. This is a complicated gene, which has been observed in 30 different *isoforms*. There are 97 versions of this gene in other organisms. These are known as *orthologues* – genes that have evolved

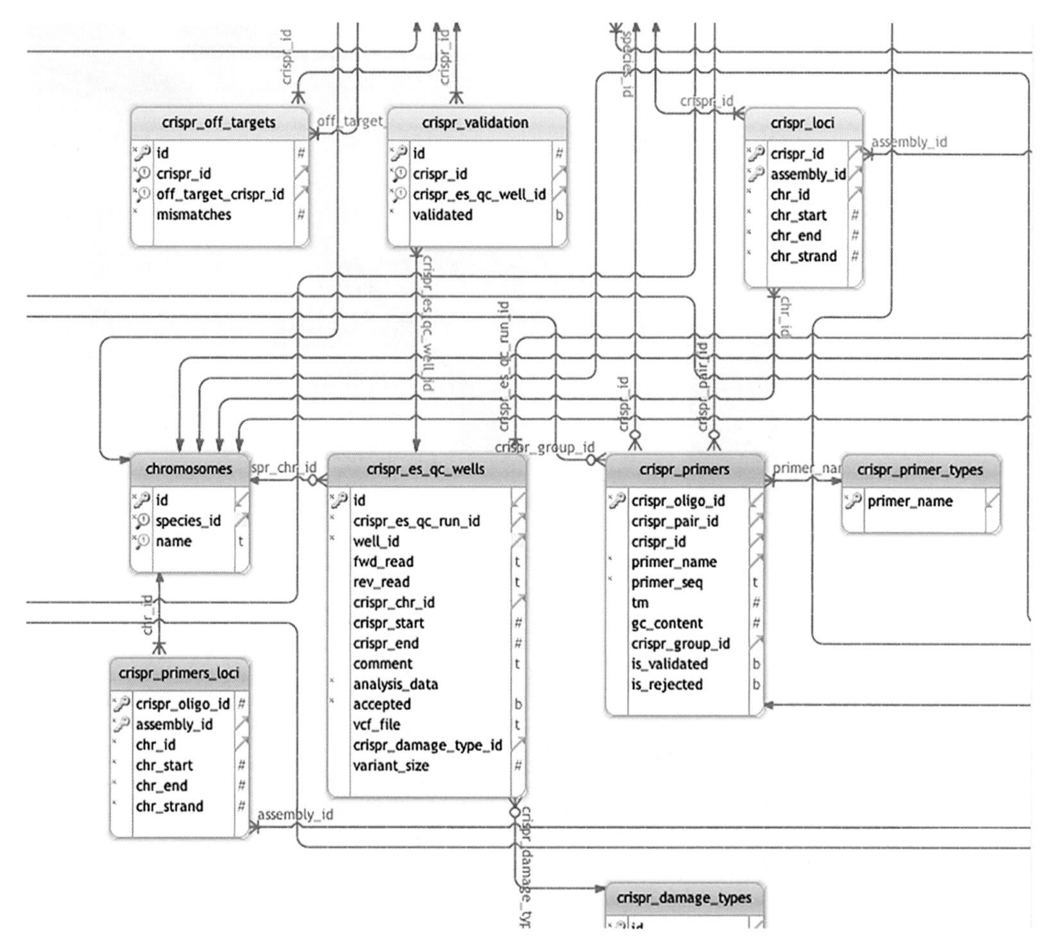

Figure 3.6 A visualisation of part of the schema of a database (actually a laboratory information system – another example of a bioinformatics application). The rectangles represent tables in the database, which is an SQL database using PostgreSQL (www.postgresql.org) as the database engine. No data are shown here as this is the metadata (data that describes data) that defines the logical relationships between attributes of the data. The lines indicate relationships explicitly defined by foreign keys (attributes in one table that point to and depend upon attributes in another table). It is these relationships that enable complex processes to be modelled in the database and make for effective storage and querying of the data.

from a common ancestral gene over time as populations also evolved to become distinct species (termed speciation).

3.4.2 Sequence Comparison

We compare sequences (whether DNA, cDNA, RNA or protein) on a regular basis. We do this to discover common regions, understand overall structures, predict functionality of unknown regions from similarity with known ones, understand evolutionary relationships, etc. *Sequence alignments* using sequences of a similar length and showing a high degree of sequence similarity are easier to interpret. Given a sequence to investigate, we would run a database search (normally using one of the Basic Local Alignment Search Tool (BLAST) suite of programs, viz. www.ensembl.org/Homo_sapiens/Tools/Blast) to find related sequences. Comparing a shorter sequence (perhaps of a few hundred bases) with a much longer sequence – say a chromosome millions of bases long – is a different task and uses specially configured database search tools.

Figure 3.7 A page from the Open Targets Platform (www.targetvalidation.org) showing the results of a search that started out as 'breast cancer genes' and led to 'hereditary breast cancer'. The top two gene target symbols are BRCA2 and BRCA1. This gives a useful way into assessing the extent of information available and then the analyst can drill down further and explore additional resources.

Whichever tools we use, the concepts of *local and global alignment* (see Section 3.2.3) become important. In genomic bioinformatics, we find ourselves comparing DNA with DNA much of the time. In other areas of bioinformatics, we may use protein/protein comparisons more. The basic rules for comparing sequences are the same in either case, but the alphabet for proteins is more extensive than for DNA. DNA comparison programs tend to be optimised for finding short stretches of identical sequence and then stringing these regions together. Protein level comparisons will normally use tables of similarity scores ('scoring matrices') to assess regions of similarity and scoring priority.

3.4.2.1 Similarity and Homology

When considering lists of hits and the relevance of results, it is important to take into consideration both the statistical significance (denoted by a score in the BLAST output) and also the biological context in which the result will be used. This is why multiple sequence alignment and some understanding of sequence similarity are important tools for the bioinformatician. It is easy to jump to conclusions about family relationships (*homology*) when parts of two sequences happen to share some sequence similarity. Language is important here.

3.4.2.2 The Dotplot

A useful visualisation tool is the dotplot. Here, two sequences (either DNA or protein) are laid out one on the X-axis and the other on the Y-axis of Cartesian space. We end up with a rectangle of cells that are marked if the XY positions have identical bases/residues. Thus we assess each cell in turn and visualise the result. More sophisticated algorithms introduce a sliding window, with a number of positions contributing to the score at

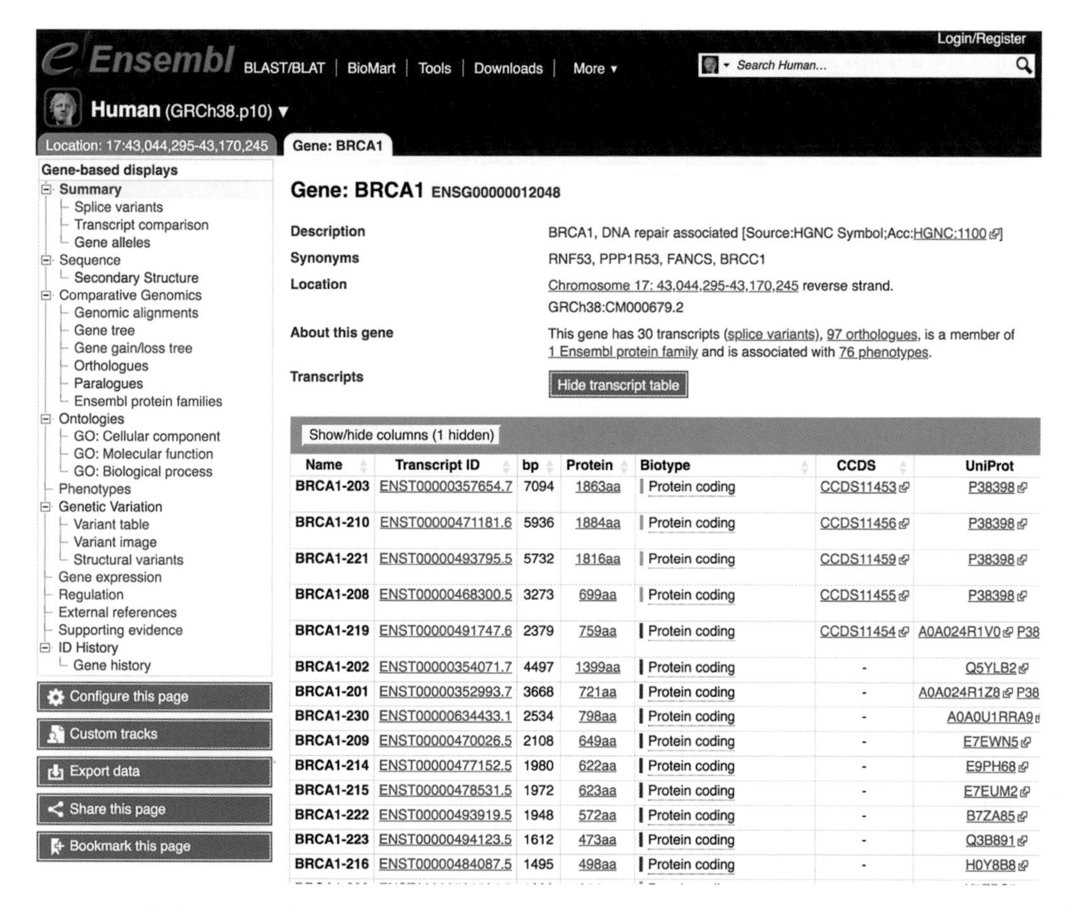

Figure 3.8 The beginning of the Ensembl genome browser page for the human BRCA1 gene. Note the navigation aid at the left of the page giving access to more detail. Of particular interest are the number of splice variants observed for this gene. There are four synonyms in the literature. The Ensembl accession code for this entry is ENSG000000012048. Databases assign their own accession codes.

any particular cell. A self/self sequence comparison is shown in Figure 3.10. Dotplots are very useful when reviewing sequence data at the genomic level following WGS as they clearly indicate *structural variants* (see Figure 3.11).

3.4.2.3 Local and Global Alignment

When the overall, global alignment quality between two sequences is important the Needleman–Wunsch algorithm is used. It uses a dynamic programming technique to find its way through a matrix – much like the dotplot matrix. This technique tries to find the best alignment for all the residues in both sequences. However, often it is the local alignment of important features that we are most concerned with and the Smith–Waterman alignment technique is used. For database searching, it is necessary to optimise the implementation of the local alignment search in order to get results back to the user in a reasonable time. This requirement led to the development of BLAST, which remains the most popular means of searching a database with a query sequence (see the detailed discussion in [5]).

3.4.2.4 Multiple Sequence Alignment

Often we want to align several sequences that may be related to each other structurally or functionally. The multiple sequence alignment (using automated tools such as Clustal Omega [6]), often supported by some

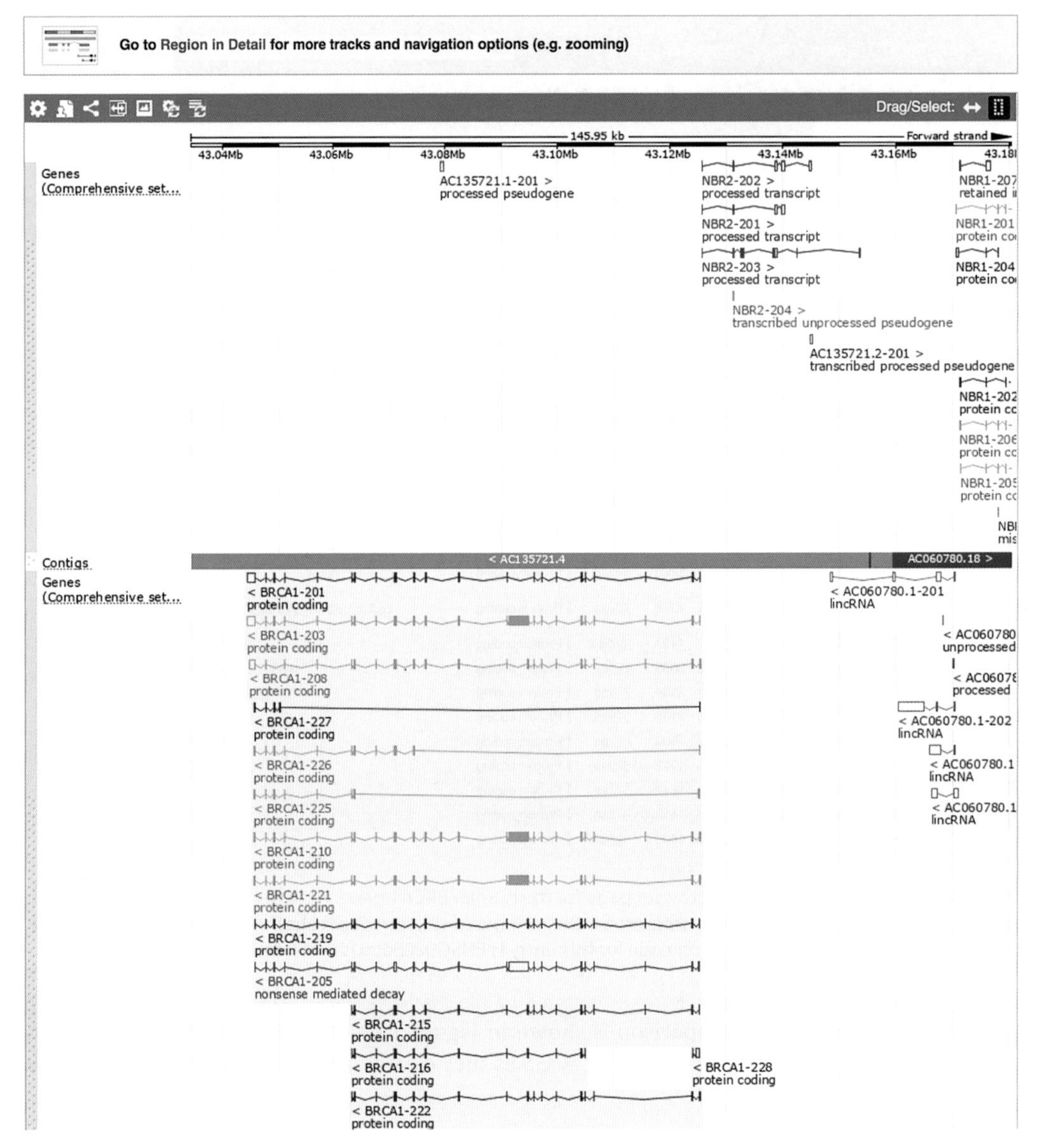

Figure 3.9 The section of the BRCA1 entry in Ensembl that displays an interactive scrollable and clickable browser for homing in on specific regions of the genome around the BRCA1 gene and its transcripts. The display in the lower section of the figure shows the structure of introns and exons (defined in the main text) that make up many of the BRCA1 splice variants. This gene is on the minus strand, which is why it is shown below the blue dividing bar. Plus strand genes would appear above the blue bar.

form of knowledge-led editing in an interactive system (e.g. Cinema [7]), is an important technique. This approach relates more than two sequences to one another. Ultimately it leads to development of discriminators and profiles for seeking out and classifying sequences, see InterPro for an integrated resource [8]. Multiple sequence alignment is an important part of the process of phylogenetic analysis, which helps to chart the evolutionary history of sequences and their functions and conserved features. In DNA sequence analysis using short read techniques, multiple sequence alignment has its equivalent in the aligned read file (VCF – see Section 3.4.2.5).

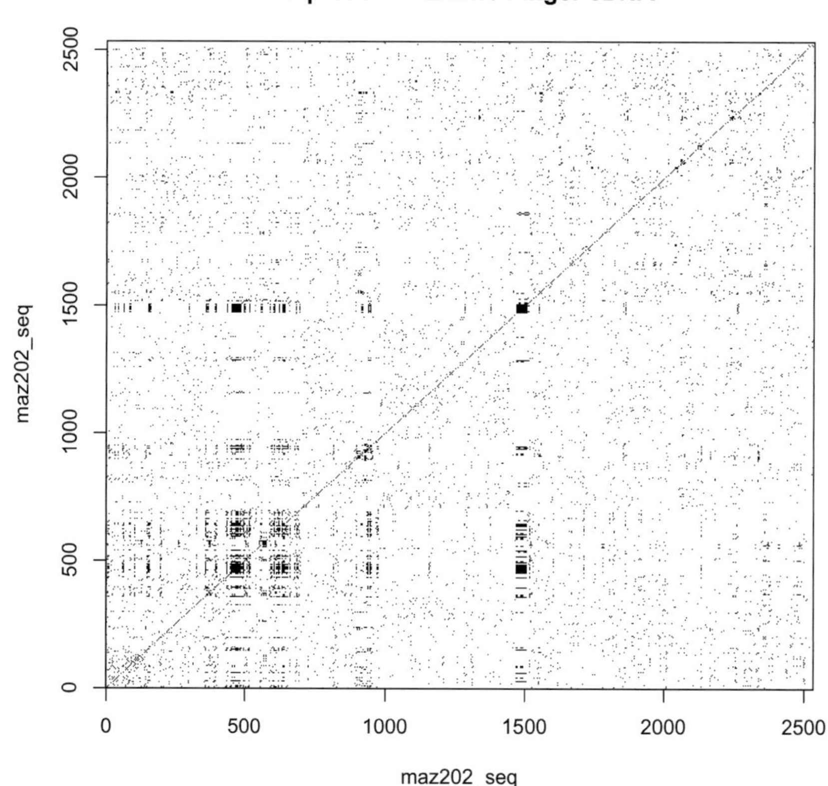

Figure 3.10 Dotplot of the DNA sequence of the Human Zinc finger transcription factor. This is a self/self comparison so the dotplot is square with the major diagonal (bottom left to top right) being completed because each base is identical with its counterpart in the other sequence. Regions of similarity within the sequence are identifiable by clusters of dots forming defined dark areas off the major diagonal. The plot is symmetrical, the upper left triangle being identical to the lower right triangle. If two different sequences were being used, there would be no such symmetry. This plot was generated by the author using the R package 'seqinr', which provides the dotplot function and allows variation of the window length (set to 5) and the number of matches per window (4) for a dot to show up in the plot. Window averaging is required for DNA sequences because of the inherently poor signal to noise ratio in DNA sequences. This is because of the four letter alphabet of DNA. The accession code analysed is: ENST00000322945.10 MAZ-202 cdna:protein_coding.

3.4.2.5 Short Read Aligners – Variant Call Format (VCF)

Short sequences generated by DNA sequencing systems are known as *short reads*. When these short reads are generated, they are made available in a number of formats. A common format is known as the *fastq* format and is often generated by sequencing systems. This is a basic sequence format in which the sequence is listed on one line and an associated line of the same length is used to denote the quality of the base call at each point in the sequence. The sequence is called from the reads by *base calling software*. Several base callers exist and parameters can be set to determine how many calls of a particular base at a specific location are required to gain a specific quality score. The variant call format is a file format that enables aligned reads to be summarised and interchanged between different programs for assembly and processing.

Sequence files generated by next generation sequencing (NGS) systems can be very large. Fastq is not a particularly efficient means of storing the called data. The Samtools project (samtools.sourceforge.net) provides

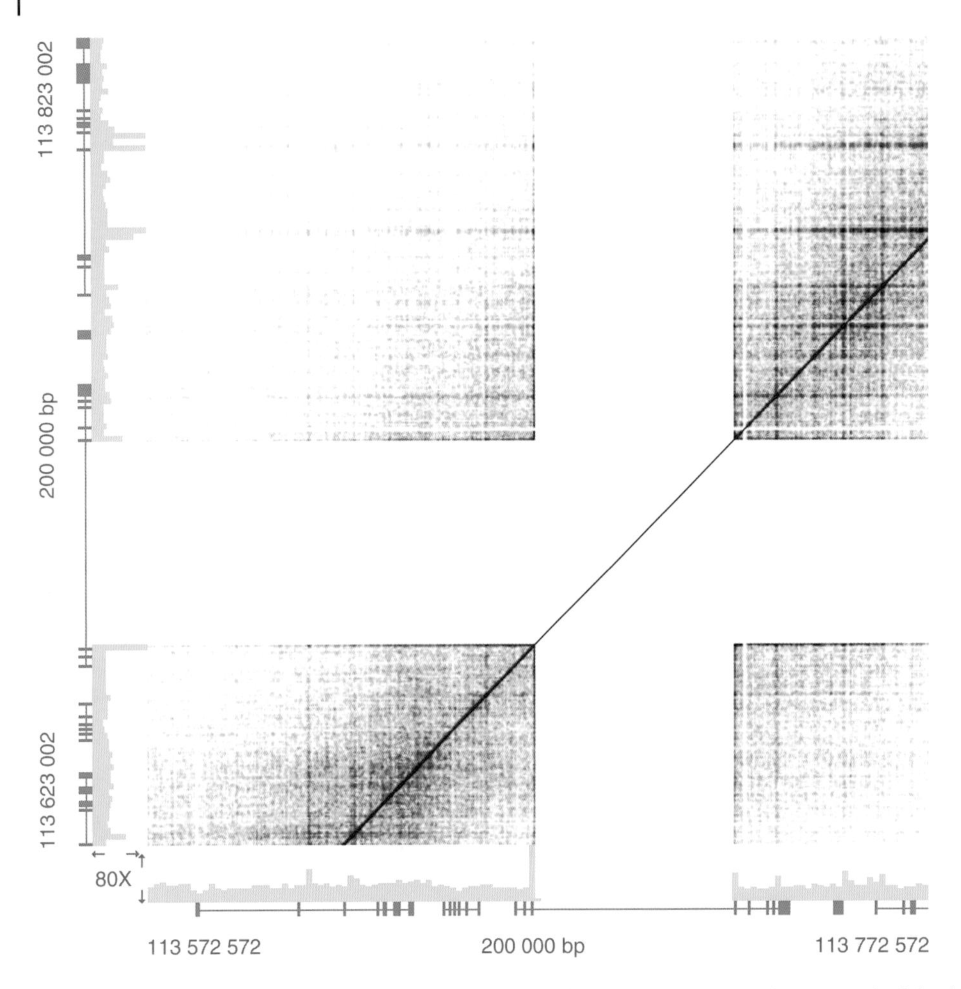

Figure 3.11 A matrix view (dotplot) of a structural variant from a chromium 10X dataset in which both haplotypes have been sequenced. There is a deletion on one of the alleles – hence the large amount of white space off the diagonal. This technology gives longer reads than the standard NGS approach through the concept of linked-reads (see text). The tools for analysing the data (10× Longranger) are provided by 10× Genomics (www.10xgenomics.com). This emphasises the power of relatively simple tools (the dotplot) in bioinformatics analysis and visualisation.

many tools for compressing and converting between various DNA sequence formats. The SAM (sequence alignment/map) format is a generic format for storing large nucleotide sequence alignments [9].

3.4.2.6 Genome Aligners – BWA

The current raft of sequencing technologies are mainly based on sequencing many short sequences (~100 to 200 or possibly 300 bases) of DNA known as *short reads.* New technologies focused on generating much longer read lengths of individual DNA molecules are in progress, but current technology makes it quick and relatively cheap to generate sequencing libraries and sequence gigabases of short read DNA. The drawback is that the large amount of sequence is broken up into these very short reads, which must then be compiled into longer stretches. Alignment is a time-consuming process, especially when dealing with the billions of reads produced by the latest sequencing technologies. The Burrows Wheeler algorithm (BWA) is an indexing scheme that was developed for compressing data very efficiently. When dealing with billions of bases of a genomic sequence, compression becomes a useful paradigm in making analysis tractable. BigBWA [10] is an example of an implementation of BWA for genomic alignment from NGS data. It combines a technology called

Hadoop – an implementation of the MapReduce programming model for distributing a big data problem across available resources. In the map phase the reads are split into subsets that are processed using the BWA technique; in the reduce phase the multiple output files are reduced into one unique solution.

Short read lengths restrict the resolution of complex regions containing a highly repetitive sequence, heterozygous sequences (sequences that exhibit differences on both alleles of a diploid species), structural variations and genome assembly. The human and mouse genomes were sequenced originally using the much longer reads generated by the Sanger sequencing method, which produced *contig* sizes of 50–80 Mb (megabases). Short read contigs are typically an order of magnitude smaller at between 10 and 100 kb (kilobases). A contig can be thought of as a block of aligned sequence. When sequence lengths are short, it is likely that there will be a gap between some of the reads that cannot be spanned by another read. At this point the contig is broken and a new one is started. The ordering of contigs is unknown without a longer read to align and join them up. NGS technologies, however, provide *depth of coverage* as opposed to the average sequence read of Sanger sequencing. This means that the quality of bases in a read and contributing to a contig can be assessed more accurately.

3.4.2.7 Long Read Technologies

In the previous section we came up against the issue of short read NGS contigs that cannot be linked together to form a longer contig. A longer read can be used to link such contigs. Methods of generating longer contigs can be achieved by generating longer reads in the first place. The single molecule real time (SMRT) system can produce reads of 10 000 bases (cf. 100–300 bases for short read NGS). Aligning such reads to form contigs can help with *structural variation* and *haplotype* calling (see www.mlo-online.com/long-read-sequencing).

The alternative NGS method is to link together short reads based on *molecular barcodes* that indicate which long stretch of DNA a group of short reads originates from. The structural variation plot in Figure 3.11 shows the analysis based on a 10× chromium WGS dataset.

3.5 Applications

3.5.1 CRISPR/Cas9 and Off-Targets

Cas9 is an enzyme (a protein) that can induce a double-stranded break in DNA. It was discovered in the *Streptococcus pyogenes* bacterium where it assists the bacterium to remember that it has seen an invading DNA virus before. It does this by incorporating the invading viral DNA into its own genome and by using sequence complementarity to recognise and cut the DNA of the virus should it infect the bacterium again. It was subsequently discovered that the Cas9 protein was responsible for this sequence-specific recognition and that this could be reprogrammed to recognise desired genomic target sites. Importantly, it could also be transferred to work in mammalian cells to cut double-stranded DNA at a defined location. The era of *precision genomics* was moved on significantly.

The repair of the double-strand break by the cellular machinery is most frequently achieved by the non-homologous end joining (NHEJ) process. This process is somewhat error prone and leads to short insertion or deletion events at the cut site. Less frequently, a mechanism known as homology directed repair (HDR) is used by the cell. This relies on copying the correct sequence from a template DNA (naturally the sister chromatid). This can be exploited to introduce defined changes and insertions of longer stretches of DNA into the genome. The results of experiments directed at specific genes use bioinformatics techniques to design primers and to prioritise CRISPR sites. The experiment can be evaluated by NGS genotyping to find out what the effect of the DNA repair has been.

The Cas9 protein scans the genome and stutters over pairs of G bases. The protein can be programmed to latch on to a specific genomic sequence of around 20 bases ending in 'NGG' (where 'N' is any base). The programming tool is a 20-base guide RNA that is provided along with the Cas9 protein. There are about 300

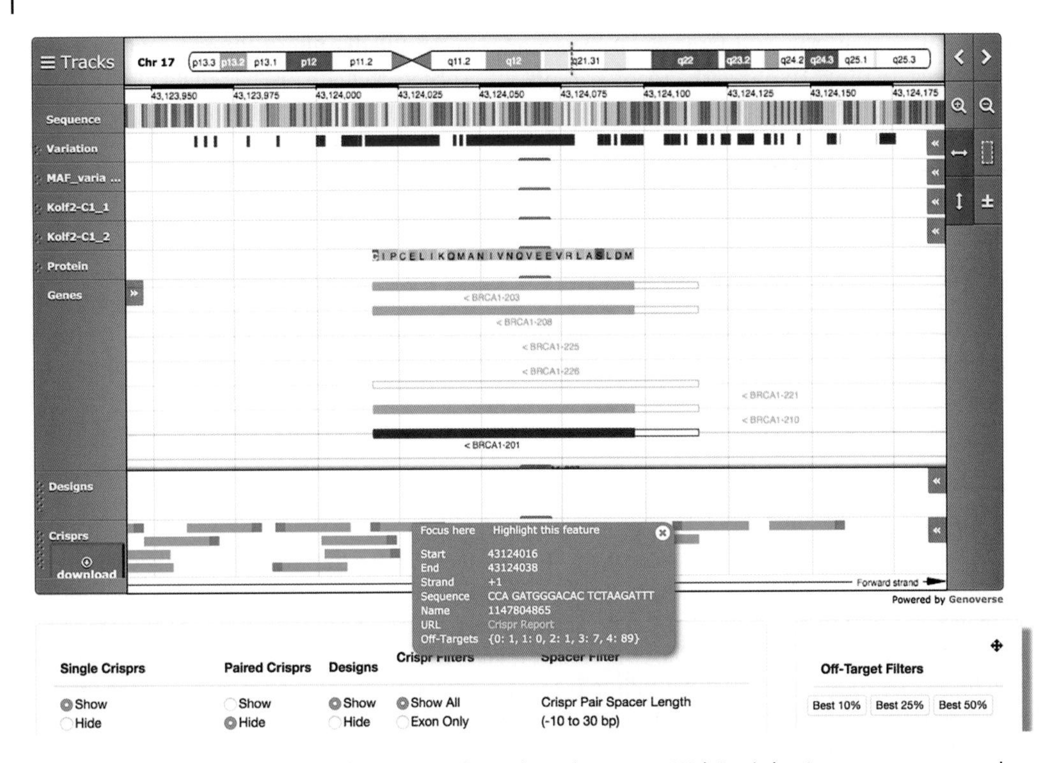

Figure 3.12 An example of a bioinformatics tool (a web application, or WebApp) that integrates a genome browser (wtsi-web.github.io/Genoverse) along with many other tools to present useful information to the scientists designing CRISPR/Cas9 precision genome editing experiments. The guide sequences (here called 'Crisprs') are shown in green in the Crispr track. The PAM (protospacer adjacent motif) site (NGG or CCN when the sequence is reversed and complemented on the other strand) is coloured in blue. The PAM site is not part of the Crispr guide but is shown in blue in the Crispr track for convenience. Whether the Crispr is PAM-right or PAM-left tells us whether we are dealing with NGG or CCN PAM sites. It is necessary because WGE stores all its Crispr location data on the plus strand of the DNA duplex. The blue pop-up window shows summary information for a Crispr object that has been clicked with the mouse. Several tracks show no data for this region.

million 'NGG' sites in the human genome. Furthermore, there can be off-target cutting effects where there are one or more mismatches between the RNA guide and its target.

Figure 3.12 illustrates the genome browser interface of the WGE Crispr design tool (www.sanger.ac.uk/htgt/wge). This tool is a bioinformatics tool that assists in the selection of sites for CRISPR/Cas9 experiments. It supports many different features, including interactive searching for CRISPR/Cas9 sites in human and mouse genomes.

The WGE implementation consists of three logical parts but users are normally only aware of the web application. This provides the context for searching and the visualisation of the search results. WGE also links to (i) a relational database (actually, PostgreSQL) that provides a persistent store of CRISPR and off-target information and (ii) a CRISPR-Analyser tool that runs on a separate server that provides fast indexed lookups to all the off-targets in the human and mouse genomes with up to 4 mismatches to the 20-base guide RNA query. The CRISPR-Analyser provides its results principally in the form of an *off-target string*. The off-target string shown in Figure 3.12 is {0: 1, 1: 0, 2: 2: 1, 3: 7, 4: 89}. This means that there is one sequence in the human genome that matches the CRISPR guide exactly (this is the query or *on-target sequence*). There are no hits with just 1 mismatch in the guide sequence; 1 hit with 2 mismatches, 7 hits with 3 mismatches and 89 hits with 4 mismatches. The off-target string is a helpful data structure that is used to help prioritise CRISPR guides for experimental design [11].

For the bioinformaticians, it is crucial to understand these basic mechanisms and to come up with computational solutions – either by programming new systems or by using tools written by others that can form part of a bioinformatics pipeline.

In researching, designing and implementing the WGE system many decisions had to be made about what exactly would constitute a 'Crispr' object in the system. Considerations of space for storage, rapidity of access to on-target and off-target data and summarisation of information all had to be carefully balanced to create a responsive system. Even the fine details of assigning IDs to Crisprs became critical (i.e. they cannot change once assigned) when a feed of Crispr locations was provided to the Ensembl genome browser.

The source code for WGE and the CRISPR-Analyser is available on GitHub at github.com/htgt/WGE and github.com/htgt/CRISPR-Analyser. The WGE code is written mainly in Perl and Javascript, and the CRISPR-Analyser in C++.

3.5.2 Drug Discovery and Target Discovery

Bioinformatics has contributed significantly to drug discovery (discovery of new medicines) since the early 1990s when sequencing programs in pharmaceutical companies were introduced to trawl short-read sequences, called expressed sequence tags (ESTs), for new variants of known genes. Drug targets are typically proteins that are amenable to having their function modulated by small molecule ligands. Such ligands are known as drugs (medicines). Now that we have more advanced sequencing capabilities, *transcriptomics* has become a major research endeavour. Projects underway are aimed at sequencing tissues (e.g. GTeX consortium) and single cells (e.g. HCA – Human Cell Atlas) from a wide variety of sources and mapping their transcripts to build up a profile of the genes that are expressed in individual cells across whole organisms.

The Human Cell Atlas (www.humancellatlas.org) is a major international collaboration in this area whose mission is 'to create comprehensive reference maps of all human cells – the fundamental units of life – as a basis for both understanding human health and diagnosing, monitoring, and treating disease'.

The GTeX Genotype Tissue Expression consortium (www.gtexportal.org) catalogues genetic variation and the influence of that variation on gene expression within and between all major tissues in the human body. By building up a database and associated tissue bank of all major tissues from 1000 individuals, bioinformaticians have a rich resource of genetic variation, gene expression, histological and clinical data to mine.

Bioinformatics techniques are intimately involved in discovering new drugs and repurposing known drugs for new uses. Bioinformatics has been used to help understand the mechanism of drug action and through *phylogenetics* and the elucidation of evolutionary pathways has been able to clarify disputed molecular mechanisms [12]. Often highly divergent species are involved in the analysis and new approaches for improving phylogenetic methods are applied in the context of *phylogenomics* [13] – i.e. phylogenetics applied at the genomic scale.

3.5.3 Transcriptomics

When we are interested in looking at the expression of the complement of genes in a cell, we need to sequence the transcripts that have been transcribed from the genomic DNA into messenger RNA. In drug discovery, understanding the genes that are actually transcribed (and later expressed) is critical to discovery of new therapeutic targets. Generally, small molecule drugs will be targeted at specific proteins, rather than DNA or RNA. Discovering all the proteins expressed by a cell is done using biophysical techniques, including advanced mass spectrometry and the tools of proteomics. It can be effective simply to look at mRNA transcripts, a technique that also provides the ability to count transcripts and to see the difference in transcription levels between cells in healthy and diseased states, or at different stages in development.

The current RNA-Seq technology that can be used to analyse RNA transcripts relies on transcribing the RNA transcripts back to DNA for sequencing, so it is in fact a DNA sequencing technique. This involves both copying (reverse transcription) and amplification (PCR) steps. From a bioinformatics perspective this creates problems because we have to assume that the reverse transcription and PCR processes have amplified the transcripts uniformly in generating the cDNA library, which may not always be the case. Transcript counting depends upon this assumption.

Newer native RNA-Seq technologies (for example, Oxford Nanopore Technologies) remove the necessity for reverse transcription to cDNA and enable single longer molecules of RNA to be sequenced. This has the added advantage of sequencing through splice sites and making alternatively spliced transcripts easier to count [14]. This ability to derive reliable transcript counts is important because experiments often involve comparing gene expression between different conditions, such as drug treated compared to non-drug treated.

Using these and associated genomics strategies, there is great potential to improve drug targeting and creation of diagnostic tools with increased sensitivity. The role of bioinformatics is central in optimising algorithms for alignment, quality metrics, standards for variant calling and interpretation as well as making the results available globally. This will ultimately foster a deeper understanding of disease and lead to greater therapeutic precision [15].

3.6 Concluding Remarks

Bioinformatics is a scientific discipline that requires understanding of molecular biology and computational information systems. Some facility with programming is valuable for bringing together tools into workflows or pipelines. However, it is also true that many bioinformaticians focus entirely on analysis and development of new approaches while delegating the coding aspects to those skilled in the art. There are many potential applications and we have focused on computational genomics here along with a small selection of tools that can help navigate around the extensive and highly interconnected world of genomic data. It is worth exploring further avenues of bioinformatics research whether in the journal *Bioinformatics* (for computational techniques and software), *Nature Biotechnology* (for technologies related to bioinformatics), or *Nature* and *Science* journals where important work on genomics is regularly published.

References

1 International Genome Sequencing Consortium (2001). Initial sequencing and analysis of the human genome. *Nature* 409: 860–921.
2 Ledford, H. (2016). AstraZeneca launches project to sequence 2 million genomes. *Nature* 532: 427.
3 Zook, J., Chapman, B., Wang, J. et al. (2014). Integrating human sequence data sets provides a resource of benchmark SNP and indel genotype calls. *Nat. Biotechnol.* 32: 246–251.
4 Livingstone, C.D. and Barton, G.J. (1993). Protein sequence alignments: a strategy for the hierarchical analysis of residue conservation. *Comput. Appl. Bio. Sci.* 9: 745–756.
5 Attwood, T.K. and Parry-Smith, D.J. (1998). *Introduction to Bioinformatics*. London: Pearson.
6 Sievers, F., Wilm, A., Dineen, D.G. et al. (2011). Fast, scalable generation of high-quality protein multiple sequence alignments using Clustal Omega. *Mol. Syst. Biol.* 7: 539.
7 Parry-Smith, D.J., Payne, A.W.R., Michie, A.D., and Attwood, T.K. (1997). CINEMA - a novel Colour INteractive Editor for Multiple Alignments. *Gene* 211 (2): GC45–GC56.
8 Finn, R.D., Attwood, T.K., Babbitt, P.C. et al. (2016). InterPro in 2017 – beyond protein family and domain annotations. *Nucleic Acids Res.* 45 (D1): D190–D199.
9 Li, H., Handsaker, B., Wysoker, A. et al. and 1000 Genome Project Data Processing Subgroup (2009). The sequence alignment/map (SAM) format and SAMtools. *Bioinformatics* 25: 2078–2079.
10 Abuín, J.M., Pichel, J.C., Pena, T.F., and Amigo, J. (2015). BigBWA: approaching the burrows–wheeler aligner to big data technologies. *Bioinformatics* 31 (24): 4003–4005.

11 Hodgkins, A., Farne, A., Perera, S. et al. (2015). WGE: a CRISPR database for genome engineering. *Bioinformatics* 31 (18): 3078–3080.
12 Xuhua, X. (2017). Bioinformatics and drug discovery. *Curr. Top. Med. Chem.* 17 (15): 1709–1726.
13 Parry-Smith, D.J. (2003). Bioinformatics: its role in drug discovery. In: *Burger's Medicinal Chemistry and Drug Discovery*, 6e, vol. 1 Drug Discovery (ed. D. Abraham). Hoboken, New Jersey, USA: Wiley.
14 Hussain, S. (2018). Native RNA-sequencing throws its hat into the transcriptomics ring. *Trends Biochem. Sci.* 43 (4): 225–227.
15 Ashley, E.A. (2016). Towards precision medicine. *Nat. Rev. Genet.* 17 (9): 507–522.

Further Reading

Lodish, H. et al. (2013). *Molecular Cell Biology*. UK: Macmillan Higher Education.
Parrington, J. (2015). *The Deeper Genome*. Oxford: OUP.
Rodriguez-Ezpeleta, N., Hakenberg, M., and Aransay, A.M. (eds.) (2012). *Bioinformatics for High Throughput Sequencing*. New York: Springer.
Wang, X. (2016). *Next Generation Sequencing Data Analysis*. Florida: CRC Press.
Watson, J. et al. (2014). *Molecular Biology of the Gene Seventh Edition*. New York: Cold Spring Harbour Press.

Websites

www.ukbiobank.ac.uk/2018/04/whole-genome-sequencing-will-transform-the-research-landscape-for-a-wide-range-of-diseases
www.nature.com/news/human-genome-project-twenty-five-years-of-big-biology-1.18436
en.wikipedia.org/wiki/List_of_sequence_alignment_software#Short-Read_Sequence_Alignment
www.ensembl.org
www.ucsc.edu
www.targetvalidation.org
www.ensembl.org/Homo_sapiens/Tools/Blast
samtools.sourceforge.net
www.mlo-online.com/long-read-sequencing
www.10xgenomics.com
www.sanger.ac.uk/htgt/wge
www.humancellatlas.org
www.gtexportal.org

4

Gene Cloning for the Analysis of Gene Expression

Huw B. Thomas and Raymond T. O'Keefe

Faculty of Biology, Medicine and Health, University of Manchester, Oxford Road, Manchester, M13 9PT, UK

4.1 Identifying Target Sequence

Identifying and defining a target sequence is an important first step in expression studies since it forms the basis for all subsequent experiments. Several well-established and fully curated databases are available online that may be used to provide the necessary nucleotide and/or protein sequence(s). These include fully comprehensive multiorganism online resources such as the National Center for Biotechnology Information (NCBI), UCSC Genome Browser and Ensembl, or more bespoke, organism-specific databases such as those available for yeast (*Saccharomyces* Genome Database [SGD], PomBase), Zebrafish (ZFIN) or *Drosophila* (Flybase). Links to each of these databases can be found at the end of this chapter. Multiorganism depositories are useful for identifying orthologous genes for cross-species expression studies, whilst single-species databases are often updated with greater regularity.

Selection of the correct sequence, and therefore the correct database source, will depend on the specifics of the intended experiment. For example, analysis of splicing consequences of a particular exon using a mini-gene splicing assay will require not only the sequence of the exon in question but also the upstream and downstream intron sequences and thus would be obtained from a genomic sequence database. Alternatively, cloning of a protein-coding gene into an expression vector would require the full complementary DNA (cDNA) sequence of the open reading frame made from messenger RNA. Often both sequences are available from the same depository, but it is important to make the distinction between all available sequences for a gene/transcript/isoform of interest.

4.2 *In Silico* Design

For cloning experiments, it is often helpful or even necessary to visualise the vector and intended insert sequence to aid with the design of the cloning experiment. Several software packages are available that enable *in silico* design and annotation of a recombinant DNA construct such as Benchling, SnapGene®, pDraw and UGENE. These programs allow the identification of potential restriction sites, the design of suitable primer sequences and the confirmation of all components in a cloning vector such as antibiotic resistance or auxotrophic markers. Some of the more sophisticated cloning software provides the means to carry out a step-wise assembly of the vector construct such that multistep cloning experiments can be modelled and, most importantly, tracked, thus providing an accurate historical record of the cloning steps carried out. Additionally, the SnapGene software suite can be used in the design of individual steps of cloning, for example calculating volumes of components for polymerase chain reaction (PCR) or predicting the results of a restriction digest experiment of a recombinant vector. All of these features simplify the experimental process and seek to reduce the likelihood of human error.

Biomolecular and Bioanalytical Techniques: Theory, Methodology and Applications, First Edition. Edited by Vasudevan Ramesh.

4.3 Primer Design

A useful feature of cloning software suites is the ability to design bespoke PCR primers for producing a cloned DNA fragment for a cloning experiment (Figure 4.1). Whilst this method of primer design can expedite the overall design process, the software wizards often adhere to strict default parameters and do not provide much flexibility in design options, such as primer length. In such cases, manual design may be more suitable. Many different aspects and features must be considered when designing primers; some of the more important considerations will be covered below.

The aim of the designing process is to choose primer sequences that will accurately amplify the target sequence from the template material with high fidelity whilst having a sequence length of approximately 20 nt and an optimal annealing temperature falling between 55 and 65 °C (although examples outside this temperature range are sometimes used), with no more than a 5 °C difference between annealing temperatures for each primer [1].

4.3.1 Utility Sequences

PCR amplification of a target sequence that precedes a cloning experiment will often exploit the relative flexibility of the 5′ end of the primer sequences to introduce additional utility sequences to either end of the target sequence (Figure 4.1c). These utility sequences are designed to accommodate the downstream cloning reaction and can include adding flanking restriction enzyme sites for traditional restriction cloning, introducing an additional 15–25 nt sequence that overlaps with an acceptor vector for Gibson assembly®, providing upstream promoter sequences for phage-derived RNA polymerases or adding homologous recombination (HR) motifs for Gateway™ cloning.

4.3.2 Annealing Temperature

The optimum annealing temperature for any pair of primers is influenced by not only the GC content of the primer/template sequence but also the polymerase used in the PCR reaction. Determining the annealing temperature to use in a PCR amplification reaction is therefore not only reliant on the chosen primer sequences. A more accurate estimation of optimal annealing temperature can be made by employing an online Tm calculator such as that provided by New England Biolabs (NEB) (https://tmcalculator.neb.com), which accounts for primer length, sequence and final concentration in addition to the polymerase and enzyme buffer used when calculating the Tm. When using primers that feature a utility sequence (as covered in Section 4.3.1), this portion of the primer sequence should be omitted from the calculations as it is not complementary to the template and will not anneal.

4.3.3 Specificity

Successful PCR relies on the accurate amplification of a single product from the template material. To this end, primers should have high specificity for a single location within the template sequence. Primers are designed to be approximately 18–25 nucleotides in length, which will usually provide sufficient specificity for the target sequence and allow easy binding to the template at the appropriate annealing temperature (Figure 4.1b). Additionally, primers should have a 40–60% GC content and have at their 3′ end a G or C to promote binding and extension of the primer. An added complexity with cloning experiments is a requirement to add a utility sequence to the amplified product by way of additional non-homologous nucleotides to the 5′ end of the primers. These additional nucleotides reduce the percentage identity between primer and template and increase the likelihood the primers may have multiple binding locations within the template sequence. It is therefore recommended that during the primer design process, candidate primer sequences are analysed

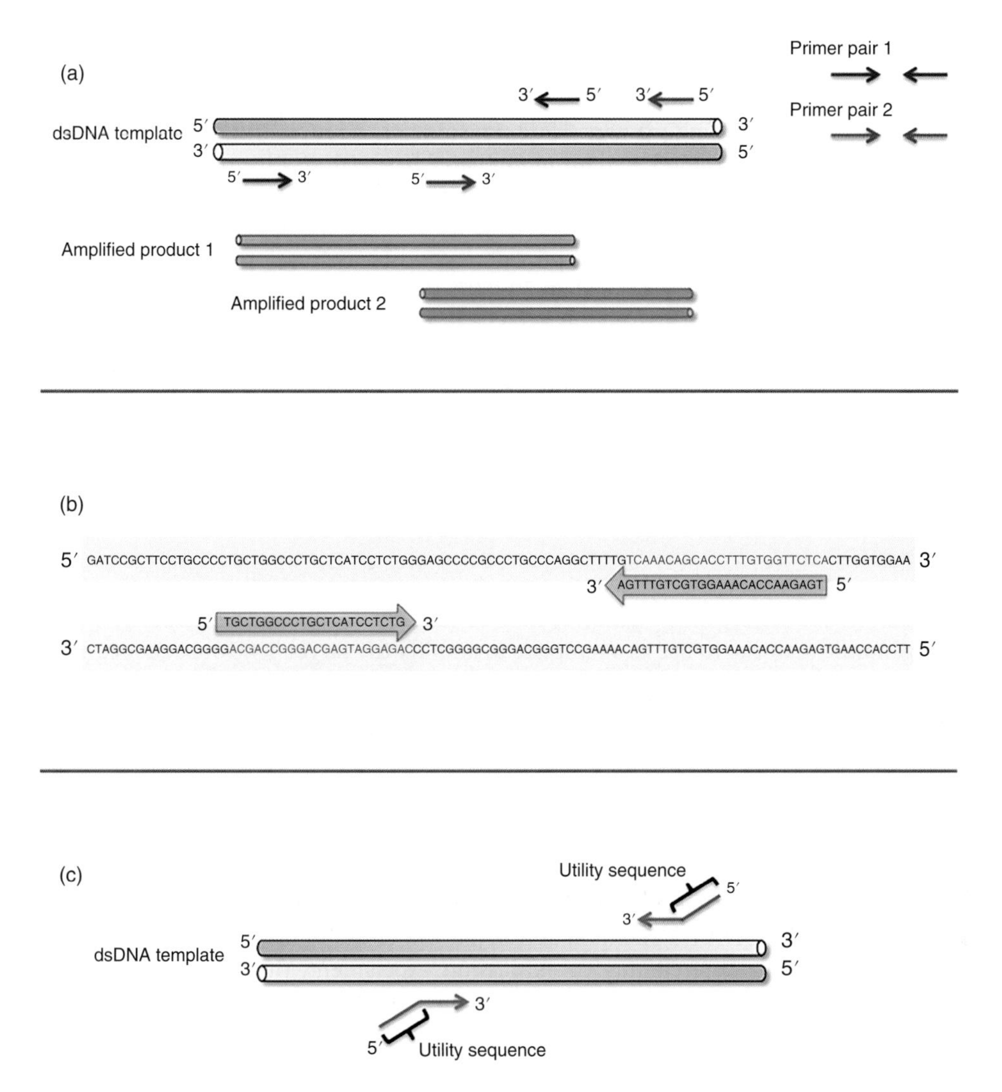

Figure 4.1 A schematic representation of a primer binding. (a) Two different DNA products can be amplified from the same double-stranded DNA template using two different pairs of primers. (b) Each primer has a complementary sequence to either the top sense strand (5′–3′) or bottom anti-sense strand (3′–5′) and align in an anti-parallel fashion. (c) An additional utility sequence (to aid downstream cloning techniques) may be incorporated into the final amplified product by inclusion of an extra sequence at the 5′ end of the primer sequences.

for sequence similarity to the template. This can be achieved using a program such as PrimerBlast (available through the NCBI website), which not only identifies potential additional binding sites for pairs of primers but also calculates the size of each product and the calculated Tm for determining an appropriate annealing temperature for the PCR amplification. Estimating the specificity of primers is especially important for protocols using a genomic DNA template for the PCR since the potential scope of binding sites is greatly increased compared to PCR amplification from a less complex template such as a donor vector.

4.3.4 Additional PCR Primer Design Considerations

Whilst the target template sequence will often be the most restricting factor, where possible it is advisable to adhere to some general rules when designing primers. Here are some of the more important considerations. Primers should not be based paired at their 3′ end for more than 2 base pairs with either themselves or another primer, as this base-pairing would provide a terminus for extension by the DNA polymerase. Ensure the 3′ end of the primers contain a GC-clamp, that is, 1–3 G or C nucleotides in the last 5 bases of the 3′ end to increase the binding potential of the primers and provide a good basis for elongation. The overall G/C content of primer sequences should not be greater than 50–60%. Avoid primer sequences that contain either multiple di-nucleotide repeating sequences (such as ATATAT) or single nucleotides repeated more than four-times consecutively (for example, GGGGG), as both of these features can contribute to mispriming. Choose primer sequences with a low secondary structure capacity, i.e. the likelihood of forming hairpin structures, which can lead to lower PCR efficiency and yields. A quantification of self-complementarity is provided by the PrimerBlast software by NCBI.

4.4 Template Preparation

For most cloning experiments of model organism genes, it may be possible (or indeed preferable) to purchase a purified sample of genomic material to use as a template to amplify the target gene or sequence prior to cloning. However, in some circumstances a suitable sample may not be available or the specifics of the study design requires the template to be prepared from a particular source, for example, a sample from a patient expressing a certain genetic variant or from a specific cell/tissue type. In these instances it is important to undertake a methodology that provides the highest quality genomic DNA to ensure success in the initial amplification of a target gene and other downstream applications.

Alternatively, it is increasingly common for cloning methods to utilise de novo synthesis of a target gene/sequence rather than PCR-based amplification from a suitable DNA template sequence [2]. Historically, this approach was costly and not a service readily available. However, an increasing number of companies now offer fully customisable synthesis of DNA fragments at a relatively low cost. There are several advantages to using de novo synthesised sequences in a cloning or expression experiment, the most significant of these being the flexibility. Since no template sequence is required, the design process is far less restrictive and sequences can be synthesised that complement the experimental design. For example, for studies involving expression of a eukaryotic protein using a bacterial system, a bespoke sequence can be designed that accurately encodes for the required protein sequence but importantly accommodates the preferred codon usage of bacteria. In instances like this, no natural template would be available for a PCR-based approach and may hinder the success of the cloning process. Another advantage of de novo synthesis is that since sequences are highly pure and their sequence verified as standard during the manufacturing process, some of the validation steps of a cloning experiment can be avoided, which can expedite the whole cloning protocol. Indeed, it is now possible to bypass any cloning procedures and design a fully bespoke vector construct, inclusive of any target genes, sequence tags, regulatory sequences and selection markers.

4.4.1 Genomic DNA Preparation

Preparation of genomic DNA from a cell/tissue sample is a relatively straightforward process involving initial cell lysis, deproteination and finally recovery of DNA by precipitation in an alcohol-based solution. Initial preparation of a sample will depend largely on the cell/tissue type as some cells are more resistant to chemical lysis and may need additional physical disruption, such as passage through a narrow gauge needle ($\leq$20 G) for whole blood cells or snap-freezing in liquid nitrogen and grinding for yeast cells, before a homogenous solution is formed. Homogenisation of the cells in the DNA lysis buffer must be achieved quickly to avoid unwanted action of cellular nucleases, which will degrade the DNA. However, the freezing of cells prior to disruption or the presence of ethylenediaminetetraacetic acid (EDTA) in the buffer will act to inhibit any DNase activity. Two important considerations of the extraction process are, firstly, during extraction, to limit the shear forces on the extracted DNA by gentle mixing and pipetting to minimise disruption of the DNA and maintain high molecular weight fragments, secondly, to remove all solvents and excess salt from the sample following extraction, which can hamper downstream processes, and, not least, to provide an accurate measurement of DNA concentration. Modern DNA extraction kits are designed to limit both the physical force and contamination levels during DNA isolation by providing optimised buffer concentrations and utilising a spin-column component for efficient purification of DNA.

Importantly, the variety of DNA extraction kits available is extensive. This variety reflects not only the range of cell types and DNA sources routinely utilised within a research context but also the scale and format of the intended DNA preparation. For example, bespoke kits are available for use on blood, urine, bacteria, bone or cultured cells. Similarly, several of these kits are available in either a single-tube (1.5 ml) scale or in a multi-well plate scale, allowing up to 96 samples to be processed concurrently. Thus it is important to choose the best available protocol/kit for the sample in question to maximise the integrity and purity of the final DNA extraction.

4.4.2 RNA to Complementary DNA (cDNA) Conversion

Following isolation of an RNA sample, it is necessary to convert the RNA to complementary DNA (cDNA) before it can be used in a PCR reaction or downstream cloning experiments. Conversion to DNA also serves to stabilise the sample, as DNA is far less susceptible to degradation than RNA. Conversion is achieved using a retroviral-derived reverse transcriptase, which transcribes the RNA to generate a complementary strand of DNA. A typical RNA to cDNA conversion reaction requires several key components. The first is a suitable RNA template, which could be a total RNA preparation or a specific RNA transcript. Other essential components include: a reverse transcriptase enzyme, dNTPs for cDNA synthesis, primers for the initiation of transcription, a suitable reaction buffer, an RNase inhibitor and finally a reducing agent such as dithiothreitol (DTT) to prevent oxidative damage to the protein components of the reaction.

The primers used are often one of three types: gene-specific for conversion of a specific gene or locus within an RNA sample; oligo-d(T), which primes at the poly-adenosine tail of a mature mRNA transcript (thereby preferentially enriching full-length mRNA within a total RNA sample); or random hexamers, which bind indiscriminately across the length of RNA template and act to prime all RNA fragments equally irrespective of whether they are full-length or not. Optional alternatives to oligo-d(T) primers are anchored oligo-d(T) primers, which are modified to include 2 random residues at their 3′ end; this increases their likelihood of annealing to the 5′ end of the poly(A)-tail and prevents priming from within the poly(A)-tail, ultimately leading to increased cDNA yields. However, oligo-d(T) priming is less suitable in circumstances where the RNA template may be degraded since any degradation of the 3′ poly(A) tail will reduce the efficiency of the priming and therefore the reverse transcription.

Several reverse transcription kits are available commercially, which include all the necessary components for a simple conversion reaction. Typically the template RNA is incubated with dNTPs and primers and heated to ensure all components are disassociated from each other before being rapidly cooled (usually by placing on ice) to aid the annealing of the primers to the template RNA. The remaining reaction components are added and the reaction proceeds at a temperature optimal for the enzyme activity. Recent bioengineering of reverse transcriptase enzymes has meant that modern kits can convert total RNA to cDNA in as little as 15 minutes. Once converted the cDNA sample may be used in a PCR reaction to amplify a gene of interest (GoI) for cloning.

4.4.3 Plasmid DNA Preparation

Plasmids are circular DNA molecules that are used as a vehicle for other DNA sequences. Plasmids can be easily and specifically maintained and propagated in bacteria as they contain sequences that allow both their replication in bacteria as well as resistance to specific antibiotics that can be used to select and grow only cells containing a desired plasmid. Many different types of plasmids have been developed and are either available commercially, directly from researchers or from a plasmid repository like Addgene (https://www.addgene.org). An important step in a cloning experiment is the isolation of plasmid DNA. This step may involve the initial isolation of a plasmid to be used in a cloning experiment or alternatively the isolation of a newly cloned recombinant construct. In either situation, careful extraction is required to ensure the sample is of high integrity but importantly is made up of plasmid DNA alone and is not contaminated by genomic DNA. Proprietary plasmid purification kit protocols are optimised to minimise the risk of contamination by genomic DNA.

4.4.4 RNA Preparation

Since RNA is much more susceptible to both physical and enzymatic degradation than DNA and that common ribonucleases are highly stable and resistant to heat denaturation, extra precautions must be taken when isolating RNA. These precautions include cleaning the work area and technical equipment (such as pipettes) with an RNase inhibitor such as RNase*Zap*™ (Thermo Fisher) and ensuring all steps of the RNA isolation protocol are carried out on ice (unless specified otherwise).

RNA isolation protocols include examples where cell disruption and RNA purification is achieved through a combination of a homogenising spin-column and a cleanup spin-column. More commonly, cells/tissues are disrupted using a phenol-based solution such as TRIzol® (Sigma) to solubilise biological material and simultaneously denature proteins. Subsequent addition of chloroform results in phase separation of the sample, where proteins are retained in the organic phase, DNA resolves to the interface and the RNA remains in an aqueous phase. An added advantage of a TRIzol isolation is that each phase constituent (RNA/DNA/protein) can be retained and individually isolated using specific reagents and methodology. However, more optimised methods are available for extraction of protein or DNA from cell/tissue samples and, as such, RNA extraction alone is the predominant application of the TRIzol reagent. The highest integrity of RNA can be achieved using a combination of phenol/chloroform extraction with TRIzol followed by a column cleanup using a proprietary kit that includes a DNase digest step to eliminate any DNA carried over.

4.5 Cloning Methods

4.5.1 Gateway Cloning

Gateway cloning is a well-established easily reversible method for cloning that relies on the rapid shuttling of a sequence from a donor vector to a destination vector utilising homologous recombination sequences present in both vectors [3]. The use of universal recombination sequence pairs (either L-R or B-P) has meant that many genes can be commercially purchased already cloned into a donor vector harbouring the necessary homologous recombination motifs. Furthermore, several destination vectors have been designed to reflect the downstream application of the cloned gene, for example vectors optimised for expression in yeast, bacteria or mammalian cells, in addition to vectors that have been designed to be used in specific circumstances such as antibody production, protein–protein interactions or RNAi assays. Cloning is carried out by incubating both vector populations in the presence of a proprietary recombinase enzyme (Clonase™) specific for either L-R of B-P homologous recombination pairs. An important benefit of this method of cloning is that recombination is irreversible but that successful recombination alters the recombination sites to that of the other possible motif pair. For example, recombination of sequences by using the L and R motifs results in the recombined sequences becoming B and P, respectively. This property provides the ability to reverse the reaction or subclone by utilising a different Clonase enzyme.

4.5.2 Gibson Cloning

Gibson assembly is a cloning technique developed in the Craig Venter Institute by Dr Daniel Gibson and colleagues [4]. Gibson cloning can be used to assemble up to five fragments in a single reaction regardless of fragment length or end compatibility. Gibson assembly has become a popular choice for molecular cloning since it requires fewer technical steps and reagents and thus is often faster and cheaper than other cloning methods. Additionally, Gibson assembly is sequence independent and can be carried out without the need for any restriction enzymes.

Gibson assembly relies on two important characteristics: firstly, all fragments feature an overlapping portion to their assigned adjacent fragment and, secondly, the simultaneous activity of three enzymes within the reaction mix.

Fragments are amplified with bespoke primers to incorporate flanking utility sequences. Here the additional sequence is a 15–25 nt overlapping sequence specific to either another fragment (where multiple fragments are to be joined together) or with the vector backbone sequence. An important advantage of Gibson cloning is that through careful design of the primer sequences, fragment sequences may be altered during the cloning protocol. For example, cloning and mutagenesis can be achieved by introducing a short indel or single nucleotide polymorphism (SNP) into the fragments. This mechanism can be useful to model the effects of small mutations within the target sequence where only a wild type template sequence is available.

During a Gibson assembly reaction, fragments are incubated with the Gibson assembly mix that contains three enzymatic components: a 5′ exonuclease, a high-fidelity DNA polymerase and a DNA ligase. The action of the exonuclease degrades a short portion of the 5′ strands on each fragment, thus revealing complementary 3′ sequences between adjoining fragments. The DNA polymerase repairs any gaps in sequence between fragments and finally each fragment is ligated together by the actions of the DNA ligase, producing a fully assembled construct. The reaction mix has been engineered such that all three enzymatic reactions can occur during a single incubation step at 50 °C.

4.5.3 Restriction Enzyme Digestion and Ligation Cloning

Restriction digest cloning takes advantage of the availability of many different restriction endonucleases that target a wide variety of short DNA sequences to cut a DNA strand. New England Biolabs (www.neb.com) is just one company that sells an extensive portfolio of restriction enzymes that can be used for cloning. Restriction enzymes can digest DNA to produce either blunt ends, 5′ overhanging ends or 3′ overhanging ends. The 5′ or 3′ overhangs are usually preferred as they add specificity and directionality in the cloning process. In the design stage it is preferable to incorporate two different restriction enzyme sites into the 5′ end of the PCR primers that will leave overhangs on each end of the DNA after digestion to allow directional cloning of the DNA into the vector. Many cloning vectors are designed with multiple restriction enzyme sites that allow the insertion of DNA sequences that have been digested with compatible restriction enzymes. A vector is first cut with two restriction enzymes and is then either gel purified or column purified to remove the unwanted piece of DNA. It is also useful to dephosphorylate the cut vector with either calf intestinal phosphatase or Antarctic phosphatase to reduce re-ligation of the vector later on. Antarctic phosphatase has the advantage of being able to be heat inactivated [5]. The DNA to be inserted into the vector is also cut with the same two restriction enzymes, or other compatible restriction enzymes, and purified to remove the unwanted pieces of DNA from the ends. Mixing the digested vector and DNA together in the presence of T4 DNA ligase inserts the DNA into the vector.

4.5.4 Ligation Independent Cloning

In ligation independent cloning, the 3′–5′ exonuclease and 5′–3′ polymerase activities of T4 DNA polymerase are utilised to produce long complementary single-stranded overhangs (10–15 nucleotides) in both the PCR product and the plasmid, which can be annealed together. PCR primers are designed to contain sequences complementary to the ends of a plasmid cut with a restriction enzyme. Specific conditions where only one nucleotide is used with the T4 DNA polymerase exonuclease activity removes nucleotides from the 3′ end of the DNA to a certain position to prepare the plasmid and PCR product with complimentary single-stranded overhangs. The plasmid and PCR product are then annealed and this mixture is transformed into bacteria, where the nick between the PCR product and the plasmid is repaired during the replication process, completing the insertion of the PCR product into the plasmid [6].

4.5.5 TA Cloning

TA cloning takes advantage of Taq DNA polymerase, which has a non-template dependent terminal transferase activity that incorporates a single deoxyadenosine (A) at the 3′ end of PCR products [7]. These PCR products with 3′A overhangs can then be incubated with a plasmid vector that has a 3′T overhang and then ligated together with DNA ligase. T overhang plasmids can be obtained commercially or can be produced by restriction enzyme digestion or by addition of T overhangs to blunt ended plasmids [8]. The disadvantage of TA cloning is that Taq DNA polymerase does not have 3′–5′ exonuclease proofreading activity so there is the possibility of introducing random mutations. If introduction of random mutation is a concern then the DNA can first be amplified with a proofreading polymerase and then 3′A's added with Taq polymerase. Also insertion of a DNA fragment by TA cloning is not directional, so potentially 50% of the resulting clones will not be in the correct orientation. A variation on TA cloning is TOPO TA cloning, where plasmids are available with the 3′T end of the plasmid fused with topoisomerase I, which catalyses the ligation of the introduced A overhang PCR product [9].

4.6 Uses for Cloned DNA Sequences

The ability to clone either the gene sequence or cDNA into a plasmid vector is essential for the analysis of gene function. A wide variety of plasmid vectors that contain certain features that facilitate gene functional analysis are either commercially available or available through Addgene (www.addgene.org). For example, to determine the subcellular localisation of a protein in living or fixed cells, cDNA sequences can be cloned in frame with sequences that code for a number of fluorescent proteins like the green fluorescent protein (GFP), enhanced GFP or mCherry. The addition of small epitope tag sequences, like the V5, FLAG, MYC or HA tag, by cloning a cDNA in frame with one of these short (8–10 amino acid) peptide epitope tags, can allow for recognition of the protein fused to the tag with a specific antibody for the tag. Tagging then allows applications like western blotting, immunoprecipitation and immunofluorescence to be carried out without the need for a specific antibody to your protein of interest. The small size of these epitope tags means they usually do not influence the function of the tagged protein. Affinity tags are another type of tag that can be fused to a protein through cloning, which allows the affinity tagged protein to be purified with a specific affinity resin. Popular affinity tags are a 6–8 Histidine tag, glutathione-S-transferase (GST), maltose binding protein (MBP) or calmodulin binding peptide (CBP).

Cloned genes can also be placed under the control of different promoter sequences that allow one to tune the expression of the cloned gene. For example, in yeast there is a selection of promoter sequences that have been engineered into plasmid vectors that provide a range of expression levels from low to high. Additionally, there are promoters that can be turned on or off by the addition of chemical inducers or inhibitors. In yeast there is the popular *GAL1* promoter that is active when cells are grown in the presence of galactose and repressed when grown in the presence of glucose. With these different promoters the effects of either under- or overexpressing a certain gene product can be determined. Another well-used system for regulating expression in mammalian cells is based on the tetracycline resistance operon in bacteria. The Tet-On and Tet-Off systems have been developed to either turn on or off the expression of a gene in the presence of tetracycline or doxycycline. In summary, these are the most common uses of a cloned gene or cDNA, but once you have a gene cloned into a plasmid vector it can then be used for many different applications.

4.7 Verifying Cloned Sequences

Following successful cloning of a gene into a vector (using any of the methods described above), constructs will need to be validated to ensure the recombinant sequence is accurate and devoid of any errors. Nucleotide level verification is usually achieved by targeted Sanger sequencing of the plasmid, but often a carefully planned restriction digest of the plasmid can be used to screen several candidate colonies to eliminate any unsuccessful clones prior to undertaking the more expensive protocol of verification by sequencing.

4.7.1 Bacterial Transformation

Transformation of the vector into a bacterial strain enables propagation of the plasmid and subsequent isolation and purification ready for verification by Sanger sequencing. In addition, some forms of cloning provide a means to positively select for successfully cloned constructs during the transformation step. For instance, homologous recombination cloning methods (Gateway) often employ a negative selection (or suicide) gene such as ccdB incorporated into the portion of the plasmid to be replaced during successful cloning. Providing the construct is transformed into a bacterial strain that is sensitive to the expression of the ccdB gene, then only constructs that have successfully replaced the ccdB gene during the cloning reaction will grow. Examples of *Escherichia coli* strains sensitive to ccdB expression include Top10 and EC100D. More commonly, transformed cells are cultured on solid media containing an antibiotic selection in which genetic resistance is provided by expression of the transformation plasmid. This resistance ensures only transformed colonies will be cultured. Antibiotics such as ampicillin, kanamycin and tetracycline are some of the more common selection markers used.

4.7.2 Mini/Midi/Maxi DNA Prep

The volume of cell culture and required mass of purified plasmid will determine what scale of preparation to use for plasmid purification. A standard mini-prep protocol will require 2–5 ml of overnight culture and yield up to 10 μg of pure plasmid DNA. Often, this amount is sufficient for both verification protocols and any downstream applications. However, larger yields may be achieved by scaling the volume of overnight culture up to 100 ml and using a midi-prep kit for up to 100 μg of plasmid or in rare circumstances a maxi-prep kit may be used to purify up to 0.5 mg of plasmid from approximately 500 ml of bacterial culture. The principles of the purification process are analogous regardless of the kit scale used but mini-prep kits offer the highest extraction efficiency whilst also having the easiest and shortest protocol.

4.7.3 Sequence Confirmation

Confirmation of plasmid constructs by Sanger sequencing is the gold-standard technique for ensuring that the cloned plasmid is accurate to the nucleotide level. Most commonly provided as an external third-party service, 'plasmid to sequence' can be achieved in as little as 24 hours. Although most service providers offer low cost sequencing and require only a small volume of purified plasmid (5–15 μl) as a starting point, the cost of the service can be reduced by additional preparation steps being carried out in-house prior to samples being sent to the sequencing provider; these steps include florescent labelling and terminator reactions.

Each sequencing reaction requires a specific sequencing primer to determine the location and direction of the sequencing reaction. Sequencing service providers offer a range of 'universal primer' sequences, such as T7, T3, M13-for and M13-rev that match sequences found in the most commonly used plasmid constructs. However, if no universal primer sequence is available or the novel fragment of DNA to be sequenced is larger than the 800–1000 bp length of a single sequencing reaction, then a single (or multiple) bespoke sequencing primer must be used. These primers can be included with the sample when sending to the sequencing provider or, alternatively, the sequencing provider may manufacture the primer prior to sequencing the sample. This primer can then be used in any subsequent sequencing reactions.

4.8 Applications of Gene Constructs

The above passages describe the necessary steps, protocol variations and common considerations when designing or implementing a study involving the cloning of a target gene into an expression vector. Once such a vector has been produced (and adequately verified), the range of applications for which a recombinant vector may be used is extremely large.

4.8.1 Recombinant Expression

A construct may be used simply to generate recombinant RNA, DNA or protein, in which case the choice of expression organism (or in some cases a particular strain) is crucial.

4.8.1.1 Bacterial System

Bacterial expression offers the fastest and easiest way to produce excess amounts of a cloned target gene. However, whilst several competent *E. coli* strains are commercially available, subtle differences in their features lend themselves to particular applications. Some of these features are designed to aid the cloning or transformation process itself, such as different antibiotic resistance or optimised plasmid replication for a high copy expression, whilst other strains have been manipulated to aid downstream expression steps. For example, the Rosetta™ and Origami™ strains have been specifically engineered for use in protein expression studies. Rosetta strains are designed to be used when expressing eukaryotic proteins requiring a 'universal' codon translation from mRNA to protein and thus are not restricted to the native codon usage of *E. coli* [10]. Likewise, Origami strains are commonly used in protein expression studies due to the mutation of their thioredoxin reductase (trxB) and glutathione reductase (gor) genes [11]. These mutations greatly enhance disulphide bond formation in the cytoplasm and leads to a much higher yield of active protein compared to comparable expressions in other bacterial host strains. Enhancement of disulphide bonds is mostly important for expression of extracellular proteins that normally exist in a more oxidising environment. Intracellular proteins (cytoplasmic/nuclear) function in a more reducing environment and do not generally have disulphide bonds.

4.8.1.2 Yeast System

For studies of eukaryotic processes, transformation of a recombinant gene vector into a yeast host enables the expression of that gene in a simple eukaryotic system. Yeast transformations are more straightforward, requiring fewer reagents and less time than transfections into higher eukaryotic cells such as human cell lines. Additionally, certain types of plasmid vectors, such as the pRS family vectors [12] are cross-compatible between bacterial and yeast systems, allowing convenient propagation in *E. coli* and maintenance in *Saccharomyces cerevisiae.*

Yeast transformation efficiency and accuracy can be aided by incorporating the genes for a step in the biosynthetic pathways for uridine (*URA3*), leucine (*LEU2*) or histidine (*HIS3*) in a plasmid, providing successfully transformed colonies the ability to grow in specific nutrient-deficient media in strains where these genes have been deleted or mutated. A huge range of plasmid constructs offering different combinations of antibiotic resistance and biosynthetic markers make for greater flexibility when designing an experiment involving transformation and exogenous expression of a recombinant gene.

4.8.1.3 Human Cell Transfection

For studies involving human or mammalian genes, transfection of a recombinant gene construct into a suitable cell line is a prerequisite for many analytical methodologies. For example, a recombinant protein may be expressed in a mammalian cell line to investigate its localisation, interacting partners or maturation process (such as transcript splicing or protein folding).

Cells used for transfection are most commonly well-established cell lines such as the human cell lines HeLa, HEK-293 or K562, or mammalian cell lines CHO and COS cells. Whilst these cell lines are relatively easy to maintain and transfect, the multiple genomic alterations they have undergone may make them poor candidates for *in vivo* studies. Fortunately, several different primary cell lines are commercially available allowing greater specificity in the choice of host cell. Importantly, commercially available transfection regents are routinely optimised to many of the most commonly used cell types, ensuring that differences between cell lines are accounted for during the transfection protocol, often meaning bespoke methodologies for different cell lines. However, transfection efficiency and complexity of culture methods can vary greatly from cell line to cell line. As such, experimental design often results in a compromise between experimental success and suitability of design.

4.8.2 *In Vitro* Production of RNA

Many plasmid cloning vectors contain, adjacent to their multicloning sites, recognition sequences for phage RNA polymerases (T7 and/or T3) that can be utilised to produce RNA by any gene cloned into the vector. The plasmid vector is usually digested with a restriction enzyme downstream of the sequence from which RNA will be produced and mixed with a phage RNA polymerase, optimised transcription buffer and the four ribonucleotides to produce an RNA by run-off transcription. The ability of phage RNA polymerase to incorporate radioactively labelled or modified nucleotides facilitates downstream detection. The RNA produced can be used for *in vitro* translation to produce proteins, for probes to use during *in situ* hybridisation and northern blotting, for structural studies or as templates for RNA processing reactions.

4.9 Case Study: Cloning of a Human Missense Variant Exon into a Minigene Splicing Vector

With the rise in population, Genome Wide Association Studies (GWAS) – which seek to genetically characterise populations by capturing and analysing many thousand individual genomes – several rare genetic diseases are increasingly being associated and linked to low frequency allele mutations in a number of human genes. In many instances these genes are poorly characterised and the functional outcomes of the single nucleotide polymorphisms (SNPs) described are as yet undefined. Whilst missense mutations in highly conserved residues often hint at a structural change to the protein leading to a change in its functional capacity (or often loss of function), examples are emerging where single missense mutations are altering a conserved splice site, a splicing factor binding motif or activating a cryptic splice site [13]. Understanding how these nucleotide variations may be affecting splicing of the pre-mRNA prior to translation can give us a greater understanding of the functional consequences of such rare genetic mutations in the pathology of their associated disease/conditions.

The following case study will describe in full the methods that may be used to identify, isolate and generate a variant-containing exon into a mini-gene splicing vector. This vector can then be used to investigate the splicing consequences of the variant exon compared to the wild type exon sequence.

4.9.1 Methods

4.9.1.1 Prerequisites

A human missense mutation to interrogate for splicing effects.
Affected gene name and exon number.
Precise mutation coordinate (genomic or cDNA) and polymorphism details (e.g. c.1305A > C).
pSpliceExpress Vector (available from Addgene – Plasmid #32485) [14].

4.9.1.2 Background

The pSpliceExpress vector construct is an example of a mini-gene vector that contains two exon sequences (in this instance derived from the rat insulin gene INS2) flanking a pair of homologous recombination (HR) motifs. The portion of the vector between the HR motifs contains the ccdB gene that allows for positive selection of vectors deficient of this portion when the construct is transformed into TOP10 competent cells. The vector is designed so that during the cloning procedure, a single fragment (usually containing a single exon with flanking intronic sequence) is homologously recombined, thus replacing the ccdB portion with an exonic sequence and producing a mini-gene construct containing three exons. This construct can then be transfected into a human or mammalian cell line, where the expressed transcripts will be subject to splicing by the endogenous splicing machinery. The effect of a single nucleotide variation between the WT mini-gene and variant mini-gene on its splicing pattern can then be compared by extracting the RNA population from the cells and carrying out reverse transcription PCR.

4.9.1.3 Identification and Amplification of Target Exon

Using the official gene name or an appropriate accession number locate the database entry for the GoI on the NCBI or Ensembl website. Save the target gene's genomic sequence (including intron portions) in FASTA format. Load the genomic sequence into SnapGene (or an equivalent sequence viewer) and using the variant coordinates, annotate the location of the polymorphism seen between WT and disease associated alleles. Save the sequence of the mini-gene vector (pSpliceExpress) into a separate file accessible by SnapGene.

4.9.1.4 *In Silico* Design

Highlight the portion of the GoI defined as 100 nt upstream (5′) of the start of exon to the site of mutation and save as a new sequence named Fragment 1. Likewise, highlight and save the portion of the GoI defined as the first nucleotide downstream of the mutation site to 100 nt downstream (3′) of the end of the exon, as Fragment 2. Use the 'Gibson assembly' software wizard available in SnapGene to design overlapping primers for the assembly of Fragment 1 and Fragment 2 into the pSpliceExpress digested vector (choose NheI and XbaI as restriction sites used to linearise the vector and 15–25 nt of overlap between fragments). See Figure 4.2. SnapGene will design two pairs of primers, one for each fragment. Rename these primers F1-for, F1-rev, F2-for and F2-rev. Primers F1-for and F2-rev will have an additional utility sequence at their 5′ ends that matches the sequence 5′ and 3′ (respectively) to the insertion site on the pSpliceExpress vector. Primers F1-rev and F2-for will overlap each other, covering the area including the mutation site. Before exporting primer sequences from SnapGene software, manually alter the sequence of the F1-rev and F2-for primers at the location that relates to the variant/mutation site. For example, if missense mutation is G > A, alter the G > A in the F2-for primer and the corresponding C > T in the F1-rev.

4.9.1.5 Fragment Amplification

PCR amplify each fragment with newly designed primers using a high fidelity DNA polymerase such as Phusion. Use F1-for and F1-rev for 'Fragment-1', F2-for and F2-rev for 'Fragment-2' and finally use F1-for and F2-rev for a single 'WT-Fragment' absent of a variant missense mutation. Use 50–250 ng of human genomic DNA as a template in a 50 μl reaction containing 1 μl dNTPs (10 μM), 1 μl forward primer (10 μM), 1 μl reverse primer (10 μM), 10 μl 5× HF buffer, 0.5 μl Phusion polymerase, made up to 50 μl with nuclease free water. Use

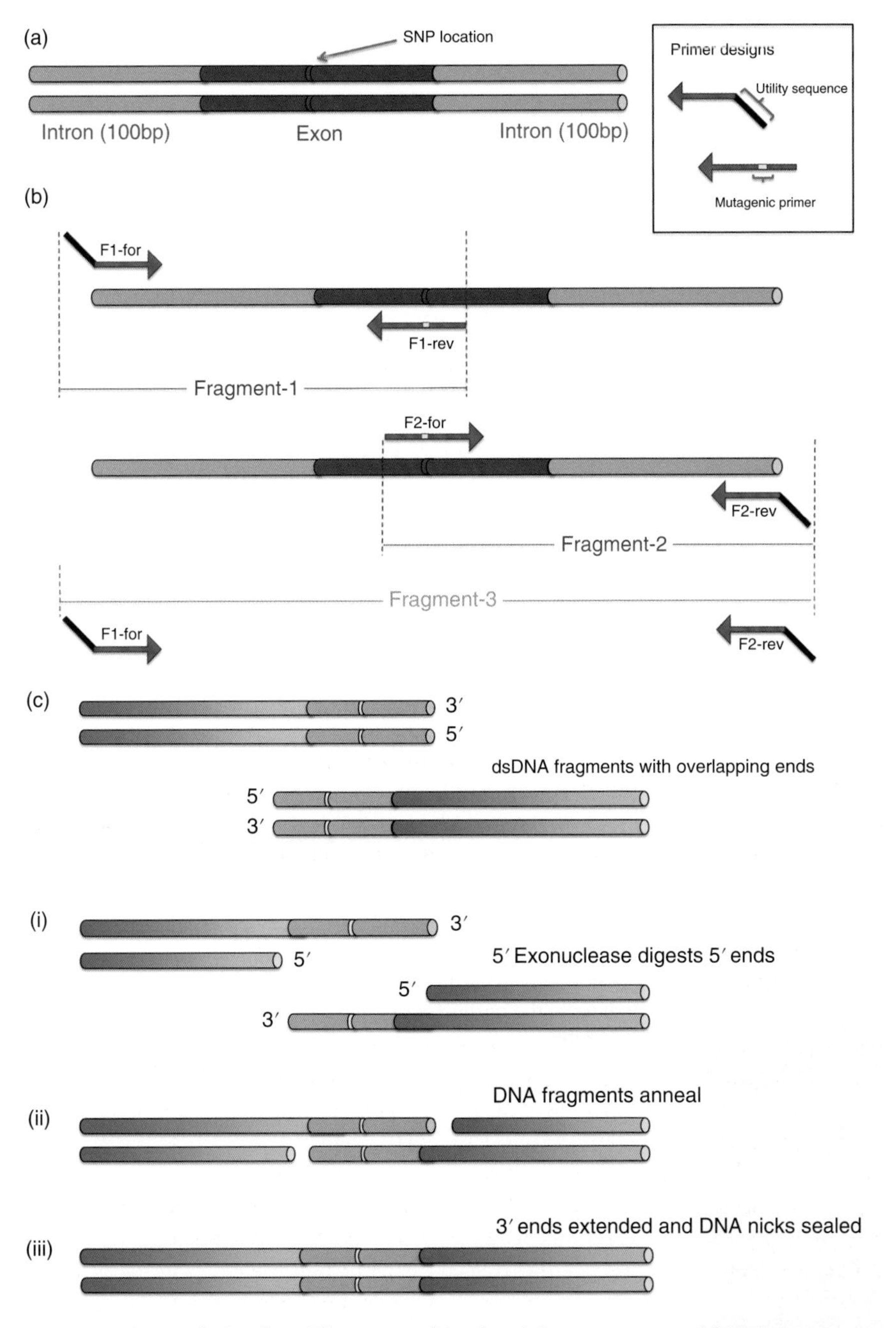

Figure 4.2 Primer design for a Gibson assembly of a mini-gene construct. (a) Mini-gene construct consisting of an exon of interest flanked by at least a 100 bp adjacent intronic sequence. Exon may contain necessary variant SNP or may need introducing during PCR using mutagenic primers. (b) Primer design for simultaneous Gibson assembly and site-specific mutagenesis. Primers F1-for and F2-rev include a utility sequence consisting of a 15–25 nt overlap with the pSpliceExpress vector. Primers F1-rev and F2-for are mutagenic and introduce a complementary SNP during amplification of Fragment 1 and Fragment 2 (see inset box). Using only primer F1-for and F2-rev primers will produce a WT Fragment 3 deficient of SNP but including overlapping sequence ends. (c) Overview of a Gibson assembly cloning method. (i) Overlapping double-stranded DNA fragments are digested at their 5′ ends, revealing complementary 3′ sequences between fragments. (ii) Fragments anneal due to complementarity. (iii) 3′ ends are extended by DNA polymerase and DNA nicks are sealed by ligase activity, producing a seamless assembled construct.

the following thermocycler settings. Initial denaturation at 98 °C for 1 minute, followed by 35 cycles of 98 °C for 30 seconds, calculated annealing temperature (55–72 °C) for 20 seconds and extension at 72 °C for 30 s/kb of amplicon, followed finally by a further extension at 72 °C for 10 minutes.

Run a small volume (5–10 μl) of each reaction on a suitable percentage agarose gel to ensure that only a single product has been amplified in each reaction. Use Primer-BLAST or SnapGene to calculate the product size for each pair of primers and check band heights against expected product size. Clean up the remaining PCR reactions using an QIAquick PCR purification kit (Qiagen) and measure the concentration of each PCR product.

Incubate 1 μg of purified pSpliceExpress vector with 1 μl each of restriction enzymes NheI and XbaI (NEB) for one hour at 37 °C in the presence of a suitable buffer, with a total reaction volume of 50 μl. Halt enzymatic activity by an incubating reaction at 65 °C for 20 minutes. Run the digested vector at 70 V for 90 minutes on a 0.8% agarose gel and excise band relating to the fragment at 4.2 kb (Figure 4.3). Purify the gel band using an QIAquick gel extraction kit (Qiagen).

Carry out the Gibson assembly by incubating digested pSpliceExpress plasmid with both Fragment-1 and Fragment-2, or with WT-Fragment alone. A total of 0.02–0.5 pmol of DNA fragments should be used with a 2–3-fold excess of fragments to vector.

Mix DNA components with 10 μl of 2× Gibson assembly Master Mix and adjust the reaction volume to 20 μl with deionised water. Incubate samples in a thermocycler at 50 °C for 60 minutes and store on ice or at −20 °C for subsequent transformation.

4.9.1.6 Bacterial Transformation

Transform Gibson assemblies into recommended competent cells provided with Gibson assembly kit (New England Biolabs) using the manufacturer's recommended protocol. Briefly, thaw an aliquot(s) of NEB 5-alpha competent *E. coli* (NEB) on ice. Add 2 μl of chilled Gibson assembly to the competent cells and mix gently (do not vortex). Incubate the mixture on ice for 30 minutes. Heat shock at 42 °C for 30 seconds and place back on ice for 2 minutes. Add 950 μl of room-temperature SOC (2% tryptone, 0.5% yeast extract, 10 mM NaCl, 2.5 mM KCl, 10 mM $MgCl_2$, 10 mM $MgSO_4$ and 20 mM glucose) media and incubate at 37 °C for at least one

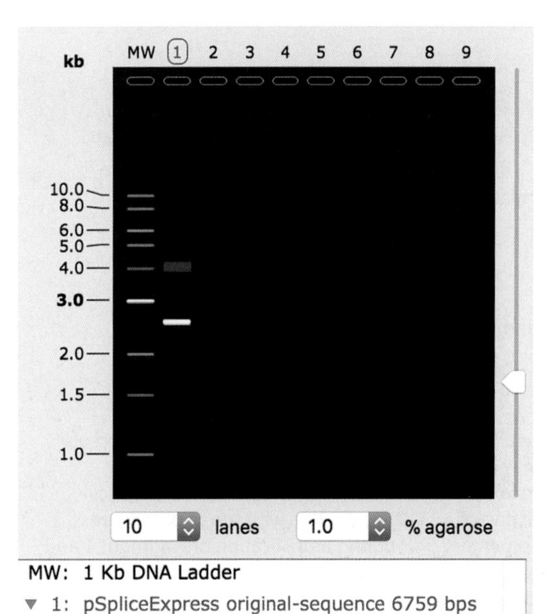

Figure 4.3 SnapGene simulated agarose gel of pSpliceExpress plasmid following digestion with restriction enzymes NheI and XbaI. The band visible at approximately 4.2 kb (shown in red) relates to the portion of plasmid needed for subsequent Gibson assembly.

hour. Shake vigorously or rotate. Warm selection plates (e.g. Carbenicillin containing Luria-Bertani [LB] agar) to 37 °C. Spread 100 µl of cells on to the plates. Briefly centrifuge the remaining 900 µl of cell culture, aspirate the media and re-suspend the pellet in 100 µl of fresh media. Spread the remaining 100 µl of cell culture on to an additional plate. This plate serves as a concentrated replicate in case transformation efficiency is low. Incubate the plates overnight at 37 °C.

4.9.1.7 Plasmid Propagation/Purification

Inoculate 5 ml of LB media (10 g Bacto-tryptone, 5 g yeast extract, 10 g NaCl per litre) containing an appropriate antibiotic (e.g. Carbenicillin) with a single colony for each sample. Incubate cultures for 12–18 hours at 37 °C shaking vigorously. Carry out a mini-prep on overnight cultures using GenElute Plasmid Miniprep Kit (Sigma) using the manufacturer's recommended protocol. Briefly, pellet 2 × 2 ml of overnight culture in a 2 ml Eppendorf tube, aspirate media following each spin and finally re-suspend in 200 µl of re-suspension solution. Lyse cells by adding 200 µl of Lysis solution. Terminate the lysis step after five minutes by adding 350 µl of neutralisation solution. Invert to mix. Centrifuge the lysate for 10 minutes at 12 000 g to pellet debris. Add 500 µl of column preparation mix to the binding column and briefly centrifuge, discard the flow-through. Transfer the cleared lysate to the binding column and centrifuge for one minute at 12 000 g, discard the flow-through. Add 750 µl of wash solution to the column, centrifuge for one minute at 12 000 g and discard the flow-through. Spin the column for an additional one minute to remove excess ethanol from the column. Transfer the column to a fresh collection tube and add 50–100 µl of elution solution to the centre of the column in a dropwise fashion. Allow to stand at room temperature for one minute. Spin the column for one minute. Measure the concentration of purified plasmid with a spectrophotometer measuring absorbance at 260 nm, using the kit elution solution as a blank. Estimate the plasmid concentration given that 1 A_{260} unit = approximately 50 µg of dsDNA. Plasmid concentrations should be 50–100 ng/µl. Confirm the sequence of cloned constructs by Sanger sequencing using the sequence 5′-CTACTCAGGAGAGCGTTCAC as a bespoke sequencing primer.

4.9.1.8 HEK Cell Transfection

Start with a 10 cm cell culture dish of HEK 293 cells in an appropriate culture media (DMEM +10% Fetal calf serum) near full confluency. Wash twice with 10 ml of phosphate buffered saline and detach cells with a trypsin/EDTA solution. Count cells using a haemocytometer or coulter counter and seed a six-well culture dish with 2.5×10^5 cells in a total volume of 1.5 ml. Incubate cells overnight at 37 °C + 5% CO_2 to reach 40–60% confluency. Per sample, combine 2.5 µl of Lipofectamine (Thermo Fisher) with 250 µl of Opti-MEM media (Thermo Fisher). Combine 200 ng of plasmid construct with 250 µl Opti-MEM media. Combine both mixes together and allow to stand for 25 minutes at room temperature. Gently add to the cells and incubate the cells for 48 hours at 37 °C + 5% CO_2.

4.9.1.9 RNA Extraction

Extract RNA from cells using an appropriate protocol such as Trizol/chloroform extraction. Briefly, aspirate the media and add 1 ml of Trizol reagent (Sigma) per well of a six-well plate. Homogenise the solution by pipetting up and down or by passing the solution several times through a 20-gauge needle. Transfer to an Eppendorf tube. Add 200 µl of chloroform and shake vigorously for 15 seconds. Allow to stand at room temperature for five minutes. Centrifuge at 4 °C for 15 minutes at 12 000 g. Carefully remove the aqueous phase into a fresh tube containing 500 µl of isopropanol. Allow to stand at room temperature for five minutes. Centrifuge at 4 °C for 10 minutes at 12 000 g. Aspirate the supernatant taking care not to disturb the pellet. Wash the pellet with 70–75% ethanol. Centrifuge at 4 °C for five minutes at 12 000 g. Remove the ethanol and allow the pellet to briefly air dry. Resuspend the pellet in 100 µl of nuclease water. Clean up the RNA using an RNeasy column kit and include the DNase digestion step. Measure the concentration of RNA using a spectrophotometer measuring absorbance at 260 nm.

4.9.1.10 cDNA Conversion

Convert all RNA samples to cDNA using the Superscript IV kit (Invitrogen). Use this step to normalise all sample concentrations by adding an equal amount of RNA into each sample reaction. Briefly, add 1 µg of

RNA (in a volume not exceeding 11 μl) to 1 μl of dNTP mix (10 mM) and 1 μl of Oligo $d(T)_{20}$ (50 μM). Adjust the volume to 13 μl with nuclease-free water. Incubate the samples at 65 °C for five minutes followed by one minute on ice. Add to the samples, 4 μl of 5× SSIV buffer, 1 μl DTT (100 mM), 1 μl of RNaseOUT RNase inhibitor and 1 μl of Superscript IV. Incubate the samples in a thermal cycler for 10 minutes at 55 °C followed by 10 minutes at 80 °C to inactivate enzyme activity. Use cDNA immediately for PCR amplification or store at −20 °C long-term.

4.9.1.11 PCR Amplification of Spliced Products

Amplify the spliced mini-gene products using primers designed to target the first and last exon of the pSplice-Express mini-gene construct. Forward primer (5′-TGCTGGCCCTGCTCATCCTCTG) and reverse primer (5′-TGGACAGGGTAGTGGTGGGCCT). Use 2 μl of cDNA sample as a template in a 25 μl reaction containing 0.5 μl of dNTPs (10 μM), 0.5 μl of forward primer (10 μM), 0.5 μl of reverse primer (10 μM), 5 μl of 5× HF buffer and 0.25 μl of Phusion polymerase, made up to 25 μl with nuclease free water. Use the following thermal cycler settings. Initial denaturation at 98 °C for 1 minute, followed by 35 cycles of 98 °C for 30 seconds, annealing/extension at 72 °C for 1 minute, followed finally by a further extension at 72 °C for 10 minutes. If the fragment is >1 kb in length then the annealing/extension step may need to be increased. Use SnapGene software to carry out a silico prediction of product sizes of the most likely splice products, e.g. endogenous vector exons only (mini-gene spliced out) or a fully spliced product (mini-gene spliced in). Run PCR products on a suitable percentage agarose gel to visualise the splicing outcome from a mini-gene assay. Possible outcomes could be exon inclusion, intron inclusion, exon skipping or activation of cryptic splice sites.

4.10 Case Study: Epitope Tagging of a Yeast Gene

When investigating the function of a novel protein, or a protein you do not have an antibody to, epitope tagging can be a quick and convenient method of providing a tool for a range of experimental techniques. Epitope tagging allows the potential for the researcher to carry out western blotting, immunolocalisation, immunoprecipitation and affinity purification with the tagged protein. Epitope tagging involves cloning of the DNA sequence containing the open reading frame of the target protein in frame with the epitope tag into a plasmid vector appropriate for the host cell type. This plasmid could then be introduced into cultured higher eukaryotic cells by transfection, into yeast cell by transformation or into bacteria cells by transformation. The following case study will describe in full the methods that may be used to clone a yeast gene into a plasmid vector containing an epitope tag that can express the tagged protein in yeast cells.

4.10.1 Methods

4.10.1.1 Yeast Genomic DNA Preparation

Genomic DNA can either be purchased or can be easily prepared from most yeast strains using the following protocol. Grow a 2 ml yeast culture in YPD (10% BactoYeast extract, 20% BactoPeptone, 20% Dextrose) broth from a single colony. Spin down 1.5 ml of cells at 2000 rpm for five minutes in an Eppendorf tube. Discard the supernatant and resuspend cells in 500 μl of nuclease free water. Spin down cells at 2000 rpm for five minutes. Add 200 μl of lysis buffer (10 mM Tris-Cl pH 8, 1 mM EDTA, 100 mM NaCl, 1% sodium dodecyl sulfate [SDS], 2% Triton-X 100), 200–300 μl of acid washed glass beads (Sigma G8772), and 200 μl of phenol:chloroform:isoamyl alcohol (25 : 24 : 1 saturated with 10 mM Tris, pH 8 and 1 mM EDTA). Vortex vigorously for three minutes. Spin for five minutes at top speed in a microfuge. Carefully remove 150 μl of the top aqueous phase to a new tube and add 300 μl of 100% ethanol and mix well. Spin for five minutes at top speed in a microfuge. Remove the supernatant and then wash the DNA pellet with 95% ethanol. Remove all the ethanol and let the DNA pellet air dry for 30 minutes. Resuspend the DNA pellet in 100 μl of Tris-Cl pH 8. Measure the concentration of the genomic DNA using a spectrophotometer measuring absorbance at 260 nm.

4.10.1.2 Amplification of Target Yeast Gene

Use the *Saccharomyces* Genome Database (SGD – https://www.yeastgenome.org) to obtain the DNA sequence of the open reading frame for your GoI. In this case study we will be using the small yeast gene *DIB1*/YPR082C as an example for epitope tagging. Search for *DIB1* on the SGD webpage; then from the *DIB1* 'Summary' page select the 'Sequence' tab where, if you scroll down the page, you will find a pull-down menu in the 'Sequence' section where you can select the DNA for the 'Coding DNA'. It is important to only download/copy the coding DNA as selecting genomic DNA, for some genes in yeast, will include intron sequences that are not needed. In the case of genes with introns, cDNA would need to be used as a template for PCR amplification instead of genomic DNA.

Load the *DIB1* coding DNA sequence into SnapGene (or an equivalent sequence viewer). Remove the stop codon (TAA) at the 3′ end of the *DIB1* coding sequence to allow in frame fusion of the *DIB1* coding sequence with the tag (Figure 4.4).

Save the sequence of a yeast epitope tagging vector into a separate file accessible by SnapGene. For example, Oxford Genetics (https://www.oxfordgenetics.com) sell a wide range of plasmid vectors for use in yeast, including plasmid vectors for epitope tagging. In this case study we will be using the pSF-TEF1-COOH-3C-FLAG vector, which has the strong yeast *TEF1* promoter sequence, a multiple cloning site with a range of unique restriction enzyme recognition sites, a sequence for the 3C Protease/PreScission protease to allow the removal of the tag following protein purification or isolation, if required, and a FLAG tag sequence to C-terminally tag the inserted open reading frame. The plasmid also contains a *URA3* marker and a 2 μm origin of replication for selection and maintenance in yeast as well as a kanamycin resistance gene and pMB1 bacteria replicon for selection and maintenance in bacteria.

Identify restriction enzyme recognition sites in the pSF-TEF1-COOH-3C-FLAG vector that are not found in the *DIB1* open reading frame sequence. In this case EcoRI and KpnI are two restriction enzyme recognition sites that are not within the *DIB1* coding sequence and are unique restriction enzyme recognition sites within the multicloning site upstream of the FLAG tag in the pSF-TEF1-COOH-3C-FLAG vector (Figure 4.5).

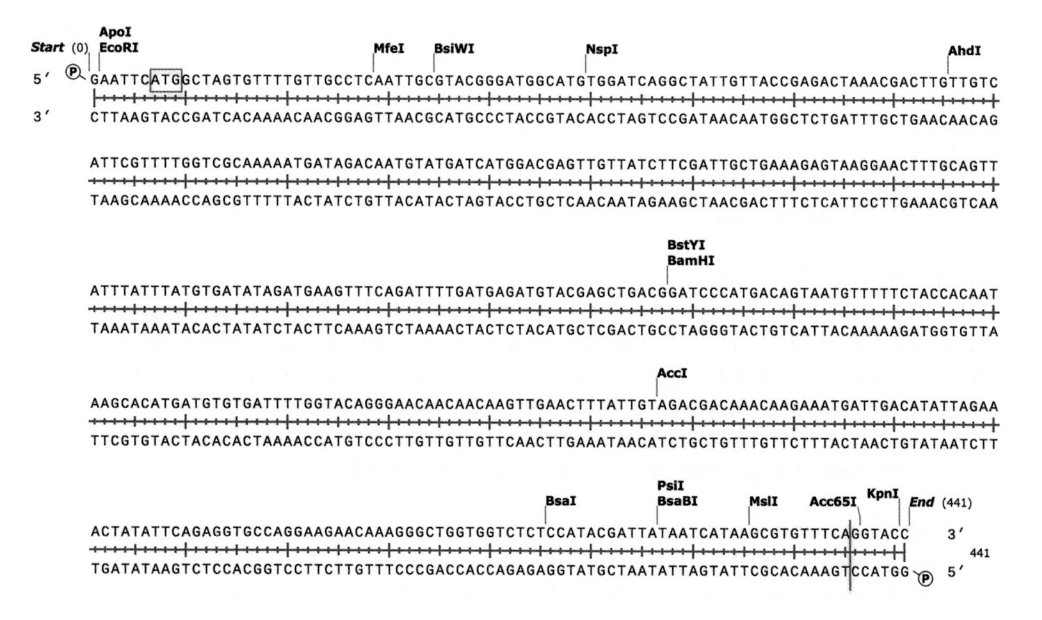

Figure 4.4 Sequence of the yeast *DIB1* gene with added EcoRI and KpnI restriction enzyme sites and the TAA stop codon removed. The Dib1 start codon is indicated with a red box and the position where the stop codon of Dib1 was removed is indicated with a red line.

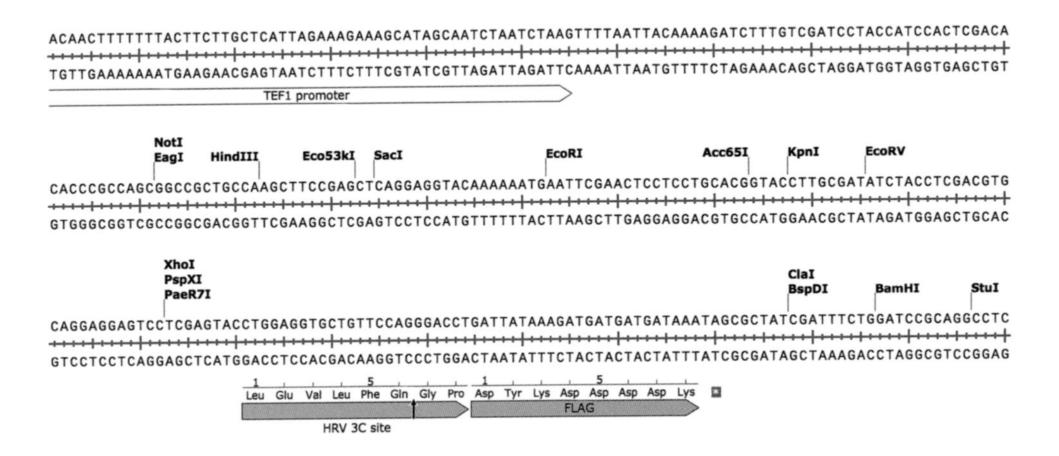

Figure 4.5 Multicloning site region of the pSF-TEF1-COOH-3C-FLAG vector.

Design PCR primers to amplify the *DIB1* coding sequence from genomic DNA. The primers should include the EcoR1 and KpnI restriction enzyme recognition sites and allow in frame fusion with the FLAG tag. Example primers for this case would be a forward primer (5′–GGGAATTCATGGCTAGTGTTTTGTTGCC with the EcoRI site underlined) and a reverse primer (5′-GGGGTACCTGAAACACGCTTATGATTATAATCG with the KpnI site underlined). Extra sequences have been added to the 5′ ends of the primers to allow a more efficient restriction digest.

Amplify the *DIB1* coding sequence with the designed primers using a high fidelity polymerase such as Phusion. Use 50–250 ng of yeast genomic DNA as a template in a 50 μl reaction containing 1 μl dNTPs (10 μM), 1 μl forward primer (1 μM), 1 μl reverse primer (10 μM), 10 μl 5× HF buffer and 0.5 μl Phusion polymerase, made up to 50 μl with nuclease free water. Use the following thermocycler settings. Initial denaturation at 98 °C for 1 minute, followed by 35 cycles of 98 °C for 30 seconds, calculated annealing temperature (55–72 °C) for 20 seconds and extension at 72 °C for 30 s/kb of amplicon, followed finally by a further extension at 72 °C for 5 minutes.

Run a small volume (5–10 μl) of the PCR on an agarose gel (at an appropriate percentage to check that only a single product has been amplified in each reaction). Use PrimerBlast or SnapGene to calculate the product size and check band heights against expected product size; in this case it should be 445 bp. Clean up the remaining PCR reaction using a column cleanup kit (i.e. Qiagen PCR purification kit) and measure the concentration of the PCR product using a spectrophotometer.

4.10.1.3 Restriction Digestion of PCR Product

Usually up to 1 μg of the cleaned-up PCR product is then digested with both EcoRI and KpnI. A typical restriction enzyme digestion reaction in this case contains up to 1 μg of PCR product, 5 μl of 10× CutSmart buffer (New England Biolabs), 10 units of EcoRI-HF (New England Biolabs) and 10 units of KpnI-HF (New England Biolabs) made up to 50 μl with nuclease free water. The reaction is then incubated at 37 °C for one hour. The restriction digest reaction is then cleaned up using a column cleanup kit (i.e. Qiagen PCR purification kit) and the concentration of the digested PCR product is measured using a spectrophotometer.

4.10.1.4 Restriction Digestion of Plasmid Vector

Usually up to 1 μg of the plasmid vector is digested with both EcoRI and KpnI. A typical restriction enzyme digestion reaction in this case contains 1 μg of pSF-TEF1-COOH-3C-FLAG, 5 μl of 10× CutSmart buffer (New England Biolabs), 10 units of EcoRI-HF (New England Biolabs) and 10 units of KpnI-HF (New England Biolabs) made up to 50 μl with nuclease free water. Add 10 units of calf intestinal phosphatase (CIP) to remove phosphates from the ends of the cut plasmid to prevent any re-ligation of singly cut plasmid. The reaction is then incubated at 37 °C for one hour. The restriction digest reaction is then cleaned up using a column cleanup kit (i.e. Qiagen PCR purification kit) and the concentration of the digested plasmid is measured using a spectrophotometer. The column purification removes the small piece of DNA liberated by the double digestion of the plasmid vector with the two enzymes.

4.10.1.5 Ligation of *DIB1* PCR Product into the pSF-TEF1-COOH-3C-FLAG Plasmid

A typical ligation reaction is carried out in a 20 µl volume consisting of 2 µl of 10× T4 DNA ligase buffer, 50 ng (~12 fmol of ~6765 bp) of restriction enzyme digested and purified plasmid vector, 10 ng (~36 fmol of ~445 bp) of restriction enzyme digested and purified *DIB1* PCR product, made up to 19 µl with nuclease free water and 1 µl (400 units) T4 DNA ligase. The molar ratio given here is 1 : 3 for the amount of vector to insert. For making a vector to insert ratio calculations the NEBioCalculator (http://nebiocalculator.neb.com) is a useful tool. A control reaction is also set up with the vector alone to determine the background of uncut/re-ligated plasmid. For ligations with compatible overhanging ends the ligation reaction is incubated at room temperature for at least 10 minutes. The resulting ligation of *DIB1* into the pSF-TEF1-COOH-3C-FLAG plasmid to give a fusion of the *DIB1* open reading frame with the FLAG tag is shown in Figure 4.6.

4.10.1.6 Transformation of Ligation Reaction

A volume of 1–10 µl from the ligation reactions is transformed into an appropriate commercially available competent bacteria strain according to the manufacturer's protocol. An example of the bacteria transformation procedure can be found in Case Study 1 (Section 4.9.1.6). Following transformation the control plate should have no or very few bacteria colonies whereas the ligation reaction with the insert should have significantly more colonies.

4.10.1.7 Plasmid Propagation/Purification

Bacteria colonies from the ligation plate should be cultured and plasmid DNA isolated to determine by DNA sequencing whether the DNA sequence has been inserted correctly and is oriented in frame with the FLAG tag. An example of a typical protocol for the plasmid propagation/purification procedure can be found in Case Study 1 (Section 4.9.1.7).

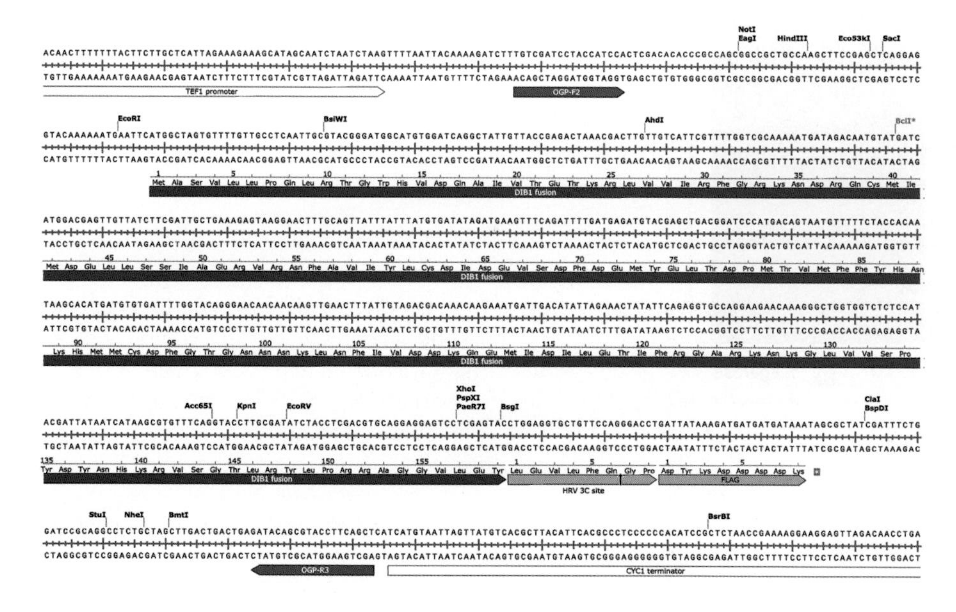

Figure 4.6 *DIB1* gene cloned into the pSF-TEF1-COOH-3C-FLAG vector in frame with the FLAG tag. Binding sites for the sequencing primers OGP-F2 and OGP-R3 are shown.

4.10.1.8 Test Restriction Digest and Sequencing

It is useful to carry out a diagnostic restriction digest of the potential plasmid constructs before sending them off for DNA sequencing. In this case a restriction digestion with the enzyme AccI will produce three bands if the *DIB1* sequence has been inserted and only two bands if it has not been inserted (Figure 4.7). Plasmids found by this test restriction digest to have the *DIB1* inserted should then be sent for DNA sequencing using the sequencing primers OGP-F2 (5′- TGTCGATCCTACCATCCA) and OGP-R3 (5′- AGCTGAAG-GTACGCTGTATC) (Figure 4.6).

4.10.1.9 Yeast Transformation

Once the plasmid sequence has been confirmed to be correct by DNA sequencing it can then be transformed into a yeast strain that contains a deletion/inactivation of the *URA3* gene to allow selection of the *DIB1*-containing pSF-TEF1-COOH-3C-FLAG plasmid, which contains the *URA3* gene. A general protocol for yeast transformation based on the Geitz method for yeast transformation is described in [15].

Inoculate the single colony into 5 ml YPD (10% BactoYeast extract, 20% BactoPeptone, 20% Dextrose) broth in a 15 ml sterile tube and incubate for 16–20 hours with shaking at 30 °C. Dilute the cells in 25 ml of YPD broth in a 50 ml sterile tube to A_{600} of 0.5 units (this amount of cells will be enough to carry out four transformation reactions). Incubate with shaking at 30 °C until $A_{600} = 2$ (approximately four hours). Pellet the cells at 2000 g for five minutes and then decant the supernatant. Re-suspend the cell pellet in 25 ml of sterile water, then pellet the cells at 2000 g for five minutes and decant the supernatant. Re-suspend the cells in 700 μl of sterile 0.1 M lithium acetate and transfer to a sterile 1.5 ml microcentrifuge tube. Pellet the cells at 10 000 g for 10 seconds and then aspirate the supernatant. Re-suspend the cells in 200 μl of 0.1 M lithium acetate. Make up a transformation mix for each transformation consisting of 36 μl of 1 M lithium acetate, 240 μl of 50% polyethylene glycol 3500, 50 μl of 2 mg/ml of salmon sperm DNA (the salmon sperm DNA is boiled for five

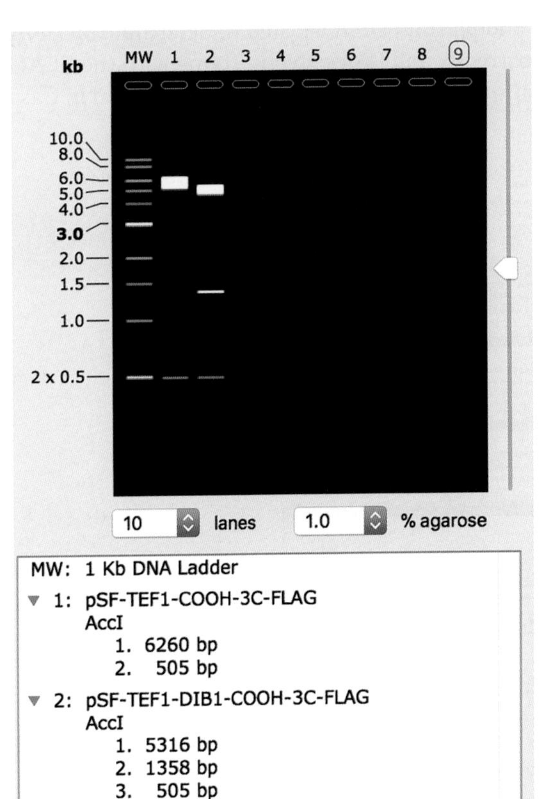

Figure 4.7 SnapGene simulated agarose gel of test restriction enzyme digests with the enzyme AccI of the pSF-TEF1-COOH-3C-FLAG vector (Lane 1) or the *DIB1* gene cloned into the pSF-TEF1-COOH-3C-FLAG vector (Lane 2). The successful cloning of *DIB1* into the pSF-TEF1-COOH-3C-FLAG produces an additional band at 1358 bp and the change in mobility of the 6260 bp band to 5316 bp.

minutes and then cooled on ice before use) and 0.2–1.0 μg of plasmid DNA in a volume of 34 μl. Be sure to also set up a negative control reaction with no plasmid DNA. Add 50 μl of cells to each transformation mix and vortex vigorously to mix. Incubate at 30 °C for 30 minutes and then incubate at 42 °C for 30 minutes. Pellet the cells at 2000 g for 15 seconds. Gently re-suspend the cells in 1 ml of sterile water and gently spread 50–200 μl of aliquots on the appropriate selective plate. In this case the appropriate selective plate would be a synthetic defined plate without URA (SD-URA). MP Biomedicals (https://www.mpbio.com) distributes a wide variety of media for yeast growth. Grow cells at 30 °C for three days to allow individual yeast colonies to form.

Transformed cells can then be used for western blotting, immunolocalisation, immunoprecipitation or affinity purification of the tagged protein.

References

1 Chuang, L.Y., Cheng, Y.H., and Yang, C.H. (2013). Specific primer design for the polymerase chain reaction. *Biotechnol. Lett.* 35 (10): 1541–1549.

2 Kosuri, S. and Church, G.M. (2014). Large-scale de novo DNA synthesis: technologies and applications. *Nat. Methods* 11 (5): 499–507.

3 Hartley, J.L., Temple, G.F., and Brasch, M.A. (2000). DNA cloning using *in vitro* site-specific recombination. *Genome Res.* 10 (11): 1788–1795.

4 Gibson, D.G. (2011). Enzymatic assembly of overlapping DNA fragments. *Method Enzymol.* 498: 349–361.

5 Rina, M., Pozidis, C., Mavromatis, K. et al. (2000). Alkaline phosphatase from the Antarctic strain TAB5 – properties and psychrophilic adaptations. *Eur. J. Biochem.* 267 (4): 1230–1238.

6 Aslanidis, C. and Dejong, P.J. (1990). Ligation-independent cloning of Pcr products (Lic-Pcr). *Nucleic Acids Res.* 18 (20): 6069–6074.

7 Clark, J.M. (1988). Novel non-templated nucleotide addition-reactions catalyzed by procaryotic and eukaryotic DNA-polymerases. *Nucleic Acids Res.* 16 (20): 9677–9686.

8 Yao, S., Hart, D.J., and An, Y.F. (2016). Recent advances in universal TA cloning methods for use in function studies. *Protein Eng. Des. Sel.* 29 (11): 550–556.

9 Shuman, S. (1994). Novel-approach to molecular-cloning and polynucleotide synthesis using vaccinia DNA topoisomerase. *J. Biol. Chem.* 269 (51): 32678–32684.

10 Novy, R., Drott, D., Yaeger, K., and Mierendorf, R. (2001). *Novations* 12: 1–3.

11 Prinz, W.A., Aslund, F., Holmgren, A., and Beckwith, J. (1997). *J. Biol. Chem.* 272: 15661–15667.

12 Sikorski, R.S. and Hieter, P. (1989). A system of shuttle vectors and yeast host strains designed for efficient manipulation of DNA in *Saccharomyces cerevisiae. Genetics* 122: 19–27.

13 Soemedi, R., Cygan, K.J., Rhine, C.L. et al. (2017). Pathogenic variants that alter protein code often disrupt splicing. *Nat. Genet.* 49 (6): 848+.

14 Kishore, S., Khanna, A., and Stamm, S. (2008). Rapid generation of splicing reporters with pSpliceExpress. *Gene* 427 (1–2): 104–110.

15 Gietz, R.D. and Woods, R.A. (2006). Yeast transformation by the LiAc/SS carrier DNA/PEG method. *Methods Mol. Biol.* 313: 107–120.

Further Reading

Green, M.R. and Sambrook, J. (2012). *Molecular Cloning: A Laboratory Manual.* Cold Spring Harbor Laboratory Press. ISBN: 978-1-936113-42-2.

Websites

NCBI (https://www.ncbi.nlm.nih.gov)
UCSC (https://genome.ucsc.edu)
Ensembl (https://www.ensembl.org/index.html)
SGD (https://www.yeastgenome.org)
PomBase (https://www.pombase.org)
ZFIN (https://zfin.org)
Flybase (http://flybase.org)
Benchling (https://benchling.com)
SnapGene (http://www.snapgene.com)
pDraw (http://www.acaclone.com)
UGENE (http://ugene.net)
NEB Tm calculator (https://tmcalculator.neb.com/#!/main)
Gateway™ cloning (https://www.thermofisher.com/uk/en/home/life-science/cloning/gateway-cloning.html)
Gibson Assembly® (https://www.neb.com/products/e2611-gibson-assembly-master-mix#Product%20Information)
Oxford Genetics (https://www.oxfordgenetics.com)
NEBioCalculator (http://nebiocalculator.neb.com)
MP Biomedicals (https://www.mpbio.com)

5

Proteomic Techniques and Their Applications

Hsueh-Fen Juan

Department of Life Science, Graduate Institute of Biomedical Electronics and Bioinformatics, National Taiwan University, Taipei, Taiwan

5.1 Significance and Background

'Proteome' was first coined by Marc Wilkins and Keith L. Williams in 1994 in the first Siena meeting on '2D Electrophoresis: from protein maps to genomes' [1] and defined in 1996 as 'the entire proteins in a cell, a tissue or an organism' [2]. The term, proteome, was not adopted immediately because the genomics researchers preferred to call the study of proteome as functional genomics. In 'A Sydney proteome story', Williams talked about proteins as 'functional genomes', but that did not really work for us, so we stayed with the name 'proteome' [3]. One year later, the term 'proteomics' was coined as the study of proteome, including protein identification, protein modification and protein–protein interaction (PPI). Proteins are functional molecules and the workhorses of the cell [3], and hence profiling proteome helps us to understand cellular processes.

In the 1980s, researchers used Edman sequencing to obtain protein sequence information for the abundant proteins after applying two-dimension gel electrophoresis (2DE) to separate proteins by size and charge. Edman sequencing was a laborious process, however, and new techniques were sought. In the 1990s, genome sequencing techniques were improved dramatically, and genome databases were built fast. Not only genome sequencing techniques but also mass spectrometry (MS) techniques enhanced proteomics progress rapidly because MS has high sensitivity for identification of proteins if the corresponding protein sequences are in the genome databases. Hence, both developments of genome sequencing and MS technologies were essential for enabling effective proteomic studies [4].

MS allows us to rapidly and reliably identify a protein in a sample on the basis of its peptide mass and becomes a vital tool for current proteomics. In 2002, John B. Fenn and Koichi Tanaka were awarded the Nobel Prize in Chemistry for their contributions in hovering through spraying and blasting, respectively, in order to study of biological macromolecules (www.nobelprize.org). John B. Fenn pioneered the use of a strong electric field for sample spraying and then producing small, charged, freely hovering ions. This method came to be called 'electrospray ionisation (ESI)'. Koichi Tanaka was the first person using an intense laser pulse to let large molecules such as proteins become released as free ions under suitable chemical environments. The phenomena were called 'soft laser desorption'.

MS always produces a very complex spectrum, which makes the analysis of the data difficult. Matthias Mann, now the director of the Max Planck Institute of Biochemistry, when a graduate student in John B. Fenn's group, quickly worked out a couple of algorithms to simplify the spectrum and make the MS data analysis easier. Matthias Mann and his colleague developed a useful bioinformatics tool, MaxQuant [5], to analyse the enormous amounts of data generated by MS-based proteomics. They provided detailed protocols about what data types can be analysed in this software and how to use it [6].

Since MS and proteomics produce a large amount of data, bioinformatics has become an essential tool. There are four major bioinformatics topics in proteomics: identification, quantification, functional analysis, and data

Biomolecular and Bioanalytical Techniques: Theory, Methodology and Applications, First Edition. Edited by Vasudevan Ramesh.

sharing [7]. First, we need tools, such as MaxQuant, to identify and quantify proteins and post-translational modification (PTM) sites from the mass spectrum data. If proteins are separated by 2DE, image analysis software tools are required to quantify the intensities of protein spots. After identification and quantification of proteins and PTM sites, we need to annotate the data with existing biological knowledge such as Gene Ontology (GO) Enrichment Analysis (http://geneontology.org/page/go-enrichment-analysis) and functional networks such as STRING (https://string-db.org). Finally, sharing the data using online repositories, such as ProteomExchange (http://www.proteomexchange.org), is also important in the proteomics community.

Proteomics has been applied to solve biological and medical problems, such as disease prognosis and diagnosis, drug-induced mechanism studies, as well as plant science and paleography. The accurate quantification in proteomics becomes an important requirement for clinical applications and biological research. In this chapter, the principles of major proteomics techniques and methods, including sample preparation, quantitation and data analysis, will be introduced and, finally, their applications will also be reviewed.

5.2 Principles of Major Proteomics Techniques

The major high throughput proteomics techniques include MS and proteome microarrays. There are two MS-based techniques: 2DE and shotgun methods (Figure 5.1). After generating high throughput data, bioinformatics becomes the key issue to obtain worthy biological explanations. Here I will also briefly introduce the bioinformatics tools and resources.

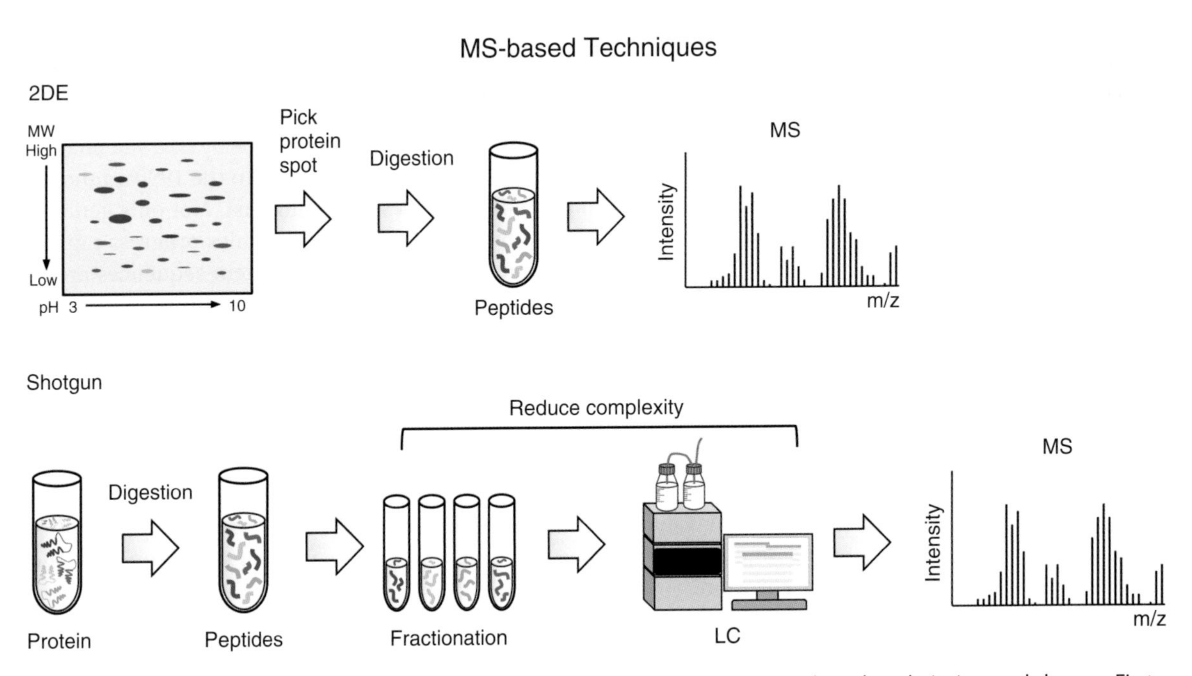

Figure 5.1 Two MS-based techniques. 2DE is a traditional method to separate proteins based on their sizes and charges. First, protein spots of interest are chosen and digested into peptides, which are loaded to MS. Shotgun proteomics is a recently developed method that directly digests proteins into peptides. To reduce the complexity of peptides, fractionation by ion exchange or reverse phase column is needed. Before being loaded to MS, liquid chromatography (LC) is used to reduce the sample complexity.

5.2.1 MS-Based Techniques

For protein identification, the proteins need to be digested into peptides, which can be sequenced by MS. Before applying the peptide samples to MS, various ionisation methods need to be considered. Although many ionisation methods exist, two are most suitable for peptides: ESI and matrix-assisted laser desorption ionisation (MALDI). MALDI is suitable for simple samples, while ESI is suitable for complex samples because it can be combined with liquid chromatography (LC-MS). Figure 5.2 illustrates these two methods.

5.2.1.1 Electrospray Ionisation (ESI)

ESI uses an electrospray at high voltage to produce ions from macromolecules such as proteins. ESI can produce multiple-charged ions, which extend the mass range of the analyser. ESI is one of the two so-called 'soft ionisation' techniques since it splits biomolecules with little fragmentation. ESI can be coupled with MS (electrospray ionisation mass spectrometry [ESI-MS]). In principle, solutes are sprayed from a capillary and droplets break down and evaporate. Peptides are multiply charged: +1, +2, +3, +4, +5, +6, etc.

In 1984, Masamichi Yamashita and John B. Fenn first reported the ESI technique [8] and later the analysis of biological macromolecules [9]. John B. Fenn was awarded the Nobel Prize in Chemistry in 2002 with Koichi Tanaka. He reviewed the history in the development of ESI MS in his laboratory [10] and described the work of Malcolm Dole, who was the first scientist to use ESI with MS in 1968 [11].

5.2.1.2 Matrix-Assisted Laser Desorption/Ionisation Time-of-Flight (MALDI-TOF)

MALDI is an ionisation method that utilises a matrix with laser energy absorbing ability to create ions from large molecules such as proteins and peptides with minimal fragmentation [12]. This method is also called a 'soft ionisation' technique suitable for samples after 2DE separation. Firstly, the sample is mixed with a matrix that absorbs ultraviolet light. Secondly, the laser is fired to irradiate the sample. Finally, the samples are ionised and then fly in the presence of an electric field to the detector. The detector records a time-of-flight spectrum, which can be equated with mass since the flight time is proportional to the sample mass.

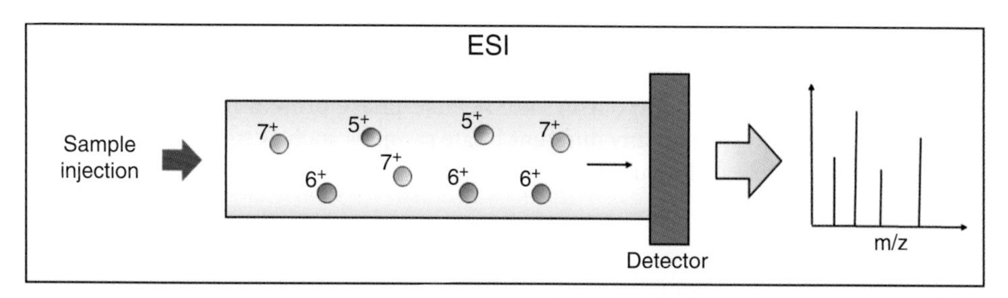

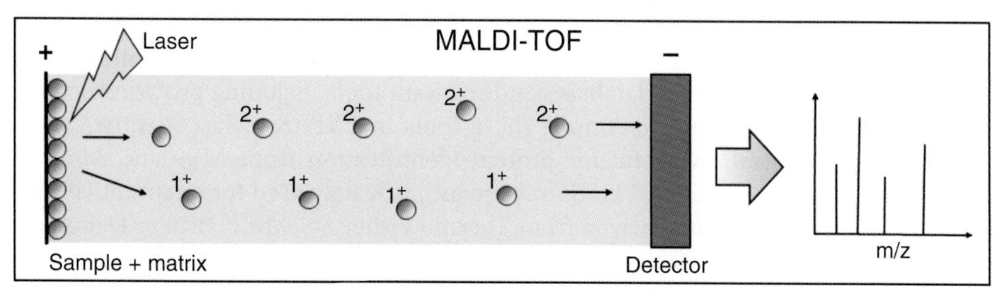

Figure 5.2 Two ionisation methods. Electrospray ionisation (ESI) is a method to produce ions from macromolecules by electrospray. Matrix-assisted laser desorption/ionisation time-of-flight (MALDI-TOF) utilises a matrix with laser energy absorbing the ability to create ions from large molecules. The ions produced from the two methods can then be detected. These two ionisation methods coupled with mass spectrometry are useful for proteomics.

Many matrixes are used now, e.g. alpha-cyano-4-hydroxy-cinnamic acid (CHCA), which is used for peptides <10 kDa, and 2,5-dihydroxybenzoic acid (DHB), which is used for proteins >10 kDa. However, in 1980, scientists were looking for a matrix suitable for large biomolecules. Fortunately, in February 1985, Koichi Tanaka mistakenly used a glycerin-UFMP (ultrafine metal powder) and found the matrix was better than cobalt UFMP to absorb photon (UV laser) and form ions. For this finding, he was awarded the Nobel Prize in Chemistry in 2002.

5.2.2 Proteome Microarray

The proteome microarray, also known as a protein microarray or protein chip, is a high throughput proteomics method used in addition to MS-based methods and has emerged as a promising approach. Proteome microarrays are miniaturised, parallel assay systems that use immobilised coded proteins of an organism in a high density format placed on a microscope slide using a contact spotter or a non-contact microarray [13, 14]. The construction of defined sets of overexpressed cloned genes for high throughput expression and purification of recombinant proteins are essential for a proteome microarray. Similar to DNA microarrays, printing proteins as microarrays is a challenging technique in the proteome microarray field. To detect proteins on a proteome chip, molecular probes are labelled with either fluorescent, affinity, photochemical or radioisotope tags [14]. Typically, the most widely used microarray labelling is fluorescence, which offers high sensitivity and can be read out on a confocal DNA microarray scanner, thereby in principle enabling the production of both qualitative and quantitative data [15].

Proteome microarrays have many applications, including protein–protein [16], protein–phospholipid [17], protein–small molecule and protein–DNA/RNA interactions, biomarker discovery and identification of enzyme (e.g. protein kinases) substrates [13–15]. In general, there are three types of proteome microarrays according to their applications: analytical, functional and reverse-phase microarrays [14]. One of the model analytical microarrays is an antibody microarray that contains antibodies immobilised on a chip and the targeted proteins can be detected either by direct labelling or using a reporter antibody in sandwich assay format similar to the enzyme-linked immunosorbent assay (ELISA), except that the reaction happens on a chip, not in solution. So-called 'functional microarrays' focus on revealing protein functions by microarrays, which are broadly applied for protein–enzyme, protein–protein, protein–lipid and protein–DNA interactions, and so on [14, 16, 17]. The third proteome microarrays are reverse-phase protein microarrays, which differ from classic protein microarrays by printing many different lysate samples such as tissues or cell lysates on the chip and then identify specific proteins with suitable probes [18].

5.2.3 Bioinformatics Tools and Resources

As noted earlier, proteomics approaches generate large amounts of data, and therefore bioinformatics tools and resources are essential for biological interpretation. Using MS-based methods, hundreds of thousands of spectrum datasets can be obtained and used to identify proteins. ExPASy is the SIB Bioinformatics Resource Portal, which provides access to scientific databases and software tools including proteomics tools and resources (https://www.expasy.org/proteomics). Among these tools in EXPASy, Mascot (http://www.matrixscience.com/search_form_select.html) is useful for protein identification from MS data. Another software, MaxQuant (http://www.biochem.mpg.de/5111795/maxquant), is widely used for quantitative proteomics with large-scale mass-spectrometric data derived from Thermo Fisher Scientific, Bruker Daltonics, AB Sciex and Agilent Technologies MS systems and supports the main labelling techniques such as stable isotope labelling with amino acids in cell culture (SILAC), di-methyl, TMT, and isobaric tags for relative and absolute quantitation (iTRAQ), as well as label-free quantification [5, 6]. Since 2015, MaxQuant has provided an updated component, termed the Viewer, which displays high resolution proteomics data by visualisation.

To explain the functions of biological molecules, network biology is core to understanding how protein–protein, DNA/RNA–protein and lipid–protein interactions occur and to determine the functions of these molecules. Many software packages can be used for network analysis including commercial, e.g. Ingenuity Pathway Analysis (IPA) and Pathway Studio, and free open tools, e.g. Cytoscape (http://www.cytoscape.org). All of these tools offer functional analysis for not only multiple gene expression but also proteomics and metabolomics experimental data. Most biological networks are generated by searching databases of curated literature-sourced interactions, such as Kyoto Encyclopedia of Genes and Genomes (KEGG, https://www.genome.jp/kegg) and IntAct (www.ebi.ac.uk/intact). Gene Set Enrichment Analysis (GSEA) is a widely used computational method for omics data analysis that determines whether a defined set of genes is over-represented and shows statistically significant differences between two biological states [19, 20]. PTM proteome analysis is sometimes different from global proteome analysis. For example, phosphoproteome data allow us to infer the phosphor motifs and upstream kinases. Hence, we developed a software application focusing on phosphoproteome data, named DynaPho (http://dynapho.jhlab.tw). The tool includes five analytical modules: phosphosite clustering, biological pathways, kinase activity, dynamics of interaction networks and the predicted kinase–substrate associations [21].

5.3 Methods for Proteomics

The methods for proteomics comprise MS-based and chip-based methods. Here I focus on MS-based methods. I will first explain how to prepare samples from cells, tissues and blood and then describe several methods for proteome quantification.

5.3.1 Sample Preparation

Samples can be taken from cells, tissues and blood. Given the recent trend of research on single-cell omics profiling, I will also describe how to isolate the single cells from cell population and tissues. Figure 5.3 briefly explains the preparation methods for different types of samples.

5.3.1.1 Cell Sample Preparation

For cell sample preparation [22, 23], in general, cells need to be cultured, harvested and then lysed using a lysis buffer containing 1% (v/v) sodium dodecyl sulfate, 50 mM Tris-HCl, 10% (v/v) glycerol and a protease inhibitor cocktail. The cells are homogenised on ice using a homogeniser for two to three minutes. The cell lysate is centrifuged at 12 500g at 4 °C for 30 minutes. The supernatant containing the protein extract is collected and the protein concentration is measured using, for example, a Pierce BCA Protein Assay Kit. To digest proteins into peptides, gel-assisted protein digestion is useful to reduce sample volume and reagents, which are sensitive for mass spectrometry [23]. Peptides can be extracted from the gel with 0.1% (v/v) trifluoroacetic acid (TFA), 50% acetonitrile (ANC)/0.1% (v/v) TFA and 100% ANC sequentially by vigorous vortexing [23]. The extracted peptide solution has to be dried using a centrifugal evaporator. Now the peptide samples are ready for identification or various tag labelling such as iTRAQ and dimethyl labelling.

For phosphoproteome experiments [23–25], the protein extract should be reduced with 10 mM dithiothreitol at room temperature for 30 minutes and carbamidomethylated with 55 mM iodoacetamide at room temperature in the dark for 30 minutes. Alkylated proteins can be digested with endopeptidase Lys-C (1 : 100 w/w) for two hours followed by sequencing grade modified trypsin (1 : 100 w/w) overnight at room temperature for two minutes. The trypsin reaction can be inactivated by acidified the peptide solution to a $pH < 3$ using TFA. To remove detergent, the acidified peptide solution should be combined with an equal volume of ethyl acetate and agitated vigorously for one minute, followed by centrifugation at 15 700g for two minutes to separate the aqueous and organic phases. The sample from the aqueous phase is dried using a centrifugal evaporator and then subjected to desalting using Styrenedivinylbenzene Empore disk membranes (SDB-XC) StageTips (catalogue no. 2340; 3 M) and eluted in a buffer containing 0.1% (v/v) TFA and 80% (v/v) acetonitrile (CAN) [26]. Before applying MS analysis, performing phosphopeptide enrichment is better to identify more

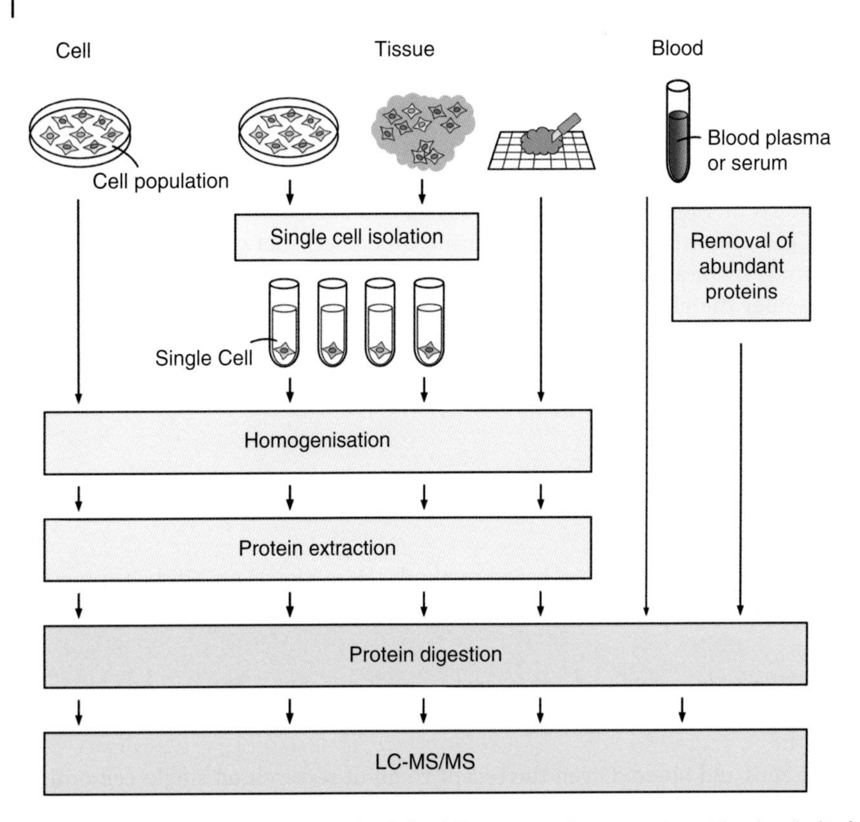

Figure 5.3 An illustration of the methods for different sample preparations. The detailed information is described in the text.

phosphopeptides. Many strategies have been developed to enrich phosphopeptides, e.g. Ni-NTA, titanium dioxide (TiO_2) enrichment and hydroxyl acid-modified metal oxide chromatography (HAMMOC) [23, 24].

Single-cell omics become important for understanding the functional differences of various cells contributing to health and disease [27]. Before performing single-cell omics profiling, cell isolation is a key issue. Many methods for single-cell isolation have been developed, including mouth pipette, serial dilution, flow-assisted cell sorting (FACS), robotic micromanipulation and microfluid platforms [28]. However, proteome profiling in single cells is a longstanding challenge since proteins cannot be amplified using the polymerase chain reaction (PCR). Recently, scientists developed a few methods to measure proteins in single cells according to conjugation of antibodies with oligonucleotides, for example, ribonucleic acid expression and protein sequencing assay (REAP-seq), which is similar to standard flow cytometry methods but with antibodies conjugated to DNA barcodes instead of fluorophores [29]. Proteins are probed using a homogeneous affinity-based proximity extension assay (PEA) using pairs of antibodies conjugated with oligonucleotides after single-cell isolation by FACS [30]. Zhu et al. developed a nanoPOTS (nanodroplet processing in one pot for trace samples) platform that used serial dilution to obtain single cells for further proteomics analysis [31]. Using this method, they claimed that they could identify >3000 proteins from as few as 10 cells [31]. The nanoPOTS method also can be applied to tissue samples.

5.3.1.2 Tissue Sample Preparation

Tissues are excised and collected for proteome and phosphoproteomic analyses. Tissue is first ground into a fine powder using a mortar and pestle in liquid nitrogen. Liquid nitrogen must be added to the mortar frequently to ensure that the tissues do not thaw during grinding [32].

For global proteomics experiments, tissue power is then suspended in lysis buffer containing 1% (v/v) safety data sheet (SDS), 50 mM Tris-HCl, 10% (v/v) glycerol, and protease inhibitor. The amount of lysis buffer added is based on the amount of tissue powder. The solution containing the tissue powder is re-suspended by pipetting until there is almost no visible pellet. The sample solution should be homogenised on ice using a homogeniser similar to cell sample preparation [23]. The following method is the same as with the cell sample preparation.

For phosphoproteomics experiments, the protein extract of each sample is denatured in an 8 M urea solution. Proteins are then reduced, carbidomethylated and diluted five times with 50 mM TEAB for trypsin digestion. Tryptic peptides are then processed by labelling and phosphopeptide enrichment, and then analysed by LC-MS/MS [24].

5.3.1.3 Blood Sample Preparation

Blood containing serum and plasma is the predominant sample used for biomarker studies [33, 34] and clinical diagnosis [35, 36]. However, MS-based plasma/serum proteomics is very challenging because of its extremely large dynamic range of protein abundances [35, 36]. Removal of abundant proteins for discovery proteomics is one method to use before applying LC-MS/MS. Many methods can be used for this purpose, including immunoaffinity-based depletion, fractionation by chromatography and electrophoresis (Figure 5.4).

Immunoaffinity-based depletion is a method using antibodies to remove abundant proteins specifically. For fractionation, chromatography such as reversed-phase (RP) or strong cation exchange (SCX) are often used to remove abundant proteins. The principles for RP and SCX chromatography are based on protein hydrophobicity and charge, respectively. Electrophoresis is used for protein separation using size (1D) or 2DE, which combines two separation methods, isoelectric point (PI) and size. Only non-abundant proteins are applied for further MS identification.

5.3.2 Quantitative Methods

To quantify proteins, two major methods can be used, gel-based and gel-free. The popular gel-based method is two-dimensional gel electrophoresis, which can be used to separate proteins by their sizes and charges, as shown in Figure 5.4. Gel-free quantitative methods comprise label-free, chemical labelling, such as dimethyl labelling [37] and iTRAQ [38], and metabolic labelling, such as SILAC [39]. Figure 5.5 briefly explains the concept for these three gel-free labelling methods.

5.3.2.1 Label-Free Quantitative Method

The label-free quantitative method can be used to quantify proteins. The advantage of the label-free method is cost-saving, but the corresponding peptide identification is not a trivial task. Therefore, powerful software is required in label-free proteomics. One of the label-free quantification methods is to calculate spectrum counts, inferring protein abundance by the number of times a peptide isobserved and the number of distinct

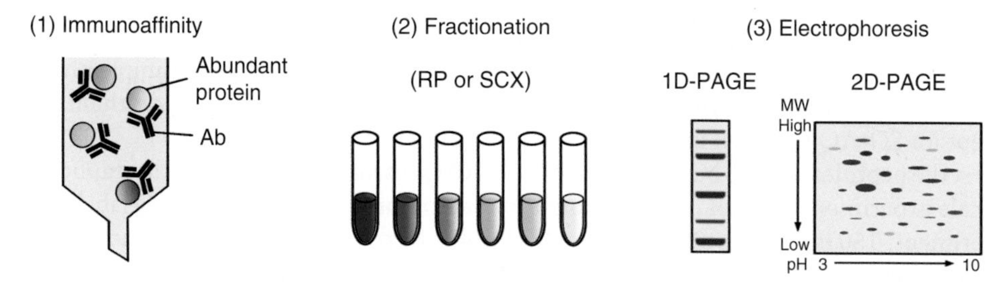

Figure 5.4 Three methods for removal of abundant proteins. Method 1, 2 and 3 indicate immunoaffinity-based depletion, fractionations by chromatography and electrophoresis, respectively. RP, reversed phase. SCX, strong cation exchange. 1D-PAGE, one-dimensional polyacrylamide gel electrophoresis. 2D-PAGE, two-dimensional polyacrylamide gel electrophoresis.

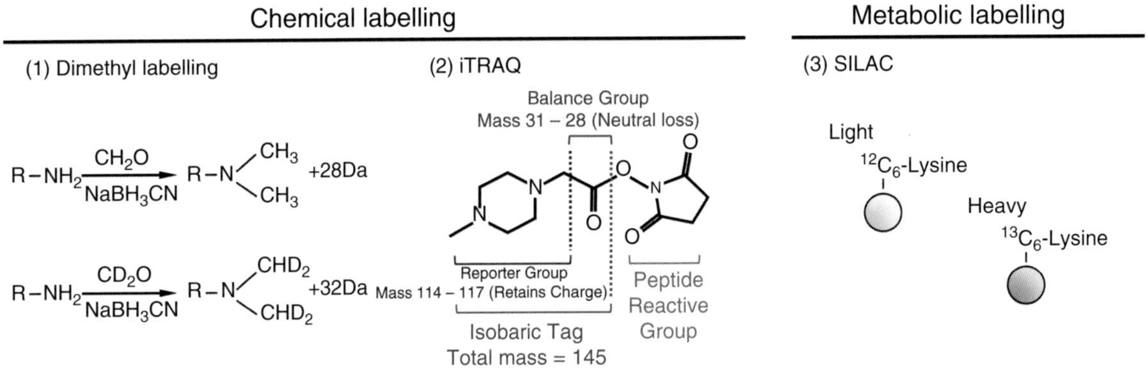

Chemical labelling

Metabolic labelling

(1) Dimethyl labelling

$R{-}NH_2 \xrightarrow[NaBH_3CN]{CH_2O} R{-}N(CH_3)_2$ +28Da

$R{-}NH_2 \xrightarrow[NaBH_3CN]{CD_2O} R{-}N(CHD_2)_2$ +32Da

(2) iTRAQ

Balance Group
Mass 31 – 28 (Neutral loss)

Reporter Group
Mass 114 – 117 (Retains Charge)

Peptide Reactive Group

Isobaric Tag
Total mass = 145

(3) SILAC

Light
$^{12}C_6$-Lysine

Heavy
$^{13}C_6$-Lysine

Figure 5.5 The concept for dimethyl labelling, iTRAQ and SILAC. The detail methods are shown and described in the following figures and text, respectively.

peptides observed from a given protein [40]. Another label-free method is quantification by comparing the intensity of mass spectrometric signal of each peptide from a given protein [40].

5.3.2.2 Dimethyl Labelling Method

Dimethyl labelling is a fast, inexpensive and easy labelling method. This method can be applied to global or PTM such as in a phosphorylation proteomic study. After trypsin digestion of proteins into peptides, dimethyl labelling can be used for comparison of two (Figures 5.5 and 5.6) or three protein samples (Figure 5.7). When labelling tryptic peptides, most peptides have a 4 or 8 Da mass difference (Figure 5.6). When the protein is cleaved after an arginine residue, only the N terminus of the peptide is labelled, so there is a 4 Da difference; when the protein is cleaved after a lysine residue, both the N terminus and the lysine residue are labelled; therefore there is an 8 Da difference.

In a previous study [24], the experimental method is described as follows.

The peptide samples are first mixed with 6 μl of 4% formaldehyde-H2 (Sigma-Aldrich) and 4% formaldehyde-D2, respectively, and then immediately 6 μl of freshly prepared 0.6 M sodium cyanoborohydride is added to each mixture. Each mixture is vigorously mixed and then reaction is allowed to proceed for 60 minutes at room temperature. Ammonium hydroxide (1%, 24 μl) is added to stop the reaction by reacting with the excess formaldehyde. Formic acid (10%, 30 μl) is further added with functions to end the labelling reaction and acidify the samples. Finally, the H- and D-labelled samples are combined at a 1 : 1 ratio and then desalted by using SDB-XC StageTips. The samples are further applied to LC-MS/MS or phosphopeptide enrichment.

5.3.2.3 iTRAQ Method

The iTRAQ method is one of the most popular chemical labelling methods. The term 'isobaric' describes the characteristic that different iTRAQ reagents (114, 115, 116 and 117) labelling different samples have equal mass. The iTRAQ reagent is amide reactive and can link to the N-terminus and lysine side chains of peptides [38]. The concept for iTRAQ is shown in Figure 5.8.

To perform iTRAQ, first the peptides need to be re-suspended in iTRAQ dissolution buffer. For the duplicate experiment and small-scale experiment, 5 mg of peptides from each sample are required for iTRAQ labelling. For the large-scale experiment, 150 mg of peptides from each sample are required [32]. Equal amounts of peptides from different samples are labelled by adding iTRAQ Reagent 114, iTRAQ Reagent 115, iTRAQ Reagent 116 or iTRAQ Reagent 117 and vortexing at room temperature for one hour. Labelled peptides are combined and dried with a centrifugal evaporator. The labelling peptide samples are now ready for LC-MS/MS analysis directly or fractionation by SCX chromatography and further applied to LC-MS/MS analysis.

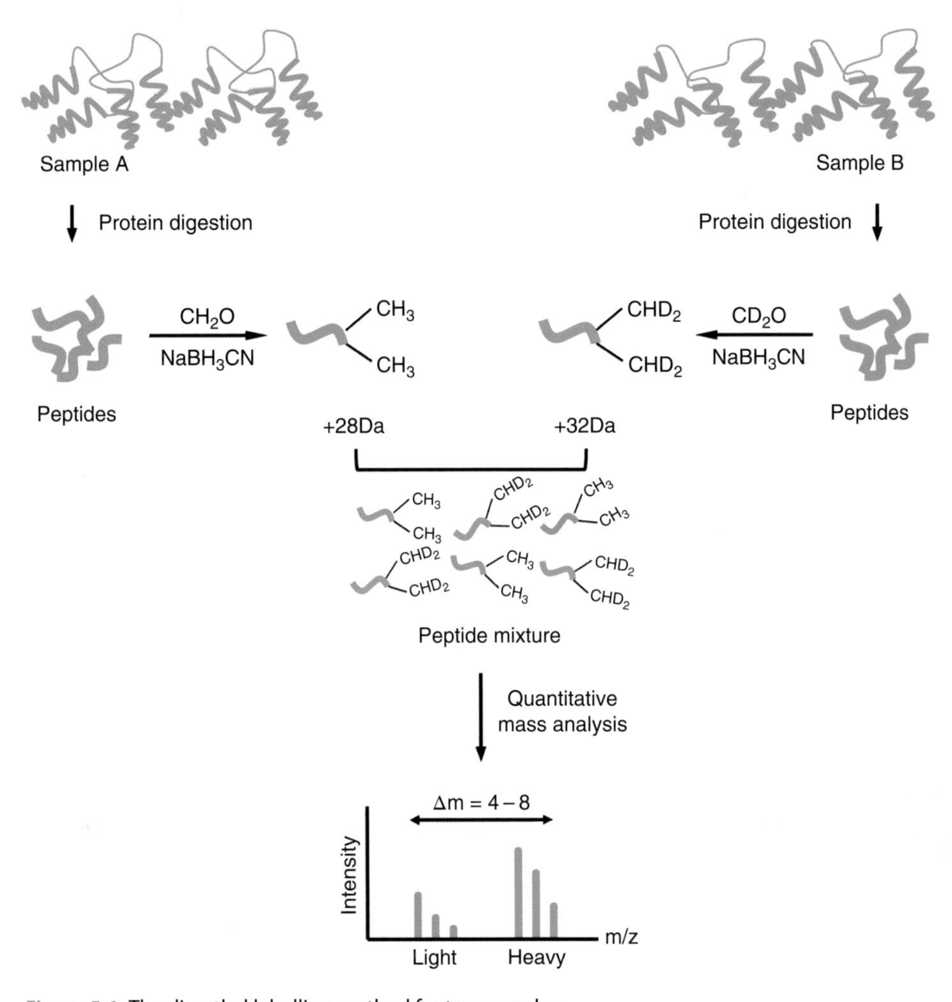

Figure 5.6 The dimethyl labelling method for two samples.

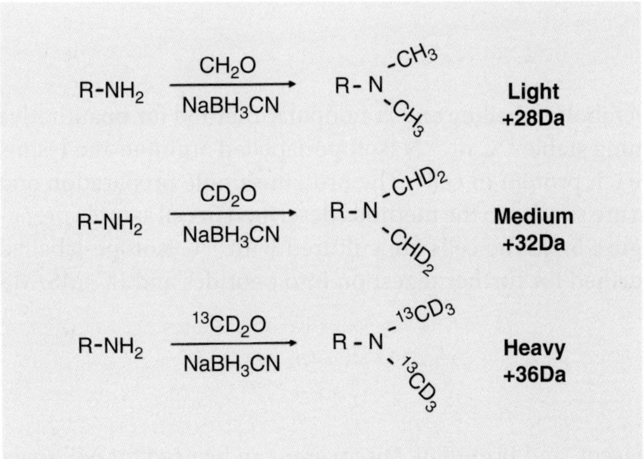

$R\text{-}NH_2 \xrightarrow[NaBH_3CN]{CH_2O} R\text{-}N(CH_3)_2$ **Light +28Da**

$R\text{-}NH_2 \xrightarrow[NaBH_3CN]{CD_2O} R\text{-}N(CHD_2)_2$ **Medium +32Da**

$R\text{-}NH_2 \xrightarrow[NaBH_3CN]{^{13}CD_2O} R\text{-}N(^{13}CD_3)_2$ **Heavy +36Da**

Figure 5.7 Triplex stable isotope dimethyl labelling can be used for three protein samples.

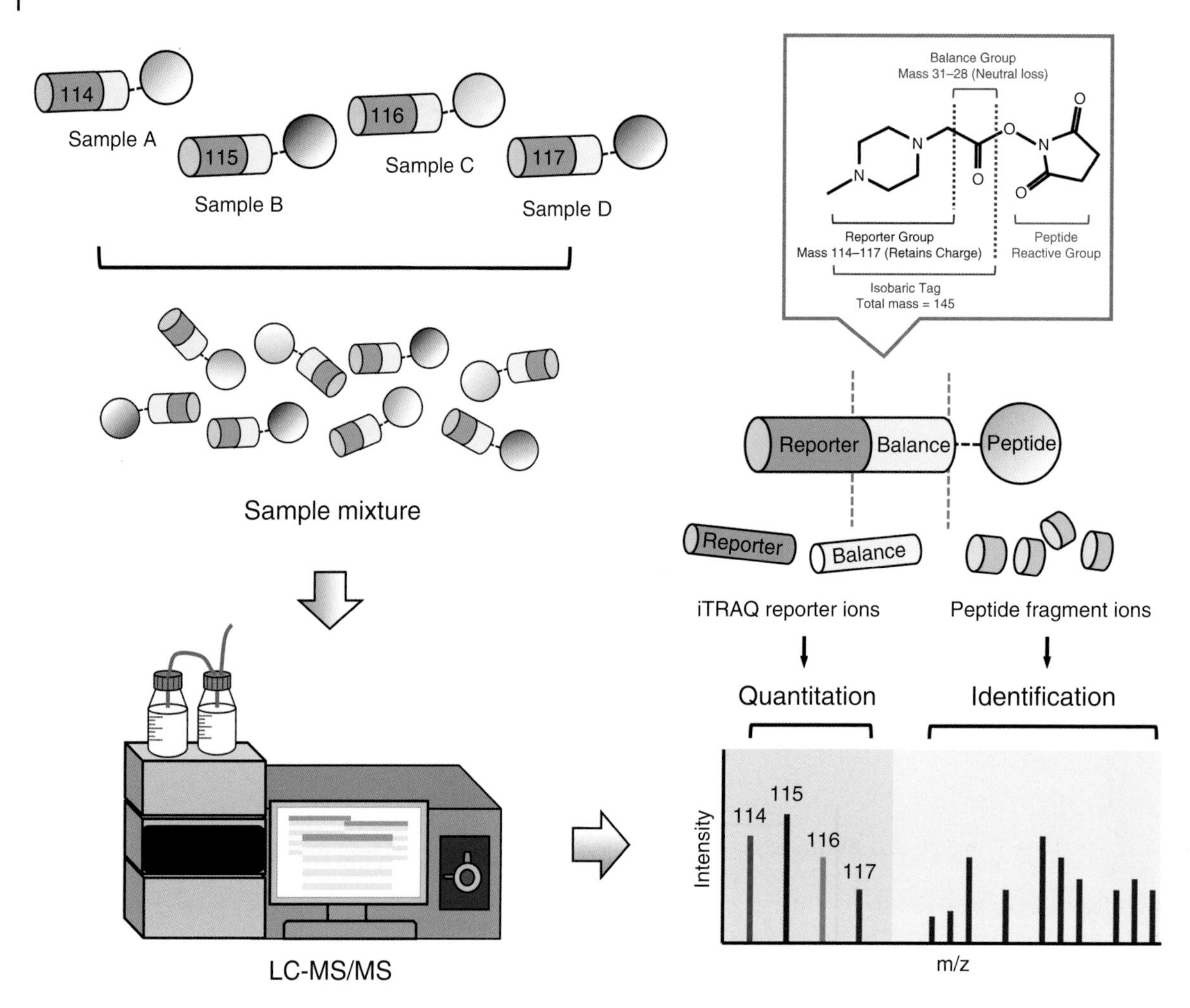

Figure 5.8 Explanation of iTRAQ. This shows that iTRAQ is used for labelling four samples, A, B, C and D. After labelling, samples are mixed together and LC-MS/MS is applied to quantify the intensity by measuring the iTRAQ report ions (114, 115, 116 and 117) and identify by peptide fragment MS data.

5.3.2.4 SILAC Method

SILAC [39] is one of the powerful approaches by metabolic labelling and is a popular method for quantitative proteomics. The cells are cultured in media containing stable ^{13}C or ^{15}N isotope-labelled arginine and lysine, so the labelled amino acids are incorporated into each protein in cells. The protein sample preparation and peptide digestion for SILAC labelling quantification are similar to the methods described in cell sample preparation above. Here, I show the SILAC method (Figure 5.9). The cells are cultured with ^{13}C isotope-labelled arginine and lysine and the labelled proteins are purified for further digestion into peptides and LC-MS/MS analysis.

5.3.3 Mass Spectrum Data Analysis

Several software packages including MaxQuant, Mascot, and Proteome Discoverer can be used for MS spectrum data to identify proteins. MaxQuant is more popular than the other three since it is free and can be run in many kinds of systems such as Microsoft Windows, Mac, and Linux. How to set the parameters when using

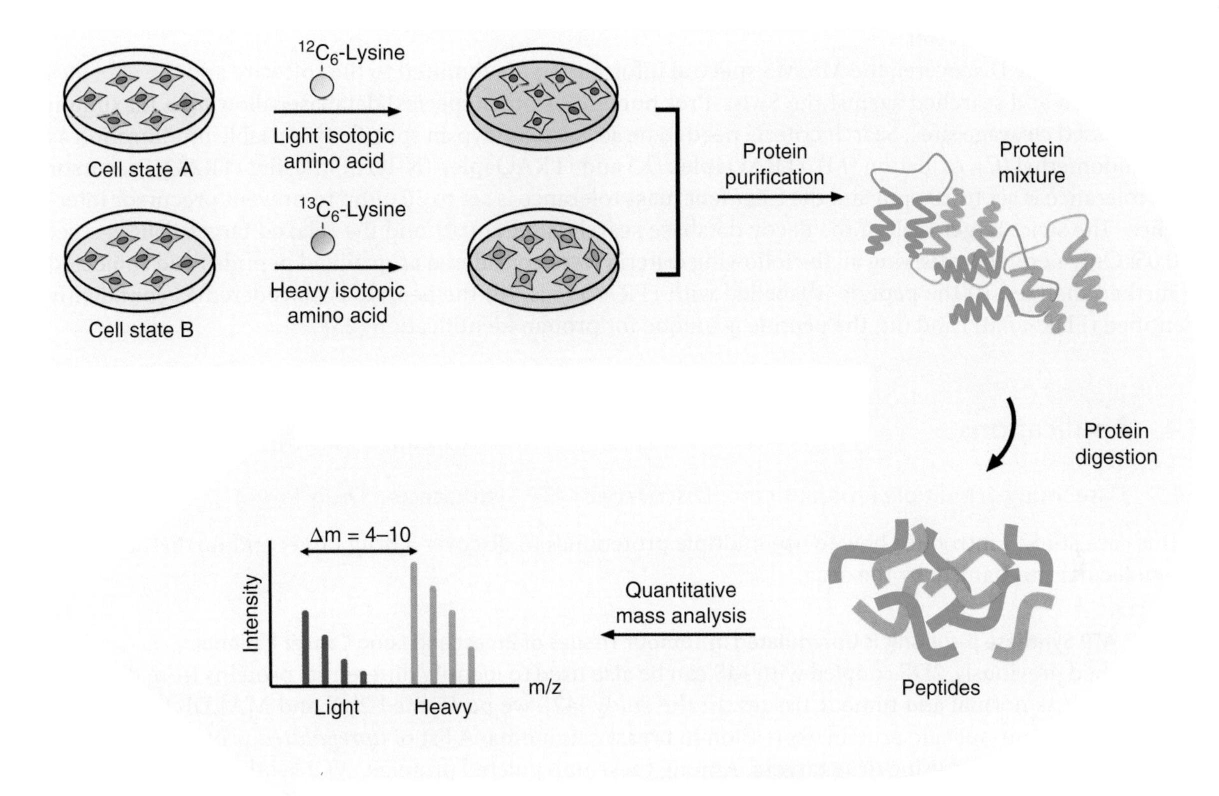

Figure 5.9 The concept for SILAC. The explanation for this method is described in the text.

the software is critical for protein identification. Here I briefly introduce the setting parameters based on our team's experience.

5.3.3.1 MaxQuant

If using MaxQuant, raw MS spectra are processed for peak detection and quantitation, and peptide identification is performed by using the Andromeda search engine and the Swiss-Prot database. Search criteria can be set, such as trypsin specificity, fixed modification of carbamidomethyl, variable modifications of oxidation and phosphorylation, and allow for up to two missed cleavages. A minimum of six amino acids in the peptide is required. The precursor mass tolerance is 3 ppm and the fragment ion tolerance is 0.5 Da. By using a decoy database strategy, peptide identification is based on the posterior error probability with a false discovery rate (FDR) of 1%. FDR is a statistical method to present the expected proportion of type I errors for multiple hypotheses testing. The detailed FDR description is described in the MaxQuant website (http://www.biochem.mpg.de/5111795/maxquant). Precursor intensities of already identified peptides are further searched and recalculated by using the 'match between runs' option in MaxQuant [24].

5.3.3.2 Mascot

If using Mascot for peptide identification, the search criteria can be set as follows: trypsin specificity allowing up to two missed cleavages, fixed modification of carbamidomethyl (C) and variable modifications of oxidation (M) and phosphorylation (ST), (Y), (D) and (H) [41]. Peptides are considered to be identified if the Mascot score yielded a confidence limit above 99% based on the significance threshold ($p < 0.01$) and if at least three successive y- or b-ions with an additional two and more y-, b- and/or precursor-origin neutral loss ions are observed, based on the error-tolerant peptide sequence tag concept [41].

5.3.3.3 Proteome Discoverer

If using Proteome Discoverer, the MS/MS spectral information is submitted to the software and the data files are combined and searched against the Swiss-Prot human (or other species) database, allowing a maximum of two missed cleavage sites. Search criteria need to be set such as trypsin specificity, variable modification as carbamidomethyl (C), oxidation (M), iTRAQ4plex (K) and iTRAQ4plex (N-term) if using iTRAQ. Precursor mass tolerance is set to 10 ppm and the fragment mass tolerance is set to 50 mmu to prevent precursor interference. The strict target FDR of the decoy database search is set at 0.01 and the relaxed target FDR was set at 0.05. Only peptides satisfying all the following criteria were considered as qualified peptides and subjected to further analyses: (i) the peptide is labelled with iTRAQ tags, (ii) the peptide is considered as confidently identified (FDR < 0.01) and (iii) the peptide is unique for protein identification [23].

5.4 Applications

5.4.1 Case Study 1: Multiple Proteomics for Discovery of ATP Synthase as a Drug Target

In this case study, I introduce how to use multiple proteomics to discover a drug target and further decipher the molecular mechanism by the drug.

5.4.1.1 ATP Synthase β-subunit is Upregulated in Tumour Tissues of Breast and Lung Cancer Patients

As described previously, 2DE coupled with MS can be also used to identify differential proteins from different samples such as normal and tumour tissues. In the study [42], we performed 2DE and MALDI-TOF-MS to identify the tumour-specific protein expression in breast carcinoma. A list of *upregulated* proteins in cancerous tissue might be promising drug targets. Among these upregulated proteins, ATP synthase β-subunit was found to be expressed at high levels in the tumour tissue. Similar results showed that ATP synthase β-subunit in lung cancer tissue was upregulated compared to adjacent normal tissues from patients.

5.4.1.2 Importance of ATP Synthase

ATP synthase is a ubiquitous multimeric protein complex that catalyses the synthesis of ATP, the common 'energy currency' of living cells. This molecular machine consists of two moieties, a transmembrane portion (Fo), the rotation of which is induced by the proton gradient, and a globular catalytic moiety (F1) that synthesises ATP [43]. In 1966, Peter D. Mitchell proposed the chemiosmotic hypothesis that a proton-motive force across the inner mitochondrial membrane is the immediate source of energy for ATP synthesis [44]. Paul D. Boyer proposed the 'binding-change hypothesis', a detailed elucidation of the mechanism by which ATP synthase catalyses the synthesis of ATP [45, 46]. John E. Walker determined the DNA sequence of the genes encoding the proteins in ATP synthase [47–49] and the first X-ray structure of F1, the soluble fraction of ATP synthase [50]. Mitchell, Boyer and Walker were awarded the Nobel Prize in Chemistry in 1978 and 1997. In 2013, Martin Karplus was awarded the Nobel Prize in Chemistry for the development of multiscale models for complex chemical systems. His research was also related to the smallest biological rotatory motor, F1-ATPase.

In general, it is localised to the mitochondrial inner membrane. Recent studies showed that ATP synthase was also found on the extracellular surface of endothelial cells in some cancer tissues, lymphocytes, hepatocytes, paraganglioma, proliferating cell lines, breast and lung cancer cells [42, 51–57]. With the property of facing outside the cell, this kind of ATP synthase is called 'ectopic ATP synthase'. Ectopic ATP synthase not only functions as an energy generator but also as proton channels and receptors for various ligands, which are involved in numerous biological processes including the mediation of intracellular pH, cholesterol homeostasis, the regulation of the proliferation and differentiation of endothelial cells, and the recognition of immune responses of tumour cells [58–60]. Ectopic ATP synthase together with the whole respiratory chain are localised on C6 glioma and lung cancer cell surfaces [56, 61]. Ectopic ATP synthase is expressed on the surfaces of various cancer cells, but not on normal or normal-like cells; therefore, researchers suggest that ectopic ATP synthase is a potential molecular target for anti-tumour and anti-angiogenesis therapies [24, 32, 42, 51, 53, 56, 57, 61–63]. Inhibition of ectopic ATP synthase showed inhibitory effects on cell proliferation in various cancer cells, suggesting the oncogenic role of ectopic ATP synthase in tumorigenesis.

5.4.1.3 2DE-Based Proteomics Reveals Ectopic ATP Synthase Inhibitor Induced Unfolded Protein Response (UPR)

To investigate the effects of targeting ectopic ATP synthase on protein expression in lung cancer, comprehensive time-course protein expression profiles using 2DE-based proteomics were analysed after ectopic ATP synthase inhibitor citreoviridin treatment [56, 57]. The amounts of protein spots were quantified using ImageMaster and proteins were identified using MS, where 49 proteins were successfully identified and further mapped to PPI networks. The constructed PPI networks were analysed by gene ontology (GO) functional enrichment. The results indicated that protein folding, negative regulation of ubiquitin-protein ligase activity involved in mitotic cell cycle and mRNA processing were major functions affected by citreoviridin. Further biological experiments such as rescued experiments and western blotting showed that citreoviridin indeed induced a reactive oxygen species (ROS) dependent unfolded protein response (UPR). The similar 2DE-based method was applied to a breast cancer study [57] and found the same results as those in lung cancer.

5.4.1.4 Dynamic Phosphoproteomics Identified a Key Phosphosite Involved in Citreoviridin-Induced Pathways

Protein phosphorylation is a reversible, ubiquitous, and fundamentally PTM and plays a significant role in a wide range of physiological processes in both eukaryotic and prokaryotic organisms and is one of the major modes of regulation in signal transduction, growth control and metabolism [24, 41, 64–66]. Dimethyl labelling of peptides, a quantitative labelling method described above, was used for this study. The phosphopeptide was enriched by HAMMOC [67, 68] and further applied to LTQ-Orbitrap XL (Thermo Scientific, Berman, Germany) equipped with a nanoACQUITY UPLC system (Waters Corp., Milford, MA). MS raw data were analysed by MaxQuant software version 1.3.0.5 [5]. Phosphorylation motif and clustering analyses were performed by Motif-X algorithm [69] and the fuzzy c-means algorithm implemented in the 'Mfuzz' package for R [70], respectively. The results showed that after treatment with citreoviridin, 41 motifs containing 33 serine (Ser) and 8 threonine (Thr) phosphosites were overrepresented. To associate the enriched motifs with specific kinases, similarities between enriched motifs and kinase recognition motifs were examined and 300 kinase recognition motifs from the PhosphoNetworks database [71] were obtained. To elucidate citreoviridin-induced signalling pathways, we also performed mathematical modelling using the time-series phosphoproteome data to construct the response network with citreoviridin treatment. The results from a clustering and constructed response network displayed the temporal relation of phosphorylated HSP90 and MAPK1 in citreoviridin treatment [24]. Additionally, site-directed mutagenesis was used to change HSP90 phosphosite Ser255 to E255 and A255 and further western blotting analysis was performed to measure the protein expression levels of the MAPK/ERK pathway. The results showed that phosphosite Ser255 of HSP90AB1 is crucial for MAPK/ERK1 signalling.

5.4.2 Case Study 2: iTRAQ Quantitative Proteomics to Elucidate Molecular Mechanisms

As described previously, iTRAQ is one of the most popular chemical labelling methods. Therefore, in case study 2, two iTRAQ quantitative proteomic applications for the study of molecular mechanisms are introduced.

5.4.2.1 Citreoviridin-Induced Molecular Mechanism in Xenograft Mice Tissue

As described in case study 1, we understand that targeting ectopic ATP synthase using citreoviridin is a potential therapeutic strategy [32]. To investigate the *in vivo* molecular mechanism by citreoviridin in lung cancer, we employed the xenograft mice model, where 5×10^6 CL1-0 cells resuspended in 0.1 ml Hanks' balanced salt solution (HBSS) were mixed with matrigel and injected subcutaneously into NOD.CB17-Prkdcscid female mice (four to five weeks old) for tumour growth. After tumour volumes reached 100 mm^3, animals were randomly assigned into two groups: those receiving intraperitoneal injection with the vehicle control (DMSO) or the ATP synthase inhibitor (citreoviridin). Tumours (control and treatment) were excised for further iTRAQ quantitative proteomics. Tumour tissues were ground into a fine powder using the method described in the previous section 'sample preparation'. Moreover, after conducting reduction, alkylation and digestion of proteins, the peptides were resuspended in iTRAQ dissolution buffer and follow the method described in the 'iTRAQ method'. Peptide samples from the control tumour (C1 and C2) and the citreoviridin treated tumour (T1 and T2) were labelled with iTRAQ 114, 115, 116 and 117 tags, respectively. For small-scale profiling, only 277 proteins with an FDR of 3.51% were identified. Therefore, we performed fractionation by SCX chromatography and obtained 2659 proteins with an FDR of 2.22%.

Due to measurement errors in the experiments and individual variations from biological replicates of samples, after protein identification, normalisation is necessary for accuracy in protein quantitation. In this study, we compared and evaluated seven different normalisation methods and found that the method 'the sum of intensities in protein quantitation' is the best. The iTRAQ signature ion intensities of peptides matching the protein were summed and the protein abundance ratio was calculated as dividing a sample's summation of intensities to another sample's summation of intensities. This is a weighted calculation because the larger intensity contributes more to the protein abundance ratio. For each protein, we calculated four protein abundance ratios, T1/C1, T2/C2, T2/C1 and T1/C2. The R value of each protein, which represents the relative abundance of the protein, was calculated with the four protein abundance ratios, T1/C1, T2/C2, T2/C1 and T1/C2. The S value of each protein, which represents the error of protein abundance ratios, was calculated by its protein abundance ratios, T1/C1 and T2/C2. Each protein had one S value and the distribution of S values can be considered as the distribution of errors. Assuming that the errors follow a normal distribution, a 1.96-fold of the standard deviation (1.96 SD) of S values is statistically significant ($P < 0.05$) and can be taken as the cut-off value. The proteins with R values larger than the cut-off value +1.96 SD of S values can be taken as upregulated proteins. On the other hand, the proteins with R values smaller than the cut-off value (1.96 SD) of S values can be taken as downregulated proteins. These proteins are considered as 'differentially expressed proteins', which can be further analysed for functional enrichment and pathway mapping. Here, in a large-scale experiment among 2659 proteins, 144 proteins were differentially expressed and applied for further analysis.

In the gene ontology biological process, DAVID Bioinformatics Resources [72, 73] were used for the functional enrichment. The top two GO biological process clusters were related to glucose metabolism and protein ubiquitination process. MetaCore was used for pathway mapping. The top pathway map enriched was the glycolysis and gluconeogenesis pathway map, which is related to glucose metabolism. Furthermore, expressions of the seven glucose-metabolism related proteins validated by western blotting were all upregulated in citreoviridin-treated tumour samples, which confirmed the results of the proteomic analysis. The results indicated that citreoviridin may reduce the glycolytic intermediates for macromolecule synthesis and further inhibit cell proliferation and tumour growth *in vivo*.

5.4.2.2 Middle Infrared Radiation-Interfered Networks

Infrared (IR) light is divided into three major regions based on wavelength: near-infrared (NIR, 0.76–1.5 μm), middle-infrared (MIR, 1.5–5.6 μm) and far-infrared (FIR, 5.6–1000 μm) [22]. NIR and FIR were used for therapy such as pain relief, muscle fatigue and inflammatory osteoarthritis [74, 75]. In the study, we found MIR inhibited the cell growth and altered the morphology of breast cancer cells. To investigate MIR-interfered networks, we performed iTRAQ quantitative proteomics in breast cancer MDA-MB-231 cells.

The experimental design was similar to that in Section 5.4.2.1. Peptides from the control were labelled with iTRAQ reagent 114 and iTRAQ reagent 115; peptides from the MIR treatment were labelled with iTRAQ reagent 116 and iTRAQ reagent 117. The same normalisation method as in Section 5.4.2.1 was used. In the study, we did not select differentially expressed proteins for functional enrichment; instead we used GSEA, which uses a running-sum statistic for the whole gene (or protein) set on a rank list of all the available expression values [19]. In MIR-induced MDA-MB-231 cells, the proteasome pathway, p53-dependent DNA damage, anaphase-promoting complex (APC) and Cdc2/CDK1-mediated degradation, mitotic pathway, tumour necrosis factor (TNF) pathway, insulin glucose pathway, and integrin and cell-to-cell were enriched, indicating that MIR might regulate cell cycle progression, induce DNA damage, alter nucleotide metabolism and affect cell adhesion related functions. We further validated the enriched functions using several biological experiments. Indeed, MIR induced cell cycle arrest at the G2/M phase and DNA damage, caused cytoskeleton rearrangement and reduced cell mobility and invasion ability of MDA-MB-231 cells.

5.5 Concluding Remarks

In the post-genome era, proteomics has become essential to understanding the functions of genes and proteins in living cells and organisms. Over the past decades, proteomics has opened our eyes to a global view and to the dynamic interactions when studying complicated biological systems. Many high-throughput and sensitive methods have been developed for proteome profiling, e.g. MS and proteome microarrays. Although data-independent acquisition (DIA) MS produces a vast amount of data, the techniques in proteomics have still a long way to go before they reach maturity.

Proteome data quality and quantity depend on sample preparation, so how to collect and prepare protein samples as well as label peptides is a critical issue. Similar to the analysis of genome data, bioinformatics is necessary for proteome data analysis. How to extract peptide sequences from mass raw data accurately and how to figure out the biological meanings, pathways and signalling, rely on bioinformatics tools and software. Therefore, the analysis tools should combine informatics techniques such as artificial intelligence and machine learning for a true understanding of the proteome data.

Proteomics can be applied to not only discovery of clinical biomarkers, drug targets and investigation of molecular mechanisms but also plant science and palaeontology. Proteomics provides the insight for disease diagnosis and therapy. The journey from a single protein to proteome to PPIs and functional modules is challenging but exciting.

Acknowledgements

This work was supported by the Ministry of Science and Technology (MOST 105-2320-B-002-057-MY3 and MOST 106-2320-B-002-053-MY3). The author thanks Dr. Chantal Ho Yin Cheung for drawing the figures and Professor Hsuan-Cheng Huang for proofreading the draft of this chapter.

References

1 Wilkins, M.R. (2009). Proteomics data mining. *Expert Rev. Proteomics* 6 (6): 599–603.
2 Wilkins, M.R., Sanchez, J.-C., Gooley, A.A. et al. (1996). Progress with proteome projects: why all proteins expressed by a genome should be identified and how to do it. *Biotechnol. Genet. Eng. Rev.* 13 (1): 19–50.
3 Williams, K.L., Gooley, A.A., Wilkins, M.R., and Packer, N.H. (2014). A Sydney proteome story. *J. Proteomics* 107: 13–23.
4 Ali-Khan, N., Zuo, X., and Speicher, D.W. (2002). Overview of proteome analysis. *Curr. Protoc. Protein Sci.* 22.1.1–22.1.19.
5 Cox, J. and Mann, M. (2008). MaxQuant enables high peptide identification rates, individualized p.p.b.-range mass accuracies and proteome-wide protein quantification. *Nat. Biotechnol.* 26: 1367–1372.

6 Tyanova, S., Temu, T., and Cox, J. (2016). The MaxQuant computational platform for mass spectrometry-based shotgun proteomics. *Nat. Protoc.* 11 (12): 2301–2319.
7 Vaudel, M., Venne, A.S., Berven, F.S. et al. (2014). Shedding light on black boxes in protein identification. *Proteomics* 14 (9): 1001–1005.
8 Yamashita, M. and Fenn, J.B. (1984). Electrospray ion source. Another variation on the free-jet theme. *J. Phys. Chem.* 88 (20): 4451–4459.
9 Fenn, J.B., Mann, M., Meng, C.K. et al. (1989). Electrospray ionization for mass spectrometry of large biomolecules. *Science* 246 (4926): 64–71.
10 Fenn, J.B. (2002). Electrospray ionization mass spectrometry: how it all began. *J. Biomol. Tech.* 13 (3): 101–118.
11 Dole, M., Mack, L.L., Hines, R.L. et al. (1968). Molecular beams of macroions. *J. Phys. Chem.* 49 (5): 2240–2249.
12 Hillenkamp, F., Karas, M., Beavis, R.C., and Chait, B.T. (1991). Matrix-assisted laser desorption/ionization mass spectrometry of biopolymers. *Anal. Chem.* 63 (24): 1193A–1203A.
13 Yang, L., Guo, S., Li, Y. et al. (2011). Protein microarrays for systems biology. *Acta Biochim. Biophys. Sin.* 43: 161–171.
14 Hall, D.A., Ptacek, J., and Snyder, M. (2007). Protein microarray technology. *Mech. Ageing Dev.* 128 (1): 161–167.
15 Duarte, J.G. and Blackburn, J.M. (2017). Advances in the development of human protein microarrays. *Expert Rev. Proteomics* 14 (7): 627–641.
16 Tsai, H.-T., and Juan, H.F. (2012). Revealing novel interacting proteins of ATP synthase by human proteome microarray. Master Thesis in National Taiwan University.
17 Lu, K.Y., Tao, S.C., Yang, T.C. et al. (2012). Profiling lipid-protein interactions using nonquenched fluorescent liposomal nanovesicles and proteome microarrays. *Mol. Cell. Proteomics* 11 (11): 1177–1190.
18 Sutandy, F.X., Qian, J., Chen, C.S., and Zhu, H. (2013). Overview of protein microarrays. *Curr. Protoc. Protein Sci.* 72: 27.1.1–27.1.16.
19 Subramanian, A., Tamayo, P., Mootha, V.K. et al. (2005). Gene set enrichment analysis: a knowledge-based approach for interpreting genome-wide expression profiles. *Proc. Natl. Acad. Sci. U.S.A.* 102 (43): 15545–15550.
20 Mootha, V.K., Lindgren, C.M., Eriksson, K.F. et al. (2003). PGC-1alpha-responsive genes involved in oxidative phosphorylation are coordinately downregulated in human diabetes. *Nat. Genet.* 34 (3): 267–273.
21 Hsu, C.L., Wang, J.K., Lu, P.C. et al. (2017). DynaPho: a web platform for inferring the dynamics of time-series phosphoproteomics. *Bioinformatics* 33 (22): 3664–3666.
22 Chang, H.Y., Li, M.H., Huang, T.C. et al. (2015). Quantitative proteomics reveals middle infrared radiation-interfered networks in breast cancer cells. *J. Proteome Res.* 14 (2): 1250–1262.
23 Cheung, C.H.Y., Hsu, C.L., Chen, K.P. et al. (2017). MCM2-regulated functional networks in lung cancer by multi-dimensional proteomic approach. *Sci. Rep.* 7 (1): 13302.
24 Hu, C.W., Hsu, C.L., Wang, Y.C. et al. (2015). Temporal phosphoproteome dynamics induced by an ATP synthase inhibitor citreoviridin. *Mol. Cell. Proteomics* 14 (12): 3284–3298.
25 Masuda, T., Tomita, M., and Ishihama, Y. (2008). Phase transfer surfactantaided trypsin digestion for membrane proteome analysis. *J. Proteome Res.* 7: 731–740.
26 Rappsilber, J., Mann, M., and Ishihama, Y. (2007). Protocol for micropurification, enrichment, pre-fractionation and storage of peptides for proteomics using StageTips. *Nat. Protoc.* 2: 1896–1906.
27 Lombard-Banek, C., Moody, S.A., and Nemes, P. (2016). Single-cell mass spectrometry for discovery proteomics: quantifying translational cell heterogeneity in the 16-cell frog (*Xenopus*) embryo. *Angew. Chem. Int. Ed.* 55 (7): 2454–2458.
28 Hu, Y., An, Q., Sheu, K. et al. (2018). Single cell multi-omics technology: methodology and application. *Front. Cell Dev. Biol.* 6: 28.

29 Peterson, V.M., Zhang, K.X., Kumar, N. et al. (2017). Multiplexed quantification of proteins and transcripts in single cells. *Nat. Biotechnol.* 35 (10): 936–939.
30 Darmanis, S., Gallant, C.J., Marinescu, V.D. et al. (2016). Simultaneous multiplexed measurement of rna and proteins in single cells. *Cell Rep.* 14 (2): 380–389.
31 Zhu, Y., Piehowski, P.D., Zhao, R. et al. (2018). Nanodroplet processing platform for deep and quantitative proteome profiling of 10–100 mammalian cells. *Nat. Commun.* 9 (1): 882.
32 Wu, Y.H., Hu, C.W., Chien, C.W. et al. (2013). Quantitative proteomic analysis of human lung tumor xenografts treated with the ectopic ATP synthase inhibitor citreoviridin. *PLoS One* 8 (8): e70642.
33 Hanash, S.M., Pitteri, S.J., and Faca, V.M. (2008). Mining the plasma proteome for cancer biomarkers. *Nature* 452 (7187): 571–579.
34 Juan, H.F., Chen, J.H., Hsu, W.T. et al. (2004). Identification of tumor-associated plasma biomarkers using proteomic techniques: from mouse to human. *Proteomics* 4 (9): 2766–2775.
35 Geyer, P.E., Kulak, N.A., Pichler, G. et al. (2016). Plasma proteome profiling to assess human health and disease. *Cell Syst.* 2 (3): 185–195.
36 Lin, L., Zheng, J., Yu, Q. et al. (2018). High throughput and accurate serum proteome profiling by integrated sample preparation technology and single-run data independent mass spectrometry analysis. *J. Proteomics* 174: 9–16.
37 Hsu, J.L., Huang, S.Y., Chow, N.H., and Chen, S.H. (2003). Stable-isotope dimethyl labelling for quantitative proteomics. *Anal. Chem.* 75 (24): 6843–6852.
38 Ross, P.L., Huang, Y.N., Marchese, J.N. et al. (2004). Multiplexed protein quantitation in *Saccharomyces cerevisiae* using amine-reactive isobaric tagging reagents. *Mol. Cell. Proteomics* 3 (12): 1154–1169.
39 Ong, S.E., Blagoev, B., Kratchmarova, I. et al. (2002). Stable isotope labelling by amino acids in cell culture, SILAC, as a simple and accurate approach to expression proteomics. *Mol. Cell. Proteomics* 1 (5): 376–386.
40 Bantscheff, M., Schirle, M., Sweetman, G. et al. (2007). Quantitative mass spectrometry in proteomics: a critical review. *Anal. Bioanal. Chem.* 389: 1017–1031.
41 Hu, C.W., Lin, M.H., Huang, H.C. et al. (2012). Phosphoproteomic analysis of *Rhodopseudomonas palustris* reveals the role of pyruvate phosphate dikinase phosphorylation in lipid production. *J. Proteome Res.* 11 (11): 5362–5375.
42 Huang, T.C., Chang, H.Y., Hsu, C.H. et al. (2008). Targeting therapy for breast carcinoma by ATP synthase inhibitor aurovertin B. *J. Proteome Res.* 7 (4): 1433–1444.
43 Pu, J. and Karplus, M. (2008). How subunit coupling produces the gamma-subunit rotary motion in F1-ATPase. *Proc. Natl. Acad. Sci. U.S.A.* 105: 1192–1197.
44 Mitchell, P. (1966). Chemiosmotic coupling in oxidative and photosynthetic phosphorylation. *Biol. Rev.* 41: 445–501.
45 Boyer, P.D., Cross, R.L., and Momsen, W. (1973). A new concept for energy coupling in oxidative phosphorylation based on a molecular explanation of the oxygen exchange reactions. *Proc. Natl. Acad. Sci. U.S.A.* 70: 2837–2839.
46 Kayalar, C., Rosing, J., and Boyer, P.D. (1977). An alternating site sequence for oxidative phosphorylation suggested by measurement of substrate binding patterns and exchange reaction inhibitions. *J. Biol. Chem.* 252: 2486–2491.
47 Gay, N.J. and Walker, J.E. (1981). The atp operon: nucleotide sequence of the region encoding the alpha-subunit of *Escherichia coli* ATP-synthase. *Nucleic Acids Res.* 9: 2187–2194.
48 Gay, N.J. and Walker, J.E. (1981). The atp operon: nucleotide sequence of the promoter and the genes for the membrane proteins, and the delta subunit of *Escherichia coli* ATP-synthase. *Nucleic Acids Res.* 9: 3919–3926.
49 Saraste, M., Gay, N.J., Eberle, A. et al. (1981). The atp operon: nucleotide sequence of the genes for the gamma, beta, and epsilon subunits of *Escherichia coli* ATP synthase. *Nucleic Acids Res.* 9: 5287–5296.

50 Abrahams, J.P., Leslie, A.G., Lutter, R., and Walker, J.E. (1994). Structure at 2.8 A resolution of F1-ATPase from bovine heart mitochondria. *Nature* 370: 621–628.

51 Das, B., Mondragon, M.O., Sadeghian, M. et al. (1994). A novel ligand in lymphocyte-mediated cytotoxicity: expression of the beta subunit of H+ transporting ATP synthase on the surface of tumour cell lines. *J. Exp. Med.* 180: 273–281.

52 Arakaki, N., Nagao, T., Niki, R. et al. (2003). Possible role of cell surface H^+-ATP synthase in the extracellular ATP synthesis and proliferation of human umbilical vein endothelial cells. *Mol. Cancer Res.* 1: 931–939.

53 Martinez, L.O., Jacquet, S., Esteve, J.P. et al. (2003). Ectopic beta-chain of ATP synthase is an apolipoprotein A-I receptor in hepatic HDL endocytosis. *Nature* 421: 75–79.

54 Burrell, H.E., Wlodarski, B., Foster, B.J. et al. (2005). Human keratinocytes release ATP and utilize three mechanisms for nucleotide interconversion at the cell surface. *J. Biol. Chem.* 280: 29667–29676.

55 Burwick, N.R., Wahl, M.L., Fang, J. et al. (2005). An inhibitor of the F1 subunit of ATP synthase (IF1) modulates the activity of angiostatin on the endothelial cell surface. *J. Biol. Chem.* 280: 1740–1745.

56 Chang, H.Y., Huang, H.C., Huang, T.C. et al. (2012). Ectopic ATP synthase blockade suppresses lung adenocarcinoma growth by activating the unfolded protein response. *Cancer Res.* 72 (18): 4696–4706.

57 Chang, H.Y., Huang, T.C., Chen, N.N. et al. (2014). Combination therapy targeting ectopic ATP synthase and 26S proteasome induces ER stress in breast cancer cells. *Cell Death Dis.* 5: e1540.

58 Chi, S.L. and Pizzo, S.V. (2006). Cell surface F1Fo ATP synthase: a new paradigm? *Ann. Med.* 38: 429–438.

59 Fu, Y. and Zhu, Y. (2010). Ectopic ATP synthase in endothelial cells: a novel cardiovascular therapeutic target. *Curr. Pharm. Des.* 16: 4074–4079.

60 Ma, Z., Cao, M., Liu, Y. et al. (2010). Mitochondrial F1Fo-ATP synthase translocates to cell surface in hepatocytes and has high activity in tumor-like acidic and hypoxic environment. *Acta Biochim. Biophys. Sin.* 42: 530–537.

61 Ravera, S., Aluigi, M.G., Calzia, D. et al. (2011). Evidence for ectopic aerobic ATP production on C6 glioma cell plasma membrane. *Cell. Mol. Neurobiol.* 31: 313–321.

62 Lu, Z.J., Song, Q.F., Jiang, S.S. et al. (2009). Identification of ATP synthase beta subunit (ATPB) on the cell surface as a non-small cell lung cancer (NSCLC) associated antigen. *BMC Cancer* 9: 16.

63 Yavlovich, A., Viard, M., Zhou, M. et al. (2012). Ectopic ATP synthase facilitates transfer of HIV-1 from antigen-presenting cells to CD4(+) target cells. *Blood* 120: 1246–1253.

64 Wang, Y.T., Tsai, C.F., Hong, T.C. et al. (2010). An informatics-assisted label-free quantitation strategy that depicts phosphoproteomic profiles in lung cancer cell invasion. *J. Proteome Res.* 9: 5582–5597.

65 Yachie, N., Saito, R., Sugiyama, N. et al. (2011). Integrative features of the yeast phosphoproteome and protein-protein interaction map. *PLoS Comput. Biol.* 7: e1001064.

66 Imami, K., Sugiyama, N., Imamura, H. et al. (2012). Temporal profiling of lapatinib-suppressed phosphorylation signals in EGFR/HER2 pathways. *Mol. Cell. Proteomics* 11: 1741–1757.

67 Kyono, Y., Sugiyama, N., Imami, K. et al. (2008). Successive and selective release of phosphorylated peptides captured by hydroxy acid-modified metal oxide chromatography. *J. Proteome Res.* 7: 4585–4593.

68 Sugiyama, N., Masuda, T., Shinoda, K. et al. (2007). Phosphopeptide enrichment by aliphatic hydroxyl acid-modified metal oxide chromatography for nano-LC-MS/MS in proteomics applications. *Mol. Cell. Proteomics* 6: 1103–1109.

69 Schwartz, D. and Gygi, S.P. (2005). An iterative statistical approach to the identification of protein phosphorylation motifs from large-scale data sets. *Nat. Biotechnol.* 23: 1391–1398.

70 Olsen, J.V., Vermeulen, M., Santamaria, A. et al. (2010). Quantitative phosphoproteomics reveals widespread full phosphorylation site occupancy during mitosis. *Sci. Signaling* 3 (104): ra3.

71 Hu, J., Rho, H.S., Newman, R.H. et al. (2014). PhosphoNetworks: a database for human phosphorylation networks. *Bioinformatics* 30: 141–142.

72 Sherman, B.T. and Lempicki, R.A. (2009). Bioinformatics enrichment tools: paths toward the comprehensive functional analysis of large gene lists. *Nucleic Acids Res.* 37: 1–13.

73 Huang, D.W., Sherman, B.T., and Lempicki, R.A. (2008). Systematic and integrative analysis of large gene lists using DAVID bioinformatics resources. *Nat. Protoc.* 4: 44–57.
74 Akhalaya, M.Y., Maksimov, G., Rubin, A. et al. (2014). Molecular action mechanisms of solar infrared radiation and heat on human skin. *Ageing Res. Rev.* 16: 1–11.
75 Inoué, S. and Kabaya, M. (1989). Biological activities caused by far-infrared radiation. *Int. J. Biometeorol.* 33: 145–150.

Further Reading

Juan, H.F. and Huang, H.C. (2012). *Systems Biology: Applications in Cancer-Related Research*. Singapore: World Scientific Publishing.
Juan, H.F. and Huang, H.C. (2018). *A Practical Guide to Cancer Systems Biology*. Singapore: World Scientific Publishing.

Website Resources

Gene Ontology (GO) Enrichment Analysis (http://geneontology.org/page/go-enrichment-analysis) STRING (https://string-db.org)
ProteomExchange (http://www.proteomexchange.org)
https://www.expasy.org/proteomics
MaxQuant (http://www.biochem.mpg.de/5111795/maxquant)
Cytoscape (http://www.cytoscape.org)
Kyoto Encyclopedia of Genes and Genomes (KEGG, https://www.genome.jp/kegg)
IntAct (www.ebi.ac.uk/intact)
DynaPho (http://dynapho.jhlab.tw)

6

Overproduction, Separation and Purification of Affinity-Tagged Proteins from *Escherichia coli*

Finbarr Hayes[1] *and Daniela Barillà*[2]

[1] *Faculty of Biology, Medicine and Health, The University of Manchester, Manchester, M13 9PL, UK*
[2] *Department of Biology, University of York, York, YO10 5DD, UK*

6.1 Introduction

Understanding protein function is a central goal of modern molecular biology. This aim typically is achieved using a combination of *in vivo* and *in vitro* approaches that together provide an integrated overview of protein features, interactions and regulation. The availability of a sufficient quantity of pure protein that can be isolated rapidly and efficiently is crucial for its biochemical, physical and structural characterisation *in vitro*. Moreover, *in vitro* studies of purified proteins are key to the development of drugs that affect proteins implicated in disease formation and progression, as well as to the optimization of proteins produced for biotechnological, bioprocessing and other industrial and commercial purposes.

The Gram-negative bacterium *Escherichia coli* remains the primary cell factory for protein production: approximately 90% of the tertiary structures that have been deposited in the Protein Data Bank have been generated from proteins that were purified from *E. coli* [1]. *E. coli* is specially suitable for protein production due to the bacterium's rapid growth kinetics in inexpensive cultivation media, high cell density, simplicity of handling and storage and ease of genetic manipulation [2]. *E. coli* has a doubling time of less than 30 minutes when growing maximally in low cost growth medium and under optimal conditions can achieve densities of more than 10^{10} cells per millilitre of culture medium. Most laboratory strains of *E. coli* are considered to be non-pathogenic, which makes handling of these strains relatively non-hazardous. Moreover, these derivatives also have been engineered to be less fit genetically than wild-type strains and therefore are less likely to survive should an inadvertent escape occur from the research environment. Nevertheless, monocultures of *E. coli* laboratory strains remain viable for extended periods at room temperature and survive for decades when stored correctly at −80 °C [3]. Finally, the sophisticated genetic manipulation systems that have been developed for *E. coli* over many years have made this bacterium the most tractable and well-studied of all model organisms [4].

Prior to the advent of molecular gene cloning techniques, proteins were purified from the multitude of other proteins and cellular components in *E. coli* using a series of sequential fractionation and chromatographic separation techniques that utilised the biochemical and biophysical features of the target protein. For example, the LacI repressor was one of the first DNA binding proteins to be isolated and characterised [5]. In one instance, purification of the repressor began with 350 l of *E. coli* culture, which was harvested and then was lysed by freezing and thawing. This step was followed by a set of enzymatic treatment, fractionation, ammonium sulphate precipitation, dialysis and chromatography steps that resulted in 350–1000 mg of purified, functional LacI protein [6]. In another example, an acetyltransferase that confers resistance to the antibiotic chloramphenicol was purified from 18 l of *Staphylcoccus aureus* culture by successive steps of lysis, centrifugation, ammonium sulphate precipitation, dialysis, ion-exchange chromatography, ultrafiltration under nitrogen, gel

Biomolecular and Bioanalytical Techniques: Theory, Methodology and Applications, First Edition. Edited by Vasudevan Ramesh.

filtration and concentration [7]. These laborious and time-consuming purification procedures, although still necessary in certain cases, largely have been supplanted by the use of affinity tags when purifying recombinant proteins from *E. coli*. Indeed, the use of tags to purify proteins from *E. coli* by affinity chromatography has revolutionised the ease with which proteins can be isolated, studied and manipulated [8].

Current techniques for protein purification typically involve, first, cloning of the cognate gene in a plasmid expression vector so that an affinity tag is fused to the target gene. The reader is referred to Chapter 4 for an in-depth description of molecular cloning strategies. The tagged gene is inserted downstream of a strong, regulatable promoter on the vector (Figure 6.1a). Controlled overexpression of the cloned, tagged gene is achieved by addition of a small molecule inducer that directly or indirectly activates the regulatable promoter. Induction is sufficiently potent that the target protein is highly overproduced and often is the principal protein observed

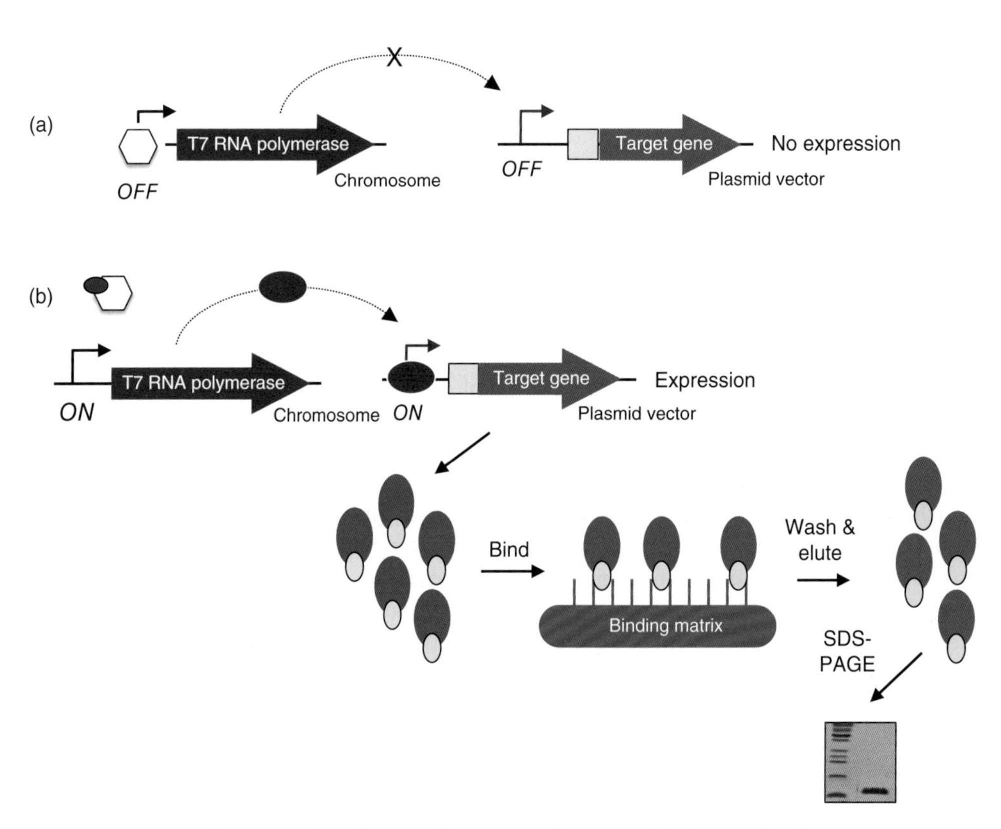

Figure 6.1 Overproduction and purification of His_6-tagged proteins. (a) The gene for the target protein (blue) is cloned in a plasmid expression vector so that the locus is fused to a sequence that encodes a His_6 affinity tag (yellow). Here the tag is at the 5′ end of the gene but may also be placed at the 3′ end depending on the choice of vector and target gene. The fusion is under the control of a promoter (bent red arrow) that is recognised by the RNA polymerase encoded by bacteriophage T7 (red). The gene for the latter is engineered into the *E. coli* chromosome and is under the control of the *lac* promoter (bent black arrow). This promoter is blocked by binding of the LacI repressor protein (white hexagon). Thus, the gene for T7 RNA polymerase is not expressed, T7 RNA polymerase is not produced and the target gene is not expressed. (b) Binding of IPTG (magenta) alters the conformation of the LacI repressor which no longer binds to the *lac* promoter. The native RNA polymerase in *E. coli* transcribes the gene for T7 RNA polymerase which binds to the promoter upstream of the target gene fusion, which then is expressed to high levels. The His_6 affinity tag allows specific capture of the target protein on a nickel matrix (green) to which other cellular proteins and components (not shown) do not bind. The matrix is washed to remove impurities and the target protein is eluted by altering the conditions on the matrix. Specifically, a solution with a high concentration of imidazole is passed over the column. Imidazole competes with the His_6-tagged protein for the nickel matrix thereby displacing the protein. The eluted protein is transferred to appropriate buffer conditions for assay and storage, and protein purity is assessed by SDS-PAGE. The identity of the protein may be verified by mass spectrometry.

when cell lysates are analysed by sodium dodecyl sulphate-polyacrylamide gel electrophoresis (SDS PAGE). The lysate is passed over a matrix for which the tagged, target protein has high affinity, whereas the plethora of other proteins and cellular material in the lysate passes unhindered through the column (Figure 6.1b). Thus, the affinity tag permits single-step enrichment of the target protein. As the target protein is bound non-covalently to the affinity matrix, the protein is released from the column by addition of an elution buffer that alters the conditions on the column. The eluted target protein often is >90% pure at this stage and may be suitable directly for functional analysis or characterisation (Figure 6.1b). If a higher degree of purity is required, one or more gel filtration or other separation techniques may be used as outlined briefly below (also see Chapter 7).

6.2 Selecting an Affinity Tag: Glutathione-S-Transferase, Maltose-Binding Protein and Hexa-Histidine Motifs

Affinity purification of tagged proteins typically begins with the overexpression of the cognate gene that is cloned under the control of a regulatable promoter in a plasmid expression vector. The cloning strategy involves the genetic tagging of the target protein with an affinity moiety that markedly eases subsequent purification on a matrix that non-covalently captures the tagged protein [9, 10]. Numerous affinity tags may be used, including glutathione-S-transferase (GST), maltose-binding protein (MBP) and hexa-histidine motifs (His_6), among others [11, 12]. Fusion proteins with these tags are purified by affinity chromatography using glutathione-sepharose, amylose and Ni^{2+} matrices, respectively. The use of His_6-tags is particularly widespread for protein capture from *E. coli* as many proteins tolerate the attachment of a string of six histidine residues (approx. 1 kDa) at the N-terminus and/or C-terminus without perturbation of protein function. GST (26 kDa) and MBP (42.5 kDa) are bulkier entities but are thought to fold independently of proteins to which they are fused. MBP is a highly soluble protein that has been used extensively to improve the solubility of fusion proteins [13, 14]. This approach can be specially useful for target proteins that are prone to aggregation when overproduced. For example, elastin-like polypeptides are intrinsically disordered peptides that are highly susceptible to aggregation above defined transition temperatures. These polypeptides were purified more readily as MBP fusions from *E. coli*. An intervening protease site allowed cleavage and subsequent separation of the functional polypeptides from the MBP-tag [15]. In another example, murine leukaemia inhibitory factor was purified as an MBP fusion protein. The fusion protein was bioactive, demonstrating that the MBP-tag did not block the functionality of the target protein [16]. The mechanism by which MBP enhances the solubility of proteins to which it is fused is uncertain [17]. Analogously, GST may improve the solubility of proteins to which it is fused and which otherwise are recalcitrant to purification. GST has a high affinity for the reduced form of glutathione (GSH). Fusions between GST and the target protein of interest can be readily immobilised on sepharose beads to which GSH is covalently attached. Other proteins and cellular materials pass through the column. The subsequent addition of an excess of free GSH displaces the fusion protein from the matrix. The fusion protein often is sufficiently pure at this stage that it is suitable for further characterisation. Moreover, the target protein may be liberated from the fusion using a protease that cleaves a linker sequence that is engineered between GST and the target protein [18]. Protease cleavage sites often are designed between the target and tag moieties to provide the option of removing the latter from the fusion protein [19]. An example of the production and purification of a GST fusion protein is provided below.

Purification of proteins with His_6-tags is achieved by immobilised metal ion affinity chromatography (IMAC). This technique uses a chelating matrix that is impregnated with soft metal ions such as Cu^{2+} or Co^{2+}, but most commonly Ni^{2+} (Figure 6.1b). Protein surfaces contain electron-donating groups, notably the imidazole side chain of histidine, which specifically recognise the non-coordinated sites of the metal ions with high affinity. Thus, proteins that are rich in histidine residues are non-covalently immobilised on matrices that contain Ni^{2+} ions. However, the interaction between the histidine side chains and the metal can be reversed by washing with a buffer that contains a high concentration of free imidazole, thereby eluting

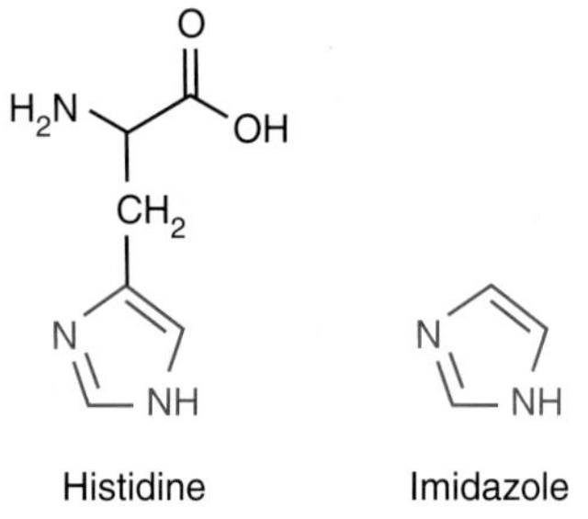

O
H_2N
OH
CH_2
N
NH
N
NH
Histidine
Imidazole

Figure 6.2 Illustration of the imidazole side chain (right) in histidine (left), which, as a canonical amino acid, additionally possesses α-amino and carboxylic acid groups.

the protein (Figure 6.1b). Imidazole competes effectively with histidine for the nickel matrix as histidine possesses an imidazole side chain (Figure 6.2). In short, a protein with electron-donating groups, in particular histidine, can be purified by reversible interactions with a metal complex [20, 21].

As most proteins do not contain a sufficient number of contiguous histidine residues for efficient IMAC, His_6-tags are engineered at the C- or N-terminus of the target protein by cloning the corresponding gene in-frame with six consecutive histidine codons in specialised expression vectors. The correct frame is achieved by using the polymerase chain reaction (PCR) and specially designed primers (Figure 6.3a), which generate a

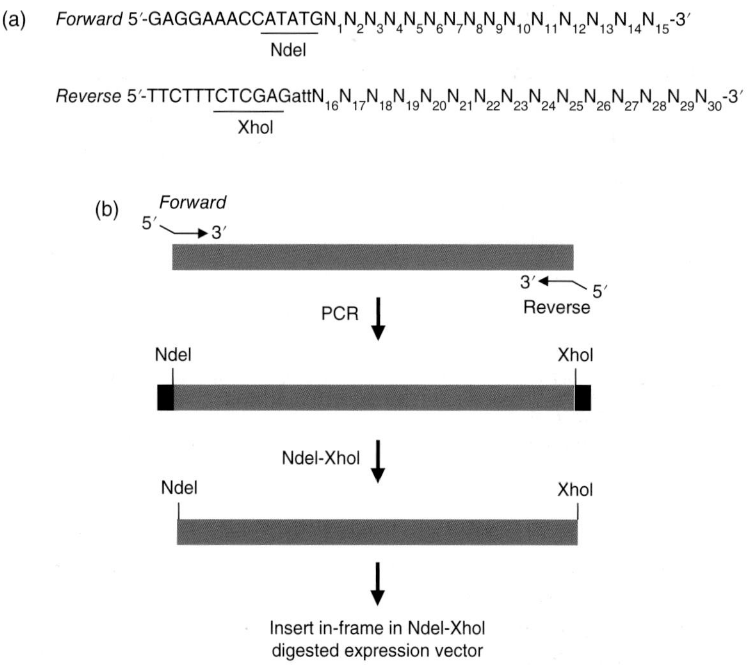

(a)
Forward 5′-GAGGAAACCATATGN1N2N3N4N5N6N7N8N9N10N11N12N13N14N15-3′
NdeI
Reverse 5′-TTCTTTCTCGAGattN16N17N18N19N20N21N22N23N24N25N26N27N28N29N30-3′
XhoI
(b)
Forward
5′
3′
3′
5′
Reverse
PCR
NdeI
XhoI
NdeI-XhoI
NdeI
XhoI
Insert in-frame in NdeI-XhoI
digested expression vector

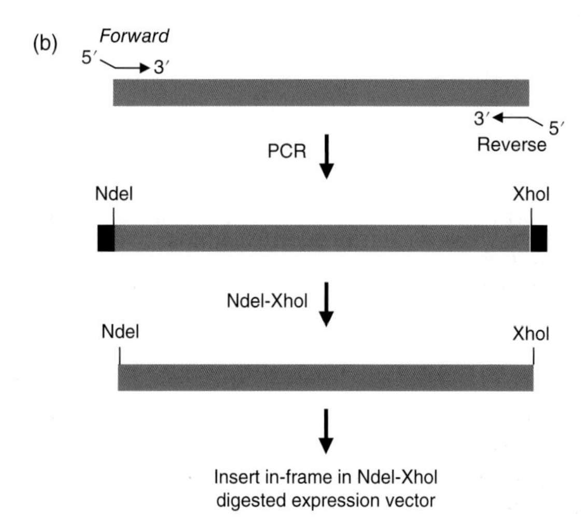

Figure 6.3 An example of target gene cloning in an expression vector. (a) Example primers for PCR amplification of a target gene. The *forward* primer anneals to the 5′ end of the target gene via residues N_1–N_{15} that represent the five codons that follow the ATG start codon of the gene. The eight bases at the 5′ end generate a tail that allows the NdeI restriction enzyme to bind stably to its recognition sequence in the final PCR product. NdeI cleaves the sequence 5′-CATATG-3′ that fortuitously includes an ATG start codon. The *reverse* primer anneals to the target gene via residues N_{16}–N_{30} that represent the five codons at the 3′ end of the gene. The lower case letters indicate the translational stop codon. The six bases at the 5′ end of the reverse primer generate a tail that allows the XhoI enzyme to bind stably to its recognition sequence in the final PCR product. (b) The *forward* and *reverse* primers are used to amplify the target gene from an appropriate template, e.g. genomic DNA or a plasmid that harbours the gene. The PCR product is digested with NdeI-XhoI and ligated into an expression vector that is digested with the same enzymes. The resultant recombinant plasmid contains an in-frame transcriptional fusion of the cloned gene and the desired tag.

product that, when cleaved with appropriate restriction enzymes, is inserted into the expression vector that is digested with the same enzymes (Figure 6.3b). The initial His_6-tags vectors [22] were improved by the addition of expression signals derived from bacteriophage T7 [23] to produce a series of plasmid derivatives (pET vectors) and host strains that are used widely for protein production in *E. coli*. The host strains were engineered to include a chromosomal gene for T7 RNA polymerase. This polymerase initiates transcription from a distinctive promoter sequence introduced into the pET vectors that is not recognised by *E. coli* RNA polymerase (Figure 6.4). Thus, the use of T7 expression signals allows for tightly regulatable and specific expression of the target gene cloned downstream of the T7 promoter. The pET vector series is available under the Novagen brand from Merck Biosciences/EMD Biosciences, Inc.

An extensive array of expression vectors in addition to the pET series is available for production of recombinant proteins in *E. coli* [11]. These vectors include, for example, plasmids that replicate using diverse replicons that provide different copy numbers and with different antibiotic resistance markers [24–26]. Other expression vectors employ arabinose-responsive promoters [27, 28] or permit protein fusions with GST, MBP and other solubility tags [29–34]. Expression vectors also have been developed that, for example, utilise recombination-based cloning techniques [35, 36] or ligation-independent cloning strategies [37]. General approaches for molecular cloning are described in Chapter 4.

6.3 The pET Vector Series: Archetypal Expression Vectors in *E. coli*

As the pET series is among the most commonly used expression vectors and illustrates key principles that are used for high-level gene expression in *E. coli*, the features of a typical pET vector are presented here. The pET33b(+) vector encodes kanamycin-resistance to allow selection of *E. coli* transformants (Figure 6.4). Other pET vectors possess different resistance genes which permit either the cotransformation of more than one plasmid with different markers into the expression host and/or the utilisation of *E. coli* hosts that display a similar chromosomal resistance as that encoded by the vector. Replication of pET33b(+) occurs via the pMB1 origin which provides a copy number of approx. 15–20 plasmids per cell. In practice this means that the vector is easy to isolate from small volumes of *E. coli* K-12 laboratory strains using standard plasmid purification procedures. The plasmid, like many other pET vectors, also contains the replication origin of bacteriophage f1. This element allows the production of a single-stranded form of the plasmid when *E. coli* is infected with the appropriate helper bacteriophage. Single-stranded DNA can be used for the sequencing of cloned genes although this feature largely has been surpassed by routine sequencing of double-stranded plasmid DNA [38].

The *lacI* gene on pET33b(+) encodes the LacI transcriptional repressor (Figure 6.4). This protein recognises the *lac* operator site on the plasmid which prevents leaky expression of genes inserted downstream of the T7 promoter. LacI also binds upstream of a chromosomally-located gene for T7 RNA polymerase in *E. coli* BL21-type strains thereby blocking production of this polymerase. Repression is relieved by addition of isopropyl β-D-1-thiogalactopyranoside (IPTG) to the growth medium. IPTG is a non-hydrolysable analogue of allolactose which is the natural inducer of the *lac* operon in *E. coli*. IPTG binds to and alters the conformation of the LacI repressor which consequently no longer interacts with either the plasmid or chromosomal *lac* operator sites. Therefore, the chromosomal gene for T7 RNA polymerase is expressed, and the polymerase binds to the T7 promoter on pET33b(+) and induces expression of the cloned downstream gene (Figure 6.1).

The His_6-tag may be engineered either at the amino- or carboxy-terminal end of the target protein in the case of the pET33b(+) vector (Figure 6.4). For the former the target gene is cloned between the NdeI and XhoI restriction enzyme sites with a translation stop codon placed after the XhoI site. The stop codon is introduced during PCR amplification of the target gene (Figure 6.3). This strategy results in a target protein with a 25 residue peptide at the amino-terminus (MGSSHHHHHHSSGLVPR↓GSRRASVH). If required, 17 of these amino acids, including the His_6-tag, may be removed from the purified protein by treatment with the thrombin protease that cleaves between the arginine and glycine residues in the sequence Leu-Val-Pro-Arg-Gly-Ser as outlined below [19, 39]. If the His_6-tag is required at the carboxy-terminal end of the target protein, the

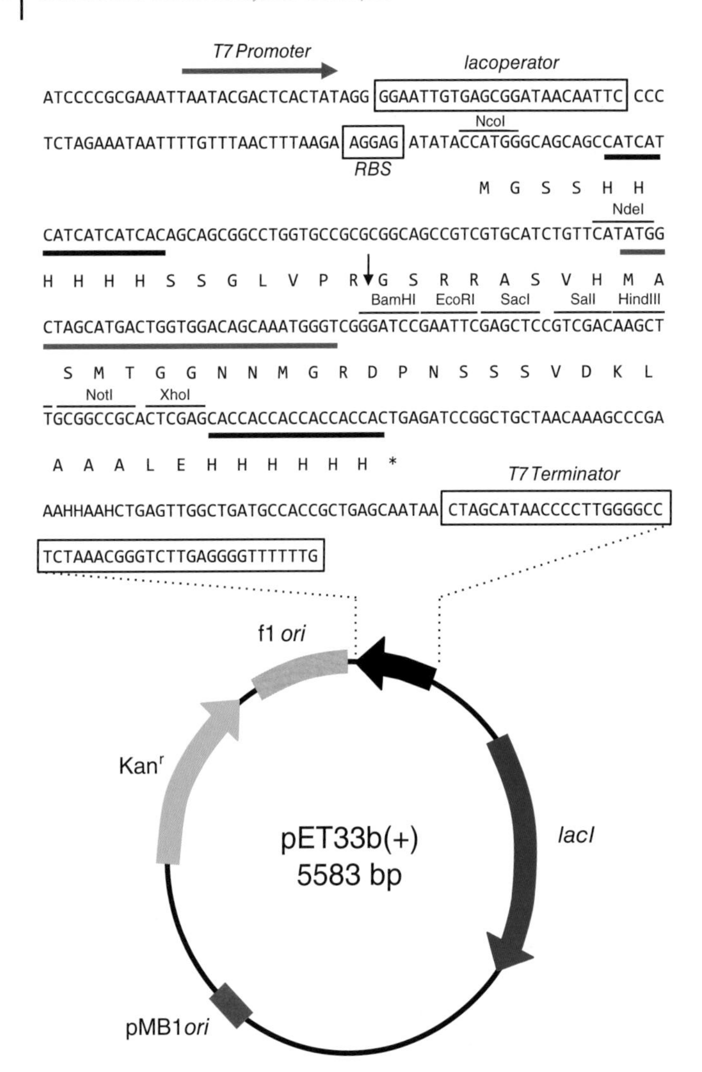

Figure 6.4 Organisation of an example expression vector. The pET33b(+) vector (Novagen) facilitates the overproduction of proteins with N- or C-terminal His_6-tags. The former may be cleaved from the purified protein. A different tag also may be included that allows detection by antibodies that recognise the tag. The vector includes the pMB1 plasmid replication origin (red arc) which allows replication at a moderate copy number in *E. coli*. The replication origin of bacteriophage f1 (grey arc) permits the production of single-stranded DNA species that are useful for sequencing purposes, although this feature has been superseded by routine sequencing of double-stranded DNA. A gene for kanamycin resistance (green arrow) allows plasmid selection. The *lacI* gene (blue arrow) encodes the LacI repressor protein that binds to *lac* operator sites both on the plasmid (boxed) and upstream of a chromosomally located gene for T7 RNA polymerase. Thus, leaky expression of genes cloned downstream of the T7 promoter in pET33b(+) is repressed and expression of the chromosomal gene for T7 RNA polymerase is also blocked. Repression is relieved by addition of IPTG to the growth medium. IPTG alters the conformation of the Lac repressor, which no longer binds to the plasmid or chromosomal *lac* operator sites that thereby allow expression of the gene for T7 RNA polymerase. The polymerase binds to the T7 promoter (red arrow) and induces expression of cloned downstream genes. The DNA sequence above the pET33b(+) vector map illustrates additional salient features involved in the expression of cloned genes. The ribosome binding site (RBS) and T7 transcriptional terminator are boxed. His_6 affinity tags are indicated by horizontal black lines. The T7 tag (horizontal red line) is an 11 amino acid sequence from T7 bacteriophage gene 10 permits detection of fusion proteins by using antibodies against the tag. Selected restriction enzyme sites are marked. These sites are used to clone the gene of interest by different strategies. For example, insertion of a target gene in-frame between the NdeI and XhoI sites produces a fusion protein that harbours a His_6-tag at the N-terminus. The tag allows affinity purification of the corresponding fusion protein. The His_6-tag subsequently may be removed from the protein by digestion with thrombin serine protease that cleaves at the site indicated by the vertical arrow (see Figure 6.7). In this example, cloning into the XhoI site introduces a stop codon that avoids addition of a His_6-tag at the C-terminus of the fusion protein. In another example, the target gene is cloned between NcoI and XhoI sites so that the target protein is produced with a non-cleavable C-terminal His_6-tag. In contrast, use of the BamHI or nearby sites allows for a gene fusion that includes the gene 10 tag.

cognate gene is cloned between the NcoI restriction enzyme site and any of the sites that are located between BamHI and XhoI in pET33b(+) (Figure 6.4). The XhoI site is used frequently as it ensures that all of the intervening sequences between NcoI and XhoI, including the 25 codons that specify the amino-terminal His_6-tag, thrombin cleavage site and the T7 tag, are removed. The T7 tag is an 11 amino acid motif derived from T7 bacteriophage gene 10 which allows detection in Western blots of fusion proteins with an amino-terminal His_6-tag by using commercially available antibodies against the T7 tag. However, anti-His-tag antibodies also are available which allow the detection of fusion proteins with His_6-tags irrespective of whether the tag is located at the amino- or carboxy-terminal. Note that a stop codon is *not* introduced after the XhoI site when a carboxy-terminal His_6-tag is required. Instead the stop codon at the 3′ end of the six histidine codons will terminate translation. Because the T7 promoter drives very effective gene expression [40], the pET33b(+) vector contains a strong transcriptional terminator that inhibits undesirable read-through by T7 RNA polymerase into other vector sequences (Figure 6.4).

6.4 IMAC of a His_6-Tagged Protein: Example Methodology with the ParF DNA Segregation Protein

The ParF protein (22 kDa) that is encoded by a bacterial multidrug resistance plasmid provides a valuable illustration of the step-by-step procedures for purification of a His_6-tagged protein in *E. coli*. ParF is representative of a widespread class of ATPases that mediate the stable segregation of plasmids during bacterial cell division [41]. The function of these and analogous NTPases is vital for the maintenance of antibiotic resistance and other plasmids in bacterial populations [42]. The *parF* gene was cloned as a PCR product between NdeI and XhoI sites in the pET22b(+) vector (Figure 6.3), thereby generating a protein fusion in which the His_6-tag is situated at the carboxy-terminal of ParF [43].

6.4.1 Testing for ParF Induction and Solubility

Different proteins exhibit different characteristics when overproduced in *E. coli*. Therefore, it is crucial to determine the optimal expression conditions when a target protein is to be produced for the first time. For example, preliminary, small-scale experiments initially were performed in which the optimal concentration of IPTG and the most appropriate time and temperature for *parF* induction were assessed. 0.1–5 mM IPTG and induction times from one to five hours were tested, as well as the expression at 30 and 37 °C. Induction at lower temperatures also may be tested as the formation of insoluble inclusion bodies (see below) may be ameliorated by the reduction in the rate of protein synthesis that occurs when the post-induction temperature is lowered [44]. Trials for *parF* expression were conducted in *E. coli* BL21(DE3), which harbours the gene for T7 RNA polymerase under control of a modified *lac* promoter. *E. coli* BL21(DE3) and its derivatives are widely used strains for induction with the pET vector series. Among these derivatives are BL21(DE3) strains that contain the pLysS plasmid. This plasmid constitutively produces low levels of T7 lysozyme, which reduces the expression of recombinant genes cloned in pET22b(+) and analogous vectors by inhibiting basal levels of T7 RNA polymerase. The use of expression strains that are pre-transformed with pLysS is especially useful when expressing recombinant genes whose products are toxic in *E. coli* as induction of these genes is repressed more effectively than without pLysS. Analysis of induced samples by SDS-PAGE revealed that induction in the BL21(DE3) strain with 1 mM IPTG for three hours at 30 °C produced the highest concentration of the approx. 23 kDa ParF-His_6 protein from the pET22b(+)-*parF* plasmid.

Proteins vary in their solubility following overproduction. Indeed it has been estimated that up to 30% of recombinant proteins are insoluble when overproduced in *E. coli* [45]. These proteins typically form inclusion bodies, which are densely packed, denatured, insoluble protein aggregates [46]. Although proteins can be recovered from inclusion bodies by denaturation and refolding of pelleted material in the presence of denaturing agents such as urea or guanidine hydrochloride [47], it is more convenient if the majority of the overproduced protein remains in the soluble (supernatant) fraction of a cell extract. If a target protein is insoluble under one set of induction conditions, different induction temperatures, growth media, inducer concentrations and induction times can be tested. Insertion and testing of the target gene in different expression vectors with different affinity tags also may be helpful, as may the co-expression of chaperone proteins that assist in folding of the target protein [48, 49]. Note that even if a protein is principally in the soluble fraction when overproduced, mutated versions of the same protein may behave differently and the solubility of these mutated proteins should be assessed independently. The following is a step-by-step protocol that was used to determine ParF-His_6 solubility.

1. *E. coli* BL21(DE3) was transformed with approx. 0.1 μg of the pET22b(+)-*parF* expression plasmid and the transformation mix was plated on Luria–Bertani (LB) agar plates containing ampicillin (100 μg/ml) for plasmid selection. Plates were incubated at 37 °C for 12–16 hours.
2. Three colonies from the transformation plate were inoculated in 10 ml of LB broth with ampicillin in a 125 ml conical flask.
3. Cultures were grown at 30 °C with shaking at 180 rpm until OD_{600} was 0.65–0.75 (approx. three hours).
4. A 100 μl aliquot of the culture was removed to a 1.5 ml microcentrifuge tube and centrifuged at approx. 10 000*g* for one minute. The supernatant was discarded and the cell pellet was resuspended in 20 μl SDS-PAGE loading buffer (50 mM Tris-HCl, pH 6.8, 100 mM dithiothreitol (DTT), 2% SDS, 0.1% bromophenol blue, 10% glycerol).
5. The sample was denatured by boiling for three minutes and was stored at −20 °C. This sample is the uninduced control for later SDS-PAGE analysis.
6. The remaining culture was induced with 1 mM IPTG at 30 °C, shaking at 180 rpm for three hours.
7. A 100 μl aliquot was removed and treated as in step 4 to check by SDS-PAGE for proper induction. This sample is the induced control.
8. The remaining induced culture was harvested by centrifugation at approx. 5000*g* for five minutes at 4 °C. The supernatant was discarded and the cell pellet was analysed immediately or was stored at −80 °C for subsequent use.
9. The pellet was thawed at room temperature and resuspended in 400 μl of binding buffer (20 mM Tris-HCl, pH 7.9, 500 mM NaCl, 5 mM imidazole, 10% glycerol).
10. 10 μl of lysozyme (10 mg/ml) and 10 μl of phenylmethylsulphonyl fluoride (100 mM) were added to assist cell lysis and inhibit protease activity, respectively. The sample was incubated at 30 °C for 15 minutes.
11. The sample was sonicated on ice six times for 15 seconds each with pause intervals of 30 seconds. Sonication causes cell lysis and also shears DNA, which aids in subsequent handling of the lysate. Pause intervals between sonication bursts avoid overheating of the sample.
12. The lysate was centrifuged at approx. 4000*g* for 60 minutes at 4 °C.
13. The supernatant was decanted and the pellet was resuspended in 400 μl of binding buffer.
14. 5 μl of 4× SDS-PAGE loading buffer were added to both 15 μl supernatant and 15 μl of pellet, and both fractions were denatured by boiling for three minutes.
15. The uninduced (step 5) and induced (step 7) controls as well as the supernatant and pellet fractions (step 14) were analysed by SDS-PAGE. The relative amounts of the 23 kDa induced protein in the supernatant and pellet fractions were inspected to assess the solubility of the overproduced ParF-His_6 protein. In this case, >90% of the protein was detectable in the supernatant fraction, indicating that ParF is highly soluble under these induction conditions.

6.4.2 Identifying the Correct Imidazole Concentration for ParF Protein Elution

Different proteins elute at different concentrations of imidazole during IMAC. To avoid perturbation of the protein conformation or quaternary structure, the lowest imidazole concentration possible should be used during elution. To identify this concentration, the protein first can be purified from a 10 ml culture using an imidazole gradient to ascertain how much imidazole should be used in the wash buffer and in the elution buffer during larger-scale purifications. To determine the most appropriate imidazole concentration for ParF-His$_6$ elution from an Ni^{2+} column, steps 1 to 12 in the preceding induction protocol were followed to produce a cleared *E. coli* lysate containing overproduced ParF-His$_6$. This lysate was then treated as follows.

1. A 15 µl aliquot of the cleared lysate was removed to a microcentrifuge tube, 5 µl of 2× SDS-PAGE loading buffer were added and the sample was stored at −20 °C for later use. This sample was the induced, cleared extract control.
2. 200 µl settled bed volumes of commercially available Ni^{2+} resin can be handled in a 1.5 ml microcentrifuge tube. Non-stick tubes were used so that the resin did not adhere to the tube walls. The following steps were done at 4 °C, although it may not be strictly necessary here as protein activity was not a concern during these trials.
3. 400 µl of Ni^{2+} resin slurry were transferred to a 1.5 ml non-stick microcentrifuge tube, which then was centrifuged at approx. 500*g* for 4 °C for five minutes. It is important to centrifuge at 400–1000*g* only as higher velocities may break the beads. The supernatant was discarded.
4. The following sequence of washes was used to charge and equilibrate the resin with Ni^{2+}. For each step the appropriate buffer was added, the tube was inverted several times to mix and then was centrifuged at approx. 500*g* at 4 °C for one minute.
   (a) Twice with 400 µl sterile deionised water
   (b) Three times with 400 µl 1× charge buffer ($NiSO_4$ solution stored at 8× concentration [400 mM $NiSO_4$])
   (c) Twice with 400 µl binding buffer (20 mM Tris, pH 7.9, 500 mM NaCl, 5 mM imidazole, 10% glycerol).

For each step, 400 µl volumes were added with a 1 ml pipette tip, but were removed with a 200 µl tip in order not to disturb the resin.

5. The charged resin was stored at 4 °C with 400 µl 1× binding buffer until ready for use. The resin is usable for many weeks when stored under these conditions.
6. The cleared extract (step 1) was added to the resin. The sample was mixed gently by inversion several times and incubated for one hour at 4 °C on a rotary device to allow binding of ParF-His$_6$ to the resin.
7. The sample was centrifuged at approx. 500*g* for 4 °C for one minute.
8. 15 µl of the supernatant were retained, 5 µl of 2× loading buffer were added and the sample was stored at −20 °C. This sample was the ParF-His$_6$-depleted extract.
9. The following sequence of steps was used to wash and elute the ParF protein from the resin. Concentrations of imidazole from 100 to 400 mM were used successively to determine the minimum concentration that was sufficient for elution of the protein. For each step the appropriate buffer was added, the tube was inverted several times to mix, incubated for five minutes at 4 °C on a rotary device and then centrifuged at approx. 500*g* for 4 °C for one minute.
   (a) Three times with 600 µl of binding buffer
   (b) Twice with 600 µl of wash buffer (20 mM Tris, pH 7.9, 500 mM NaCl, 85 mM imidazole, 10% glycerol)
   (c) The bound protein was eluted twice with 300 µl of elution buffer (20 mM Tris-HCl, pH 7.9, 500 mM NaCl, 10% glycerol) containing in turn 100, 200, 300 or 400 mM imidazole.
10. 15 µl of each fraction were retained, 5 µl of 2× loading buffer were added and the samples were analysed immediately by SDS-PAGE or were stored at −20 °C for later processing.
11. Visual comparison by SDS-PAGE of ParF-His$_6$ concentrations in the cleared extract control (step 1), the depleted extract (step 8) and the wash and elution fractions (step 9) revealed that 300 mM was the optimal concentration of imidazole required for elution of the protein from the Ni^{2+} resin.

6.4.3 Optimised Protocol for ParF Purification

After determination of the optimal induction and imidazole elution conditions for ParF-His$_6$ in IMAC as described above, the protocol below was employed to purify the protein on a larger scale. Although protein purification trials with small culture volumes generally are scalable to larger volumes, caution should be exercised. An aliquot of the large-scale culture after induction always should be analysed by SDS-PAGE to verify correct overproduction before proceeding to purification of the target protein. In the event that expression at a larger scale does not replicate fully the expression that was optimised using smaller culture volumes, the induction time, inducer concentration and other parameters may need to be modified to improve the overproduction that is obtained when employing larger culture volumes. It is important also to transform the expression plasmid freshly into the *E. coli* host strain for each induction procedure. Long-term maintenance of the plasmid in the expression strain, either on agar plates or as frozen stock cultures, is not recommended. *E. coli* BL21(DE3) and other strains that are commonly used for high level gene expression are proficient in homologous recombination that may induce plasmid rearrangements and also possess endonuclease activity that potentially results in plasmid loss. Instead the plasmid should be stored at −80 °C in a standard *E. coli* K-12 strain that is used for molecular cloning. These strains have been mutated both to be deficient in homologous recombination and in endonuclease activity and provide a stable environment in which to avoid genetic alterations to the expression plasmid. The strain that contains the plasmid should be streaked for single colonies on appropriate selective agar plates at 37 °C. The strain subsequently can be propagated as a broth culture from which plasmid DNA may be prepared freshly using standard isolation procedures (see Chapter 4).

1. *E. coli* BL21(DE3) was transformed with the pET22b(+)-*parF* expression plasmid, selecting on LB plates containing ampicillin (100 μg/ml) at 37 °C for 12–16 hours.
2. Approximately five colonies from this transformation were inoculated in 6 ml of LB broth with ampicillin in a 125 ml conical flask.
3. The culture was grown at 30 °C, shaking at 180 rpm, until OD_{600} was 0.65–0.75 (approx. three hours).
4. The 6 ml culture was used to inoculate 600 ml of LB with ampicillin in a 2 l flask.
5. The culture was incubated at 30 °C, shaking at 180 rpm, until OD_{600} was 0.65–0.75 (approx. three hours).
6. A 100 μl aliquot was removed to a microcentrifuge tube, centrifuged at approx. 10 000*g* for one minute, the supernatant was discarded and the cell pellet was resuspended in 20 μl 2× SDS-PAGE loading buffer. The sample was denatured and stored at −20 °C as the uninduced control.
7. The remaining culture was induced with 1 mM IPTG with incubation at 30 °C, shaking at 180 rpm, for three hours.
8. A 100 μl aliquot was removed to check for correct overproduction (induction control) (Figure 6.5a, lane O).
9. The induced culture was dispensed in 100 ml aliquots and centrifuged at approx. 5000*g* for five minutes. The pellets were stored at −80 °C. Cell pellets retained for many months at −80 °C can be used for protein recovery in the case of ParF. The long-term viability of other recombinant proteins under these conditions may vary.
10. A 100 ml cell pellet was thawed at room temperature.
11. The pellet was resuspended in 15 ml of 1× binding buffer and the following were added: 150 μl of soybean trypsin inhibitor (1 mg/ml), 150 μl of lysozyme (10 mg/ml) and 150 μl of phenylmethylsulphonyl fluoride (100 mM).
12. The sample was incubated for 15 minutes at 30 °C, an additional 150 μl of lysozyme solution were added and incubation was continued for another 15 minutes.
13. The sample was sonicated on ice 5 times for 30 seconds each, with pause intervals of 1 minute, to disrupt the cells.
14. The lysate was centrifuge at approx. 5000*g* for 40 minutes at 4 °C. This low-speed centrifugation step maintains ParF-His_6 in the supernatant. Other overproduced proteins may be more robust and may tolerate higher centrifugation speeds.
15. The supernatant extract was collected and loaded on to a pre-equilibrated, commercially available Ni^{2+} column (2.5 ml settled bed resin) at 4 °C.

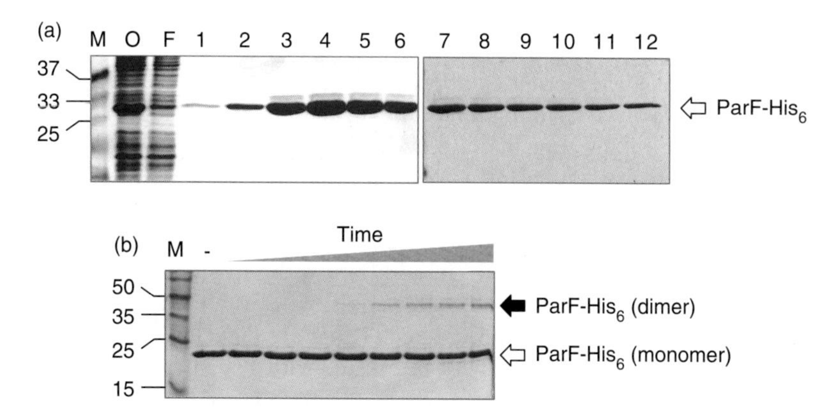

Figure 6.5 Purification and chemical cross-linking of ParF-His$_6$. (a) The ParF-His$_6$ protein was overproduced in *E. coli* BL21(DE3) as described in the text and aliquots from the purification steps were analysed by SDS-PAGE and stained with Coomassie Blue. Lane M, protein size markers; lane O, overproduced ParF-His$_6$; lane F, flowthrough after binding of ParF-His$_6$ to the Ni^{2+} column; lanes 1–12, fractions collected after release of ParF-His$_6$ from the column with elution buffer. The open arrow indicates the position of ParF-His$_6$ in the overproduced and elution fractions. (b) Chemical cross-linking of purified ParF-His$_6$. A series of reactions was assembled that contained ParF-His$_6$ (2.5 μg), ATP (1 mM), and DMP (10 mM). The reactions were incubated at 23 °C for 0, 0.5, 1, 3, 5, 15, 30, 60, and 120 minutes (left to right). Reactions were quenched by addition of 33 mM Tris-HCl, pH 7.5 and an equal volume of SDS-PAGE loading buffer, placed on ice, and then analysed by SDS-PAGE and stained with Coomassie Blue. Open and closed arrows indicate the positions of monomeric and dimeric ParF-His$_6$, respectively. Lane M, protein size markers. Selected size markers (kDa) are indicated on both panels.

16. The loading–elution circle was closed and ParF-His$_6$ was loaded for 90 minutes using a peristaltic pump at a flowrate = 3.0.
17. The column was washed first with 30 ml of binding buffer followed by 70 ml of wash buffer (20 mM Tris, pH 7.9, 500 mM NaCl, 85 mM imidazole, 10% glycerol); 50–100 μl volumes of this flowthrough were retained in each case for later SDS-PAGE analysis (Figure 6.5a, lane F).
18. ParF-His$_6$ was eluted with 12 ml of elution buffer using the optimal imidazole concentration that was determined above (20 mM Tris-HCl, pH 7.9, 500 mM NaCl, 10% glycerol, 300 mM imidazole).
19. DTT was added to 2 mM final concentration to all the elution fractions to stabilise the protein. DTT prevents the oxidation of proteins in reducing environments and inhibits the formation of undesirable disulfide bonds. Although ParF-His$_6$ lacks cysteine residues that might form disulfide bonds, the protein proved to be more stable with DTT in the buffer. The effects of DTT on other recombinant proteins needs to be assessed empirically.
20. The Bradford assay [50] and SDS-PAGE were performed to identify the elution fractions in which ParF-His$_6$ was most concentrated (Figure 6.5a, lanes 1–12). Protein concentrations also can be determined using commercially available kits. Fractions 3–6 were retained in this case. The protein was estimated to be >90% pure (Figure 6.5).
21. The selected fractions were exchanged into a storage buffer (30 mM Tris, pH 7.5–8.0, 100 mM KCl, 10% glycerol, 2 mM DTT) by a buffer exchange using 5 ml HiTrap desalting columns pre-packed with Sephadex® G-25 Superfine (GE Life Sciences). The storage buffer was used to equilibrate the column and to elute the protein. Alternatively, successive dialysis into the storage buffer was performed.
22. Purified ParF-His$_6$ was aliquoted in 100 μl volumes, flash-frozen with liquid nitrogen and stored at −80 °C. Each 100 ml volume of induced culture (step 9) produced several milligrams of purified protein. The protein was stable for many months under these storage conditions. The stability of other recombinant proteins needs to be assessed by trial-and-error.

6.4.4 Demonstrating the Functionality of Purified ParF-His_6 Protein

ParF is a monomer when bound to ADP but forms higher-order complexes in the presence of ATP [51]. One example assay of the purified ParF-His_6 protein involved examination of the formation of these higher-order species after covalent cross-linking with the cross-linking reagent dimethyl pimelimidate (DMP). DMP reacts with amine-containing moieties such as those in proteins to form covalent amidine linkages. These linkages are insensitive to SDS-PAGE and thus treatment with DMP permits the visualisation of dimeric and other oligomeric forms of proteins. Aliquots of purified ParF-His_6 were incubated with ATP and DMP for up to 120 minutes and the reactions were analysed by SDS-PAGE. In addition to the monomeric form of the protein that was evident at approx. 23 kDa, a dimeric product became visible as the incubation time increased (Figure 6.5b). Thus, purified ParF-His_6 forms higher-order species that are detectable by chemical cross-linking. The ParF-His_6 protein also displayed ATPase activity, interacted with a partner protein and exhibited other properties *in vitro*, which verified that the purified protein was functional [51].

6.5 Production and Purification of a GST-Tagged Protein: Example Methodology with the C-Terminal Domain of Yeast RNA Polymerase II

RNA polymerase II is one of three RNA polymerases in eucaryotic cells. This multisubunit complex is responsible for the bulk of gene transcription and therefore is a fulcrum for regulation of gene expression. The C-terminal domain (CTD) of the largest subunit of RNA polymerase II interacts with numerous RNA processing factors and also is a site for phosphorylation that profoundly influences the domain's function [52]. CTD contains repeats of a heptapeptide motif, which in the yeast *Saccharomyces cerevisiae* comprises 26 copies of the consensus sequence Tyr-Ser-Pro-Thr-Ser-Pro-Ser [53, 54]. Yeast CTD has been purified by various strategies including as a GST tagged protein in *E. coli* [55].

The gene segment that encodes CTD was amplified by PCR from yeast genomic DNA and inserted in the pGEX-4T-1 vector (GE Healthcare Life Sciences) using EcoRI and XhoI restriction enzymes [55] (Figure 6.6a). The primers used for amplification were designed analogously to those illustrated in Figure 6.3a. As outlined above for the pET vector series, the cloning generated an in-frame N-terminal fusion of the genes for GST and CTD. Unlike the pET vector series, expression of the fused gene from pGEX-4T-1-CTD is driven by the *tac* promoter, which is a hybrid between the *lac* and *trp* promoters that is stronger than either of the parental promoters [56]. The pGEX-4T-1 vector includes the *lacI^q* gene, which overproduces the Lac repressor. As described above, IPTG inhibits binding of the repressor at the *tac* promoter, which therefore is available for recognition by *E. coli* RNA polymerase that directs the high-level expression of the fusion gene for GST-CTD.

Production of the GST-CTD protein in *E. coli* was optimised as described above for the ParF-His_6 protein: the most suitable incubation temperature, IPTG concentration, induction time and solubility characteristics were determined by tests with small-scale volumes of cultures that harboured the pGEX-4T-1-CTD expression plasmid. Samples were analysed by SDS-PAGE to visualise production of the GST-CTD protein fusion. Unlike the pET system detailed above, the pGEX-4T-1 expression vector does not involve regulation by T7 RNA polymerase and therefore does not require specialised strains such as *E. coli* BL21(DE3). Therefore different laboratory derivatives of *E. coli* K-12 were assessed for GST-CTD production from pGEX-4T-1-CTD. The following is the optimised protocol for production and purification of the GST-CTD fusion protein and is a modified version of the procedure described elsewhere [57].

1. *E. coli* DH5α was transformed with the pGEX-4T-1-CTD expression plasmid (0.1 μg) with selection on LB plates containing ampicillin (100 μg/ml) at 37 °C for 12–16 hours.
2. Approximately five colonies from this transformation were inoculated in 6 ml of LB broth with ampicillin in a 125 ml conical flask.
3. The culture was grown at 30 °C, shaking at 180 rpm, until OD_{600} was approx. 0.6 (approx. four hours).
4. The 6 ml pre-culture was used to inoculate 1 l of LB with ampicillin in a 3 l flask.

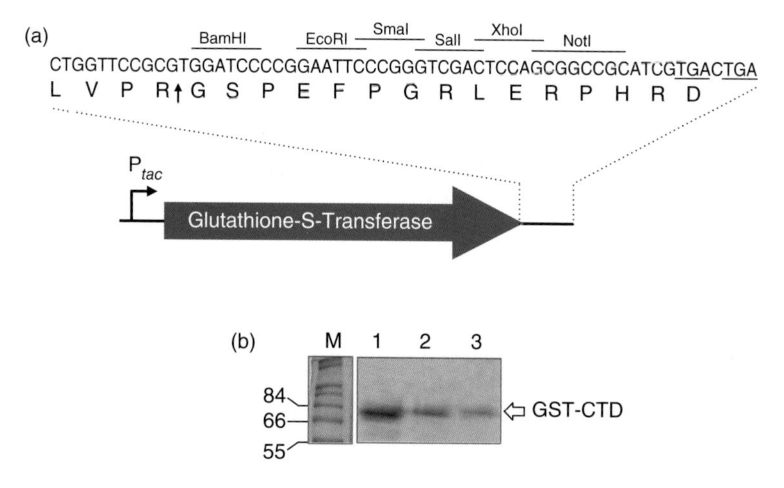

Figure 6.6 Cloning and purification of GST-CTD. (a) Organisation of the multiple cloning site in vector pGEX-4T-1. This vector (GE Healthcare Life Sciences) facilitates the overproduction of proteins with an N-terminal GST tag. The gene for GST (blue) is followed in-frame by the expanded DNA sequence. Selected restriction enzyme sites are marked. These sites are used to clone the gene of interest as an in-frame PCR product between, for example, the EcoRI and XhoI sites to produce a fusion protein that harbours the GST tag at the N-terminus. The tag allows affinity purification of the corresponding fusion protein. If desired, the GST tag subsequently may be removed from the protein by digestion with thrombin serine protease that cleaves at the site indicated by the vertical arrow. Stop codons are underlined. Expression of the fusion gene is driven by the *tac* promoter. (b) SDS-PAGE analysis of purified GST-CTD. The GST-CTD fusion protein was purified as detailed in the text. Aliquots of the elution fractions 1, 2, and 3 were analysed by SDS-PAGE and stained with Coomassie Blue. Lane M, protein size markers. Selected size markers (kDa) are indicated.

5. The culture was incubated at 30 °C, shaking at 180 rpm, until OD_{600} was approx. 0.6 (approx. four hours).
6. A 100 μl aliquot was removed to a microcentrifuge tube, centrifuged at approx. 10 000*g* for one minute, the supernatant was discarded and the cell pellet was resuspended in 20 μl 2× SDS-PAGE loading buffer. The sample was denatured and stored at −20 °C as the uninduced control for subsequent SDS-PAGE analysis.
7. The remaining culture was induced with 1 mM IPTG with incubation at 30 °C, shaking at 180 rpm, for 12 hours.
8. A 100 μl aliquot was removed to check for correct induction (induction control) in SDS-PAGE analysis.
9. The induced culture was dispensed in 500 ml aliquots and centrifuged at approx. 5000*g* for five minutes. The pellets were stored at −80 °C.
10. A 500 ml cell pellet was thawed on ice (approx. one hour).
11. The pellet was resuspended in 20 ml of TZ buffer (50 mM Tris-HCl, pH 7.9, 12.5 mM $MgCl_2$, 0.5 mM ethylenediaminetetraacetic acid (EDTA), 100 mM KCl, 20% glycerol, 1 mM β-mercaptoethanol, 10 μM $ZnCl_2$).
12. The protease inhibitors leupeptin, phenylmethylsulphonyl fluoride and soybean trypsin inhibitor were added to the resuspended cell pellet to final concentrations each of 1 μg/ml, along with the detergent Triton X-100 (1%) and 1 ml of lysozyme (20 mg/ml).
13. The sample was incubated on ice for 30 minutes and then was sonicated on ice 6 times for 30 seconds each with pause intervals of one minute.
14. The lysate was centrifuged at approx. 8000 *g* for 60 minutes at 4 °C.
15. The supernatant extract (approx. 40 ml) was collected and mixed with one bed volume of GSH-sepharose resin. The resin was prepared by centrifuging 1.33 ml of resin slurry, resuspending in 10 ml of TZ buffer and washing twice with this buffer.
16. The sample was incubated with gentle agitation for 30 minutes at room temperature to permit binding of the GST-CTD protein to the resin.
17. The lysate–resin mix was centrifuged for five minutes and the supernatant was removed and discarded.

18. The resin with the bound GST-CTD was washed with 20 ml of 50 mM Tris-HCl, pH 7.9, 1 M NaCl, 1 mM β-mercaptoethanol, and soybean trypsin inhibitor, leupeptin and phenylmethylsulphonyl fluoride at final concentrations of 1 μg/ml.
19. The sample was centrifuged and washed with 20 ml of phosphate buffered saline (137 mM NaCl, 10 mM Na_2HPO_4, 1.8 mM KH_2PO_4, 2.7 mM KCl, pH 7.4) that contained Triton X-100 (1%), 1 mM β-mercaptoethanol and the protease inhibitors described in the preceding step.
20. GST-CTD was eluted from the glutathione sepharose resin with 2 ml of 50 mM Tris-HCl, pH 7.5 that contained 15 mM reduced glutathione with incubation at room temperature for 10 minutes.
21. The sample was centrifuged and approx. 1.5 ml of the supernatant was collected (fraction 1).
22. Steps 20 and 21 were repeated twice to produce fractions 2 and 3.
23. Fractions 1–3 were dialysed against 500 ml of TZ buffer, aliquoted in 100 μl volumes, flash-frozen with liquid nitrogen and stored at −80 °C.
24. The Bradford assay and SDS-PAGE were performed to determine protein concentration and purity (Figure 6.6b), respectively. The protein was estimated to be >90% pure.

6.5.1 Demonstrating the Functionality of Purified GST-CTD Protein

The interaction of CTD with a plethora of RNA processing factors is detectable both *in vivo* and *in vitro* using a variety of experimental approaches. These interactions are modulated by the phosphorylation state of the CTD [52]. The GST-CTD purified above was phosphorylated *in vitro*. Both the phosphorylated and unphosphorylated forms were tested for interactions with proteins in yeast whole-cell extracts using affinity chromatography assays (also known as pull-down assays). These experiments revealed diverse interacting partners with the CTD including the Pcf11 protein that links mRNA 3′-end processing, transcriptional elongation and termination of transcription in *S. cerevisiae* [55]. As the tag was not cleaved from the GST-CTD fusion protein in this case, the tagged protein was demonstrated to be functional in pull-down assays and the GST tag did not interfere detectably with the interaction of CTD with partner proteins. An important negative control experiment in these assays was to test that purified GST alone did not interact with the proteins to which GST-CTD is bound.

6.6 Further Purification of Tagged Proteins

The preceding examples illustrated cases in which His_6 and GST affinity tags did not perturb the behaviours of recombinant proteins *in vitro*. Nevertheless, the potential impact of the tag on the function and structure of the purified protein should always be considered. Although bulky tags such as GST and MBP are considered to fold independently of the target protein to which they are fused, these tags are more likely to affect protein behaviour than compact tags such as His_6, which generally are considered innocuous [58]. Nevertheless, there are numerous reports that His_6-tags also can affect protein activity [59–61]. Therefore the production, purification and comparison of proteins with N- or C-terminal tags may be appropriate. In practice the behaviour of a protein with one of these tags, e.g. a C-terminal tag, may be examined initially and the second tagged version of the protein, e.g. with an N-terminal tag, may be characterised only if the first protein either is not produced properly or displays aberrant properties *in vitro*. Alternatively, certain expression vectors are designed specifically to allow proteolytic cleavage of His_6-tag and other tags after protein purification (Figure 6.4) [19]. In this case the tagless protein and cleaved tag are separated by passing the reaction through an Ni^{2+} column to which the tag and uncleaved protein will bind but through which the tagless protein will pass unhindered (Figure 6.7). Co-purified proteins that have affinity for the His_6 matrix will also be retained on the column. The cleavage reaction typically leaves a 'stub' of a few amino acids between the cleavage site and the target protein. For example, cloning of a target gene between NdeI and XhoI restriction sites in pET33b(+) generates a fusion protein that harbours the amino acid sequence MGSS<u>HHHHHH</u>SSGLVPR↓GSRRASVH at the N-terminus

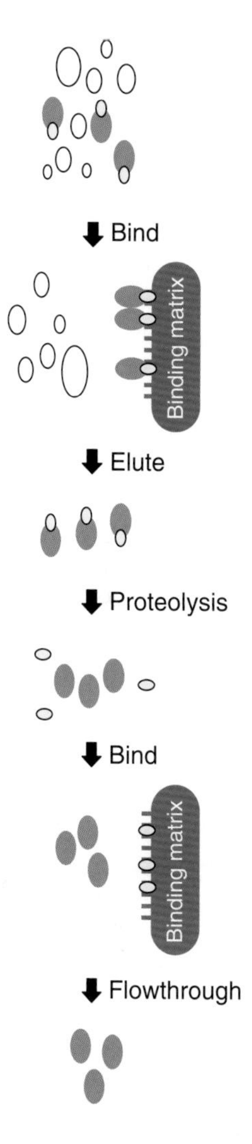

Figure 6.7 Separation and purification of a tagless target protein. The overproduced fusion protein comprises the target protein (blue) with a His_6 or other affinity tag (yellow). The tagged protein is purified from other proteins and cellular components (open symbols) in *E. coli* by IMAC or equivalent affinity chromatography. The fusion protein binds non-covalently to the affinity matrix (green) whereas other proteins pass through the column. The tagged protein is eluted and purified by altering the buffer conditions on the matrix, e.g. with a high concentration of imidazole in the case of His_6-tagged proteins. The purified protein is treated with an appropriate protease (not shown) that cleaves a linker sequence that is engineered between the target protein and the tag. The cleavage reaction is applied to a fresh affinity column to which the liberated tag binds whereas the purified tagless target protein emerges in the flowthrough. The commercially available protease may be engineered to possess the same tag as the target fusion protein so that it also is captured on the matrix.

(Figure 6.4). Digestion of the purified protein with the thrombin protease removes 17 residues, including the His_6 motif, but leaves the eight amino acid tail GSRRASVH. Ion exchange chromatography also may be used to purify the untagged protein from the protease cleavage reaction. Strategies have been developed for purification of overproduced tagless proteins by single-step procedures, although these approaches have yet to gain widespread popularity [62, 63].

The example purification procedures for ParF-His_6 and GST-CTD described above typically generate proteins that are >90% pure based on inspection by SDS-PAGE. This level of purity often is sufficient for *in vitro* analyses. However, if the presence of contaminating proteins is a concern, a mutated non-functional version of the tagged protein can be purified by an identical procedure and tested in the relevant assay(s) in parallel with the wild-type protein. The mutated protein can be generated by site-directed mutagenesis of a codon, which specifies an amino acid that is known to be key for protein function [64]. If the mutated protein displays no activity then the activity observed with the wild-type protein can be attributed more confidently to the latter and not to contaminating proteins. Alternatively, the wild-type protein can be purified further using additional chromatographic techniques, which are discussed in detail in Chapter 7. Principal among these techniques are ion exchange chromatography that utilises the reversible interaction between a charged protein and a chromatographic matrix that possesses an opposite charge, gel filtration chromatography that separates protein species based on molecular size and hydrophobic interaction chromatography that purifies proteins based on hydrophobicity [65–67].

6.7 Alternative Hosts for Protein Production

Although *E. coli* will continue to be the workhorse for recombinant protein production for the foreseeable future, expression platforms also have been developed for protein production in other bacterial species [68]. However, one noteworthy drawback with protein purification from *E. coli* and other bacteria is the absence of extensive post-translational modification processes in these hosts. Proteins in eucaryotic species often are subject to glycosylation, phosphorylation, acetylation and other modifications after translation [69]. Post-translational modifications may modulate the structure and/or function of eucaryotic proteins. Proteins that lack these modifications when produced and purified from *E. coli* may not exhibit all of the characteristics of the equivalent modified versions. Moreover, numerous eucaryotic proteins and protein complexes have also proven to be recalcitrant to overproduction in *E. coli*. However, several alternative hosts have been developed for the purification of post-translationally modified proteins as well as mammalian and other recombinant proteins that are refractory to production in bacteria [1, 70–73]. *Pichia pastoris* is a methylotrophic yeast that can utilise methanol as a sole carbon and energy source. The genes involved in methanol utilisation are highly inducible, which led to the development of methanol-inducible expression systems in this host [74]. Recent developments in genome analysis of *P. pastoris*, along with promoter engineering, the optimization of codon usage and gene dosage, and enhancements to protein secretion and methanol metabolic pathways, make this organism an attractive host for the production of heterologous proteins [75].

The yeast *S. cerevisiae* is generally recognised as safe, is genetically tractable and is used extensively for the industrial production of certain biochemicals and biofuels due to its tolerance of alcohols and harsh conditions [76, 77]. In addition, a range of expression plasmids that possess inducible or constitutive promoters and different selectable markers have been developed for recombinant protein production in *S. cerevisiae* [78]. Glycosylation is one of the most important post-translational modifications of eucaryotic proteins. Glycosylation patterns in *P. pastoris* and *S. cerevisiae* differ both from each other and from those that occur in human cells. Therefore, caution needs to be exerted in choosing which of these hosts is more appropriate to produce a recombinant target protein [79]. This issue has been alleviated in part by the engineering of *P. pastoris* derivatives that mimic the glycosylation patterns in human cells [80].

Baculovirus is an insect virus that has been adapted for protein production in cell lines. Notably, baculovirus-infected insect cell lines have been used extensively for the production of a range of recombinant proteins, including proteins that do not fold properly in bacterial expression systems, glycoproteins, membrane proteins, vaccines and multiprotein complexes [81]. Baculovirus expression systems avail particularly of the viral polyhedrin and p10 promoters, which are the primary promoters that are active during the late phase of virus infection. Baculovirus also has been engineered with mammalian promoters for the production of recombinant proteins in mammalian cell lines [82].

6.8 Concluding Remarks

The production and simplified purification of tagged recombinant proteins from *E. coli* has been central to the success of molecular biology in recent decades [83]. Proteins now are purified routinely for fundamental functional studies, as reagents and targets for biomedical and pharmaceutical use, and as industrial enzymes. Continued efforts to improve further the protein production capacity of *E. coli* centre on the enhancement of expression vectors, host strains and induction systems [84, 85], as well as the use of genomic, proteomic, and metabolic engineering strategies to modulate gene expression and control [86, 87]. A wide array of sophisticated expression systems that incorporate diverse plasmid vectors, different affinity purification tags and tightly regulated promoters are available for use in *E. coli* [11]. The breadth of expression platforms may seem daunting when embarking on a first venture into recombinant protein production and purification. However, many of the available expression systems have been tried, tested and honed by a plethora of researchers over many years using multifarious proteins from sundry organisms [88]. The simplest start point for a researcher who wishes to delve into the realm of protein production potentially is to use a His_6-tag vector of the pET or a related series that has been proven to be robust for the expression of a variety of disparate genes. Investigation of the relevant literature concerned with overproduction, separation and purification of proteins that are related to the protein of interest is also highly recommended as *nihil novi sub sole.*

Acknowledgements

Work in the laboratory of DB is supported by the Biotechnology and Biological Sciences Research Council.

References

1 Nettleship, J.E., Assenberg, R., Diprose, J.M. et al. (2010). Recent advances in the production of proteins in insect and mammalian cells for structural biology. *J. Struct. Biol.* 172: 55–65.
2 Huang, C.J., Lin, H., and Yang, X. (2012). Industrial production of recombinant therapeutics in *Escherichia coli* and its recent advancements. *J. Ind. Microbiol. Biotechnol.* 39: 383–399.
3 Son, M.S. and Taylor, R.K. (2012). Growth and maintenance of *Escherichia coli* laboratory strains. *Curr. Protoc. Microbiol.* Chapter 5:Unit 5A.4.
4 Blount, Z.D. (2015). The unexhausted potential of *E. coli*. *eLife* 4: e05826.
5 Gilbert, W. and Müller-Hill, B. (1966). Isolation of the Lac repressor. *Proc. Natl. Acad. Sci. U.S.A.* 56: 1891–1898.
6 Rosenberg, J.M., Khallai, O.B., Kopka, M.L. et al. (1977). Lac repressor purification without inactivation of DNA binding activity. *Nucleic Acids Res.* 4: 567–572.
7 Winshell, E. and Shaw, W.V. (1969). Kinetics of induction and purification of chloramphenicol acetyltransferase from chloramphenicol-resistant *Staphylococcus aureus*. *J. Bacteriol.* 98: 1248–1257.
8 Amarasinghe, C. and Jin, J.P. (2015). The use of affinity tags to overcome obstacles in recombinant protein expression and purification. *Protein Pept. Lett.* 22: 885–892.
9 Raran-Kurussi, S. and Waugh, D.S. (2017). Expression and purification of recombinant proteins in *Escherichia coli* with a His_6 or dual His_6-MBP tag. *Methods Mol. Biol.* 1607: 1–15.
10 Wood, D.W. (2014). New trends and affinity tag designs for recombinant protein purification. *Curr. Opin. Struct. Biol.* 26: 54–61.
11 Rosano, G.L. and Ceccarelli, E.A. (2014). Recombinant protein expression in *Escherichia coli*: advances and challenges. *Front. Microbiol.* 5: 172.
12 Zhao, X., Li, G., and Liang, S. (2013). Several affinity tags commonly used in chromatographic purification. *J. Anal. Methods Chem.* 2013: 581093.
13 Lebendiker, M. and Danieli, T. (2017). Purification of proteins fused to maltose-binding protein. *Methods Mol. Biol.* 1485: 257–273.
14 Waugh, D.S. (2016). The remarkable solubility-enhancing power of *Escherichia coli* maltose-binding protein. *Postepy Biochem.* 62: 377–382.
15 Bataille, L., Dieryck, W., Hocquellet, A. et al. (2015). Expression and purification of short hydrophobic elastin-like polypeptides with maltose-binding protein as a solubility tag. *Protein Expression Purif.* 110: 165–171.

16 Guo, Y., Yu, M., Jing, N., and Zhang, S. (2018). Production of soluble bioactive mouse leukemia inhibitory factor from *Escherichia coli* using MBP tag. *Protein Expression Purif.* 150: 86–91.
17 Raran-Kurussi, S. and Waugh, D.S. (2012). The ability to enhance the solubility of its fusion partners is an intrinsic property of maltose-binding protein but their folding is either spontaneous or chaperone-mediated. *PLoS One* 7: e49589.
18 Harper, S. and Speicher, D.W. (2011). Purification of proteins fused to glutathione S-transferase. *Methods Mol. Biol.* 681: 259–280.
19 Waugh, D.S. (2011). An overview of enzymatic reagents for the removal of affinity tags. *Protein Expression Purif.* 80: 283–293.
20 Block, H., Maertens, B., Spriestersbach, A. et al. (2009). Immobilized-metal affinity chromatography (IMAC): a review. *Methods Enzymol.* 463: 439–473.
21 Porath, J., Carlsson, J., Olsson, I., and Belfrage, G. (1975). Metal chelate affinity chromatography, a new approach to protein fractionation. *Nature* 258: 598–599.
22 Hochuli, E., Bannwarth, W., Döbeli, H. et al. (1988). Genetic approach to facilitate purification of recombinant proteins with a novel metal chelate adsorbent. *Nat. Biotechnol.* 6: 1321–1325.
23 Studier, F.W., Rosenberg, A.H., Dunn, J.J., and Dubendorff, J.W. (1990). Use of T7 RNA polymerase to direct expression of cloned genes. *Methods Enzymol.* 185: 60–89.
24 Bartosik, A.A., Markowska, A., Szarlak, J. et al. (2012). Novel broad-host-range vehicles for cloning and shuffling of gene cassettes. *J. Microbiol. Methods* 88: 53–62.
25 Santos, P.M., Di Bartolo, I., Blatny, J.M. et al. (2001). New broad-host-range promoter probe vectors based on the plasmid RK2 replicon. *FEMS Microbiol. Lett.* 195: 91–96.
26 Scott, H.N., Laible, P.D., and Hanson, D.K. (2003). Sequences of versatile broad-host-range vectors of the RK2 family. *Plasmid* 50: 74–79.
27 Chakravartty, V. and Cronan, J.E. (2015). A series of medium and high copy number arabinose-inducible *Escherichia coli* expression vectors compatible with pBR322 and pACYC184. *Plasmid* 81: 21–26.
28 Guzman, L.M., Belin, D., Carson, M.J., and Beckwith, J. (1995). Tight regulation, modulation, and high-level expression by vectors containing the arabinose P_{BAD} promoter. *J. Bacteriol.* 177: 4121–4130.
29 Bedouelle, H. and Duplay, P. (1988). Production in *Escherichia coli* and one-step purification of bifunctional hybrid proteins which bind maltose. *Eur. J. Biochem.* 171: 541–549.
30 Bird, L.E. (2011). High throughput construction and small scale expression screening of multi-tag vectors in *Escherichia coli*. *Methods* 55: 29–37.
31 Cabrita, L.D., Dai, W., and Bottomley, S.P. (2006). A family of *E. coli* expression vectors for laboratory scale and high throughput soluble protein production. *BMC Biotechnol.* 6: 12.
32 Correa, A., Ortega, C., Obal, G. et al. (2014). Generation of a vector suite for protein solubility screening. *Front. Microbiol.* 5: 67.
33 Guan, C., Li, P., Riggs, P.D., and Inouye, H. (1988). Vectors that facilitate the expression and purification of foreign peptides in *Escherichia coli* by fusion to maltose-binding protein. *Gene* 67: 21–30.
34 Smith, D.B. and Johnson, K.S. (1988). Single-step purification of polypeptides expressed in *Escherichia coli* as fusions with glutathione S-transferase. *Gene* 67: 31–40.
35 Jia, B. and Jeon, C.O. (2016). High-throughput recombinant protein expression in *Escherichia coli*: current status and future perspectives. *Open Biol.* 6: 160196.
36 Salim, L., Feger, C., and Busso, D. (2016). Construction of a compatible Gateway-based co-expression vector set for expressing multiprotein complexes in *E. coli*. *Anal. Biochem.* 512: 110–113.
37 Schmid-Burgk, J.L., Schmidt, T., Kaiser, V. et al. (2013). A ligation-independent cloning technique for high-throughput assembly of transcription activator-like effector genes. *Nat. Biotechnol.* 31: 76–81.
38 Shendure, J., Balasubramanian, S., Church, G.M. et al. (2017). DNA sequencing at 40: past, present and future. *Nature* 550: 345–353.
39 Jenny, R.J., Mann, K.G., and Lundblad, R.L. (2003). A critical review of the methods for cleavage of fusion proteins with thrombin and factor Xa. *Protein Expression Purif.* 31: 1–11.

40 Tang, G.Q., Bandwar, R.P., and Patel, S.S. (2005). Extended upstream A-T sequence increases T7 promoter strength. *J. Biol. Chem.* 280: 40707–40713.
41 McLeod, B.N., Allison-Gamble, G.E., Barge, M.T. et al. (2017). A three-dimensional ParF meshwork assembles through the nucleoid to mediate plasmid segregation. *Nucleic Acids Res.* 45: 3158–3171.
42 Hayes, F. and Barillà, D. (2010). Extrachromosomal components of the nucleoid: recent developments in deciphering the molecular basis of plasmid segregation. In: *Bacterial Chromatin* (ed. C.J. Dorman and R.T. Dame), 49–70. Dordrecht: Springer Publishing.
43 Barillà, D. and Hayes, F. (2003). Architecture of the ParF*ParG protein complex involved in prokaryotic DNA segregation. *Mol. Microbiol.* 49: 487–499.
44 Papaneophytou, C.P. and Kontopidis, G. (2014). Statistical approaches to maximize recombinant protein expression in *Escherichia coli*: a general review. *Protein Expression Purif.* 94: 22–32.
45 Leibly, D.J., Nguyen, T.N., Kao, L.T. et al. (2012). Stabilizing additives added during cell lysis aid in the solubilization of recombinant proteins. *PLoS One* 7: e52482.
46 Ramón, A., Señorale-Pose, M., and Marín, M. (2014). Inclusion bodies: not that bad…. *Front. Microbiol.* 5: 56.
47 Singh, A., Upadhyay, V., Upadhyay, A.K. et al. (2015). Protein recovery from inclusion bodies of *Escherichia coli* using mild solubilization process. *Microb. Cell Fact.* 14: 41.
48 Correa, A. and Oppezzo, P. (2015). Overcoming the solubility problem in *E. coli*: available approaches for recombinant protein production. *Methods Mol. Biol.* 1258: 27–44.
49 Saccardo, P., Corchero, J.L., and Ferrer-Miralles, N. (2016). Tools to cope with difficult-to-express proteins. *Appl. Microbiol. Biotechnol.* 100: 4347–4355.
50 Bradford, M.M. (1976). A rapid and sensitive method for the quantitation of microgram quantities of protein utilizing the principle of protein-dye binding. *Anal. Biochem.* 72: 248–254.
51 Barillà, D., Rosenberg, M.F., Nobbmann, U., and Hayes, F. (2005). Bacterial DNA segregation dynamics mediated by the polymerizing protein ParF. *EMBO J.* 24: 1453–1464.
52 Buratowski, S. (2009). Progression through the RNA polymerase II CTD cycle. *Mol. Cell* 36: 541–546.
53 Babokhov, M., Mosaheb, M.M., Baker, R.W., and Fuchs, S.M. (2018). Repeat-specific functions for the C-terminal domain of RNA polymerase II in budding yeast. *G3* 8: 1593–1601.
54 Hsin, J.P. and Manley, J.L. (2012). The RNA polymerase II CTD coordinates transcription and RNA processing. *Genes Dev.* 26: 2119–2137.
55 Barillà, D., Lee, B.A., and Proudfoot, N.J. (2001). Cleavage/polyadenylation factor IA associates with the carboxyl-terminal domain of RNA polymerase II in *Saccharomyces cerevisiae*. *Proc. Natl. Acad. Sci. U.S.A.* 98: 445–450.
56 de Boer, H.A., Comstock, L.J., and Vasser, M. (1983). The *tac* promoter: a functional hybrid derived from the *trp* and *lac* promoters. *Proc. Natl. Acad. Sci. U.S.A.* 80: 21–25.
57 Peterson, S.R., Dvir, A., Anderson, C.W., and Dynan, W.S. (1992). DNA binding provides a signal for phosphorylation of the RNA polymerase II heptapeptide repeats. *Genes Dev.* 6: 426–438.
58 Carson, M., Johnson, D.H., McDonald, H. et al. (2007). His-tag impact on structure. *Acta Crystallogr. D Biol. Crystallogr.* 63: 295–301.
59 Booth, W.T., Schlachter, C.R., Pote, S. et al. (2018). Impact of an N-terminal polyhistidine tag on protein thermal stability. *ACS Omega* 3: 760–768.
60 Mohanty, A.K. and Wiener, M.C. (2004). Membrane protein expression and production: effects of polyhistidine tag length and position. *Protein Expression Purif.* 33: 311–325.
61 Sabaty, M., Grosse, S., Adryanczyk, G. et al. (2013). Detrimental effect of the 6 His C-terminal tag on YedY enzymatic activity and influence of the TAT signal sequence on YedY synthesis. *BMC Biochem.* 14: 28.
62 Cooper, M.A., Taris, J.E., Shi, C., and Wood, D.W. (2018). A convenient split-intein tag method for the purification of tagless target proteins. *Curr. Protoc. Protein Sci.* 91: 5.29.1–5.29.23.
63 Guan, D. and Chen, Z. (2014). Challenges and recent advances in affinity purification of tag-free proteins. *Biotechnol. Lett.* 36: 1391–1406.

64 Peracchi, A. (2001). Enzyme catalysis: removing chemically 'essential' residues by site-directed mutagenesis. *Trends Biochem. Sci.* 26: 497–503.

65 O'Fágáin, C., Cummins, P.M., and O'Connor, B.F. (2011). Gel-filtration chromatography. *Methods Mol. Biol.* 681: 25–33.

66 Jungbauer, A. and Hahn, R. (2009). Ion-exchange chromatography. *Methods Enzymol.* 463: 349–371.

67 McCue, J.T. (2009). Theory and use of hydrophobic interaction chromatography in protein purification applications. *Methods Enzymol.* 463: 405–414.

68 Gómez, S., López-Estepa, M., Fernández, F.J., and Vega, M.C. (2016). Protein complex production in alternative prokaryotic hosts. *Adv. Exp. Med. Biol.* 896: 115–133.

69 Khoury, G.A., Baliban, R.C., and Floudas, C.A. (2011). Proteome-wide post-translational modification statistics: frequency analysis and curation of the swiss-prot database. *Sci. Rep.* 1: 90.

70 Contreras-Gómez, A., Sánchez-Mirón, A., García-Camacho, F. et al. (2014). Protein production using the baculovirus-insect cell expression system. *Biotechnol. Prog.* 30: 1–18.

71 Fernández, F.J. and Vega, M.C. (2016). Choose a suitable expression host: a survey of available protein production platforms. *Adv. Exp. Med. Biol.* 896: 15–24.

72 Gasser, B., Prielhofer, R., Marx, H. et al. (2013). *Pichia pastoris*: protein production host and model organism for biomedical research. *Future Microbiol.* 8: 191–208.

73 Wang, G., Huang, M., and Nielsen, J. (2017). Exploring the potential of *Saccharomyces cerevisiae* for biopharmaceutical protein production. *Curr. Opin. Biotechnol.* 48: 77–84.

74 Zahrl, R.J., Peña, D.A., Mattanovich, D., and Gasser, B. (2017). Systems biotechnology for protein production in *Pichia pastoris*. *FEMS Yeast Res.* 17: fox068.

75 Byrne, B. (2015). *Pichia pastoris* as an expression host for membrane protein structural biology. *Curr. Opin. Struct. Biol.* 32: 9–17.

76 Kutyna, D.R. and Borneman, A.R. (2018). Heterologous production of flavour and aroma compounds in *Saccharomyces cerevisiae*. *Genes* 9: E326.

77 Turner, T.L., Kim, H., Kong, I.I. et al. (2018). Engineering and evolution of *Saccharomyces cerevisiae* to produce biofuels and chemicals. *Adv. Biochem. Eng. Biotechnol.* 162: 175–215.

78 Darby, R.A., Cartwright, S.P., Dilworth, M.V., and Bill, R.M. (2012). Which yeast species shall I choose? *Saccharomyces cerevisiae* versus *Pichia pastoris*. *Methods Mol. Biol.* 866: 11–23.

79 Vieira Gomes, A.M., Souza Carmo, T., Silva Carvalho, L. et al. (2018). Comparison of yeasts as hosts for recombinant protein production. *Microorganisms* 6: E38.

80 Hamilton, S.R., Davidson, R.C., Sethuraman, N. et al. (2006). Humanization of yeast to produce complex terminally sialylated glycoproteins. *Science* 313: 1441–1443.

81 Kost, T.A. and Kemp, C.W. (2016). Fundamentals of baculovirus expression and applications. *Adv. Exp. Med. Biol.* 896: 187–197.

82 Mansouri, M. and Berger, P. (2018). Baculovirus for gene delivery to mammalian cells: past, present and future. *Plasmid* 98: 1–7.

83 Gileadi, O. (2017). Recombinant protein expression in *E. coli*: a historical perspective. *Methods Mol. Biol.* 1586: 3–10.

84 Gupta, S.K. and Shukla, P. (2016). Advanced technologies for improved expression of recombinant proteins in bacteria: perspectives and applications. *Crit. Rev. Biotechnol.* 36: 1089–1098.

85 Schlegel, S., Genevaux, P., and de Gier, J.W. (2017). Isolating *Escherichia coli* strains for recombinant protein production. *Cell. Mol. Life Sci.* 74: 891–908.

86 Liu, M., Feng, X., Ding, Y. et al. (2015). Metabolic engineering of *Escherichia coli* to improve recombinant protein production. *Appl. Microbiol. Biotechnol.* 99: 10367–10377.

87 Mahalik, S., Sharma, A.K., and Mukherjee, K.J. (2014). Genome engineering for improved recombinant protein expression in *Escherichia coli*. *Microb. Cell Fact.* 13: 177.

88 Konczal, J. and Gray, C.H. (2017). Streamlining workflow and automation to accelerate laboratory scale protein production. *Protein Expression Purif.* 133: 160–169.

Further Reading

Bonner, P.L.R. (2018). *Protein Purification*, 2e. CRC Press Inc.

Janson, J.C. (ed.) (2011). *Protein Purification: Principles, High Resolution Methods, and Applications*, 3e. Wiley.

Rosenberg, I.M. (2005). *Protein Analysis and Purification: Benchtop Techniques*. Birkhäuser.

Simpson, R.J., Adams, P.D., and Golemis, E.A. (eds.) (2009). *Basic Methods in Protein Purification and Analysis: A Laboratory Manual*. Cold Spring Harbor Laboratory Press.

Scopes, R.K. (2010). *Protein Purification: Principles and Practice*. New York: Springer.

7

Chromatography: Separation Techniques in Biology

W John Lough and Mark Carlile

School of Pharmacy and Pharmaceutical Sciences, University of Sunderland, Chester Road, Sunderland, SR1 3SD, UK

7.1 Introduction to Chromatographic Separation

Chromatography is a term that is used to describe a group of separation techniques that utilise the distribution of molecules between a stationary phase and a mobile phase [1]. Separation is achieved by exploitation of a molecule's physicochemical properties that promote it to interact with the stationary phase or stay within the mobile phase. Interactions with the stationary phase promote a slower passage through the separation system. It is this distribution and the modulation of the distribution between these phases that makes chromatography a very powerful and versatile separation technique. Different types of molecule have different distribution constants and hence come off the column at different times. Thus, they are separated from one another before being measured. Importantly, this means that quantitative determinations based on chromatography can be optimised to give good specificity. In an analytical method validation to demonstrate that an analytical method is fit for its intended purpose, accuracy, precision, linearity, limit of quantitation and limit of detection may be evaluated, but it is the test for specificity that is often the most challenging of the validation tests [2]. Therein lies perhaps the main reason for the very widespread use of chromatography in biology.

The field of 'biology' is vast and complex. Within this there are a large and very diverse number of applications for which the power of chromatography may be effectively exploited. Consequently, the entire range of manifestations of 'chromatography' can find some use in biology. However, especially in the purification and analysis of intact, large biological molecules, liquid column chromatography (Figure 7.1) is by far the most common form of chromatographic separation used [3, 4]. However, the application of gas-phase chromatographic separation is also becoming more common, especially for the analysis and profiling of cellular metabolism [5].

The most common forms of liquid chromatographic separation (Table 7.1) are referred to as modes of chromatography, which differ in terms of their stationary phase chemistry and physical properties and the associated mobile phase (one stationary phase can be used for more than one mode of chromatography). The stationary phase is commonly regarded as the solid matrix (usually particles/beads) and its associated surface chemistry. However, this is almost always porous and the 'stationary phase' might better be regarded as the porous matrix and the stagnant mobile phase therein (making constituting up to 90% of the total 'stationary phase'). The solid matrix is typically composed of regular beads of <1–100 μm in diameter. The beads are either xerogels (e.g. Sephadex™: able to shrink and swell dependent on the mobile phase) and aerogels (e.g. silica, porous glass: volume is independent of the mobile phase). Silica finds greater use for modern quantitative analysis (cf. preparative/purification) where there has been adaptation and modification of technology common in high performance liquid chromatography (HPLC) of small molecules. The choice of composition of the mobile phase is dependent on the mode of the separation method and the nature of the samples applied.

Biomolecular and Bioanalytical Techniques: Theory, Methodology and Applications, First Edition. Edited by Vasudevan Ramesh.

Mobile phase reservoirs: need to be **above** the pumps to generate a siphon effect; sintered filter on end of inlet tube; should be labelled with content, prepared by…., date of preparation, date of expiry; in modern systems dissolved gases are removed post-reservoir by inline membrane degassing (amber glass may be recommended in some cases but clear glass allows visual inspection for particulates, cloudiness *etc* (Mel Euerby, Chromatographic Society "Grass Roots" course, 2016)

Detector: in the past this might have been a variable wavelength UV detector set at a single wavelength but more commonly now a diode-array UV detector capable of capturing many wavelengths simultaneously; also an alternative detector (*e.g* fluorescence, electrochemical, evaporative light scattering detector) might be used on its own or in series with a low volume connector; **for small biological molecules at low concentrations in complex mixtures mass spectrometric detection is almost mandatory.** The detector output may be collected by a PC which controls the entire system and/or be fed directly into a laboratory information management system (LIMS)

Column compartment (usually temperature controlled): the column is connected to the injector and detector by zero dead volume connecting tubing; this compartment may also contain more than one column with a switching valve to direct flow to one or the other columns *e.g.* during an overnight sequence of different methods

Autosampler: gives accurate, precise injection volumes into the high pressure system; sample solutions may be injected individually or as part of a sequence for an automated run; modern systems may contain in the order of 100 vials of ~2 mL volume making injections at <20 μL with control of depth of needle immersion in vial, speed of solution withdrawal and injection (useful for viscous solutions) and a range of rinse sequences

Pumping system: need to deliver pulse-free, continuous, high pressure flow of mobile phase to the top of the column; can be anything from a binary pumping system with high pressure mixing after the pumps to a quaternary solvent system with low pressure mixing before the pump(s); for HPLC pressures up to in the order of 400 bar are delivered and often over 1000 bar for ultra-high-performance liquid chromatography (uHPLC)

Figure 7.1 A typical modern system for high-performance liquid chromatography (HPLC). The instrument shown is also suitable for uHPLC in which higher pressures are used in obtaining good efficiency in short run times on columns containing sub-2-μm particles. Older systems had clearly identifiable pumps, detectors, etc., but, to reduce their laboratory footprint, modules of newer systems are stacked. Systems used for chromatography of biological samples may be biocompatible (e.g. an inert titanium/PEEK flow path not only reduces interactions of proteins with metallic surfaces but is also suitable for use with harsh salt conditions and de-salting prior to MS detection).

7.1.1 The Theory of Chromatographic Separation

The mobile phase flow and analyte separation when applied to chromatography can be explained by several main mathematical relationships and physical parameters. Essentially a chromatogram, the signal response product of a chromatographic separation, is a plot of analyte concentration versus eluent volume or, less fundamentally, time (Figure 7.2). The retention time (t_r) or volume for a particular analyte is an important parameter for characterising the separation. Retention time is dependent on the volumetric flowrate. However, the retention time also depends on the dimensions of the column used. Therefore, the retention time is usually reported as a normalised retention quantity known as the capacity factor (or retention factor) (k). Selectivity (or separation factor) (α) is the ratio of the retention (capacity) factors (k) of two peaks. The efficiency (N) is reported as the number of theoretical plates for a column. Traditionally, a 'plate' (the name is taken from the 'plates' used in fractional distillation towers used in the petroleum industry) is defined as a hypothetical separation layer within the column in which the two phases establish equilibrium with each other. The higher the number of plates the more efficient the separation will be. The concept of a layer in which equilibrium is established is a rather abstract one in that chromatography is a dynamic process and equilibrium is never reached. For simplicity it may be useful to note that analytes elute from the chromatography column as a peak with a Gaussian distribution and the plate height is related to the variance of that peak. The Gaussian distribution arises from the band broadening processes that take place in the column, giving rise to random error in the time taken for molecules of a single analyte to elute from the column. Resolution (R) is defined as the separation of two analyte peaks, taking into account the individual peak widths. The main mathematical relationships used in

Table 7.1 Different modes of liquid chromatographic separation commonly used in the analysis of biological samples.

Physicochemical property	Mode of chromatography	Pressure format	References
Size and shape	Gel filtration/size exclusion separation for proteins and nucleic acids	HIGH LOW	[6–8]
Net charge	Anion and cation exchange chromatography for proteins. Water softening and demineralisation. pH quenching	HIGH LOW	[4, 9, 10]
Isoelectric point	Chromatofocusing of proteins	MEDIUM	[11, 12]
Hydrophobicity	Hydrophobic interaction and reversed-phase chromatography of proteins, nucleic acids and small molecule drugs	HIGH LOW	[13, 14]
Biological function	Affinity chromatography for proteins and nucleic acids. Immobilised ligand chromatography for proteins and nucleic acids. Immunoadsorption chromatography for proteins	HIGH LOW	[15–17]
Carbohydrate content	Lectin affinity chromatography	HIGH	[18]
Free thiols	Chemisorption (covalent) chromatography applied to proteins	HIGH	[19, 20]
Metal binding	Immobilised metal ion affinity chromatography for proteins	HIGH LOW	[21, 22]
Chiral	Typically separation of pharmaceutical drug enantiomers but also biomolecules (e.g. *d*, *l*- amino acids were amongst the first analytes to be enantio-resolved and there is a revival of interest in the determination of trace d- enantiomer content in natural amino acids)	HIGH	[23–27]
Ion pair	Used to separate charged substances using reversed-phase HPLC. Works through the addition of mobile phase additives. Used to separate small molecule drug, metabolites, and nucleic acids	HIGH	[28–30]

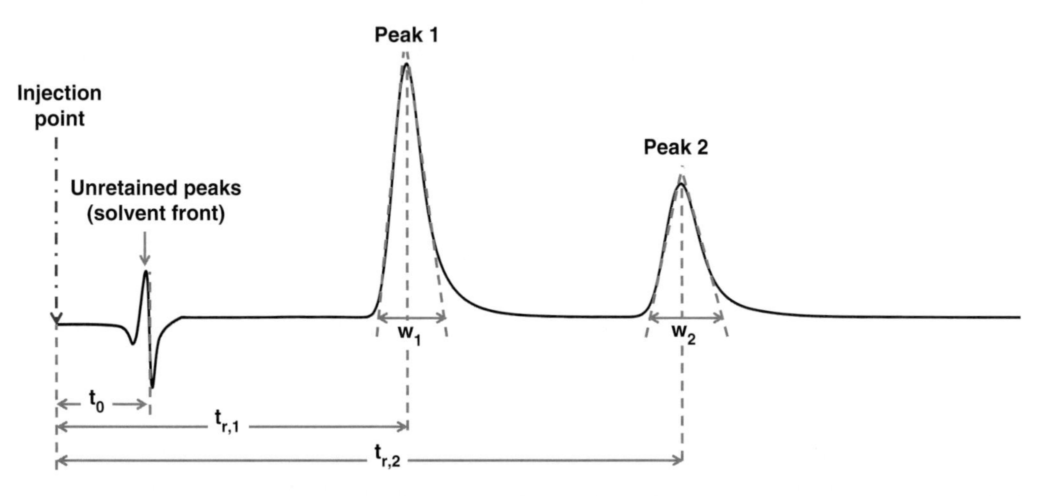

Figure 7.2 Chromatographic separation of a mixture of two analytes showing the measurements that need to be made in order to calculate capacity factor, selectivity, efficiency and resolution.

chromatographic separation are shown below:

$$\text{Capacity factor } k = \frac{t_2 - t_1}{t_0}, \quad \text{Selectivity } \alpha = \frac{k_2}{k_1}, \quad \text{Efficiency } N = 16\left(\frac{t}{w}\right)^2,$$

$$\text{Resolution } R_s = \frac{(t_2 - t_1)}{\tfrac{1}{2}\,(w_1 + w_2)}$$

where

k is the retention factor
t_r is the retention time for a particular analyte (e.g. $r = 1$ for the first peak)
t_0 is the column void volume (solvent front)
N is the number of theoretical plates
w_n is the width of an individual peak
R_s is the resolution of two individual peaks

The above equation for resolution is used for the actual measurement of resolution from a chromatogram. This should not be confused with an alternative equation for resolution, which may be obtained by substituting the other three equations into the equation for resolution.

The Purnell equation [31] describes the resolution of two similar peaks under chromatographic separation taking into account the selectivity, capacity and plate number:

$$\underbrace{R_s}_{\text{Resolution}} = \underbrace{\frac{1}{4}\sqrt{N}}_{\text{Efficiency factor}}\;\underbrace{\frac{k}{(1+k)}}_{\text{Retention factor}}\;\underbrace{\frac{(\alpha - 1)}{\alpha}}_{\text{Selectivity factor}}$$

The Purnell equation is an extremely useful relationship that can be used to gain a deeper insight into what qualitatively is obvious. Better resolution may be had by having narrower peaks (better efficiency) and by having a larger ratio of capacity factors (selectivity). As the Purnell equation also indicates, better resolution may be had by increasing the capacity factor. However, not only is this at the expense of increasing analysis time but there is little benefit in terms of increasing resolution for $k > 10$.

The separation of two analytes (Figure 7.2) is only possible if their retention times differ by virtue of having different distribution coefficients between the mobile phase and stationary phase. While the principal goal in this case is to obtain reliable baseline resolution in as short a time as possible, an associated goal is to obtain this not just through good peak efficiency but also through good peak symmetry; i.e. there may be deviations from the anticipated symmetrical Gaussian peak shape in the form of fronting and/or tailing. There are several main causes for non-Gaussian peak shapes, which include: column stationary phase deterioration, column under- or overpacking, mobile phase impurities or loss of liquid handling control and/or efficiency [32, 33]. Peak width is usually measured at both the baseline inflection points and at half of the maximum signal height (Figure 7.2). By monitoring peak width over repeated separations, both column and system performance may be determined. For example, if a column is still being conditioned with the sample there will be a reduction in peak width over time. Also of note is the fact that as the concentration of an analyte in the sample increases there is usually a consequent broadening of the peak (mass overloading). Other factors leading to an increased peak width are dilute sample injections over long periods of time (usual in preparative protein chromatography and volume overloading), very shallow or elongated gradient elution profiles, mobile phase impurities or incorrect make-up and/or problems with liquid handling.

Gradient elution is in contrast to isocratic elution. When eluting analytes in chromatographic separations, there are three distinct types of elution (Figure 7.3) available. *Isocratic elution* utilises one mobile phase consistently across the whole separation. Isocratic elution is used in size exclusion chromatography but is rarely used when other modes of liquid chromatography that are used for the separation of large biologicals. This applies in particular to preparative liquid chromatography. *Gradient elution* utilises two (or very occasionally more) mobile phases and over a period of time gradually mixes the phases so that the composition changes

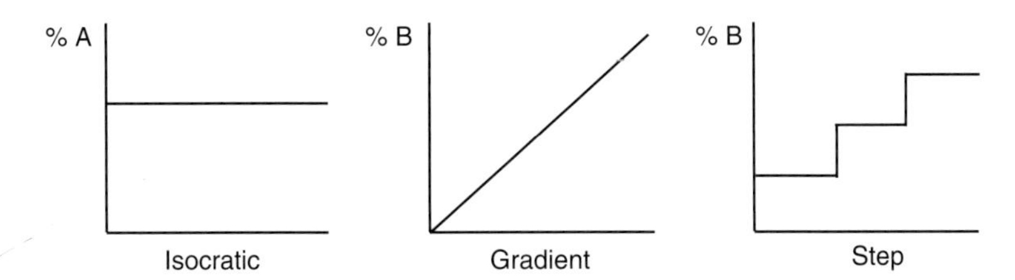

Figure 7.3 Standard elution profiles used in liquid chromatography.

from the initial composition to a new composition. For the majority of liquid separations, the gradient is linear. *Step elution,* like gradient elution, utilises two (or more) mobile phases, but rather than gradually changing the composition it uses a series of 'step-up' transitions to achieve a new mobile phase composition. In the analysis of small molecules a gradient is commonly used to shorten run times by bringing strongly retained sample constituents off the column more quickly. In preparative work on large biologicals the application of a gradient is more commonly used to facilitate separation and aid resolution of closely related analytes. Step elution is used to generate fast but crude 'cuts' of adsorbed species from the column. It is common to use step elution after the mobile phase composition needed for a specific analyte elution has been characterised via gradient elution.

In considering the theory of chromatography in the context of biological applications it must be remembered that it is not just about manipulating retention, selectivity and efficiency to obtain resolution of large molecules. Biological samples, particularly those containing small molecules, can be very complex, say several hundreds of thousands of components compared to maybe tens of components in a small molecule drug sample. A real problem can be that, even with very narrow peaks, there are simply too many sample constituents for them all to be resolved in a given run time. In other words, there is insufficient peak capacity. Temperature, flow rate, use of smaller particles, etc., can be used to increase peak capacity. However, for a step improvement it is necessary to use multidimensional chromatography where the peak capacity is the product of the peak capacity of each dimension. For this reason, comprehensive 2D-LC (very small volumes of eluent from the first column are all chromatographed on a second column with different selectivity in real time) is attracting great interest [34]. In the meantime, in the quantitative determination of complex biological samples it is almost the norm to add another dimension when using a single column through the use of a mass spectrometer as the detector. Mass spectrometry in its own right is a powerful tool in the analysis of biologicals (Chapter 9) but while the focus here is more on the actual chromatography of biological molecules, including for preparative purposes, it cannot be emphasised enough that liquid chromatography in combination with mass spectrometric detection is very powerful for the characterisation and quantitative determination of biologicals.

7.1.2 Chromatography for Biological Samples

Biological samples are, by their very nature, complex, containing impurities that are analyte-related (minority) and analyte-unrelated (majority). Analytical and preparative separation methods need to be robust in terms of their reproducibility and, for preparative work, throughput (multiple injections/loadings), while maintaining a selectivity for a physicochemical property for the analyte. Chromatography as a separation technique can provide this selectivity requirement even when applied to heterogeneous sample mixtures from a variety of sources.

Chromatography can also be used in-line with other preparative and analytical techniques since it can provide the analyte in an eluent that is suitable for further downstream separation and analysis (for example spectrophotometry, electrophoresis, circular dichroism, nuclear magnetic resonance [NMR] and mass spectrometry).

However, the dynamic nature of biological samples can be a technical challenge when applying any form of separation technique. Ensuring that the sample applied to the chromatographic technique is indeed representative of the true biological system from which the sample was taken needs to be considered at all stages of sample preparation through to data analysis. When applied correctly, chromatography can be non-destructive, modulated in terms of physical properties (pH, temperature, conductivity), have efficient sample processing times, and take advantage of computation integration and processing methods to circumvent many of these challenges.

7.1.3 Biological Sample Preparation for Chromatography

While there are chromatographic applications for isolating [35] and characterising intact cells [36], in just about all cases for chromatographic separation of biological samples some initial form of cellular disruption is required [37]. Disruption methods will release various compartmentalised hydrolases (phosphatases, glycosidases and proteases) that can alter the composition of the lysate [38]. In all modern experimental approaches, it is essential to preserve sample composition from the point of generation to analysis. Indeed, managing sample complexity can be an added challenge when working with biological material [39, 40]. Even the most sophisticated modern instrument cannot produce meaningful data without some form of rational sample preparation. Biological samples, irrespective of their origin, will contain a heterogeneous mixture of small and macromolecular species (Table 7.2), all of which will behave differently when applied to chromatographic separation.

At the start of the sample preparation phase, there are several important considerations that need to be taken into account since they impact the sample preparation strategy:

Measurement: The application of a simple measurement like UV-spectrophotometry may still incur problems if the sample does not have a chromophore or if the chromophore is masked by sample impurities leading to coelution and/or signal broadening. The application of more sophisticated measurement techniques like mass spectrometry [5] or NMR spectroscopy [50] may require the removal of specific sample constituents prior to their use, not to mention other associated technical considerations (Table 7.3).

Quantitation: Accurate quantitation is needed for all chromatographic separations. The level of quantitation can range from basic peak area determination to more complex comparisons with internal standards to generate accurate titre values for the analyte and in some cases the main impurity peaks.

What is the sample composition and how much is known about it? The organic versus inorganic, solid versus soluble sample composition is by far the most important property of the sample to consider prior to chromatographic separation. Of course, it goes without saying that if the sample is solid or a mixture of solid and liquid then exactly how subsampling is carried out is a major consideration. Failure to account for sample composition can lead to poor chromatographic separation and even failure of the chromatographic media and liquid pumping systems. The sample volume and viscosity will affect the solubility of many analytes and this may change as a sample is stored based on the environmental conditions.

Only after addressing these considerations can an appropriate strategy for sample pre-treatment be determined.

7.1.4 Extraction and Cleanup Methods for Chromatographic Sample Preparation

Through sample extraction, it is possible to change the sample composition to suit the chromatographic separation method and/or reduce the sample complexity quite significantly. Common sample extraction methods used for biological samples prior to chromatographic separation (Table 7.4) and subsequent sample cleanup techniques that can be applied to an extracted sample (Table 7.5) can, in many cases, be applied in series to yield a sample that is 'clean enough' to be further processed via chromatographic separation.

Table 7.2 General consideration and techniques for chromatographic separation of biological samples.

Sample type	Description	Technical considerations	
Protein and peptides	Large bio-macromolecule consisting of one or more chains of amino acids joined via peptide bonds. Proteins are found in all forms of life and have structural and catalytic functions. Peptides are short chains of 2 (di-peptide) to 20–30 amino acids in length. The structure of a protein directs its function and so maintenance of the 3D conformation of a protein is important when preparing a sample. Proteins can be found in both soluble and insoluble forms depending on their function.	• The amino acid sequence of a protein dictates its physicochemical properties. Proteins and peptides are sensitive to extremes of temperature and pH and need to be buffered to stay in solution. Each protein will have an isoelectric point at which it will have an equal number of positive and negative charges on its constituent amino acids and will precipitate out of solution. • Proteins are susceptible to denaturation when exposed to chaotropes and/or reducing agents. Increasing concentrations (or prolonged exposure) of oxidising agents may lead to changes in the hydrophobicity profile of the protein and eventual unfolding. • Recombinant proteins have a tendency to aggregate if generated using strong induction conditions that overwhelm the protein folding capacity of the expression system (example T7 promoter induction in *E. coli* strains). • UV spectrophotometric analysis of proteins is routinely carried out at 280 nm. • Proteins are stable when frozen.	[41–43]
Nucleic acids and oligonu-cleotides	Large bio-macromolecule consisting of one or two strands of nucleotides joined by phosphodiester bonds. Nucleic acids carry the genetic information for all known forms of life. *DNA:* Is a stable duplex of two strands of deoxyribonucleotides (*A*denine, *T*hymine, *G*uanine and *C*ytosine). The stable structure is formed by the stacking of the nucleotide bases and the exclusion of water to the phosphate backbone on the outside of the duplex. *RNA*: Can be found in both single-stranded and duplex conformations. RNA consists of strands of ribonucleotides and the base *T*hymine is replaced with *U*racil. *Oligonucleotides* are short double- or single-stranded nucleic acids that can be composed of ribo- or deoxyribonucleotides or unnatural nucleotide monomers.	• The nucleotide sequence carries information that can be turned into a protein sequence or directs the formation of a structural, regulatory or catalytic RNA. Care must be taken not to modify both the base sequence or the structure of any isolated samples. • DNA has a maximal absorbance at 260 nm and has hypochromic properties, wherein, as it denatures, the absorption at 260 nm increases. • DNA is a relatively stiff molecule with a high axial ratio leading to an increase in sample viscosity at high concentrations. • In strong acids and high temperature nucleic acids are hydrolysed to their constituent nucleotides. More dilute acid treatment leads to the breakage of glycosidic bonds in the purine nucleotides, leading to the sample becoming apurinic. • Addition of alkali to DNA will change the tautomeric state of the bases and results in denaturation through loss of hydrogen bonding. Addition of alkali to RNA will lead to hydrolysis of the nucleotide strand. • As with all polymers, nucleic acids can be separated based upon their size and indeed sizing is used routinely for characterising both DNA and RNA preparations. • Phenol-chloroform extraction of nucleic acids will isolate DNA and/or RNA while denaturing any bound or associated proteins. Ethanol can be used to precipitate-out nucleic acids from the aqueous layer of the extraction, when the Na^+ ion concentration is >0.3 M, therein concentrating them from relatively dilute samples. There are numerous proprietary methods available for isolation of nucleic acids.	

(Continued)

Table 7.2 (Continued)

Sample type	Description	Technical considerations	
		• Nucleic acids are stable when frozen. Ideally DNA should be stored in a lyophilised form. • DNA can be a major contaminant in protein isolation and poses an obstacle to sample processing due to its viscosity.	
Lipids	A diverse family of biological macromolecules with mostly hydrophobic properties. Lipids can be long chained 'fatty' acids or shorter hydrocarbon chains with polar phosphate, saccharide groups or even proteins attached. Lipids are a major energy storage molecule in the cell and the main constituent of cell membranes. Steroids are also part of the lipid family.	• Due to their hydrophobicity, lipids are poorly soluble in aqueous solutions but very soluble in organic solvents (e.g. acetone, chloroform, ether). Lipids added to aqueous buffer will form micelles. • When working with lipids in the lab it is usually important to dissociate the lipid from other cellular components (usually protein). • Care must be taken not to hydrolyse or oxidise the lipids in the sample by controlling the level and activity of co-purified hydrolase enzymes and reactive oxygen species. • Lipids are stable when frozen but temperatures below −30 °C need to be ensured to stop acylation and hydrolysis. Slow thawing can have similar effects on lipid samples.	[44]
Carbohydrates	Carbohydrates are a diverse family of biomolecules containing carbon, hydrogen and oxygen. The monomeric unit of carbohydrates is the saccharide. Carbohydrates are commonly referred to as sugars. Within biological systems, carbohydrates are used for their energy storage and structural properties.	• As with proteins and lipids, carbohydrates can be found in both soluble and insoluble forms. As with proteins and lipids, care must be taken to not change the carbohydrate analyte profile of the sample by chemical and/or enzymatic hydrolysis and microbial spoilage. • Exhaustive ethanol partition and extraction has been applied to carbohydrate isolation, but there is a technical movement to derivatisation and direct analysis. • Elevated temperatures will promote hydrolysis of carbohydrate samples and it is common to use temperatures in excess of 40 °C when carrying out an isolation. Other methods utilise acid hydrolysis to extract carbohydrates for analysis. • Carbohydrates are stable when frozen.	[45–47]
Clinical samples	Clinical (patient-derived) samples will contain a mixture of all biological macromolecules and bio-inorganic analytes in different proportions dependent on the source. It should be noted that the generation of clinical samples usually indicates some form of perturbed physiological process associated with the patient and the sample. Common clinical samples are: • Whole blood and blood fractions • Tissue biopsies • Urine samples • Tissue swabs • DNA isolates	• Maintaining sample integrity is essential when dealing with any clinical isolate so as to preserve any biomarkers and generate a true representation of a pathological state. It is difficult to generate a holistic analytical profile for all analytes in a clinical sample, but it is sometimes the subtle interactions and fluctuations between analytes that provide the most insight into a biological system or process. • Controlling and accounting for variability in clinical analysis is important. *Patient-derived variability*: will be present due to genetic makeup, normal biological variability and environmental and physiological factors specific to the patient. *Pre-analytical variability*: will be derived from sample preparation and storage. *Analytical variability*: will arise from changes in analytical method performance – can be accounted for through the use of technical standards and replicates.	[39, 48, 49]

Table 7.3 Detection methods used with chromatographic separations.

Detection classification	Description and technical considerations	References
General detectors	*UV/visual light absorbance*: Visible light detection (400–750 nm) and the more commonly used ultraviolet detection (100–400 nm; e.g. proteins absorb strongly at 280 nm by virtue of aromatic amino acids such as tryptophan) are a routine detection method used by biological sample analysis. The detector measures the absorbance of light by chromophores in the sample. For most applications UV/Vis detection obeys the Beer Lambert law but may show a non-linear relationship at high analyte concentrations. Many detectors will allow multiple wavelengths to be monitored. Diode array detectors allow scanning of wavelength ranges and can generate 3D plotting of the detection signal. Picogram per ml detection levels are achievable.	[51, 52] [53]
	Refractive index: Involves measuring the change in refractive index of the column eluate. As an analyte passes through the detector the refractive index will change compared to the mobile phase. The greater the refractive index the greater the difference between the sample and mobile phase. Thus, the sensitivity will be higher for the higher difference in refractive index between the sample and mobile phase. The refractive index is said to be the only universal detection method for chromatography. Used in the analysis of carbohydrates and small molecules. Samples do not need a chromophore.	[54, 55]
	Evaporative light scattering: Causes vaporisation of the eluate combined with quantification of the analyte by light scattering. The eluate emerging from the column is combined with a flow of air or nitrogen to form an aerosol – the eluent is evaporated from the aerosol by passage through an evaporator and the emerging dry particles of analyte are irradiated with a light source and the scattered light detected by a photodiode. The intensity of the scattered light is determined by the quantity of analyte present and its particle size. Samples do not need a chromophore.	
Selective detectors	*Fluorescence*: Can be applied to samples that fluoresce – a major limitation of its use (although many samples can be derivatised to induce fluorescence). Fluorescent detection is extremely sensitive down to the femtogram per ml level.	[42, 56, 57]
	Conductivity: Measures electrical impedance of the column eluate, which will change as analytes elute from the column. Its application is not common for biological samples but can be developed for protein separation. However, the sensitivity for protein is relatively low at the microgram per ml level.	[58]
	Electrochemical: Selective detection of electroactive analytes. Amperometric and coulometric forms of detection can be used. Analytes in the eluate flow through the flow cell, wherein molecules at the electrode surface undergo either an oxidation or a reduction, resulting in a current flow between the two electrodes. Sensitivity can be at the nanogram per ml level.	
	Flame photometric: Applied in gas chromatographic separation for looking at residual metals such as tin, boron, arsenic and chromium as well as for sulphur and phosphorous containing compounds. Samples are carried in a hydrogen/air flame. Detection is via chemiluminescene at specific wavelengths, which when passed into a photomultiplier gives an electrical signal that can then be measured. Sensitivity can be at the picogram per ml level for some applications.	

(Continued)

Table 7.3 (Continued)

Detection classification	Description and technical considerations	References
Structure determining detectors (hyphenated analysis)	Combining sample separation with identification and/or structural deduction is a very powerful analytical method combination that is applied in the analysis of many biological samples.	
	Mass spectrometry (MS): The analyte is introduced into a mass spectrometer after elution wherein its mass is determined. The mobile phase is mostly removed before the sample is introduced to the spectrometer. Detection can be via by total ion current (TIC) or selected ion monitoring (SIM). Mass spectrometry is able to distinguish analytes in overlapping peaks via ion monitoring as long as the analytes have a unique molecular ion or fragment ion. Mass spectrometry is very sensitive down to the femtogram per ml range for most analytes.	[59, 60]
	Fourier transform inferred (FTIR) spectroscopy: Infrared spectroscopy arises because different analytes produce different spectral fingerprints when exposed to infrared radiation. The Fourier transform converts the detector output to an interpretable spectrum that provides structural insights. Sensitivity is in the nanogram per ml range.	[61, 62]
	Nuclear magnetic resonance (NMR) spectroscopy: The sample is transferred directly into the NMR in the eluate solvent. The sample can be transferred during the chromatographic separation and the NMR spectra are then acquired either in on-flow mode (continuously, while the chromatography is running), alternatively the NMR spectra are acquired under static conditions, requiring 'parking' of the analytes before analysis. Successful application has been seen with pharmaceutical analysis. Selectivity is in the femtogram per ml range.	[63]

7.1.5 Protein Separation

As alluded to earlier, applications of chromatography to 'biology' are many and varied. However, the analysis of proteins has always been of paramount importance. Biochemists have been studying proteins since the early nineteenth century and the efficient chromatographic separation techniques in use today were developed in the early 1940s and 1950s and porous organic polymers were developed as part of the Manhattan Project (to develop the first atomic bomb!). Tiselius developed the basic chromatographic principles of chromatographic frontal analysis (continuous feed of the sample solution; the first constituent that breaks through may be obtained pure) in the early 1940s and applied a refined set of these techniques to the analysis of ovalbumin, serum albumin and immunoglobulins in the 1950s. In the subsequent six decades many different forms of chromatographic separation for protein preparation (purification) and analysis (concentration and structural determination) have been developed.

7.2 General Considerations for Protein Separation by Chromatography

The first consideration for protein separation by chromatography is to be aware of problems specific to proteins that are not encountered in liquid chromatography of small molecules. The most obvious of these is size. In recent research, very small particles with no pores have been used. That notwithstanding, large pore particles (usually greater than 300 Å) must be used. The size is also responsible for slow diffusion in the mobile phase and slow mass transfer in and out of the stationary phase. The former can be an advantage at very low flowrates but the latter is more prevalent at normal flowrates and contributes to poor efficiency. Proteins may be denatured and so in preparative work it is prudent to work at close to natural conditions using aqueous buffers at pH not too far away from 7.4. It is best to avoid polar organic solvents such as methanol and acetonitrile. The use of very non-polar stationary phases is also inadvisable given that proteins can denature the

Table 7.4 Sample extraction methods commonly used prior to chromatographic separation of biological samples.

Extraction method	Description and technical considerations	References
Solid–liquid extraction	The sample is placed in a sealed container and solvent is added that dissolves/extracts/leaches the analyte of interest. The solution is then separated from the solid by filtration or centrifugation. Particularly useful for extracting small molecules from biological samples.	[64–66]
Liquid–liquid extraction	Also known as solvent extraction or partitioning. Analytes in the sample are separated or portioned between two immiscible liquids (usually polar water and a non-polar organic solvent) based on their relative solubility in each of the liquids. Particularly useful for extracting small molecules, pharmaceutical compounds and secondary metabolites.	[67, 68]
Homogenisation	*Biological tissue:* Sample is placed in a blender or a mechanical homogeniser containing a solvent or aqueous buffer. The sample is homogenised to a homogeneous finely 'chopped' state. The liquid or suspended solids can be removed (filtration/centrifugation) for further workup. *Microbial cultures:* The culture is first harvested via centrifugation to yield a cell pellet. The pellet is then resuspended in a cold aqueous buffer and before mechanical homogenisation at high pressure (> 600 bar/8700 psi). Homogenisation is aided through the use of detergents. It is usual to carry out 3–5 passes through the homogeniser using optical density and/or microscopy to monitor cell lysis.	[69–71]
Sonication	The use of sound waves to vigorously agitate particles in a solution. The sound waves can be applied via the use of a probe immersed into the sample or through sound waves being applied to a sample container immersed in an ultrasonic bath. Particularly useful for resuspending solid samples (e.g. protein aggregates) and lysing small bacterial culture volumes, although care must be taken to not denature protein and nucleic acid analytes of interest. Also, can be used to generate emulsions and nano-emulsions with more homogeneous characteristics.	[72–75]

surface of these or even may adsorb irreversibly. Lastly, large proteins are heterogeneous and, in a small way, this contributes to band broadening.

Proteins have four main physicochemical properties that can be exploited by four different modes of chromatographic separation (Tables 7.1 and 7.2):

7.2.1 Size/Shape – Exploited by Size Exclusion Chromatography (SEC)

In size exclusion chromatography (SEC) (Figure 7.4a and b) the molecular sieving properties of porous materials are used to separate proteins based upon their size and shape. Traditionally in SEC, a column of microparticulate cross-linked copolymers (styrene or divinylbenzene) and with a narrow range of pore sizes is in equilibrium with the mobile phase. More recently, it is common to use silica-based particles typically with an immobilised diol surface for protein SEC. Large analytes are excluded from the pores and pass through the spaces between the particles and are the first analytes to elute from the column. Smaller analytes are distributed between the mobile phase inside and outside the particles and will therefore take longer to pass through the column. The smallest analytes are the last to elute from the column. The mobile phase within a particle is available to an analyte that is able to penetrate the particle. The distribution of an analyte in a SEC column is determined by the total volume of mobile phase inside and outside the particles. For a given type of particle, the distribution coefficient (K_d) of a particular analyte between the inner and outer mobile phase is a function of its molecular size. If the analyte is large and completely excluded from the mobile phase within the particle, $K_d = 0$. If the analyte is sufficiently small to gain complete access to the inner mobile phase, $K_d = 1$. However, a variation in pore size will mean that the K_d values vary between 0 and 1. It is this complete variation of K_d between these two limits that makes it possible to separate analytes within a narrow molecular

Table 7.5 Sample 'clean-up' methods used prior to chromatographic separation.

Clean-up method	Description and technical considerations	References
Filtration	Sample is passed through a paper or membrane filter (or SPE cartridge) to remove suspended particulates; 0.2 μm filtration is recommended for all samples prior to chromatographic separation to ensure sample homogeneity and to protect the stationary phase. Various filter modalities are available, ranging from charged membranes to depth filters designed to remove specific particulate sizes. Note that filter flushing may be needed to recover valuable samples from filter media in some cases. Ultrafiltration can be used to concentrate samples and also remove specific impurities of a set MW by using filter membranes with pores of a specific molecular weight cut-off (MWCO). This technique is available in centrifugation-tube format.	[76]
Centrifugation and sedimentation	*Centrifugation:* Sample is separated by centrifugal force to remove particulates to yield a clarified supernatant and a particulate pellet. Used to prepare cell-free samples and to clarify protein and nucleic acid preparations. Often used as an alternative to filtration for smaller biological samples. *Sedimentation:* The sample is allowed to settle under gravity undisturbed. Sedimentation is a slow process and the rate of sedimentation is dependent on Stokes's radius.	[77]
Solid-phase extraction (SPE)	A form of chromatography itself – in SPE the sample is passed through a small disposable cartridge or column filled with a stationary phase that is able to adsorb the analyte of interest while allowing contaminating analytes to pass through with little or no adsorption. Used frequently in an ion exchange (IEX) 'charged' format to remove contaminating proteins from test solutions.	[1, 28, 78]
Lyophilisation	Also known as freeze drying. An aqueous sample is frozen followed by controlled water removal via sublimation under vacuum. Particularly useful for thermally sensitive analytes (e.g. proteins) and samples containing volatile compounds (which will be removed during the drying). Can be used to concentrate an analyte from a large volume for subsequent reconstitution in a chromatography-appropriate buffer or solution.	[10, 79, 80]
Evaporation	Liquid is removed from the sample by gentle heating at atmospheric pressure, usually with flowing air (or under N_2 gas). A vacuum can be applied to remove liquids with a low volatility. Generally used on small samples under centrifugal force to retain sample analytes.	[81, 82]
Dialysis	Analytes move between two aqueous liquid phases across a semi-permeable membrane. Analytes transfer from one liquid to the other, based on differential concentration. Adaptations to this technique utilise membranes that have specific MWCO pores for more controlled analyte processing. Generally used to remove proteins from samples and for buffer exchange.	[83, 84]
Precipitation	Analytes can be 'brought-out' of solution by modulation of their physicochemical properties. A common method for preparing proteins is to change the pH of the protein solution. As the pH approaches the isoelectric point of the protein it will precipitate out of solution and can be sedimented and harvested via centrifugation. An alternative method for protein precipitation is through the addition of ammonium sulphate, which will remove water from around the protein and cause it to precipitate out of solution. Both of these methods can be used to remove impurities while maintaining the analyte of interest in solution.	[85, 86]

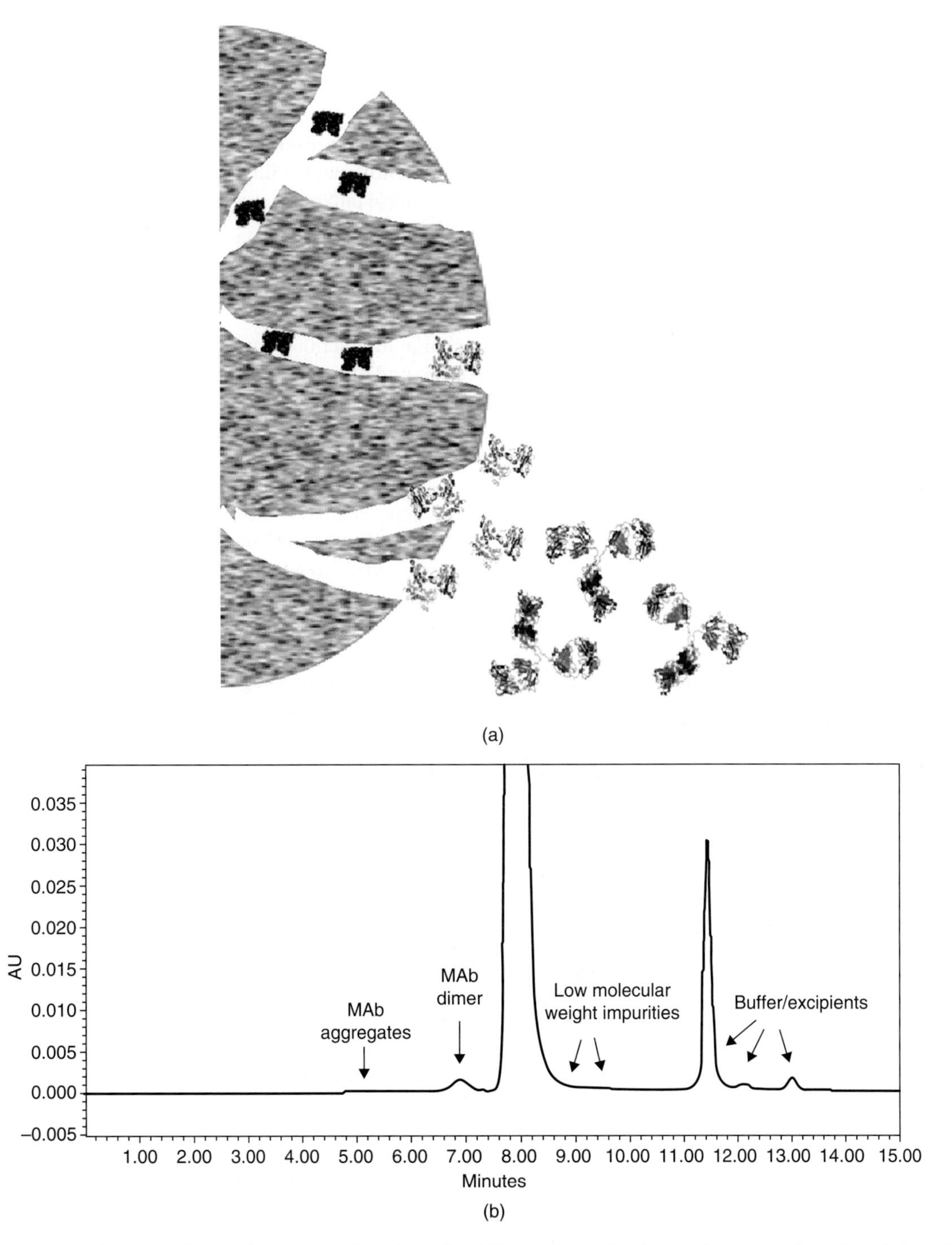

Figure 7.4 (a) Size exclusion chromatography – there should be no adsorption/desorption on and from the solid matrix. Differences in residence times in the column arise from the 'size' of the molecules, small molecules totally permeating the pores of the particles where flow is diffusion controlled and hence taking longer to traverse the column. They elute at t_0 with all the separations taking place *before* t_0. (b) Monitoring aggregation and impurities of monoclonal antibodies using size exclusion chromatography. Experimental conditions: column – Agilent Bio SEC-3300 Å, 7.8 × 300 mm; mobile phase: 150 mM phosphate, pH 7; flowrate: 1.0 ml/min; temperature: ambient; sample: monoclonal antibody (10 μl, 5 mg/ml).

size range on a given particle type. It should be noted that since the protein separations take place before t_0, SEC columns are usually longer and wider than those for other modes of chromatography.

The sizing properties of SEC make it a very powerful technique for characterising proteins in their native conformation, but SEC can also be applied for desalting protein preparations and concentration of particular analytes. The application of SEC has also been successful for separating nucleic acids [6, 87], viral particles [88], large molecular weight formulation products [89] and multicomponent cellular assemblies [90]. For all of the applications, good knowledge of common SEC stationary phase polymers and the associated technical considerations (Table 7.6) is important.

7.2.2 Charge – Exploited by Ion Exchange Chromatography (IEX)

This form of chromatographic separation relies on charge–charge interactions between the analyte and the stationary phase. Ion exchange chromatography (IEX) (Figure 7.5a and b) is a very versatile form of chromatography that can be applied over a wide range of biomolecules, synthetic polymers and small molecules. Both cation and anion exchanger stationary phases have a high binding capacity and their retention and resolution can be modulated through the ionic strength and pH of the mobile phase. The exchangers are either acidic (cation exchanger, negatively charged) or basic (anion exchanger, positively charged). Analytes are exchanged on the stationary phase through ionic interactions (a positively charged analyte interacting with a negatively charged stationary phase functional group, and vice versa). IEX can be carried out using either a strong or weak exchanger, wherein the strength of the exchanger refers to the extent to which the ionisation state of the functional groups varies with pH. It is *not* the strength of the interaction between the stationary phase and the analyte. The selection of either a weak or strong exchanger (Table 7.6) depends on the nature of the separation.

Table 7.6 Chromatographic stationary phase description and technical considerations for the preparation and analysis of proteins.

Mode	Stationary phase description and technical considerations
Size exclusion chromatography (SEC)	*Low pressure stationary phase polymers:* Dextran (Sephadex, Sephacryl), Agarose (Sepharose) and Polyacrylamide (Bio-gels) *High pressure stationary phase polymers:* Polyvinyl chloride (Fractogel) and Dextran (Superdex) SEC resins are prone to compression and sometimes difficult to pack. For some applications, fine resolution is only obtained with excessively long columns (30–100 cm), which can be prohibitive on many systems. SEC columns should be cleaned and stored in 200 mM sodium acetate, 20% v/v ethanol. 0.5 M NaOH can be used for brief robust cleaning if needed.
Ion exchange chromatography (IEX)	*Anion exchangers:* *Strong:* Trimethylaminomethyl: -O-$CH_2N^+(CH_3)_3$ (cellulose matrix) Triethylaminoethyl: -O-$CH_2CH_2N^+(CH_2CH_3)_3$ (dextran or cellulose matrix) *Weak:* Aminoethyl: -O-$CH_2CH_2N^+H_3$ on an agarose matrix Diethylaminoethyl: -O-$CH_2CH_2N^+(CH_2CH_3)_2H$ (cellulose matrix) *Cation exchangers:* *Strong:* Sulpho: $-SO_3^-$, O-$CH_2SO_3^-$, O-$CH_2CH_2CH_2SO_3^-$ (cellulose, dextran, or polystyrene matrix) *Weak:* Carboxymethyl: O-CH_2COO^- (cellulose or dextran matrix) It is important that IEX columns are adequately equilibrated prior to sample application so as to bind and prepare all available functional groups with a counter ion (for proteins this would be Na^+Cl-) and to apply a set mobile phase pH across the entire column. IEX can be used in gradient or step elution modes and elution can be via pH changes or counterion concentration changes. A final high ionic strength wash after the elution step will allow the column to be cleaned and regenerated for subsequent reuse. 0.5 M NaOH can be used for brief robust cleaning. IEX columns should be stored in 20% v/v ethanol between uses.

(Continued)

Table 7.6 (Continued)

Mode	Stationary phase description and technical considerations
Hydrophobic interaction chromatography (HIC)	*Common HIC and RPC stationary phase chemistries:* Hydrophilic → Hydrophobic $-(OCH_2CH_2)_nOH$ (Ether); Polypropylene glycol; Phenyl; $-O-CH_2CH_2CH_3$ (Butyl); $-O-CH_2CH_2CH_2CH_2CH_2CH_3$ (Hexyl); $-O-(CH_2)_{17}CH_3$ (C18) The performance of HIC separation is mostly dependent on the properties of the protein and the stationary phase chemistry. The stronger the ligand hydrophobicity and/or the more surface hydrophobicity on the protein the stronger the interaction. However, these interactions are also affected by the presence of contaminating analytes, detergents, pH and the ligand density/degree of substitution. The dynamic binding of the analyte and the subsequent resolution is determined by the flowrate, salt concentration and mobile phase properties. Complete elution of bound species can be a problem if a very hydrophobic ligand is matched with a high salt concentration in the injection sample. Selecting a less hydrophobic ligand or a less substituted matrix can help solve this problem. Temperature can have an effect on some complex sample separations also. Columns must not be left with bound analytes and can be fully eluted and cleaned by 1 M NaOH and are usually stored in 20% v/v ethanol between uses.
Affinity chromatography	*Established affinity chromatography applications:* • Purification of kinase and dehydrogenase enzymes through the immobilisation of nucleotides (5′-AMP and 2′5′-ADP) • Purification of heparin binding proteins (growth factors, DNA, lipoproteins, membrane receptors) via the immobilisation of heparin • Isolation and identification of DNA binding proteins through the immobilisation of DNA (specifically promoter sequences and regulatory elements) • Isolation of fatty acid binding proteins through the immobilisation of fatty acids • Isolation of immunoglobulins through the immobilisation of Protein A and Protein G ligands • Isolation of mRNA through the immobilisation of poly-Thymine • Isolation of recombinant proteins through the engineering of fusion protein partners and metal affinity binding motifs Elution of the bound protein (or other analyte) needs to be complete to ensure performance upon reuse. Elution is usually through the application of high ionic strength buffers (1.5 M KCl or 2 M NaCl) or changes to the pH of the mobile phase that change the surface charge of the protein and allow it to dissociate from the column. Cleaning of the resin can be carried out using detergents, chaotropes in low or high pH buffers containing 1.5 M NaCl. Strong acids and bases should not be applied to affinity resins so as to maintain the ligand integrity. Storage should be in 20% v/v ethanol.

If the separation is using a wide pH range (gradient pH-based elution) then a strong exchanger is preferable as it ensures that retention via ion exchange will take place at both extremes of pH. Weak anion exchangers should be used when the binding capacity of the stationary phase is known to change as the pH of the mobile phase varies. In most IEX method development exercises it is advisable to select and screen strong exchangers first before moving to a weak exchanger to, if required, introduce an additional element of selectivity.

For protein separation via IEX the isoelectric point (pI, wherein the protein has no overall net charge) of the protein of interest should be obtained as this will allow an appropriate exchanger and mobile phase to be selected. This is particularly important if a pH gradient is being used instead of a buffer gradient. At a pH lower than the pI, the protein will have a net positive charge (suitable for cation exchange), while at a pH above the pI the protein will have a net negative charge (suitable for anion exchange). However, the stability of the protein as pH varies and the physicochemical properties of any putative impurities in the sample should also be considered so as to ensure that chromatograms obtained are representative of the initial sample. IEX may also be successfully employed for the separation of nucleic acids [91], drugs and impurities [10].

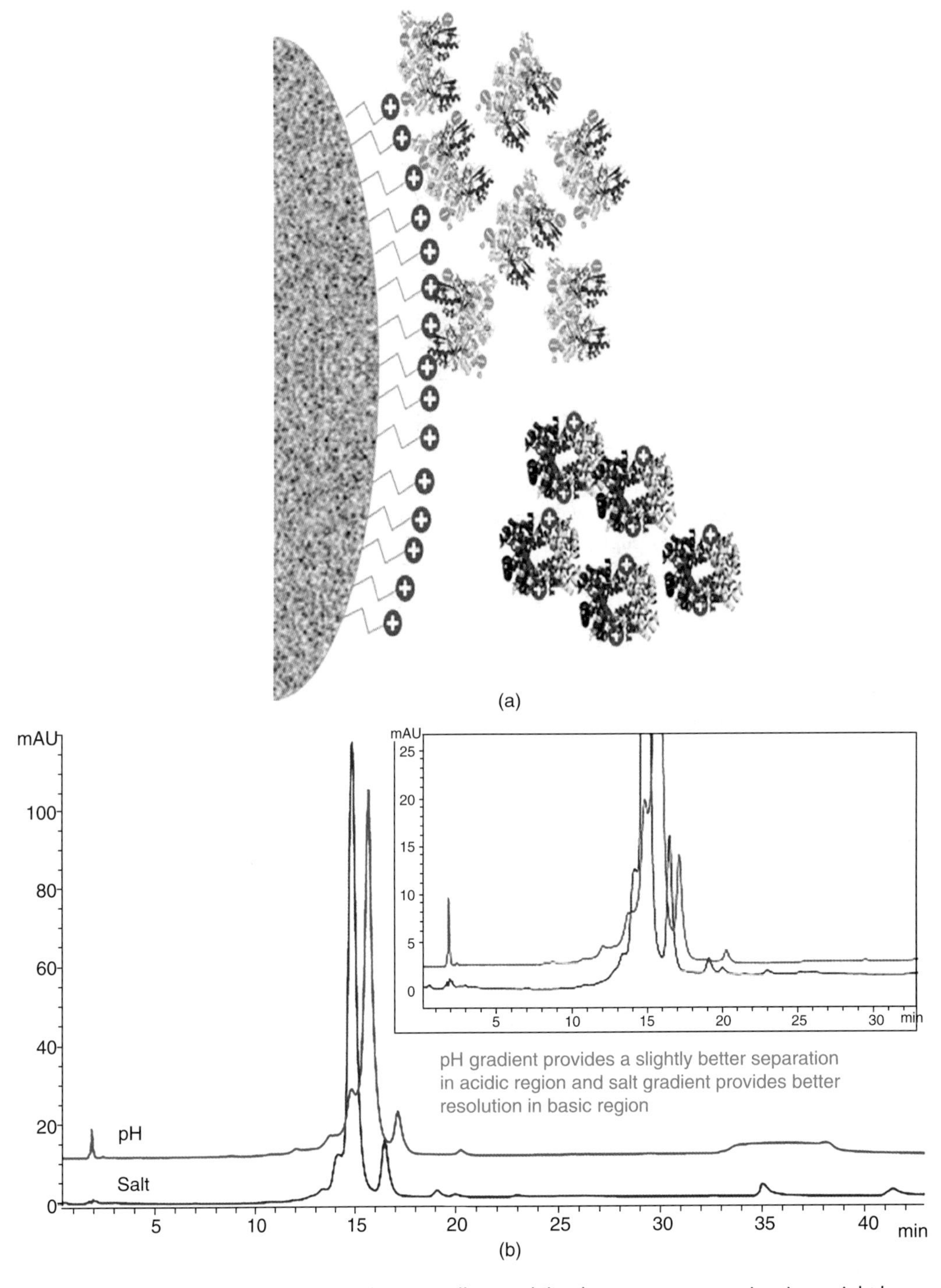

Figure 7.5 (a) Ion-exchange chromatography – typically immobilised quaternary ammonium ions might be used for a strong anion exchanger and immobilised sulphonic acid groups for a strong anion exchanger. Retention of a particular protein will depend on its pI and the mobile phase pH, which can be used to manipulate selectivity. Retention can also be manipulated by the ionic strength of the aqueous buffer mobile phase, increasing numbers of competing ions and reducing the retention of analyte ions. For gradient elution the mobile phase changes from one with low ionic strength to one with high ionic strength. (b) Comparison of salt gradient *v* pH gradient in ion-exchange chromatography showing that the outcomes are complementary. Column: Proteomix SCX-NP5, 5 μm, 4.6 × 250 mm PEEK; mobile phase: a 10 mM phosphate, pH 7.0 B: A + 0.5 M NaCl; flowrate: 1.0 ml/min; detection: UV, 280 nm column temperature: 25 °C; samples: NIST mAb injection, 60 μg (pI 9.18, in 12.5 mM histidine, pH 6.0).

7.2.3 Hydrophobicity – Exploited by Hydrophobic Interaction Chromatography (HIC) and Reversed-Phase Chromatography (RPC)

Hydrophobic interaction chromatography (HIC) (Figure 7.6a and b) has been applied for protein separation for over 30 years [92, 93]. Separation is based on the adsorption of hydrophobic areas on the surface of an analyte on to the stationary phase (or, in some models, partition into a less polar area of mobile phase close to the stationary phase). For proteins the majority of hydrophobic residues are buried in the centre of the protein protected from the aqueous environment of the surface; indeed, it is the formation of this hydrophobic core that drives protein folding in globular proteins. Hydrophilic amino acids on the protein surface are covered with an ordered layer of water. Hydrophobic 'patches' on a protein surface can be exposed by treating the protein with salt ions (usually ammonium sulphate), which preferentially take up the ordered water molecules and allow the hydrophobic areas to be exploited for capture and separation activities. Interactions are via weak van der Waals forces between the protein and the stationary phase. However, protein–protein interactions are possible via the exposed hydrophobic areas and may cause aggregation of the protein and eventual precipitation. HIC can be applied in both low pressure protein capture (selective initial isolation step) and high pressure analysis formats. For low pressure chromatography, HIC makes an ideal intermediate or polishing step (further treatment of the 'captured' protein) in protein purification. The binding of proteins to an HIC stationary phase is normally in high salt concentrations and elution proceeds via the gradual decrease in salt concentration. This gradual change in salt concentration will cause a rehydration of the protein surface and a loss of interaction with the stationary phase. Changing the mobile phase pH can also be used to elute HIC separations, but the resulting separation can be quite variable for complex samples as this changes the separation to a more mixed mode (HIC:IEX) format and requires specialised stationary phase chemistries, which must have good batch-to-batch reproducibility. When applied in an HPLC format, HIC can be used to separate very closely related analytes using very shallow gradients and small resin beads or particle sizes ($\leq$5 μm). The stationary phases commonly used for HIC are naturally hydrophobic, containing groups such as butyl, octyl and phenyl attached to a matrix, which promotes more protein–matrix interaction rather than protein–protein interactions. This quality of HIC makes it very versatile and able to accommodate high loading capacities. Common stationary phases include Phenyl Sepharose and Phenyl-5PW for low pressure HIC and Bio-Gel TSK Phenyl and Spherogel TSK Phenyl for HIC HPLC.

HIC is very similar to reversed-phase chromatography (RPC, also commonly abbreviated as RPLC or RP-HPLC). However, in RPC the mobile phase will contain a polar-organic constituent, such as methanol or acetonitrile, that may promote sample denaturation, especially for proteins. RPC tends not to be suitable for protein purification, except in very special circumstances [79], but is a very versatile and commonly used application for protein analysis. RPC has a lower binding capacity than HIC. While HIC is commonly used for protein separation it has also been applied to lipids [94], carbohydrates [95] and nucleic acids [96]. Common HIC and RPC stationary phase chemistries (Table 7.6) can be used in either HIC or RPC but generally more hydrophobic chemistries are used in RPC with polar organic solvents in the mobile phase and less hydrophobic chemistries are used in HIC with highly or completely aqueous buffer containing mobile phases.

RPC is very much the most common mode of chromatography for small molecules [97]. Therefore, as far as proteins are concerned, RPC comes into its own in the elucidation of protein structure where it is used for peptide analysis of protein tryptic digests (Section 7.4).

7.2.4 Biological Function and/or Structure – Exploited by Affinity Chromatography

Affinity chromatography (Figure 7.7a and b) exploits the biological function or specific structure of an analyte for separation making it extremely specific and capable of purification to high purity from very complex sample mixtures [17]. It is this specificity that makes affinity chromatography very expensive to use. Affinity chromatography is probably best characterised by the purification of enzymes by immobilising substrates/ligands

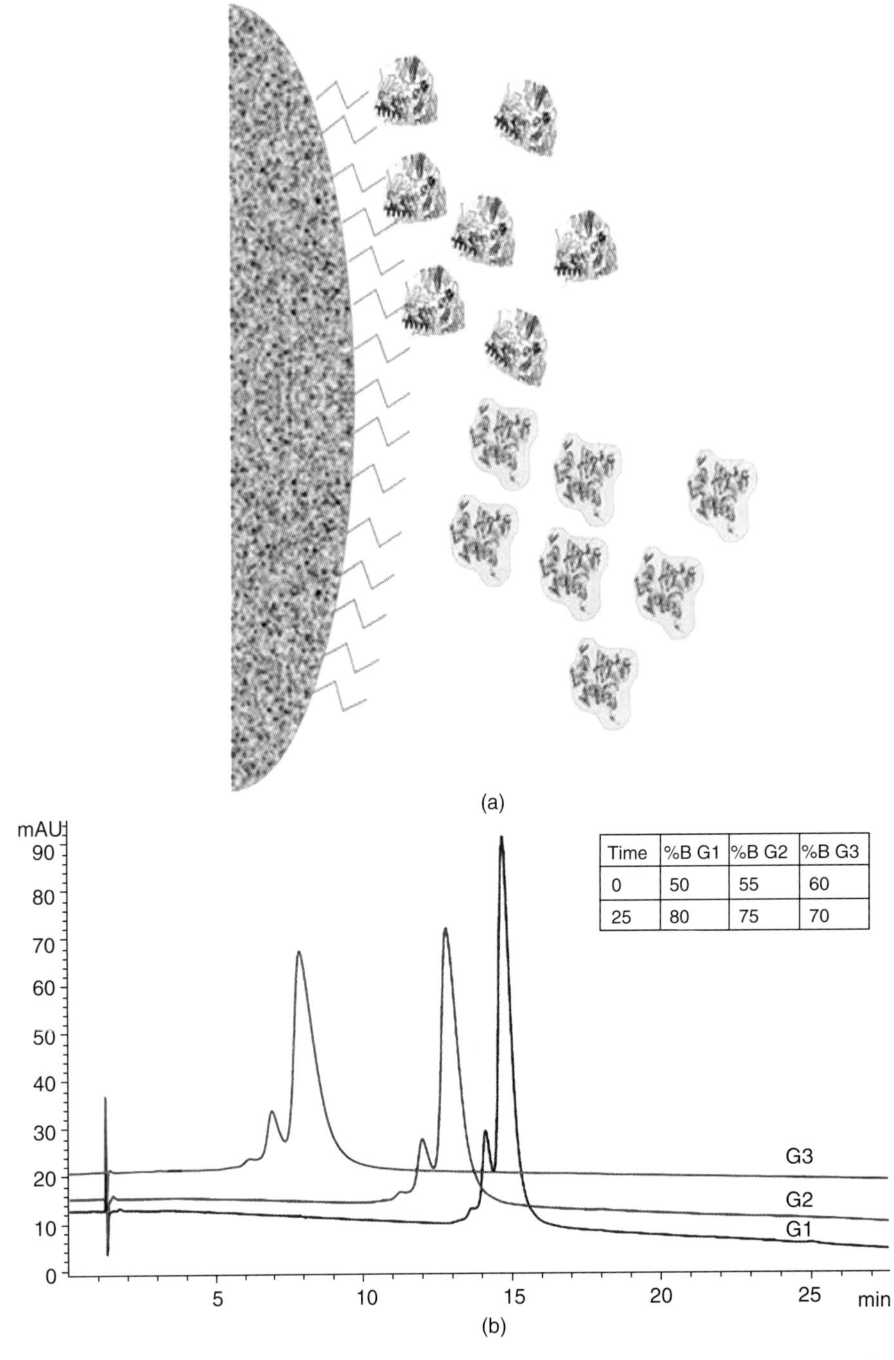

Time	%B G1	%B G2	%B G3
0	50	55	60
25	80	75	70

Figure 7.6 (a) Hydrophobic interaction chromatography – as in reversed-phase chromatography, retention is based on hydrophobic interaction/partition but in this case the stationary phase is less non-polar so less retentive so that an aqueous mobile phase with little or no polar–organic component must be used to obtain retention, thereby avoiding denaturing the protein analytes. Retention can be manipulated by the ionic strength of the aqueous buffer mobile phase, increasing numbers of competing ions effectively, making the mobile phase more polar and driving hydrophobic areas of the protein on to the stationary phase. Therefore, for gradient elution the mobile phase changes from one with high ionic strength to one with low ionic strength. (b) Gradient optimisation for hydrophobic interaction chromatography (HIC). Column: Proteomix HIC Butyl-NP5, 5 µm, NP 4.6 × 100 mm; mobile phase: A: 2 M ammonium sulphate, in 100 mM sodium phosphate pH 7.0; B: 100 mM sodium phosphate pH 7.0; flowrate: 0.5 ml/min. Detector: UV, 280 nm; column temperature: 30 °C; sample: NIST mAb 10 mg/ml (pI 9.18, in 12.5 mM histidine, pH 6.0); injection volume: 2 µl.

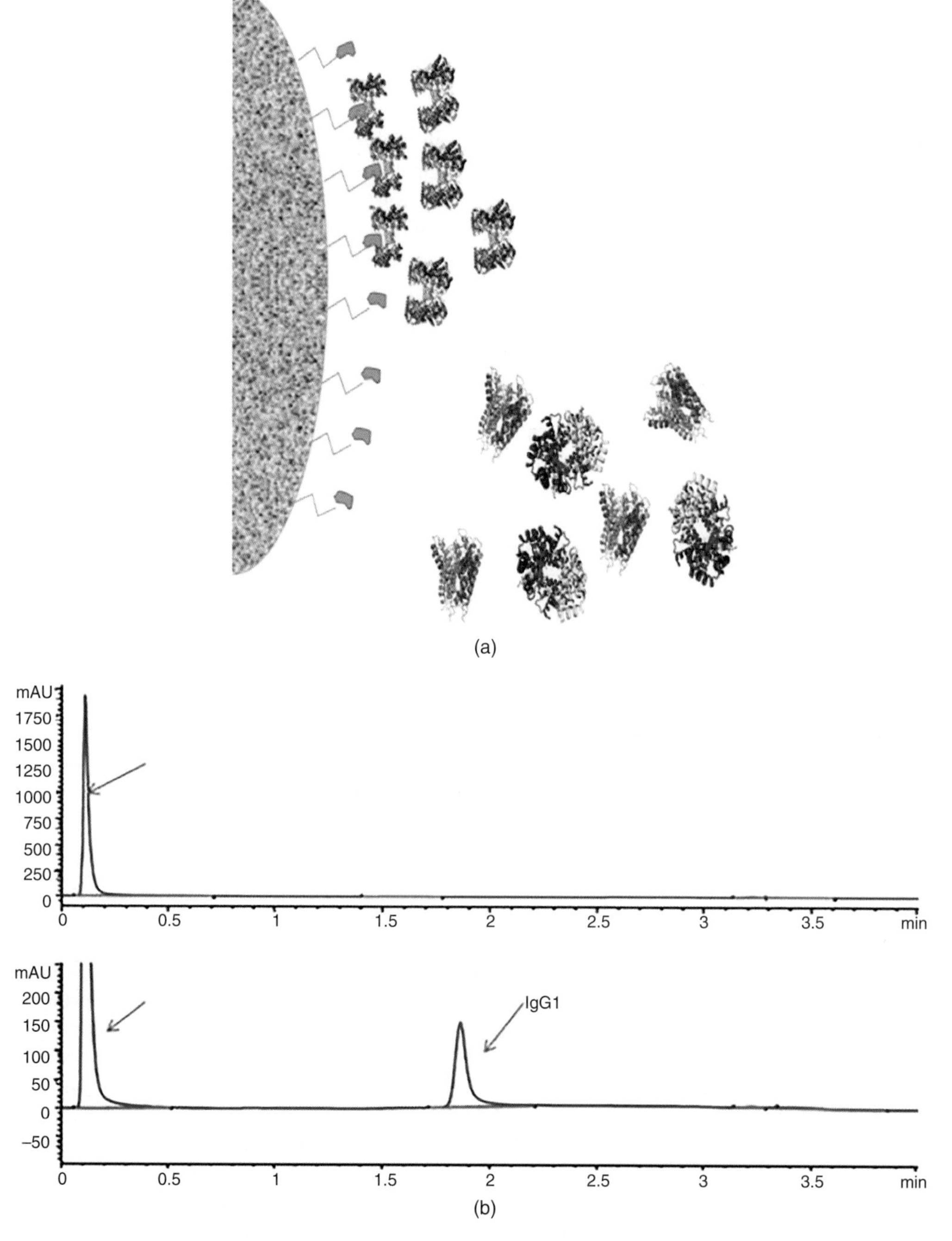

Figure 7.7 (a) Affinity chromatography – interaction between the targeted protein and the immobilised ligand is highly selective, so much so that low efficiency arising from slow mass transfer of the protein on and off the ligand is usually not an issue. It is important that the immobilisation of the ligand is such that there are no non-specific interactions with the bridging groups. (b) Specificity of Bio-monolith Protein A to IgG1. Experimental conditions: column – Agilent Bio-monolith Protein A; mobile phase: A (binding buffer) – sodium phosphate buffer, 20 mM, pH 8.0, B (eluting buffer) – citric acid, 0.1 M, pH 2.8; gradient: 0% B from 0 to 0.5 min (for washing), 100% B from 0.5–1.5- 2 min (for eluting), 0% B from 2.1–4 min (for equilibrating); sample: IgG1 (2.5 mg/ml) and *E. coli* lysate (10 mg/ml); injection: 5 µg; flowrate: 1.0 ml/min; detection: UV at 280 DAD from Agilent Bio-inert 1260 LC system.

(Table 7.6) on to chromatography stationary phase matrices. The enzymes would bind to the immobilised substrate once introduced to the column [98, 99]. The interaction between the stationary phase ligand and the analyte needs to be reversible so that the column can be reused and so the analytes can be eluted. Elution is usually carried out through the application of free-ligand changes in pH or salt gradients. Affinity chromatography is now used as the primary capture step for purifying immunoglobulin molecules used for therapeutic applications, but is also used for separating nucleic acids, membrane-bound proteins and even whole cells. The use of affinity methods requires a detailed knowledge of the biological function and to some extent structure of the analyte of interest. The stationary phases need to be relatively inert in terms of their binding characteristics but possess suitable chemistry to allow attachment of the ligand. In this way retention is exclusively by the affinity interaction and resolution is not lost under the influence of non-specific interactions. Ligands are usually added along with a spacer of up to 10 units (usually hydrocarbon chains but can contain carbonyl or amine groups) in length to allow correct orientation of the ligand and stop steric hindrance during binding to closely immobilised ligands. The matrix is usually an agarose or dextran base. Unlike other forms of chromatographic separation, affinity chromatography can be carried out in both batch-resin and packed-column formats.

7.3 Engineering Proteins for Streamlined Chromatographic Separations

For the isolation of native proteins from biological tissue or clinical samples, the chromatography may be limited by the physicochemical properties of the protein of interest. However, with the advent of recombinant DNA technologies, proteins can now be engineered to be more easily purified via chromatographic methods. There are several options available when trying to engineer a protein for a more simplified purification. There are three main approaches for engineering proteins for more efficient purification.

The addition of a purification tag: Purification tags are small amino acid sequences (<20 residues) that can be used in conjunction with affinity chromatography for purification. Commonly used purification tags are:

Immobilised metal affinity chromatography (IMAC) associated with the surface properties of a target protein. IMAC (also known as metal chelate chromatography) utilises immobilised divalent cations (Ni^{2+}, Co^{2+}, Zn^{2+}, Cu^{2+}) that interact via coordinate bonds with the imidazole groups of histidine, indole groups of tryptophan and to a lesser extent thiol groups of cysteines. Proteins harbouring these residues on their surface will be adsorbed to the stationary phase and can be selectively eluted by altering the pH of the mobile phase, the addition of a chelating agent (usually ethylenediaminetetraacetic acid [EDTA]) or a solution of a free functional group (imidazole, indole). The molecular biology techniques used to generate His-tagged proteins are generated by molecular biology techniques (Section 7.2.3). Most IMAC resins have a cellulose base and incorporate a spacer arm between the metal ion and the resin bead. Proteins can be engineered via recombinant DNA technologies [100, 101] to have these tags at either their N- or C-terminal. The incorporation of 6× histidine residues is the most common form of tag used in IMAC separations, as cysteine incorporation tends to lead to protein aggregation and tryptophan incorporation can cause significant changes to the protein structure. Due to its specificity, IMAC can be used on very 'dirty' samples (crude cell extracts) and still achieve target protein purities of >95% from a single separation.

Introduction of a FLAG-Tag or S-Tag on to the N-terminal of a protein allows immunoaffinity chromatography to be used for protein purification. The FLAG tag (Asp-Tyr-Lys-Asp-Asp-Asp-Asp-Lys) is an 8-reside hydrophilic amino acid sequence that is recognised by an immobilised monoclonal antibody on the stationary phase. The binding on the FLAG tag to the stationary phase is Ca^{2+}-dependent. Elution is carried out via the introduction of EDTA to the column and subsequent chelation of the Ca^{2+}. The FLAG tag can be removed through treatment of the eluate with enterokinase [102]. Similarly, the 15-amino acid residue S-tag (Lys-Glu-Thr-Ala-Ala-Ala-Lys-Phe-Glu-Arg-Gln-His-Met-Asp-Ser) derived from pancreatic ribonuclease A, can be used for protein purification, usually in batch preparations [103]. Removal of the S-tag is via subtilisin treatment.

Both of these small proprietary tag-based purification systems are very useful for small preparations of recombinant proteins, but as the size of the preparation increases the financial costs may become prohibitive and the digestion product isolation can become more complex.

Polyarginine C-terminal tags have been used in conjunction with cation exchange chromatography for protein purification. The arginine tag is sequentially removed through the application of carboxypeptidase B [104].

The generation of a fusion protein: There are several systems (Chapter 6.2) available for the expression of fusion proteins that, like the small affinity tags discussed above, utilise recombinant DNA technologies. The fusion protein systems that have been characterised the most are glutathione S-transferase (GST) and maltose binding protein (MBP). Both of these systems are supplied as proprietary kits [105].

Both systems fuse the GST or MBP protein to the N- or C-terminal of the protein of interest and then use immobilised ligands (Glutathione for GST fusions; amylose for MBP fusions) on agarose-based stationary phases. Both systems can allow on-column cleavage or elution via pH changes or the addition of free ligand and subsequent cleavage. Thrombin is used for GST cleavage and Factor Xa for MBP cleavage. If the digestion is carried out after the initial affinity capture step, then a follow-on IEX separation is usually employed to separate the fusion protein from the protein of interest.

The generation of a fusion protein not only allows for efficient purification of target proteins but also confers stability to otherwise unstable proteins [106] and can enhance solubility for aggregation-prone protein preparations [107].

Manipulating the expression format of the protein: Recombinant DNA technologies have also allowed users to manipulate the expression format and titre of recombinant proteins in order to aid subsequent purifications. The use of strong induction promoters (the bacteriophage T7 promoter in *E.coli* [108] or the cytomegalovirus (CMV) promoter in mammalian cell cultures [109]) for expressing recombinant proteins has facilitated the generation of aggregated misfolded protein known as inclusion bodies [110]. These inclusion bodies can be efficiently isolated through cellular homogenisation, homogenate centrifugation and detergent washing to yield protein preparations that are >95% pure for the protein of interest. These washed inclusion bodies can then be solubilised and refolded [111] to generate soluble proteins that are easily captured through the application of IEX chromatography. This expression and purification scheme is applied to the generation of many bacterially expressed recombinant protein therapeutics that are generated on a commercial scale [112].

Another application of recombinant DNA technologies, has allowed recombinant expressed proteins to be sorted and/or secreted to the periplasm (bacterial cells), cell membrane or growth media (bacterial, plant, insect and mammalian cells) for a more simplified purification strategy. As proteins are generated, specific sorting signals are built in to the amino acid sequence; these signals will ensure that a protein is sorted to the correct site of activity. These signals can be manipulated for engineering protein sorting [113, 114]. Isolation of proteins from the periplasm of bacterial cells is carried out via an osmotic shock of the cells [115], while extracellular signals will ensure that expressed proteins are transported across the cell membrane into the growth media. As in the application of the strong promoter systems described above, sorting signals are much used in the generation of commercial-scale protein therapeutics.

7.4 Example Chromatographic Separations of Biological Samples

7.4.1 On-Column Protein Folding

Rationale: As described earlier (Section 7.3), the rapid expression of proteins in *E.coli* cell cultures can lead to the formation of inclusion bodies. Historically, these protein aggregates were seen as 'lost' products and efforts were made to try to express the protein in a more soluble format. However, the relatively high purity obtained through the generation and isolation of inclusion bodies has more recently made them more attractive for protein purification. There are many solubilisation and refolding protocols available for application to many proteins expressed as inclusions bodies. However, for some proteins the solubilisation and refolding procedures lead to a loss in yield and/or a reduction in product quality. On-column refolding can in many circumstances circumvent these problems through stabilisation of the protein during the molecular shaping into its final conformation. In addition, on-column refolding allows processing at much smaller volumes (refolding is usually carried out in volumes that are 20-fold larger than the solubilisation and therefore necessitate large column loadings of dilute protein samples) and aids in the analysis and preparation of proteins generated via high throughput screening (HTS) technologies [116]. During method development activities combining the refolding and primary capture chromatography steps can simplify an overall purifications scheme. However, the on-column refolding step has been employed as part of the primary capture chromatography step and after an initial chromatographic capture and crude purification.

Technical considerations: The application of on-column protein refolding is dependent on the protein of interest, the available time, money and chromatographic hardware available. For inexperienced users it may be prudent to first define a test-tube refolding protocol and then capture the column step for a refolding product and attempt to combine the two steps. Due to its high capacity and robust stationary phase chemistry, IEX is an ideal capture and separation mode for refold samples. The stationary phase will be exposed to chaotropes (e.g. 8 M urea or 5 M guanadine-HCl), reducing agents (e.g. 10 mM dithiothreitol [DTT]), oxidising agents (e.g. 5 mM cystamine-HCl) as well as extremes of pH, and as such will need to be properly cleaned, regenerated and stored after use. Both strong and weak exchangers can be applied and it is common to screen several exchangers for optimisation. Proteins should be loaded on to the column at concentrations of $\leq$10 mg/ml. The solubilisation sample should be carried out with 'clean' inclusion bodies that are devoid of extraction buffer components, e.g. detergents. Protein folding can be a spontaneous process both *in vivo* and *in vitro* in test-tube preparations, but for immobilised protein it is advisable to slowly introduce the refolding environment through the application of a gradient. Elution is carried out using high salt buffers (0.5–1 M NaCl) or a change in the mobile phase pH.

Simplified method: A baseline AIEX protocol for on-column refolding.

7.4.2 On-Column Protein Refolding via Anion Exchange Chromatography

i) Wash the inclusion body preparation to achieve a clean 'white' protein mass – this can be done with water or aqueous buffer/1 M urea mix.
ii) Store inclusion bodies as a 25–50% w/v slurry in the washing solution. This slurry can be stored at −80 °C prior to chromatographic processing.
iii) Prepare the solubilisation buffer (50 mM Tris-HCl, 8 M urea, 10 mM DTT, pH 9.0) and then slowly add the inclusion body slurry to a final concentration of 10 mg/ml protein. Allow the solubilisation to mix for 30–60 minutes. Note that if a cation exchange chromatography (CIEX) column is used then the base buffer should be modified (e.g. sodium acetate).
iv) Equilibrate an anion exchange column with five column volumes of solubilisation buffer and then load the solubilisation mixture on to the column. Wash the column with five column volumes of solubilisation buffer.
v) Refold the bound protein by applying the refold buffer (50 mM Tris-HCl, pH 9.0) from 0 to 100% over 10 column volumes.
vi) Elute the column using a step elution with 50 mM Tris-HCl, 0.5 M NaCl, pH 9.0.
vii) The column should be cleaned and stored after use.

For an example on-column refolding of recombinant lysozyme (Figure 7.8a), showing the associated size exclusion separation of key elution samples from the on-column refolding (Figure 7.8a), the diethylaminoethanol (DEAE) elution of the refolded lysozyme is resolved into conjoined two peaks. The peak 1 refolded sample shows a shift in retention time (16 minutes) compared to the solubilised load material

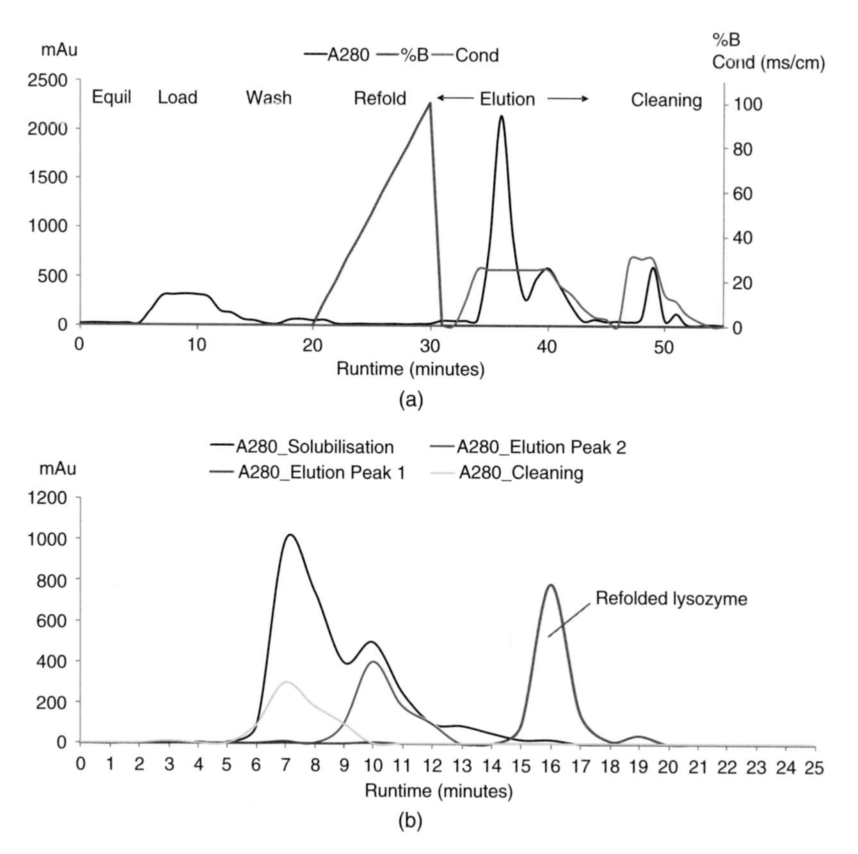

Figure 7.8 On-column refolding. Recombinant lysozyme inclusion bodies (expressed in *E.coli*) were washed three time in purified water and then stored as a 56% w/v slurry at −80 °C. After thawing, 100 mg of slurry was added to solubilisation buffer (50 mM Tris-HCl, 8 M urea, 10 mM DTT, pH 9.5) and stirred for 30 minutes for solubilisation. (a) The solubilisation was loaded on to a 1 ml DEAE FF column that had been pre-equilibrated in solubilisation buffer. Then load washing was performed with solubilisation buffer. Refolding was carried out by introducing the refolding buffer (50 mM Tris-HCl, pH 9.0) over 10 column volumes (0–100% B). The refolding protein was then eluted using 50 mM Tris–HCl, 0.5 M NaCl, pH 9.0 as a step elution for 10 column volumes. After elution, the column was cleaned with 0.5 M NaOH. All steps were carried out at 1 ml/min and the protein was monitored using A_{280}. Elution fractions are highlighted as shaded regions. (b) Key fractions from the DEAE on-column refold were diluted to 25 ml and separated using size exclusion chromatography (column: 5 ml of Bio-Gel P-6 gel filtration column) 0 at 1 ml per minute.

(main peak max. at 7 minutes). The solubilisation sample resolves into two conjoined peaks (peak max. at 7 and 10 minutes, respectively). The DEAE elution peak 2 contains higher molecular weight material (lower retention time) with a similar retention time peak max. of 10 minutes to some of the solubilised material. The DEAE column cleaning step contains high molecular weight material similar to the solubilisation sample (peak max. at 7 minutes). The conclusions from this on-column refolding experiment are that lysozyme can be captured post-solubilisation and refolded using on-column methods. The post peak material seen on the DEAE column contains higher molecular weight material that could be aggregates or misfolded protein. The cleaning step removes material that is likely to be non-refolded protein.

Associated analytical technologies: Assaying for protein structure can be a relatively complex undertaking. However, there are some techniques (circular dichroism, protein NMR or hydrogen-deuterium exchange mass-spectroscopy) that will provide detailed information on the folding dynamics and crucially the final 3D conformation of the protein. However, there are simpler analytical technologies that can be employed to

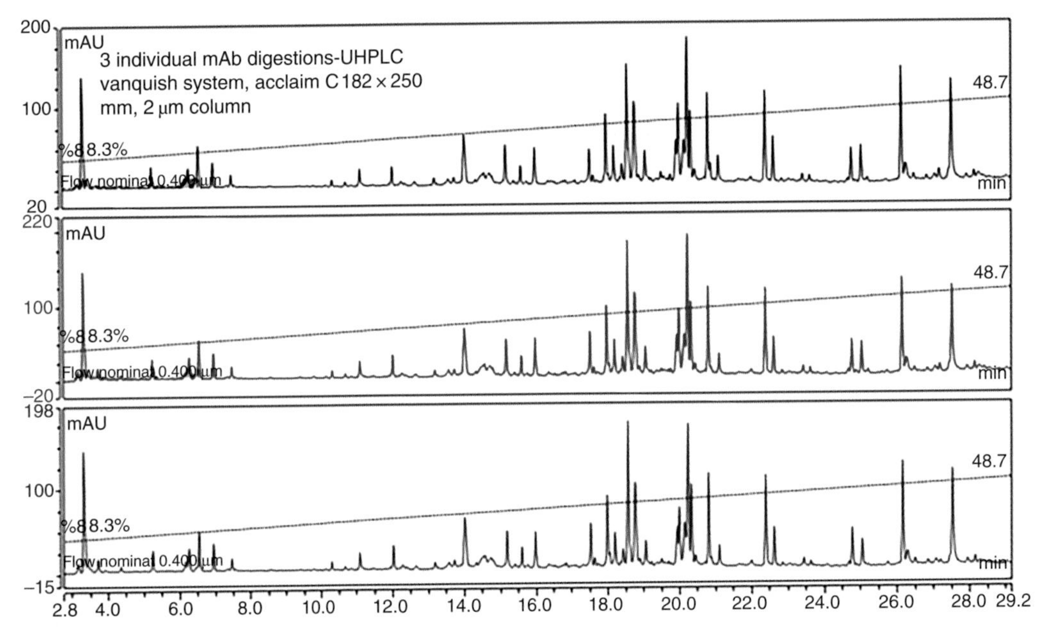

Figure 7.9 Reproducibility of tryptic digestions using an immobilised enzyme kit system. Three individual mAb Digestion–ns - UHPLC Vanquish System, Acclaim C18 2 × 250 mm, 2 μm column. Source: Courtesy of Thermo Fisher Scientific.

monitor the changes in protein conformation, for example analytical ultracentrifugation and size exclusion chromatography.

7.4.3 Protein Structure Elucidation

Even after isolating an intact protein, it may still be difficult to characterise the purified product. However, such a 'top-down' study of a large molecule can be complemented by a 'bottom-up' approach to protein analysis. This could involve, for example, papain digestion of a monoclonal antibody, but this results in a few still relatively large fragments. On the other hand, deglycosylation of monoclonal antibodies allows glycan analysis by hydrophilic interaction liquid chromatography (HILIC) liquid chromatography–mass spectrometry (LC-MS) to study the sugars responsible for molecular recognition, HILIC being a relatively recently popularised mode of LC involving a polar stationary phase and a mobile phase containing a high proportion of a polar organic solvent and a low proportion of aqueous buffer that is suitable for polar analytes and is characterised by good MS sensitivity [117]. This type of analysis is very valuable as it is post-translational modifications such as glycosylation that endow proteins with many of their important properties. This is particularly the case in the study of monoclonal antibodies and 'biosimilars' used as biopharmaceutical drugs.

However, the most common form of the 'bottom-up' approach is to carry out a tryptic digest of the protein and subsequently characterise the product peptides by LC (Figure 7.9) or, more frequently, LC-MS or even 2D-LC-MS. This chromatography approach is now widely used and the chromatography used has been perfected to the point that the actual digestion is a significant variable. To this end, manufacturers are now introducing kits to carry out solid phase digestion to reduce the variability caused by time, temperature and autocatalytic digestion of the enzyme.

An even greater challenge is to determine low levels of biopharmaceutical drugs in biological fluids. This can be and is carried out on the intact protein but even here the bottom-up approach can be applied. A signature product peptide may be found for a particular biopharmaceutical and levels of that signature peptide may be monitored as a surrogate for the biopharmaceutical.

7.5 Other Applications of Chromatography for Biological Sample Preparation and Analysis

While the content of this chapter has been focused mainly on the separation of proteins, advances in chromatographic hardware and column chemistry have allowed liquid and gas chromatography to be applied to many biologically relevant fields. This is not new. For many years catecholamines in biological fluids have been studied, e.g. by HPLC with electrochemical detection. Prostanoid levels have been monitored in the study of analgesic drugs and corticosteroids in the study of depression. Gas chromatography, after chemical derivatisation of analytes, can be used in the field of metabolomics and supercritical fluid chromatography has been used effectively with MS detection in the field of lipidomics. However, more recent illustrative examples include:

- The purification of viral associated particles for vaccine production and characterisation [118].
- Cell membrane biology and receptor characterisation [119].
- The linkage of chromatography with mass spectroscopy (LC-MS/GC-MS) or NMR (LC-NMR) has enabled the study of metabolic fluxes and perturbations in diseased cells [5, 120] and engineered expression strains [121].
- Clinical sample analysis has also benefitted from hyphenated techniques: lipid quantification [122], bile acid analysis [123] and prenatal screening [124].
- The separation of gene expression products for pathogen identification and characterisation is also a relatively recent application of chromatography that has enabled a more efficient and cost-effective approach to electrophoretic separation [125].

In summary, the application of chromatography to biological research has enabled biologists to gain a detailed insight into their sample composition and confidence in their data that was once only available to synthetic and analytical chemists.

That is not to say that chromatography in biology is a mature field. The pace of biomedical research shows no signs of abating. In the field of biopharmaceuticals the need for innovation is a constant and, going beyond large biological molecules as drugs, research into drug–antibody complexes and 'new modalities' such as oligonucleotides as drugs will throw up new challenges to the chromatographer.

References

1 Campíns-Falcó, P., Sevillano-Cabeza, A., Herráez-Hernández, R. et al. (2012). Solid-phase extraction and clean-up procedures in pharmaceutical analysis. In: *Encyclopedia of Analytical Chemistry*, 1–17. Chichester, UK: Wiley.

2 Shabir, G.A., John Lough, W., Arain, S.A., and Bradshaw, T.K. (2007). Evaluation and application of best practice in analytical method validation. *J. Liq. Chromatogr. Relat. Technol.* 30: 311–333.

3 Hansen, S.H. and Pedersen-Bjergaard, S. (2015). Analysis of small-molecule drugs in biological fluids. In: *Bioanalysis of Pharmaceuticals*, 207–260. Wiley.

4 Samsonov, G.V., Pisarev, O.A., and Melenevsky, A.T. (1993). Chromatographic purification and superpurification of biologically active compounds using heteroreticular and composite ion exchange resins at low pressure. *Pure Appl. Chem.* 65: 2287–2290.

5 Pasikanti, K.K., Ho, P.C., and Chan, E.C.Y. (2008). Gas chromatography/mass spectrometry in metabolic profiling of biological fluids. *J. Chromatogr. B* 871: 202–211.

6 Latulippe, D.R. and Zydney, A.L. (2009). Size exclusion chromatography of plasmid DNA isoforms. *J. Chromatogr. A* 1216: 6295–6302.

7 Hong, P., Koza, S., and Bouvier, E.S.P. (2012). Size-exclusion chromatography for the analysis of protein biotherapeutics and their aggregates. *J. Liq. Chromatogr. Relat. Technol.* 35: 2923–2950.

8 Ó'Fágáin, C., Cummins, P.M., and O'Connor, B.F. (2017). Gel-filtration chromatography. In: *Methods in Molecular Biology*, 15–25. Springer.
9 Budelier, K. and Schorr, J. (2001). Purification of DNA by anion-exchange chromatography. In: *Current Protocols in Molecular Biology*, Unit 2.1B. Hoboken, NJ, USA: Wiley.
10 Bolto, B., Dixon, D., and Eldridge, R. (2004). Ion exchange for the removal of natural organic matter. *React. Funct. Polym.* 60: 171–182.
11 Burness, A.T.H. and Pardoe, I.U. (1983). Chromatofocusing of sialoglycoproteins. *J. Chromatogr. A* 259: 423–432.
12 Mohammad, J. (2010). Chromatofocusing. *Cold Spring Harb. Protoc.* 2010: pdb.top67.
13 Canene-Adams, K. (2013). Reverse-phase HPLC analysis and purification of small molecules. *Methods Enzymol.* 533: 291–301.
14 Cummins, P.M. and O'Connor, B.F. (2011). Hydrophobic interaction chromatography. *Methods Mol. Biol.* 681: 431–437.
15 Chockalingam, P.S., Jurado, L.A., and Jarrett, H.W. (2001). DNA affinity chromatography. *Mol. Biotechnol.* 19: 189–199.
16 Urh, M., Simpson, D., and Zhao, K. (2009). Chapter 26: Affinity chromatography. In: *Methods in Enzymology*, 417–438. Elsevier.
17 Wiederschain, G.Y. (2007). Handbook of affinity chromatography. *Biochemist* 72: 793–793. Portland Press.
18 West, I. and Goldring, O. (1994). Lectin affinity chromatography. *Mol. Biotechnol.* 2: 147–155.
19 Norsko, J.K. (1990). Chemisorption on metal surfaces. *Reports Prog. Phys.* 53: 1253–1295.
20 Toyo'oka, T. (2009). Recent advances in separation and detection methods for thiol compounds in biological samples. *J. Chromatogr. B Anal. Technol. Biomed. Life Sci.* 877: 3318–3330.
21 Porath, J. (1992). Immobilized metal ion affinity chromatography. *Protein Expr. Purif.* 3: 263–281.
22 Tsai, C.F., Wang, Y.T., Lin, P.Y., and Chen, Y.J. (2011). Phosphoproteomics by highly selective IMAC protocol. *NeuroMethods* 57: 181–196.
23 Gilbert, M.T. (1987). Chiral chromatography. In: *High Performance Liquid Chromatography*, 291–312. Elsevier.
24 Lam, S. (1989). Chiral liquid chromatography. In: *Chiral Liquid Chromatography* (ed. W. Lough), 83–101. Blackie Publishing Group.
25 Lämmerhofer, M. (2012). Chiral separations and enantioselectivity. *J. Chromatogr. A* 1269: 1–2.
26 Asnin, L. (2012). Adsorption models in chiral chromatography. *J. Chromatogr. A* 1269: 3–25.
27 Fiori, M., Scintu, M.F., and Addis, M. (2013). Characterization of the lipid fraction in lamb meat: comparison of different lipid extraction methods. *Food Anal. Methods* 6: 1648–1656.
28 Perrone, P.A. and Brown, P.R. (1984). Ion-pair chromatography of nucleotides. *J. Chromatogr. A* 317: 301–310.
29 Cecchi, T. and Passamonti, P. (2009). Retention mechanism for ion-pair chromatography with chaotropic reagents. *J. Chromatogr. A* 1216: 1789–1797.
30 Dolan, J.W., Kirkland, J.J., and Snyder, L.R. (2010). *Introduction to Modern Liquid Chromatography*, 3e. Wiley.
31 Purnell, J.H. (1966). The correlation of separating power and efficiency of gas-chromatographic columns. *J. Chem. Soc.* 1268–1274.
32 Miyabe, K. and Guiochon, G. (1999). Peak tailing and column radial heterogeneity in linear chromatography. *J. Chromatogr. A* 830: 263–274.
33 Miyabe, K., Matsumoto, Y., Niwa, Y. et al. (2009). An estimation of the column efficiency made by analyzing tailing peak profiles. *J. Chromatogr. A* 1216: 8319–8330.
34 Stoll, D.R. (2010). Recent progress in online, comprehensive two-dimensional high-performance liquid chromatography for non-proteomic applications. *Anal. Bioanal. Chem.* 397: 979–986.
35 Anspach, F.B., Curbelo, D., Hartmann, R. et al. (1999). Expanded-bed chromatography in primary protein purification. *J. Chromatogr. A* 865: 129–144.

36 Co, C.C., Ho, C.-C., and Kumar, G. (2012). Motility-based cell sorting by planar cell chromatography. *Anal. Chem.* 84: 10160–10164.
37 Moldoveanu, S. and David, V. (2015). *Modern Sample Preparation for Chromatography*, 1–453. Elsevier.
38 Kolars, J.C., Lown, K.S., Schmiedlin-Ren, P. et al. (1994). CYP3A gene expression in human gut epithelium. *Pharmacogenetics* 4: 247–259.
39 Holland, N.T., Smith, M.T., Eskenazi, B., and Bastaki, M. (2003). Biological sample collection and processing for molecular epidemiological studies. *Mutat. Res.* 543: 217–234.
40 Note S (2001) Biological Sample Preparation 331.1. Molecules 1–12.
41 Stadtman, E.R. and Levine, R.L. (2006). Protein oxidation. *Ann. NY Acad. Sci.* 899: 191–208.
42 Laurence, J.S. and Middaugh, C.R. (2010). Fundamental structures and behaviors of proteins. In: *Aggregation of Therapeutic Proteins*, 1–61. Wiley.
43 Konermann, L. (2012). Protein Unfolding and Denaturants. In: *eLS*. Wiley.
44 Christie, W. and Han, X. (2012). *Lipid Analysis: Isolation, Separation, Identification and Lipidomic*, 2e. Pergamon.
45 Dell, A. and Reason, A.J. (1993). Carbohydrate analysis. *Curr. Opin. Biotechnol.* 4: 52–56.
46 Saulnier, L., Vigouroux, J., and Thibault, J.-F. (1995). Isolation and partial characterization of feruloylated oligosaccharides from maize bran. *Carbohydr. Res.* 272: 241–253.
47 Stick, R.V. (2001). Carbohydrate-based vaccines. In: *Carbohydrates*, 233–237. Elsevier.
48 Yu, C., Cohen, L.H., and Arbor, A. (2004). Tissue sample preparation — not the same old grind. *Sample Prep. Perspect.* 2–6.
49 Guthrie, J.W. (2012). General considerations when dealing with biological fluid samples. In: *Comprehensive Sampling and Sample Preparation*, 1–19. Elsevier.
50 Kralicek, A.V. and Ozawa, K. (2011). Advances in cell-free protein synthesis for NMR sample preparation. In: *Biomolecular NMR spectroscopy*, 3–19.
51 Penquer, A. (2009). Review of the available uv detectors. *EAS Publ. Ser.* 37: 193–198.
52 Rutan, S. (1993). Diode array detection in HPLC. *J. Chromatogr. A* 657: 457.
53 Krattiger, B., Bruin, G.J., and Bruno, A. (1994). Refractive index detector. *Anal. Chem.* 66: 11A–15A.
54 Kohler, M., Haerdi, W., Christen, P., and Veuthey, J.L. (1997). The evaporative light scattering detector: some applications in pharmaceutical analysis. *Trends Anal. Chem.* 16: 475–484.
55 Mathews, B.T., Higginson, P.D., Lyons, R. et al. (2004). Improving quantitative measurements for the evaporative light scattering detector. *Chromatographia* 60: 625–633.
56 Geng, X., Shi, M., Ning, H. et al. (2018). A compact and low-cost laser induced fluorescence detector with silicon based photodetector assembly for capillary flow systems. *Talanta* 182: 279–284.
57 Luci, G., Intorre, L., Ferruzzi, G. et al. (2018). Determination of ochratoxin a in tissues of wild boar (sus scrofa L.) by enzymatic digestion (ED) coupled to high-performance liquid chromatography with a fluorescence detector (HPLC-FLD). *Mycotoxin. Res.* 34 (1): 1–8.
58 McMinn, D.G. (1992). The flame ionization detector. *Detect. Capill. Chromatogr.* 23: 7–21.
59 Cohen, S., Gagnieu, M.-C., Lefebvre, I., and Guitton, J. (2013). LC-MS bioanalysis of nucleotides. In: *Handbook of LC-MS Bioanalysis*, 559–572.
60 Yan, Z., Cheng, C., and Liu, S. (2012). Applications of mass spectrometry in analyses of steroid hormones. In: *LC-MS in Drug Bioanalysis*, 251–286. Boston, MA: Springer.
61 Mohamed, M.A., Jaafar, J., Ismail, A.F. et al. (2017). Fourier transform infrared (FTIR) spectroscopy. In: *Membrane Characterization*, 3–29. Elsevier.
62 Subramanian, A. and Rodriguez-Saona, L. (2009). Fourier transform infrared (FTIR) spectroscopy. In: *Infrared Spectroscopy for Food Quality Analysis and Control*, 145–178. Elsevier.
63 Silva Elipe, M.V. (2011). *LC-NMR and Other Hyphenated NMR Techniques*, 1–220. Wiley.
64 Simeonov, E., Tsibranska, I., and Minchev, A. (1999). Solid–liquid extraction from plants — experimental kinetics and modelling. *Chem. Eng. J.* 73: 255–259.

65 Bucić-Kojić, A., Planinić, M., Tomas, S. et al. (2007). Study of solid–liquid extraction kinetics of total polyphenols from grape seeds. *J. Food Eng.* 81: 236–242.

66 Majors, R.E. (2013). Novel sorbents for solid-liquid extraction. *LC GC Eur.* 26: 685–691.

67 Golumbic, C. (1951). Liquid-liquid extraction analysis. *Anal. Chem.* 23: 1210–1217.

68 Tedder, D.W. (2014). Liquid-liquid extraction. In: *Albr'ght's Chemical Engineering Handbook. CRC Press,,* 709–738.

69 Frank, K., Köhler, K., and Schuchmann, H.P. (2012). Stability of anthocyanins in high pressure homogenisation. *Food Chem.* 130: 716–719.

70 Ouzineb, K., Lord, C., Lesauze, N. et al. (2006). Homogenisation devices for the production of miniemulsions. *Chem. Eng. Sci.* 61: 2994–3000.

71 Shao, S., Guo, T., Gross, V. et al. (2016). Reproducible tissue homogenization and protein extraction for quantitative proteomics using micropestle-assisted pressure-cycling technology. *J. Proteome Res.* 15: 1821–1829.

72 Oldenburg, K., Pooler, D., Scudder, K. et al. (2005). High throughput sonication: evaluation for compound solubilization. *Comb. Chem. High Throughput Screen.* 8: 499–512.

73 Mahdi Jafari, S., He, Y., and Bhandari, B. (2006). Nano-emulsion production by sonication and microfluidization – a comparison. *Int. J. Food Prop.* 9: 475–485.

74 Stepanskiy, L.G. (2012). Sonication-induced unfolding proteins. *J. Theor. Biol.* 298: 77–81.

75 Campaniello, D., Bevilacqua, A., Sinigaglia, M., and Corbo, M.R. (2016). Using homogenization, sonication and thermo-sonication to inactivate fungi. *PeerJ* 4: e2020.

76 Scheer, L. (2009). The use of syringe filters for analytical sample preparation: including HPLC and dissolution testing. *Filtration* 9: 308–309.

77 Fabry, T.L. (1987). Mechanism of erythrocyte aggregation and sedimentation. *Blood* 70: 1572–1576.

78 Pawliszyn, J. (2012). Theory of solid-phase microextraction. In: *Handbook of Solid Phase Microextraction*, 13–59. Elsevier.

79 Mollerup, I., Jensen, S.W., Larsen, P. et al. (2002). Insulin, purification. *Encycl. Bioprocess Technol.* 992: 1219–1222.

80 Teagarden, D.L., Speaker, S.M., Maetin, S.W.H., and Österberg, T. (2010). Practical considerations for freeze-drying in dual chamber package systems. In: *Freeze Drying/Lyophilisation of Pharmaceutical and Biological Products*, 494–526. CRC Press.

81 Burtis, C.A. (1990). Sample evaporation and its impact on the operating performance of an automated selective-access analytical system. *Clin. Chem.* 36: 544–546.

82 Walker, G.M. and Beebe, D.J. (2002). Evaporation-driven microfluidic sample concentration. In: *2nd Annual International IEEE-EMBS Special Topic Conference on Microtechnologies in Medicine and Biol–gy - Proceedings*, 523–526.

83 Boutelle, M.G. and Fillenz, M. (1996). Clinical microdialysis: I role of on-line measurement and quantitative microdialysis. In: *Clinical Aspects of Microdialysis*, 13–20.

84 Lange, E.C.M. (2013). *Microdialysis in Drug Development*, 13–33. New York, NY: Springer.

85 Baduel, C., Mueller, J.F., Tsai, H., and Gomez Ramos, M.J. (2015). Development of sample extraction and clean-up strategies for target and non-target analysis of environmental contaminants in biological matrices. *J. Chromatogr. A* 1426: 33–47.

86 Ng, L.L. (1983). Sample preparation by salts precipitation and quantitation by high-performance liquid chromatography with uv detection of selected drugs in biological fluids. *J. Chromatogr. A* 257: 345–353.

87 Ellegren, H. and Låås, T. (1989). Size-exclusion chromatography of DNA restriction fragments. Fragment length determinations and a comparison with the behaviour of proteins in size-exclusion chromatography. *J. Chromatogr. A* 467: 217–226.

88 Vajda, J., Weber, D., Brekel, D. et al. (2016). Size distribution analysis of influenza virus particles using size exclusion chromatography. *J. Chromatogr. A* 1465: 117–125.

89 Ruysschaert, T., Marque, A., Duteyrat, J.L. et al. (2005). Liposome retention in size exclusion chromatography. *BMC Biotechnol.* 5: 11.

90 Benedikter, B.J., Bouwman, F.G., Vajen, T. et al. (2017). Ultrafiltration combined with size exclusion chromatography efficiently isolates extracellular vesicles from cell culture media for compositional and functional studies. *Sci. Rep.* 7: 15297.

91 Westman, E., Eriksson, S., Låås, T. et al. (1987). Separation of DNA restriction fragments by ion-exchange chromatography on FPLC columns mono P and mono Q. *Anal. Biochem.* 166: 158–171.

92 Kennedy, R.M. (1990). Hydrophobic chromatography. *Methods Enzymol.* 182: 339–343.

93 Queiroz, J.A., Tomaz, C.T., and Cabral, J.M.S. (2008). Hydrophobic interaction chromatography of proteins. *J. Chromatogr. A* 1205: 46–59.

94 Fischer, W. (1996). Molecular analysis of lipid macroamphiphiles by hydrophobic interaction chromatography. *J. Microbiol. Methods* 7025: 129–144.

95 El Rassi, Z. (1995). Reversed-phase and hydrophobic interaction chromatography of carbohydrates and glycoconjugates. *J. Chromatogr. Libr.* 58: 41–101.

96 Savard, J.M. and Schneider, J.W. (2007). Sequence-specific purification of DNA oligomers in hydrophobic interaction chromatography using peptide nucleic acid amphiphiles: extended dynamic range. *Biotechnol. Bioeng.* 97: 367–376.

97 Lough, W. (2000). Reversed-phase liquid chromatography. In: *Encyclopaedia of Analytical Chemistry*, 11442–11450. Wiley.

98 Cherkasov, I.A. (1972). Affinity chromatography of enzymes. *Russ. Chem. Rev.* 41: 891–903.

99 Farooqui, A.A. (1980). Purification of enzymes by heparin-sepharose affinity chromatography. *J. Chromatogr. A* 184: 335–345.

100 Cheng, Y., Liu, D., Feng, Y., and Jing, G. (2003). An efficient fusion expression system for protein and peptide overexpression in *Escherichia coli* and NMR sample preparation. *Protein Pept. Lett.* 10: 175–181.

101 Loughran, S.T., Loughran, N.B., Ryan, B.J. et al. (2006). Modified his-tag fusion vector for enhanced protein purification by immobilized metal affinity chromatography. *Anal. Biochem.* 355: 148–150.

102 Einhauer, A. and Jungbauer, A. (2001). The FLAGTM peptide, a versatile fusion tag for the purification of recombinant proteins. *J. Biochem. Biophys. Methods* 49: 455–465.

103 Raines, R.T., McCormick, M., Van Oosbree, T.R., and Mierendorf, R.C. (2000). The S.Tag fusion system for protein purification. *Methods Enzymol.* 326: 362–376.

104 Fuchs, S.M. and Raines, R.T. (2009). Polyarginine as a multifunctional fusion tag. *Protein Sci.* 14: 1538–1544.

105 GE Healthcare (2009). Glutathione S-transferase (GST) gene fusion system. *GST Gene Fusion Syst.* 1–8.

106 Costa, S., Almeida, A., Castro, A., and Domingues, L. (2014). Fusion tags for protein solubility, purification and immunogenicity in Escherichia coli: the novel Fh8 system. *Front. Microbiol.* 5: 63.

107 Rabhi-Essafi, I., Sadok, A., Khalaf, N., and Fathallah, D.M. (2007). A strategy for high-level expression of soluble and functional human interferon alpha as a GST-fusion protein in *E. coli*. *Protein Eng. Des. Sel.* 20: 201–209.

108 Tabor, S. (2001). Expression using the T7 RNA polymerase/promoter system. In: *Current Protocols in Molecular Biology*. Wiley, Chapter 16: Unit 16.2.

109 Moritz, B., Becker, P.B., and Göpfert, U. (2015). CMV promoter mutants with a reduced propensity to productivity loss in CHO cells. *Sci. Rep.* 5: 16952.

110 Ramón, A., Señorale-Pose, M., and Marín, M. (2014). Inclusion bodies: not that bad.... *Front. Microbiol.* 5 (56): 1–6.

111 Singh, A., Upadhyay, V., and Panda, A.K. (2015). Solubilization and refolding of inclusion body proteins. In: *Insoluble Proteins: Methods and Protocols*, 283–291.

112 Panda, A.K. (2003). Bioprocessing of therapeutic proteins from the inclusion bodies of. *Adv. Biochem. Eng. Biotechnol.* 85: 43–93.

113 Haryadi, R., Ho, S., Kok, Y.J. et al. (2015). Optimization of heavy chain and light chain signal peptides for high level expression of therapeutic antibodies in CHO cells. *PLoS One* 10: e0116878.

114 Ng, D.T.W. and Sarkar, C.A. (2013). Engineering signal peptides for enhanced protein secretion from Lactococcus lactis. *Appl. Environ. Microbiol.* 79: 347–356.

115 Rathore, A.S., Bilbrey, R.E., and Steinmeyer, D.E. (2003). Optimization of an osmotic shock procedure for isolation of a protein product expressed in E. coli. *Biotechnol. Prog.* 19: 1541–1546.

116 Oganesyan, N., Kim, S.H., and Kim, R. (2005). On-column protein refolding for crystallization. *J. Struct. Funct. Genom.* 6: 177–182.

117 McCalley, D.V. (2017). Understanding and manipulating the separation in hydrophilic interaction liquid chromatography. *J. Chromatogr. A* 1523: 49–71.

118 Cazares, L.H., Ward, M.D., Brueggemann, E.E. et al. (2016). Development of a liquid chromatography high resolution mass spectrometry method for the quantitation of viral envelope glycoprotein in Ebola virus-like particle vaccine preparations. *Clin. Proteomics* 13: 18.

119 Caculitan, N.G., Kai, H., Liu, E.Y. et al. (2014). Size-based chromatography of signaling clusters in a living cell membrane. *Nano Lett.* 14: 2293–2298.

120 Watson, G.W., Wickramasekara, S., Maier, C.S. et al. (2015). Assessment of global proteome in LNCaP cells by 2D-RP/RP LC-MS/MS following sulforaphane exposure. *EuPA Open Proteom.* 9: 34–40.

121 Albrecht, S., Kaisermayer, C., Reinhart, D. et al. (2018). Multiple reaction monitoring targeted LC-MS analysis of potential cell death marker proteins for increased bioprocess control. *Anal. Bioanal. Chem.* 410: 3197–3207.

122 Blanchard, V., Ramin-Mangata, S., Billon-Crossouard, S. et al. (2018). Kinetics of plasma apolipoprotein E isoforms by LC-MS/MS: a pilot study. *J. Lipid Res.* 59: 892–900.

123 Krautbauer, S. and Liebisch, G. (2018). LC-MS/MS analysis of bile acids. In: *Methods in Molecular Biology*, vol. 1730, 103–110. Springer.

124 López Uriarte, G.A., Burciaga Flores, C.H., Torres de la Cruz, V.M. et al. (2018). Proteomic profile of serum of pregnant women carring a fetus with down syndrome using nano uplc Q-tof ms/ms technology. *J. Matern. Neonatal. Med.* 31: 1483–1489.

125 Ueda, S., Yamaguchi, M., Eguchi, K., and Iwase, M. (2016). Identification of Cereulide-producing Bacillus cereus by nucleic acid chromatography and reverse transcription real-time PCR. *Biocontrol Sci.* 21: 45–50.

Further Reading

Conductivity E, Tds S (2004) 760. Electrical Conductivity/Salinity Fact Sheet. Water Resour 0:2–6.

Dong, M.W. (2006). *Modern HPLC for Practising Scientists*. Wiley.

Gooding, K.M. and Regnier, F.E. (2002). *HPLC of Biological Macro- Molecules, Revised and Expanded*. CRC Press; Taylor and Francis Group.

Grigore, M.-N., Ivanescu, L., and Toma, C. (2014). *Halophytes: An Integrative Anatomical Study*. Springer International Publishing.

Snyder, L.R. (1992). Chapter 1, Theory of chromatography. In: *Journal of the Chromatography Library*, A1–A68.

Snyder, L.R., Joseph, J.K., and Dolan, J.W. (2010). Ionic samples: reversed-phase, ion-pair, and ion-exchange chromatography. In: *Introduction to Modern Liquid Chromatography*, 303–360. Hoboken, NJ, USA: Wiley.

Striegel, A.M., Yau, W.W., Kirkland, J.J., and Bly, D.D. (2009). *Modern Size Exclusion Chromatography: Practice of Gel Permeation and Gel Filtration Chromatography*, 2e. Wiley.

Timms, J.F. and Cutillas, P.R. (2010). Overview of quantitative LC–MS techniques for proteomics and activitomics. *Methods Mol. Biol.* 658: 19–45.

Walls, D. and Loughran, S. (2011). *Protein Chromatography: Methods and Protocols*. Springer.

8

Synthetic Methodology in Chemical Biology

Richard C. Brewster and Stephen Wallace

Institute of Quantitative Biology, Biochemistry and Biotechnology, School of Biological Sciences, University of Edinburgh, Edinburgh, ,UK

8.1 Introduction

Chemical biology is a diverse field encompassing a wide range of techniques and processes. One challenging area in this field is the study of biomolecules in their native environments to determine structure, function and dynamics.

This chapter looks to introduce some of the different approaches to biomolecule modification in this field. The first section will look at *in vitro* methods, starting with an introduction to peptide synthesis and how these peptides can be ligated to produce synthetic proteins. We will then consider some of the different chemical reactions for modifying endogenous amino acids on proteins *in vitro*.

The second section will discuss the use of bioorthogonal reactions for protein modification *in vivo*. We will look at how design principles can be used to improve reaction kinetics *in vivo* and how the judicious choice of a reaction manifold can lead to improved properties in single cells and whole animals. We will then provide an overview of different methods for the introduction of new functional groups into living cells using both metabolic and synthetic biology approaches.

Finally, a case study will consider the field of histone post-translational modifications (PTMs). Three very different methods for incorporating unnatural amino acid (UAA) modifications into a histone protein are considered and their differences and complementarities discussed.

8.2 Peptide Synthesis

Peptides play a fundamental role in essentially all physiological and biochemical processes. They form the basis of a number of pharmaceutical agents and their unique chemistry has facilitated the study of many biological systems. By definition, peptides are short chains of amino acids linked by amide (peptide) bonds. Generally they vary in length from 2 to 50 amino acids, with anything longer than this generally denoted as a protein. Nature has evolved remarkably efficient methods for constructing polypeptide chains as proteins are required for almost all cellular functions. As dictated by the Central Dogma – '*DNA makes RNA, makes protein*' – proteins are synthesised *in vivo* by first decoding genomic DNA into mRNA via transcription, which is then translated into a polypeptide at the ribosome. Here, tRNA–amino acid pairs (called aminoacyl-tRNAs) are sequentially polymerised according to their cognate messenger RNA in a rapid and sequence-specific manner.

In the field of biochemistry, methods have been optimised in recent years for producing recombinant proteins in cells using modern molecular biology and this can yield large amounts of a specific protein in a range of microbial hosts.

Biomolecular and Bioanalytical Techniques: Theory, Methodology and Applications, First Edition. Edited by Vasudevan Ramesh.

However, producing recombinant peptides is a much greater challenge and often requires extensive optimisation, especially for shorter peptides (<100 amino acids in length) as these are often degraded by proteases in the host cell [1]. It is also difficult to incorporate unnatural modifications using recombinant techniques, although advances in this field are discussed in section 8.6.

The requirement for peptides of any length, sequence and bearing unnatural modifications from both industrial and research fields has led to the development of extremely robust methods for peptide synthesis in the field of synthetic organic chemistry.

8.3 Amide Bond Synthesis

The basic reaction behind peptide synthesis is deceptively simple, an electrophilic carboxylic acid on one amino acid undergoes nucleophilic attack by an amine from another amino acid, forming an amide bond and eliminating water. The energy barrier for this process, however, is extremely high and temperatures >100 °C are required for the uncatalysed reaction (Figure 8.1) [2]. This low reactivity can be attributed to the stability of carboxylate and ammonium salts – the predominant structure of amino acids in water – which are significantly less electrophilic and nucleophilic, respectively.

Early methods to synthesise amide bonds first converted the carboxylic acid to an acid chloride, which can then react with amines in the presence of excess base. Acid chlorides are highly reactive and therefore not very stable, and most methods to synthesise them require harsh conditions. Most modern amide bond formation reactions now use coupling reagents, as shown in Figure 8.2. Typically the hydroxyl group reacts with a carbodiimide to form an *O*-acylurea, an extremely unstable intermediate. Nucleophilic addition of the amine gives the amide product and urea. The highly unstable nature of the *O*-acylurea can, however, cause different reaction products, including the *N*-acylurea and epimerisation of the α-position of the L-amino acid (Figure 8.3). Several additives have been developed that can stabilise the *O*-acylurea as an activated ester, which reduces this unwanted epimerization process. HOBt (1-hydroxy-1*H*-benzotriazole) is one such reagent and has been shown to significantly reduce side reactions and epimerisation during amide bond formation. Many modern amide coupling methods now include the carbodiimide and HOBt motif in the same molecule (see HATU and PyBOP) (Figure 8.2).

Now assume we want to synthesise a simple dipeptide and were to mix two amino acids together with a coupling reagent. The result may produce a small amount of the dipeptide, but there would be a multitude of different length peptide chains with a random sequence of amino acids as each of our amino acids can react with another equivalent of itself (Figure 8.4).

This is where the concept of protecting group chemistry is important. If the amine on one amino acid is protected and the carboxylic acid on the other is protected, then the product of the reaction will be a single dipeptide. If orthogonal protecting groups are used on the *C*- and *N*-terminii, then one can be selectively deprotected and another amino acid with the corresponding *C*/*N* protecting groups in place can be added to extend the peptide. Any functional group that is nucleophilic or electrophilic can interfere with the amide coupling reactions and so must also be orthogonally protected to prevent side reactions and be resistant to the conditions used to deprotect the *C*-/*N*-terminus. This includes the side chains of the naturally occurring

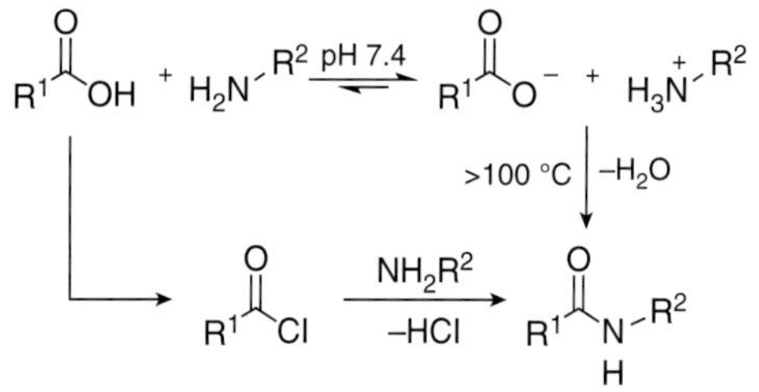

Figure 8.1 Amide bond synthesis by thermal reaction or by carboxylic acid activation via an acyl chloride.

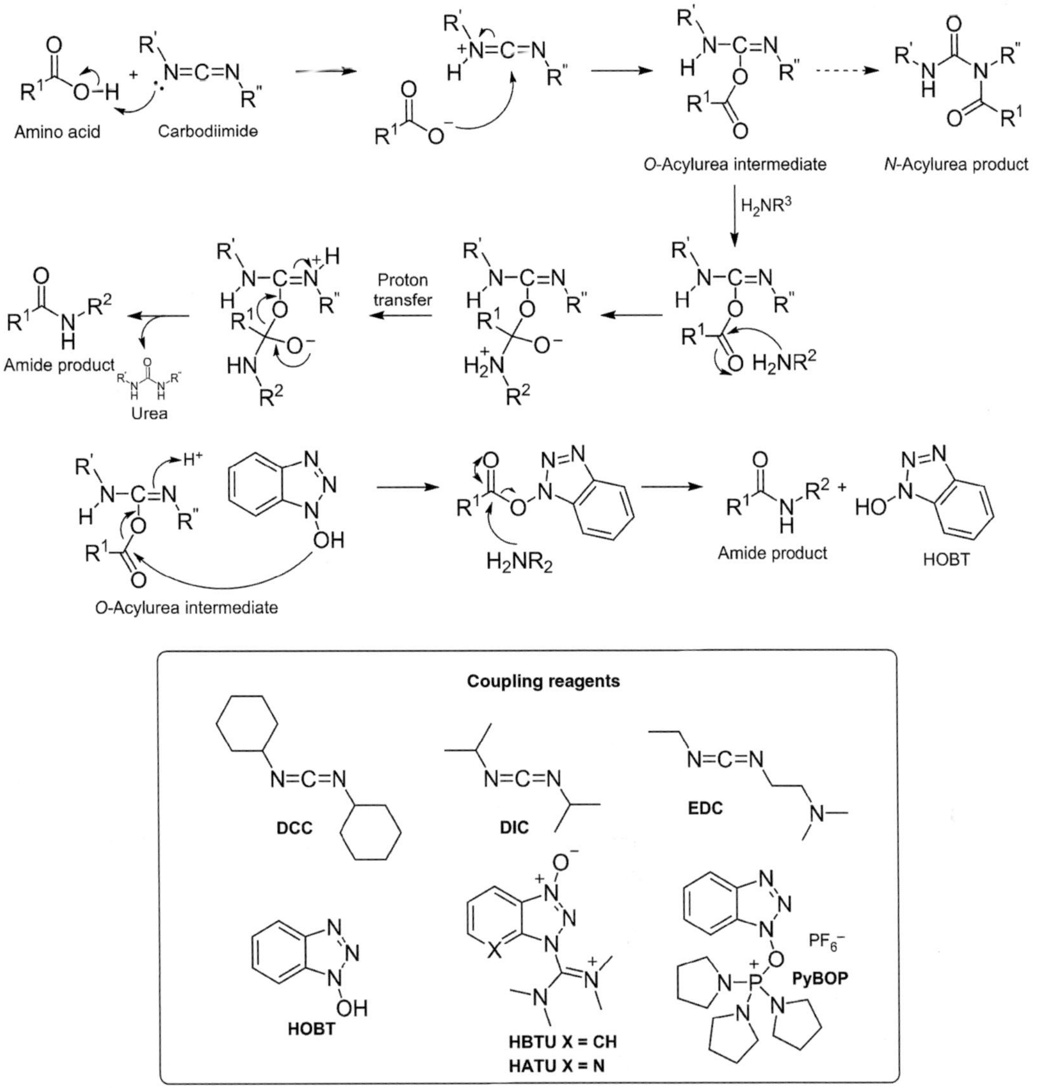

Figure 8.2 Common coupling reagents used in synthetic organic chemistry for amide bond synthesis and the mechanism of carbodiimide-mediated coupling reactions.

amino acids; e.g. lysine has an amine and aspartic/glutamic acid have a carboxylic acid that can participate in amide bond formation reactions and so must be protected.

8.3.1 Solid Phase Peptide Synthesis (SPPS)

Each step in a peptide synthesis requires purification, so that coupling or deprotection reagents and by- products can be removed to allow the next step of the reaction to proceed. This can be particularly time-consuming and generates large volumes of chemical waste. Realising this limitation, the field of 'solid phase peptide synthesis' (SPPS) was developed by Bruce Merrifield in 1963, for which he won a Nobel Prize in 1984 [3]. Using this approach, the first amino acid is coupled through its *C*-terminus to an insoluble solid (polymer) support or 'resin' and, instead of a separate purification step, excess amino acid and coupling reagent can simply be

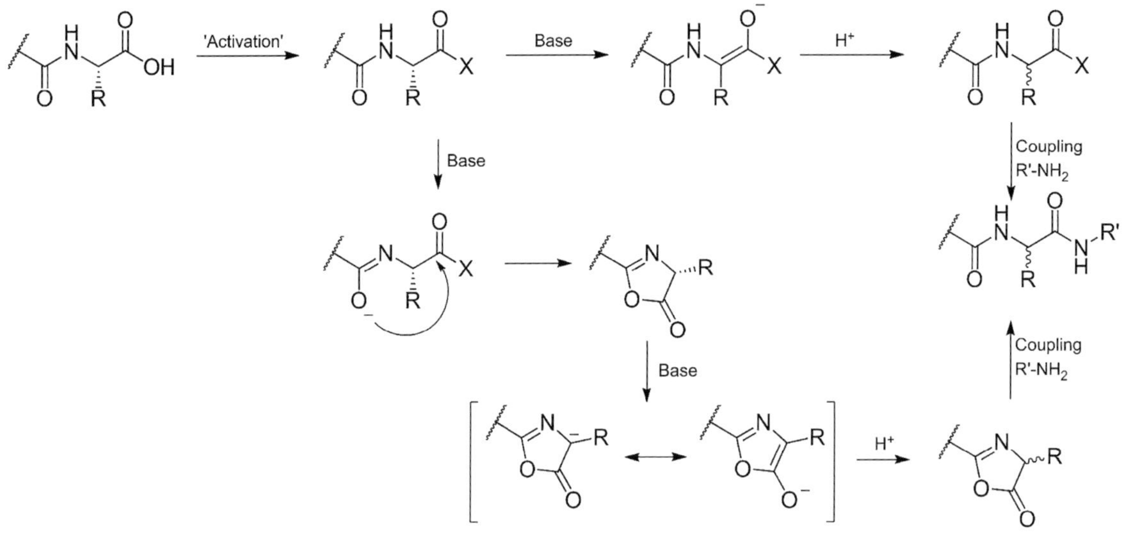

Figure 8.3 Mechanisms of based catalysed racemization of amino acid residues in SPPS.

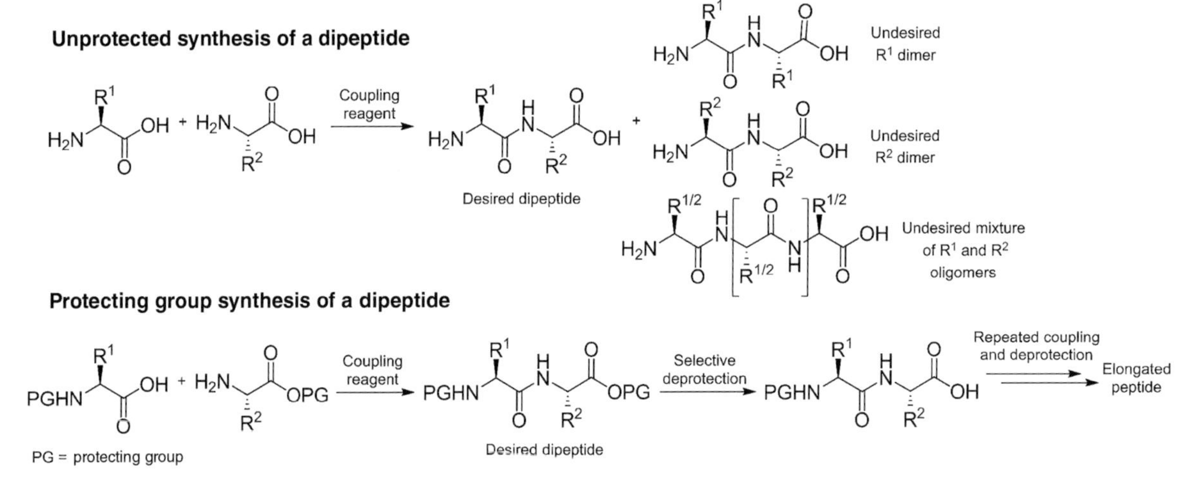

Figure 8.4 Synthesis of a simple dipeptide on unprotected amino acids will lead to a complex mixture of products formed due to competing side reactions from other reactive functional groups in the molecule. If these other functional groups are protected then only one product is formed, giving a higher yielding reaction. If one of the protecting groups can be selectively removed then the peptide can be elongated.

washed away leaving the amino acid functionalised solid support. The *N*-terminus can then be deprotected in the solid phase, the reagents washed away and another amino acid added using a coupling reagent to extend the peptide. The cycle is then simply repeated until the desired amino acid sequence has been synthesised. The peptide can then be deprotected and cleaved from the resin concurrently (normally with trifluoroacetic acid [TFA] or hydrofluoric acid [HF]) and then further purified, if required, by reverse phase high performance liquid chromatography (RP-HPLC).

Introduction of the Wang linker (*para*-alkoxybenzyl alcohol) on to Merrifield's resin increases the lability of the peptide to acid cleavage, which reduced side reactions observed during the acid-catalysed cleavage step during SPPS (Figure 8.5). Other linkers now exist that allow a more acid labile cleavage, e.g. 2-chlorotrityl

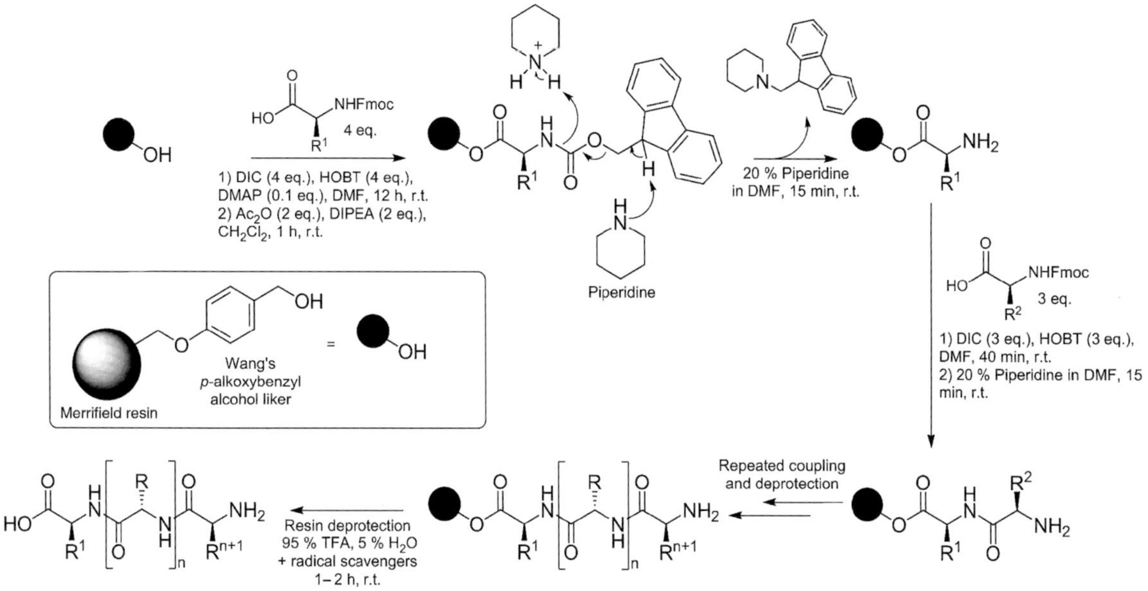

Figure 8.5 SPPS using a Wang functionalised resin and Fmoc protecting groups. The first amino acid is coupled to the solid support via an esterification reaction. Fmoc deprotection of the amine using piperidine allows subsequent amide couplings and this procedure is repeated until the target peptide is made. Cleavage from the resin is achieved using TFA, which also removes side chain protecting groups giving the unprotected peptide product.

linker. The ability to use more acid-labile linkers allows for peptides to be removed from the solid support still fully protected, which can be useful for further functionalisation, e.g. cyclisation. Primary amide incorporation at the *C*-terminus can be achieved using a Rink amide linker; this modification can increase peptide stability to proteases and alter the charge of the peptide.

In general, SPPS has many advantages over solution phase synthesis. For example, excess reagent can be used at each step to drive reactions to completion and then simply washed away, bypassing any laborious liquid–liquid separation, solvent evaporation and chromatography steps. SPPS is also easily automated and peptide synthesisers are available that can synthesise peptide sequences rapidly with minimal user input.

Two amine protecting group strategies now dominate SPPS, *tert*-butyloxycarbonyl (Boc) or fluorenylmethoxycarbonyl (Fmoc). Boc groups are acid labile protecting groups and can be rapidly cleaved using 25–50% (TFA) in DCM. The Fmoc protecting group is base labile and is normally cleaved using 20% piperidine in DMF. For each method the amino acid side chain protecting groups must be stable under Fmoc or Boc deprotection conditions. Fmoc is used more commonly than Boc as the deprotection conditions are milder, leading to fewer side reactions, and the final resin cleavage uses TFA instead of HF, which is extremely toxic.

The key to SPPS is good conversion for each step. Consider the synthesis of a 10 amino acid peptide, which requires a total of 18 reaction steps. If each of these reactions yields 90% of the product, the final yield will be ~15%. If this is improved to 95% for each reaction, then the overall yield will be 40%. Similarly, a 99% yield for each reaction will result in an overall yield of 83%. These differences may seem small, but their impact on the overall efficiency of polypeptide synthesis via SPPS can be dramatic. It is for this reason that optimisation of coupling reagents and protecting groups is so important in the field of peptide synthesis.

Typical Procedure for Fmoc SPPS

1. Wang resin (1 g, 1 mmol/g loading) is swollen in CH_2Cl_2/DMF (10 ml; 8 : 2) for 1.5 hours and drained.
2. Fmoc amino acid (4 mmol) is dissolved in the minimum amount of DMF with HOBT (4 mmol) and DMAP (0.4 mmol), and DIC (4 mmol) is added. The mixture is added to the resin, which is mixed for 12–20 hours.

3. The resin is drained and washed with DMF (2 × 10 ml), CH_2Cl_2 (2 × 10 ml) and MeOH (2 × 10 ml), dried and 10–20 mg are weighed into a flask. 20% Piperidine in DMF (1 ml) is added and mixed for 30 minutes, the solution is filtered off and absorbance is measured at 301 nm. Fmoc deprotection product is quantified using the extinction coefficient ($\varepsilon = 7800\ dm^3/mol\ cm$). If the loading percentage is acceptable the resin is capped using Ac_2O (20 mmol) and DIPEA (20 mmol) in DMF (3 ml). If loading is not acceptable, step 2 is repeated.
4. The resin is drained and washed with DMF (2 × 10 ml), CH_2Cl_2 (2 × 10 ml) and DMF (10 ml). A solution of 20% piperidine in DMF is added (5 ml) and mixed for 5 minutes. The solution is drained and washed as before and 20% piperidine (5 ml) is added and mixed for 10 minutes, drained and washed again.
5. Fmoc-amino acid (3 mmol) and HOBt (3 mmol) are dissolved in the minimum amount of DMF and DIC is added. The solution is added to the resin and mixed for 40 minutes at room temperature (r.t.). The resin is drained and washed and Fmoc deprotection is done by repeating step 4.
6. A cleavage cocktail is prepared based on 95% TFA and 5% H_2O (additional scavengers are added depending on the side chain protecting groups used), which are added to the resin and mixed for 1–2 hours. The resin is drained into a flask and washed with TFA (3 × 3 ml). The peptide is precipitated with cold ether and collected by vacuum filtration, purity is checked by high-performance liquid chromatography (HPLC) and purified if required.

For more information about SPPS several comprehensive reviews have been written [4]. Further information on amino acid protecting groups used can be found in the review by Isidro-Llobet et al. [5] and on coupling reagents in the review by El-Faham and Albericio [6].

The major advantages of SPPS is that it is quick, predictable and scalable. Peptides can be readily synthesised on a milligram-to-gram scale in most laboratories, or up to multikilogram scale industrially [7]. The iterative nature of this method also means that peptides can be easily modified to include unnatural motifs or functional groups that cannot be introduced using cellular methods. Common modifications of this type include acetylation of the *N*-terminus and primary amide synthesis at the *C*-terminus, cyclic peptides, incorporation of UAAs, amino acids bearing PTMs or peptides bearing fluorophores, ligands or other small organic compounds.

However, while SPPS has made peptide synthesis routine for most labs, this technique is extremely wasteful and environmentally irresponsible. Excess chemical reagents are commonly used at each step to increase coupling efficiencies and the solvents and reagents used are often toxic and non-renewable. While the cost of reagents has reduced in recent years due to mass scale production to meet a growing demand for this technology in industry, SPPS remains a cost ineffective way of synthesising polypeptides compared to recombinant DNA methods [8].

There is also a limitation on the size of peptides than can be synthesised on resin, which is normally limited to about 50 amino acids, although through optimisation of conditions, resin and loading values this can be increased to >100 amino acids for certain sequences. However, purification of these large compounds can also be challenging. Although these improvements have enabled the synthesis of small proteins (e.g. RNAse A, 124 amino acids) [9], development of other techniques such as native chemical ligation (NCL) has since been favoured.

8.3.2 Native Chemical Ligation (NCL)

SPPS has made the synthesis of peptides very tractable, but protein synthesis remains a challenge using this technique. Conceptually, if we could couple multiple peptides together in an ordered fashion, a protein could be synthesised *in vitro* from entirely synthetic peptide precursors. One technique for achieving this is NCL [10]. Using this method, unprotected peptides, one bearing a thioester at the *C*-terminus and the other containing a cysteine residue at the *N*-terminus, can be ligated. The reaction proceeds via a transthioesterification followed by an *S*-to-*N* acyl shift giving the amide product as shown in Figure 8.6. Sequential NCL coupling

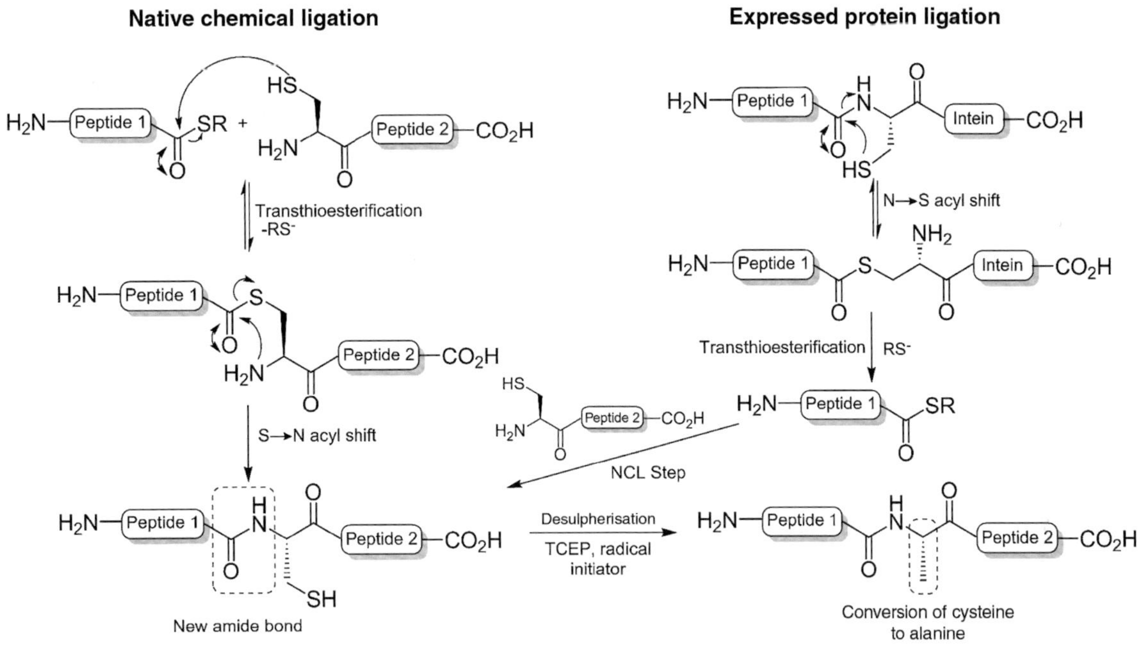

Figure 8.6 Mechanism of the native chemical ligation (NCL) reaction on an unprotected peptide/protein. Transthioesterification is followed by an *S*-to-*N* acyl shift to afford the new amide bond. Expressed protein ligation (EPL) relies on intein excision (an intein is an intervening protein, which is a natural protein segment that undergoes 'protein splicing', where a section protein will remove itself through generation of a thioester [10]), where an *N*-to-*S* acyl shift gives the thioester product [11]. Addition of a free thiol, typically mercaptophenylacetic acid (MPAA), causes transthioesterification, giving a protein thioester, which can undergo an NCL reaction with a synthetic peptide.

reactions can lead to the synthesis of large proteins bearing multiple synthetic modifications. As the cysteine residue may not be required in the final peptide/protein sequence, desulphurisation of cysteine can be achieved in a separate step using triscarboxyethyl phosphine (TCEP) and a radical initiator, converting the cysteine into an alanine residue. The drawback here is that if a cysteine residue is required in the sequence it must be protected.

NCL produces a single ligated peptide product and therefore can be considered as highly chemoselective. Purification by RP-HPLC or size exclusion chromatography means that products are normally separable from starting materials. However, the applicability of this reaction has been limited by slow reaction rates and difficulties in synthesising thioesters, although the development of new reagents in recent years has significantly improved these issues [12].

A significant enhancement in NCL reactivity comes from increasing the reactivity of the thioester. Mercaptophenyl acetic acid (MPAA) is often added to the reaction mixture, causing an additional transthioesterification to occur. The intermediate aryl-thioester (which are often too reactive to isolate) reacts much faster than an alkyl thioester, increasing the overall rate of reaction. A drawback to the use of excess MPAA, however, is that it can complicate purification and MPAA must be removed before desulphurisation [13].

Trifluoroethanethiol (TFET) has been proposed as an alternative to MPAA as the electron withdrawing nature of the CF_3 group gives a pK_a of 7.3, allowing efficient transthioesterification at pH 6.8. The benefit of TFET is that it is volatile (bp = 35 °C) and so can be easily removed, allowing desulphurisation and ligation to be done in a 'one-pot' reaction, as shown in the example in Figure 8.7 for the synthesis of a tick-derived protein Chimadanin, where two sequential ligations and desulphurisation are all done as 'one-pot'. Of particular note

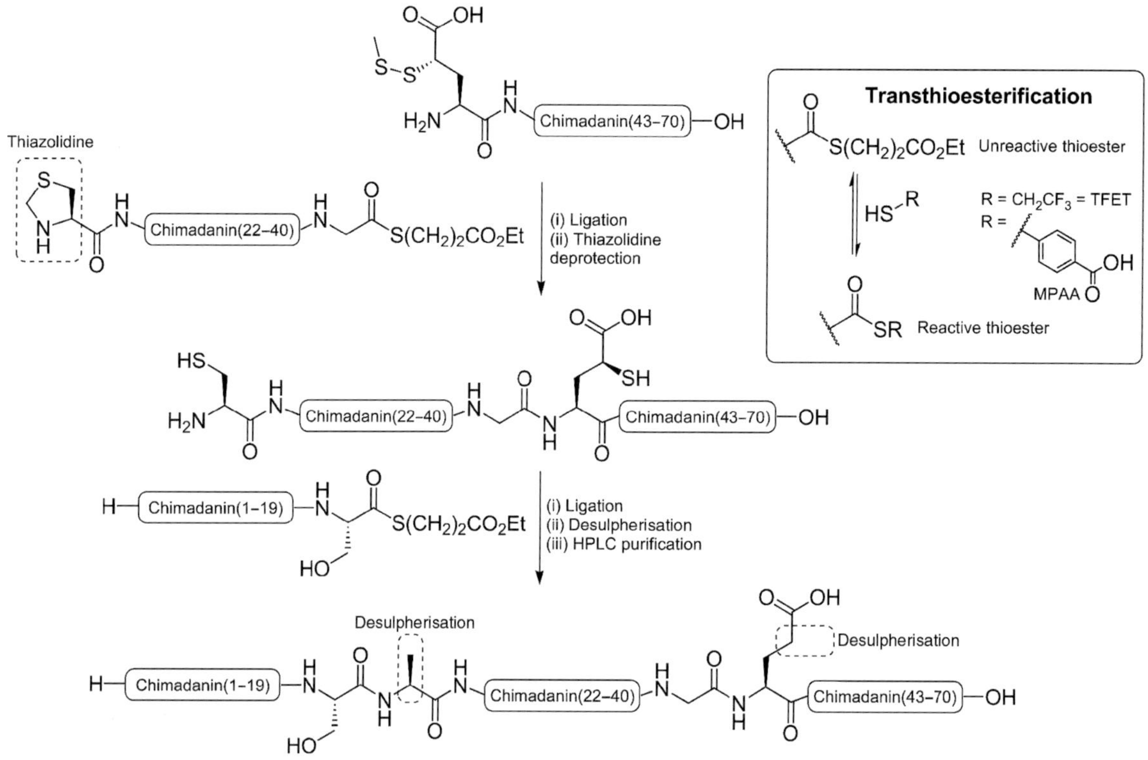

Figure 8.7 The synthesis by Payne et al. of Chimadanin using an NCL strategy [13].

is the author's use of a modified glutamic acid bearing a thiol. This approach allowed the first ligation site to be at Glu rather than Ala [13].

Example procedure for Chimadanin synthesis by NCL [13]: (*NCL is a more complex technique than some of the examples given, so a general protocol is not really applicable. The procedure below is given to give readers an idea of the experimental method using the example shown in* Figure 8.7.)

i) *Ligation.* A solution of peptide Chimadanin (43–70) (3.8 mg, 1.13 μmol, 1.2 eq.) in ligation buffer (370 μl; 6 M Gn·HCl, 100 mM NaPi, 25 mM TCEP, pH 6.8) was added to peptide Chimadanin (22–40) (2.6 mg, 0.94 μmol, 1.0 eq.) with TFET (7.5 μl, 2 vol.%). The pH was readjusted to 6.8 with NaOH (3 M) and the solution incubated for 2 hours at 30 °C. Analysis by HPLC-MS confirmed complete conversion.

ii) *Thiazolidine deprotection.* A solution of $MeONH_2$·HCl (390 μl; 0.2 M in 6 M Gn·HCl, 100 mM NaPi pH 3.4) was added to the reaction mixture and incubated for three hours at 30 °C. Analysis by HPLC-MS confirmed complete conversion.

iii) *Ligation.* The pH of the reaction mixture was adjusted to 7.0 using NaOH (3 M) followed by addition of peptide Chimadanin (1–19) (3.2 mg, 1.2 μmol, 1.3 eq. in 6 M Gn·HCl, 100 mM NaPi) and TFET (18 μl, 2 vol.%) and incubated for 18 hours at 30 °C. Analysis by HPLC-MS confirmed complete conversion.

iv) *Desulphurisation.* A solution of TCEP (1 M) and glutathione (100 mM) in buffer (925 μl; 6 M Gn·HCl, 100 mM NaPi) was added to give a solution containing 500 mM TCEP, 40 mM glutathione and 0.5 mM Chimadanin intermediate. The solution was adjusted to pH 6.2 and degassed by sparging with $Ar_{(g)}$ for 10 minutes, which also removed TFET. VA-044 was added to give a concentration of 20 mM and the mixture was incubated at 37 °C for 5 hours. Analysis by HPLC-MS confirmed complete conversion. The product was purified by reversed-phase semipreparative HPLC, giving Chimadanin (3.1 mg, 35% yield).

NCL is an extremely powerful strategy for synthesising proteins bearing UAAs and PTMs. It has been used to synthesise an enzyme using entirely D-amino acids to create a protein mirror image [14], for the synthesis of glycoproteins [15] and for protein synthesis of up to 304 amino acids [16]. In these examples NCL has been successfully used to access protein targets that are not yet possible to synthesise using purely biological methods.

One way of combining the benefits of biological and chemical protein synthesis methods is through the use of expressed protein ligation (EPL), where a protein expressed using recombinant DNA techniques can be further extended using synthetic peptide fragments [17]. The protein is first expressed bearing a thioester at the *C*-terminus through a separate intein excision step. The thioester is then reacted with a peptide bearing a cysteine residue at the *N*-terminus, which in turn induces the desired transthioesterification and *S*-/*N*-acyl shift, as for NCL. Alternatively, the protein can be genetically modified to contain an additional *N*-terminal Cys residue that can then react with a peptide thioester in an analogous fashion [18].

The major advantage of EPL is that the amount of peptide synthesis and NCL reactions can be significantly reduced, enabling larger proteins/polypeptides to be constructed in a more efficient manner.

8.3.3 Reactions of Endogenous Amino Acids

In the first section we focused on how synthetic chemistry can be used to generate short peptides sequences and then how these can be coupled to synthesise larger proteins. The major benefit of this approach to protein synthesis is that a large amount of chemical divergence can be incorporated site selectively into the protein/peptide chain. The disadvantages are that the process is expensive, time consuming and requires multiple purification steps. This in turn generates large volumes of chemical waste. Protein synthesis can also be achieved via plasmid-based overexpression in microorganisms such as *Escherichia coli* and *Saccharomyces cerevisiae.*

In addition to these approaches, the functional groups on amino acids can also be reacted selectively to introduce synthetic modifications. Although these reactions can be selective for a particular amino acid, they generally do not give site selectivity; e.g. if a lysine-selective reagent is used, the reagent could react with any Lys residue in the protein, not just at one particular site. Some selectivity can be obtained if a particular residue is solvent exposed or if the surrounding amino acids in the protein's tertiary structure increase the reactivity of one particular residue, but these effects are not generalisable and vary a lot between proteins and residues [19].

Within this field, a wide range of reagents exist to target almost all canonical amino acids (apart from the aliphatic side chains, and Ser and Thr), which have been recently reviewed [19, 20].

The most common compounds that are attached to proteins (represented by 'R' in Figures 8.8 and 8.9) using these techniques are fluorophores or biotin. The addition of a fluorophore to a protein has many benefits for protein imaging using microscopy. Biotinylation is commonly used for affinity purification, a technique that takes advantage of the extremely high binding affinity ($K_D > 10$ M [12]) of biotin to the proteins avidin and streptavidin.

Some of the most common commercially available reagents for amino acid conjugation are shown in Figure 8.6; these reagents are typically focused on Lys and Cys, the most nucleophilic amino acids.

The electrophilic NHS esters and the water-soluble sulpho-NHS derivative are reactive towards primary amines, giving an amide bond product. The reaction is selective for Lys and *N*-terminal primary amines in proteins and is pH dependent, as the nucleophilic free amine ($-NH_2$) rather than the non-nucleophilic ammonium ion ($-NH_3^+$) is required for a reaction to occur. Due to the pK_a of Lys (10.5), a pH 7–9 is generally required for this reaction to occur. It should be noted that the *N*-terminus of peptide chains has a lower pK_a, typically around 8, which allows for some selectivity over Lys [21].

Typical procedure for lysine modification with NHS esters: To protein (50–100 μM) in amine-free buffer (100 mM, pH 7–9) is added a 10-fold molar excess of NHS-ester (5–50 mM dissolved in water or DMSO). The solution is mixed and incubated at room temperature for 4–6 hours or at 4 °C overnight. The labelled protein is normally purified by size exclusion chromatography or buffer exchange.

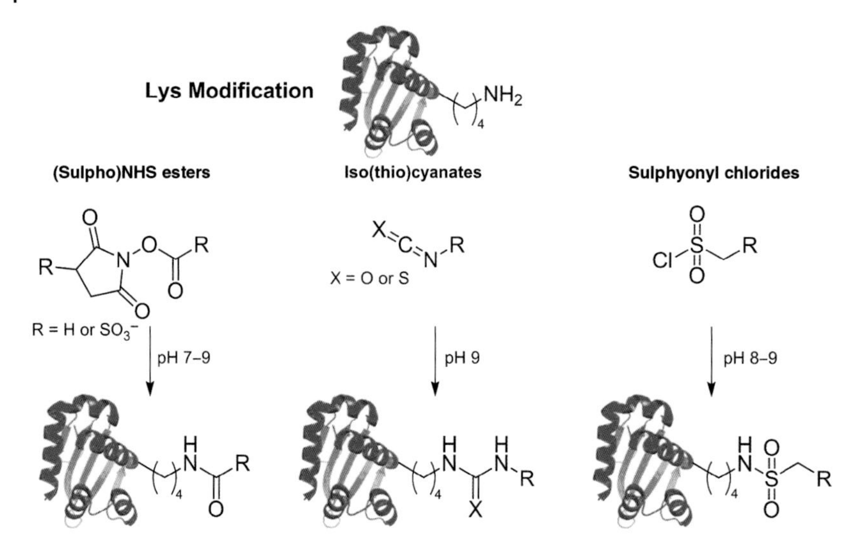

Figure 8.8 Common commercial reagents for selective reaction of Lys on proteins and peptides.

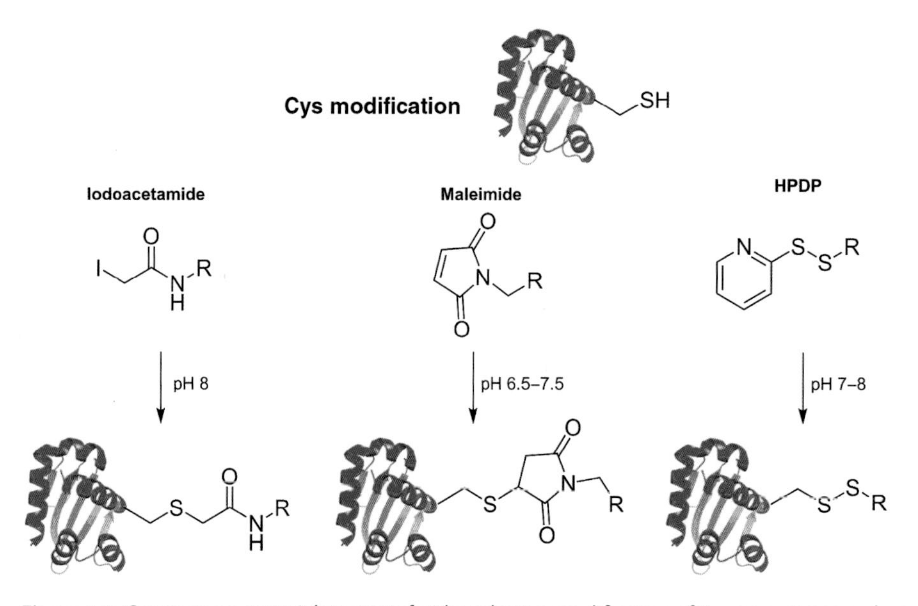

Figure 8.9 Common commercial reagents for the selective modification of Cys on proteins and peptides.

L-cysteine (Cys) is another amino acid that can be targeted for protein bioconjugation. Cys is the most nucleophilic amino acid, especially when deprotonated ($pK_a = 8.5$). The low abundance of Cys in proteins (<2%) can enable protein modification via site-directed mutagenesis, where a point mutation can be introduced to give a recombinant protein with a single Cys at a chosen location. Modification at Cys is normally achieved using an alkylating reagent, Michael addition reaction or via a dehydroalanine (Dha) intermediate. All of these methods show excellent selectivity at carefully controlled pH [22].

Typical procedure (maleimide and iodoacetamide): To protein (50–100 μM) in buffer (100 mM phosphate or Tris, must be free of thiols) at pH 7 (maleimides) or pH 8 (iodoacetamide) is added TCEP (10-fold molar excess) and the solution is incubated in the dark for one hour. The reagent is dissolved in water or DMSO (1–10 mM) and a 10–20-fold molar excess is added. The solution is incubated for one hour at room temperature in the dark or overnight at 4 °C. The reaction can be quenched by addition of excess thiol (e.g. β-mercaptoethanol or glutathione [100 eq.]) and purified by buffer exchange or gel filtration.

Typical procedure (HPDP): Protein must be reduced using TCEP or dithioreitol (DTT) and thoroughly desalted to remove any excess reagent. HPDP reagent is dissolved in DMF to give a concentration of ~4 mM and 100 μl of this is added to the reduced protein (10–50 μM) in buffer (1 ml 1 mM EDTA in phosphate-buffered saline [PBS], pH 7–8), giving a final concentration of HPDP of 0.4 mM. The solution is mixed well and incubated at room temperature for two hours. Excess reagent is removed by desalting or gel filtration.

Most Cys selective reactions are unreactive towards disulphides, which can form between two Cys residues via oxidation in air. For this reason, the first step of any Cys reaction normally involves a reduction step. Reduction is normally achieved through incubation of the protein with DTT or *tris*(2-carboxyethyl)phosphine (TCEP). DTT must normally be removed before the conjugation as it contains free thiols that can react and reduce the effective concentration of the reagent. For this reason, the use of TCEP is recommended for most procedures.

Iodoacetamide has been used to alkylate Cys residues since 1935 and is still used as a technique to 'cap' reactive Cys side chains in proteomic analysis, such as tryptic digestion. Iodoacetamides react rapidly and selectively at pH 8.0 via an S_N2 mechanism.

Maleimides are more reactive than iodoacetamides and react through a Michael-type addition to the maleimide alkene. The highly electrophilic nature of these reagents means that the deprotonation of Cys-SH is not required for the reaction to occur, so these reagents can be used at physiological pH (6.5–7.5). For both reagents the rate of reaction will increase at higher pH, but this is often accompanied by increased side reactions with other amino acids.

HPDP reagents are 'activated' disulphides that react with free thiols at near-neutral pH to give a disulphide linked product. The disulphide bond is cleavable by reduction with DTT or TCEP, so these reagents are favoured for applications where the modification can be later removed.

8.3.4 Dehydroalanine (Dha) Procedure

Dehydroalanine (Dha) can also be used to introduce UAAs and PTM mimics into proteins using a two-step 'tag and modify' approach. The chemistry of Dha is an interesting example of *umpolung* chemistry, where a nucleophilic sulphur atom is eliminated using a reagent such as 2,5-dibromohexanediamide to afford an electrophilic alkene [23] (Figure 8.10). This effectively reverses the electronic properties of the amino acid. Conjugation to Dha is then achieved by nucleophilic addition to the alkene using thiol-based reagents or by means of radical addition using alkyl halides [24].

Although the reactions of Dha are extremely powerful for creating modified proteins, the stereocentre at the α-position of the parent amino acid is epimerised and therefore the products are formed as an approximately 1 : 1 mixture of diastereomers [24]. The stereoablative nature of this reaction can limit the utility of this reaction in some experiments. The Dha procedure is given as follows:

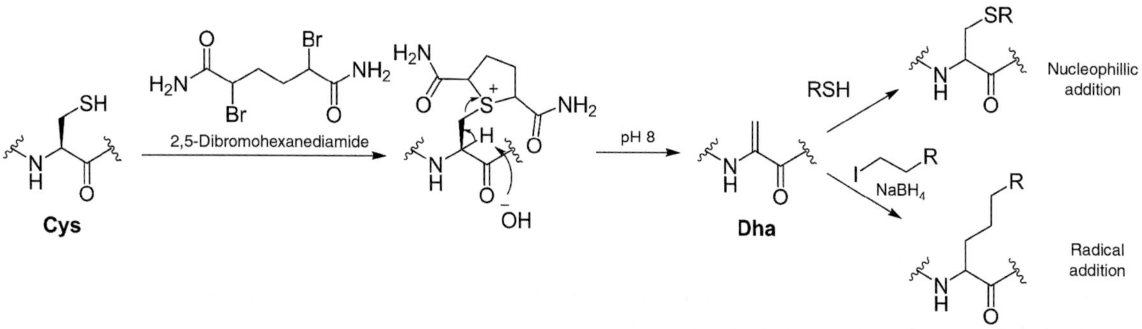

Figure 8.10 Formation of dehydroalanine (Dha) from Cys converts a good nucleophile into an excellent electrophile. Dha can be selectively reacted with nucleophiles, e.g. thiols, or alkyl iodides in a radical addition reaction.

1. A cysteine containing protein (60 μM) in PBS (10 mM; pH 8) is reduced using DTT (300-fold molar excess) by shaking at room temperature for one hour. DTT is removed using size exclusion chromatography or buffer exchange.
2. Dibromohexanediamide (150-fold molar excess) is weighed into an Eppendorf and reduced protein solution is added. The reaction is shaken at 37 °C for four hours or until the reaction is complete by HPLC-MS.
3. Dibromohexanediamide is precipitated by centrifugation (1 minute, 16 k × *g*), giving the Dha product.
4. For nucleophilic addition, a thiol (twofold molar excess) is added to the Dha protein solution, which is incubated at 37 °C with shaking. Additional thiol (twofold molar excess) is added every 20 minutes 10 times. The reaction is incubated for 2–10 hours, with completion determined by HPLC-MS and the product is purified by dialysis of size exclusion chromatography.

8.4 Bioorthogonal Chemistry

All of the techniques mentioned so far in this chapter are limited to the reactions of functional groups present in the 20 canonical amino acids. This has led to the development of methods to specifically modify proteins *in vitro*; however, for *in vivo* modifications achieving selectivity for one protein is extremely difficult. This is largely due to the density and functional complexity of the cell interior, where one must first intercept the correct protein but then also achieve selectivity over other small molecule cofactors, lipids and metabolites.

This challenge has led to the development of the field of bioorthogonal chemistry. First introduced by Bertozzi et al. in 2003, a bioorthogonal reaction can be defined as a non-enzymatic reaction that occurs *in vivo* but yet *neither interacts nor interferes with the underlying biological system.* The rates of these reactions must be fast enough under physiological conditions to enable rapid bioconjugation kinetics and the substrate(s) and product(s) of the reaction must also be non-toxic to the cell [25].

Research in this field has developed rapidly in recent years and it is now possible to perform bioorthogonal reactions in a range of cellular environments at reaction rates approaching that of native enzymes. Despite this success, the selection of the 'right' bioorthogonal reaction for a given application remains a challenging process of elimination. For example, the second-order rate constants of these reactions generally range from 10^{-2} to 10^{5} M^{-1} s^{-1}. For a 1 μM reaction this equates to a half-life ranging from 10 seconds to 3 years. Increasing the rate of reactions is achieved by increasing the reactivity of the reaction components, and this often makes them more difficult to prepare and therefore more expensive. In general, higher levels of reactivity are also accompanied by an increased susceptibility to nucleophilic attack by free thiols in the cell. These liabilities make many of the most sophisticated bioorthogonal chemistries unsuitable for use in multicellular organisms and whole animals (e.g. live cell imaging in nematodes and mice) where factors such as tissue penetration, circulatory and immunological effects are important considerations. Slower reactions use more stable reagents but in order to achieve appreciable reaction kinetics more reagent must be used, which can increase toxicity effects. Navigating the balance between reactivity, deactivation and toxicity is the art of bioorthogonal reaction selection.

By and large, pericyclic reactions have seen the most success in this field. The majority of the reactions developed have focused on the use of azides. Azides are yet to be discovered in nature and possess weak electrophilic properties and a dipolar moment, meaning they are ideal for cyclisation reactions. They are also stable under biological conditions. Due to their small size, azides can also be readily incorporated into cells through metabolic, genetic or activity based pathways. Alkynes are also popular diene- and dipolarophiles due of their appropriate reactivity and compact size.

8.4.1 The Staudinger Ligation

One of the first bioorthogonal reactions to be widely used is the Staudinger ligation reaction between an azide and a phosphine [26]. The mechanism (Figure 8.11) follows the classic Staudinger reduction of azides, where

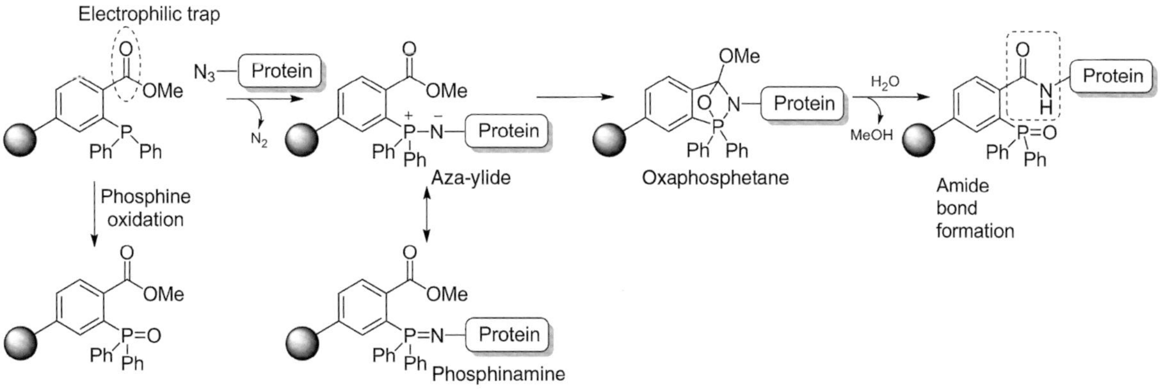

Figure 8.11 Mechanism for Staudinger ligation reaction of azides and phosphines bearing an electrophilic trap. Initial Staudinger reduction of the azide releases nitrogen gas and forms the aza-ylide, which undergoes cyclisation giving the oxaphosphetane which is hydrolysed forming the amide bond.

the azide reacts with a phosphine followed by elimination of N_2. During the reaction the aza-ylide can undergo hydrolysis, yielding the free amine or, through displacement of a proximal ester, intramolecular cyclisation to give the oxaphosphetane, which then hydrolyses to give an amide product with the phosphine oxide attached.

The reaction was modified by Bertozzi et al. and Raines et al. in 2000 to give the traceless Staudinger ligation reaction [27]. In this variant of the classic Staudinger ligation the product does not contain a phosphine oxide, which is instead eliminated during the reaction mechanism instead of methanol. Staudinger ligation reactions have been used extensively to couple fluorophores to biomolecules, but the reaction has been limited by unacceptably slow reaction kinetics [28]. Attempts to increase the rate of the ligation reaction have been made by adding electron donors to the phosphine, but this has been found to also increase the rate of phosphine oxidation (the major side reaction in the Staudinger ligation), leading to poor product conversions.

Typical procedure for Staudinger protein labelling
Azido protein (50–500 µM) is mixed with phosphine probe (10-fold molar excess) in buffer (50–100 mM Tris or PBS; pH 7–8, 6 M guanidine-HCl). Reaction mixtures are incubated for 15 hours at 37 °C and purified by gel chromatography or buffer exchange.

8.5 The Copper-Catalysed Azide-Alkyne Cycloaddition Reaction (CuAAC)

The azide-alkyne 1,3-dipolar cycloaddition reaction is an extremely popular method for bioconjugation. The thermally initiated Huisgen reaction was discovered in 1893 and showed that heating azides and alkynes at high temperatures yielded a mixture of 1,4- and 1,5-triazoles. Despite the reaction being exothermic (i.e. thermodynamically favourable), the activation energy between terminal alkynes and azides is too high for the reaction to occur at room temperature, significantly limiting its potential as a bioorthogonal reaction [29].

Meldal et al. and Sharpless et al. simultaneously discovered that Cu(I) could be used to catalyse the reaction, allowing the reaction to proceed at room temperature; this is now known as the copper-catalysed alkyne-azide cycloaddition (CuAAC) reaction [29]. This extremely powerful reaction gives exclusively 1,4-triazoles and epitomises the concept of 'click' chemistry – reactions that are high yielding, simple to perform, wide in scope, conducted in benign solvents and create by-products that are easily removed. One of the key drivers in the uptake of CuAAC chemistry is that both the azide and alkyne are small, stable and easy to introduce into a wide variety of chemical probes and biomolecules. This has led to the availability of a wide range of commercially available CuAAC reagents, making this reaction a first port-of-call for many applications in chemical biology.

CuAAC reactions are typically performed in water in the presence of an organic solvent (if required) to aid the solubility of reagents. The reactions are sensitive to air since oxygen can readily oxidise Cu(I) to Cu(II), producing reactive oxygen species (ROSs), which are toxic to cells. Sodium ascorbate is a soft reducing agent that is compatible with most proteins and is often used to reduce any Cu(II), giving Cu(I) and dehydroascorbic acid [30].

The mechanism of the CuAAC reaction has been somewhat debated in recent years, but Fokin et al. published the widely accepted mechanism in 2013 (shown in Figure 8.12) which involves two moles of Cu(I) species in the catalytic cycle [31].

Although CuAAC chemistry is orthogonal to many biological reactions and is widely used for coupling to peptides and proteins, it cannot be described as bioorthogonal as the Cu(I) catalyst is extremely toxic to cells. Cu(I) is readily chelated by many proteins, which can alter their structure and function and can also produce

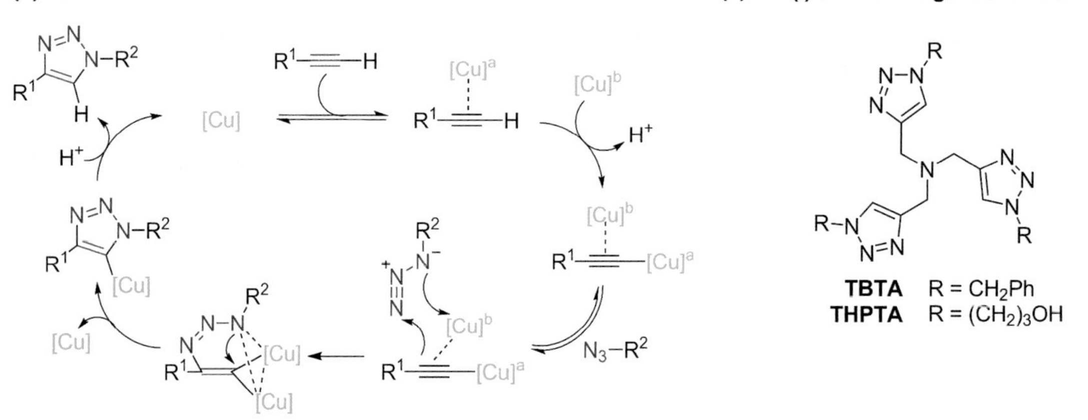

Figure 8.12 The copper-catalysed azide-alkyne cycloaddition (CuAAC) reaction and its mechanism. (a) The thermal Huisgen reaction producing a mix of 1,4- and 1,5-triazoles; (b) CuAAC reaction, which is solvent and functional group tolerant; (c) picolyl azide chelation of copper, which allows for lower use of Cu(I) concentration; (d) mechanism of the CuAAC reaction [31]; (e) common Cu(I) chelating ligands to reduce toxicity, side reactions and increase the rate of reaction [30].

ROSs. Chelating ligands such as TBTA stabilise the Cu(I) and limit interactions with proteins, or Click-iT® Plus reagents chelate Cu(I) giving enhanced rates and lowering the Cu(I) concentration – both solutions have been shown to mitigate toxicity. However, the best solution to this problem is to remove the need for copper entirely. This has been achieved in recent years through the elegant design of new reagents that contain a pre-installed molecular strain.

Typical procedure for chelation assisted CuAAC [32]

1. A biomolecule bearing an azide or alkyne (2–100 μM) is dissolved in degassed PBS buffer (100 mM) and alkyne/azide reagent is added in 2–10-fold molar excess.
2. $CuSO_4$ (2.5 μl; 20 mM in H_2O) and chelating ligand THPTA (5 μl; 50 mM in H_2O) are pre-mixed and added to the protein solution followed by freshly prepared sodium ascorbate (25 μl; 5 mM in H_2O).
3. The solution is mixed, the reaction sealed to prevent oxygen diffusion and left at room temperature for one hour. The reaction can be quenched by addition of excess ethylenediamine tetraacetic acid (EDTA) solution.
4. Copper removal can be achieved by dialysis with EDTA.

8.5.1 The Strain-Promoted Azide-Alkyne Cycloaddition Reaction (SPAAC)

In 1953 it was discovered that cyclooctynes react with azides at room temperature to form triazoles. The reason for this enhanced reactivity over aliphatic alkynes is increased ring strain, which promotes the alkyne-azide cyclisation in the absence of a metal catalyst. For cyclooctyne this has been measured at ~18 kcal/mol compared to cyclooctane at 12.1 kcal/mol [33]. In general, ring strain is inversely proportional to the size of the ring. However, cyclooctyne is the smallest stable cyclic alkyne that is known to exist at ambient temperature.

Bertozzi et al. first demonstrated the potential of cyclooctynes for cell and protein labelling in what is now known as the 'strain promoted azide-alkyne cycloaddition' (SPAAC) reaction, shown in Figure 8.13 [34]. Bertozzi went on to demonstrate that cells decorated with sialic acid azides in their membrane glycoproteins via the metabolic incorporation of peracetylated ManAz could be labelled with a biotin-cyclooctyne reagent and imaged using an avidin bound fluorophore with no toxic effects to the cells [31].

This paper showcased the potential of SPAAC chemistry, but the non-optimised cyclooctyne reagent used showed extremely slow reaction kinetics. This has led to the development of new ultra-fast SPAAC reagents where reactivity has been enhanced by reagent design. Dibenzoazacyclooctynes (DIBACs) and bicyclo-[*6.1.0*]-nonynes (BCNs) (Figure 8.14) are two classes of reagents that are now widely used for SPAAC reactions that are 130 and 58 times more reactive, respectively, than the cyclooctyne used by Bertozzi et al. [35]. However, as mentioned previously, an increase in reaction rate is generally accompanied by decreased stability *in vivo*.

Another limitation of these cyclooctyne reagents is the hydrophobic nature of the cyclooctyne ring, especially the dibenzyl DIBAC compound, which has led to complications in their use in protein labelling. In particular, these reagents have been shown to partition in the hydrophobic environment of the cell membrane, inhibiting their diffusion into the cytosol and making intracellular targets difficult to visualise during labelling experiments. For this reason, even though DIFO and BCN have slower rates of reaction, they are sometimes preferred to DIBAC.

8.5.2 Tetrazine Ligations

Although the reactions of azides for biological labelling have provided selectivity and allowed genetic incorporation of reactive handles, they have become the limiting factor in the search for ever faster reactions. Other reactive groups such as nitrones and syndones have also been shown to react with cyclooctynes, but are also limited by their slow reaction kinetics.

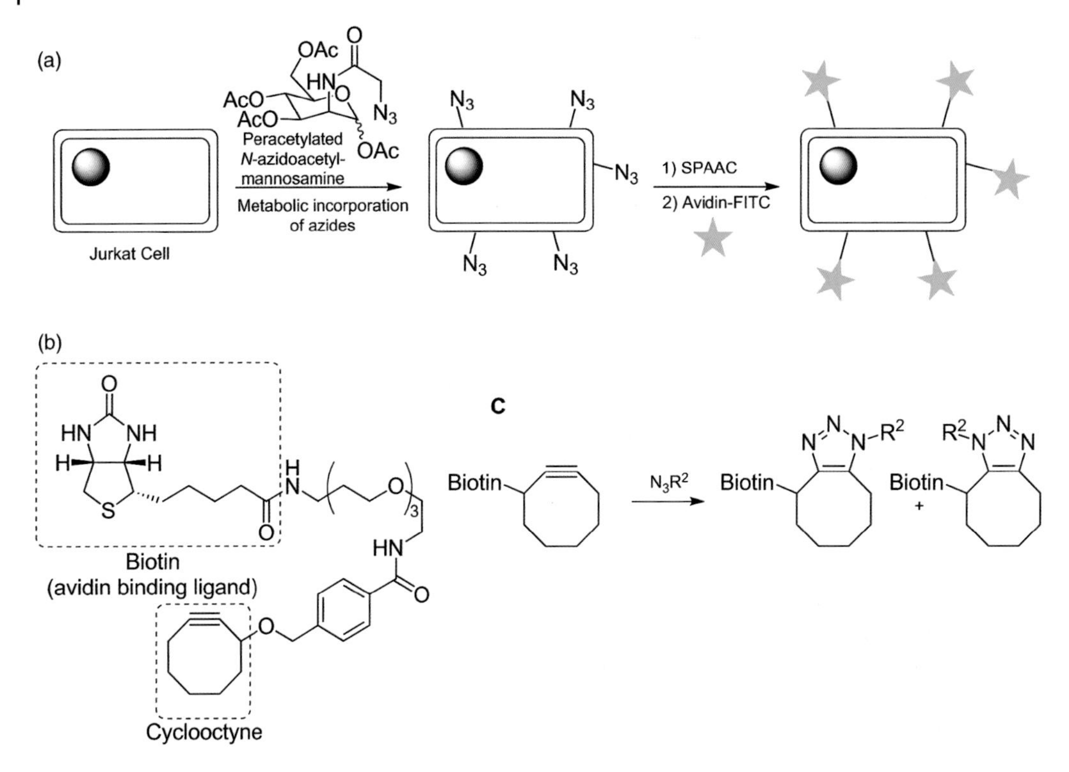

Figure 8.13 Seminal example of SPAAC reaction on live cells. (a) Azides are incorporated into cell surface glycoproteins by incubation of Jurkat cells with peractylated *N*-azidoacetylmannosamine. SPAAC reaction with a biotin cyclooctyne probe (b) conjugates the biotin probe through a 1,2,3-triazole (c). Addition of fluorescently tagged Avidin binds to the biotin molecule, fixing the fluorophore to the cell surface [34].

	Cyclooctyne	DIFO	BCN	DIBAC
Rate of reaction ($\times 10^{-3}$ M^{-1} s^{-1})	2.4[a]	76[a]	140[b]	310[b]

Figure 8.14 Bertozzi et al.'s cyclooctyne compound and a selection of commercial cyclooctynes and their rate of reaction with BnN_3. Conditions: (a) CD_3CN; (b) CD_3CN:D_2O (3 : 1) [35].

However, 1,2,4,5-tetrazines also react with strained alkynes and alkenes via an inverse electron demand Diels-Alder (IEDDA) cycloaddition (Figure 8.15a) and these typically have much faster rates than any other bioorthogonal reaction. Cycloadditions with BCNK range between 3 and 1245 M^{-1} s^{-1}, 10–100 times faster than the DIBAC SPAAC reaction with azides [36] (Figure 8.15b). Reactions with transcyclooctene (TCO) derivatives are truly exceptional, with second-order rate constants reported up to 10^6 M^{-1} s^{-1}. The rate of cycloaddition is dependent on a number of factors, including the electronic nature of the substituents on the tetrazine, the ring strain of the dienophile, sterics and solvent choice (Figure 8.15) [37].

Once again, the high reactivity of *trans*-cyclooctene is accompanied by high levels of instability and short half-lives in biological serum (3.26 hours for TCO). Although the exact mechanism of its deactivation has yet

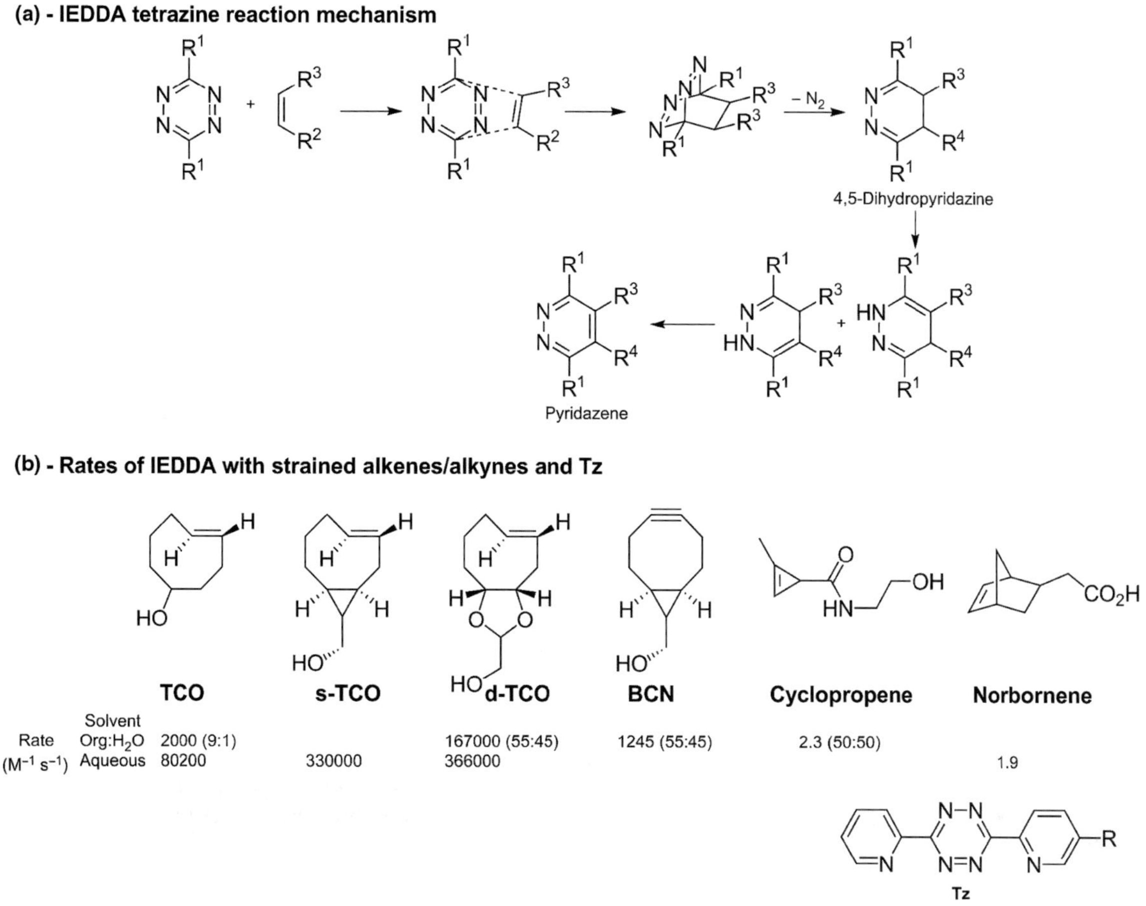

(a) - IEDDA tetrazine reaction mechanism

4,5-Dihydropyridazine

Pyridazene

(b) - Rates of IEDDA with strained alkenes/alkynes and Tz

Rate ($M^{-1} s^{-1}$)	Solvent	TCO	s-TCO	d-TCO	BCN	Cyclopropene	Norbornene
	$Org:H_2O$	2000 (9:1)		167000 (55:45)	1245 (55:45)	2.3 (50:50)	
	Aqueous	80200	330000	366000			1.9

Tz

Figure 8.15 The IEDDA reaction of tetrazines. (a) Mechanism of the tetrazine IEDDA reaction. (b) Rates of reaction for selected alkynes and alkenes with tetrazine Tz [37].

to be confirmed, it has been shown that TCO readily isomerises to the less reactive *cis*-cyclooctene *in vivo*. Further increasing the ring strain of TCO by fusing a cyclopropane (s-TCO) or dioxolane (d-TCO) increases the rate of reaction even more and, in the case of d-TCO, also increases the stability of this reagent in serum for up to four days. For biological labelling studies in more complex environments, cyclopropenes have recently emerged as suitable dienophiles for the IEDDA reaction with tetrazines. Despite their slower reaction kinetics these reagents perform well in whole animals, while also retaining their bioorthogonal properties.

Overall, the exceptional properties of the IEDDA reaction means that this bioorthogonal reaction is the state-of-the art in the field and has found widespread use in the field of chemical biology, especially in situations where high labelling efficiencies and low reagent toxicity *in vivo* are of paramount importance.

8.6 Unnatural Amino Acid Incorporation

One of the major advancements that has allowed the concept of bioorthogonal chemistry to flourish is the ability to genetically incorporate unnatural functional groups into living cells. There are three main methods that are generally used to achieve this: metabolic incorporation and incorporation of UAAs through selective pressure incorporation in auxotrophic cells, or by genetic code expansion using orthogonal translation.

Incorporation of UAAs can be achieved using amino acids that are similar in structure to natural amino acids. Using this approach, incorporation can simply be achieved via the addition of the UAA to the culture medium and relies on the promiscuity of the cognate aminoacyl-tRNA synthetase. However, the efficiency of this system is greatly improved by making the cells auxotrophic for one or more amino acids, as the activation of UAAs by aa-tRNAs is much slower than for natural amino acids.

This was first demonstrated using selenomethionine (Se-Met) for use in X-ray crystallography due to its increased diffraction properties relative to L-Met [38]. This technique has been predominantly used in recombinant protein expression, where cells are first grown in the presence of all 20 natural amino acids and then transferred into media containing 19 amino acids and the UAA before protein expression is induced. However, the survival rate of auxotrophic cells cultured exclusively in the presence of UAA is typically very low, as the proteome-wide incorporation of the UAA at high concentration can lead to toxicity effects. For this reason, labelling efficiencies are generally very low using this approach.

The substrate scope has been improved by targeting the inherent promiscuity of the methionyl-tRNA synthetase (Met-RS), which has allowed the metabolic incorporation of azidohomoalanine (Aha), homopropargylglycine (Hpg) and homoallylglycine (Hag) UAAs (Figure 8.16). These UAAs have been applied in a range of bioconjugation experiments. The metabolic incorporation of unnatural Trp, Phe and Leu-derived amino acids have also been achieved, but this has required a degree of protein engineering to alter the natural specificity of the respective tRNA synthetase (Figure 8.16) [39].

Perhaps the most elegant approach to UAA incorporation is via genetic code expansion. In this approach, the amber STOP codon (TAG) is repurposed to direct the incorporation of a UAA using an amber-decoding tRNA (CUA) and an orthogonal aminoacyl tRNA synthetase (aaRS) (Figure 8.17). Identification of a suitable orthogonal aaRS can be achieved from non-common amino acid use in microorganisms (e.g. the pyrrolysyl aaRS from *Methanosarcina mazei*) or via directed evolution. Suppression of a target amber codon via this approach therefore allows for the site-specific incorporation of a UAA to any residue in a protein in a genetically directed manner. Although protein yields in these experiments are often lower than via more traditional methods, this technique has undeniable value. Since its initial development in *E. coli*, this technology has since been expanded for use in a variety of microorganisms, mammalian cells, and more recently in whole animals [40].

While expansion of the genetic code has allowed UAAs to be incorporated into proteinaceous material in a variety of organisms, the stochastic labelling of other cellular structures (e.g. oligosaccharides and fatty acids)

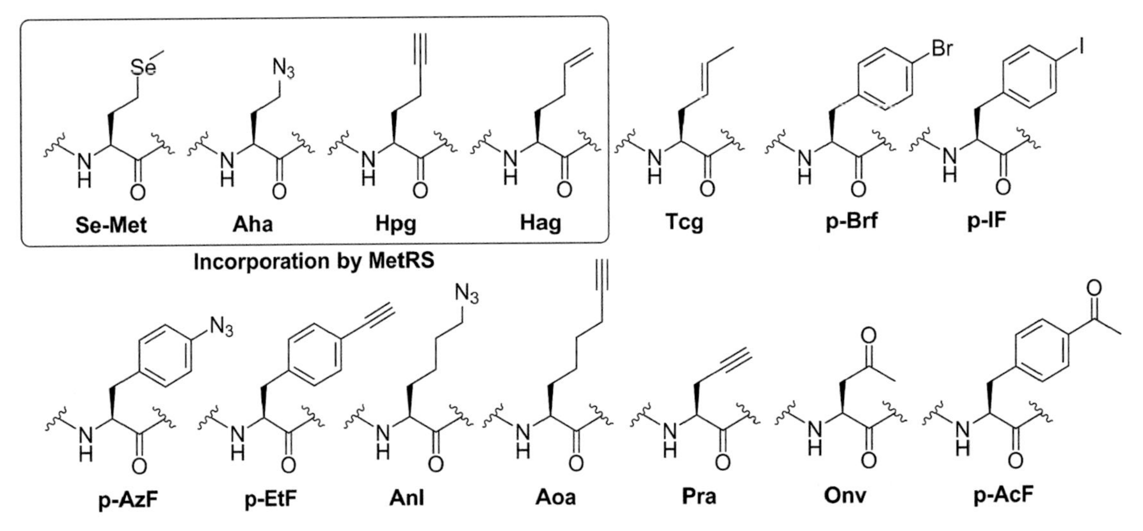

Figure 8.16 UAAs that can be incorporated using selective pressure incorporation. Se-Met, Aha, Hpg, and Hag can all be incorporated using native MetRS [39].

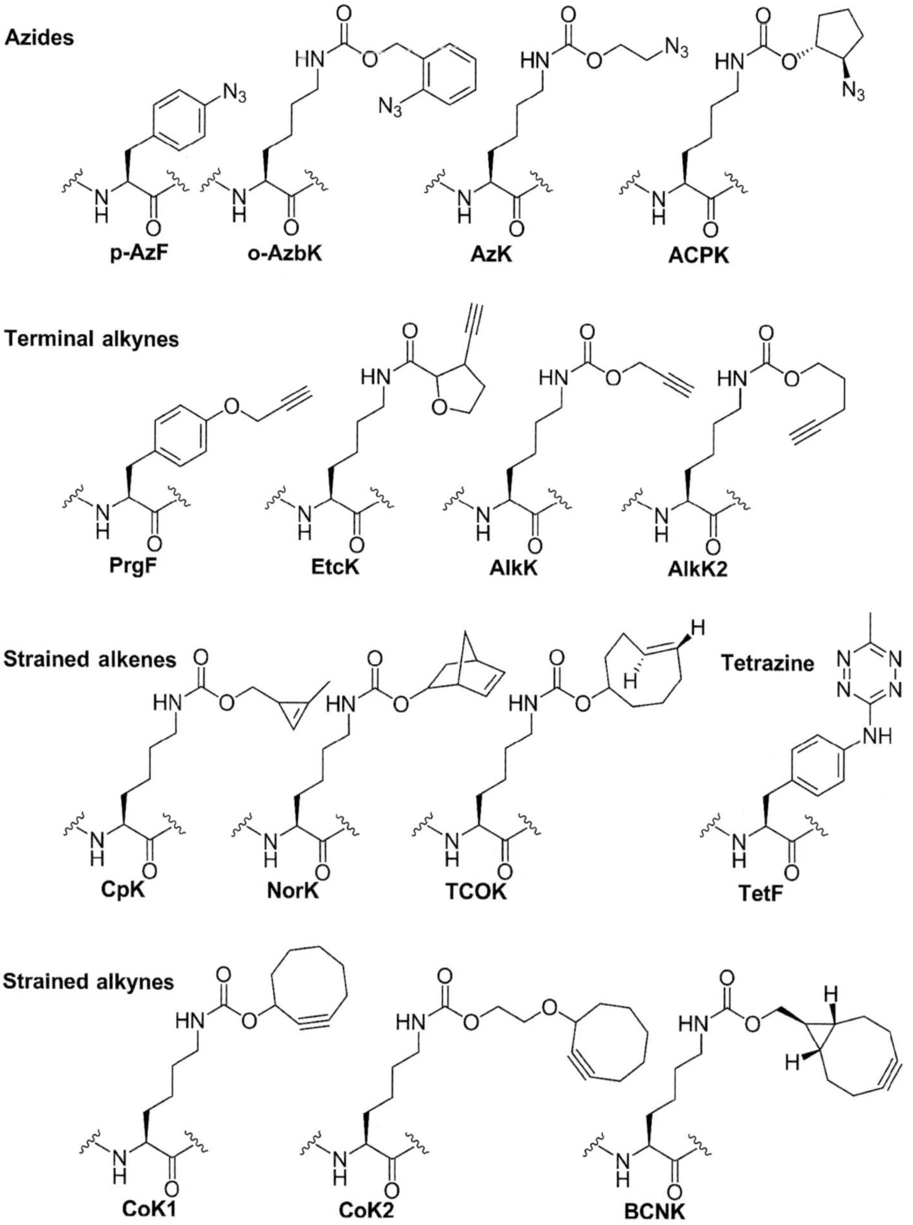

Azides

p-AzF o-AzbK AzK ACPK

Terminal alkynes

PrgF EtcK AlkK AlkK2

Strained alkenes

CpK NorK TCOK

Tetrazine

TetF

Strained alkynes

CoK1 CoK2 BCNK

Figure 8.17 Unnatural amino acids that can be introduced into proteins through: (a) selection pressure incorporation or (b) amber stop codon techniques [37, 39].

requires a different approach. Metabolic incorporation of metabolite mimics bearing unnatural functional groups has been shown to be an excellent way to achieve this. By incubating a cell in the presence of an unnatural, yet structurally related, metabolite these structures can be incorporated into the cell metabolome via native metabolic incorporation.

Cell-surface glycans were some of the first substrates to be the target of this approach. Bertozzi et al. first demonstrated the incorporation azido sialic acid (Sia) by incubation of Jurkat and HeLa cells with peracetylated *N*-azidoacetylmannosamine (ManAz; Figure 8.18 and Figure 8.13) through a de novo biosynthesis

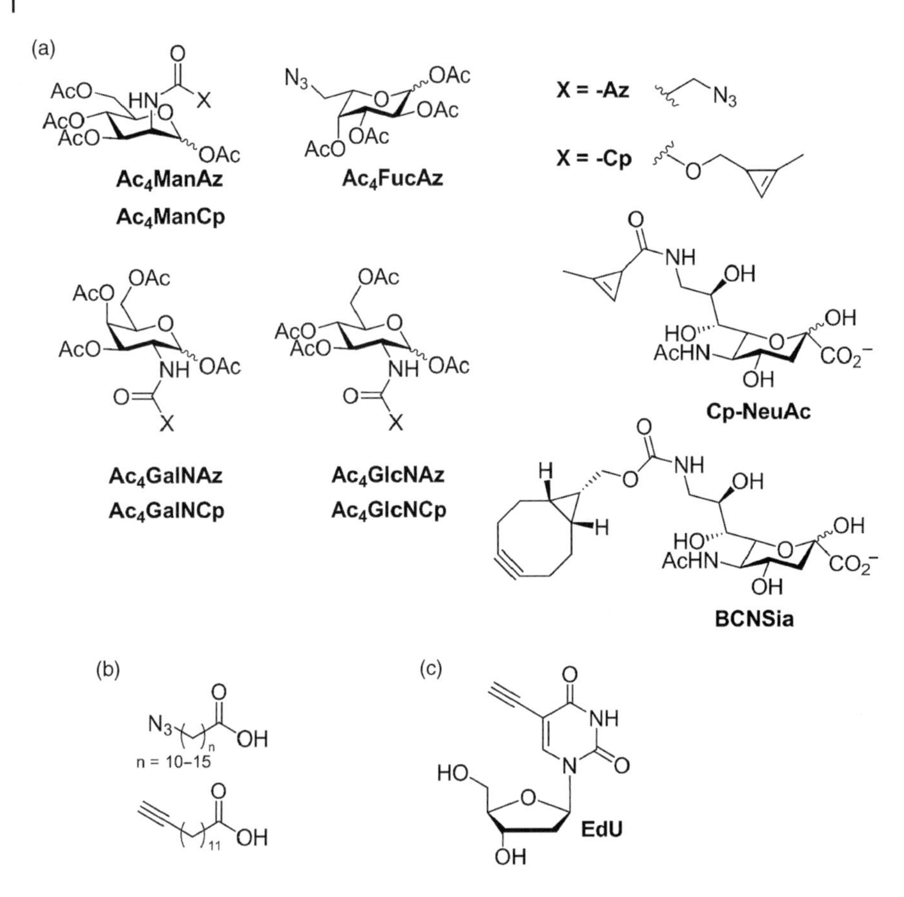

Figure 8.18 Unnatural metabolites for (a) incorporation of azido and cyclopropyl groups into carbohydrates and incorporation of BCN into sialic acid; (b) incorporation of azide and alkyne groups into fatty acids; (c) incorporation of an alkyne into DNA via EdU.

pathway [23]. Further work has since incorporated azido derivatives of *N*-acetylglucosamine (GlcNAc) and *N*-acetyl galactosamine (GalNAc), GlcNAz and GalNAz, respectively, using an appropriate salvage pathway in mammalian cells. Prescher et al. recently expanded the versatility of this approach by incorporation of various cyclopropene labelled glycans for IEDDA coupling [41]. Surprisingly, Bertozzi et al. have shown that larger groups such as BCN could also be incorporated using this approach [42].

Finally, lipid biosynthesis pathways and nucleic acids have also been targeted for incorporation of unnatural functional groups. Myristoylation and palmitoylation have been studied through the metabolic incorporation of azide and alkyne-modified fatty acids, and DNA has also been labelled via the metabolic incorporation of 5-ethynyl-2′-deoxyuridine (EdU) in cells.

8.7 Case Studies

8.7.1 Post-Translational Modification of Histone Tails

Histones are the proteins responsible for packaging DNA inside the nucleus of eukaryotic cells. The histone is composed of four core components called H2A, H2B, H3 and H4, which form the histone octamer, the major repeating unit of chromatin. PTMs are covalent modifications to the side chains of amino acids that alter the chemical properties. There are around 20 confirmed PTMs in histones focused on Lys residues, although Ser, Thr, Arg, His and Glu are also modified.

Histone PTMs are thought to act by altering interactions between amino acid side chains on histones and DNA. ε-NH_2 Lys acetylation is one of the most commonly observed histone PTMs and can be installed on histones by the enzyme histone acetyltransferase (HAT) or removed by histone deacetylases (HDACs). Acetylation converts a positively charged amine residue (the ε-NH_2 of Lys is protonated at the physiological pH) to a neutral amide. This change in charge state is thought to reduce the interaction between DNA and the histone, causing the DNA to be coiled less tightly and leaving it more available for transcription.

The ability to synthesise synthetic or semi-synthetic histones bearing PTMs at specific sites has greatly facilitated understanding the effect PTMs have and whether PTMs at specific sites act as markers for protein–protein interactions.

This section will compare three methods to generate histone PTMs: through alkylation of cysteine residues, use of NCL to create synthetic proteins and through amber stop codon techniques.

8.7.1.1 Cysteine Alkylation

ThiaLys (shown in Figure 8.19) is a lysine mimic obtained by alkylation of cysteine using 2-bromoethylamine. Substitution of a carbon atom for a thiol causes a slight bond lengthening (+0.28 Å) and shift in pK_a (−1.1) compared to Lys, but the mimic is recognised by the lysine specific protease trypsin. Shokat et al. employed a cysteine alkylation strategy to synthesise lysine methylation mimics in histone proteins [43]. Histones are the proteins that package DNA and modification of Lys residues in histones are thought to control DNA transcription, replication and repair.

A point mutation of Lys to Cys allowed introduction of a single Cys residue into the histones, which were expressed recombinantly in *E. coli*. Incubation of the histones with bromoethylamines bearing a methylation mark gave synthetic methyl lysine products that have facilitated the study of the role Lys methylation plays. A similar procedure has been used successfully to generate dimethyl and trimethyl Lys mimics [43].

Procedure: Lyophilised histones (5–10 mg) were dissolved in alkylation buffer (900 μl, 1 M HEPES pH 7.8, 4 M guanidinium chloride, 10 mM D/L-methionine). DTT (20 μl, 1 M) was added to the solution and incubated at 37 °C for one hour. Bromoethylamine (100 μl, 1 M in H_2O) was added and incubated for two hours in the dark at room temperature. DTT (10 μl, 1 M in H_2O) was added followed by bromoethylamine (50 μl, 1 M in

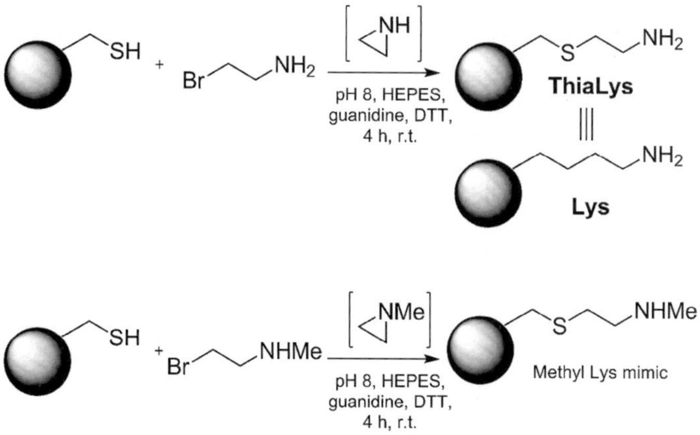

SH + Br NH2 [NH] pH 8, HEPES, guanidine, DTT, 4 h, r.t. → S NH2 ThiaLys ≡ NH2 Lys

SH + Br NHMe [NMe] pH 8, HEPES, guanidine, DTT, 4 h, r.t. → S NHMe Methyl Lys mimic

Figure 8.19 Alkylation of Cys to give ThiaLys, a lysine mimic, using 2-bromoethyl amine, which reacts with Cys via an aziridine intermediate. By using (methyl)-2-bromoethylamine methyl Lys mimics were synthesised, facilitating the study into the role of Lys methylation has in histone–DNA interactions.

H_2O) and the reaction incubated for a further two hours at room temperature. The reaction was quenched by addition of β-mercaptoethanol (50 μl) and purified using a PD-10 size exclusion column and lyophilised.

8.7.1.2 Native Chemical Ligation (NCL)

NCL and EPL are extremely powerful tools to synthesise modified histones as most histone modifications are found at the *C*- or *N*-terminus. For NCL, sequential ligation and *N*-terminal deprotection strategies have been used to introduce histone PTMs. For the synthesis of histone H2A containing a phosphonated tyrosine, Brik et al. used the method shown in Figure 8.20 [44]. Three peptides were prepared using Fmoc SPPS and the sequences were designed so that each ligation site was at the position of an Ala residue in the native sequence. By doing this the Cys left after ligation could be desulphurised to Ala, leaving the native sequence. The *N*-terminal of peptide H2A(49-86) was protected as a thiazolidine, which can be removed in the presence of $[Pd(allyl)Cl]_2$ and $MgCl_2$. The *C*-terminus was activated using an *N*-acyl-benzimidazolinone group to increase the rate of ligation.

Procedure: Ligation of fragments H2A(49-86) and H2A(88-130) was achieved in the presence of 20 eq. 4-mercaptophenylacetic acid (MPAA; a catalyst) and 10 eq. of TCEP at pH 7.3 for three hours. Thiazolidine deprotection was achieved using 100 eq. $MgCl_2$ and 15 eq. $[Pd(allyl)Cl]_2$ at 37 °C in one hour. Addition of

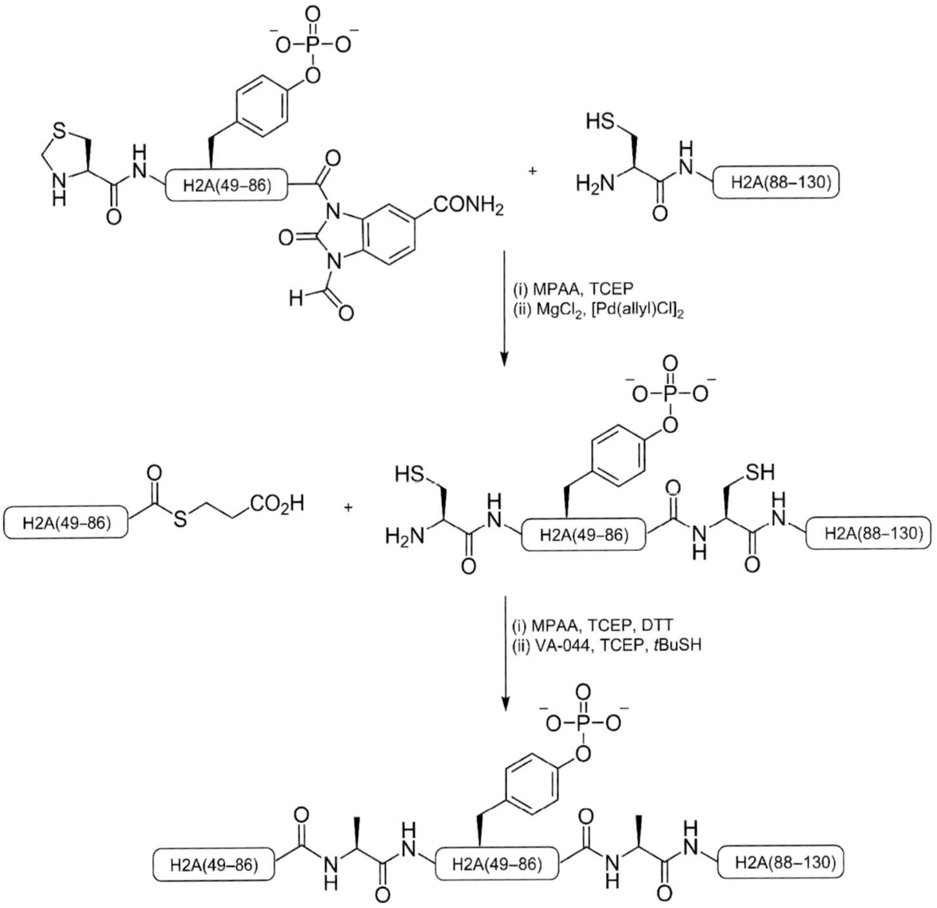

Figure 8.20 One pot sequential NCL reaction, followed by desulphurisation to give a synthetic histone H2AY57p from three peptide fragments prepared using Fmoc SPPS [44].

the third peptide fragment H2A(1-48) with 75 eq. MPAA and 37.5 eq. TCEP gave complete ligation after a further seven hours. The reaction was quenched with 75 eq. DTT and dialysed overnight. Desulphurization using 100 eq. VA-044 and TCEP for six hours gave the product, which was purified by RP-HPLC, giving the protein in a 16% yield for the four steps. For a detailed protocol see Brik et al. [44].

8.7.1.3 Genetically Encoded Methyl-Lys Incorporation

The incorporation of an UAA using amber stop codon techniques relies on having a tRNA-synthetase/$tRNA_{CUA}$ pair that are specific for the UAA. A major challenge for the incorporation of Me-Lys is that although a synthetase that could accept Me-Lys is possible, achieving discrimination between Me-Lys and Lys is difficult since Lys is smaller in size and similarly charged. To circumvent this issue, Chin et al. developed a two-step method to incorporate Me-Lys into histones (Figure 8.21). N^{ε}-tert-butyloxycarbonyl-L-lysine has been previously shown to be incorporated into myoglobin in *E. coli* using an evolved mutant of the pyrrolysyl-tRNA synthetase/$tRNA_{CUA}$ pair [44]. Using this technique Chin et al. were able to incorporate N^{ε}-tert-butyloxycarbonyl-N^{ε}-methyl-L-lysine (BocMeLys) into an His-tagged H3K9BocMe, which was purified using an Ni-affinity column. Boc groups are acid labile and treatment of the protein with 1% TFA for 4 hours at 37 °C gave quantitative removal or 'deprotection' of the Boc protecting group yielding H3K9me1, which was confirmed by electrospray ionisation mass spectrometry (ESI-MS) and western blot using an anti-H3K9me1 antibody.

Procedure: Expression of Histone H3K9Boc-me1 was achieved by transformation of *E. coli* Bl21(DE3) cells with pBKPylS (which encodes the *Methanosarcina berkeri* pyrrolysyl-tRNA synthetase, *Mb*PylRS) and pCDF-PylT-H3K9TAG (which encodes histone H3 bearing an amber codon at position 9 and an *N*-terminal His6-tag followed by a TEV protease cleavage site sequence, as well as $Mb\text{tRNA}_{CUA}$ on an *lpp* promoter and *rrnC* terminator, where the plasmid has a spectinomycin resistance marker). Cells were recovered in 1 ml of SOC media for one hour at 37 °C, before incubation (16 hours, 37 °C, 250 rpm) in 100 ml of 2 × TY containing kanamycin (50 μg/ml) and spectinomycin (70 μg/ml); 25 ml of this overnight culture was used to inoculate 500 ml of 2 × TY supplemented with kanamycin (25 μg/ml), spectinomycin (35 μg/ml) and 2 mM of BocMeLys. Cells were grown (37 °C, 250 rpm) and protein expression was induced at OD600 ~0.9 by addition of isopropyl β-D-1-thiogalactopyranoside (IPTG) to a final concentration of 1 mM. After five hours of induction, cells were harvested and resuspended in 50 ml of 1 × PBS containing 1 mM DTT, lysozyme (1 mg/ml), DNaseI (100 μg/ml), 1 mM PMSF and Roche protease inhibitor cocktail. The cells were disrupted by sonication. The cell lysates were centrifuged at 17 000 rpm for 20 minutes at 4 °C. The supernatant was discarded and the pellet was retained as the insoluble fraction. The pellet was resuspended in 25 ml of 1 × PBS supplemented with 1 mM DTT and 1% Triton-X, and centrifuged at 17 000 rpm for 20 minutes at 4 °C. The pellet was subsequently resuspended in 25 ml of 1 × PBS containing 1 mM DTT and centrifuged at 17 000 rpm for 20 minutes at 4 °C. The insoluble fraction was incubated

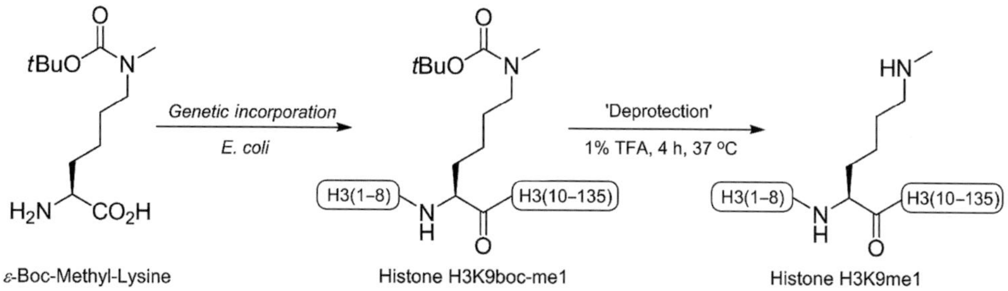

Figure 8.21 Genetic incorporation of a Boc protected methyl lysine analogue using the amber stop codon technique. After protein expression and purification, the Boc group is removed using a low concentration of strong acid, giving native H3K9me1 [45].

in 350 μl of DMSO for 30 minutes at room temperature and dissolved in 25 ml of 20 mM Tris-HCl buffer (pH 8.0) containing 6 M guanidinium hydrochloride and 1 mM DTT. The solution was incubated with vigorous shaking at 37 °C for one hour and centrifuged at 17 000 rpm for 20 minutes at 4 °C. The supernatant was equilibrated with 1 ml of 50% Ni-NTA beads (Qiagen) for one hour at room temperature. The beads were collected by centrifugation at 2400 rpm for five minutes. The beads were washed with 15 ml of 100 mM sodium phosphate buffer (pH 6.2) containing 8 M urea and 1 M DTT. The protein was eluted with 20 mM sodium acetate buffer (pH 4.5) supplemented with 7 M urea, 200 mM NaCl and 1 mM DTT in 500 μl fractions. The fractions of the purified proteins were analysed by 4–12% sodium dodecyl sulphate-polyacrylamide gel electrophoresis (SDS-PAGE). The protein-containing fractions were combined, dialysed overnight in 1 mM DTT solution and stored at −20 °C. Heterochromatin protein 1 homologue beta (HP1b) from mouse, cloned into pET-16 (Novagen) expression vector, was expressed in *E. coli* C41(DE3) and purified by Ni affinity, anion exchange chromatography and gel filtration. For the preparation of monomethylated histones, the protein H3K9boc-me1 (40 nmol) was incubated with shaking (800 rpm) in 1 ml of 1% TFA for four hours at 37 °C to produce H3K9me1. The protein was rebuffered to 1 mM DTT (1.5 ml) using a sephadex G25 column. The hexahistidine tag was removed by incubating with TEV protease (1.5 mg/ml, 100 μl) in 50 mM Tris buffer (pH 7.4) for five hours at 30 °C and overnight dialysis in 1 mM DTT.

8.8 Conclusion

This case study has showcased three different techniques that can be used to produce a histone containing a PTM. The alkylation method has the benefit of being the simplest of the three methods to perform, and it can be easily performed in conjunction with the introduction of a Cys residue via site-directed mutagenesis. These can be alkylated using commercial reagents without the requirement for specialist chemical handling equipment and knowledge. It also provides a method to generate three different states of Lys methylation. The modification is a mimic of Me-Lys, however, and the true impact this can have on results generated from these compounds can vary for each individual application.

UAA incorporation is an extremely powerful technique and, in this case, produced a 'native' H39Kme1 protein, although it should be noted that the strong acid required to remove the Boc group would not be tolerated by all proteins. This method allows for the Me-Lys to be expressed at any point in the sequence, but no examples yet exist to introduce dimethyl- or trimethyl-Lys residues using this technique.

NCL is ultimately the most flexible technique of the three as almost any PTM can be incorporated into any position. As this technique develops, ligation reactions are becoming higher yielding and faster, and new techniques in SPPS have allowed milder Fmoc strategies to be used that have avoided the use of extremely toxic hydrofluoric acid-mediated deprotection steps. However, each sequence is different and this technique can require extensive optimisation of reaction conditions to achieve acceptable results.

8.8.1 Overall Conclusion

Overall, this chapter explores a variety of modern methodologies that have been developed in recent years to synthesise, modify and intercept polypeptides and proteins *in vitro* and in whole organisms. Through years of innovation in the fields of synthetic organic chemistry, chemical biology and synthetic biology, bioconjugation methods now exist that enable the learned practitioner to deploy a suitable methodology to almost any given experimental question. With these tools available and now well understood, the field of chemical biology looks set to remain a powerful approach to exploring the biological sciences.

References

1 (a) Li, Y. (2011). *Protein Expression Purif.* 80: 260–267. (b) Palm, C., Jayamanne, M., Kjellander, M., and Hällbrink, M. (2007). *Biochim. Biophys. Acta Biomembr.* 1768: 1769–1766.
2 Marcia de Figueiredo, R., Suppo, J.-S., and Campagne, J.-M. (2016). *Chem. Rev.* 116: 12029–12122.
3 Merrifield, R.B. (1963). *J. Am. Chem. Soc.* 85: 2149–2154.
4 (a) Behrendt, R., White, P., and Offer, J. (2016). *J. Pept. Sci.* 22: 4–27. (b) Paradis-Bas, M., Tulla-Puche, J., and Albericio, F. (2016). *Chem. Soc. Rev.* 45: 631–654. (c) Palomo, J.M. (2014). *RSC Adv.* 4: 32658–32672. (d) Mäde, V., Els-Heindl, S., and Beck-Sickinger, A.G. (2014). *Beilstein J. Org. Chem.* 10: 1197–1212.
5 Isidro-Llobet, A., Álvarez, M., and Albericio, F. (2009). *Chem. Rev.* 109: 2455–2504.
6 El-Faham, A. and Albericio, F. (2011). *Chem. Rev.* 111: 6557–6602.
7 Bray, B.L. (2003). *Nat. Rev. Drug Discovery* 2: 587–593.
8 Guzman, F., Barberis, S., and Illanes, A. (2007). *Electron. J. Biotechnol.* 10 (2, 279–314.).
9 Gutte, B. and Merrifield, R.B. (1971). *J. Biol. Chem.* 246: 1922–1941.
10 Dawson, P.E., Muir, T.W., Clark-Lewis, I., and Kent, S. (1994). *Science* 266: 776–779.
11 Shah, N.H. and Muir, T.W. (2014). *Chem. Sci.* 5: 446–461.
12 Malins, L.R. and Payne, R.J. (2014). *Curr. Opin. Chem. Biol.* 22: 70–78.
13 Thompson, R.E., Liu, X., Alonso-Garcia, N. et al. (2014). *J. Am. Chem. Soc.* 136: 8161–8164.
14 Milton, R.C. and Kent, S.B. (1992). *Science* 256: 1445–1448.
15 Unverzagt, C. and Kajihara, Y. (2013). *Chem. Soc. Rev.* 42: 4408–4420.
16 Kumar, K.S.A., Bavikar, S.N., Spasser, L. et al. (2011). *Angew. Chem. Int. Ed.* 50: 6137–6141.
17 Muir, T.W., Sondhi, D., and Cole, P.A. (1998). *Proc. Natl Acad. Sci. USA* 95: 6705–6710.
18 Berrade, L. and Camarero, J.A. (2009). *Cell. Mol. Life Sci.* 66: 3909–3922.
19 Koniev, O. and Wagner, A. (2015). *Chem. Soc. Rev.* 44: 5495–5551.
20 deGruyter, J.N., Malins, L.R., and Baran, P.S. (2017). *Biochemistry* 56: 3863–3873.
21 Chen, D., Disotaur, M.M., Xiong, C. et al. (2017). *Chem. Sci.* 8: 2717–2722.
22 Chalker, J.M., Bernardes, G.J.L., Lin, Y.A., and Davis, B.G. (2009). *Chem. Asian J.* 4: 630–640.
23 Chalker, J.M., Gunnoo, S.B., Boutureira, O. et al. (2011). *Chem. Sci.* 2: 1666–1676.
24 Wright, T.H., Bower, B.J., Chalker, J.M. et al. (2016). *Science* 354: aag1465.
25 Sletten, E.M. and Bertozzi, C.R. (2009). *Angew. Chem. Int. Ed.* 48: 6974–6998.
26 Saxon, E. and Bertozzi, C.R. (2000). *Science* 287: 2007–2010.
27 (a) Saxon, E., Armstrong, J.I., and Bertozzi, C.R. (2000). *Org. Lett.* 2: 2141–2143. (b) Nilsson, B.L., Kiessling, L.L., and Raines, R.T. (2000). *Org. Lett.* 2: 1939–1941.
28 Schilling, C.I., Jung, N., Biskup, M. et al. (2011). *Chem. Soc. Rev.* 40: 4840–4871.
29 Meldal, M. and Tornøe, C.W. (2008). *Chem. Rev.* 108: 2952–3015.
30 McKay, C.S. and Finn, M.G. (2014). *Chem. Biol.* 21: 1075.
31 Worrell, B.T., Malik, J.A., and Fokin, V.V. (2013). *Science* 340: 457–460.
32 Presloski, S.I., Hong, V.P., and Finn, M.G. (2011). *Curr. Protoc. Chem. Biol.* 3: 153–162.
33 Bach, R.D. (2009). *J. Am. Chem. Soc.* 131: 5233–5243.
34 Agard, N.J., Prescher, J.A., and Bertozzi, C.R. (2004). *J. Am. Chem. Soc.* 126: 15046–15047.
35 Dommerholt, J., Rutjes, F.P.J.T., and van Delft, F.L. (2016). *Top. Curr. Chem.* 374: 16.
36 (a) Chen, W., Wang, D., Dai, C. et al. (2012). *Chem. Commun.* 48: 1736–1738. (b) Lang, K., Davis, L., Wallace, S. et al. (2012). *J. Am. Chem. Soc.* 134: 10317–10320.
37 Oliveira, B.L., Guoa, Z., and Bernardes, G.J.L. (2017). *Chem. Soc. Rev.* 46: 4895–4950.
38 Barton, W.A., Tzvetkova-Robev, D., Erdjument-Bromage, H. et al. (2009). *Protein Sci.* 15: 2008–2013.
39 Lang, K. and Chin, J.W. (2014). *Chem. Rev.* 114: 4764–4806.

40 (a) Schmied, W.H., Elsässer, S.J., Uttamapinant, C., and Chin, J.W. (2014). *J. Am. Chem. Soc.* 136: 15577–15583. (b) Davis, L. and Greiss, S. (2018). Genetic Encoding of Unnatural Amino Acids in *C. elegans*. In: *Noncanonical Amino Acids*, Methods in Molecular Biology, vol. 1728 (ed. E. Lemke). New York, NY: Humana Press.
41 Patterson, D.M., Nazarova, L.A., Xie, B. et al. (2012). *J. Am. Chem. Soc.* 134 (45): 18638–18643.
42 Agarwal, P., Beahm, B.J., Shieh, P., and Bertozzi, C.R. (2015). *Angew. Chem. Int. Ed. Engl* 54 (39): 11504–11510.
43 Simon, M.D., Chu, F., Racki, L.R. et al. (2007). *Cell* 128: 1003–1012.
44 Maity, S.K., Jbara, M., Mann, G. et al. (2017). *Nat. Protoc.* 12: 2293–2322.
45 Nguyen, D.P., Garcia Alai, M.M., Kapadnis, P.B. et al. (2009). *J. Am. Chem. Soc.* 131: 14194–14195.

Further Reading

Algar, W.R., Dawson, P., and Medintz, I.L. (2017). *Chemoselective and Bioorthogonal Ligation Reactions: Concepts and Applications*. Wiley VCH.
Dobson, C.M., Gerrard, J.A., and Pratt, A.J. (2002). *Foundations of Chemical Biology*. USA: Oxford University Press.
Hermanson, G.T. (2013). *Bioconjugate Techniques*. Academic Press.
Jones, J. (2002). *Amino Acid and Peptide Synthesis*. USA: Oxford University Press.

9

Reaction Chemical Kinetics in Biology

Nicholas J. Harmer[1] *and Mirella Vivoli Vega*[2]

[1] *Living Systems Institute, University of Exeter, Stocker Road, Exeter, EX4 4QD, UK*
[2] *Department of Biomedical Experimental and Clinical Sciences, University of Florence, Viale Morgagni 50, 50134 Florence, Italy*

9.1 Significance

The function of many proteins is to act as catalysts for biological reactions. These protein catalysts, *enzymes*, generally speed up the rate of one or at most a few reactions, with a limited range of potential substrates. There is a general scientific interest in quantitatively understanding how enzymes alter the rate of reactions. Quantifying how reaction rates change underpins our understanding of cellular physiology; for industrial usage, it is critical to know how fast reactions will take place and there is an increasing interest in using synthetic biology to engineer new biochemistry into organisms. Enzyme assays have been foundational to our understanding of biology in the past and will contribute to an even broader range of applications in the next wave of biological sciences.

9.1.1 Examples of Enzyme Use (Clinical and Biotechnology)

Defects in enzymes underlie many human genetic conditions. Two of the most common human homozygous genetic defects in enzymes are mutations in the enzymes phenylalanine 4-hydroxylase (PAH; Figure 9.1a) [4, 5] and glucose-6-phosphate dehydrogenase (G6PD; Figure 9.1b) [6]. PAH hydroxylates phenylalanine to tyrosine as part of the phenylalanine catabolism pathway (KEGG pathway map 00360 [7]). Reduction in the rate of this enzyme leads to an accumulation of phenylalanine in the blood (hyperphenylalaninemia). This leads to the clinical symptoms of phenylketonuria (PKU). Prevention of PKU is now achieved by testing newborns for elevated phenylalanine via a heel prick test [4, 5]. Understanding of the PAH enzyme's structure and enzymology have allowed PAH function to be substantially restored in approximately 20% of patients by use of the modified *cofactor*[1] sapropterin dihydrochloride [8] that promotes enzyme folding and function. G6PD oxidises glucose-6-phosphate as one of the two NADPH generating steps in the pentose phosphate pathway (Figure 9.1b). Over 200 disease causing mutations are currently known [9]. G6PD deficiency is generally treated by management: the genetic defects are generally sufficiently minor to be asymptomatic in the absence of haemolytic or oxidative stressors [6]. Testing for G6PD levels is routine before a selection of drugs are prescribed, as these drugs act as such stressors [10, 11]. Firm diagnosis of G6DH deficiency is achieved through an enzyme assay for activity levels in red blood cells [12]. An accurate quantitative assay is essential as disease penetrance and symptoms correspond closely to residual enzyme activity.

1 *Cofactors* are metal ions or small molecules (e.g. haem, NAD^+) that form part of the enzyme and are required for the function of the enzyme. Some are catalytic (e.g. metal ions), while others must be regenerated by other cellular systems (e.g. glutathione).

Biomolecular and Bioanalytical Techniques: Theory, Methodology and Applications, First Edition. Edited by Vasudevan Ramesh.

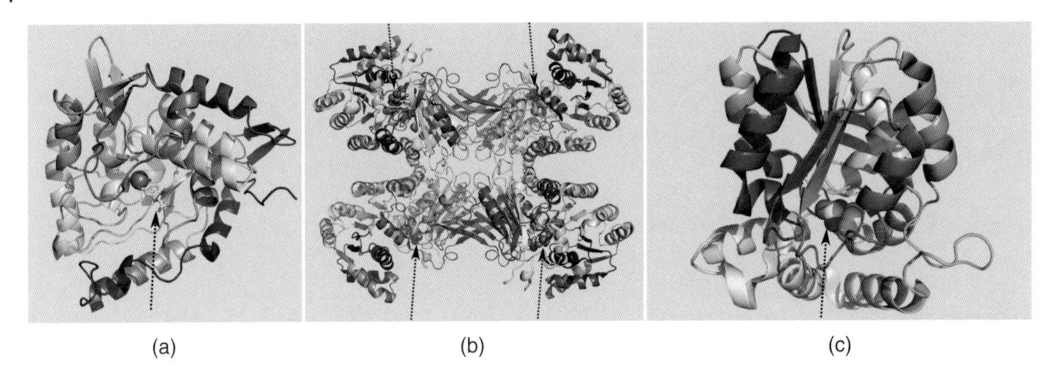

Figure 9.1 Examples of enzymes important to health and industry. Enzymes are shown as cartoon (rainbow colours: N-terminus blue, C-terminus red); ligands are shown as spheres or sticks (carbon: yellow; oxygen: red; nitrogen: blue; iron: brown). (a) Phenylalanine 4-hydroxylase catalytic domain, with the (catalytic) iron and inhibitor L-DOPA (3,4-dihydroxyphenylalanine; sticks, dashed arrow) highlighted. (b) Glucose 6-phosphate dehydrogenase tetramer, with glucose 6-phosphate (dashed arrows). (c) γ-lactamase monomer, with a substrate analogue (dashed arrow). Images generated using PyMOL from RCSB PDB IDs: 6PAH [1], 5UKW [2], 1HL7 [3].

Another key use of enzymes is in the manufacturing of high value chemicals, such as drug molecules and perfumes [13, 14]. Enzymes offer many advantages for chemical synthesis, including high specificity, high reaction yields, strong stereo- and regio-selectivity, and the potential to combine several reaction steps [15, 16]. Enzymes can be readily incorporated into very small reactors for syntheses on a micro or nanoscale [17], facilitating efficient synthesis at all scales. For example, γ-lactamases (Figure 9.1c) allow the enantiomeric resolution of the bicyclic lactam 2-azabicyclo[2.2.1]hept-5-en-one, a key building block for antivirals (e.g. carbovir) [18]. The major drawback that has restricted the use of enzymes on a wider scale has been their high specificity. Together with the limited tolerance to heating and solvents that most enzymes display, this makes finding suitable enzymes for industrial biotransformations difficult. These challenges are being overcome by a combination of directed evolution, protein design and semi-rational mutagenesis. The capacity to rapidly test many protein variants is the major bottleneck of these methods. Obtaining the right industrial properties requires elegantly designed enzyme assays to provide sensitivity and throughput at the very high levels required.

9.2 Overview of Kinetics and Its Application to Biology

Enzyme kinetics represents one branch of the broader field of chemical kinetics [19]. This field includes investigations of the relationship between the concentration of a substrate in a reaction and the rate of the reaction. For a chemical reaction, the *rate* (the change in concentration of the reactants or products) is generally related to the concentration of the reactants by a simple power law. Depending on the nature of the reaction, this may be zero order (substrate concentration does not affect the reaction), first order (rate is proportional to substrate concentration) or second order (rate is proportional to the square of substrate concentration; Figure 9.2). Higher order reactions are rarer, but do occur [19]. For reactions with multiple substrates, the reaction rate will be zero, first or second order with respect to each substrate: again, the overall order of the reaction is usually second order or less.

Enzyme kinetics have a greater level of complications. These arise because the enzyme catalyst is a highly complex molecule and undergoes structural changes on a very rapid (nanosecond to microsecond) timescale. The enzyme reaction usually occurs on a microsecond to second timescale and includes many steps (including substrate binding, the formation of intermediates and the release of products). Consequently, the process by which the enzyme speeds up the reaction must be considered to understand the nature of the catalytic process and the relevant kinetic parameters.

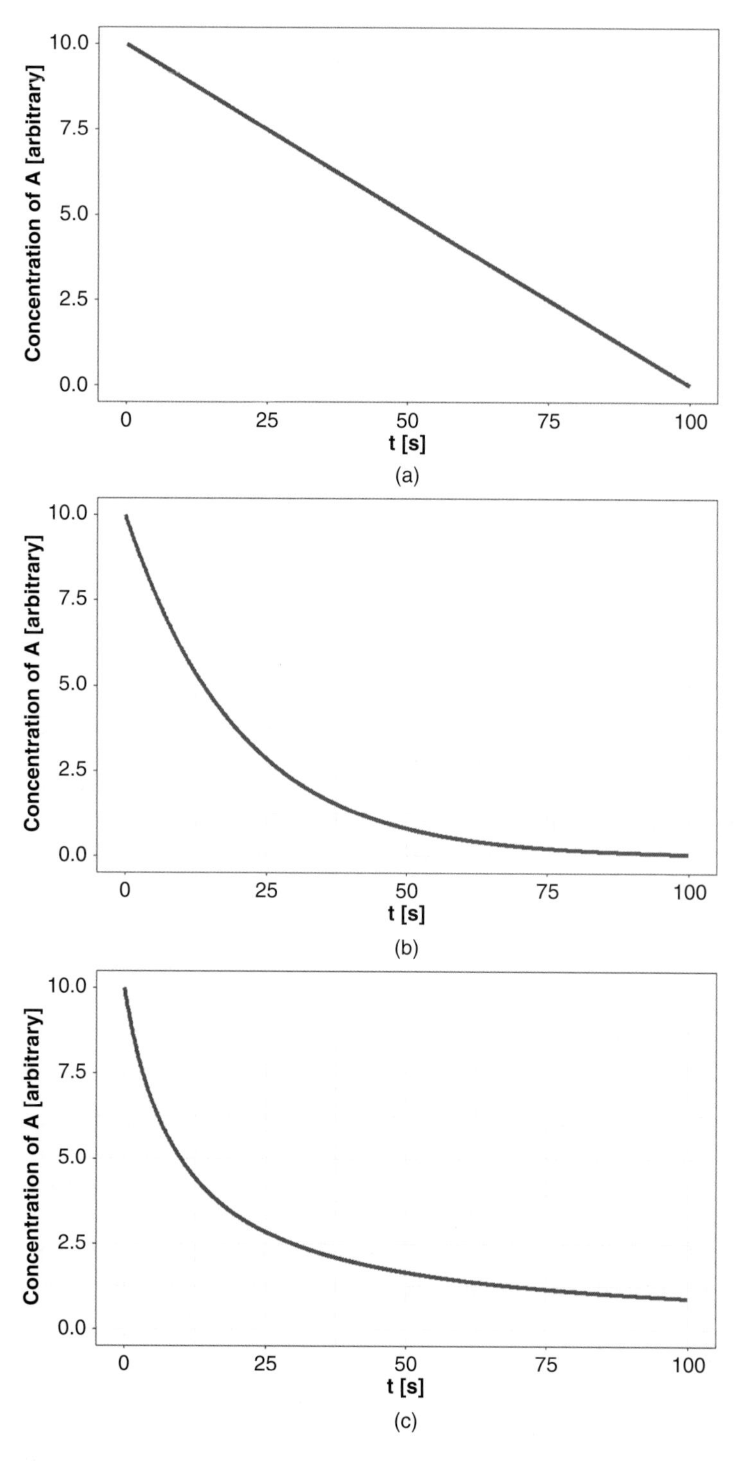

Figure 9.2 Reactions with different orders with respect to the substrate. In each case, the concentration of substrate *A* is shown reducing from an arbitrary starting point, over an arbitrary time period. (a) Zero-order reaction. Here, the reduction in concentration of *A* shows no dependence on the substrate concentration. (b) First-order reaction. In this case, the rate of the reaction (and so consumption of *A*) is directly proportional to the concentration of *A*. The rate slows considerably as *A* is consumed. (c) Second-order reaction. The rate of reaction is related to the square of the concentration of *A* (e.g. two molecules of *A* combining to form some product). The reaction initially proceeds rapidly, but falls significantly as the concentration of *A* approaches zero.

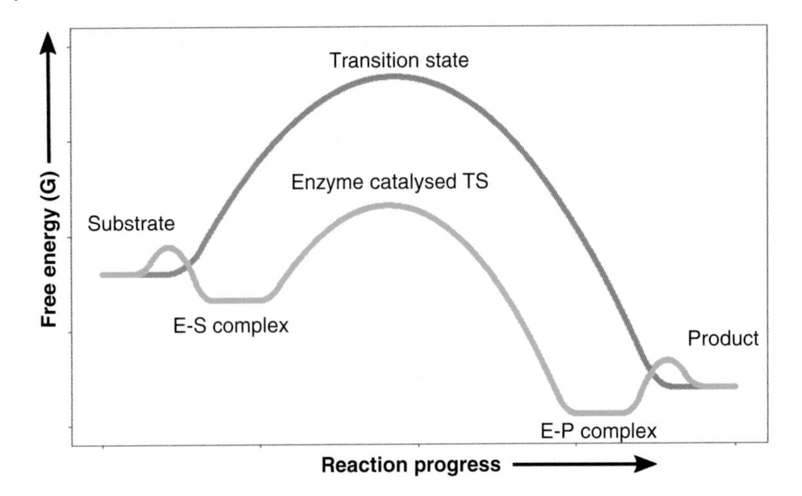

Figure 9.3 Schematic of enzyme catalysis. The uncatalysed reaction (red) has to overcome a large free energy barrier to reach the transition state (TS). A simple example enzyme catalysed reaction (blue) provides a reaction route at a lower free energy. The free energy barrier is lower, allowing more molecules to react. The reaction is more complicated, with enzyme-substrate (E-S) and enzyme-product (E-P) complexes.

The nature of enzyme catalysis is to provide a reaction pathway with a lower free energy barrier than the uncatalysed reaction (Figure 9.3). Most enzyme-catalysed reactions take place at a very slow rate in the absence of catalysis. All reactions must pass through one or more *transition state(s)* that are maxima in free energy. The rate of the reaction will be determined by the proportion of substrate molecules that are able to access this state. In order to access the enzyme-catalysed intermediate state, the substrate must first bind to the enzyme. As binding of substrate to the enzyme is usually a fast, reversible event, the rate of these reactions is far too rapid to be determined except with specialist instruments designed to collect rate data on millisecond timescales (see *Advanced Methods*, Section 9.6).

9.2.1 Steady-State Kinetics

Most enzymology studies instead use *steady-state kinetics*. These are measured on a seconds to minutes timescale and so can be studied using widely available instruments. The assumption of a steady state (see below) allows the simplification of the reaction scheme. The resulting kinetic equation is familiar to most biological sciences and bioorganic chemistry students from an early part of their studies.

The *Michaelis–Menten equation* (Eq. (9.1)) is derived using a model of an enzyme with a single substrate and a single product. There is a single step between each of the following states: free enzyme and substrate; enzyme–substrate complex; enzyme–product complex; and free enzyme and product. The transitions between each of these states is determined by a rate constant in either direction (Figure 9.4). To determine all of these rate constants would require a large quantity of data, including data at millisecond timescales, as the binding and release of both substrate and product from the enzyme is likely to take place on such timescales.

$$v = \frac{V_{max}\,[S]}{K_M\,[S]}, \tag{9.1}$$

where V is the initial reaction rate, V_{max} is the reaction rate at infinite substrate concentration, $[S]$ is the substrate concentration and K_M is the Michaelis constant.

Deriving the Michaelis–Menten equation relies on assumptions to simplify the scheme. The reaction is assumed to be unidirectional (i.e. $k_{-2} = 0$) and the product release is assumed to be fast and irreversible ($k_3 = \infty$, $k_{-3} = 0$). These assumptions are most reasonable when the product concentration is zero or low, and simplify the reaction scheme considerably (Figure 9.4b).

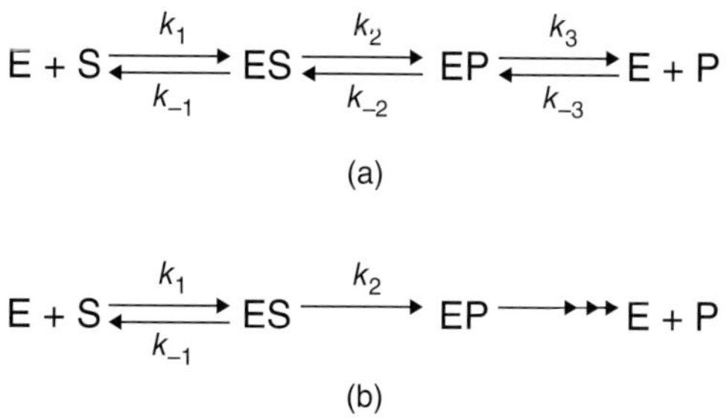

Figure 9.4 Reaction scheme for an enzyme (E) catalysed reaction with one substrate (S) and one product (P). (a) Each reaction is modelled as reversible in the general case. Each step has a rate constant (k_1, k_{-1}, etc.). (b) Simplified scheme. The conversion from the enzyme–substrate complex to an enzyme-product complex is considered irreversible and the release of product is considered to take place considerably faster than this conversion. This considerably reduces the complexity of the system, simplifying experimental determination.

$$\mathrm{E + S} \underset{k_{-1}}{\overset{k_1}{\rightleftharpoons}} \mathrm{ES} \underset{k_{-2}}{\overset{k_2}{\rightleftharpoons}} \mathrm{EP} \underset{k_{-3}}{\overset{k_3}{\rightleftharpoons}} \mathrm{E + P}$$

(a)

$$\mathrm{E + S} \underset{k_{-1}}{\overset{k_1}{\rightleftharpoons}} \mathrm{ES} \overset{k_2}{\longrightarrow} \mathrm{EP} \longrightarrow \mathrm{E + P}$$

(b)

Secondly, the *steady state approximation*, introduced by Briggs and Haldane [20], is applied. It is assumed that as the reaction progresses, it will rapidly reach a state where the rate of formation and breakdown of the enzyme–substrate complex (the remaining intermediate in the reaction) will become equal. The consequence of this is that there is no change in the concentration of the enzyme–substrate complex (Eq. (9.2)). Modelling a generic reaction demonstrates that this rapidly becomes true and that this is true for reactions across a wide range of reasonable values for k_1 and k_{-1} (Figure 9.5). The steady-state approximation is given as

$$k_1\,[E][S] = k_{-1}\,[ES] + k_2\,[ES]. \tag{9.2}$$

In most experiments, the substrate concentration will considerably exceed the enzyme concentration and so the formation of the enzyme–substrate complex will have no real effect on the substrate concentration. However, there will be a significant effect on the enzyme concentration. The enzyme concentration will no longer be the initial concentration added to the experiment:

$$[E] = [E_0] - [ES]. \tag{9.3}$$

Combining (9.2) and (9.3):

$$k_1\,[S]\,([E_0] - [ES]) = k_{-1}\,[ES] + k_2\,[ES]. \tag{9.4}$$

Rearranging (9.4):

$$k_1\,[S]\,[E_0] = k_{-1}\,[ES] + k_2\,[ES] + k_1[S][ES], \tag{9.5}$$

$$k_1\,[S]\,[E_0] = (k_{-1} + k_2 + k_1[S])\,[ES], \tag{9.6}$$

$$\frac{k_1\,[S]\,[E_0]}{k_{-1} + k_2 + k_1[S]} = [ES], \tag{9.7}$$

$$\frac{[S]\,[E_0]}{\frac{k_{-1}+k_2}{k_1} + [S]} = [ES]. \tag{9.8}$$

Equations (9.3) to (9.8) are rearrangements of the steady-state approximation to define $[ES]$.

In a simplified scheme for the reaction (Figure 9.3b), the observed rate of the reaction will be equal to the rate at which product P is generated:

$$\frac{\mathrm{d}[P]}{\mathrm{d}t} = k_2\,[ES] = v. \tag{9.9}$$

Equation (9.9) is a statement of the rate of the reaction.

Substituting the enzyme–substrate complex determination from (9.8) into (9.9) gives

$$v = \frac{k_2\,[E_0][S]}{\frac{k_{-1}+k_2}{k_1} + [S]}. \tag{9.10}$$

Equation (9.10) is the derived Michaelis–Menten equation following Briggs and Haldane [20].

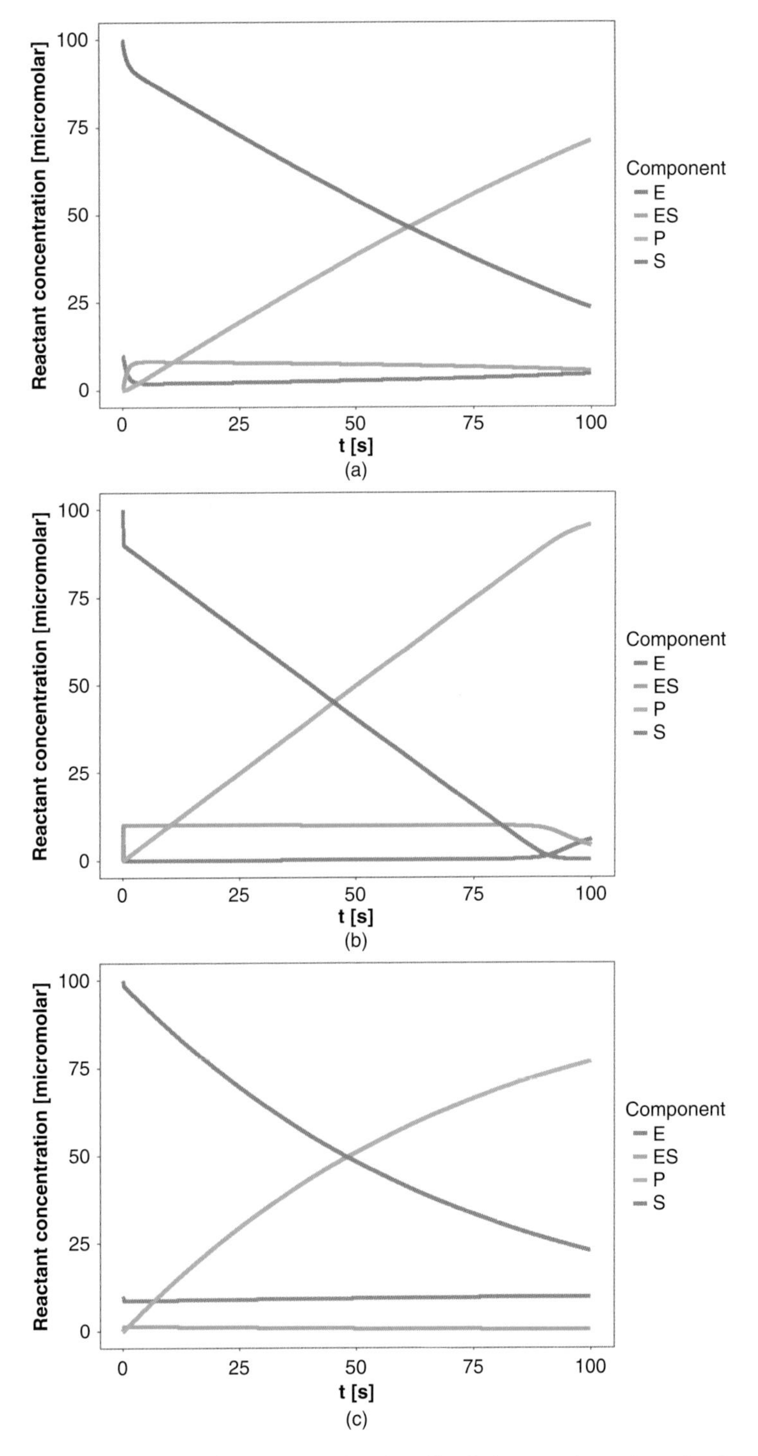

Figure 9.5 The steady state approximation. Simulations are shown for a simple unisubstrate enzyme obeying the approximation in Figure 9.3b. A high enzyme concentration is used so that the ES complex concentration is on the same scale as the substrate and product. Three scenarios are illustrated to demonstrate that the steady-state approximation holds across a range of K_M. (a) Approximating average K_M, the steady state holds with a slow drop in concentration after the initial phase of the reaction. (b) Approximating low K_M (high k_1), the steady state is maintained until the reaction is almost complete. (c) Approximating high K_M (low k_1; higher k_{-1}), the ES complex concentration reduces once the initial phase (10–20% of substrate exhausted) is complete.

We now define $V_{max} = k_2 [E_0]$ and $K_M = (k_{-1} + k_2)/k_1$. Substituting these into Eq. (9.10) recovers Eq. (9.1), the form in which this is most commonly presented.

By considering how the Michaelis–Menten equation is derived, important points regarding an enzyme that follows these equations can be observed. Firstly, the rates obtained are only valid during the early parts of the reaction, as the assumptions made include the fact that there is no product. This is particularly the case for reversible reactions, as the reverse reaction will start to take place at a significant rate as product levels increase. Consequently, while this is not always practical, it is ideal to determine the rate of the reaction during the first 10% of substrate usage. Secondly, the steady-state approximation implies that firstly there will be a lag phase before the reaction reaches its maximum rate and that the rate of the reaction will then remain similar (in the absence of product effects) over a significant timeframe (Figure 9.5). Therefore, for many reactions, a certain 'dead-time' between the experimental setup and measurement can be tolerated. Careful choice of the amount of enzyme used will allow the reaction to proceed sufficiently slowly to allow the time required for measurements to be started and collection of data points to be made.

There are several other treatments that use similar assumptions but a different mathematical scheme. These also arrive at the same general equation. These alternative treatments are often used in more complicated situations (e.g. with multiple substrates) as they make the determination of rate equations more convenient. These treatments are reviewed in detail in more specialist texts [19, 21, 22].

9.2.2 Beyond the Michaelis–Menten Equation

9.2.2.1 Enzyme Cooperativity

There are several common scenarios where modification of the Michaelis–Menten equation is necessary. An important example is *cooperative enzymes* [23]. Cooperative enzymes show responses to ligands that are either sigmoidal (positive cooperativity; Figure 9.6a) or are hyperbolic (like Michaelis–Menten enzymes), with a more pronounced plateau at high ligand concentrations (negative cooperativity; Figure 9.6b). Both positive and negative cooperativity are common [24]. It is also likely that more enzymes are cooperative than is generally appreciated, with many being considered as Michaelis–Menten due to a scarcity or lack of sensitivity in the data (e.g. [25]).

Cooperativity was traditionally understood in the context of multimeric enzymes. In such enzymes, binding of substrate to one active site makes the binding of substrate to another active site more likely (positive cooperativity) or less likely (negative cooperativity; Figure 9.6). There are also cases of enzymes that display both positive and negative cooperativity [25–28]. Cooperative kinetics – especially positive cooperativity – is also observed in many monomeric enzymes [29]. Such enzymes generally adopt a low activity state in the absence of substrate, which switches slowly into a high activity state (where substrate binding occurs). Cooperative kinetics should therefore be considered for monomeric as well as multimeric enzymes.

Steady-state kinetics can also be studied for cooperative enzymes. There are several models available for understanding cooperative kinetics [30]. Classically, the two dominant hypotheses were the cooperative (Monod–Wyman–Changeux) and sequential (Koshland–Némethy–Filmer) models [31, 32]. Many enzymes display features of both models, and these likely represent two extremes of a spectrum of enzyme behaviour [19, 33]. As both models require several additional fast kinetic steps or equilibria, cooperative enzymes are usually fitted instead to a variation of the *Hill equation*, applied to Michaelis–Menten kinetics [19] (Eq. (9.11)). Positively cooperative enzymes will show $h > 1$, while negatively cooperative enzymes will show $h < 1$:

$$v = \frac{V_{max} [S]^h}{(K_{½}^h + [S]^h)}, \tag{9.11}$$

where $K_{½}$ is the substrate concentration at half-maximal enzyme rate and h is the Hill coefficient. When $h = 1$, the equation reverts to the standard Michaelis–Menten equation.

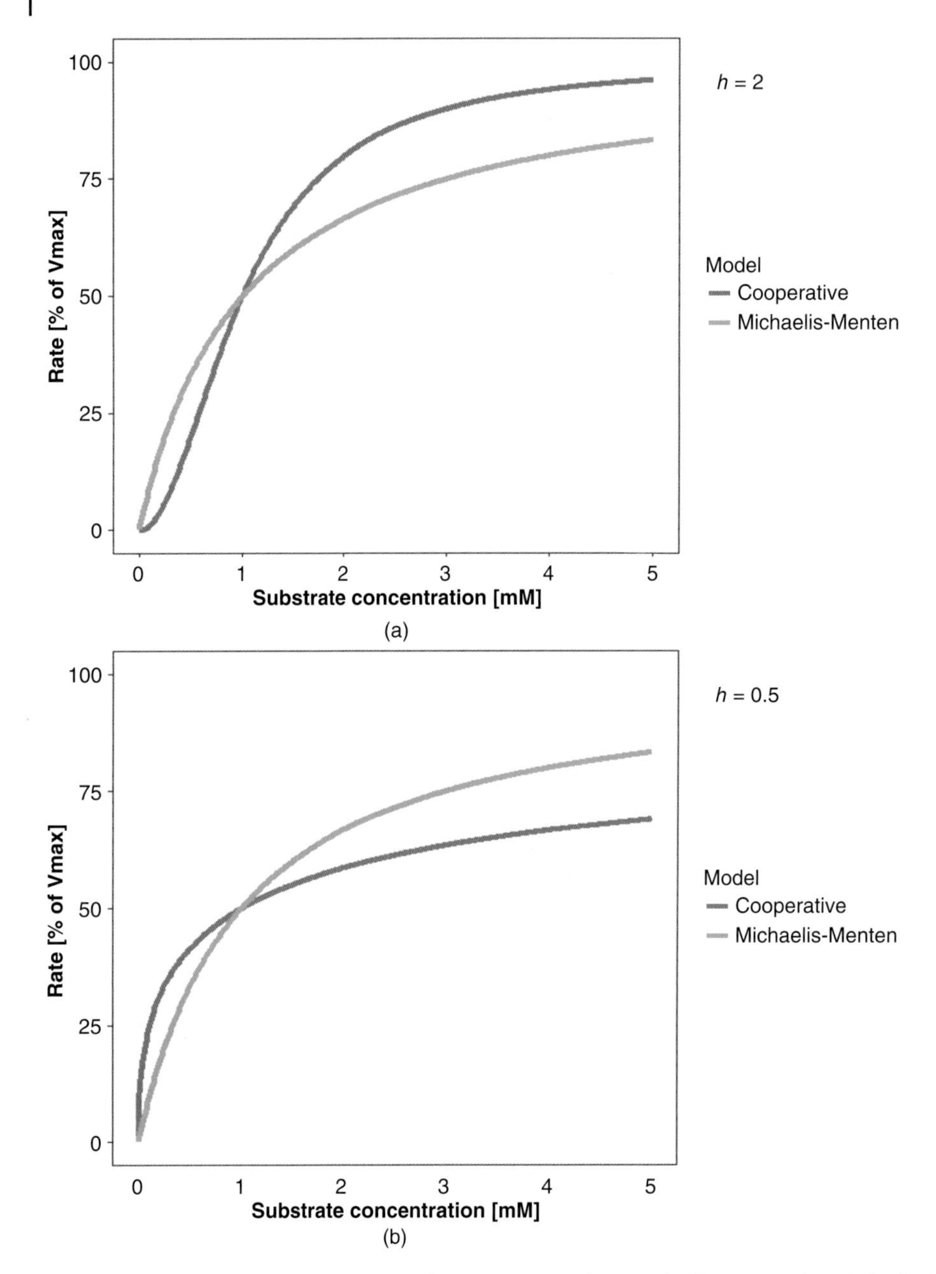

Figure 9.6 Cooperative enzymes. Figures show cooperative kinetics (red), compared to Michaelis–Menten kinetics (blue). The compared kinetics have $K_M/K_{½} = 1$ mM. (a) Positive cooperativity ($h = 2$). The rate at low substrate concentrations is considerably lower than the Michaelis–Menten case. (b) Negative cooperativity ($h = 0.5$). The approach to saturation is considerably slower than the Michaelis–Menten model.

9.2.2.2 Multiple Substrates

Enzymes that have more than one substrate – the majority of enzymes – also deviate from Michaelis–Menten kinetics. With multiple substrates (and usually multiple products), the reaction schemes become more complicated (Figure 9.7). Each addition to the reaction will result in extra kinetic steps that must be modelled. Even in the simplest case (sequential bisubstrate enzyme – e.g. benzaldehyde lyase [34, 35]), this results in two extra

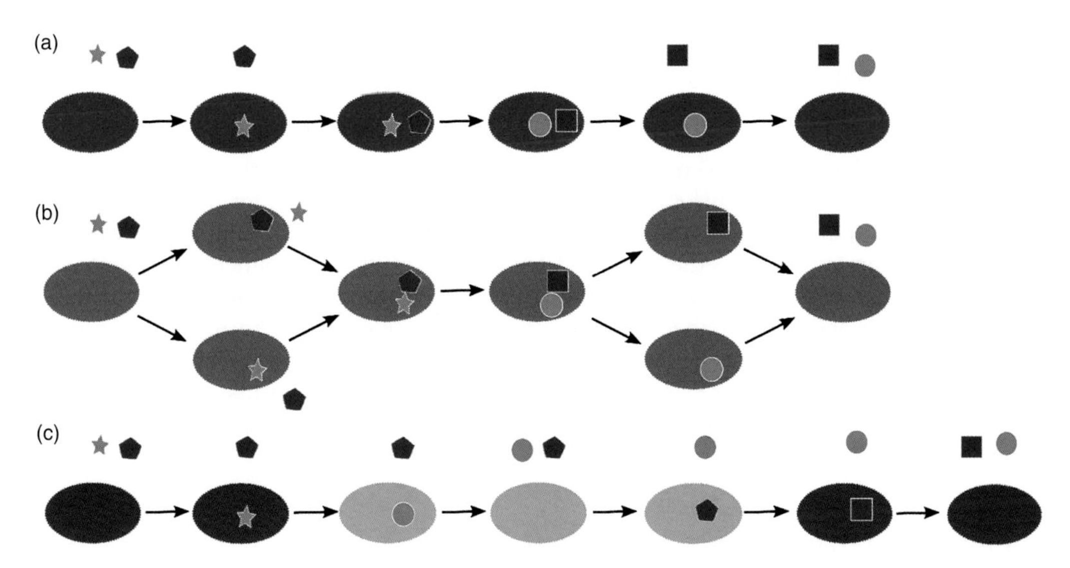

Figure 9.7 Bisubstrate reaction mechanisms. Approximately 60% of enzymes have two substrates and two products. The substrates and products bind and release in three main schemes. Enzymes are shown as red/pink/maroon/green ovals; substrates as yellow stars/blue pentagons; products as yellow circles/blue squares. (a) Ordered sequential bisubstrate mechanism. One substrate (yellow) must bind before the second (blue). After reaction, the product corresponding to the second substrate to bind is released first. (b) Random sequential bisubstrate mechanism. Either substrate can bind to the enzyme first; after the reaction, either product can be released first. (c) 'Ping-pong' mechanism. One substrate will bind to the enzyme, and react, producing one product and a modified enzyme (green). After the first product is released, the second substrate binds. This then reacts to form the second product, with the enzyme returned to its original state.

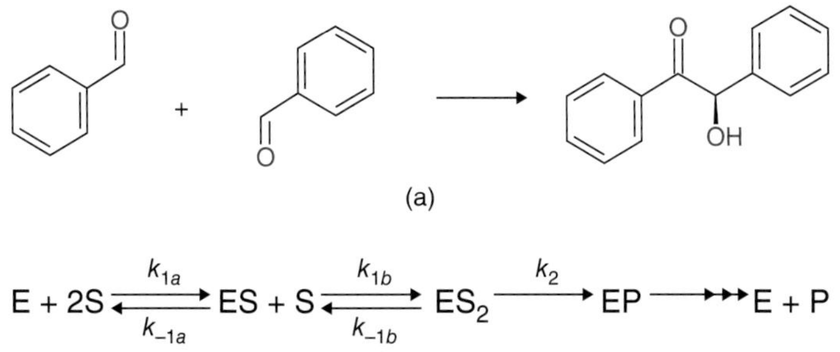

O + O ⟶ O OH

(a)

$$E + 2S \underset{k_{-1a}}{\overset{k_{1a}}{\rightleftharpoons}} ES + S \underset{k_{-1b}}{\overset{k_{1b}}{\rightleftharpoons}} ES_2 \xrightarrow{k_2} EP \longrightarrow E + P$$

(b)

Figure 9.8 Forward reaction for a simple bisubstrate enzyme. The enzyme benzaldehyde lyase [34, 35] combines two molecules of benzaldehyde to form the product *R*-benzoin (a). As the two substrates are identical, the kinetics are simpler. Nevertheless, there are two additional rate constants in the reaction (b).

kinetic constants (Figure 9.8). Many of these more complicated schemes can also be simplified using similar assumptions to those in the Michaelis–Menten equation, and the steady-state approximation. For example, in the case of the sequential bisubstrate reaction (Figure 9.8), the corresponding rate equation is [21, 22]

$$v = \frac{V_{max}[A][B]}{K_i^A K_M^B + K_M^B[A] + K_M^A[B] + [A][B]}. \tag{9.12}$$

This equation describes the general case for a steady-state ordered mechanism [21]. In cases where the formation of the EA complex is in equilibrium (i.e. $k_{-1a} \gg k_2$; equilibrium ordered mechanism), the $K_M{}^A[B]$ in the denominator is removed.

It should be clear from this rate equation that it is considerably more challenging to determine all of the necessary parameters. The equation now contains four parameters (V_{max}, $K_M{}^A$, $K_M{}^B$ and $K_i{}^A$) and two variables (for the two substrate concentrations). A general approach is to set the concentration of one substrate to a high (saturating) level, where the substrate concentration is expected to be well in excess of K_M (and so the $[S]/(K_M + [S])$ term for that substrate approximates to 1). The *apparent* K_M ($K_{M\,app}$) of the other substrate in these conditions can then be determined. This K_M will be an amalgam of the true K_M for this substrate and the $K_i{}^A K_M{}^B$ term. The *case study* demonstrates the determination of these apparent K_M values for an example substrate. A more involved experiment allows the accurate determination of all four parameters (see *protocol*).

Enzymes that have more than two substrates can also be investigated using similar schemes. As the number of substrates increases, so does the number of parameters and consequently it becomes increasingly important to have a clear understanding of the likely reaction mechanism. The same caveats regarding the Michaelis–Menten equation apply even more so in these more complicated mechanisms. This is so particularly for the effects of product inhibition, as the greater number of products make it likely that this will be significant considerably earlier in the reaction.

9.3 Determination of Enzyme Kinetic Mechanisms

9.3.1 Michaelis–Menten Parameters

Investigation of the Michaelis–Menten parameters of a newly isolated enzyme for a presumed substrate is generally an incremental process. The design of a well-designed enzyme assay is discussed in detail in key textbooks [19, 21]. Once this is available, obtaining reliable, high quality data will take several scouting experiments to obtain the right conditions. Acquiring data at the right enzyme and substrate concentrations is essential to determine the kinetic parameters with high confidence. This is particularly important when attempting to differentiate between mechanisms.

The first prerequisite is to determine an appropriate concentration of enzyme to use. An ideal enzyme concentration will allow an accurate determination of the rate (ideally while the first 10% of substrate is consumed) at high substrate concentrations. This is essential for an accurate determination of V_{max}. However, the enzyme concentration used should also permit determination of the reaction rate when this rate is 5–10% of the high substrate case. This is important for collecting data at substrate concentrations well below K_M, and so for an accurate determination of K_M, and also for identifying positively cooperative enzymes (which are typically identified by unexpectedly low rates at low substrate concentrations). These criteria are generally easier to achieve with continuous, rather than stopped, assays [19, 21]. It is important in when designing this experiment to consider the effects of the enzyme's buffers and carriers (especially glycerol) on the reaction when determining the highest practical concentration. Experiments should include a no enzyme control, to account for non-enzymatic changes in the signal (e.g. natural slow breakdown of NADH at 37 °C). An example of a refined experiment is shown in Figure 9.9. Note that the units for enzyme reactions are conventionally given as units of time and concentration (ideally k_{cat} in per second and K_M in mM or μM). Extensive recommendations are available at https://www.beilstein-strenda-db.org/strenda/public/guidelines.xhtml [36].

Ideal features from a refined experiment would show:

- A graph of enzyme concentration against rate (Figure 9.9b) shows that at the enzyme concentration selected, the rate of reaction is proportional to the enzyme concentration.
- The reaction rate can be accurately measured, ideally in the first 10% of substrate usage, for the selected enzyme concentration (i.e. for a continuous assay, a linear section of at least five points can be measured; for a stopped assay, the rate of reaction is consistent over at least two of the time periods tested).
- An accurate measurement could be taken with the rate at 5–10% of the observed rate at the enzyme concentration selected.

Examples where these criteria would not be met are shown in Figure 9.10. In many cases, it may not be possible to fulfil all of these criteria for a given enzyme reaction. In this case, the experimenter must select

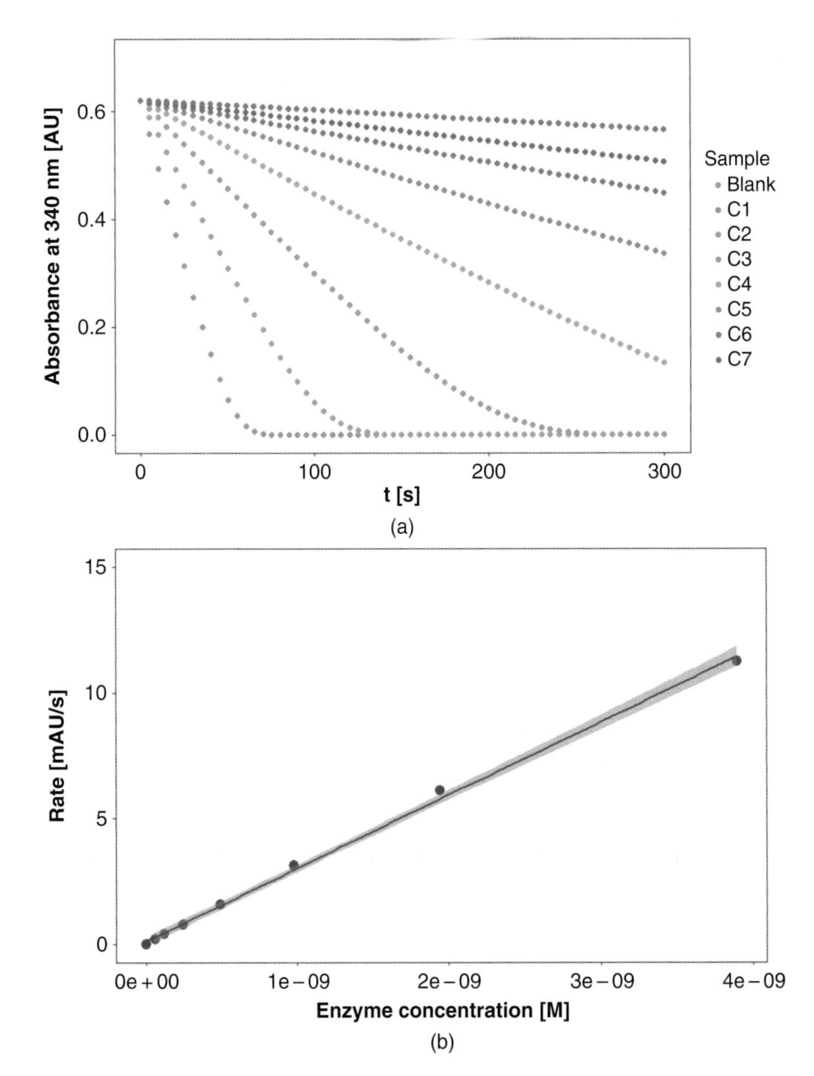

Figure 9.9 Example of an optimised experiment to determine ideal enzyme concentrations. Data are modelled for lactate dehydrogenase from rabbit muscle, using the reverse reaction: pyruvate + NADH → lactate + NAD^+. The reaction is modelled as irreversible. Kinetic parameters: $K_M{}^{pyruvate} = 150$ μM; $K_M{}^{NADH} = 12$ μM; $K_i{}^{pyruvate} = 75$ μM; $k_{cat} = 6.7 \times 10^8$ per second; background rate of NADH degradation 0.03 μM/s. The model uses a volume of 200 μL, with initial concentrations of 1 mM pyruvate, 100 μM NADH. Enzyme is modelled at seven concentrations, with a maximum concentration of 3.9 nM (equivalent to ~0.03 unit in 200 μL – C1), with twofold dilutions between the concentrations to a minimum of 61 pM (C7). A control with no enzyme is used to determine the background rate (Blank). NADH concentrations are converted to the absorbance at 340 nm (AU: absorbance unit) that would be detected using the extinction coefficient $\varepsilon = 6220$ 1/M/cm. Data are shown for every five seconds, corresponding to detection with a good plate reader. (a) High concentrations of the enzyme (C1, C2) lead to the reaction consuming over 20% of the substrate before many data points can be collected and so would not be appropriate concentrations (especially as fewer data can be collected when using more samples). Concentrations C4 and C5 consume approximately 10–20% of the substrate in 100 seconds and are clearly still within the linear phase: these would allow good data analysis when the substrate concentrations and rate are lower in later experiments. Concentrations C6 and especially C7 show a detectable rate, but when the substrate concentrations are reduced, the rate may not be distinguishable from background. This might result in cooperative behaviour being missed or incorrectly modelled. (b) The rate of reaction is plotted against enzyme concentration. The rate of reaction is linearly related to the enzyme concentration, indicating that enzyme is limiting across this concentration range (note that the highest enzyme concentration shows a slightly lower rate than predicted).

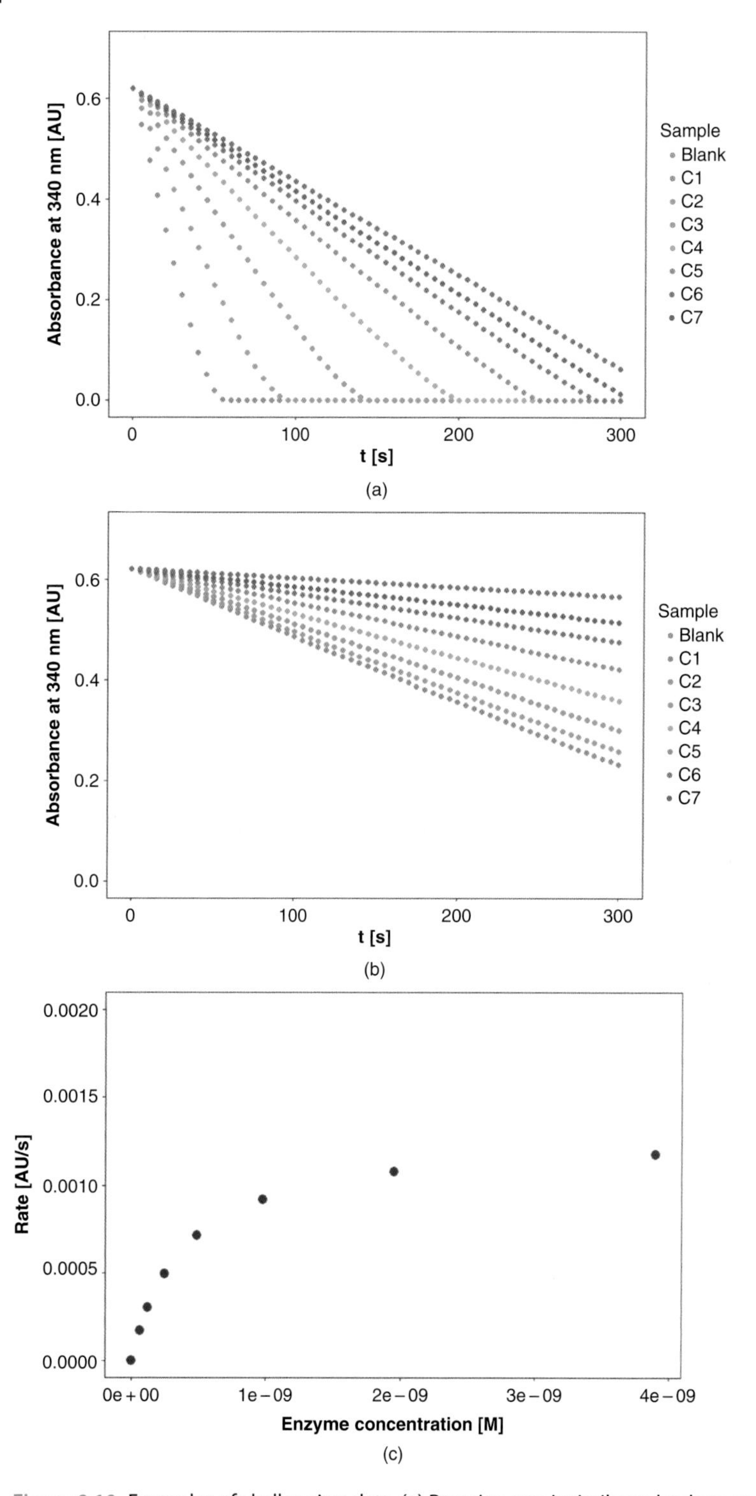

Figure 9.10 Examples of challenging data. (a) Reaction rate is similar to background rate. Here, it will be necessary to use a higher enzyme concentration and continue the reaction until more substrate is exhausted. (b) Enzyme is not limiting at test concentrations. Here the conditions of the reaction make the enzyme no longer the sole determinant of rate. The graph of rate versus enzyme concentration (*C*) does not follow a straight line. Enzyme concentrations at the lower end of the concentration range will have to be used. This may make data collection challenging as the observed rates will be small. A compromise concentration, where the enzyme is still just limiting (most likely around C5) but the rate is still sufficiently above background will have to be chosen.

which criterion to compromise. Making this choice will require consideration of the scientific goals of the enzyme assay.

Once the enzyme concentration has been established, the next stage in determining the kinetic parameters for a monosubstrate enzyme is to obtain a broad estimate of K_M. After the data have been collected (see the detailed protocol below), they are fitted to the Michaelis–Menten equation using a statistical software package. Examples of such packages are *R* (conveniently implemented using RStudio), Graphpad, KaleidaGraph and SPSS. Although linear transformations such as the *Lineweaver-Burk plot* (an example is shown in Figure 9.15) were used before such packages were available, they can amplify errors considerably and should not be used [37]. This will generally provide one of the following outcomes:

- The data fit reasonably well to the Michaelis–Menten equation, with the estimated K_M well within the range of substrate concentrations tested. In this case it is appropriate to move to more detailed investigations.
- The data appear to follow a straight line passing through the origin. This indicates that the data do not cover values above K_M sufficiently well. This generally indicates that the substrate is not an ideal substrate for this enzyme. This is often observed in situations where the initial substrate concentration has been selected based on a homologue and the enzyme studied does not prefer this substrate. If it is possible to increase the highest substrate concentration, repeat the experiment starting with the highest practicable concentration. In cases where a saturating substrate concentration cannot be reached, it is possible to determine k_{cat}/K_M (e.g. see [38, 39]).
- The enzyme shows a high, similar rate in all conditions. This indicates that the substrate range is all above K_M. In this case, the experiment should be repeated, starting at one of the lower substrate concentrations and using fourfold dilutions.
- The data fit poorly to the Michaelis–Menten equation. There are several ways in which this might be the case. In this situation, it is best to fit the data to a model for cooperativity (Eq. (9.11)) or substrate inhibition (Eq. (9.13)) as appropriate. The estimated parameters should be used to perform a more detailed experiment, using the wider substrate range discussed. This detailed experiment must be undertaken with the expectation that the data analysis will require one of the more complicated scenarios discussed later.

Examples of an initial substrate range determination in these scenarios is shown in Figure 9.11.

Once an estimate of K_M has been determined in this manner, more detailed experiments should be performed. It is highly important to have a good range of substrate concentrations both above and below K_M. Substrate concentrations considerably in excess of K_M are necessary to obtain a good estimate of V_{max}. This is essential not only as V_{max} is one of the two parameters to be fitted but also as the K_M value is highly dependent on V_{max} (as, by definition, it is the substrate concentration with a half-maximal rate). Substrate concentrations below K_M are also essential, as these are required to obtain a good estimate of K_M. These low substrate concentration points are also important for identifying cooperative enzymes: substantial deviation from the Michaelis–Menten model in these low substrate concentration points is the best indicator of cooperativity. Therefore, even though these data can be more challenging to obtain accurately, they offer considerable value in understanding the enzyme. These data should be distributed geometrically around K_M – that is, each substrate concentration should be an *n*-fold dilution of the previous substrate concentration. Although arithmetic distributions (i.e. 0.25×, 0.5×, 0.75×, 1×, 1.25×, 1.5×, 1.75× K_M) are superficially more attractive, these will give little information if the estimate of K_M is inaccurate. Once these data have been obtained, they should again be fitted to the Michaelis–Menten equation, as illustrated above.

Examples of such a refinement experiment is shown in Figure 9.12. Likely outcomes from these experiments are:

- The data show a good fit to the Michaelis–Menten equation, with the determined K_M well within the concentrations tested (at least three non-zero data points above and below K_M; Figure 9.12b). In this case, the result may be sufficient for the experimental requirements. If a more accurate determination of the kinetic parameters is required, the experiment should be further refined, as discussed below.

- The data fit reasonably well to the Michaelis–Menten equation, but the refined value of K_M has altered by at least twofold from the initial estimate (Figure 9.12c). In this case, a further refinement of the experiment will be required, as discussed below.
- The data show significant deviations from the Michaelis–Menten equation (Figures 9.12d to f). In these cases, alternative experimental treatments are required. These are detailed in the following sections.

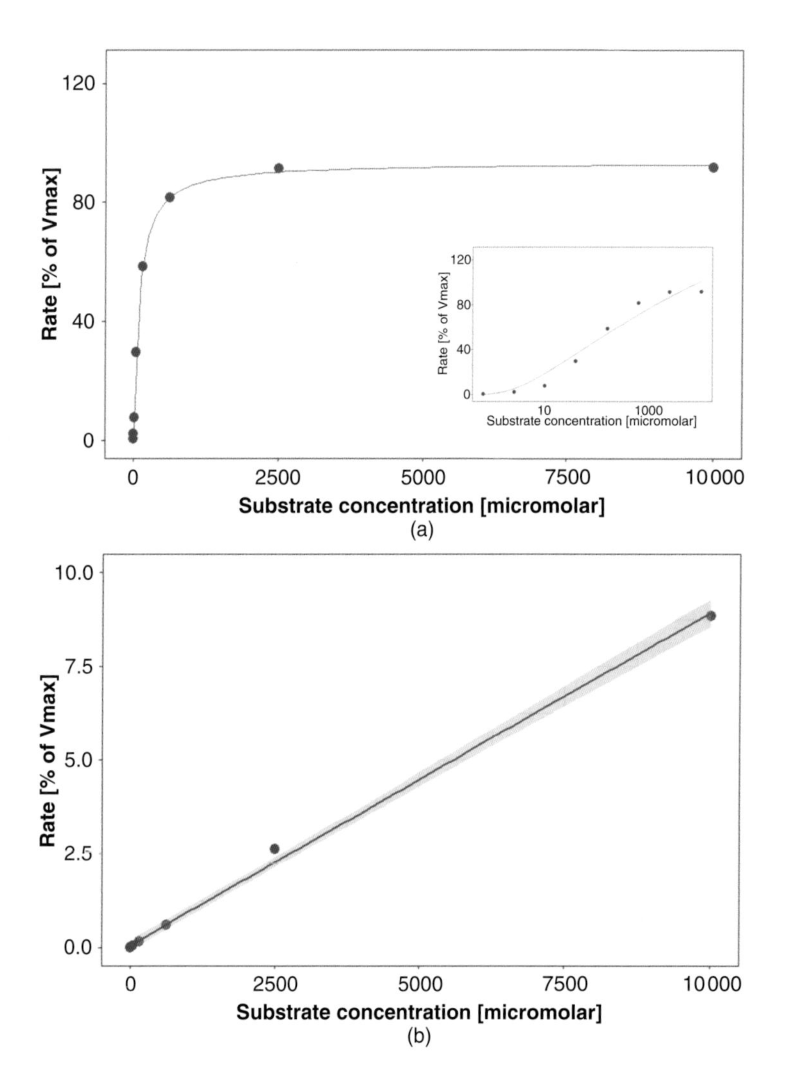

Figure 9.11 Examples of results from an initial substrate concentration experiment. All figures show simulated data, with random errors added for verisimilitude. (a) Effective determination of an estimate of K_M. The data fit well to the Michaelis–Menten equation, with points distributed well above and below the estimated K_M (~100 μM). The inset shows the *x*-axis on a log scale, to highlight the fact that the fit at low concentrations is good. (b) Substrate concentrations used are too low. The data fit best to a straight line, indicating that the substrate concentration is not yet approaching K_M. Higher substrate concentrations would be required in this instance. (c) Substrate concentrations used are too high. Almost all of the substrate concentrations are above K_M, leading to a poor estimate of K_M. This is made clearer using a log scale for the *x*-axis (inset). A repeated experiment, starting at a lower substrate concentration (likely removing the top three concentrations) would be necessary. (d) Results are not monotonic (here substrate inhibition is modelled). The data consequently are a poor fit to the Michaelis–Menten equation. The data should be fitted to the substrate inhibition model. Follow-up experiments would require more data to model the additional parameters required.

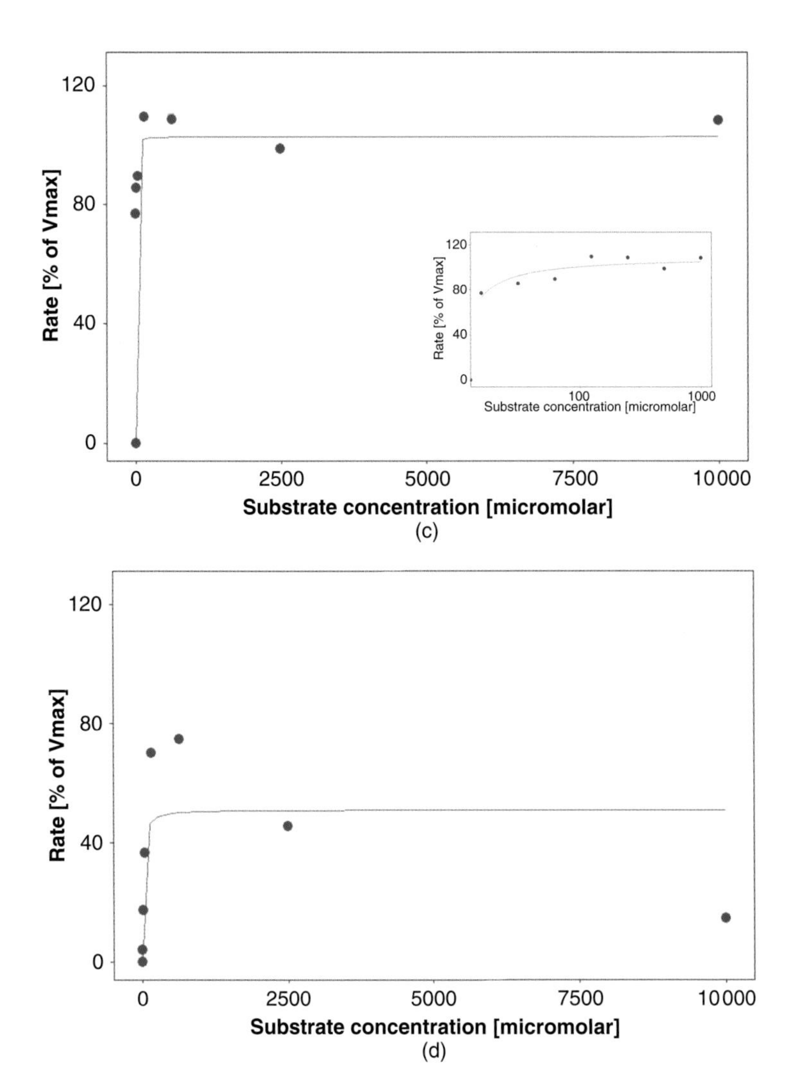

Figure 9.11 (*Continued*)

Where the experiment requires further more refinement, it is often sensible to collect more data points. Where the data are not determined sufficiently accurately, the next step would be to look for accuracy improvements in the enzyme assay.

9.3.2 Cooperative Enzymes

When the data for a refined enzyme assay remain monotonic (i.e. as substrate concentration increases, the rate always increases), but show deviations from the Michaelis–Menten equation, the most likely explanation is that the enzyme is showing cooperative behaviour (Figure 9.12d and e). Firstly, fit the data to a cooperative model rather than the Michaelis–Menten equation (Figure 9.13). This will provide estimates of the kinetic parameters V_{max}, $K_{½}$[2] and h for the reaction. These parameters will likely be poorly determined as there are

2 For cooperative enzymes, the parameter $K_{½}$ is used rather than K_M. The $K_{½}$ is the substrate concentration where the enzyme activity is 50% of maximum. $K_{½}$ is used to reflect that this parameter will be raised to a power in the cooperative equation.

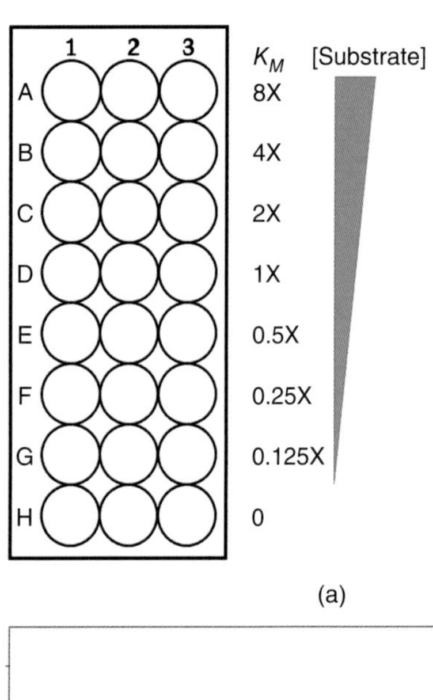

(a)

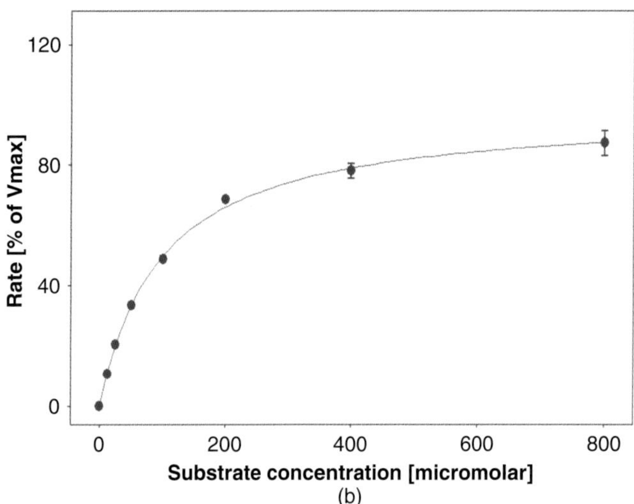

(b)

Figure 9.12 Experimental determination of refined estimates of V_{max} and K_M. Examples are simulated as triplicate experiments with random errors added (normal distribution, SD = 5%) for verisimilitude. (a) Example plate layout for detailed experiment. (b) Example of a successful experiment (expected $V_{max} = 100\%$, $K_M = 100$ μM). The data obtained give $V_{max} = 98 \pm 2\%$, $K_M = 96 \pm 6$ μM. Errors in K_M of less than 10% of the observed value indicate an accurate experiment. (c) Example of an experiment where the estimated K_M was inaccurate (expected $V_{max} = 100\%$, $K_M = 500$ μM). The data obtained give $V_{max} = 100 \pm 3\%$, $K_M = 500 \pm 30$ μM. Although the values obtained are accurate, with only one data point at a substrate concentration above K_M this cannot be trusted. (d) Example of a positively cooperative enzyme (expected $V_{max} = 100\%$, $K_M = 100$ μM, $h = 2$). The data fit poorly to the Michaelis–Menten equation (red line; observed $V_{max} = 125 \pm 7\%$, $K_M = 170 \pm 30$ μM), but fit well to the cooperative enzyme equation (Eq. (9.11); blue line; observed $V_{max} = 98 \pm 2\%$, $K_{½} = 98 \pm 3$ μM, $h = 1.9 \pm 0.1$). Note that the major deviation is at a low substrate concentration. (e) Example of a negatively cooperative enzyme (expected $V_{max} = 100\%$, $K_M = 100$ μM, $h = 0.5$). The data fit poorly to the Michaelis–Menten equation (red line; observed $V_{max} = 71 \pm 2\%$, $K_M = 30 \pm 3$ μM), but fit better to the cooperative enzyme equation (blue line; observed $V_{max} = 100 \pm 10\%$, $K_{½} = 90 \pm 50$ μM, $h = 0.50 \pm 0.06$). Note that there are deviations from Michaelis–Menten at a low substrate concentration, and also at high concentrations. (f) Example of substrate inhibition (expected $V_{max} = 100\%$, $K_M = 100$ μM, $K_i = 400$ μM). The data fit poorly to the Michaelis–Menten equation (red line; observed $V_{max} = 41 \pm 2\%$, $K_M = 21 \pm 6$ μM), but fit clearly better to the substrate inhibition equation (Eq. (9.13); blue line; observed $V_{max} = 90 \pm 10\%$, $K_M = 80 \pm 20$ μM, $K_i = 500 \pm 100$ μM). Note that there are deviations from Michaelis–Menten at high substrate concentrations, but at low concentrations the fit is reasonable.

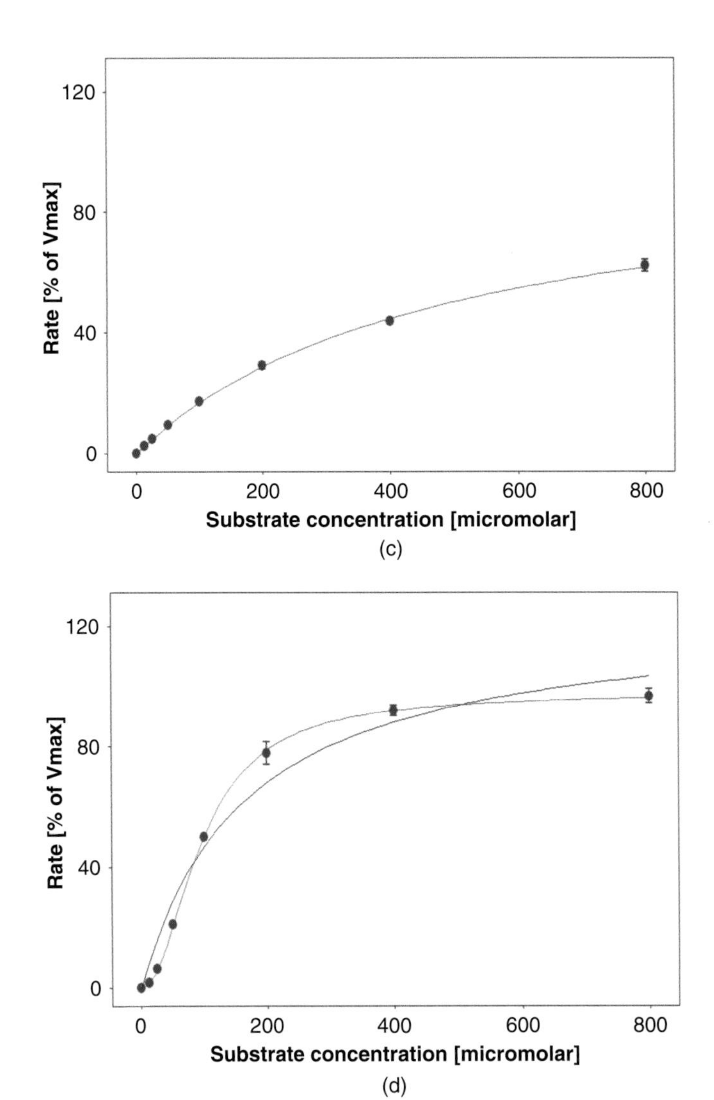

Figure 9.12 (*Continued*)

insufficient data to fit so many parameters. A following experiment should be performed using a larger number of data points (12–16) and refined substrate concentrations. For positively cooperative enzymes, collecting data up to 5× $K_{½}$ is sufficient. With positive cooperativity, the reaction approaches V_{max} more rapidly than for a Michaelis–Menten enzyme and so a tight focus around $K_{½}$ is most appropriate. For negatively cooperative enzymes, a wider range (up to 10× or even 20× $K_{½}$) is appropriate. Negatively cooperative enzymes approach V_{max} at much higher substrate concentrations than Michaelis–Menten enzymes and show low rates only at quite low multiples of $K_{½}$. As described below, these data should be distributed geometrically around $K_{½}$. These experiments should provide an accurate determination of the kinetic parameters. As discussed in the case of Michaelis–Menten enzymes, if there is a desire to determine the parameters more accurately, more data points can be obtained; however, it is also likely that the biggest improvements in data quality will come from improving the enzyme assay itself.

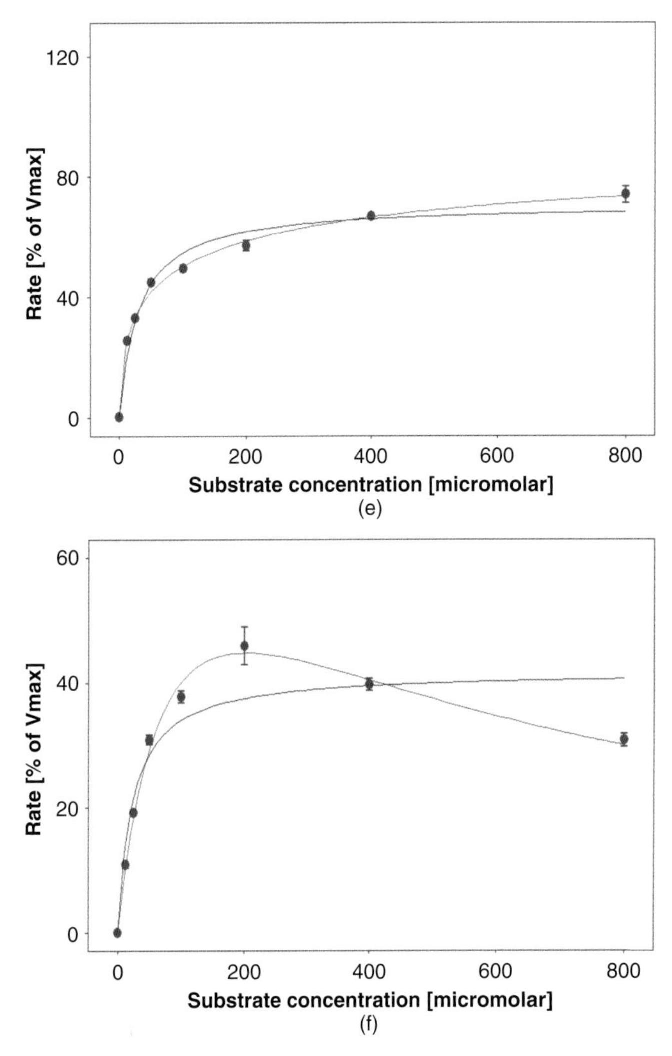

Figure 9.12 *(Continued)*

9.3.3 Substrate Inhibition

Approximately 20% of enzymes show inhibition by their own substrates [40]. This often occurs in enzymes with at least two substrates, where one substrate can bind to site that the second substrate should occupy in addition to the catalytic binding site; alternatively, the substrate can bind at an allosteric site. Enzymes that display substrate inhibition show a classical effect where, above a certain substrate concentration, the rate of reaction starts to fall in an approximately hyperbolic fashion (Figure 9.14a). The Michaelis–Menten equation is modified by an additional term that causes the K_M to inflate with increasing substrate concentration as the *inhibition constant* K_i is approached and exceeded [41]:

$$v = \frac{V_{\max}[S]}{K_M + \left([S]\left(1 + \frac{[S]}{K_i}\right)\right)}, \qquad (9.13)$$

where v is the initial reaction rate, V_{max} is the reaction rate at infinite substrate concentration, $[S]$ is the substrate concentration, K_M is the Michaelis constant and K_i is the substrate inhibition constant.

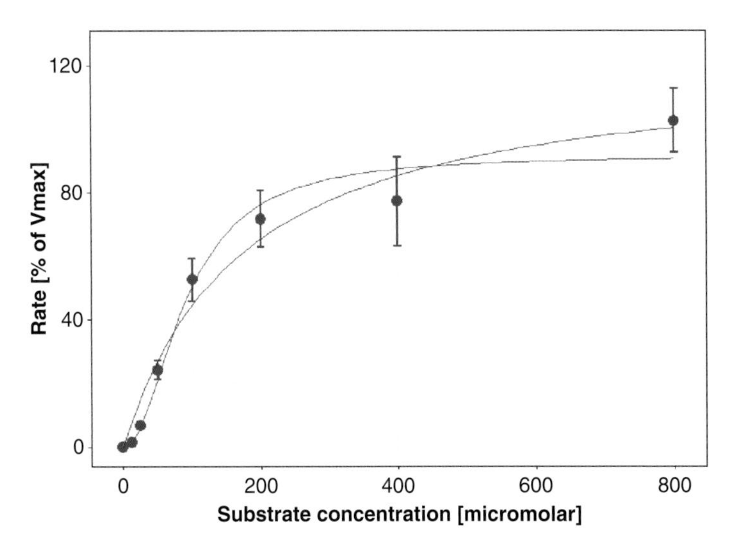

Figure 9.13 Example of a cooperative enzyme with significant errors. Example is simulated as triplicate experiments (expected $V_{max} = 100\%$, $K_{½} = 100\ \mu M$, $h = 2$) with random errors added (normal distribution, SD = 20%). The data fit poorly to the Michaelis–Menten equation (red line; observed $V_{max} = 120 \pm 10\%$, $K_M = 170 \pm 50\ \mu M$), but fit well to the cooperative enzyme equation (Eq. (9.11); blue line; observed $V_{max} = 100 \pm 10\%$, $K_{½} = 110 \pm 30\ \mu M$, $h = 1.4 \pm 0.4$).

Where substrate inhibition is detected or suspected, experiments should be performed with a minimum of 12 data points in triplicate (including a negative control). These should be distributed to have at least three substrate concentrations below the lower of K_M and K_i (ideally reaching at least three times less than the lower of these), at least three substrate concentrations above the greater of K_M and K_i (ideally reaching at least three times greater than the higher of these) and substrate concentrations between the two constants if possible (Figure 9.14b).

9.3.4 Determination of the Mechanism of a Bisubstrate Enzyme

Bisubstrate enzymes adopt three common general mechanisms (random sequential binding, ordered sequential binding and ping-pong mechanisms; Figure 9.7). Distinguishing which of these mechanisms an enzyme adopts, and the order of substrate binding if appropriate, adds a great deal of value to understanding the enzyme. This is particularly the case when the goals of a project require the modulation of enzyme activity (e.g. in a drug development program or when seeking to understand the role of an enzyme *in vivo* using known inhibitors). By selecting the right substrate binding site to target for the enzyme mechanism and the experimental conditions, a stronger effect can be achieved. For example, kinetic analysis of dihydrofolate reductase (DHFR), a target for antimicrobials and anticancer drugs, indicated that release of the tetrahydrofolate product was the limiting step in the reaction [42]. Consequently, drugs that target DHFR generally bind to the substrate/product folate binding site.

The first stage in determining the mechanism is to perform a kinetic parameter determination for one substrate at different concentrations of the second. A Lineweaver–Burk plot of the resulting data will show one of two patterns. Parallel fitted straight lines correspond to a ping-pong mechanism, while fitted lines that meet at a point correspond to sequential bisubstrate models (Figure 9.15). It does not matter which of the two substrates is considered to be the 'first' or 'second' substrate for this experiment: the results are identical regardless of which is chosen (demonstrated in Figure 9.16 from Eqs. (9.12) and (9.14), it should be apparent that the substrates are equivalent in the general cases). For an accurate determination of the correct model, the data should then be fitted to the competing models using suitable statistical software [21, 22]. This will

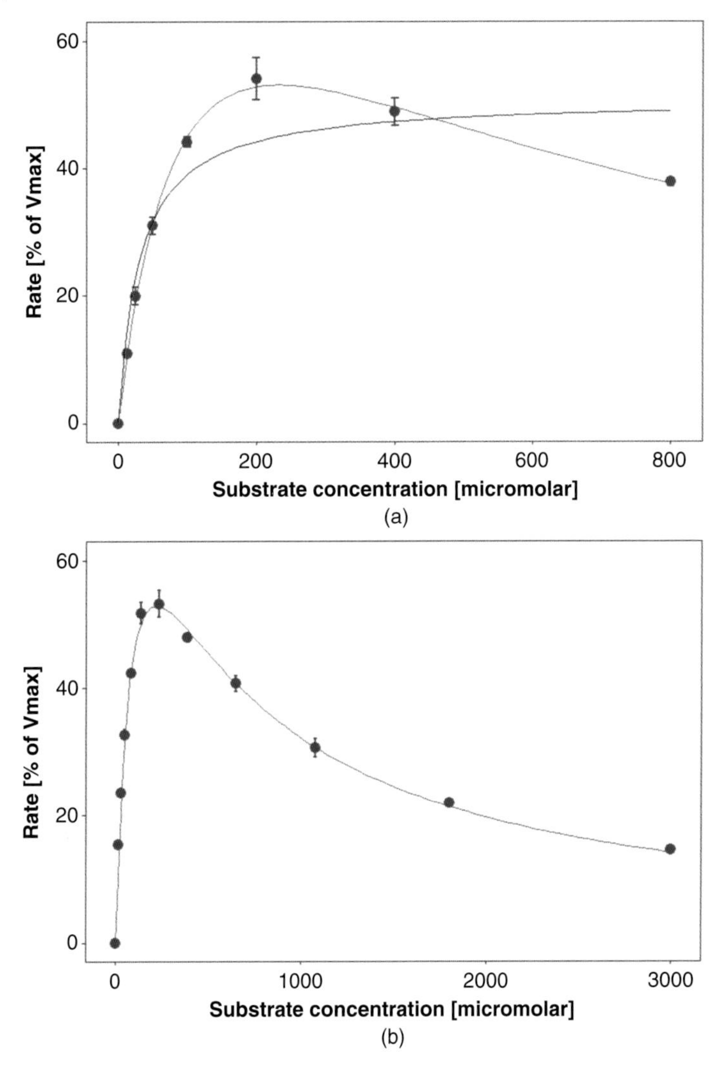

Figure 9.14 Experimental determination of refined estimates of V_{max} and K_M for an enzyme showing substrate inhibition. Examples are simulated as triplicate experiments with random errors added (normal distribution, SD = 5%) for verisimilitude. (a) Example of an initial experiment showing substrate inhibition (expected $V_{max} = 100\%$, $K_M = 100\ \mu M$, $K_i = 500\ \mu M$). The data obtained show a clear deviation from the Michaelis–Menten equation (red line; $V_{max} = 51 \pm 2\%$, $K_M = 30 \pm 7\ \mu M$). Note that the deviation is not readily apparent from the parameter estimates obtained (which might represent a result with high errors and a less accurate initial estimate of K_M) and that it is necessary to examine the graph. The data fit well to Eq. (9.13) (blue line; $V_{max} = 100 \pm 10\%$, $K_M = 110 \pm 20\ \mu M$; $K_i = 500 \pm 100\ \mu M$). Note that while the fitted parameters are a good estimate of the modelled data, they are not obviously more accurate than the Michaelis–Menten equation parameters, as three parameters have been fitted to eight substrate concentrations (underdetermination). (b) A more detailed experiment with 12 data points (here chosen as 0, 18, 30, 50, 84, 140, 233, 389, 648, 1080, 1800, 3000 μM substrate). Accurate estimates are obtained for all three parameters (observed $V_{max} = 99 \pm 4\%$, $K_M = 97 \pm 8\ \mu M$, $K_i = 500 \pm 40\ \mu M$).

also provide accurate determination of all of the kinetic parameters:

$$v = \frac{V_{max}[A][B]}{K_M^B[A] + K_M^A[B] + [A][B]}, \tag{9.14}$$

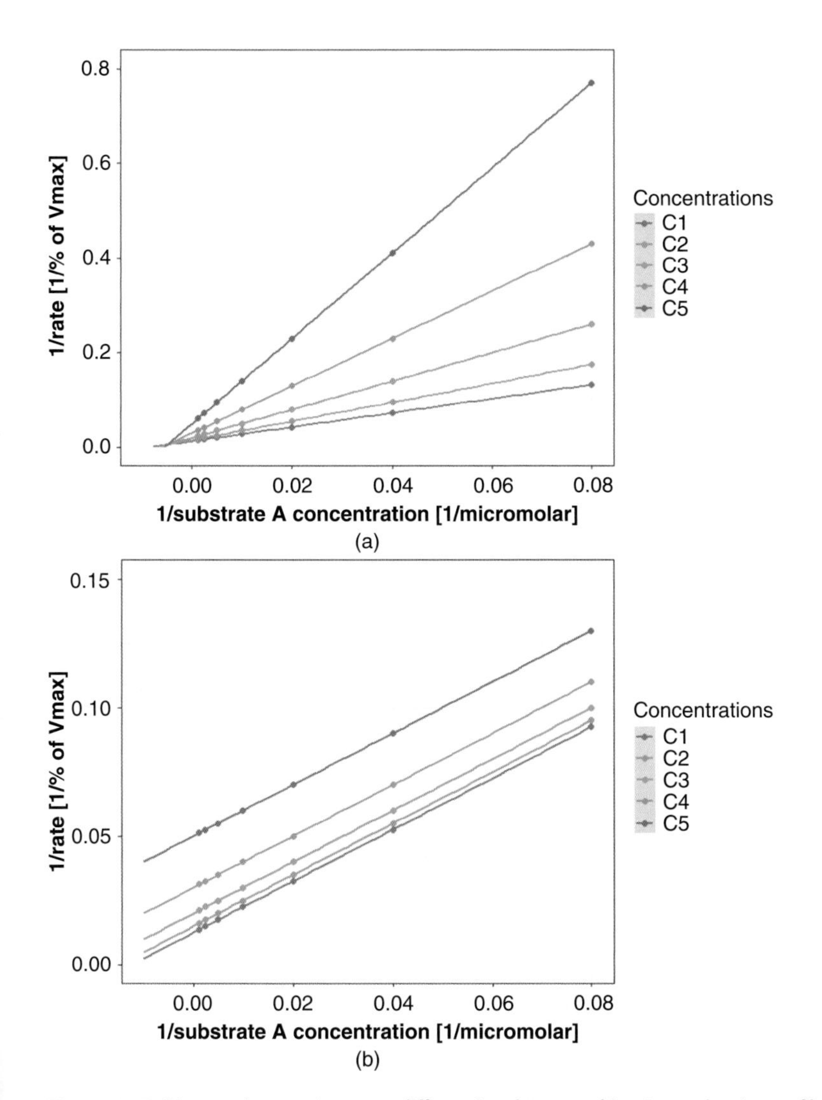

Figure 9.15 Diagnostic experiment to differentiate between kinetic mechanisms of bisubstrate enzymes. Enzymes are modelled with parameters of $V_{max} = 100\%$, $K_M{}^A = 100\ \mu M$, $K_M{}^B = 300\ \mu M$, $K_i{}^A = 200\ \mu M$. In each case, five parallel experiments at different concentrations of substrate B are performed (C1, C2, C3, C4 and C5: 4×, 2×, 1×, 0.5×, 0.25× $K_M{}^B$), testing eight concentrations of A. Lineweaver–Burk plots of perfect data are shown. (a) sequential mechanism; (b) ping-pong mechanism.

$$v = \frac{V_{\max}[A][B]}{K_i^A K_M^B + K_M^B[A] + K_M^A[B] + [A][B]} \qquad (9.12).$$

These equations for testing bisubstrate mechanisms correspond to the sequential and ping-pong mechanisms, respectively.

The quality of fit to the two models can then be compared statistically to determine the likelihood that each model is correct. In this case an appropriate test is the Akaike Information Criterion test [43, 44]. This test uses likelihood-based methods to distinguish which of two (or more) theoretical models are more likely to give the data observed, and is readily implemented by major statistical packages. The experiment should be designed to collect sufficient data to give the desired statistical confidence in the preferred model (Figure 9.17).

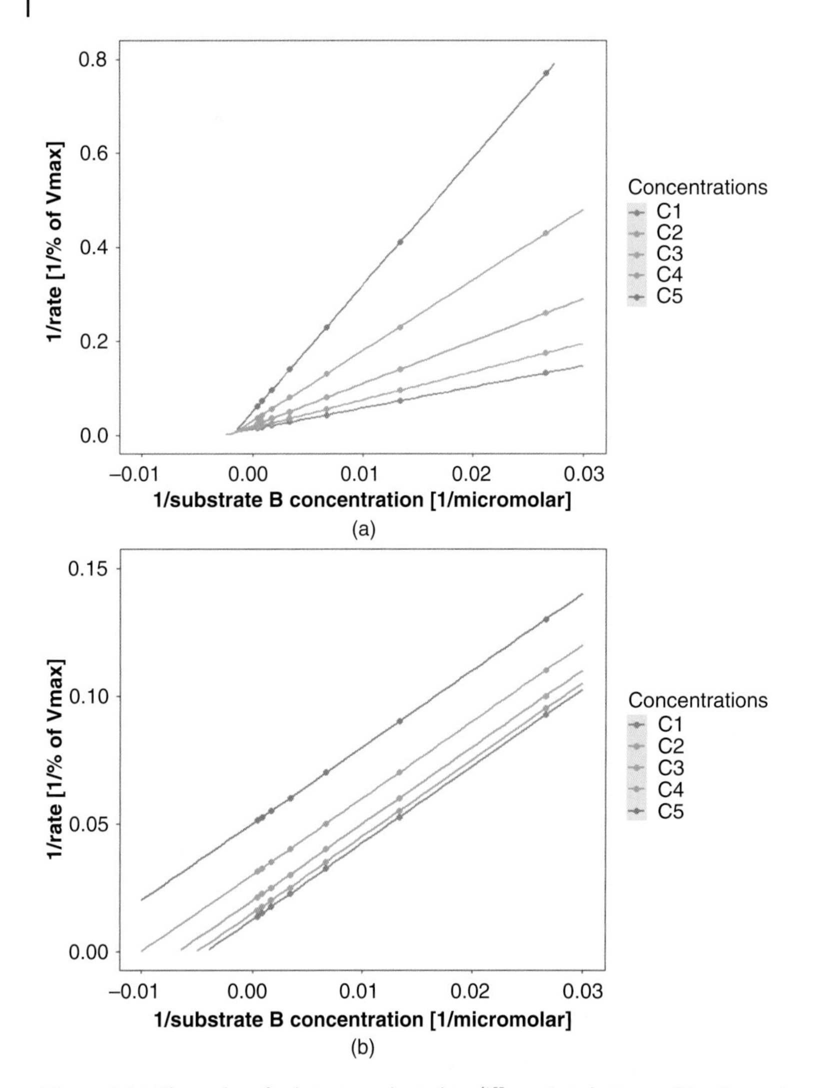

Figure 9.16 The order of substrates selected to differentiate between kinetic mechanisms of bisubstrate enzymes does not matter (cf. Figure 9.15). Enzymes are modelled with parameters of V_{max} = 100%, $K_M{}^A$ = 100 μM, $K_M{}^B$ = 300 μM, $K_i{}^A$ = 200 μM). In each case, five parallel experiments at different concentrations of substrate A are performed (C1, C2, C3, C4 and C5 correspond to 4×, 2×, 1×, 0.5×, 0.25 $K_M{}^A$), each exploring eight concentrations of B. Lineweaver–Burk (double reciprocal) plots of perfect data are shown. (a) Sequential mechanism; (b) ping-pong mechanism.

This experiment can identify a ping-pong mechanism. However, it cannot distinguish between the random equilibrium sequential and ordered sequential mechanisms. To distinguish between these two mechanisms, a detailed product inhibition study must be performed, using each product as the inhibitor with both substrates. The product inhibition equations for the two reactions types are beyond the scope of this chapter (for detail see Chapter 6 of [21] and Chapter 6 of [19]). The experiments should be performed using one substrate at either $K_{M\,app}$ or saturating levels (see Table 9.1) and varying the other substrate above and below $K_{M\,app}$. The products should be tested using at least five product concentrations, ideally covering a little product, the $K_{M\,app}$ of the cognate substrate and the 'saturating' concentration of the cognate substrate (see the *protocol* below). An example of such an experiment is shown in Figure 9.18. The range of useful experiments and the expected outcomes for the different mechanisms are listed in Table 9.1.

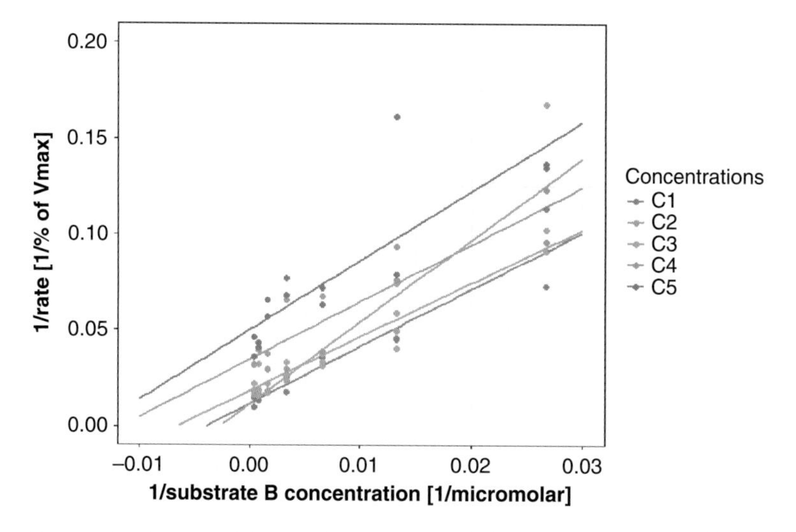

Figure 9.17 Errors in data make interpretation challenging. Data are modelled with the same parameters as Figure 9.16b, but with errors added (normal distributions with SD of 20% proportional error, 10% of V_{max} absolute error) to simulate the increased errors seen at low substrate concentrations. Data are simulated in duplicate: diagnosing the correct mechanism would be challenging.

Table 9.1 Experimental parameters for determining the mechanism of a bisubstrate enzyme using substrate inhibition.

Substrate varied	Product inhibitor used	Constant substrate and concentration	Inhibition pattern		
			Ordered 'steady state'	Ordered equilibrium	Rapid random equilibrium
A	Q	B, $K_M{}^B$	Competitive[a)]	Competitive	Competitive
B	Q	A, $K_M{}^A$	Non-competitive	Competitive	Competitive[b)]
A	P	B, $K_M{}^B$	Non-competitive	No inhibition	Competitive[b)]
A	P	B, 10–50× $K_M{}^B$	Uncompetitive	No inhibition	Competitive[c)]
B	P	A, $K_M{}^A$	Non-competitive	No inhibition	Competitive

A more detailed description is given in [21], which also provides details for more unusual bisubstrate enzymes.

a) In rare cases where there is isomerisation of the enzyme between states, this can show non-competitive inhibition.

b) Where the product binds only with the other substrate present, this can manifest as other types of inhibition.

c) In cases where only the EBQ complex is a dead-end complex, no inhibition will be seen.

9.4 Technique/Protocol: Determination of Michaelis–Menten Parameters for a Bisubstrate Enzyme and Use of Product Inhibition to Determine Mechanism

All of the experiments shown in this protocol will be illustrated using a model glucose-6-phosphate dehydrogenase enzyme from *Leuconostoc mesenteroides* (Figure 9.19), which can readily be obtained commercially. This enzyme has well-established kinetics and shows an ordered 'steady-state' sequential mechanism [49, 50]. This will be modelled to show no substrate inhibition or cooperativity and product inhibition by NADPH, as has been experimentally established [48]. When analysing the data, these should be considered as discussed above. The enzyme will be considered to have two substrates (*A*: $NADP^+$ and *B*: glucose-6-phosphate) and two products (*P*: 6-phospho-D-glucono-1,5-lactone and *Q*: NADPH).

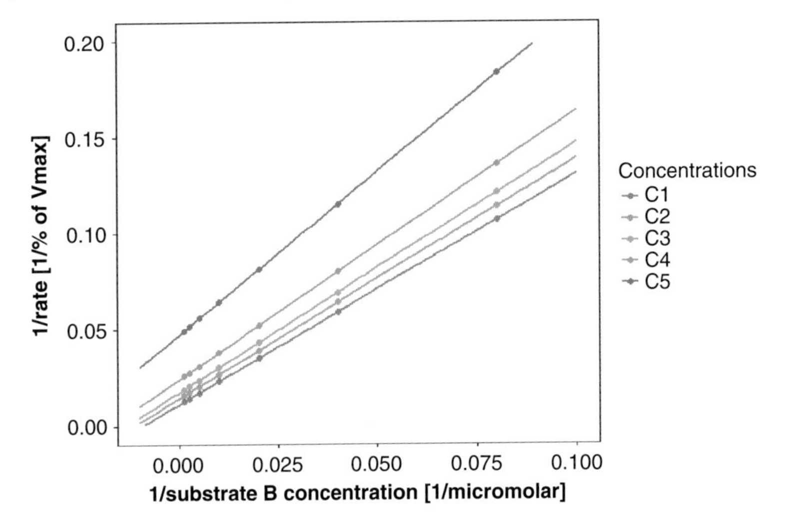

Figure 9.18 Example of a product inhibition experiment. Data are modelled for a steady-state ordered mechanism, using product P as the inhibitor, varying substrate A, with substrate B at 10× K_M. Parameters are as for Figure 9.17b, with K_M and K_i for the products set at double the substrate K_M. The resulting Lineweaver–Burk plot is nearly but not perfectly parallel. Akaike's Information Criterion clearly identifies this as uncompetitive inhibition ($P < 10^{-10}$).

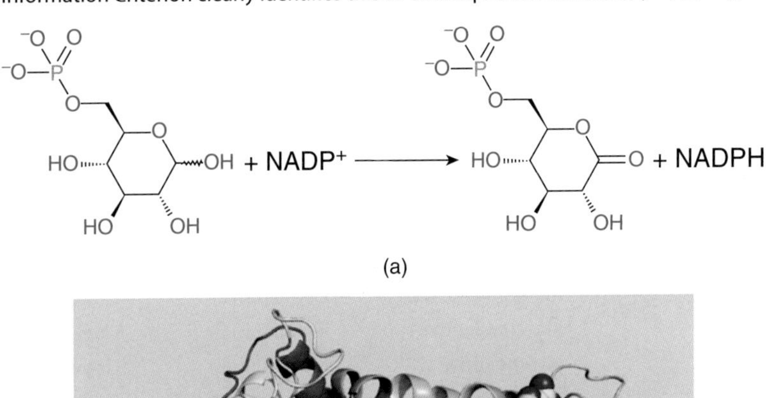

(a)

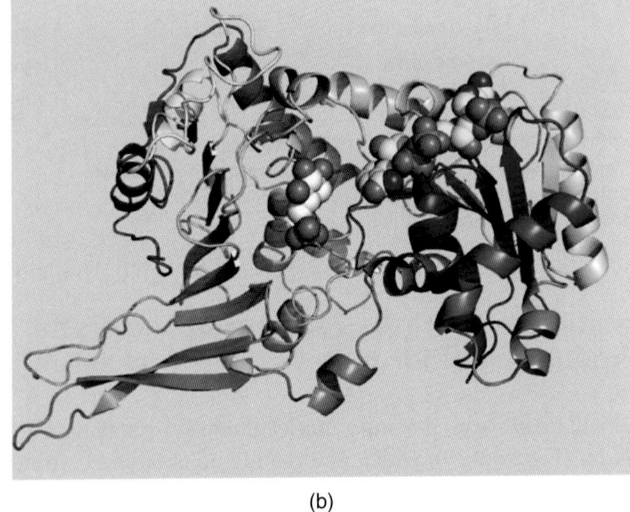

(b)

Figure 9.19 Glucose-6-phosphate dehydrogenase (G6PD) reaction. (a) Overview of G6PD reaction. (b) Structure of *Leuconostoc mesenteroides* G6PD in complex with glucose-6-phosphate (spheres, carbons white) and NADP$^+$ (spheres, carbons yellow). Figure generated using PyMOL from PDB IDs 1H9A [45] and 1E77 [46] using PyMOL. Approximate kinetic parameters for *L. mesenteroides* G6PD are [47, 48]: K_M^{NADP+}: 5.7 μM, $K_M^{glucose\text{-}6\text{-}phosphate}$: 81 μM, K_i^{NADP+}: 5.07 μM, K_i^{NADPH}: 37.6 μM, k_{cat}: 288 per second.

9.4.1 Method Requirements

- Suitable enzyme assay for determining the initial reaction rate at different substrate and product concentrations.
- Sufficient enzyme and substrates for up to 600 reactions, with each substrate at saturating concentrations.
- Good statistical software (e.g. Graphpad Prism, RStudio, KaleidaGraph, SPSS).

9.4.2 Procedure

9.4.2.1 Determine Suitable Concentrations of the Enzyme to be Used

1. Define an initial estimate of K_M for each component. This should be done using any literature available for substrates A and B with enzymes of the class tested. Where no such literature exists, define K_M as one-tenth of the highest concentration of the relevant substrate that can be practically used.
2. Perform six to eight rate experiments taking single readings with different concentrations of enzyme. Experiments should be performed using substrates A and B at 10× the estimated K_M (step 1). The enzyme concentrations used should start at the highest practical concentration of the enzyme, followed by four- to 10-fold dilutions (depending on the concentrations used and any prior knowledge; where there is no prior knowledge, use 10-fold dilutions). A no enzyme control should be included, using only carrier (Figure 9.20a and b).
3. Plot the rate of the reaction observed in step 2 against the enzyme concentration (Figure 9.20c). The enzyme should be used only in a range where the rate is proportional to the enzyme concentration.
4. Examine the data from step 2. Select a concentration that meets the criteria from step 3: where sufficient data can be obtained to get an accurate rate measurement in the first 10% of substrate use for a continuous assay[3]; where the rate is consistent over at least two time periods when using a stopped assay; and where 5–10% of the observed rate could still be measured above the background level for the reaction.
5. Perform eight rate experiments in triplicate using different concentrations of the enzyme. Perform these at 8×, 4×, 2×, 1×, 0.5×, 0.25× and 0.125× the estimated enzyme concentration determined in step 4, with a no enzyme control (Figure 9.21).
6. Repeat steps 3 and 4 with these more detailed data. Select an enzyme concentration that meets the criteria discussed in these steps as well as possible (Figure 9.9).

9.4.2.2 Determine Estimates of K_M for both Substrates[4]

7. Perform eight rate experiments taking single readings for substrate A. Use the enzyme at the concentration determined in step 6. Use substrate B at 10× the estimated K_M. For substrate A, choose the concentrations based on the quality of the K_M estimate in step 1. Where the K_M was estimated from the literature, perform the experiments with substrate A concentrations of 8×, 4×, 2×, 1×, 0.5×, 0.25× and 0.125× the estimated K_M, with a no substrate control. Where the highest practicable concentration of substrate A is being used, use fourfold dilutions of this (i.e. 1×, 0.25×, 0.0625×, 0.0156×, 0.004×, 0.001× and 0.00025×) with no substrate control (Figure 9.22).
8. Fit these data to the Michaelis–Menten equation using a statistical package. The data will generally show one of the following outcomes (Figure 9.11):
   - The data fit reasonably well to the Michaelis–Menten equation, with the estimated K_M well within the range of substrate concentrations tested. Proceed to step 9.

---

3 In many cases, this can prove impossible to achieve. Increasing this range to 20% of substrate use is still valid (see Figure 9.5), but carries a greater risk of confounding product inhibition effects.

4 When determining the substrate concentrations in these sections, keep in mind that one of the substrates or products will be measured to determine the reaction rate. This may well affect the concentrations of substrate that can be practically used. This may make accurate determination of K_M for one substrate impossible and use of saturating levels of substrate impractical. In this case the $K_{M\,app}$ for the other substrate can be determined at a concentration as close to ideal as is practicable.

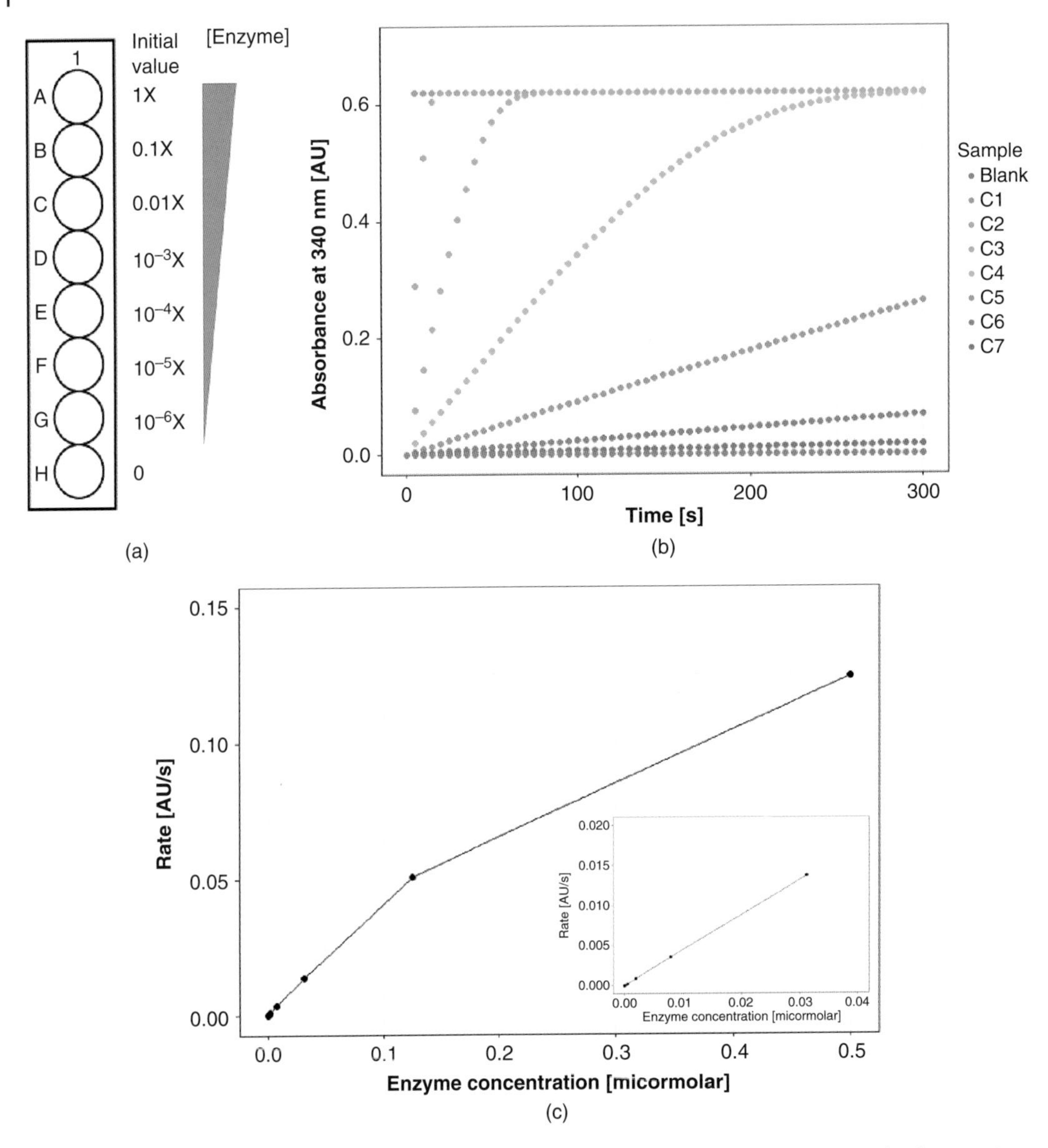

Figure 9.20 Design of the initial experiment to determine enzyme concentration. (a) Plate setup for the experiment where the likely enzyme concentration is unknown. For row A, the highest practical concentration of enzyme should be used. (b) Example with *L. mesenteroides* glucose-6-phosphate dehydrogenase. Here, fourfold dilutions of enzyme are shown from a starting concentration of 500 nM (C1–C7; initial level calculated from the enzyme unit definition). The assay oxidises $NADP^+$ to NADH, which is measured at 340 nm (AU: absorbance unit). (c) Plot of the initial reaction rate against enzyme concentration shows that at high enzyme concentrations, the rate is no longer proportional to enzyme concentration. The inset shows that linearity is maintained up to C3. The data suggest that an enzyme concentration between C4 and C5 would be most appropriate to give enough data to measure and a measureable rate at low substrate concentration.

- The data appear to follow a straight line passing through the origin. If it is possible to increase the concentration of substrate *A*, repeat step 8 with the experiment starting with the highest practicable concentration. Where this has already been done, it will not be possible to determine kinetic parameters for substrate *A*. An alternative is discussed above. Proceed to step 9 for substrate *B*, using the highest practicable concentration of substrate *A*.

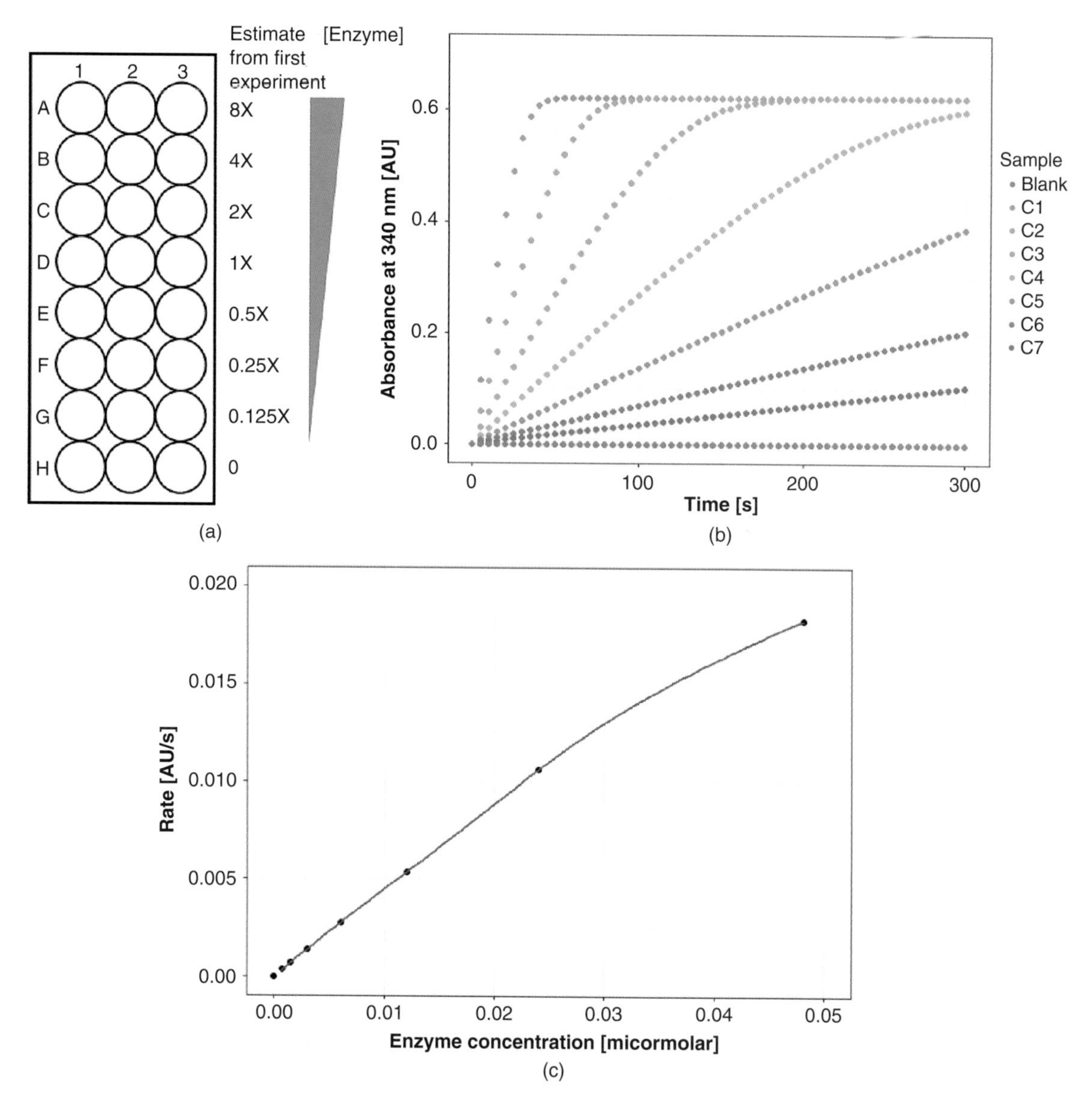

Figure 9.21 Design of the refined experiment to determine enzyme concentration. (a) Plate setup for the experiment, using the result from an initial experiment to guide enzyme concentrations. (b) Example with *L. mesenteroides* glucose-6-phosphate dehydrogenase. Here, twofold dilutions of enzyme are shown from a starting concentration of 48 nM (C1–C7; 48 nM represents 8× the suggested value of 6 nM from Figure 9.20). The assay is designed as in Figure 9.20. (c) Plot of the initial reaction rate against enzyme concentration shows that the high enzyme concentrations start to lose proportionality to enzyme concentration. Here, an enzyme concentration around C4 (6–8 nM) would be most appropriate to give enough data to measure and a measureable rate at low substrate concentration.

- The enzyme shows a high, similar rate in all conditions. Repeat step 8, starting at one of the lower substrate *A* concentrations and using fourfold dilutions.
- The data fit poorly to the Michaelis–Menten equation. Fit the data to an appropriate more complicated equation as described above. Proceed to step 9; the method should be modified as was discussed in the earlier theoretical section according to the scenario that the data correspond to.

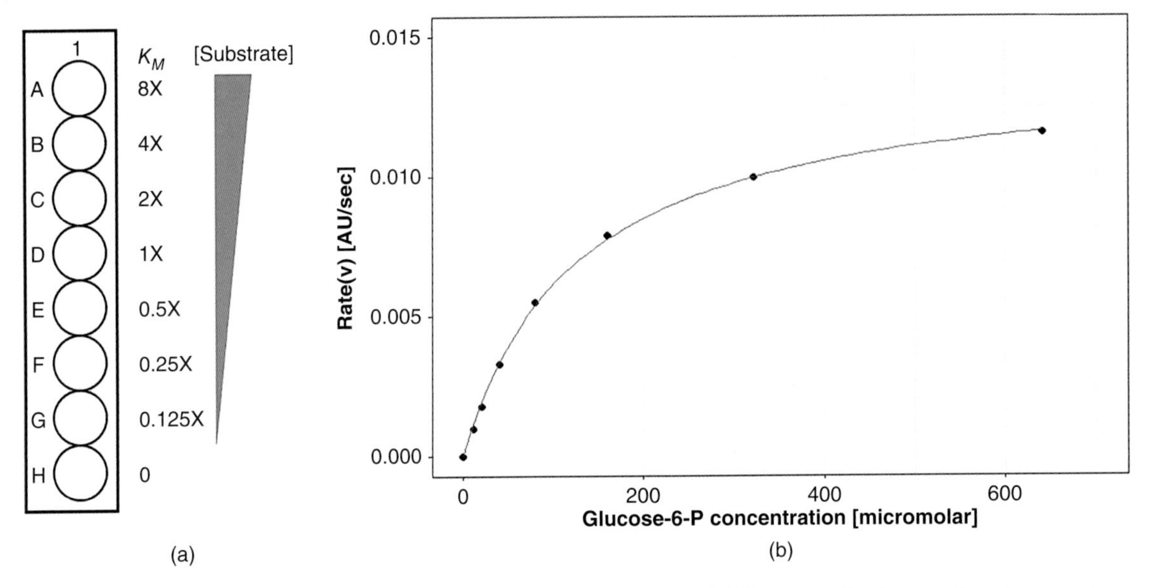

Figure 9.22 Design of the initial experiment to determine the enzyme rate. (a) Plate setup for the experiment, here using literature values to guide substrate concentrations. (b) Example with *L. mesenteroides* glucose-6-phosphate dehydrogenase, varying the concentration of glucose-6-phosphate (AU: absorbance unit). Here, twofold dilutions of substrate are shown from a starting concentration of 640 μM (C1–C7; 640 μM represents 8× the literature value of 81 μM). The model includes random errors in substrate concentrations added and measurements to simulate experimental data. The fit to the Michaelis–Menten equation gives $K_{M\,app}{}^{G6P}$ of 120 ± 10 μM.

9. Repeat step 7, varying the concentration of substrate *B*. Use a concentration of substrate *A* that is 10× the K_M estimated in step 8. Use the highest practicable concentration of substrate *A* if 10× K_M is greater than this.
10. Repeat step 8 for the analysis of the data collected in step 9.
11. Compare the estimate of K_M for substrate *B* used in step 7 with the refined estimate from step 10. If the refined estimate is no more than twice the original estimate, or the highest practicable concentration of substrate *B* was used, proceed to step 13. Otherwise, repeat the estimate of K_M for substrate *A* as described in step 7, using the refined estimate of K_M for substrate *B* from step 10.
12. Compare the estimate of K_M for substrate *A* used in step 9 with the refined estimate from step 10. If the refined estimate is no more than twice the original estimate or the highest practicable concentration of substrate *A* was used, proceed to step 13. Otherwise, repeat the estimate of K_M for substrate *B* as described in step 9, using the refined estimate of K_M for substrate *A* from step 11. Return to step 11 to consider whether another iteration is required.

9.4.2.3 Determine the Kinetic Parameters for both Substrates Accurately

13. Perform eight[5] rate experiments in triplicate using different concentrations of substrate *A*. Perform these using twofold dilutions from 8× the estimated K_M determined in steps 7–12 (i.e. 8×, 4×, 2×, 1.0×, 0.50×, 0.25X× 0.125× K_M) with a no substrate control. Use a substrate *B* concentration of 10× K_M estimated in steps 7–12, or the highest practicable concentration of this substrate if this is lower (Figure 9.23).
14. Fit these data to the Michaelis–Menten equation using a statistical package. Inspect the output carefully for deviations from the expected shape, especially at low substrate concentrations if these are determined with good confidence.

5 More points, with a subtly different concentration range, may be necessary if the kinetics do not follow the Michaelis–Menten kinetics as discussed above.

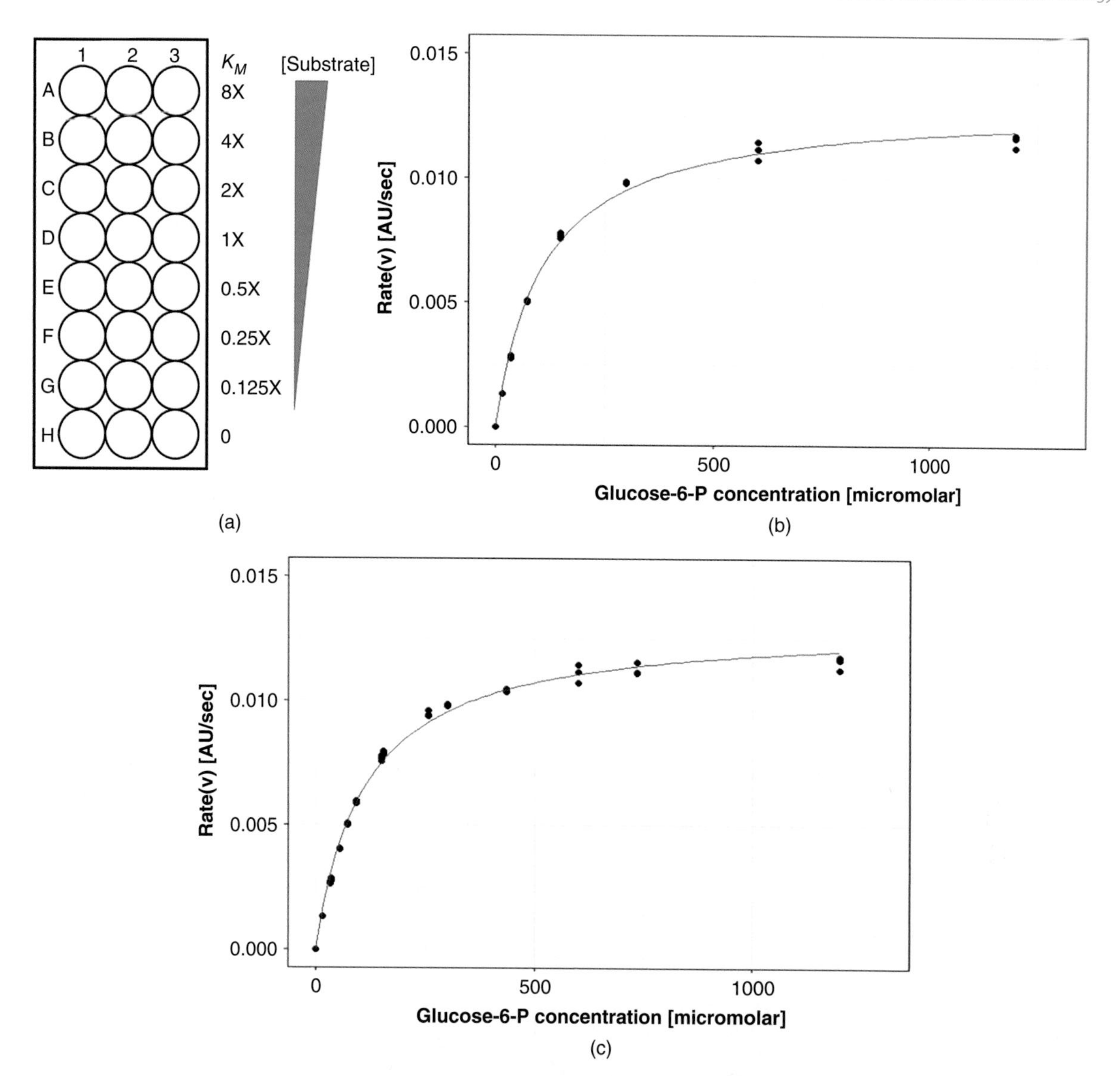

Figure 9.23 Design of refined experiment to determine enzyme kinetic parameters. (a) Plate setup for the experiment, using values from an initial experiment to guide substrate concentrations. (b) Example with *L. mesenteroides* glucose-6-phosphate dehydrogenase, varying the concentration of glucose-6-phosphate. Here, twofold dilutions of substrate are shown from a starting concentration of 1200 μM (C1–C7; 1200 μM represents 8× the estimated value of $K_{M\,app}$ from an initial experiment). The model includes random errors in substrate concentrations added and measurements to simulate experimental data. The fit to the Michaelis–Menten equation gives $K_{M\,app}^{G6P} = 110 \pm 5$ μM, $V_{max} = 13.0 \pm 0.02$ mAU/s. (c) Similar simulation to *B*, where additional data using a smaller substrate range for increased accuracy have been added. Addition data use 1.7-fold from a starting concentration of 750 μM (5× the estimated value of $K_{M\,app}$ from an initial experiment). This simulation gives $K_{M\,app}^{G6P} = 111 \pm 3$ μM, $V_{max} = 13.1 \pm 0.01$ mAU/s. Note that the errors are reduced. Ten repetitions of these simulations show that the results obtained are typical.

15. Repeat step 13, varying the concentrations of substrate B. Use a substrate A concentration of 10× K_M estimated in steps 7–12 or the highest practicable concentration of this substrate if this is lower.
16. Fit these data to the Michaelis–Menten equation as in step 14.
17. If the determined values of K_M differ substantially from the estimates used (here – change by at least twofold), repeat steps 13 and/or 15 using the determined K_M as a new estimate. Iterate until the determined values are consistent with the experimental design. Where the determined K_M increased at least twofold, it will be necessary to repeat the experiments to determine the kinetic parameters of the other substrate.
18. If the accuracy of the determined kinetic parameters is sufficient for the experimental need, proceed to step 19. Otherwise, consider the approaches outlined above for increasing the accuracy of the determined kinetic parameters. A first approach is to perform an additional experiment using a tighter range of substrate concentrations (e.g. using 1.7-fold dilutions from 5× the estimated K_M: 5×, 2.9×, 1.7×, 1.0×, 0.60×, 0.35×, 0.21X×K_M) and combining these data with the data collected in step 13 (Figure 9.23c).

9.4.2.4 Determining the Mechanism of the Bisubstrate Enzyme

19. Perform a series of rate experiments in triplicate using different concentrations of substrates A and B. Perform these at sets of eight concentrations of substrate A, using twofold dilutions from 8× the K_M determined in step 14 as above with a no substrate control. Perform five sets of experiments, using substrate B concentrations of 0.25×, 0.5×, 1×, 2×, 4× the K_M determined in step 16 (Figure 9.24). The outcome of this experiment is independent of whether substrate A or B has more concentrations tested, so it is perfectly valid to choose either substrate as A for the purpose of this assay.
20. Fit the data obtained to the models for ping-pong and sequential reaction mechanisms (Figure 9.25a). If the data fit better to the ping-pong model, this identifies the mechanism as a ping-pong mechanism. If the data fit better to the sequential model, proceed to step 21 to determine which type of sequential mechanism is preferred. In comparing the models, consider how strong the discrimination is. The selected model

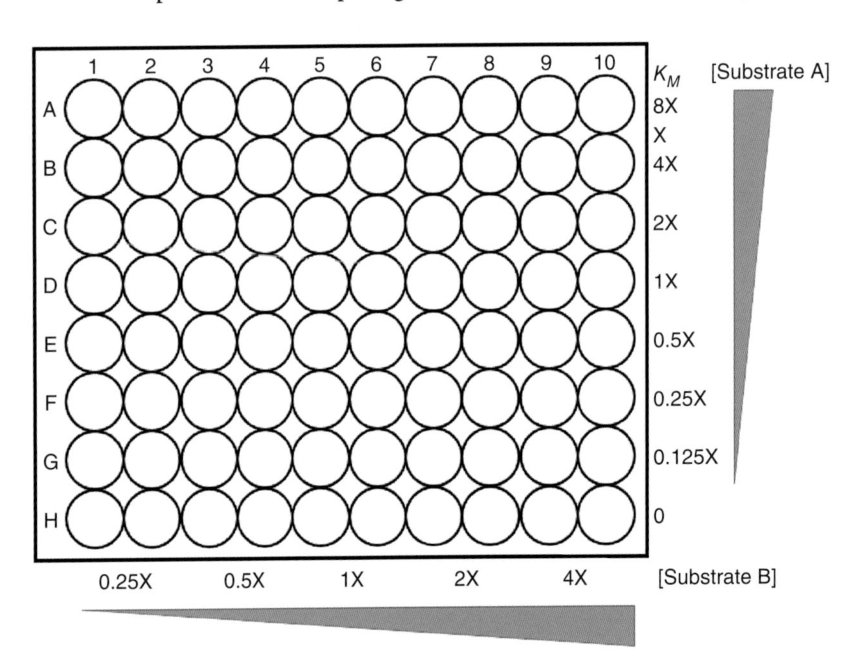

Figure 9.24 Design of the experiment to determine the bisubstrate enzyme mechanism. Suggested plate setup for the experiment, using values from earlier experiments to guide substrate concentrations. For continuous reactions, the plate will likely need to be split into several tests for convenience.

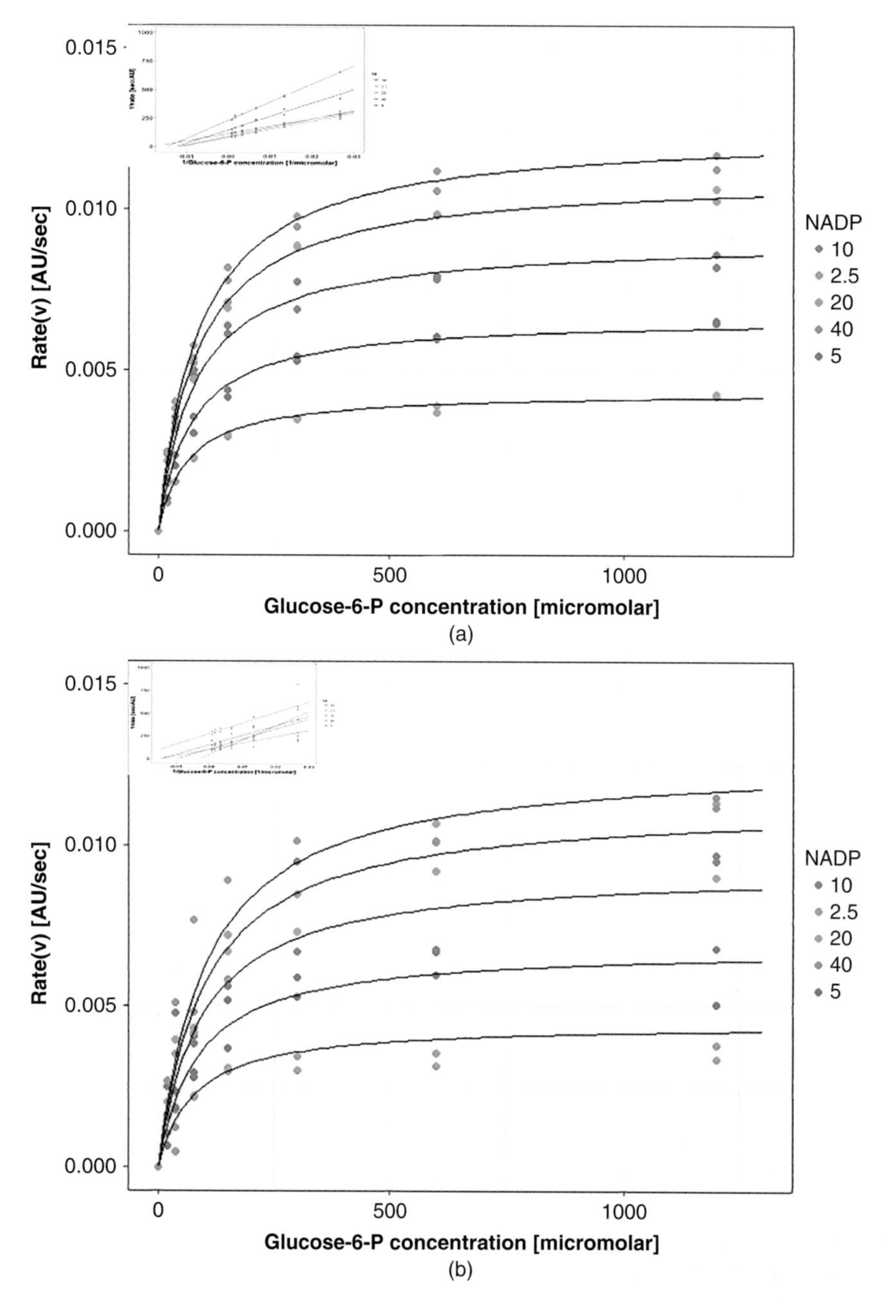

Figure 9.25 Experiments to determine bisubstrate enzyme mechanism. Examples with *L. mesenteroides* glucose-6-phosphate dehydrogenase, varying the concentration of glucose-6-phosphate and $NADP^+$. Here, twofold dilutions of glucose-6-phosphate are shown from a starting concentration of 1200 μM (C1–C7; 1200 μM represents 8× the estimated value of $K_{M\,app}$ from an initial experiment) and five twofold dilutions of $NADP^+$ from a starting concentration of 40 μM (40 μM represents 4× the estimated value of $K_{M\,app}$; this would be practically challenging to achieve as the $K_{M\,app}$ is low, and the reaction would proceed to completion rapidly). Models include random errors in substrate concentrations added and measurements to simulate experimental data. (a) Model data fit better to the sequential model than the ping-pong model (consistent with the method used to generate the model; Akaike's information criterion gives $p = 4 \times 10^{-9}$). Inset: Lineweaver–Burk plot of the same data. (b) Model data with increased errors. These data give an equivocal differentiation between the models (Akaike's information criterion: $p > 0.2$; implemented using RStudio). Inset: Lineweaver–Burk plot of the same data is not discriminating.

should show a significantly better fit than the alternative model. Where the discrimination is equivocal, it may be necessary to collect a wider range of data (Figure 9.25b).

9.4.2.5 Determining which Sequential Mechanism Is Adopted by a Bisubstrate Enzyme

21. Perform a series of rate experiments in duplicate using different concentrations of substrate A and one product. Perform these at sets of eight concentrations of substrate A, using twofold dilutions from 8× the $K_{M\,app}$ determined in step 14 as above with a no substrate control. Perform five sets of experiments, using substrate B concentration of the $K_{M\,app}$ determined in step 16 and the relevant product at 0.25×, 0.5×, 1×, 2×, 4× the $K_{M\,app}$ of the respective substrate determined in steps 14 and 16 (Figure 9.26).
22. Repeat step 21 with the second product, if it is possible to obtain this.
23. Repeat steps 21 and 22, using substrate B at eight different concentrations with constant substrate A.
24. Fit the data obtained for each product with each substrate to the competitive, uncompetitive and non-competitive models of inhibition (Figure 9.26b). Select the model that provides the best fit. In comparing the models, consider how strong the discrimination is. The selected model should show a significantly better fit than the alternative model. Where the discrimination is equivocal, it may be necessary to collect a wider range of data.
25. Compare the obtained results to Table 9.1. This should indicate whether the enzyme adopts a rapid equilibrium random sequential mechanism or an ordered sequential mechanism.

9.5 Case Study: Determination of Michaelis–Menten Parameters for a Bisubstrate Enzyme

Polysaccharides (PS) are macromolecules formed from chains of sugar monosaccharides linked through condensation of hydroxyls on pairs of sugars (Figure 9.27) [51]. These macromolecules have a wide range of important roles in cellular behaviour and homeostasis [52]. PS are essential for the biology of many bacteria, in particular helping to modulate the immune system [53, 54]. Consequently, PS are commonly used as vaccines for infectious disease [55–57] and the enzymes that synthesise them have been proposed as targets for next generation antimicrobials [58, 59]. Understanding mechanism and kinetic parameters for such enzymes is important for this drug design: compounds binding to the site of the first substrate in an ordered sequential mechanism will have greater effects than those binding to the second substrate, for example.

WcbL is a kinase from the tropical pathogen *Burkholderia pseudomallei* that catalyses the 1-phosphorylation of D-*manno*-heptose-7-phosphate (Figure 9.28) [60]. This forms part of the GDP-heptose biosynthesis pathway for formation of *B. pseudomallei*'s capsular polysaccharide [61–63]. Vivoli et al. solved the structure of the enzyme and characterised it biochemically [60]. Initial tests on the enzymes gave estimates for K_M as approximately 50 and 200 μM for D-*manno*-heptose-7-phosphate and ATP, respectively. However, a more detailed assay of the kinetics of ATP revealed a clear deviation from Michaelis–Menten kinetics (Figure 9.29a). Fitting these data to a cooperative model shows a Hill coefficient of 1.9 ± 0.1 (indicating near perfect cooperativity for the dimeric enzyme). This result required a second, more detailed assay to determine: consequently, the estimate of $K_{½}$ for the final experiment was well determined and close to the final, fitted result. The kinetics of the other substrate, D-*manno*-heptose-7-phosphate, also showed cooperativity (Figure 9.29b). In this case, the cooperativity is lower ($h = 1.5 \pm 0.1$). Furthermore, as this substrate shows, a low $K_{½}$ measuring the rates accurately at substrate concentrations lower than $K_{½}$ is challenging for the assay used. The deviation from Michaelis–Menten is far more subtle in this case and might well have been missed with a less accurate assay (as was indeed the case for the first enzyme in this pathway) [25, 62].

Crystal structures of WcbL identified the binding sites for the two substrates (Figure 9.30). The location of the binding sites suggested that the mechanism might be an ordered sequential reaction: the sugar–phosphate substrate binds deeply into the active site, while ATP binds over the active site and apparently blocks access to the sugar binding site. The enzyme was therefore tested using product inhibition. Here, one product

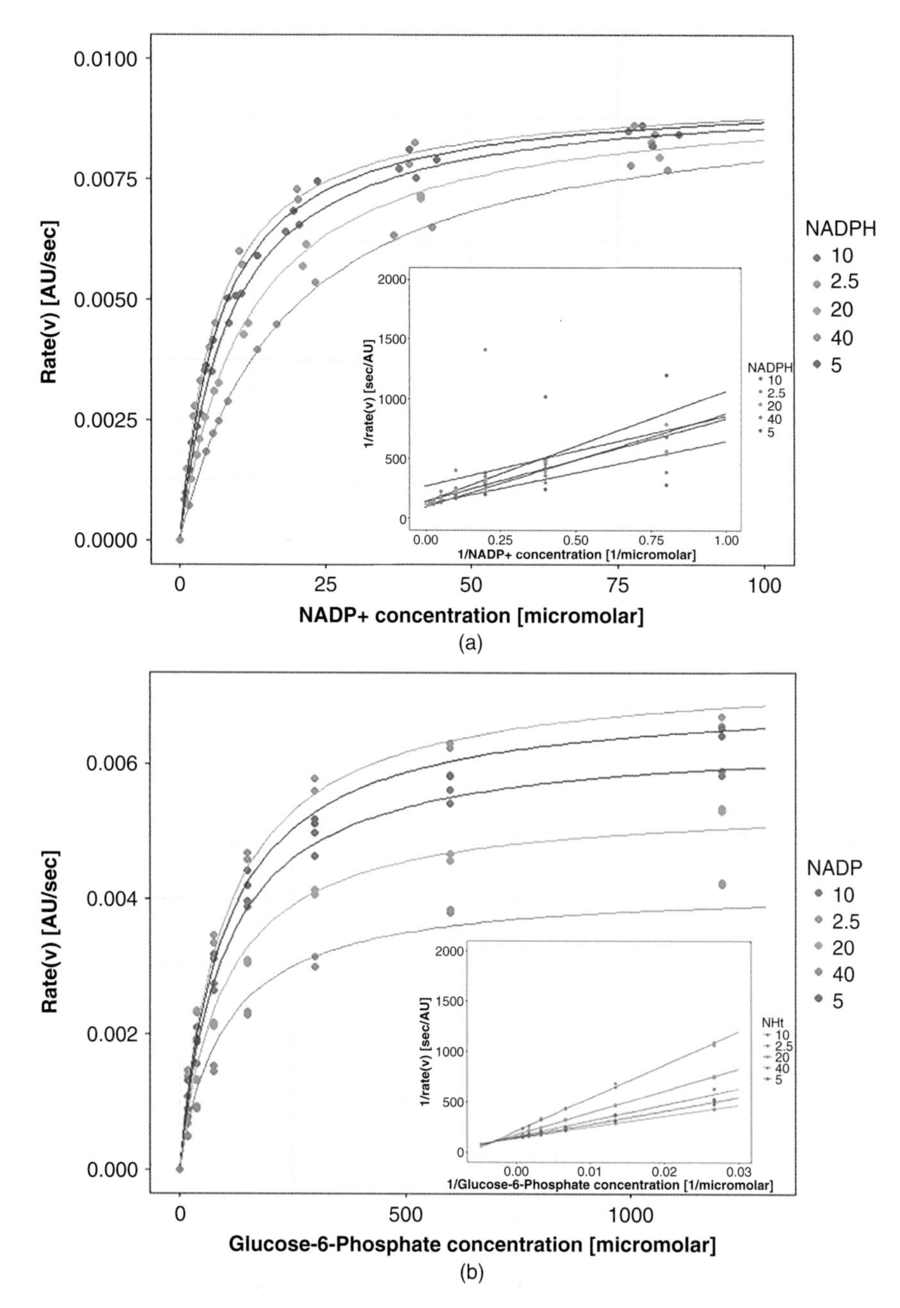

Figure 9.26 Experiments to determine the bisubstrate enzyme mechanism. Examples with *L. mesenteroides* glucose-6-phosphate dehydrogenase, using NADPH as a product inhibitor (the second product is insufficiently stable to use experimentally). Models include random errors in substrate concentrations added and measurements to simulate experimental data. (a) Experiments varying the concentration of $NADP^+$ at constant glucose-6-phosphate (150 μM). Twofold dilutions of $NADP^+$ are shown from a starting concentration of 80 μM and five twofold dilution of NADPH from a starting concentration of 40 μM. The model fits best to competitive inhibition (Akaike's Information Criterion; $p < 10^{-30}$). (b) Experiments varying the concentration of glucose-6-phosphate at constant $NADP^+$ (10 μM). Twofold dilutions of glucose-6-phosphate are shown from a starting concentration of 1200 μM and five twofold dilutions of NADPH from a starting concentration of 40 μM. The model fits best to non-competitve inhibition (Akaike's Information Criterion; $p < 10^{-11}$; implemented in RStudio). The corresponding Lineweaver–Burk plots are shown inset for reference.

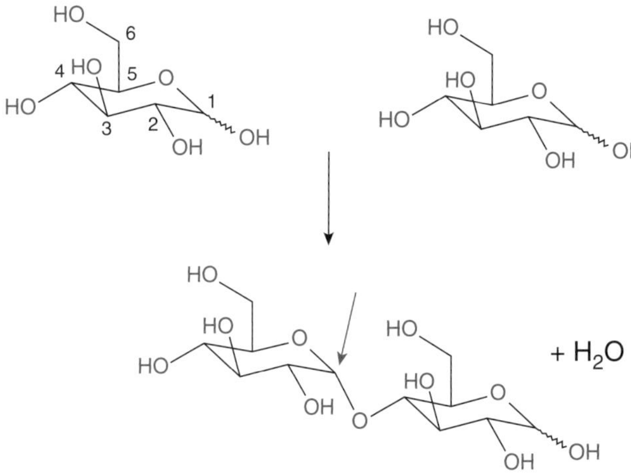

Figure 9.27 Basic structure of polysaccharides. Top left: a monosaccharide unit (glucose shown) has the generic formula $C_xH_{2x}O_x$. The most common have $x = 5–7$. Most sugars form rings as shown. The stereochemistry at positions 2–5 is fixed; position 1 has interchangeable stereochemistry in water. Two sugars can be polymerised by a condensation reaction, producing a disaccharide (bottom). Note that on polymerisation, the centre indicated by the red arrow now has fixed stereochemistry.

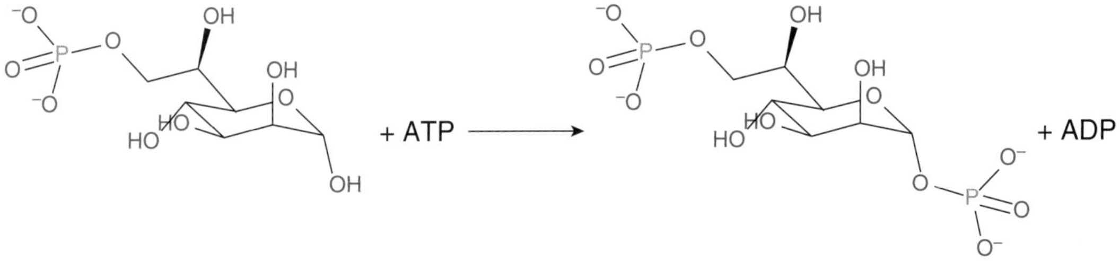

Figure 9.28 Overview of the WcbL reaction. The seven carbon sugars are used by bacteria in their surface polysaccharides.

(D-*manno*-heptose-1,7-bisphosphate) was not readily available. Consequently, only the other product (ADP) was tested. This showed non-competitive inhibition when varying the non-cognate sugar substrate and uncompetitive inhibition when varying ATP (at sugar concentrations well above K_M). In both cases, there was strong statistical support both for inhibition and for the selected mechanism over competitive inhibition. These data indicate an ordered, steady-state bisubstrate reaction (Table 9.1) and that ADP is expected to be the first product released (Figure 9.7a, blue square); no inhibition would be expected for an equilibrium ordered reaction and competitive inhibition for a rapid random equilibrium mechanism.

The experimental data on WcbL therefore provide strong evidence for the enzyme mechanism and for cooperativity in the enzyme. Similar mechanisms have been observed in other sugar kinases (e.g. [64, 65]). The kinetic and structural data here correlate very well, giving strong confidence in the conclusions of the kinetic studies.

9.6 More Advanced Methods

This chapter has described the determination of the steady-state kinetics of common enzymes. This is sufficient for many applications. Where additional insight is required, more advanced methods can

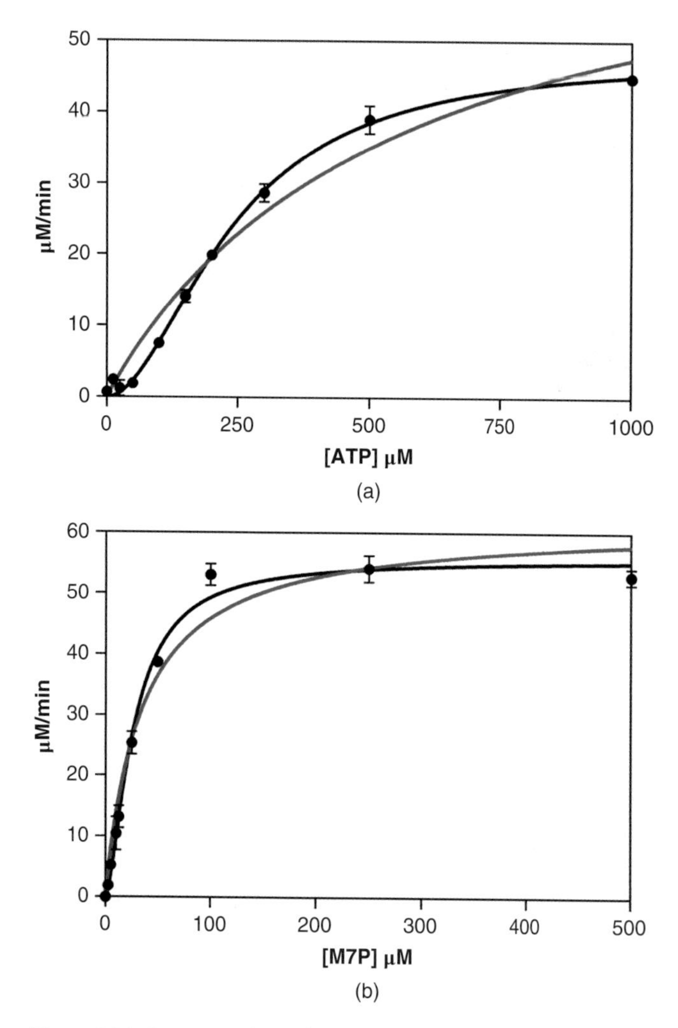

Figure 9.29 Enzyme activity of WcbL. (a) Activity of WcbL with ATP. Here, activity was tested at 10 ATP concentrations. The data show a clear deviation from Michaelis–Menten kinetics (red line). The Hill value h determined is 1.9 ± 0.1, indicating a high degree of cooperativity for a dimeric enzyme. (b) Activity of WcbL with the heptose substrate (M7P). This again shows cooperativity, although not to the same extent ($h = 1.5 \pm 0.1$).

be used. *Pre-steady state kinetics* uses instruments providing millisecond resolution (e.g. stopped-flow or quenched-flow). The time resolution and amount of data obtained allows determination of individual kinetic parameters rather than derivative parameters such as K_M [66, 67], allowing deep insight into mechanisms [68]. *Kinetic isotope effects* are observed when the isotope of one atom involved in the reaction is altered [19, 66]. The reaction rate generally slows when the mass of an atom involved in the reaction increases. Such reactions allow the experimenter to distinguish between possible reaction mechanisms, and potentially to identify rate-limited steps in reactions (e.g. [69]). Finally, structural methods (e.g. with free-electron lasers) are providing greater insight by following reactions with high time resolution after activating the enzyme (e.g. [70, 71]). These methods are currently only available for photoactivatable enzymes; further developments will increase the range of systems suitable for these studies.

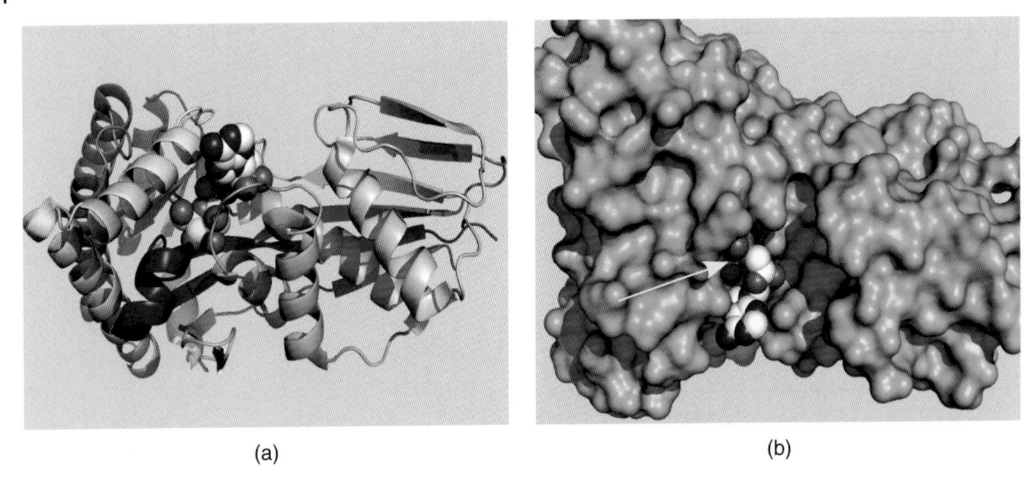

Figure 9.30 WcbL binds to its substrates in an ordered manner. The structure of WcbL bound to substrate analogues reveals that the binding of ATP occludes the sugar binding site. (a) Cartoon representation of a WcbL monomer bound to both substrates. Mannose (phosphoheptose analogue) is shown as spheres with yellow carbons; AMP-PNP (ATP analogue) is shown as spheres with white carbons. (b) Surface representation of WcbL bound to substrates. ATP binds over the sugar binding pocket. One surface of the sugar can be observed in the occluded pocket (yellow arrow). The figure was prepared using PyMOL, based on PDB IDs 4UT4 and 4UTG [60].

9.7 Concluding Remarks

Biology requires the coordination of many enzyme reactions to operate efficiently. Today's medicine and biotechnology exploit our current understanding of enzymes to provide interventions and solutions. The increasing capabilities of microfluidics and robotics offer the potential to generate enzyme data at a significantly higher rate. This offers the opportunity to study enzymes in far greater detail, obtaining high quality data in smaller volumes at higher rates. These kinetic data can be used to optimise assays and in particular build enzyme cascades with ideal properties for a desired application. A thorough understanding of the properties of enzymes will underpin their use in diverse areas such as biotechnology (e.g. sensing), manufacturing (e.g. biocatalytic production of fine chemicals) and healthcare (e.g. diagnostic tests).

References

1 Erlandsen, H., Flatmark, T., Stevens, R.C., and Hough, E. (1998). Crystallographic analysis of the human phenylalanine hydroxylase catalytic domain with bound catechol inhibitors at 2.0 A resolution. *Biochemistry* 37 (45): 15638–15646.
2 Ranzani, A.T. and Cordeiro, A.T. (2017). Mutations in the tetramer interface of human glucose-6-phosphate dehydrogenase reveals kinetic differences between oligomeric states. *FEBS Lett.* 591 (9): 1278–1284.
3 Line, K., Isupov, M.N., and Littlechild, J.A. (2004). The crystal structure of a (−) gamma-lactamase from an Aureobacterium species reveals a tetrahedral intermediate in the active site. *J. Mol. Biol.* 338 (3): 519–532.
4 Blau, N., van Spronsen, F.J., and Levy, H.L. (2010). Phenylketonuria. *Lancet* 376 (9750): 1417–1427.
5 Mitchell, J.J., Trakadis, Y.J., and Scriver, C.R. (2011). Phenylalanine hydroxylase deficiency. *Genet. Med.* 13 (8): 697–707.
6 Cappellini, M.D. and Fiorelli, G. (2008). Glucose-6-phosphate dehydrogenase deficiency. *Lancet* 371 (9606): 64–74.

7 Kanehisa, M., Furumichi, M., Tanabe, M. et al. (2017). KEGG: new perspectives on genomes, pathways, diseases and drugs. *Nucleic Acids Res.* 45 (D1): D353–D361.

8 Blau, N., Belanger-Quintana, A., Demirkol, M. et al. (2009). Optimizing the use of sapropterin (BH(4)) in the management of phenylketonuria. *Mol. Genet. Metab.* 96 (4): 158–163.

9 Minucci, A., Moradkhani, K., Hwang, M.J. et al. (2012). Glucose-6-phosphate dehydrogenase (G6PD) mutations database: review of the 'old' and update of the new mutations. *Blood Cells Mol. Dis.* 48 (3): 154–165.

10 Luzzatto, L. and Seneca, E. (2014). G6PD deficiency: a classic example of pharmacogenetics with on-going clinical implications. *Br. J. Haematol.* 164 (4): 469–480.

11 Youngster, I., Arcavi, L., Schechmaster, R. et al. (2010). Medications and glucose-6-phosphate dehydrogenase deficiency: an evidence-based review. *Drug Saf.* 33 (9): 713–726.

12 Minucci, A., Giardina, B., Zuppi, C., and Capoluongo, E. (2009). Glucose-6-phosphate dehydrogenase laboratory assay: how, when, and why? *IUBMB Life* 61 (1): 27–34.

13 Bornscheuer, U.T., Huisman, G.W., Kazlauskas, R.J. et al. (2012). Engineering the third wave of biocatalysis. *Nature* 485 (7397): 185–194.

14 Reetz, M.T. (2013). Biocatalysis in organic chemistry and biotechnology: past, present, and future. *J. Am. Chem. Soc.* 135 (34): 12480–12496.

15 Bommarius, A.S. (2015). Biocatalysis: a status report. *Annu. Rev. Chem. Biomol. Eng.* 6: 319–345.

16 Wohlgemuth, R. (2010). Biocatalysis – key to sustainable industrial chemistry. *Curr. Opin. Biotechnol.* 21 (6): 713–724.

17 Wohlgemuth, R., Plazl, I., Znidarsic-Plazl, P. et al. (2015). Microscale technology and biocatalytic processes: opportunities and challenges for synthesis. *Trends Biotechnol.* 33 (5): 302–314.

18 Taylor, S.J.C., Mccague, R., Wisdom, R. et al. (1993). Development of the biocatalytic resolution of 2-Azabicyclo[2.2.1]Hept-5-En-3-one as an entry to single-enantiomer carbocyclic nucleosides. *Tetrahedron-Asymmetry* 4 (6): 1117–1128.

19 Cornish-Bowden, A. (2012). *Fundamentals of Enzyme Kinetics*. 4th, completely revised and greatly enlarged edition. ed., 498. Weinheim: Wiley-Blackwell.

20 Briggs, G.E. and Haldane, J.B. (1925). A note on the kinetics of enzyme action. *Biochem. J.* 19 (2): 338–339.

21 Cook, P.F. and Cleland, W.W. (2007). *Enzyme Kinetics and Mechanism*. London ; New York: Garland Science. 404 p.

22 Leskovac, V. (2003). *Comprehensive Enzyme Kinetics*. New York: Kluwer Academic/Plenum Pub. 438 p.

23 Koshland, D.E. Jr. and Hamadani, K. (2002). Proteomics and models for enzyme cooperativity. *J. Biol. Chem.* 277 (49): 46841–46844.

24 Koshland, D.E. Jr. (1996). The structural basis of negative cooperativity: receptors and enzymes. *Curr. Opin. Struct. Biol.* 6 (6): 757–761.

25 Vivoli, M., Pang, J., and Harmer, N.J. (2017). A half-site multimeric enzyme achieves its cooperativity without conformational changes. *Sci. Rep.* 7 (1): 16529.

26 Bloom, C.R., Kaarsholm, N.C., Ha, J., and Dunn, M.F. (1997). Half-site reactivity, negative cooperativity, and positive cooperativity: quantitative considerations of a plausible model. *Biochemistry* 36 (42): 12759–12765.

27 Cook, R.A. and Koshland, D.E. Jr. (1970). Positive and negative cooperativity in yeast glyceraldehyde 3-phosphate dehydrogenase. *Biochemistry* 9 (17): 3337–3342.

28 Ferrari, M.E., Fang, J., and Lohman, T.M. (1997). A mutation in *E. coli* SSB protein (W54S) alters intra-tetramer negative cooperativity and inter-tetramer positive cooperativity for single-stranded DNA binding. *Biophys. Chem.* 64 (1–3): 235–251.

29 Porter, C.M. and Miller, B.G. (2012). Cooperativity in monomeric enzymes with single ligand-binding sites. *Bioorg. Chem.* 43: 44–50.

30 Cornish-Bowden, A. (2014). Understanding allosteric and cooperative interactions in enzymes. *FEBS J.* 281 (2): 621–632.

31 Koshland, D.E. Jr. Nemethy, G., and Filmer, D. (1966). Comparison of experimental binding data and theoretical models in proteins containing subunits. *Biochemistry* 5 (1): 365–385.

32 Monod, J., Wyman, J., and Changeux, J.P. (1965). On the nature of allosteric transitions: a plausible model. *J. Mol. Biol.* 12: 88–118.

33 Cui, Q. and Karplus, M. (2008). Allostery and cooperativity revisited. *Protein Sci.* 17 (8): 1295–1307.

34 Domínguez de María, P., Stillger, T., Pohl, M. et al. (2006). Preparative enantioselective synthesis of benzoins and (R)-2-hydroxy-1-phenylpropanone using benzaldehyde lyase. *J. Mol. Catal. B Enzym.* 38: 43–47.

35 Muller, C.R., Perez-Sanchez, M., and Dominguez de Maria, P. (2013). Benzaldehyde lyase-catalyzed diastereoselective C–C bond formation by simultaneous carboligation and kinetic resolution. *Org. Biomol. Chem.* 11 (12): 2000–2004.

36 Apweiler, R., Armstrong, R., Bairoch, A. et al. (2010). A large-scale protein-function database. *Nat. Chem. Biol.* 6 (11): 785.

37 Lineweaver, H., Burk, D., and Deming, W.E. (1934). The dissociation constant of nitrogen-nitrogenase in Azotobacter. *J. Am. Chem. Soc.* 56: 225–230.

38 Harrison, R.K. and Stein, R.L. (1990). Substrate specificities of the peptidyl prolyl cis-trans isomerase activities of Cyclophilin and Fk-506 binding-protein – evidence for the existence of a family of distinct enzymes. *Biochemistry* 29 (16): 3813–3816.

39 Norville, I.H., Harmer, N.J., Harding, S.V. et al. (2011). A *Burkholderia pseudomallei* macrophage infectivity potentiator-like protein has Rapamycin-inhibitable peptidylprolyl isomerase activity and pleiotropic effects on virulence. *Infect. Immun.* 79 (11): 4299–4307.

40 Reed, M.C., Lieb, A., and Nijhout, H.F. (2010). The biological significance of substrate inhibition: a mechanism with diverse functions. *Bioessays* 32 (5): 422–429.

41 Copeland, R.A. (2000). *Enzymes : A Practical Introduction to Structure, Mechanism, and Data Analysis*, 2e. New York: Wiley. 397p.

42 Fierke, C.A., Johnson, K.A., and Benkovic, S.J. (1987). Construction and evaluation of the kinetic scheme associated with dihydrofolate reductase from *Escherichia coli*. *Biochemistry* 26 (13): 4085–4092.

43 Burnham, K.P. and Anderson, D.R. (2004). Multimodel inference – understanding AIC and BIC in model selection. *Sociol. Method Res.* 33 (2): 261–304.

44 Burnham, K.P., Anderson, D.R., and Burnham, K.P. (2002). *Model Selection and Multimodel Inference : A Practical Information-Theoretic Approach*, 2e. New York: Springer. 488p.

45 Naylor, C.E., Gover, S., Basak, A.K. et al. (2001). NADP(+) and NAD(+) binding to the dual coenzyme specific enzyme *Leuconostoc mesenteroides* glucose 6-phosphate dehydrogenase: different interdomain hinge angles are seen in different binary and ternary complexes. *Acta Crystallogr. D Biol. Crystallogr.* 57: 635–648.

46 Cosgrove, M.S., Cover, S., Naylor, C.E. et al. (2000). An examination of the role of Asp-177 in the His-Asp catalytic dyad of *Leuconostoc mesenteroides* glucose 6-phosphate dehydrogenase: X-ray structure and pH dependence of kinetic parameters of the D177N mutant enzyme. *Biochemistry* 39 (49): 15002–15011.

47 Levy, H.R. (1989). Glucose-6-phosphate dehydrogenase from *Leuconostoc mesenteroides*. *Biochem. Soc. Trans.* 17 (2): 313–315.

48 Olive, C., Geroch, M.E., and Levy, H.R. (1971). Glucose 6-phosphate dehydrogenase from *Leuconostoc mesenteroides*. Kinetic studies. *J. Biol. Chem.* 246 (7): 2047–2057.

49 Cosgrove, M.S., Naylor, C., Paludan, S. et al. (1998). On the mechanism of the reaction catalyzed by glucose 6-phosphate dehydrogenase. *Biochemistry* 37 (9): 2759–2767.

50 Levy, H.R., Christoff, M., Ingulli, J., and Ho, E.M.L. (1983). Glucose-6-phosphate-dehydrogenase from Leuconostoc-Mesenteroides – revised kinetic mechanism and kinetics of Atp inhibition. *Arch. Biochem. Biophys.* 222 (2): 473–488.

51 Varki, A. and Kornfeld, S. (2015). Historical background and overview. In: *Essentials of Glycobiology* (ed. A. Varki, R.D. Cummings, J.D. Esko, et al.), 1–18. NY: Cold Spring Harbor.
52 Varki, A. and Gagneux, P. (2015). Biological functions of Glycans. In: *Essentials of Glycobiology* (ed. A. Varki, R.D. Cummings, J.D. Esko, et al.), 77–88. NY: Cold Spring Harbor.
53 Mazmanian, S.K. and Kasper, D.L. (2006). The love-hate relationship between bacterial polysaccharides and the host immune system. *Nat. Rev. Immunol.* 6 (11): 849–858.
54 Neff, C.P., Rhodes, M.E., Arnolds, K.L. et al. (2016). Diverse intestinal bacteria contain putative zwitterionic capsular polysaccharides with anti-inflammatory properties. *Cell Host Microbe* 20 (4): 535–547.
55 Balmer, P., Borrow, R., and Miller, E. (2002). Impact of meningococcal C conjugate vaccine in the UK. *J. Med. Microbiol.* 51 (9): 717–722.
56 Jackson, L.A., Gurtman, A., van Cleeff, M. et al. (2013). Immunogenicity and safety of a 13-valent pneumococcal conjugate vaccine compared to a 23-valent pneumococcal polysaccharide vaccine in pneumococcal vaccine-naive adults. *Vaccine* 31 (35): 3577–3584.
57 Peltola, H. (2000). Worldwide *Haemophilus influenzae* type b disease at the beginning of the 21st century: global analysis of the disease burden 25 years after the use of the polysaccharide vaccine and a decade after the advent of conjugates. *Clin. Microbiol. Rev.* 13 (2): 302–317.
58 Taylor, P.L. and Wright, G.D. (2008). Novel approaches to discovery of antibacterial agents. *Anim. Health Res. Rev.* 9 (2): 237–246.
59 Tedaldi, L. and Wagner, G.K. (2014). Beyond substrate analogues: new inhibitor chemotypes for glycosyltransferases. *Medchemcomm* 5 (8): 1106–1125.
60 Vivoli, M., Isupov, M.N., Nicholas, R. et al. (2015). Unraveling the *B. pseudomallei* Heptokinase WcbL: from structure to drug discovery. *Chem. Biol.* 22 (12): 1622–1632.
61 Cuccui, J., Milne, T.S., Harmer, N. et al. (2012). Characterization of the *Burkholderia pseudomallei* K96243 capsular polysaccharide I coding region. *Infect. Immun.* 80 (3): 1209–1221.
62 Harmer, N.J. (2010). The structure of sedoheptulose-7-phosphate isomerase from *Burkholderia pseudomallei* reveals a zinc binding site at the heart of the active site. *J. Mol. Biol.* 400 (3): 379–392.
63 Reckseidler, S.L., DeShazer, D., Sokol, P.A., and Woods, D.E. (2001). Detection of bacterial virulence genes by subtractive hybridization: identification of capsular polysaccharide of *Burkholderia pseudomallei* as a major virulence determinant. *Infect. Immun.* 69 (1): 34–44.
64 Imriskova, I., Arreguin-Espinosa, R., Guzman, S. et al. (2005). Biochemical characterization of the glucose kinase from *Streptomyces coelicolor* compared to *Streptomyces peucetius* var. caesius. *Res. Microbiol.* 156 (3): 361–366.
65 Rivas-Pardo, J.A., Herrera-Morande, A., Castro-Fernandez, V. et al. (2013). Crystal structure, SAXS and kinetic mechanism of hyperthermophilic ADP-dependent glucokinase from *Thermococcus litoralis* reveal a conserved mechanism for catalysis. *PLoS One* 8 (6): e66687.
66 Fersht, A. (2017). *Structure and Mechanism in Protein Science : A Guide to Enzyme Catalysis and Protein Folding*. New Jersey: World Scientific. 631pp.
67 Johnson, K. (2003). *Kinetic Analysis of Macromolecules : A Practical Approach*. Oxford: Oxford University Press. 256pp.
68 Kellinger, M.W. and Johnson, K.A. (2010). Nucleotide-dependent conformational change governs specificity and analog discrimination by HIV reverse transcriptase. *Proc. Natl Acad. Sci. USA* 107 (17): 7734–7739.
69 Rankin, J.A., Mauban, R.C., Fellner, M. et al. (2018). Lactate Racemase nickel-pincer cofactor operates by a proton-coupled hydride transfer mechanism. *Biochemistry* 57 (23): 3244–3251.
70 Horrell, S., Kekilli, D., Sen, K. et al. (2018). Enzyme catalysis captured using multiple structures from one crystal at varying temperatures. *IUCrJ* 5 (Pt 3): 283–292.
71 Spence, J. and Lattman, E. (2016). Imaging enzyme kinetics at atomic resolution. *IUCrJ* 3: 228–229.

10

Mass Spectrometry and Its Applications

Blagojce Jovcevski and Tara L. Pukala

School of Physical Sciences, University of Adelaide, Adelaide, South Australia 5005, Australia

10.1 Significance

Cellular function is underpinned by the complex interaction of a range of biomolecules, from small compounds including lipids, sugars and metabolites to macromolecules such as proteins, DNA and large carbohydrates. In order to gain deep knowledge of a biological function, it is not only imperative to identify the biomolecules involved but also to understand their structures at both elemental and three-dimensional levels. It is also critical to have the capability to quantify their abundance and identify intermolecular associations with spatial and temporal control. Given these considerations, mass spectrometry (MS) has developed over recent decades to become a central technique in bioanalytical chemistry. Currently it is arguably the most sensitive, precise and rapid method for structural characterisation of analytes, particularly biomolecules, and unique amongst the structural biology methods in that it can report on all levels of biomolecular structure and dynamics. Given the analytical advantages of the technology, the use of mass spectrometers today is almost ubiquitous in analytical, commercial, research and academic laboratories. For example, they are used at airports to screen for traces of explosives, in clinical laboratories for diagnostics using plasma or urine and to provide quality assessments for environment, food and water monitoring, amongst many other applications. Furthermore, the range of MS applications in biology only promises to grow as the technology continues to improve.

A view of the history of MS demonstrates its foundations in physics, developments in chemistry and applications in biology, highlighting the interdisciplinary concepts and practices of the field. The foundations of MS can be attributed to Joseph J. Thompson more than a century ago, with his observations of cathode rays and their contribution to the understanding of atomic structure; he was awarded the 1906 Nobel Prize in Physics for the discovery of the electron [1]. The first mass spectrometer, called at the time a parabola spectrograph, was developed by Thompson and his then assistant, Francis Aston, in the early 1900s. This was utilised for the detection of elemental isotopes, work that led to the awarding of the Nobel Prize to Aston in 1922, who remarkably discovered 212 of the 287 naturally occurring isotopes [2, 3]. These early single-sector spectrometers, however, were superseded by double-sector and tandem double-sector instruments, which provided enhanced analytical capacity. Strong foundations in MS theory and instrument design have since enabled the further development of spectrometers capable of meeting the demands of chemists and biologists.

The first applications of MS in biology emerged in the 1940s, with analysis of heavy stable isotope tracers used to study processes such as CO_2 production in animals [4]. The 1950s and early 1960s principally saw efforts to apply MS in the measurement of the molecular weight of small organic molecules, such as natural products, to verify their structure. Moreover, this renewed interest in the use of MS for molecular characterisation led to the realisation that detailed mechanistic understanding of gas-phase ion fragmentation could be used to further determine structures *de novo*, underpinning future developments in tandem MS. However,

Biomolecular and Bioanalytical Techniques: Theory, Methodology and Applications, First Edition. Edited by Vasudevan Ramesh.

despite advances in mass accuracy and resolution, extended mass range and improved analyte quantitation resulting from instrument development, it was not until implementation of soft ionisation methods in the 1980's that biological MS came to the forefront.

Methods for ion generation have evolved dramatically from the early emission of positive ions to 'classical' methods including electron ionisation (EI) and chemical ionisation (CI), amongst others. Although these early ionisation methods established MS as an analytical technique, they were not suitable for the direct analysis of large, polar, thermally labile molecules relevant to biology, such as proteins and DNA. Mass spectral analysis of biomolecules only became feasible with the advent of ion desorption methods, the earliest of which included field and plasma desorption and fast atom bombardment methods, the latter of which remains in limited use today. However, the revolutionary application of MS to biological systems is largely attributed to the development of 'soft' ionisation methods such as electrospray ionisation (ESI) [5] and matrix-assisted laser desorption ionisation (MALDI) in the mid to late 1980s [6, 7]. The new and robust capability to analyse biomolecules afforded by ESI and MALDI drove improved processes for peptide and protein sequence analysis. The first protein sequence was studied by MS in 1990 [8], which started the development of peptide mass fingerprinting techniques to sequence proteins. The ability to identify proteins was further enhanced by the development of software programs to search peptide mass spectral data against online databases of amino acid sequences, demonstrating an early realisation of the synergies between MS and computational data analysis.

In conjunction with developments in MS instrumentation and ionisation methods, and in particular the coupling of MS with other separation methods such as chromatography, it has become increasingly possible to investigate large numbers of biomolecules in a high-throughput fashion. Consequently, there has been an escalating drive to use MS to study the total complement of a given molecule class with the goal of elucidating the biological state of the whole system. This has led to the development of 'omics' applications, focusing on the analysis of one group of biomolecules. For example, comprehensive investigation of the 'omics' cascade [9] from genomics, transcriptomics and proteomics to metabolomics has had an enormous impact in the emergent field of systems biology. For very complex groups of molecules subomics have arisen, for example in the post-translational protein modification field, which has given rise to areas such as phosphoproteomics and glycomics. Progress in these disparate areas depends upon overcoming the common challenge of interpreting the large datasets generated and, despite interim successes, many data interpretation problems in MS are still challenging, particularly due to their interdisciplinary nature.

MS has traditionally been utilised to provide structural information at a primary level, but it is gaining increasing use to probe the three-dimensional conformations of biomolecules and can inform on important biomolecular interactions that underpin cellular function. Notably, experimental evidence demonstrates that the conformation of ions in the gas phase are often not significantly modified in the absence of solvation has emerged in recent years [10] and has validated MS as a structural biology tool through the development of native MS approaches. Furthermore, the observation that non-covalent interactions are largely retained in such native MS experiments means that binding interactions can be directly observed [11] and considerations such as the stoichiometry, stability and dynamic assembly of higher order complexes can be probed. Consequently, integrative approaches that combine a range of MS-based data (often with other biophysical analyses) are becoming commonplace in development of structural models of biomolecules [12].

While qualitative analysis by MS has played a significant role in many important discoveries, modern science increasingly relies on quantitative data. Several factors affect the performance of MS with regards to quantitative parameters such as range of detection, accuracy and reproducibility. These range from instrumental factors that affect ion transmission and detection and ionisation considerations such as ionisation efficiency, which are often related to physical properties of the analytes or composition of the sample. These challenges mean that only a relatively small minority of studies have attempted to provide a comprehensive quantitative description of the biological system under investigation. Nevertheless, the use of MS in quantitative analysis exploits its discriminating selectivity and sensitivity as a detector, allowing a signal to be ascribed with high certainty to a particular analyte, even at low concentrations.

10.2 Theories and Principles of Biomolecular Mass Spectrometry

At its most fundamental level, MS involves measurement of the mass-to-charge (m/z) ratio of an ion, from which molecular structure can be inferred. The versatility of this method arises from the fact that MS analysis can provide elemental, isotopic and molecular level detail for organic and inorganic samples, offering both qualitative and quantitative insights. Analytes can span from single atoms to biomolecular assemblies in the megadalton size range and can be sampled from the gas, liquid or solid state.

Every MS experiment encompasses at a minimum three essential components, namely, (i) generation of gaseous ions from the sample, (ii) separation of analytes according to m/z and (iii) detection of the relative abundance of the ions. The mass spectrometer in its most basic form is comprised of an ion source, mass analyser and detector, responsible for each of these aspects, respectively, although many variations on the design of these components exist (Figure 10.1). The operation of the mass spectrometer also requires a collision-free path for the ions (with the exception of dedicated collision cells) and hence a vacuum system is required to maintain varied low operating pressures at different stages throughout the instrument. Additional segments may be included for sample introduction, including online pre-fractionation such as by chromatography or ion resolution (e.g. ion mobility, IM). In many configurations, tandem MS analyses are feasible whereby target

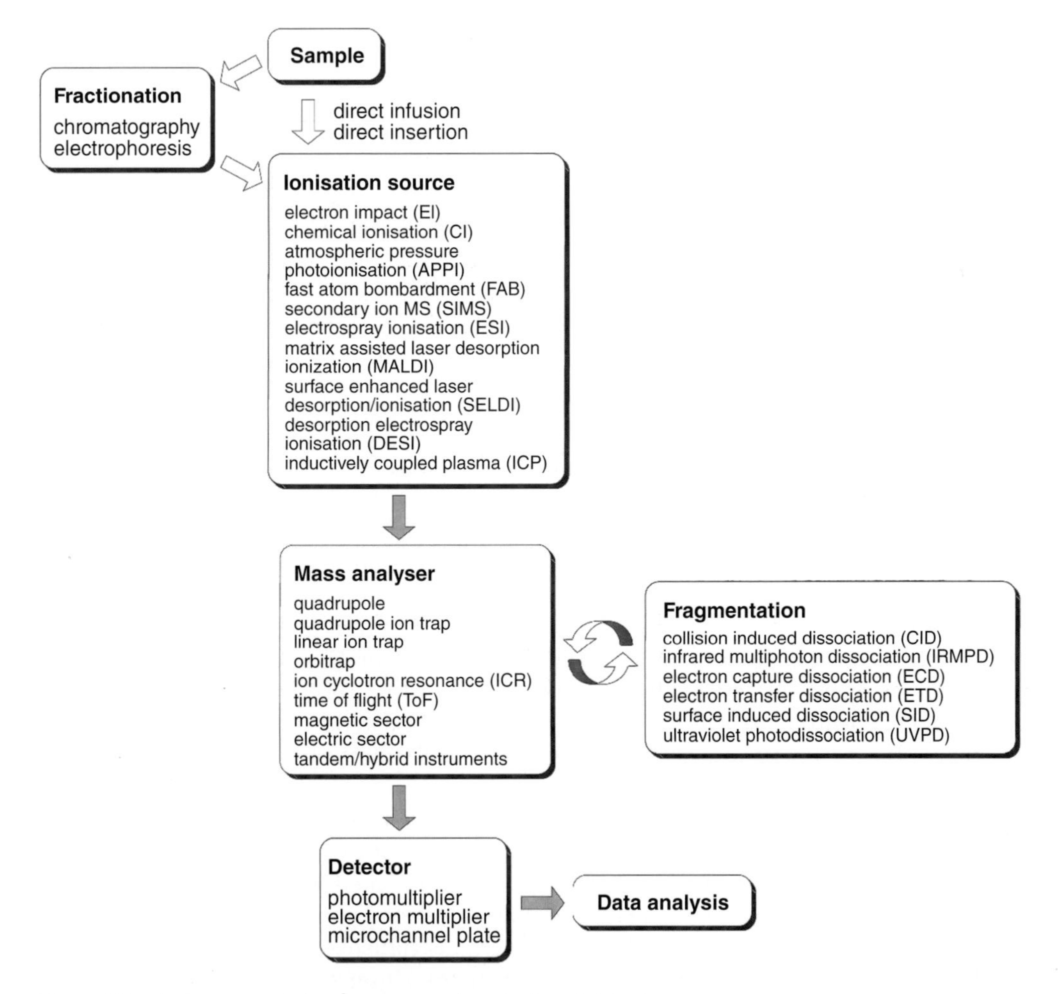

Figure 10.1 Basic components of a mass spectrometric experiment.

ions of a defined *m/z* are selected and subsequently activated for fragmentation, often by collision with an inert gas such as argon (e.g. collision-induced dissociation, CID). The multiple fragment ions are then analysed in the second-stage mass analyser to give information on structural features of the precursor ion. Finally, the data are recorded and interrogated computationally, with advances in computing power and informatics further driving applications in structural and systems biology.

MS provides both quantitative and qualitative data in a relatively high throughput manner with unrivalled accuracy, irrespective of the analyte starting state. Although thousands of different analytes can potentially be detected by MS from a single sample, most biomolecular analyses tend to examine only one class of molecules in a given experiment to gain insight into a particular biological system. Given the varying analytical challenges each class of compounds presents, the MS instrumentation employed for a particular analysis is often chosen to maximise the information that can be afforded, balancing factors such as sensitivity, dynamic range, mass accuracy, resolution and speed of analyses. Consequently, a solid understanding of the fundamental principles of these stages of the MS experiment, in particular the choice of ionisation and mass analysis method, is often critical to optimal and relevant outcomes.

10.2.1 Gas-Phase Ion Generation of Biomolecules

To attain quantitative and qualitative data with extreme precision and accuracy by MS, efficient ionisation techniques are first required to generate analytes suitable for transmission into the mass analyser. Unless they are already in the gas phase, analytes need to be first vaporised, which can occur by a variety of methods including exposure to heat, high electric fields, laser irradiation and/or bombardment with atoms or ions. Analyte ionisation may occur before, during and after transfer to the gas phase, and can give rise to ions with either a positive or negative charge that can be selected by controlling the polarity of the applied electric field at the ion source. Selection of an appropriate ionisation method is critical and can often be the most significant determinant of a successful MS analysis.

Since the vapour pressure of biopolymers is negligible and biomacromolecules are often susceptible to degradation at high temperature, ion desorption techniques are principally required for the ionisation of analytes in biomolecular MS. Such ionisation approaches include ESI (and its variants such as nanoelectrospray and desorption electrospray ionisation, DESI) and MALDI. The type of analyte under investigation or the question to be answered will dictate which ionisation type is the most applicable and practical. While an incredible number of alternative ionisation approaches have been reported, for practicality here we focus only on those that are commonly applied in contemporary biomolecular analysis.

10.2.1.1 Electron Ionisation

Electron ionisation (EI, formerly known as electron impact ionisation or electron bombardment ionisation) was one of the earliest ionisation methods coupled with MS [13] and is still commonly used today for analysis of samples, which can be volatilised, such as metabolites and natural products. In an EI source, electrons produced by heating a wire filament are accelerated to 70 eV perpendicular to the flow of gas phase analyte. The close passage of highly energetic electrons at low pressure (ca. 10^{-5} to 10^{-6} Torr) causes large fluctuations in the electric field around the neutral molecules and induces electronic excitation with the expulsion of an electron from the analyte. This generates predominantly singly charged radical cations, which are directed towards the mass analyser.

EI is described as a hard ionisation method since there is sufficient transfer of energy to induce unimolecular bond dissociation reactions of molecular ions, giving rise to fragments of lower *m/z* before mass analysis. These fragmentation pathways often follow predictable cleavage reactions dependent on chemical structure, and therefore the fragmentation pattern can be interpreted to convey structural information about the analyte. The process of ionisation and fragmentation is reproducible between instruments and mass spectral libraries are widely available to undertake identification of EI amenable analytes. However, compounds that contain particularly labile bonds are likely to dissociate to such an extent that no molecular ions are observed and therefore molecular mass determination is not directly possible. Such an approach is also not suited to polar, involatile and thermally labile molecules, although chemical derivatisation can be used to some extent to reduce polarity, increase volatility and in some cases direct characteristic fragmentation [14].

10.2.1.2 Inductively Coupled Plasma

The inductively coupled plasma (ICP) is an ionisation source that fully decomposes a sample into its constituent elements and transforms those elements into ions. The source consists of an ICP torch, comprising concentric quartz tubes that contain the sample aerosol and argon support gas. An oscillating current is produced in the source by application of radiofrequency (RF) energy to an induction coil that wraps around the tubes. When a spark is applied to the argon flowing through the ICP torch, electrons are stripped from the argon atoms, forming argon ions. These ions are caught in the oscillating fields and collide with other argon atoms, forming an argon discharge or plasma. The plasma attains a temperature of 6000–8000 K, which is sufficient to desolvate and dissociate the molecules and then remove an electron from the constituent atoms, thereby forming principally singly charged positive ions of the elements contained in the analyte. Following ionisation, ions are detected by the mass spectrometer, which can differentiate between elemental isotopes, commonly using quadrupole mass analysers, although magnetic sector and time-of-flight (ToF) analysers also play a role due to their high resolving power.

While perhaps not among the more broadly utilised ionisation methods coupled to MS, we highlight ICP briefly here due to the unique and widespread applications it enables through evaluation of elemental composition (see Section 10.4 on Applications for examples). Inductively coupled plasma mass spectrometry (ICP-MS) is one of the leading tools for the determination of elements and isotopes, and has several advantages, such as the ability to sample from a variety of matrices with high sensitivity (concentrations below 1 part per trillion), a wide dynamic range and the possibility of spatial resolution.

10.2.1.3 Electrospray Ionisation

In ESI, gas-phase ions are generated from molecules in solution by passing the sample through a capillary at a low flow rate, typically in the nl/min to μl/min range. The tip of the capillary is held at a high voltage (typically 1.5–4 kV) [5] with respect to a counterelectrode and positive or negative ions are selected by controlling the polarity of this applied field. Charge accumulation at the capillary terminus causes the liquid surface to adopt a conical shape known as a 'Taylor cone', the tip of which is drawn into an elongated filament that becomes unstable as the charge density increases [15, 16]. At a given point, this instability results in a spray of finely divided and electrically charged droplets. As the charged droplets move across both an electric field gradient and a pressure gradient towards the counterelectrode and high vacuum components of the mass spectrometer, their size decreases as the solvent evaporates. Typically, desolvation is assisted by either a weak counterflow of hot nitrogen gas or passage of the ions through a heated capillary. With evaporation of the solvent, charge concentration in the droplets increases to the point where Coulomb repulsion overcomes surface tension (the Rayleigh limit), ultimately resulting in fission of the droplets and a repeated size decrease and desolvation (Figure 10.2a). Two models are widely proposed for the formation of isolated gaseous ions from ESI, namely the ion evaporation model (IEM) or the charged residue model (CRM) [15]. The IEM proposes that ions are directly emitted from the small droplets due to electrostatic repulsion as the droplet radius decreases. In contrast, the CRM suggests repeated droplet fission continues to the point where no further evaporation can occur. In all likelihood both mechanisms are probable, with low molecular weight analytes following the IEM and the CRM being applicable to large globular species such as proteins [17]. A chain ejection model (CEM) has also been proposed for disordered polymers [18].

ESI is a very 'soft' ionisation method whereby little energy is provided to the analyte in the process of transfer to the gas phase. It is largely for this reason that electrospray ionisation mass spectrometry (ESI-MS) is an essential technique in the study of biomolecules, which are often labile and otherwise susceptible to fragmentation. Furthermore, the non-covalent interactions often important in the formation of biomolecular assemblies can be representatively transferred into the gas phase, allowing detection of binding interactions and

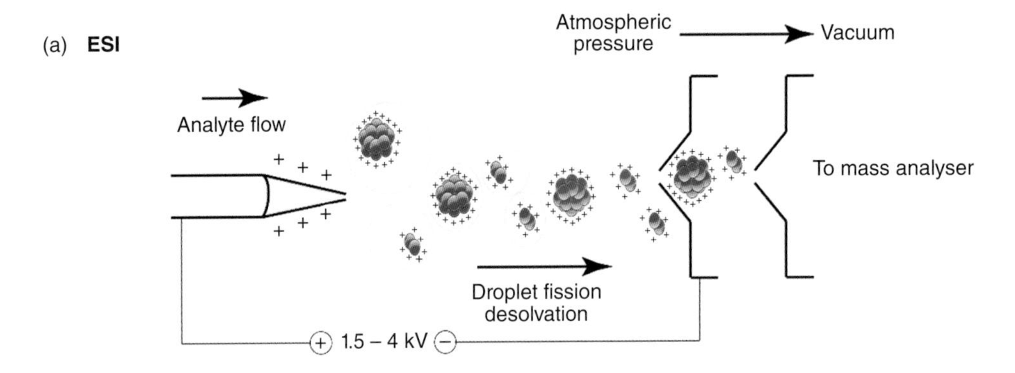

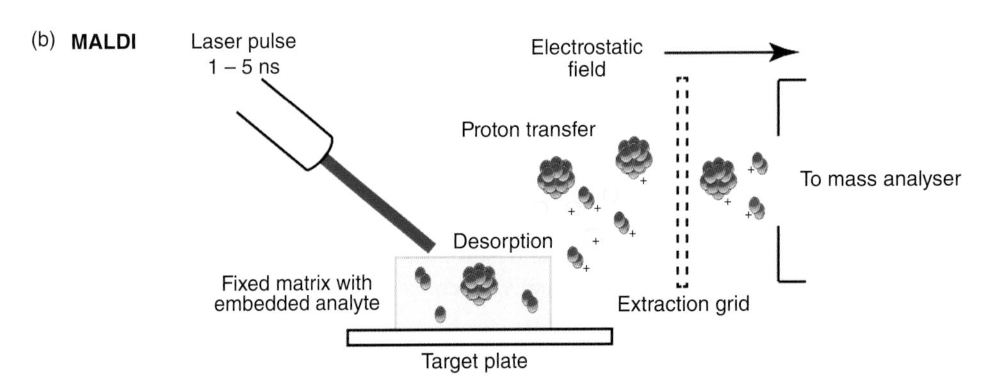

Figure 10.2 Common methods for ionisation of biomolecules. (a) Electrospray ionisation (ESI). Analyte is introduced from solution, with the flow passing through an electrospray needle that has a high potential difference (with respect to a counterelectrode). This results in the spray of charged droplets from the needle with a surface charge of the same polarity as the needle. Solvent evaporation occurs as the droplets are directed towards the counterelectrode, giving rise to repeated droplet shrinkage and droplet fission until free gas-phase ions result. (b) Matrix-assisted laser desorption ionisation (MALDI). A laser is focused on to the surface of the matrix–analyte mixture. The matrix chromophore absorbs the laser irradiation, resulting in rapid vibrational excitation and localised desorption. Ejected clusters consisting of analyte molecules surrounded by matrix and salt ions. Proton transfer and cation attachment taking place in the desorbed matrix–analyte plume gives rise to ionised analytes. The matrix molecules evaporate from the clusters to leave free analyte ions in the gas phase.

complex stoichiometries [19, 20]. ESI-MS is also a sensitive method; low concentration of analyte (micromolar to femtomolar) is adequate. Finally, another useful feature of the ESI process is the formation of multiple charges on macromolecular ions, where these charges are statistically distributed across the ionisable sites of the analyte, giving rise to multiple peaks in the mass spectrum at differing m/z ratios (Figure 10.3a). Since the mass of large biopolymers typically falls outside the range of many mass analysers, increasing the charge state decreases the measured m/z ratio, effectively extending the mass range of the analyser to accommodate species up to the megaDalton range. The molecular weight of an ion can be determined from the ESI mass spectrum by analysis of any two adjacent charge state signals along the m/z axes, by solving simultaneous equations given the peaks have the same mass and of related charge state (i.e. z and $z+1$).

Solvent volatility and acid/base properties are important considerations for ESI and, typically, polar solvents such as methanol, acetonitrile or acetonitrile:formic acid:water mixtures are used to facilitate the ESI process. The analyte can be introduced to the source in solution either from a syringe pump or commonly as the eluent flow from liquid chromatography (LC), allowing direct coupling of ESI-MS to online separation methods. In the case of protein analysis, these specific solvent conditions induce protein unfolding/denaturation, where the unfolding allows the ionisation of all basic residues (in positive mode), which aids in accurate

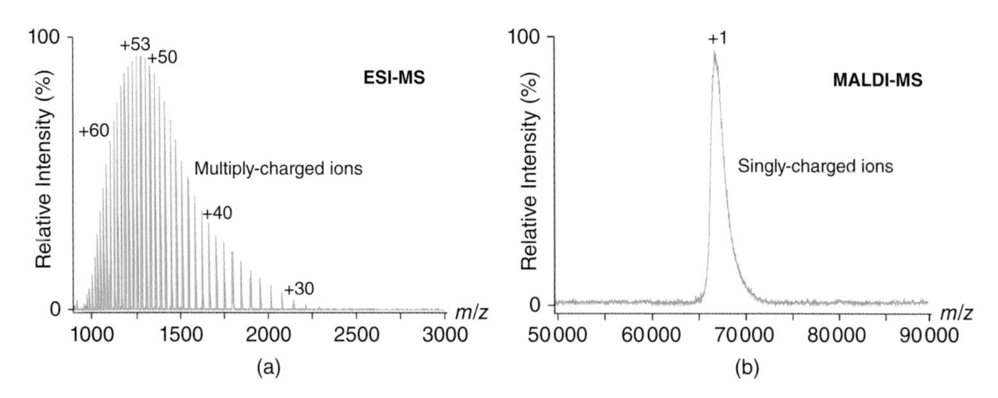

Figure 10.3 Representative mass spectra of an intact protein. (a) ESI mass spectrum of the 66.5 kDa protein bovine serum albumin from denaturing conditions, showing a series of multiply charged protein ion peaks and the corresponding singly charged ion observed in the (b) MALDI mass spectrum of the same protein. Selected charged states are indicated on the spectra. Source: Data courtesy of Parul Mittal (presently at The University of Adelaide, Australia).

mass determination (Figure 10.3). To retain physiologically relevant structural information, ESI can also be performed using other non-denaturing volatile buffers such as ammonium acetate or ammonium bicarbonate [21]. Most classes of biological polymers are amenable to ESI ionisation, though proteins are usually analysed in the positive ion mode while oligonucleotides, for example, tend to be preferably ionised in the negative ion mode given the abundance of anionic phosphate groups.

A variant of ESI using lower flowrates is nanoESI, introduced by Wilm and Mann in 1994 [16]. In this mode of operation, the sample is typically sprayed from a borosilicate glass capillary that has been pulled to a fine tip, and a conductive metal coating (such as gold or platinum) on the outside of the glass capillary or a thin platinum wire introduced from the large end of the capillary is used to supply the electric potential. These experiments require much less analyte solution and, unlike the syringe-forced flow in normal ESI, the pull exercised by the applied electric field causes, for non-viscous solutions, a self-flow controlled by the capillary tip diameter. Due to the very small capillary tip orifices (1–5 μm in diameter) and the absence of external pumping, the primary droplet sizes are much smaller, improving desolvation processes and decreasing sample consumption. Furthermore, the tolerance to salt contamination from buffers or additives common in samples from biological preparations is increased in nanoESI [22].

Developments in biomolecule ionisation have also seen moves towards ambient ionisation approaches, which allow the ionisation of untreated samples in an open environment. In 2004 DESI, the first of now almost 30 ambient ionisation methods for MS, was first described [23]. The DESI experiment directs an electrospray plume at an acute angle (approximately 45°) towards analyte deposited on an insulating surface a few millimetres away. The splashed droplets then carry desorbed, ionised analytes into the atmospheric pressure interface for sampling by the mass spectrometer. Therefore, DESI represents a combination of ESI and desorption mechanisms. Recent studies have shown that data obtained from DESI experiments are comparable to those obtained in nanoESI experiments [24], though with the key advantage of the relatively short time it takes to analyse samples, with minimal sample preparation. The DESI approach also allows biomolecular interactions to be observed 'on the fly' as various analytes such as lipids, ligands and cofactors may be applied to the surface and allowed to react on the millisecond scale with other molecules, such as proteins, from the spray source. Finally, DESI can be used to investigate samples *in situ*, specifically to observe both spatial and temporal distributions [25]. However, some disadvantages to DESI, including low spatial resolution (200 μm), lower mass acquisition range (up to 2000 *m/z* compared to MALDI reaching 50 000 *m/z*) and lower ionisation efficiency have limited its more widespread use.

10.2.1.4 Matrix-Assisted Laser Desorption Ionisation

MALDI is a solid-state desorption method that produces ions by subjecting the analyte, which is dispersed in a solid matrix and deposited on a surface, to laser irradiation. Despite the widespread applications of MALDI, exact details of the desorption and ionisation mechanisms remain unclear. However, at a fundamental level, the energy of the laser is transferred largely to thermal energy, which results in the transfer of a localised region of the material to the vapour phase in the form of an expanding plume containing a mixture of neutral and ionised matrix and analyte species. The ions generated are then guided to a mass analyser (Figure 10.2b).

As can be seen in the example shown in Figure 10.3b, the extent of charging of macromolecular ions produced by MALDI is significantly decreased compared with ESI and hence the large majority of ions are typically singly charged. While this simplifies the mass spectrum, it has the effect of requiring mass analysers with an extended *m/z* range and typically results in reduced resolution and mass accuracy. However, once set up, ions are easy to generate and the method is highly sensitive (in the femtomolar to attomolar range) and more tolerant to salt contamination. MALDI plates can also have from hundreds to thousands of wells for the sample-matrix mixtures to be placed, facilitating high throughput analysis. Like ESI, MALDI is a 'soft' ionisation method, although increased laser flux often gives rise to deposition of excessive energy, resulting in some fragmentation. Consequently, MALDI is less commonly used to characterise higher order protein structures compared to ESI-based approaches.

The ability to measure a mass spectrum from a discrete region of the sample gives rise to the possibility of spatial analysis by MALDI. By acquiring MS data over a two-dimensional array (over thousands of pixels), the spatial distribution of molecules can be deciphered and the ion image of each biomolecule can be later reconstructed and overlapped with surface images. For example, MALDI-imaging MS has developed as a standalone field for analysis of various biomolecules directly from intact thin tissue sections [26].

10.2.2 Mass Analysers

Once ions have been generated and transferred to vacuum regions of the mass spectrometer, they can be selectively and specifically manipulated using an array of MS approaches. In all cases the underlying principles of measurement are the same, which relate to the ability of magnetic and/or electric fields to influence the motion of charged atoms and molecules. As described by Newton's second law of motion, the acceleration of an object as a result of a net force is directly proportional to (and in the direction of) the magnitude of the force and inversely proportional to the mass of the object. Consequently, the trajectory of two identically moving and equally charged particles will be altered to different extents under the influence of an electric and/or magnetic field if they have a different mass. Similarly, the particle's charge is important as, according to the Lorentz force law, it relates directly to the extent of the force developed by the applied field. Therefore, by controlling the applied forces to manipulate particle trajectories, packets of ions with the same *m/z* ratio are separated in space or time from other ions and selectively detected. It is possible to deconvolve these ion motions to infer the mass-to-charge ratio (*m/z*) of distinct particles, forming the basis of the mass measurement. Results are usually presented in the form of a mass spectrum, with the *m/z* ratio plotted on the *x*-axis and the signal intensity plotted on the *y*-axis.

After selecting an appropriate ionisation method, the next major consideration in an MS experiment is the choice of mass analyser. There is a wide range of mass analysers available, differing in their compatibility with various ion sources. For example, continuous mass analysers that maintain a continuous flow of ions are compatible with ESI, while pulsed mass analysers are compatible with MALDI or otherwise require accumulation of ions for pulsed injection to be compatible with continuous ionisation sources. Analysers also vary in their ability to investigate ions of certain types, as well as in differing analytical considerations, such as sensitivity, accuracy, mass range and duty cycle, amongst others (Table 10.1). Sector, ToF, quadrupole, ion trap and ion cyclotron resonance (ICR) mass analysers evolved in parallel and are now the most commonly used spectrometers for the study of biomolecules [27]. The operating principles of these instruments are described in more detail here.

Table 10.1 Comparison of features for different mass analysers commonly used in biological mass spectrometry.

	Sector[a]	Time of flight	Quadrupole filter	Quadrupole ion trap	FT ICR	Orbitrap
Resolution	High – very high	Low – high	Low – medium	Low – high	Highest	Very high
Accuracy	High	High	Low	Low – medium	Very high	Very high
m/*z* range	High	Very high	Low	Low – medium	Medium	Low – very high[b]
Sensitivity	Medium	High	High	High	Medium	High
Dynamic range	High	Medium	High	Low – medium	Medium	High
Quantification	Very good	Medium – good	Good – very good	Poor	Medium	Good

a) Double focusing.
b) Dependent on the instrument model.

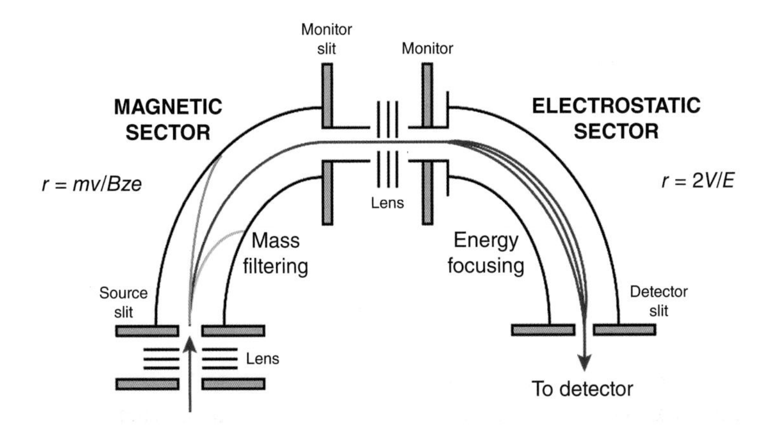

Figure 10.4 Schematic representation of a double focusing sector mass analyser with BE geometry. Ions extracted from the source are accelerated by a potential (*V*) and enter the sector analyser with velocity *v*. Electrostatic (E) or magnetic (B) fields cause the ion trajectory to be curved, with a path of radius *r*. Trajectories of ions of lower *m/z* are influenced more than those of high *m/z*.

10.2.2.1 Sector Analysers

The first commercial mass spectrometers available in the 1940s employed magnetic sector mass analysers based largely on the prototypes developed by Arthur Dempster [28] and remain widely utilised in the MS to date. In these instruments, ions leaving the source are focused and accelerated to a high velocity before passing through a magnetic sector. Here a magnetic field is applied perpendicular to the ion motion, causing the ions to be directed on a curved trajectory (Figure 10.4). If all ions are accelerated to the same kinetic energy prior to introduction to the magnetic field, then ions of a specific m/z will have a unique radius of curvature (r) (for a constant velocity v and magnetic field strength B), as given by Eq. (10.1), where e is the unit of elementary charge:

$$\frac{m}{z} = \frac{Ber}{v}. \tag{10.1}$$

Since ions of higher m/z ratio are deflected to a lesser extent than those of lower m/z, scanning the magnetic field enables ions of different m/z to be focused on to an exit slit for selective transmission of a particular ion or, alternatively, a detector array can be used for simultaneous recording of multiple m/z ions with varying spatial location.

Later developments in magnetic sector analysers led to significant improvements in resolution and other performance characteristics. Perhaps the most significant of these was the production of a double-focusing instrument [28], which adds a second sector with an electrostatic field ('E' sector), either before or after the magnetic sector ('E' sector), known as EB or BE geometries, respectively. A mass spectrum is usually obtained by scanning the magnetic sector over a desired m/z range, while the electrostatic field of the E sector is kept constant to allow passage of ions whose kinetic energy-to-charge ratio is equal to eV (where V is the acceleration voltage). Since mass resolution in sector instruments is dependent on both the spatial dispersion of the incoming ion beam as well as the ion kinetic energy spread, the magnetic sector filters ions with differing momentum and the electrostatic mass analysers are efficient ion kinetic energy filters. Hybrid instruments combining these two analysers therefore enable double focusing, and hence offer extremely high resolution.

Sector analysers are well suited to monitoring selected ions at a single m/z with high resolution, particularly useful in quantitative applications. Additionally, the presence of a second analyser allows MS/MS experiments to be performed. In this case, ion fragmentation experiments occur in a high collisional energy regime, allowing access to fragments that are often not otherwise detected.

10.2.2.2 Time-of-Flight Analysers

ToF mass analysers were conceptualised more than 50 years ago, with the first report of a 'pulsed mass spectrometer with time dispersion' in 1946 [29]. They are based on very simple physical principles, whereby in the ToF analyser an accelerating potential (V) will give an ion of charge (z) an energy of zV, which can be equated to kinetic energy (E_{kin}) according to

$$E_{kin} = zeV = \frac{mv^2}{2}, \tag{10.2}$$

where m and v are the mass and velocity of the ion, respectively. Since velocity can be related to distance (d) and time (t), the above equation can be rewritten as shown below:

$$\frac{m}{z} = \frac{2eVt^2}{d^2}. \tag{10.3}$$

If the ions travel a fixed distance to a detector through a field free region, those of larger m/z will have a longer ToF. Therefore, the differences in the time taken for ions to arrive at the detector from the pusher are determined and converted to a measure of m/z in the mass spectrum (Figure 10.5). With no theoretical upper mass limit, ToF mass analysers are ideally suited to the study of biomacromolecules, particularly coupled with MALDI.

In what is termed the 'linear mode', ions extracted from the source are unidirectionally accelerated by application of an electrostatic field in short pulses, moving into a drift space containing no field (Figure 10.5a). A limitation of this configuration is that not all ions have the exact same initial position and velocity, and therefore a spread in arrival time results, leading to the formation of broad, low amplitude signals and hence limiting resolution and sensitivity. One approach to improve resolution in ToF mass analysers involves orthogonal acceleration of ions. Here ions are typically sampled from an ion beam from a continuous ion source. The applied electric field is designed to generate a pulsed force that is exclusively at right angles to the initial axis of the ion beam. As the beam is nearly parallel, the ions in it have zero average velocity in the direction of this force prior to its application, therefore reducing the complication of initial velocity dispersion [30]. Correction of the initial energy distribution can also be achieved using a reflectron device consisting of a series of electric fields that repulse the ions back along the flight tube, usually at a slightly displaced angle (Figure 10.5b). Here, for two ions with identical m/z but different velocities, the faster ion will penetrate deeper into the decelerating region of the reflectron and hence have a longer flight path, compensating for its greater velocity. Firstly, the resultant increase in ion path lengths leads to greater separation between packets of ions and, secondly, the ions are thus refocused, thereby greatly increasing the resolution of the measurement.

The advantages offered by ToF instruments including a very high duty cycle, high transmission efficiency, extended m/z range, fast repetition rates and compatibility with pulsed ionisation sources make them very popular analysers suitable for a wide variety of applications in the biological sciences.

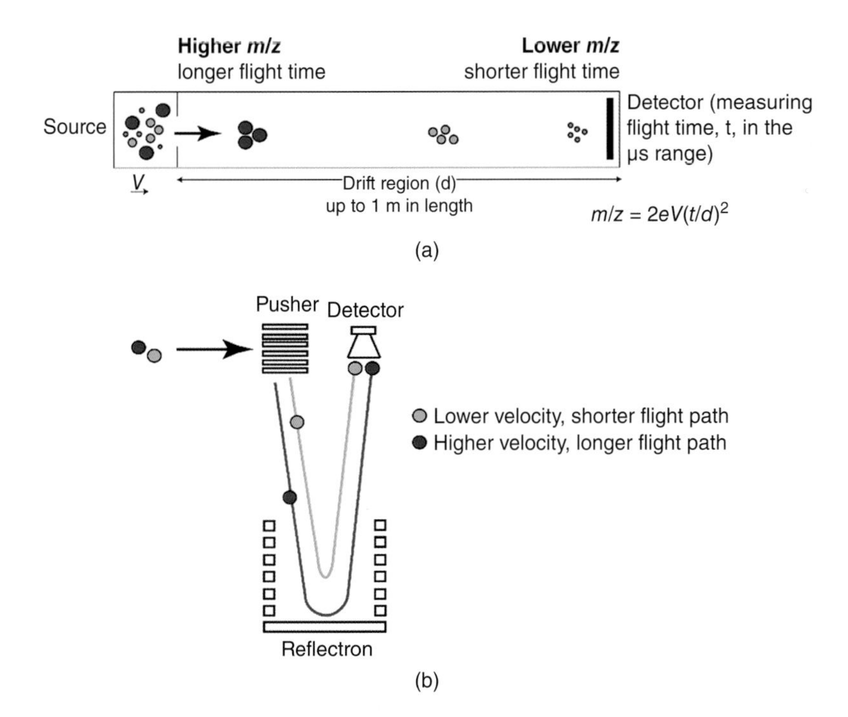

Figure 10.5 Schematic representation of the time-of-flight mass analyser. (a) Linear mode. Ions extracted from the source are accelerated over a defined electric potential *V* and separated in time according to *m/z* as they traverse the flight tube drift region. (b) Orthogonal acceleration ToF operated in the reflector mode. The pusher applies a pulsed electric potential at right angles to the ion beam and ions with different initial potential energies are refocused by mass in the reflectron to increase resolution.

10.2.2.3 Quadrupole Mass Filters

'Mass filter' devices achieve ion separation due to their ability to selectively maintain stable trajectories for ions of certain *m/z* ratios, while others become unstable. One such example, the quadrupole mass analyser, therefore can be considered as a 'tunable' mass filter that transmits ions within a narrow *m/z* range (typically a 1 *m/z* transmission window). The quadrupole analyser is comprised of four precisely parallel rods with a hyperbolic cross-section, to which a direct current (DC) and alternating RF potentials are applied to produce oscillating electric fields [31]. Ions are introduced in a continuous beam by means of a low accelerating potential along the central axis. The ions are sequentially repelled and attracted by the pairs of rods due to the oscillating fields, and hence they oscillate in the *yz*- and *xz*-planes as they travel through the quadrupole filter. Only ions of a particular *m/z* are able to traverse the quadrupole region and all other ions are radially ejected or hit the rods (Figure 10.6a).

The ion path transforms to a complex spiral-like propagation, according to the Mathieu equation [32], regardless of the ion's initial velocity or initial position. Stable solutions of Mathieu's equation are usually simplified by defining the a and q terms, where a is proportional to the DC and q is proportional to the RF. These solutions are commonly presented as 'stability diagrams' (Figure 10.6b), highlighting the limited number of combinations of a and q that lead to stable trajectories for ions of a defined *m/z* to traverse the quadrupole analyser. The quadrupole is operated with a fixed ratio of DC and RF voltages, which determines the resolution of the device. Varying the voltages while keeping this 'working' ratio fixed gives a scan line that passes consecutively through respective regions of stability for ions of differing *m/z*. In this way, a mass spectrum can be obtained for ions with *m/z* ranging from small to large. Another important consideration from the stability diagram is that the quadrupole will allow passage of all ions if the DC component is set to zero. This

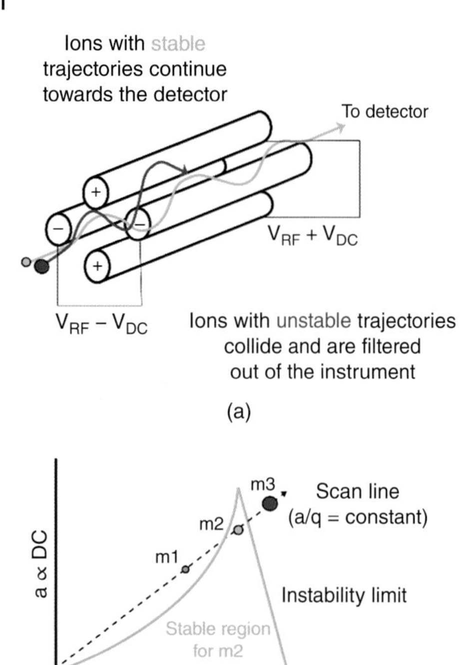

Figure 10.6 Schematic representation of a quadrupole mass filter. (a) A quadrupolar electric field is generated using four parallel electrodes that are connected diagonally. Examples of stable and unstable ion trajectories are shown. (b) Stability diagram for the quadrupole mass filter.

is commonly termed the 'rf only mode', whereby the quadrupole simply operates as an ion guide rather than an ion filter.

The *m/z* range of a typical quadrupole mass spectrometer is limited to *m/z* 2000, though it is possible to increase this by variation in the applied electric fields, albeit at a cost in resolution. The scan rate of a typical quadrupole mass spectrometer is high enough to allow direct coupling to chromatographic separation and this type of mass analyser is extensively used in tandem MS experiments (described later).

10.2.2.4 Quadrupole Ion Traps

The notion of utilising a quadrupolar field to trap and store ions arose as a natural extension of quadrupole mass filter development and introduction of the 3D ion trap represents an important development in quadrupole technology. A quadrupole ion trap device confines ions by the formation of a potential well, when appropriate potentials are applied to three electrodes (two end-caps and one ring electrode) of hyperbolic cross-sections. The internal volume of a typical 3D ion trap is approximately 1 cm^3.

Initially, ions are pulsed into the analyser under the influence of a field that maintains them in a stable oscillation (produced by an appropriate low RF potential applied to the central ring electrode), and they are therefore 'trapped' (Figure 10.7). The amplitude and frequency of the applied RF fields determine the mass range of trapped ions. A helium bath gas is used to stabilise the ion trajectories in the trap by acting as an energy sink, keeping the ions in tight orbits in the centre of the trap. Filling the ion trap in practice leads to enhanced full-spectrum sensitivity compared to the linear quadrupole mass analyser, but too many ions adversely affects mass resolution and accuracy due to space-charge effects. To acquire a mass spectrum, the fundamental RF voltage applied to the ring electrode is ramped and resonant absorption of energy progressively increases the amplitude of oscillations of the ions, destabilising the ion trajectories. Ions of increasing *m/z* are therefore sequentially ejected from the trapping volume to the detector, and based on the frequency being used at the time of detection, the *m/z* of the ion can be calculated.

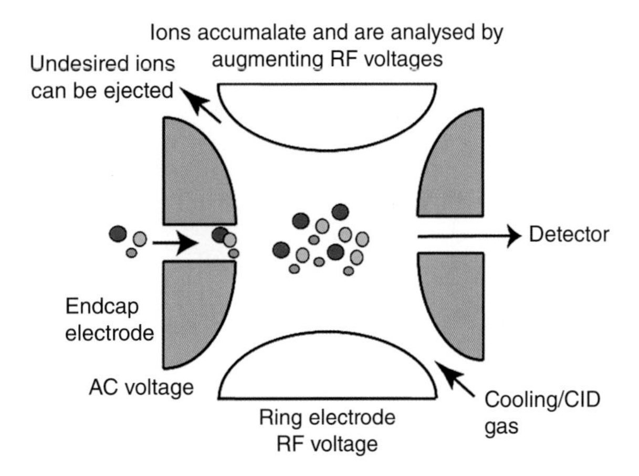

Figure 10.7 Schematic representation of a quadrupole ion trap.

A quadrupole ion trap instrument has a limited mass range and lower resolution compared with a ToF analyser, but transmission can be enhanced when proper ion populations are provided by pre-accumulation. Importantly, in a similar fashion to the quadrupole filter, superimposing RF and DC potentials on the ring electrode allows for storage of ions of a particular *m/z*. Once the ion of interest has been isolated, they can be induced to undergo energetic collisions with the bath gas and the increase in ion internal energy can give rise to fragmentation. Scanning the fragment ions gives rise to an MS/MS spectrum, or alternatively a specific fragment can be selected for a further round of fragmentation and structure interrogation, in what is known as multistage tandem MS or simply MS^n experiments. Typically, MS^5 can be achieved based on the ion capacity of the trap, as the ion population is depleted with each subsequent MS stage.

More recently, 2D or linear versions of the ion traps have been introduced, which creates a cylindrical space for the ion cloud by a radially confining 2D RF field. Such analysers provide much greater storage capacity and avoid the limitations imposed by space-charge distortion.

10.2.2.5 Ion Cyclotron Resonance

Ion confinement to a limited volume can also be achieved by using a combination of electrostatic and magnetic fields, and gives rise to instruments that currently provide highest mass accuracy in the form of a Fourier transform–ion cyclotron resonance (FT-ICR) MS. In an FT-ICR instrument, a strong magnet surrounds an ICR cell consisting of three pairs of electrodes (trapping plates, excitation plates and receiver plates). In the magnetic field, ions of a given *m/z* describe cyclotron motions with a radius *r* perpendicular to the magnetic field lines (Figure 10.8a). The principles of mass measurement in ion cyclotron resonance mass spectrometry (ICR-MS) relate to the ion cyclotron resonance frequency (f) of each ion as it rotates in a magnetic field (B). A spectrum is obtained by scanning the magnetic field of an electromagnet to bring ions of different *m/z* to resonate, based on

$$\frac{m}{z} = \frac{eB}{2\pi f}. \tag{10.4}$$

For a population of ions whose orbiting motions are in phase, an image current will be induced on detector plates with a characteristic angular frequency. Conversion from the time domain to the frequency domain through Fourier transformation allows the cyclotron frequency of each population to be determined and corresponding *m/z* values measured.

Since resolution is linearly increased with increasing magnetic field strength, large magnets up to 25 T are employed. Furthermore, the method is non-destructive, so the ions can be monitored for longer periods to increase resolution. Finally, frequency is a physical parameter that can be measured very accurately and, consequently, FT-ICR-MS offers the advantage of being able to detect ions with exceedingly high resolution, typically exceeding 10^5 (defined as $m/\Delta m_x$) in commercial instruments and far outperforming other

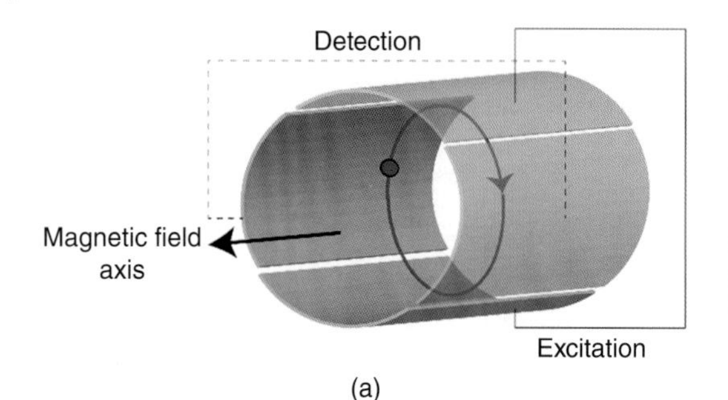

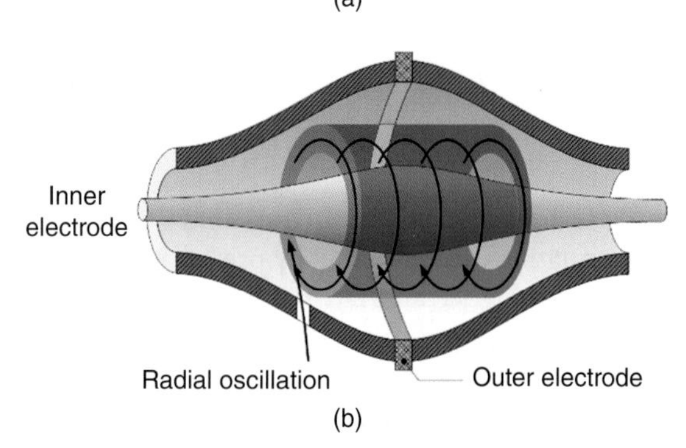

Figure 10.8 Schematic representation of high resolution mass analysers. (a) A cylindrical ion cyclotron resonance mass analyser. The ICR cell formed by three pairs of electrodes is aligned with the bore of the magnet so that the magnetic field is coaxial with the trapping axis (two trapping electrodes not shown). (b) Orbitrap mass analyser. An outer barrel-like electrode and a coaxial inner spindle-like electrode traps ions in an orbital motion around the spindle. In both cases, orbiting ions are shown in red and the image current from the trapped ions is detected.

mass analysers excluding the orbitrap. While image current detection is generally less sensitive that other ion counting methods used in other instruments, data acquisition can be carried out with the same ion population over an extended period of time, thereby enhancing the signal-to-noise ratio to the point that FT-ICR has been used to trap and detect individual macromolecular ions [33].

Fragmentation of the ions isolated in the ICR cell can be induced by a variety of means, including collision-induced dissociation, photoactivation and electron-based methods. Since FT-ICR MS is a trapping-based method, multiple stages of fragmentation can be performed in sequence, often utilising a combination of activation techniques to improve structure interrogation.

10.2.2.6 Orbitrap

A more recent major development in MS instrumentation took place with the invention of the orbitrap [34, 35]. This instrument utilises an axially symmetric spindle electrode and two cup-shaped outer electrodes facing each other to create a 'quadrologarithmic' potential. Ions are injected tangentially between the central and outer electrodes, and an applied radial electric field leads to circular movement of ions around the central spindle. In addition, an axial electric field initiates harmonic axial oscillations along the central electrode (Figure 10.8b). Mass spectra are generated in a manner similar to FT-ICR MS whereby the image current resulting from the dynamically trapped ions is measured and converted from the time domain to the frequency domain by Fourier transform. The frequency of axial oscillations is proportional to $(m/z)^{-1/2}$ and thus the frequency domain spectrum can be converted to a mass spectrum.

The high accuracy with which the axial frequency can be defined in this instrument, and the fact that this frequency is essentially independent of the energy and spatial arrangement of the ions, provides exceptional performance benefits. Ultrahigh resolution measurements in excess of 10^5 $(m/\Delta m_x)$ are possible, depending on the acquisition time, and at a resolution in this order of magnitude, a mass accuracy within 1 ppm can be achieved. Coupled with the fact that the orbitrap is significantly smaller and does not require cryogenic cooling as compared with FT-ICR MS, this instrument is a widely attractive choice for laboratories requiring high resolution MS capabilities.

10.2.3 Ion Detection

The final stage of the MS experiment is to produce a mass spectrum describing the relative abundance of ions for any given m/z. For this, the role of the detector is to convert a measure of the incoming ions to a signal that can be registered electronically and transferred to the computational system of the instrumentation for analysis. Ion detection may be accomplished in a variety of ways and is dependent on the instrument being used. For example, for instruments that produce and transmit a continuous ion beam, such as discussed for FT-ICR and orbitrap instruments, ions arriving at the detector represent an electrical current (and hence the term MS is used rather that mass spectroscopy). Faraday cup collectors are predominantly used for detection in isotope ratio measurements and some inorganic MS. Otherwise, the most commonly employed detectors are the electron multiplier and the microchannel plate [36]. In both cases, ions striking a metal plate give rise to a cascade of secondary electron emissions that result in amplification of the signal, which can then be recorded as a function of m/z or at selected m/z values.

10.2.4 Hybrid Instrumentation

The combination of upstream separation methods with mass spectrometric detection is a well-developed field and has enabled analysis and identification of molecules from increasingly complex mixtures and biological matrices. For example, gas chromatography provides a direct means for sample separation and introduction to the mass spectrometer for volatile, low molecular weight compounds [37]. Most commercial mass spectrometers now allow for integration of LC coupled to ESI sources, and particularly in the field of proteomics, reverse phase columns are commonly utilised for pre-fractionation of analytes. Capillary electrophoresis has also been successfully integrated with ESI in biological applications, although this is technically more difficult due to the need for high voltages in electrophoresis separation.

A further ion separation method that has attracted significant attention in recent years for biomolecule analysis relies on the differing mobility of ions of varied three-dimensional structure through a buffer gas in the presence of a low electric field (E). In these low field conditions where diffusion processes dominate, the velocity (v) of an ion is inversely proportional to its collisional cross-section (CCS) according to

$$v = \frac{3ze}{16N \times CCS}\left(\frac{2\pi}{\mu k_B T}\right)^{1/2}.E \tag{10.5}$$

where z is the ion charge, μ is the reduced mass of the ion–gas pair ($\mu = mM/(m+M)$, where m and M are the ion and gas particle masses), k_B is Boltzmann's constant and N is the number density of the buffer gas. Therefore, measuring the drift time of an ion through a gas cell of intermediate pressure and defined distance yields the CCS (typically measured in $Å^2$) of a molecule of interest. It also provides an orthogonal dimension of separation to m/z that can resolve structurally different ion populations such as configurational or conformational isomers.

The combination of multiple analysers in a single MS instrument has been utilised with great success. For example, typical triple quadrupole instruments consist of three identical quadrupoles placed in series, utilising the central quadrupole as a collision cell and ion guide [38]. In this geometry, various types of tandem MS experiments are possible that are particularly useful for studies requiring quantitation [39]. Furthermore, an increasing number of hybrid mass spectrometers are emerging that combine more than one type of analyser, often capitalising on the advantages of each component. For example, the combination of a quadrupole mass filter, ion mobility cell and a ToF mass analyser has been particularly popular (Figure 10.9) and offers tandem MS capabilities, extended mass range, high resolution and fast analysis.

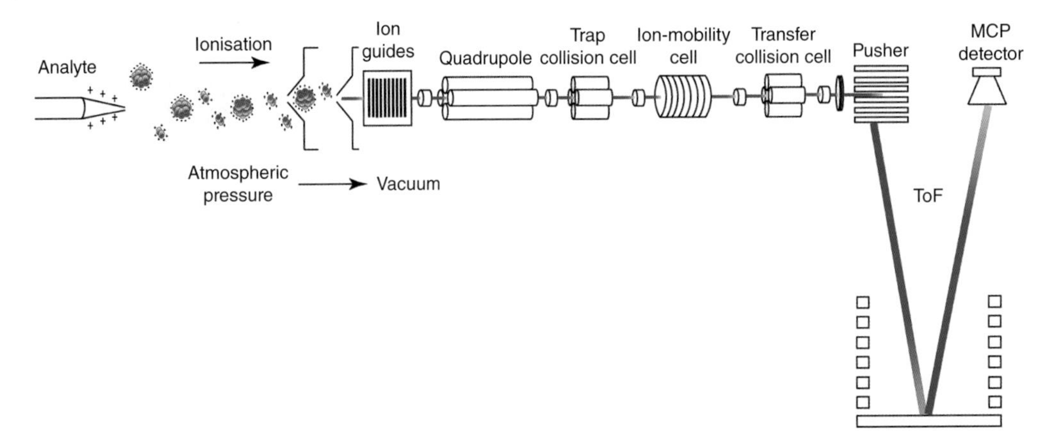

Figure 10.9 Schematic representation of a hybrid IM-Q-ToF mass spectrometer. Analytes undergo ionisation and transfer into the gas phase, where the ions are focused through a series of ion guides to the quadrupole. Selected ions are then transmitted to the collision cell where they can undergo collisions with inert gases to induce fragmentation. Ions may also have the possibility of passing through an ion mobility cell for mobility/conformation studies. All ions are subsequently transferred to the pusher, from which they are accelerated towards the MCP detector where their time-offlight (ToF) is recorded and transformed into a mass spectrum.

10.3 Techniques and Methodology in Biomolecular Mass Spectrometry

MS is a highly adaptable technology, with many variations on the basic mass measurement experiment available, utilising the ability to manipulate ions in the gas phase to provide molecular insights from kinetics and thermodynamics to structural organisation. Furthermore, a range of different solution-phase chemistries can be employed to probe the structural features and functionality of biomolecules, using the mass spectrometer as a reporter of the chemistry undertaken. Consequently, it is not possible to explore all these methodological approaches in detail here. Rather, we have chosen key examples of techniques and methods commonly utilised to provide additional information regarding the structure and function of biomolecules, with a particular focus on peptide and protein analysis.

10.3.1 Tandem Mass Spectrometry

While molecular weight information is a useful descriptor of a compound, it is not sufficient in most analyses to provide unambiguous identification, even for relatively small analytes at high mass accuracy. Furthermore, even if the elemental composition can be deduced, the molecular weight gives no indication of the structure of the molecule and therefore additional methods are required to this end. Tandem MS (also known as MS/MS or MS^n) is commonly used to further probe the structure of an analyte based on the manner in which that ion fragments with increasing activation energy. Due to the unambiguous nature of MS/MS in terms of ion peak assignment, it is a vital tool for the characterisation of biomolecules that cannot be achieved using MS on its own. In an MS/MS experiment, ions of interest at a selected *m/z* (precursor ions) from an initial mass spectral acquisition are isolated and fragmented, and the *m/z* of the resulting product ions are then measured. The sites of fragmentation, and hence the fragment ions observed, are often predictably and characteristically dependent on the structure of the molecule and therefore can be interpreted to provide details of molecular structure. This approach is widely applied in both organic and biomolecular MS and is particularly well suited for the sequencing of biopolymers [40].

The primary sequence determination of peptides and proteins by MS represents an excellent example of the power of tandem MS [40]. Since most proteins and peptides are linear polymers, cleavage of a single covalent bond along the backbone generates a fragment ion (or two complementary fragment ions) classified as shown

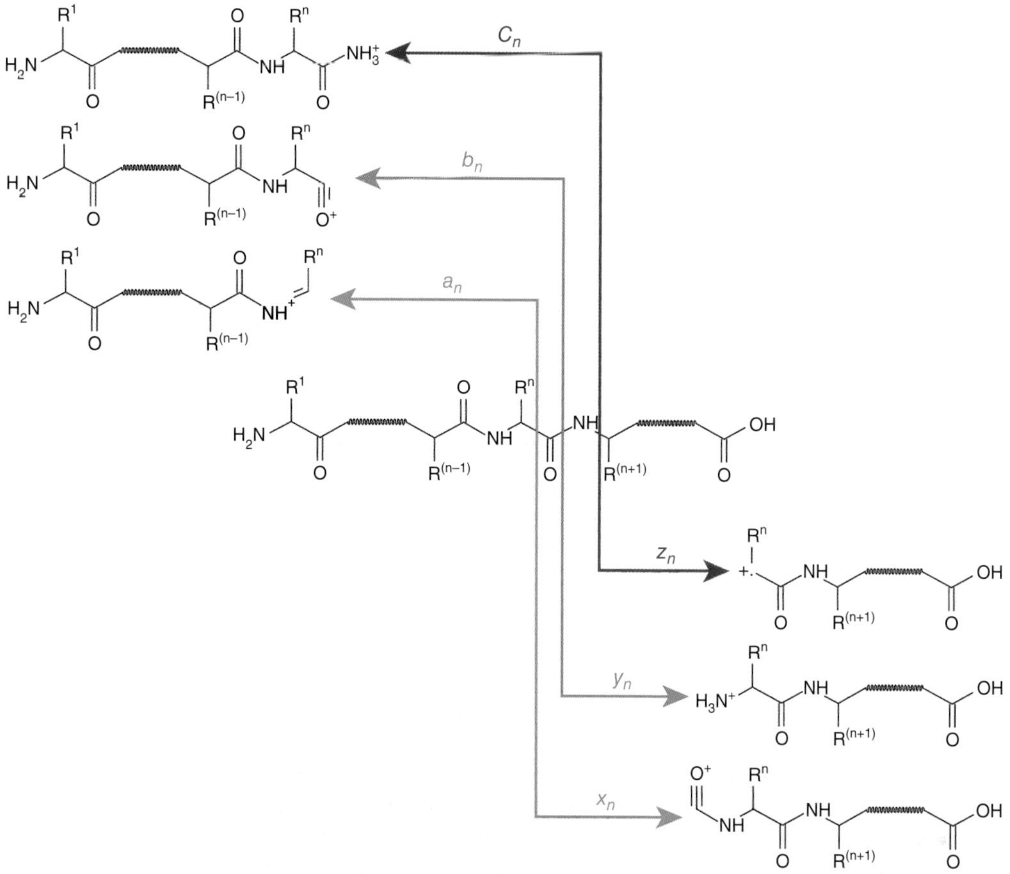

Figure 10.10 Nomenclature for common peptide ion fragments important in protein sequencing [41].

in Figure 10.10 [41], depending on the position of the bond that is cleaved and whether the charge is retained on the N- or C-terminal portion of the peptide. As a result of the heterogeneous nature of fragmentation, an array of peptide bonds are cleaved to give a series of fragment ions with distinctive mass differences, allowing for the identification of sequential amino acid residues. Cyclic and disulphide-linked polypeptides are a special case, whereby a single bond cleavage does not necessarily produce distinct (physically separated) fragments.

Tandem MS can be performed in an ion trap-type instrument, whereby all ions are first ejected except those at the m/z of interest. The ions are then excited to induce fragmentation and the fragments are sequentially ejected for mass measurement. Using such an instrument, multiple cycles of selection and fragmentation are possible to reveal greater structural detail, giving rise to MS^n experiments, which is primarily limited by ion abundances. Alternatively, MS/MS analysis can be achieved through the coupling of two distinct mass analysers, separated by some activation regime. For example, in the hybrid Q-ToF instrument shown in Figure 10.9, the quadrupole is used for ion selection, which is then separated from the ToF mass analyser by a pressurised collision cell to induce fragmentation. Other instrument configurations such as triple quadrupole and ToF/ToF instruments are also common to enable tandem analyses.

10.3.2 Gas-Phase Ion Activation

There are many advantages to manipulating the energy of ions in the gas phase, including the opportunity to induce fragmentation in a highly informative manner for structure determination, as described for the tandem

MS analysis above. In addition, slightly increasing the ion internal energy can help improve spectral resolution by a process that has been termed 'collisional cleaning'. This process essentially removes excess buffer, salts or other contaminants that may be non-specifically bound to the analyte, increasing the signal-to-noise ratio and enhancing peak shape, without inducing dissociation or unfolding of the biomolecules being analysed, thereby indirectly improving the accuracy of the molecular weight measurement [42].

10.3.2.1 Collision-Induced Dissociation

There are various means of increasing the internal energy of the ions in the gas phase to induce the dissociation of both non-covalent and covalent bonds in order to provide structural information. Perhaps the most commonly applied method is collisional activation, whereby a fraction of the ion kinetic energy is converted to vibrational excitation upon collision with a neutral gas molecule, typically in either a dedicated collision cell or inside an ion trap. This technique is referred to as collision-induced dissociation (CID). Under a low energy CID regime (tens of eV), multiple collisions are required in order to accumulate enough energy to induce bond dissociation and therefore is typically a slow process, with ion activation times in the low millisecond range. Alternatively, the high energy regime (keV), as utilised for example in MALDI-ToF-ToF analysis, mainly relies on single events and thus is a very fast event on the order of nanoseconds.

Gas-phase activation has been widely utilised to study the topological arrangements of non-covalent assemblies such as multisubunit protein complexes. CID typically results in the ejection of highly charged monomers and complementary 'stripped oligomers' (i.e. n–1-mers) for many protein complexes [19, 43, 44]. The asymmetric charge partitioning (by which the lower mass monomer takes a disproportionally large amount of charge compared to the larger stripped complex) has been attributed to the unfolding of the monomer subunits during dissociation, which allows higher occupancy of charge. The dissociation pathway of these assemblies can therefore be observed and quantified to provide information regarding binding interactions, such as the propensity for subunits to disassemble (i.e. deriving binding affinities and strength of binding interfaces) and can assist in determining the abundance of various subunits present within a biomolecular assembly (i.e. stoichiometries) from a single experiment. More recently, gas-phase activation approaches have assisted in the study of structure and dynamics for membrane proteins [45–47]. In these experiments, the protein assemblies are electrosprayed directly from detergents, micelles or nanodiscs, and ion activation is utilised to gently release the protein from the detergent in the gas phase such that native structures and interactions are maintained [45, 48]. Given the difficulty in studying the structure of membrane proteins and their critical importance in cells, this represents an exciting direction for biological MS.

10.3.2.2 Other Activation Methods

In principle, any exothermic process can be utilised to induce ion activation. For example, photoexcitation of ions in the gas phase often results in fragmentation, with excitation by infrared photons (using CO_2 lasers at 10.6 μm wavelengths), particularly useful for large biomolecules in a technique known as IR multiphoton dissociation. In the case of ultraviolet photodissociation (UVPD), a close match between the energy of the excitation photon and certain chemical bonds can lead to highly selective fragmentation processes. For example, the maximum absorption of a disulphide bond occurs at a wavelength close to 150 nm, which allows for selective cleavage of thiol linkages in disulphide-bonded peptide ions upon irradiation using a 157 nm laser [49].

Gas-phase electron transfer processes underpin other ion fragmentation methods, namely electron capture dissociation (ECD) and electron transfer dissociation (ETD). In these cases, dissociation is extremely fast, providing a unique capability to preserve chemically labile groups such as glycan modifications [50] and phosphorylation sites [51], which are often important in the analysis of post-translational modifications of proteins.

One of the major factors limiting the fragmentation yield by CID is the low efficiency of energy conversion. Another manner by which dissociation pathways can be manipulated in the gas phase that overcomes this issue is by surface-induced dissociation (SID). In this approach the ions are directed towards a solid surface within the mass spectrometer, giving rise to fast and highly energetic collisions. This obviates the restriction of dissociation by slow, multistep, low barrier processes seen in CID, and therefore often gives rise to complementary, structurally informative fragmentation of large biomolecules [52–54]. For example, in the case of protein complexes, distinct subcomplexes that are representative of the subunit architecture of the native conformation can be observed by SID [55]. These types of dissociation experiments help provide a more accurate picture of the structural interactions that biomolecules exhibit, which cannot be readily determined using bulk-averaged experiments.

10.3.3 Solution Phase Chemistry

Various forms of solution-phase chemistry are used to introduce covalent modifications in biomolecules that can be localised by MS analysis to derive information regarding structure and molecular dynamics. Here we limit discussion to three key areas, particularly directed at protein structure investigation, namely covalent labelling, chemical cross-linking mass spectrometry (CXL-MS) and hydrogen–deuterium exchange mass spectrometry (HDX-MS). Each provide a key structural constraint that can be combined in an integrated MS-based approach to develop models of protein structure, particularly for cases not amenable to traditional high resolution structural biology methods such as nuclear magnetic resonance spectroscopy and X-ray crystallography.

10.3.3.1 Covalent Labelling

Covalent labelling involves the introduction of irreversible modifications at reactive sites in a biomolecule to obtain information about solvent exposed regions. This provides conceptually similar information to hydrogen–deuterium exchange (HDX) methods, although the irreversibility of the chemistry involved means sample handling and processing are more similar to standard proteomics methods. A range of different labelling strategies is possible, depending on the type of chemistry involved, and can be limited to a few amino acids with specific labelling reagents or broadly distributed to improve spatial resolution. The simultaneous modification of many residue types is possible by oxidative labelling, typically with hydroxyl radicals, which is also frequently described as ‘footprinting’ (Figure 10.11) [56, 57]. Proteins that have been covalently modified are most commonly analysed by a proteomics workflow, regardless of the actual labelling chemistry used, where enzymatic digestion gives rise to peptide mixtures for liquid chromatography mass spectrometry (LC-MS) analysis and sequencing by MS/MS. Modified amino acids are identified by their specific mass shifts compared to unmodified residues with the help of database search software and can be incorporated into structural models as a measure of surface accessibility [56].

10.3.3.2 Chemical Cross-Linking

Chemical cross-linking (CXL) of biomolecules is an extremely useful analytical approach capable of probing inter- and intramolecular distances within and across biomolecules, both *in vitro* and *in vivo*. The CXL method for obtaining structural information on proteins is based on the formation of a covalent bond between two spatially proximate residues. This bond formation can be within a single or between two polypeptide chains and provides an upper distance constraint for incorporation into structural models [58–60].

Conventional CXL reagents typically contain two reactive sites connected through a spacer or linker region, generally an alkyl chain. More sophisticated designs include features to address analytical challenges, such as incorporation of additional reactive groups or isotopic labels, MS cleavable sites and affinity purification tags. Most commonly, the reactive groups of the CXL reagents target primary amino groups of lysine residues and the N-terminus using N-hydroxysuccinimidyl or sulfosuccinimidyl esters. However, despite these esters exhibiting fast conjugation, they are also susceptible to rapid hydrolysis in aqueous solutions with half-lives on a scale of tens of minutes under physiological conditions (pH 7; 25–37 °C). More targeted CXL reactions can be carried out by linking cysteine residues using maleimide functional groups, as the relative abundance of cysteine (<2%) in proteins/peptides provides higher specificity.

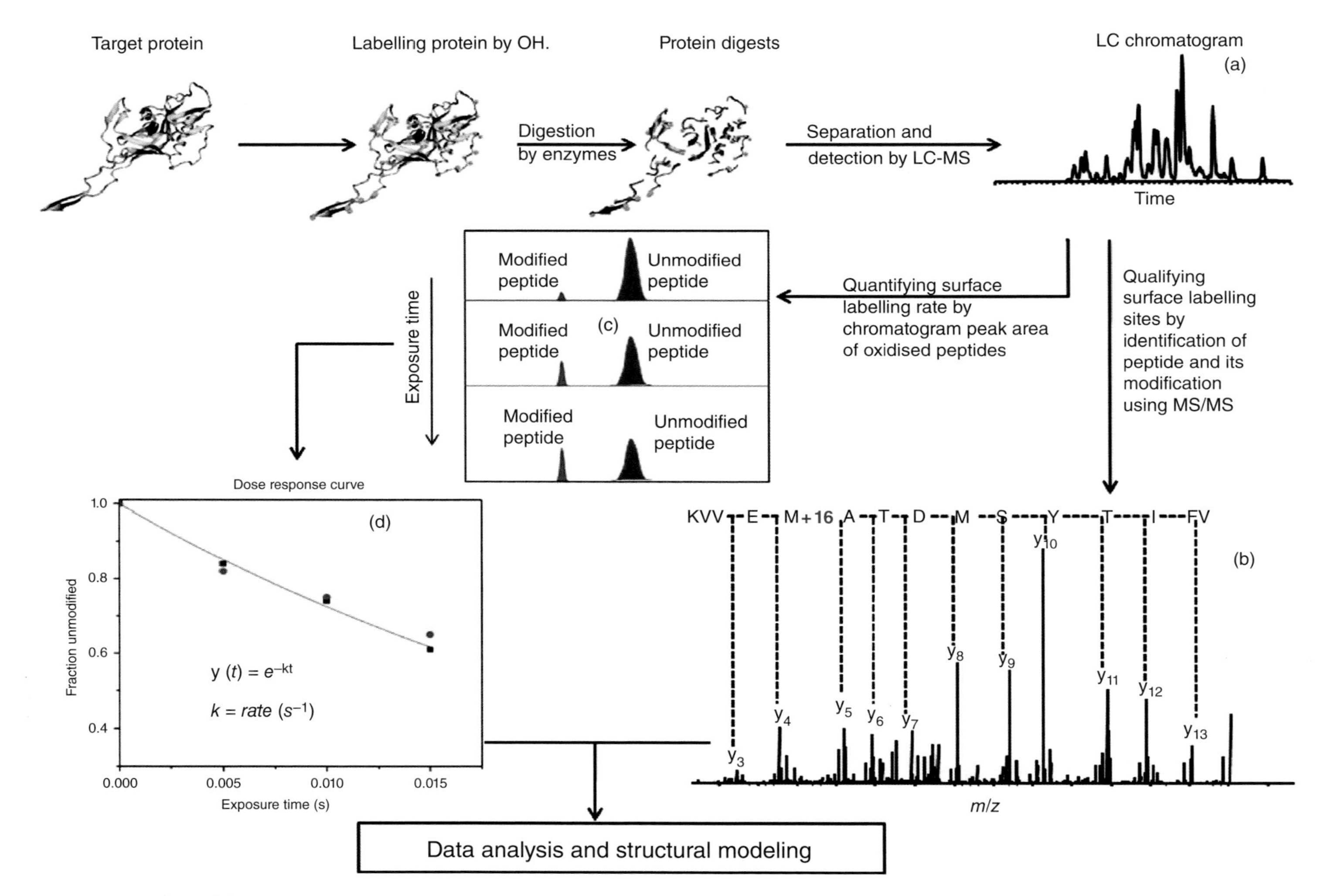

Figure 10.11 Example workflow using chemical labelling to identify protein modification sites. (a) Target proteins are labelled using hydroxyl radicals and digested into peptides prior to LC separation. (b) Peptide fingerprinting using tandem MS identifies both the peptide sequence and position of covalent modification (in this case oxidation). The abundance (c) and rate of oxidation (d) of both modified and unmodified peptides can be quantified as a function of exposure time to labelling to provide information regarding surface exposure of particular residues. Source: Reprinted with permission from [56]. Copyright (2011) American Chemical Society.

Following CXL, proteins of interest are typically enzymatically digested with trypsin and peptides are separated and identified by LC-MS/MS. Cross-linked peptides represent just a small fraction of the total peptide components, as the resulting mixture consists of a majority of unmodified or modified, but not cross-linked, peptides (termed 'dead-end' or 'mono-links'). Based on their precursor mass and tandem MS fragmentation patterns, putative cross-linked peptides and their site-specific residue linkages can be assigned. Consequently, unambiguous charge state information for both precursor and fragment ions and accurate mass measurements are important for interpretation of CXL-MS data. Dedicated analysis software for interpretation of CXL-MS data has also been developed to assist with cross-link identification. This approach has been extremely useful in identifying conformational changes in macromolecular assemblies, for example upon post-translational modification (Figure 10.12).

10.3.3.3 Hydrogen–Deuterium Exchange

HDX reactions occurring in solution and monitored by ESI-MS form the basis of investigation of structure and conformational dynamics for biomolecules, in particular proteins. This methodology takes advantage of the fact that acidic hydrogens in the protein can readily exchange with deuterium from D_2O to produce a mass shift (approximately 1.006 Da per D atom) that can be detected by MS. The exchange of backbone amide hydrogens occurs on a timescale well suited to MS analysis, and are largely dependent on the higher-order structure, or specifically the solvation and H-bonding patterns of the analytes. Consequently, increased or decreased rates of isotopic exchange can correlate with structural differences and thus HDX-MS is a versatile tool for monitoring dynamic conformational changes, for example in the case of protein folding/misfolding [61, 62] or protein– ligand [63, 64] and protein–protein interactions [65].

While HDX is facile in solution, the exchange process is reversible. Therefore, it is vital that the deuterated form of the protein at a selected time point is preserved/stabilised, which is usually achieved by quenching the solution to a much lower temperature (typically 2–4 °C) and pH (approximately 2.5). In most applications of HDX, labelled proteins are then enzymatically digested into peptides using pepsin, which is active under acidic conditions, and the amount of deuterium incorporated at the peptide level can be determined by the mass shift from the unmodified peptide, often assisted by dedicated software packages. Due to the lack of proteolytic specificity of pepsin, peptide identification is achieved using LC-MS/MS. To increase the spatial resolution from the peptide level to the individual amino acid level, data from overlapping peptides are combined. Differential rates of HDX in peptide regions are often mapped on to high resolution protein structures where available to highlight regions of conformational diversity [63].

An example of the HDX data can be seen in Figure 10.13 for prolyl oligopeptidase (PREP), a highly conserved proteolytic enzyme involved in numerous processes including brain function (memory and learning) and neuropathology. In this work, HDX-MS provided the first near-residue resolution analysis of global dynamics in the presence or absence of inhibitor bound in the active site, providing a clear structural basis for the activation mechanism [63].

10.3.4 Experimental Considerations for Biomolecular Mass Spectrometry

The analyses of biomolecules (e.g. drugs, lipids, proteins, etc.) for accurate mass measurement is enabled by high quality sample preparation, particularly when isolating analytes from complex mixtures such as crude extracts and cell lysates [66]. This section focuses on sample isolation and preparation approaches in the context of proteins, whereby these principles can also be applied to the characterisation of DNA/RNA, carbohydrates and other biomolecules. The majority of sample preparation techniques applied for the MS analyses of proteins involve the use of protein precipitation, using ammonium sulphate, to separate this class of compounds from cellular debris. Further separation and purification is achieved using fast protein liquid chromatography (FPLC), where proteins of interest can be separated from complex mixtures by properties such as binding affinity, hydrophobicity, size and charge. It is important to consider the extraction and purification

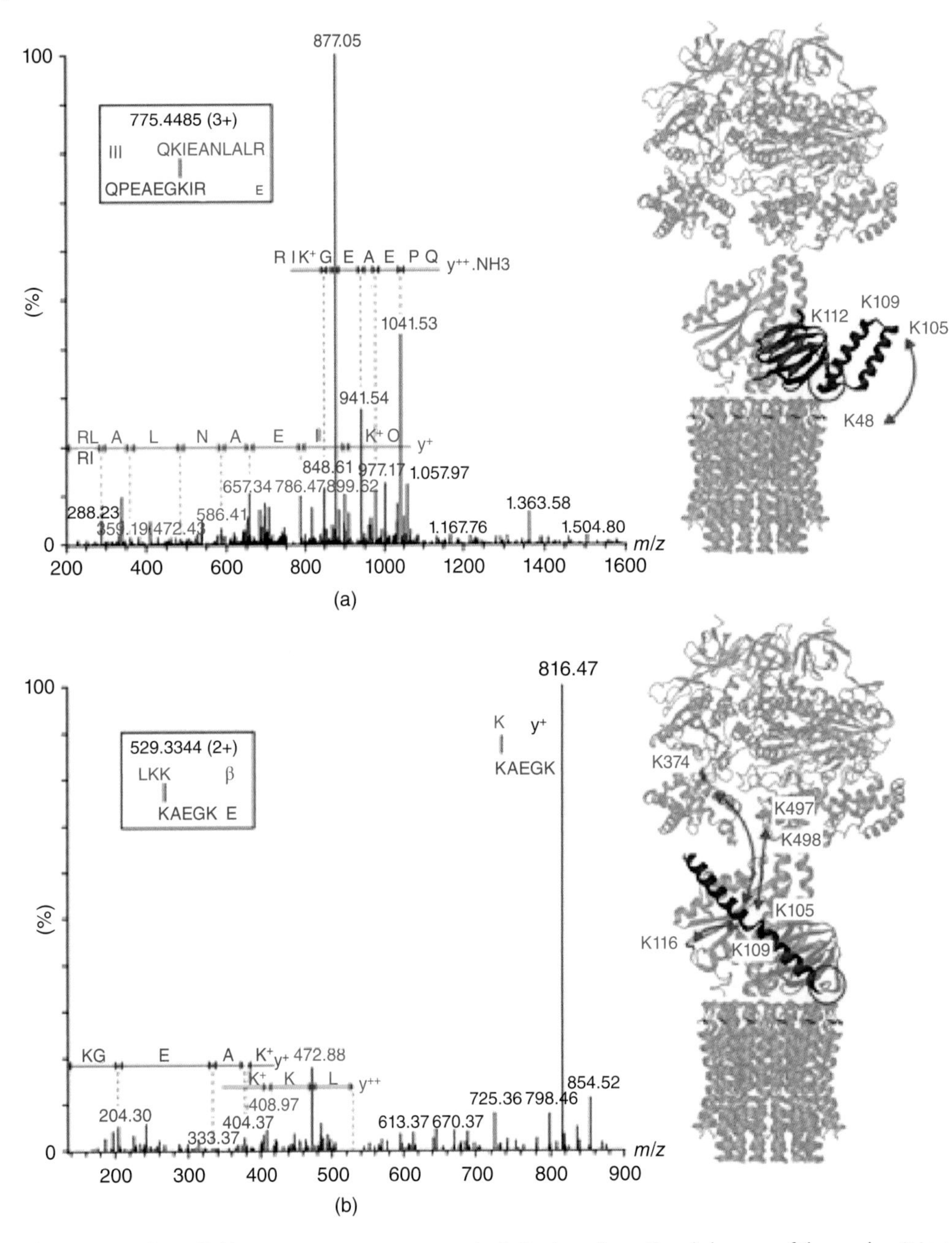

Figure 10.12 Cross-linking mass spectrometry reveals distinct conformational changes of the ε subunit in cATPase. Using a bis(sulfosuccinimidyl)suberate (BS3) lysine XL agent, the ε subunit exhibits a compact conformation (a) when bound to subunit III in the membrane ring (blue). XL-MS also shows an extended conformation (b) where the β-sheet is bound to the F_1 head (blue). Source: Figure adapted from [60] under the terms of the Creative Commons CC BY license.

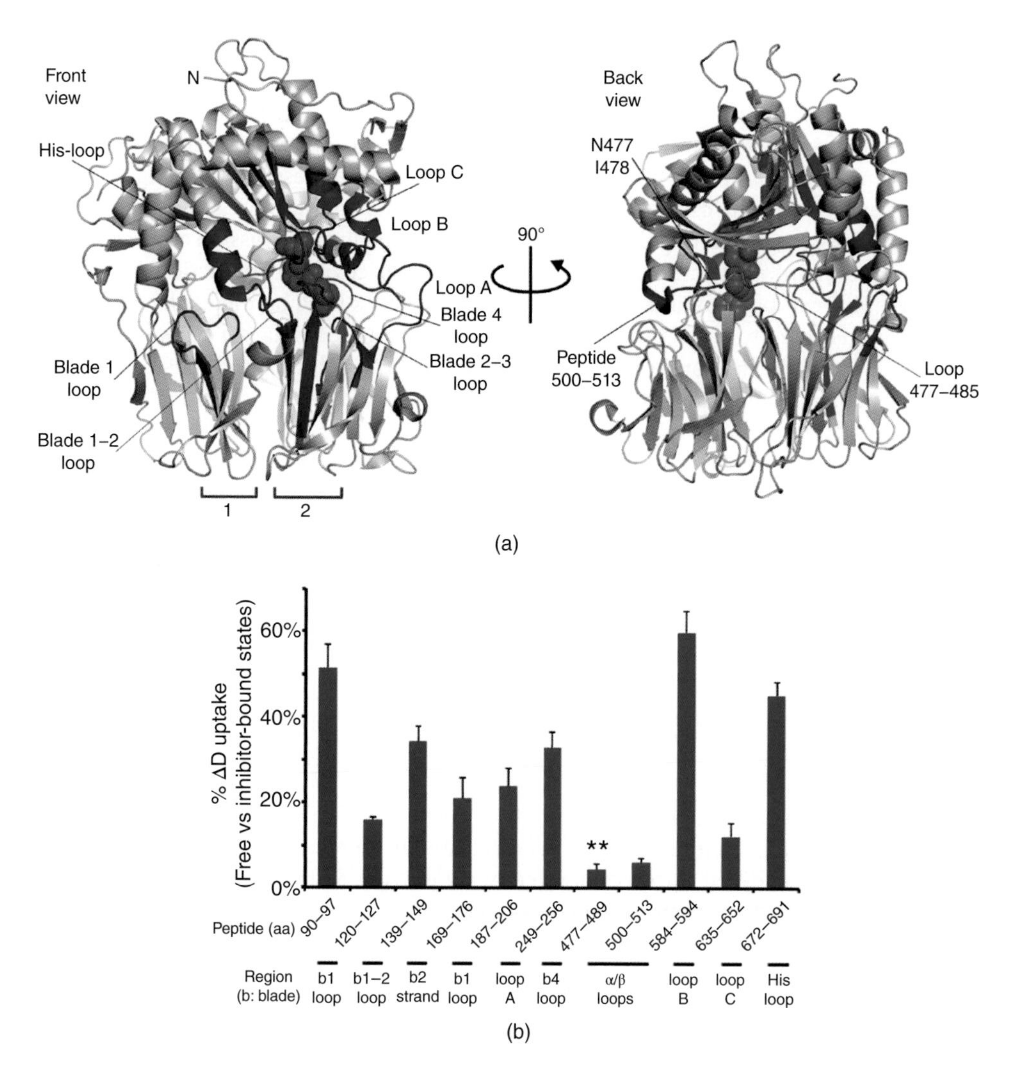

Figure 10.13 Change in the dynamics of human prolyl oligopeptidase (PREP) in response to inhibitor binding, probed by comparative H/D exchange mass spectrometry [63]. (a) Map of the PREP regions that show protection towards H/D exchange at ≥100 seconds of incubation in deuterated buffer upon inhibitor binding. Regions with differences are highlighted in blue. (b) Difference in deuterium (D) uptake (%) of human PREP in the free versus the inhibitor-bound state at isotope labelling times of ≥100 seconds. The structural regions of each peptide are indicated below. Source: Figure adapted from [63] under the terms of the Creative Commons CC BY license.

process carefully in order to maximise the yield of purified product and minimise the possibilities of misfolding or aggregation that can hamper downstream characterisation by MS.

Samples for MS analyses need to be prepared in suitable buffer systems depending on the type of experiments being performed. For accurate mass determination of biomolecules, denaturing solvents such as methanol (20% v/v) or formic acid and acetonitrile (1 : 40% v/v) are used, which maximise sensitivity and mass resolution. In the case where the native structures and interactions of biomolecules are of interest, analytes must be exchanged into volatile buffers such as ammonium acetate and ammonium bicarbonate. It is important to note that solutions containing sodium and chloride ions are not suitable for MS analyses due to their deleterious influence on the spectra due to ionisation suppression and/or adduct formation (by Na^+ ions). Typically, protein sample concentrations needed for native MS analyses are on the order of

0.1–50 μM, whereas lower concentrations are suitable for non-native/denatured samples, though this is of course somewhat dependent on the MS instrumentation used.

The instrument parameters employed during an MS experiment will determine what is observed in a spectrum, which can be manipulated in a variety of ways by subtle adjustments at various stages of the acquisition. Overall, optimisation of instrument parameters is an extensive and paramount process for attaining accurate and meaningful MS data. For example, in the case of native MS where it is important to preserve the solution state of the biomolecules, parameters associated with desolvation, ionisation, ion transmission and activation must be adjusted to prevent protein unfolding and/or dissociation of the non-covalent interactions responsible for maintaining structure. This can be dependent on the size and stability of the biomolecular assembly, whereby harsher conditions are required to aid in the desolvation and transmission of larger complexes compared to those that are smaller. Typically, the conditions that aid in high mass resolution and sensitivity come at the expense of unwanted structural changes and therefore a balance needs to be determined and due caution exercised when interpreting MS data.

Determining the mass of a macromolecular ion is now relatively straightforward and software development has advanced such that mass analysis and deconvolution have become automated and robust with increasing accuracy. Deconvolution of complicated spectra where multiple charge state series are present with a considerable degree of overlap (e.g. polydisperse proteins) has traditionally been performed manually and is extremely time consuming and can be subject to significant error and misinterpretation. In the light of this, research groups have developed custom software that is freely accessible and shown to be quite effective in accurate deconvolution [67].

10.4 Applications

The range of MS applications in chemistry and biology is incredibly extensive. While it is clearly not possible to cover all of these applications in detail, the following examples selected from recent literature demonstrate the power of MS analysis broadly considered in three basic contexts moving from small to large analytes, namely elemental and isotope analysis, small organic molecule MS and studies of macromolecular structures.

10.4.1 Elemental Analysis

Accurate determination of elements in various sample types is essential for many fields, including health, medicine and environmental science, and from research to industry. ICP-MS has undergone significant development over the last decade and while early adopters of the technology were largely from the geochemical field due to the superior ability of this technology to detect rare earth elements, it has been increasingly applied in other areas including the life sciences.

One key feature of ICP-MS is the ability to measure nearly 70% of the elements in the periodic table, offering a unique insight into the metal components of biological samples in particular. As a recent example, ICP-MS was utilised in a high-throughput fashion to reveal assimilated metals and metalloproteins from the biomass of a prototypical microbe (*Pyrococcus furiosus*) [68]. Metal ion cofactors afford proteins with unique reactivity and have significant effects on protein stability and, consequently, metalloproteins are critical to most biological processes. However, it remains difficult to predict in an organism from the genome sequence alone as metal coordination sites are diverse and poorly recognised. In this example, shifting from a protein-based purification to a metal-based identification allowed the determination of all metals assimilated by the organism from its environment and identification of its metalloproteins on a genome-wide scale. This study enabled the detection of metals known to be utilised by the organism (cobalt, iron, nickel, tungsten and zinc) and identified others that were originally thought not to be assimilated (lead, manganese, molybdenum, uranium and vanadium). Purification of eight of the 158 unexpected metal containing components yielded four novel nickel- and molybdenum-containing proteins, while another four purified proteins contained substoichiometric amounts of misincorporated lead and uranium. This application demonstrates that metalloproteomes are much more extensive and diverse than previously recognised and promise to provide key insights in cell biology.

In addition to detection of particular elements, the interrogation of specific isotopes of an element can also give useful biological insight. Because the isotope ratios of elements such as carbon, hydrogen, oxygen, sulphur and nitrogen can become locally enriched or depleted through a variety of kinetic and thermodynamic factors, measurement of the isotope ratios can be used to differentiate between samples that otherwise share identical chemical compositions. Isotope ratio determinations are usually achieved using magnetic sector instruments and are used in a variety of applications, including geological dating, and have become an analytical standard in a wide range of fields from forensics to food science. For example, since technology was first reported in 1994 to distinguish between endogenous and exogenous (synthetic) testosterone and its metabolites [69], doping control laboratories have utilised analysis of the carbon isotope ratios of endogenous steroids by gas chromatography mass spectrometry (GC-MS) to distinguish between naturally elevated steroid profiles and their illicit administration. One such high profile case relates to the 2006 Tour de France winner, Floyd Landis. Soon after the race conclusion it was reported that the urine specimen obtained from Landis was found to be positive for synthetic testosterone, where the final confirmation of this exogenous substance was determined by measuring the $^{13}C/^{12}C$ stable isotope ratios in four metabolites of testosterone via isotope ratio MS. Although Landis denied the doping, the Court of Arbitration for Sport announced a unanimous finding against him. Interestingly, some reservations have since been bought to light about the quality of the MS analysis [70], highlighting the importance of rigorous interpretation of MS data.

10.4.2 Mass Spectrometry of Organic Compounds

MS is also the tool of choice for small-molecule analysis in areas such as natural products discovery and particularly metabolomics, which aims to provide a comprehensive assessment of a wide range of endogenous metabolites in a given biological sample. Most simply, based on the measured *m/z*, characteristic fragment ions and their peak intensities, the formula and chemical structure of organic molecules can be determined manually and/or by comparison with reference spectral libraries. However, the molecular diversity in small organic compounds is incredibly expansive; for compounds comprised of only carbon, hydrogen, oxygen, nitrogen, sulphur and phosphorus, with a molecular weight less than 2000 Da and that are subject to the constraints of the 'seven golden rules of metabolomics', more than 2 billion compounds are possible and ~600 million of those are highly probable [71]. The Human Metabolome Database contains records for more than 42 000 metabolites, from peptides to sugars to enzyme cofactors, although the total is likely to be significantly higher. To probe the complexities of biological systems, huge numbers of molecules must therefore be inventoried, offering an extreme challenge in development of MS-based approaches to address the needs in sensitivity, selectivity, accuracy, dynamic range and resolution.

The number of MS-based metabolomics studies have increased exponentially over the last decade. To date such approaches have been applied to track metabolic pathways and measure dynamic flux in the metabolome, for example in response to drugs, toxins and various disease states, amongst others. Even individual cells within the same population may differ dramatically, with cell-to-cell heterogeneity stemming from differences in cell lifecycle, environmental influences and stochastic factors, and can provide new insight into differing phenotypes. Detecting metabolic changes at such low levels requires sensitive analysis, ideally suited to MS approaches. For example, using state-of-the-art MS-based metabolomics approaches and microarray technology, it is possible to profile individual cells. One recent ground-breaking study utilised such a platform to present examples of biological insight at the single-cell level, including metabolite–metabolite correlations, and visualisation of coexisting subpopulations within a genetically identical sample – one characterised by low levels of the metabolite fructose 1,6-bisphosphate and another with high levels [72].

Advances in ambient ionisation methods for imaging, such as DESI and MALDI, have opened new avenues for real-time detection of small organic molecules for the characterisation of biological specimens such as tissue samples, and shows immense utility for clinical applications [73]. One example is in surgical intervention

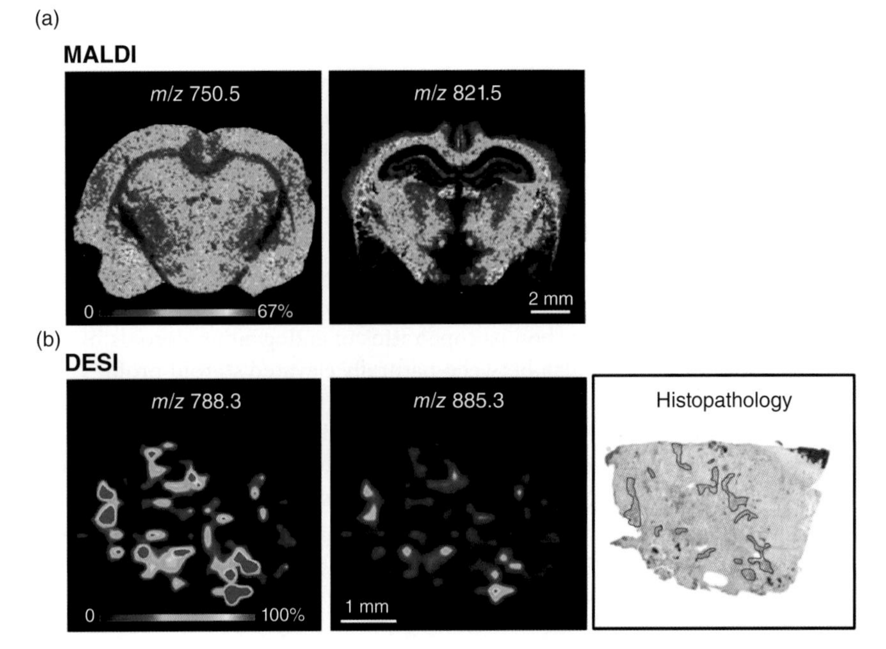

Figure 10.14 MALDI and DESI imaging of biological tissue samples. (a) MALDI-MS imaging of rat brain cryosections showing the distribution of various lipids using a novel negative ion MALDI matrix (4-phenyl-α-cyanocinnamic acid amide). The distribution of phosphatidyl ethanolamines (*m/z* 750.5, left panel) and phosphatidyl glycerols (*m/z* 821.5, right panel) were able to be visualised on this novel matrix compared to other matrices. Source: Figure adapted with permission from [74, 75]. Copyright (2011) American Chemical Society. (b) DESI-MS ion images showing the distribution of *m/z* 788.3 (left panel) and *m/z* 885.3 (middle panel), corresponding to a characteristic meningioma lipid profile. The histopathology image is shown on the right, with the main regions containing meningioma cells indicated with red lines. The distribution of meningioma cells observed by microscopy correlates with the distribution of characteristic ion signals from the DESI-MS images. Source: Reproduced with permission from PNAS.

for cancer therapy, where the main goal is to maximise tumour resection while preserving healthy tissue. However, existing techniques do not afford the molecular information needed to define tumour boundaries. While histopathology has long been held as the gold standard in imaging of tumours, diagnostic information is only accessible on the timeframe of hours and is limited to a few samples. Widespread studies have demonstrated the diagnostic potential of imaging disease tissues with examples including bladder, kidney, prostate and brain cancers. In the latter case, pioneering work to rapidly analyse and classify brain tumours based on lipid profiles acquired by desorption electrospray ionisation mass spectrometry (DESI-MS) has been shown to discriminate gliomas and meningiomas from surgical specimens who underwent brain tumour resection (Figure 10.14). The samples analysed included tumours of different histological grades and tumour cell concentrations. The molecular diagnosis derived from DESI-MS imaging correlated exceedingly well to histopathology, demonstrating the ability of this approach to providing rapid diagnosis and tumour margin assessment in near-real time [74]. Overall, MALDI and, to a lesser extent DESI, imaging approaches have been able to provide significant insights and potential in both research and diagnostic settings.

10.4.3 Mass Spectrometry for Macromolecules

All processes that underpin biological activity in living organisms are ultimately reliant on macromolecules such as oligonucleotides, polysaccharides and proteins. Consequently, elucidating the structures, interactions and dynamics of these macromolecules is an essential pursuit in the understanding of health, disease and biological activity. For these reasons, the development of ionisation methods that enabled the gas-phase

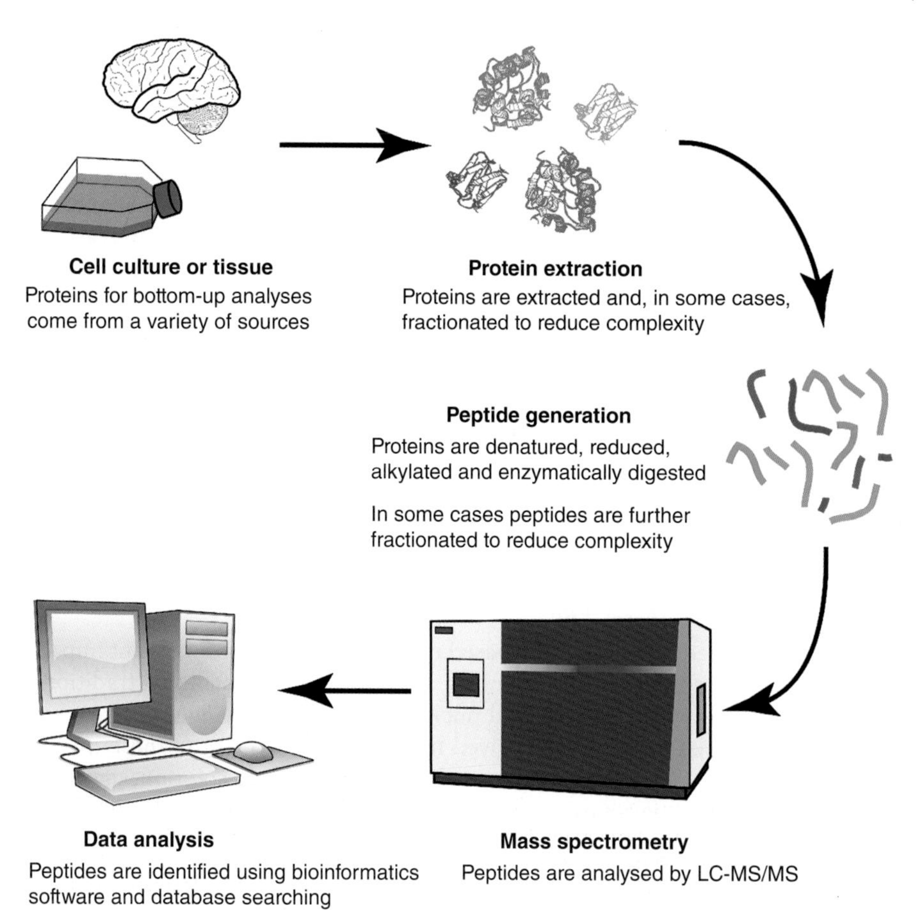

Figure 10.15 Basic workflow of a 'bottom-up' proteomics experiment for protein identification. Proteins are extracted and subjected to proteolytic digestion. The resulting peptides are usually separated using one or more dimensions of liquid chromatography, interfaced to a mass spectrometer using electrospray ionisation. Peptides are analysed by MS and tandem MS methods and identified using bioinformatics approaches by matching to protein sequence databases.

generation of large ions revolutionised the discipline of biological MS, with the field of proteomics arguably the quintessential application of this technology in the modern era.

The term 'proteomics' was first used to describe the large-scale characterisation of the entire protein complement of a cell, tissue or organism [76], and has more recently come to be directly associated with the MS identification of proteins in a biological sample. In a generic proteomics experiment, the proteins are first isolated and often fractionated to define the 'subproteome' for analysis. Since MS detection of proteins is less sensitive than that of peptides, and mass measurement and fragmentation of intact proteins are often insufficient for unambiguous identification, the proteins are typically analysed by a 'bottom-up' proteomics approach (Figure 10.15), whereby the protein is enzymatically digested, often by trypsin. Following separation of the peptides, usually by online LC coupled to ESI, mass spectra of the eluting peptides are recorded along with a series of MS/MS experiments for abundant precursor ions. The MS and MS/MS data are commonly searched against a protein sequence database to allow the identity of the constituent peptides and therefore the initial proteins (or partial gene products) to be determined.

Determination of post-translational modifications of proteins, such as the carbohydrate portion of glycoproteins or occupancy of phosphorylation sites provides another analytical challenge to researchers interested in the proteomic description of the cell, especially when only minimal amounts of sample are available. Reasons for this include decreases in ionisation and fragmentation propensities of modified peptides and, in particular, the isomeric structural heterogeneity and their very low abundance. Characterisation of oligosaccharides is more challenging than that of proteins and oligonucleotides due to the isomeric nature of the sugar subunit and its ability to form heterogeneous, branched structures. However, it is possible to determine the structures of oligosaccharides using tandem MS in an analogous fashion to peptides. For a complete proteomics analysis, it is not only necessary to define the identity of the protein components, but also the relative abundances of the proteins as a function of time, environmental stimulus or biological state. Consequently, it is increasingly desired to add a quantitative dimension to proteomics experiments, for example, quantifying the abundance of particular peptides under various conditions (e.g. healthy and diseased states). This is commonly achieved utilising stable-isotope dilution, in which pairs of chemically identical analytes of different isotopic composition are differentiated in the mass spectrometer by their mass difference and the signal intensities for such analyte pairs is used to give an accurate representation of the relative abundance. Stable isotope tags can be introduced to proteins via metabolic labelling, enzymatic transfer or chemical bioconjugation using isotope coded mass tags [77].

Illustrating the power and importance of proteomics, the genome sequencing project of the malaria parasite *Plasmodium falciparum* has recently been complemented by large-scale proteomics efforts. A large number of proteins were identified in the sexual and non-sexual human stages of the parasite, and relative quantitation was achieved between stages. From this work, a set of more than 200 proteins has been identified as possible site-specific drug or vaccine targets for future research [78].

The analysis of protein complexes is another area where MS-based methods have had a significant impact, particularly since such massive, dynamically interacting ensembles are often not amenable to traditional structural biology techniques such as X-ray crystallography and nuclear magnetic resonance spectroscopy. It has been known for more than 20 years that information derived from the mass spectra of these macromolecules, particularly for ESI-MS analysis of proteins, can be correlated with their solution structural properties. For example, the extent of multiple charging on proteins is found to differ substantially using ESI from non-denaturing or denaturing conditions, and is attributed to conformational changes taking place in solution that are reflected in the number of accessible ionisable sites on the protein [79]. Consequently, the ESI charge state distribution can be used as a measure of a three-dimensional macromolecular structure. The soft nature of ESI has also been shown to maintain the structural integrity of protein–protein and protein–ligand interactions throughout the ionisation process, enabling the study of higher order structures and binding associations through 'native' MS. The overwhelming majority of structural MS studies have focused on proteins, although similar principles are also applicable to other macromolecular complexes.

The combination of ion mobility (IM) with MS has been used extensively to determine the structures of compounds from small molecules such as drugs and explosives to subunit architecture of large macromolecular assemblies such as molecular chaperones [80, 81] and ATPases [82]. IM-MS is also capable of discerning between structural isomers such as L- and D-isoforms of peptides [83], heterogeneous glycans of identical molecular weight [84], differing unsaturation sites in lipids [85] or conformationally different protein states amongst many other examples, offering unrivalled separation of conformers and unique capabilities for the analysis of a 3D structure (Figure 10.16). IM-MS is commonly used in conjunction with gas-phase dissociation methods to extensively probe the structural dynamics of proteins [86]. More recently, it has been utilised to provide modelling constraint data, in conjunction with complementary biophysical techniques, to determining the structure and conformation of macromolecular complexes such as integral membrane proteins [87] and viruses [88]. Overall, IM-MS is a key component of the MS toolbox for elucidating the structure and dynamics of biomolecules as a means of providing a structural rationale for their function.

One of the most straightforward applications of native MS is the determination of subunit stoichiometry in a biological complex, which, with few exceptions, shows excellent consensus between that derived by other methods. This has led to successful analysis of an extensive range of non-covalent biomolecular complexes, including impressive examples such as the intact ribosome [89]. With the introduction of commercially available IM-MS instrumentation, this technology has achieved a rapid rise in popularity to study the overall

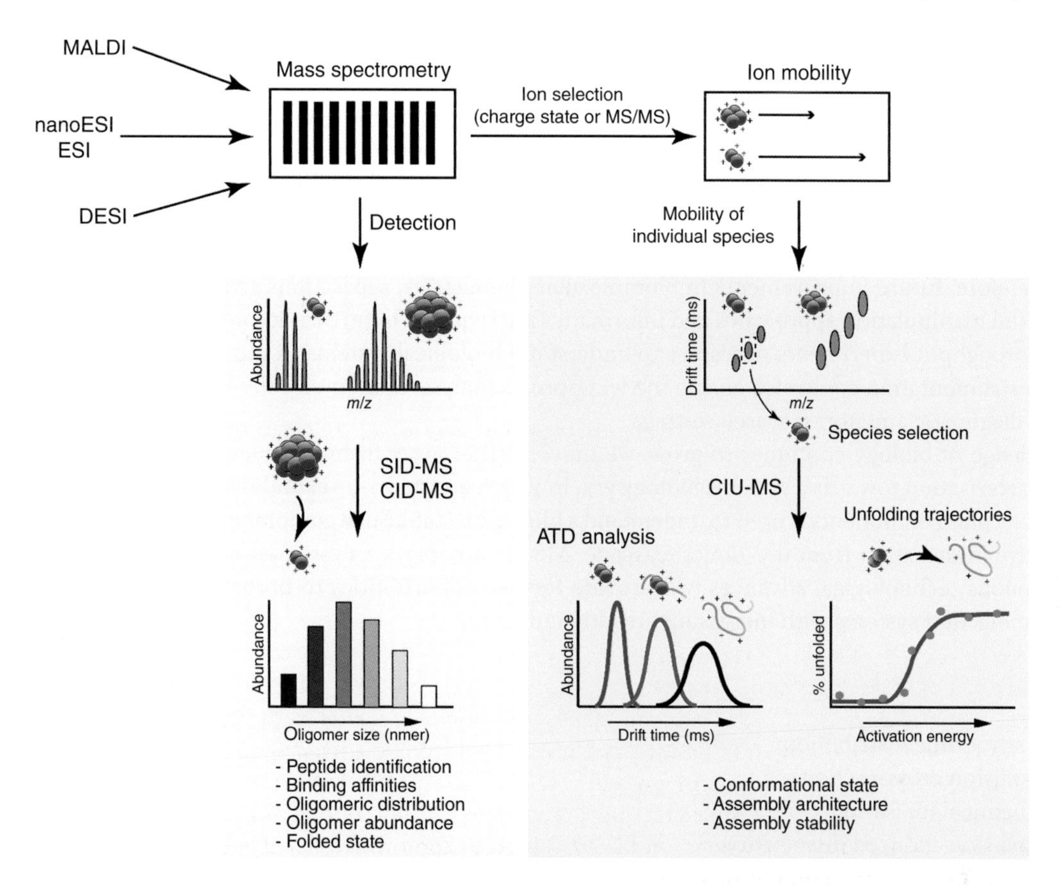

Figure 10.16 Example workflow for the investigation of structure and dynamics for proteins by mass spectrometry. In order to preserve the native-like state, proteins are prepared in a volatile buffer and are transferred to the gas phase using soft ionisation methods. Ions are detected based on their *m/z*, giving identity and stoichiometry. Activation methods such as CID and SID allow investigation of the binding interactions within a protein complex (blue box). IM-MS can provide information regarding the conformational state of protein assemblies by comparing the arrival time distribution (ATD) and corresponding CCS of different species. The stability of individual protein species can also be analysed by IM-MS by observing their unfolding dynamics as a result of collision-induced unfolding (CIU) (orange box).

structure of biomolecular assemblies, monitor changing protein conformations or reduce sample complexity by complementing the mass measurement with another gas-phase separation dimension. Combined with gas-phase activation methods, the stoichiometry, conformation, binding interactions and dynamics of a structural assembly can be potentially determined from a single MS analysis (Figure 10.16). This can certainly be further complemented by utilising an integrated approach including other structural constraints from MS-based methods such as HDX, CXL and chemical labelling [90], as well as other biophysical techniques. The benefits of an increasingly integrated structural biology approach are effectively illustrated by the Mediator complex, a multiprotein assembly that functions as a transcriptional coactivator in all eukaryotes, but is refractory to traditional structural biology methods due to its conformational flexibility and diverse composition. The middle module of the Mediator complex was first investigated by numerous MS-based experiments (native MS, tandem-MS and IM-MS), which, in combination with light scattering, small-angle X-ray scattering and pull-down assays, allowed for determination of its overall subunit topology [91]. This putative model of the complex was further refined using CXL-MS [92]. Similarly, a structure of the Mediator head module was determined by X-ray crystallography and CXL-MS [93] and, most recently, a model of the full Mediator-Pol II pre-initiation complex comprising 52 protein subunits has been proposed largely from MS-based analyses [94].

10.5 Concluding Remarks

Throughout its history, MS has always provided a significant contribution to the science of the time. However, despite the tremendous developments in MS technologies in recent decades, a number of existing challenges remain that currently prohibit the comprehensive analysis of the full spectrum of biomolecules. These include inadequate sensitivity and dynamic range, contributing to poor sample utilisation. Finally, automated data interpretation, particularly for small molecules, and the volume of data relevant to biological studies can be inadequate. Therefore, future improvements in biomolecular chemistries, separations science, fundamental ion generation and manipulation approaches and informatics are required to further accommodate the broad reaching high throughput experiments necessary to understand biological systems. A move to miniaturised, transportable instrumentation could also enable the widespread analysis of biomolecules in a broadly accessible fashion in diagnostic and non-research settings.

As our knowledge of biology continues to grow, we move further away from a reductionist approach in molecular characterisation towards a systems biology era, in which a systems-level understanding of interactions among molecular components is used to understand a biological state. Such comprehensive investigation requires extensive information from the 'omics' cascade. MS already plays an essential role in this pursuit, although continuous technological advances will provide further opportunities to probe the structure and function of biomolecular systems with increasing breadth and depth.

Abbreviations

ATD	arrival time distribution
CCS	collision cross-section
CI	chemical ionisation
CID	collision-induced dissociation
CIU	collision-induced unfolding
CRM	charged residue model
CXL	chemical cross-linking
DC	direct current
DESI	desorption electrospray ionisation
DNA	deoxyribonucleic acid
ECD	electron capture dissociation
EI	electron ionisation
ESI	electrospray ionisation
ETD	electron transfer dissociation
FPLC	fast protein liquid chromatography
FT-ICR	Fourier transform ion cyclotron resonance
HDX	hydrogen–deuterium exchange
ICP	inductively coupled plasma
IEM	ion evaporation model
IM	ion mobility
IRMS	isotope ratio mass spectrometry
LC	liquid chromatography
MALDI	matrix-assisted laser desorption ionisation
MS	mass spectrometry
RF	radiofrequency
SID	surface-induced dissociation

References

1 Thomson, J.J. (1913). *Rays of Positive Electricity and Their Applications to Chemical Analyses*. London: Longmans, Green and Co.

2 F.W. Aston (1919). LXXIV. A positive ray spectrograph, The London, Edinburgh, and Dublin Philosophical Magazine and Journal of Science, 38, 707–714.

3 Aston, F.W. (1919). Neon. *Nature* 104: 334.

4 Lifson, N., Gordon, G.B., Visscher, M.B., and Nier, A.O. (1949). The fate of utilized molecular oxygen and the source of the oxygen of respiratory carbon dioxide, studied with the aid of heavy oxygen. *J. Biol. Chem.* 180: 803–811.

5 Yamashita, M. and Fenn, J.B. (1984). Electrospray ion source. Another variation on the free-jet theme. *J. Phys. Chem.* 88: 4451–4459.

6 Karas, M., Bachmann, D., and Hillenkamp, F. (1985). Influence of the wavelength in high-irradiance ultraviolet laser desorption mass spectrometry of organic molecules. *Anal. Chem.* 57: 2935–2939.

7 Tanaka, K., Waki, H., Ido, Y. et al. (1988). Protein and polymer analyses up to m/z 100 000 by laser ionization time-of-flight mass spectrometry. *Rapid Commun. Mass Spectrom.* 2: 151–153.

8 Chowdhury, S.K., Katta, V., and Chait, B.T. (1990). Electrospray ionization mass spectrometric peptide mapping: a rapid, sensitive technique for protein structure analysis. *Biochem. Biophys. Res. Commun.* 167: 686–692.

9 Dettmer, K., Aronov, P.A., and Hammock, B.D. (2007). Mass spectrometry based metabolomics. *Mass Spectrom. Rev.* 26: 51–78.

10 Ruotolo, B.T., Giles, K., Campuzano, I. et al. (2005). Evidence for macromolecular protein rings in the absence of bulk water. *Science* 310: 1658–1661.

11 Loo, J.A., Benchaar, S.A., and Zhang, J. (2013). Integrating native mass spectrometry and top-down MS for defining protein interactions important in biology and medicine. *Mass Spectrom.* 2: S0013.

12 Li, H., Nguyen, H.H., Ogorzalek Loo, R.R. et al. (2018). An integrated native mass spectrometry and top-down proteomics method that connects sequence to structure and function of macromolecular complexes. *Nat. Chem.* 10: 139.

13 Mirsaleh-Kohan, N., Robertson, W.D., and Compton, R.N. (2008). Electron ionization time-of-flight mass spectrometry: historical review and current applications. *Mass Spectrom. Rev.* 27: 237–285.

14 Knapp, D.R. (1990). Chemical derivatization for mass spectrometry. In: *Methods in Enzymology*, 314–329. Academic Press.

15 Wilm, M. (2011). Principles of electrospray ionization. *Mol. Cell. Proteomics* 10.

16 Wilm, M.S. and Mann, M. (1994). Electrospray and Taylor-cone theory, doles beam of macromolecules at last. *Int. J. Mass Spectrom.* 136: 167–180.

17 Konermann, L., Ahadi, E., Rodriguez, A.D., and Vahidi, S. (2013). Unraveling the mechanism of electrospray ionization. *Anal. Chem.* 85: 2–9.

18 Konermann, L., Rodriguez, A.D., and Liu, J. (2012). On the formation of highly charged gaseous ions from unfolded proteins by electrospray ionization. *Anal. Chem.* 84: 6798–6804.

19 Aquilina, J.A., Benesch, J.L., Bateman, O.A. et al. (2003). Polydispersity of a mammalian chaperone: mass spectrometry reveals the population of oligomers in alphaB-crystallin. *Proc. Natl Acad. Sci. USA* 100: 10611–10616.

20 Hyung, S.J., Robinson, C.V., and Ruotolo, B.T. (2009). Gas-phase unfolding and disassembly reveals stability differences in ligand-bound multiprotein complexes. *Chem. Biol.* 16: 382–390.

21 Hernandez, H. and Robinson, C.V. (2007). Determining the stoichiometry and interactions of macromolecular assemblies from mass spectrometry. *Nat. Protoc.* 2: 715–726.

22 Schmidt, A., Karas, M., and Dulcks, T. (2003). Effect of different solution flow rates on analyte ion signals in nano-ESI MS, or: when does ESI turn into nano-ESI? *J. Am. Soc. Mass Spectrom.* 14: 492–500.

23 Takáts, Z., Wiseman, J.M., Gologan, B., and Cooks, R.G. (2004). Mass spectrometry sampling under ambient conditions with desorption electrospray ionization. *Science* 306: 471–473.

24 Ambrose, S., Housden, N.G., Gupta, K. et al. (2017). Native desorption electrospray ionization liberates soluble and membrane protein complexes from surfaces. *Angew. Chem. Int. Ed.* 56: 14463–14468.

25 Abbassi-Ghadi, N., Golf, O., Kumar, S. et al. (2016). Imaging of esophageal lymph node metastases by desorption electrospray ionization mass spectrometry. *Cancer Res.* 76: 5647–5656.

26 Aichler, M. and Walch, A. (2015). MALDI imaging mass spectrometry: current frontiers and perspectives in pathology research and practice. *Lab. Invest.* 95: 422–431.

27 Maher, S., Jjunju, F.P.M., and Taylor, S. (2015). 100 years of mass spectrometry: perspectives and future trends. *Rev. Mod. Phys.* 87: 113–135.

28 Dempster, J.A. (1918). A new method of positive ray analysis. *Phys. Rev.* 11: 316–325.

29 Stephens, W.E. (1946). A pulsed mass spectrometer with time dispersion. *Phys. Rev.* 69: 691.

30 Guilhaus, M., Mlynski, V., and Selby et al., D. (1997). Perfect timing: time-of-flight mass spectrometry. *Rapid Commun. Mass Spectrom.* 11: 951–962.

31 Douglas, D.J. (2009). Linear quadrupoles in mass spectrometry. *Mass Spectrom. Rev.* 28: 937–960.

32 Campana, J.E. (1980). Elementary theory of the quadrupole mass filter. *Int. J. Mass Spectrom. Ion Phys.* 33: 101–117.

33 Bruce, J.E., Cheng, X., Bakhtiar, R. et al. (1994). Trapping, detection, and mass measurement of individual ions in a fourier transform ion cyclotron resonance mass spectrometer. *J. Am. Chem. Soc.* 116: 7839–7847.

34 Makarov, A. (2000). Electrostatic axially harmonic orbital trapping: a high-performance technique of mass analysis. *Anal. Chem.* 72: 1156–1162.

35 Hu, Q., Noll, R.J., Li, H. et al. (2005). The orbitrap: a new mass spectrometer. *J. Mass Spectrom.* 40: 430–443.

36 Liu, R., Li, Q., and Smith, L.M. (2014). Detection of large ions in time-of-flight mass spectrometry: effects of ion mass and acceleration voltage on microchannel plate detector response. *J. Am. Soc. Mass Spectrom.* 25: 1374–1383.

37 Koek, M.M., Jellema, R.H., van der Greef, J. et al. (2011). Quantitative metabolomics based on gas chromatography mass spectrometry: status and perspectives. *Metabolomics* 7: 307–328.

38 Wang, E.H., Combe, P.C., and Schug, K.A. (2016). Multiple reaction monitoring for direct quantitation of intact proteins using a triple quadrupole mass spectrometer. *J. Am. Soc. Mass Spectrom.* 27: 886–896.

39 Yang, L., Amad, M., Winnik, W.M. et al. (2002). Investigation of an enhanced resolution triple quadrupole mass spectrometer for high-throughput liquid chromatography/tandem mass spectrometry assays. *Rapid Commun. Mass Spectrom.* 16: 2060–2066.

40 Medzihradszky, K.F. and Chalkley, R.J. (2015). Lessons in de novo peptide sequencing by tandem mass spectrometry. *Mass Spectrom. Rev.* 34: 43–63.

41 Biemann, K. (1988). Contributions of mass spectrometry to peptide and protein structure. *Biomed. Environ. Mass Spectrom.* 16: 99–111.

42 Benesch, J.L. (2009). Collisional activation of protein complexes: picking up the pieces. *J. Am. Soc. Mass Spectrom.* 20: 341–348.

43 Jovcevski, B., Kelly, M.A., Rote, A.P. et al. (2015). Phosphomimics destabilize Hsp27 oligomeric assemblies and enhance chaperone activity. *Chem. Biol.* 22: 186–195.

44 Benesch, J.L., Aquilina, J.A., Ruotolo, B.T. et al. (2006). Tandem mass spectrometry reveals the quaternary organization of macromolecular assemblies. *Chem. Biol.* 13: 597–605.

45 Marty, M.T., Hoi, K.K., Gault, J., and Robinson, C.V. (2016). Probing the lipid annular belt by gas-phase dissociation of membrane proteins in naanodiscs. *Angew. Chem. Int. Ed. Engl.* 55: 550–554.

46 Marty, M.T., Hoi, K.K., and Robinson, C.V. (2016). Interfacing membrane mimetics with mass spectrometry. *Acc. Chem. Res.* 49: 2459–2467.

47 Hopper, J.T., Yu, Y.T., Li, D. et al. (2013). Detergent-free mass spectrometry of membrane protein complexes. *Nat. Methods* 10: 1206–1208.

48 Marty, M.T., Zhang, H., Cui, W. et al. (2012). Native mass spectrometry characterization of intact nanodisc lipoprotein complexes. *Anal. Chem.* 84: 8957–8960.

49 Fung, Y.M., Kjeldsen, F., Silivra, O.A. et al. (2005). Facile disulfide bond cleavage in gaseous peptide and protein cations by ultraviolet photodissociation at 157 nm. *Angew. Chem. Int. Ed. Engl.* 44: 6399–6403.

50 Leymarie, N. and Zaia, J. (2012). Effective use of mass spectrometry for glycan and glycopeptide structural analysis. *Anal. Chem.* 84: 3040–3048.

51 Kim, M.S. and Pandey, A. (2012). Electron transfer dissociation mass spectrometry in proteomics. *Proteomics* 12: 530–542.

52 Ma, X., Loo, J.A., and Wysocki, V.H. (2015). Surface induced dissociation yields substructure of *Methanosarcina thermophila* 20S proteasome complexes. *Int. J. Mass Spectrom.* 377: 201–204.

53 Wysocki, V.H., Jones, C.M., Galhena, A.S., and Blackwell, A.E. (2008). Surface-induced dissociation shows potential to be more informative than collision-induced dissociation for structural studies of large systems. *J. Am. Soc. Mass Spectrom.* 19: 903–913.

54 Wysocki, V.H., Joyce, K.E., Jones, C.M., and Beardsley, R.L. (2008). Surface-induced dissociation of small molecules, peptides, and non-covalent protein complexes. *J. Am. Soc. Mass Spectrom.* 19: 190–208.

55 Zhou, M., Jones, C.M., and Wysocki, V.H. (2013). Dissecting the large noncovalent protein complex GroEL with surface-induced dissociation and ion mobility–mass spectrometry. *Anal. Chem.* 85: 8262–8267.

56 Wang, L. and Chance, M.R. (2011). Structural mass spectrometry of proteins using hydroxyl radical based protein footprinting. *Anal. Chem.* 83: 7234–7241.

57 Calabrese, A.N., Ault, J.R., Radford, S.E., and Ashcroft, A.E. (2015). Using hydroxyl radical footprinting to explore the free energy landscape of protein folding. *Methods* 89: 38–44.

58 Leitner, A., Faini, M., Stengel, F., and Aebersold, R. (2016). Crosslinking and mass spectrometry: an integrated technology to understand the structure and function of molecular machines. *Trends Biochem. Sci.* 41: 20–32.

59 Holding, A.N. (2015). XL-MS: protein cross-linking coupled with mass spectrometry. *Methods* 89: 54–63.

60 Schmidt, C., Zhou, M., Marriott, H. et al. (2013). Comparative cross-linking and mass spectrometry of an intact F-type ATPase suggest a role for phosphorylation. *Nat. Commun.* 4: 1985.

61 Del Mar, C., Greenbaum, E.A., Mayne, L. et al. (2005). Structure and properties of alpha-synuclein and other amyloids determined at the amino acid level. *Proc. Natl Acad. Sci. USA* 102: 15477–15482.

62 Xiao, Y. and Konermann, L. (2015). Protein structural dynamics at the gas/water interface examined by hydrogen exchange mass spectrometry. *Protein Sci.* 24: 1247–1256.

63 Tsirigotaki, A., Elzen, R.V., Veken, P.V. et al. (2017). Dynamics and ligand-induced conformational changes in human prolyl oligopeptidase analyzed by hydrogen/deuterium exchange mass spectrometry. *Sci. Rep.* 7: 2456.

64 Mistarz, U.H., Brown, J.M., Haselmann, K.F., and Rand, K.D. (2016). Probing the binding interfaces of protein complexes using gas-phase H/D exchange mass spectrometry. *Structure* 24: 310–318.

65 Iacob, R.E., Krystek, S.R., Huang, R.Y. et al. (2015). Hydrogen/deuterium exchange mass spectrometry applied to IL-23 interaction characteristics: potential impact for therapeutics. *Expert Rev. Proteomics* 12: 159–169.

66 Smits, A.H. and Vermeulen, M. (2016). Characterizing protein-protein interactions using mass spectrometry: challenges and opportunities. *Trends Biotechnol.* 34: 825–834.

67 Marty, M.T., Baldwin, A.J., Marklund, E.G. et al. (2015). Bayesian deconvolution of mass and ion mobility spectra: from binary interactions to polydisperse ensembles. *Anal. Chem.* 87: 4370–4376.

68 Cvetkovic, A., Menon, A.L., Thorgersen, M.P. et al. (2010). Microbial metalloproteomes are largely uncharacterized. *Nature* 466: 779.

69 Becchi, M., Aguilera, R., Farizon, Y. et al. (1994). Gas chromatography/combustion/isotope-ratio mass spectrometry analysis of urinary steroids to detect misuse of testosterone in sport. *Rapid Commun. Mass Spectrom.* 8: 304–308.

70 Blackledge, R.D. (2009). Bad science: the instrumental data in the Floyd Landis case. *Clin. Chim. Acta* 406: 8–13.

71 Aksenov, A.A., da Silva, R., Knight, R. et al. (2017). Global chemical analysis of biology by mass spectrometry. *Nat. Rev. Chem.* 1: 0054.

72 Ibáñez, A.J., Fagerer, S.R., Schmidt, A.M. et al. (2013). Mass spectrometry-based metabolomics of single yeast cells. *Proc. Natl Acad. Sci. USA* 110: 8790–8794.

73 Nemes, P. and Vertes, A. (2012). Ambient mass spectrometry for *in vivo* local analysis and in situ molecular tissue imaging. *Trends Anal. Chem.* 34: 22–34.

74 Eberlin, L.S., Norton, I., Orringer, D. et al. (2013). Ambient mass spectrometry for the intraoperative molecular diagnosis of human brain tumors. *Proc. Natl Acad. Sci. USA* 110: 1611–1616.

75 Fulop, A., Porada, M.B., Marsching, C. et al. (2013). 4-Phenyl-alpha-cyanocinnamic acid amide: screening for a negative ion matrix for MALDI-MS imaging of multiple lipid classes. *Anal. Chem.* 85: 9156–9163.

76 Anderson, N.G. and Anderson, N.L. (1996). Twenty years of two-dimensional electrophoresis: past, present and future. *Electrophoresis* 17: 443–453.

77 Aebersold, R. and Mann, M. (2003). Mass spectrometry-based proteomics. *Nature* 422: 198.

78 Lasonder, E., Ishihama, Y., Andersen, J.S. et al. (2002). Analysis of the *Plasmodium falciparum* proteome by high-accuracy mass spectrometry. *Nature* 419: 537.

79 Chowdhury, S.K., Katta, V., and Chait, B.T. (1990). Probing conformational changes in proteins by mass spectrometry. *J. Am. Chem. Soc.* 112: 9012–9013.

80 Baldwin, A.J., Lioe, H., Hilton, G.R. et al. (2011). The polydispersity of alphaB-crystallin is rationalized by an interconverting polyhedral architecture. *Structure* 19: 1855–1863.

81 Jovcevski, B., Kelly, M.A., Aquilina, J.A. et al. (2017). Evaluating the effect of phosphorylation on the structure and dynamics of Hsp27 dimers by means of ion mobility mass spectrometry. *Anal. Chem.* 89: 13275–13282.

82 Zhou, M., Politis, A., Davies, R. et al. (2014). Ion mobility-mass spectrometry of a rotary ATPase reveals ATP-induced reduction in conformational flexibility. *Nat. Chem.* 6: 208–215.

83 Fouque, K.J.D., Garabedian, A., Porter, J. et al. (2017). Fast and effective ion mobility-mass spectrometry separation of D-amino-acid-containing peptides. *Anal. Chem.* 89: 11787–11794.

84 Zhu, F.F., Trinidad, J.C., and Clemmer, D.E. (2015). Glycopeptide site heterogeneity and structural diversity determined by combined lectin affinity chromatography/IMS/CID/MS techniques. *J. Am. Soc. Mass Spectrom.* 26: 1092–1102.

85 Shvartsburg, A.A., Isaac, G., Leveque, N. et al. (2011). Separation and classification of lipids using differential ion mobility spectrometry. *J. Am. Soc. Mass Spectrom.* 22: 1146–1155.

86 Lanucara, F., Holman, S.W., Gray, C.J., and Eyers, C.E. (2014). The power of ion mobility-mass spectrometry for structural characterization and the study of conformational dynamics. *Nat. Chem.* 6: 281–294.

87 Kar, U.K., Simonian, M., and Whitelegge, J.P. (2017). Integral membrane proteins: bottom-up, top-down and structural proteomics. *Expert Rev. Proteomics* 14: 715–723.

88 Uetrecht, C. and Heck, A.J.R. (2011). Modern biomolecular mass spectrometry and its role in studying virus structure, dynamics, and assembly. *Angew. Chem. Int. Ed.* 50: 8248–8262.

89 Rostom, A.A., Fucini, P., Benjamin, D.R. et al. (2000). Detection and selective dissociation of intact ribosomes in a mass spectrometer. *Proc. Natl Acad. Sci. USA* 97: 5185–5190.

90 Hyung, S.-J. and Ruotolo, B.T. (2012). Integrating mass spectrometry of intact protein complexes into structural proteomics. *Proteomics* 12: 1547–1564.

91 Koschubs, T., Lorenzen, K., Baumli, S. et al. (2010). Preparation and topology of the mediator middle module. *Nucleic Acids Res.* 38: 3186–3195.

92 Lariviere, L., Plaschka, C., Seizl, M. et al. (2013). Model of the mediator middle module based on protein cross-linking. *Nucleic Acids Res.* 41: 9266–9273.

93 Robinson, P.J.J., Bushnell, D.A., Trnka, M.J. et al. (2012). Structure of the mediator head module bound to the carboxy-terminal domain of RNA polymerase II. *Proc. Natl. Acad. Sci. USA* 109: 17931–17935.
94 Robinson, P.J., Trnka, M.J., Bushnell, D.A. et al. Structure of a complete mediator-RNA polymerase II pre-initiation complex. *Cell* 166: 1411–1422.e16.

Further Reading

Gross, J.H. (2017). *Mass Spectrometry – A Textbook*, 3e. Heidelberg: Springer-Verlag.
Hillenkamp, F. and Peter-Katalinic, J. (eds.) (2007). *MALDI-MS: A Practical Guide to Instrumentation, Methods and Applications*. Weinheim: Wiley-VCH.
de Hoffmann, E. and Stroobant, V. (2007). *Mass Spectrometry – Principles and Applications*, 3e. Wiley.
Kaltashov, I.A. and Eyles, S.J. (2005). *Mass Spectrometry in Biophysics: Conformation and Dynamics of Biomolecules*. Wiley.
Kinter, M. and Sherman, N.E. (2000). *Protein Sequencing and Identification Using Tandem Mass Spectrometry*. New York: Wiley-Interscience.
Kool, J. and Niessen, W.M.A. (2015). *Analyzing Biomolecular Interactions by Mass Spectrometry*. Wiley.
Lehmann, W.D. (2010). *Protein Phosphorylation Analysis by Electrospray Mass Spectrometry: A Guide to Concepts and Practice*. Cambridge: The Royal Society of Chemistry.
Pottiez, G. (2015). *Mass Spectrometry: Developmental Approaches to Answer Biological Questions*. Springer.
Schalley, C.A. and Springer, A. (2009). *Mass Spectrometry and Gas-Phase Chemistry of Non-covalent Complexes*. Hoboken: Wiley Interscience.
Watson, J.T. and Sparkman, O.D. (2007). *Introduction to Mass Spectrometry – Instrumentation, Applications and Strategies for Data Interpretation*, 4e. Chichster: Wiley.

11

Applications and Complementarity of Analytical Ultracentrifugation and Light-Scattering Techniques

Chad A. Brautigam

Department of Biophysics, UT Southwestern Medical Center, 5323 Harry Hines Boulevard, Dallas, TX, 75390-8816, USA

11.1 Introduction

Analytical ultracentrifugation (AUC) and light scattering (LS) are methods that have a rich history in the characterisation of biological macromolecules [1, 2]. For example, AUC was critical to the discovery that proteins exist as discrete species [3, 4]. Both techniques were used to probe the solution properties of important, large biomolecular assemblies like haemoglobin [4] and viruses (e.g. [5, 6]). In the modern laboratory, advances in instrumentation, methodologies, and analyses have allowed these approaches to be routinely used on biomolecules as small as peptides or even sugars [7–9].

AUC and LS have many advantages that continue to fuel their popularities. Perhaps foremost among these is that they report on molecular attributes that are critical to understanding biological functioning: molar mass, size, shape, and stoichiometry. However, a practical advantage is that they do not require any kind of labelling or immobilisation; in other words, they are *solution* techniques that can be deployed on *native* biomolecules. This is a particular advantage when studying ensembles of macromolecules in the context of their interactions with one another (see Section 11.5.4). Further, AUC and LS are based on easily accessible physical and signal-processing fundamentals, and equipment to carry out both types of experiments is readily available.

The goal of this chapter is to give the reader a strong foundational knowledge of these methods. Although various fields employ AUC and LS, this chapter will focus on applications for solutions of purified biological macromolecules, particularly proteins. First, AUC will be covered, followed by LS. For each method, the information content, experimental apparatuses, and their respective theoretical fundaments will be covered. Finally, examples are shown that exemplify the power and complementarity of these approaches.

11.2 Analytical Ultracentrifugation

11.2.1 Information Content of an AUC Experiment

There are two commonly used modes for the AUC experiment. In sedimentation equilibrium (SE), the macromolecular solution is exposed to low centrifugal fields for long periods of time, resulting in a Boltzmann-type distribution of the macromolecules containing information on their molar mass(es). The other mode, called sedimentation velocity (SV), makes observations on the concentration distribution of macromolecules when subjected to high centrifugal fields for relatively short times. This chapter focuses solely on SV, which is currently more widely employed; the interested reader may find excellent background information on SE in the literature [10–12].

Biomolecular and Bioanalytical Techniques: Theory, Methodology and Applications, First Edition. Edited by Vasudevan Ramesh.

The primary information gained from an SV experiment performed on biological macromolecules is size. A simple thought experiment can illustrate this fact. A mixture of different proteins in aqueous solution will have different sizes, but each has approximately the same density and similar shape. If this solution is then subjected to a centrifugal field, the larger, more massive particles will sediment faster than the smaller ones. By observing the sedimentation over time (which is the essence of an SV experiment), one can arrive at an accurate estimate of molecular size and the translational diffusion coefficient. By the laws of hydrodynamics, these quantities can be related to hydrodynamic radius, molar mass[1] and shape.

11.2.2 The AUC Experiment

In the context of this chapter, the goal of an SV experiment is to apply a centrifugal field to a macromolecular solution and to observe the evolution of the macromolecular concentration as a function of time and radius (from the centre of rotation). To conduct an AUC experiment, the macromolecular solution to be studied is placed in a sector-shaped compartment (Figure 11.1, inset). This shape prevents the radially sedimenting particles from colliding with the chamber walls. Two such compartments are typically present in the part used for the experiment, which is called a *centrepiece* (see Figure 11.1, inset). Because dual-beam methods (see below) are used to detect the macromolecules as they sediment, one chamber is reserved for the sample and one for a reference solution that does not contain the macromolecule.[2] The centrepiece has four walls, but six are obviously needed to house the solution. When fully assembled, the centrepiece is sandwiched in a cylindrical cell between optically transparent windows, forming the fifth and sixth sides and allowing optical tracking of radial concentration profiles. The sample and the reference buffer are introduced into such assembled cells through external filling ports.

After assembly and sealing, the cells are placed in the centrifugation rotor. Modern rotors have spaces for four or eight cells, but one cell is always occupied by a counterbalance, limiting the experiments to three or seven cells per run, respectively. During the experiment, light emanating from above the spinning rotor (Figure 11.1) passes through both sectors and is detected by sensors located below the rotor.[3] These detectors

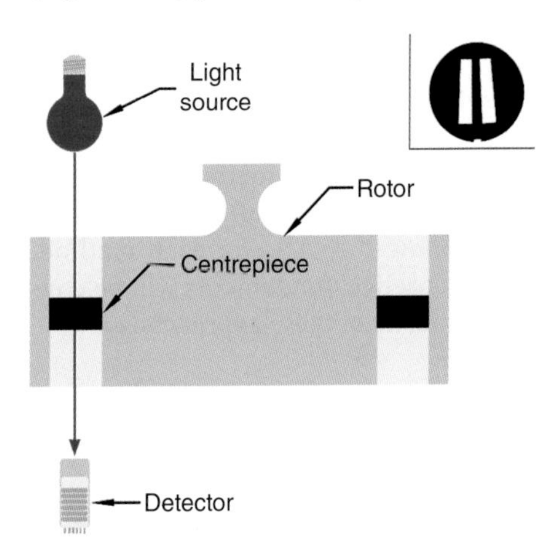

Figure 11.1 A simplified schematic view of the AUC experiment (not to scale). The rotor (centre) is shown in cross-section such that two of the cell-containing holes can be viewed. The light source (red light bulb) is positioned over the rotor and is timed to flash when the subject hole (the left one in this depiction) passes beneath it. The light passes through the sector(s) and impinges on detectors below the rotor. The location of the light source shown here is accurate for the interference optics, but for the absorbance optics it is positioned below and to the side of the rotor; light from this flash lamp passes up to the position of the light bulb, which is actually a toroidal diffraction grating. The detector is a series of lenses and mirrors culminating in a CCD camera for the interference optics, but is a photomultiplier tube masked by a radially movable slit for the absorbance optics. Shown in the inset is the centrepiece as viewed from the perspective of the light source.

1 A dimensional analysis of the equations presented later in this chapter demonstrates that the proper units for molar mass are grams per mole (g/mol). However, in keeping with the tradition in the biological sciences of presenting masses in units of Daltons (Da), the assumption is made that 1 Da = 1 g/mol; the masses are presented in Da.

2 It is not always necessary to have a reference buffer and thus a technique called *pseudoabsorbance* can be used in which both sectors are filled with samples. While this method doubles experimental throughput, it has caveats and drawbacks, and it is therefore not presented here. The interested reader is referred to [13].

3 An exception to this arrangement is found in an after-market modification to modern AUCs. This fluorescence module is essentially a confocal optical apparatus mounted over the rotor and able to move radially. The light source and detector are thus both located *over* the rotor.

are fast and radially sensitive, fulfilling the experimental requirements of collecting radial concentration profiles as a function of time.

Two light-generation and detection systems are commonly used in the modern analytical ultracentrifuge. The most widely used one flashes ultraviolet or visible light through the solution; the light passes through a pinhole (or 'slit') below the centrifuge, finally impinging on a photomultiplier tube. The slit is mounted on a rail, allowing it to be moved radially during the collection. Light is collected from both the reference and sample sectors, and the comparison of the two allows calculation of the absorbance of the sample as a function of radius according to

$$A = \log\left(\frac{I_r}{I_s}\right). \tag{11.1}$$

where A refers to absorbance, I the intensity of detected light, and the subscripts r and s depict the reference and sample sectors, respectively. Colloquially, this detection apparatus is referred to as the 'absorbance optics'.

The second light generation and detection system works on a different principle. Laser light is split into two lines that span the entire radial range of the reference and sample sectors. The light passes through the sectors and a window located below the rotor. From here, a series of lenses and mirrors recombine and focus the light on a charge-coupled device, which detects a pattern of dark and light lines that are the result, respectively, of destructive and constructive interference of the laser light. Wherever there is a path-length (i.e. refractive-index) difference, these lines will bend. Such differences occur where there are concentration gradients in the sample sector. A Fourier analysis is performed on the alternating light/dark image, and this result, as a function of radius, is reported in a digital file that can be analysed, allowing the weight concentration of macromolecules to be calculated. This system is a Rayleigh interferometer, but is often termed the 'interference optics'. Notably, a related system, called 'Schlieren optics' [14], measures the refractive-index gradients (which can be related to the concentration gradients $\mathrm{d}c/\mathrm{d}r$) and may be reintroduced in future AUC instruments [15].

There are many factors that influence the choice of detection system and the optimal experimental conditions. Concentrations, buffer components, purity, chemical compositions, sample stability, and expected outcomes, among other things, must be carefully weighed before initiating an AUC experiment. These factors are too multitudinous to treat comprehensively in this chapter, and the reader is referred elsewhere for detailed discussions on these matters [9]. For the purposes at hand, it is best to have pure macromolecular solutions in simple buffering systems (e.g. phosphate-buffered saline, PBS) at concentrations that do not exceed 1 mg/ml. Under these conditions, the experiment will usually succeed, and the choice of detection system does not bear on the final result. The absorbance optics will be used in the protocol presented below (Section 11.4.3).

11.2.3 AUC Theory and Data Analysis

Under ideal conditions (i.e. infinite dilution with no net attractive or repulsive forces between individual macromolecules), the movement of species in a centrifugal field is described by two equations. The first, the 'Svedberg equation', relates macromolecular properties that can be gleaned from an SV experiment. The second equation yields a mathematical representation of the concentration profiles observed in radial space and time, and thus it can be used to fit the SV data acquired from the optical systems directly. It is called the 'Lamm equation'.

The Svedberg equation is named for Theodor Svedberg, inventor of the analytical ultracentrifuge and recipient of the Nobel Prize in Chemistry in 1926. It can be derived from a consideration of the balance of forces on a particle that manifests during the SV experiment. For the purposes of this discussion, a one-dimensional ordinate space is established (Figure 11.2a) that has the radius from the centre of rotation, r, as its sole ordinate. Three forces are present (Figure 11.2b): (i) centrifugal force (F_{cent}), acting in the centrifugal direction (positive r); (ii) frictional force (F_{frict}), centripetally opposing the centrifugal force; and (iii) a buoyant force,

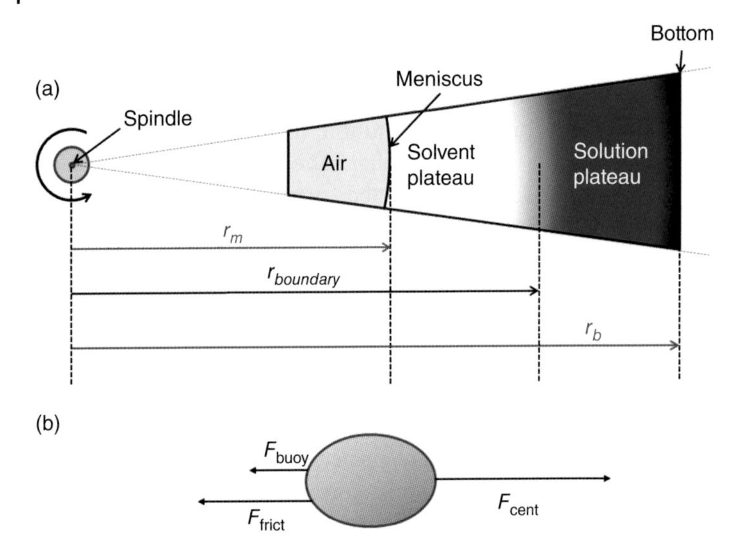

Figure 11.2 The centrifugation sector and forces. (a) The centrifugation sector (not to scale). The sector is rotated about a spindle axis, the origin for *r*. Inside the sector, solvent is shown as white and the darkness of the blue colour represents the concentration of the solute. A time approximately midway through the SV experiment is shown. (b) Forces on the sedimenting particle. The direction of sedimentation is the same as in part (a).

(F_{buoy}), which also opposes the centrifugal force. These three forces are shown, respectively, on the left-hand side of the following equation:

$$F_{cent} - F_{frict} - F_{buoy} = m\omega^2 r - vf - m\omega^2 r(\bar{v}\rho) = 0. \tag{11.2}$$

Here, m is the mass of the particle, ω is the angular velocity (in radians/s) of the rotor, v is the radial velocity of the particle, f is the frictional coefficient, $\bar{v}$ is the partial specific volume of the particle, and ρ is the density of the solution. Hydration effects on the particle are ignored for the purposes of this derivation. These forces sum to zero because they are balanced in the SV experiment. The third term on the left-hand side of the equation, according to Archimedes' principle, is the mass of the displaced solvent. This equation can be rearranged:

$$m\omega^2 r(1 - \bar{v}\rho) - vf = 0 \tag{11.3}$$

and thus

$$m(1 - \bar{v}\rho) = \frac{vf}{\omega^2 r}. \tag{11.4}$$

The $\omega^2 r$ term in Eq. (11.4) is called the 'centrifugal field' and is proportional to the centrifugal force felt by the particle. We can define a velocity per unit field as the sedimentation coefficient $s \equiv v/\omega^2 r$, so that

$$m(1 - \bar{v}\rho) = sf. \tag{11.5}$$

Note that s is a field-normalised velocity and therefore it is a property that is independent of the applied rotor speed. A number of substitutions may now be made because of the identities $m = M/N_A$ and $D = kT/f$, where M is the molar mass, N_A is Avogadro's number, D is the translational diffusion coefficient, k is Boltzmann's constant, and T is the absolute temperature. Thus

$$M(1 - \bar{v}\rho) = \frac{sTkN_A}{D} \tag{11.6}$$

and, with the knowledge that the universal gas constant R is equal to kN_A, the substitution allows formulation of the Svedberg equation:

$$M = \frac{sRT}{D(1 - \bar{v}\rho)}. \tag{11.7}$$

This equation links three macromolecular properties: the sedimentation coefficient, the diffusion coefficient, and the molar mass. The other terms are either constant or known by the experimenter. The units of s are seconds, where s values are expressed in terms of Svedbergs (with the symbol S), with 1 S equalling 10^{-13} seconds.

The other master equation of SV is the Lamm equation, first reported by Ole Lamm in 1929 [16]. Rather than concerning the properties of the sedimenting particles, the Lamm equation seeks to account for the change in particle concentration as a function of time. This ultimately results in a description of the radially dependent concentration data that are the output of an SV experiment.

Before embarking on a derivation of the Lamm equation, it is instructive to conduct a thought experiment on SV. If it is assumed that the centrifugal field can be instantaneously applied at time $t_0 = 0$ and that the solute is denser than the solvent, then at a later time $t_1 > t_0$ the material throughout the cell will have advanced in radius; material near to the bottom of the cell will build up at this impermeable barrier, and material at the meniscus will have moved centrifugally so that there is no material left at the meniscus (Figure 11.2a). In SV parlance, the meniscus has been 'cleared' and there is a 'boundary' between the solute-containing and the solute-depleted regions of the sector. The red lines in Figure 11.3 show this boundary as being infinitely sharp, but Nature (and statistical thermodynamics) does not support such a boundary. There will be a tendency for some material at the boundary to diffuse, forming a sigmoidal boundary (blue lines in Figure 11.3). Thus, two processes are observable in SV: sedimentation and diffusion.

The Lamm equation, which ultimately seeks to describe the shape and time evolution of the boundary, must therefore account for two fluxes: those due to sedimentation and diffusion. The flux due to sedimentation (j_{sed}) at a given point r in the sector is

$$j_{\text{sed}} = cv = c\omega^2 rs, \tag{11.8}$$

where c is the weight concentration of the particles. The substitution resulting in the right-hand side of Eq. (11.8) is made possible by the earlier definition of the sedimentation coefficient. The diffusion flux (j_{diff}) is simply due to Fick's Second Law:

$$j_{\text{diff}} = -D\frac{\partial c}{\partial r}. \tag{11.9}$$

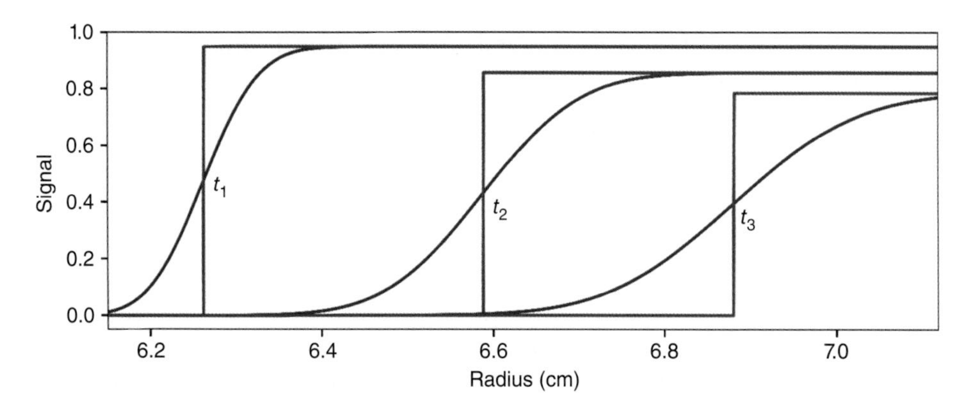

Figure 11.3 Sedimentation with and without diffusion. The red lines show what sedimentation would look like without diffusion; the blue lines show realistic diffusion. The system simulated was for a 66 000 Da protein with a sedimentation coefficient of 4.4 S at 50000 rpm with a starting concentration of 1 signal unit. Three timepoints are shown, with $t_3 > t_2 > t_1$.

Thus, the total flux is

$$j_{tot} = c\omega^2 rs - D\frac{\partial c}{\partial r}. \tag{11.10}$$

The sectorial geometry of the centrifugation chamber and the requirement for mass conservation enforces that

$$-\frac{1}{r}\left(\frac{\partial(rj_{tot})}{\partial r}\right) = \frac{\partial c}{\partial t}, \tag{11.11}$$

where t represents time. Substituting yields the Lamm equation:

$$\frac{\partial c}{\partial t} = \frac{1}{r}\frac{\partial}{\partial r}\left[rD\frac{\partial c}{\partial r} - c\omega^2 r^2 s\right]. \tag{11.12}$$

Numerous efforts have focused on solving the Lamm equation. For instance, examine two terms in an approximate analytical solution to the Lamm equation, the Faxén approximation [17]:

$$c(r,t) = \left(\frac{c_0 \mathrm{e}^{-2\omega^2 st}}{2}\right)\left\{1 - \Phi\left[\frac{r_m(\omega^2 st + \ln(r_m) - \ln(r))}{2\sqrt{Dt}}\right]\right\}. \tag{11.13}$$

Here r_m is the radial position of the meniscus, c_0 is the initial, evenly distributed concentration and Φ is the symbol for the error function. The important realisation to take away from Eq. (11.13) is that *a solution to the Lamm equation takes the form of* $c(r,t)$, *which is directly comparable to AUC data.*[4] No general, exact analytical solutions of the Lamm equation are available, but numerical solutions can be efficiently carried out with modern computers. Consequently, Lamm equation solutions, when scaled correctly, can be and are used to directly fit AUC data [19, 20].

However, fitting a Lamm equation solution or even several solutions to AUC data usually results in inaccuracies. This is because the method requires a priori knowledge of all sedimenting species, but a biological sample is rarely so well-conditioned. To overcome this obstacle, Schuck [21, 22] pioneered an approach wherein the data (depicted as $a(r,t)$) can be fitted as the sum of many such solutions scaled by a continuous distribution $c(s)$:

$$a(r,t) \approx \int_{s_{min}}^{s_{max}} c(s)L(s,D,r,t)\mathrm{d}s. \tag{11.14}$$

Here, the function L represents a Lamm equation solution in which c_0 is arbitrarily set to 1. Although the $c(s)$ distribution is the state of the art in SV data analysis, it carries two important caveats: (i) all species are assumed to have the same frictional ratio (f_r, the ratio of the experimental frictional coefficient to the frictional coefficient of an unhydrated sphere of equal volume) and (ii) all species are hydrodynamically ideal and non-interacting. In most cases, neither of these seriously detracts from the data analysis.

The integral used in the $c(s)$ analysis (Eq. (11.14)) presents a well-known mathematical difficulty (one that will be encountered again in the analysis of dynamic light scattering (DLS) data, Section 11.3.3.2). Specifically, there are many $c(s)$ distributions that will fit the data with similar statistical rigour. Further, distributions with unphysical high frequency fluctuations will be among these statistically acceptable solutions. The strategy

4 The data scans obtained by the optical systems described in this chapter are understood as moments frozen in time, where a proxy for concentration (e.g. absorbance units or interference fringes) is captured as a function of radius. A conversion constant always lies between the data and the true concentration. For the absorbance measurements, this is the extinction coefficient multiplied by the path length of the centrepiece, which is typically 1.2 cm. For interference data, a proportionality constant analogous to the extinction coefficient can be calculated; it is referred to as a 'signal increment', and can be calculated from the amino acid composition in the case of proteins [18]. It should be noted that the fringe displacement measured by Rayleigh interferometry is relative to the meniscus; thus, for accurate concentration quantitation in the methods described in this chapter, clearance of the species of interest from the meniscus is desirable.

adopted to minimise these issues in the case of AUC (and DLS) is called 'regularisation'. A full treatment of the mathematical nuances of regularisation is well beyond the scope of this chapter and the reader is referred to other literature for specifics [21, 23]. In the current context, it should suffice to realise that, to achieve the best $c(s)$ distribution, the difference between the data and the right-hand side of Eq. (11.14) must be minimised:

$$\text{Min}\left\{\left[a(r,t)-\int_{s_{\min}}^{s_{\max}} c(s)L(s,D,r,t)\mathrm{d}s\right]^2+\gamma\int\zeta(s)\mathrm{d}s\right\}. \tag{11.15}$$

If $\gamma = 0$, Eq. (11.15) reduces to the usual technique of least-squares minimisation, but this would lead to the problem delineated above, i.e. solution degeneracy and instability. Thus, with $\gamma > 0$, a term is added to the minimisation problem. In practice, the function $\zeta(s)$ is chosen to either minimise oscillations in $c(s)$ (the Tikhonov–Phillips approach) or to minimise the information content of $c(s)$ (the maximum-entropy approach). Notably, inclusion of the regularisation term guarantees that the resulting $c(s)$ distribution will not be the one that best fits the data. Therefore it is necessary to modulate γ such that the final solution meets certain statistical criteria set by the user. These concepts are robustly implemented in Schuck's AUC analysis software SEDFIT [21, 22].

The $c(s)$ distribution condenses tens of thousands or even hundreds of thousands of data points into a single, simple-to-interpret, one-dimensional plot (Figure 11.4). It relates the size (i.e. sedimentation coefficient) of the particles analysed to their normalised signal populations. The f_r is refined during the analysis and can be related to D, so use of the Svedberg equation (Eq. (11.7)) allows the determination of the molar mass of a species represented by any given peak in the distribution.

The $c(s)$ distribution is widely used to determine molar masses, oligomerisation states, and amounts of contaminants or aggregates [21, 25, 26]. There are many offshoots of the $c(s)$ distribution that may be employed in special circumstances. For example, these may provide for increased resolution [25, 27] or relief from the stricture of similar frictional ratios [28].

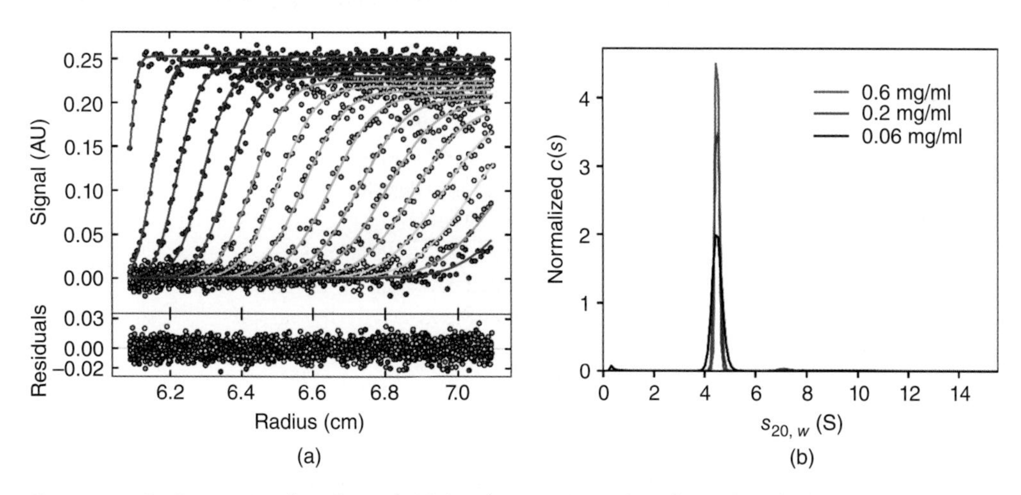

Figure 11.4 Sedimentation data, fits and $c(s)$ distributions. (a) SV data, fits and residuals. Absorbance data (circles) for the 0.2 mg/ml sample of Tp0037 (see Section 11.5.1.3) are shown as circles; each scan is coloured according to time, with early times in violet and late in red. Every fifth scan and every third data point used in the analysis are shown. A $c(s)$ distribution (see part b) was fitted to these data (respectively coloured lines). Residuals between the data and the fit lines are shown in the lower graph. Systematic noise elements [24] were subtracted from the data. (b) The $c(s)$ distributions for the Tp0037 data. The distributions were normalised by the total signal to aid visual interpretation. The $c(s)$ distribution calculated from the data in part (a) is shown in blue.

11.3 Light Scattering

11.3.1 Information in LS Experiments

Light-scattering experiments on biologically derived samples can generally be divided into two categories. The first is 'static light scattering' (SLS), which is concerned with the intensity of light scattered from an aqueous biological sample. From a biochemical perspective, the quantity usually realisable from this observation is molar mass. In the case of large particles, the radius of gyration may also be accessed. The second category is 'dynamic light scattering' (DLS). This method examines the fluctuations in light intensity from a macromolecular solution to glean information regarding the size (i.e. the hydrodynamic radius) and monodispersity of the scattering species. The hydrodynamic radii measurable by DLS in modern instrumentation varies from about 1 nm to 5 μm.

11.3.2 The LS Experiment

Despite the difference in the analytic methods, the experimental setup for both SLS and DLS experiments is similar (Figure 11.5). A polarised laser light source is shone on to the target solution, which may be contained in a cuvette or in a flow cell. Most of the light merely passes through the sample, but, in the presence of a macromolecule, some excess will be scattered. A detector or detectors at a fixed angle or angles θ in the plane of the incident light senses the scattered light and relays this information to a computer, which records the observations. It is usually useful to use a long-wavelength laser to avoid absorption in the sample; typical wavelengths range from 650 to 850 nm. Among the types of detectors used are photomultiplier tubes, photodiodes, and avalanche photodiodes. The entire LS experiment can often be accomplished in minutes, making LS much simpler and faster than AUC.

There are some limitations on the molecules that can be studied by LS. Most importantly, the samples must be free of large contaminants, such as dust, as these will dominate the signal and hence skew the results. High purity (>95%) is desirable, but the presence of impurities can be overcome using separation techniques

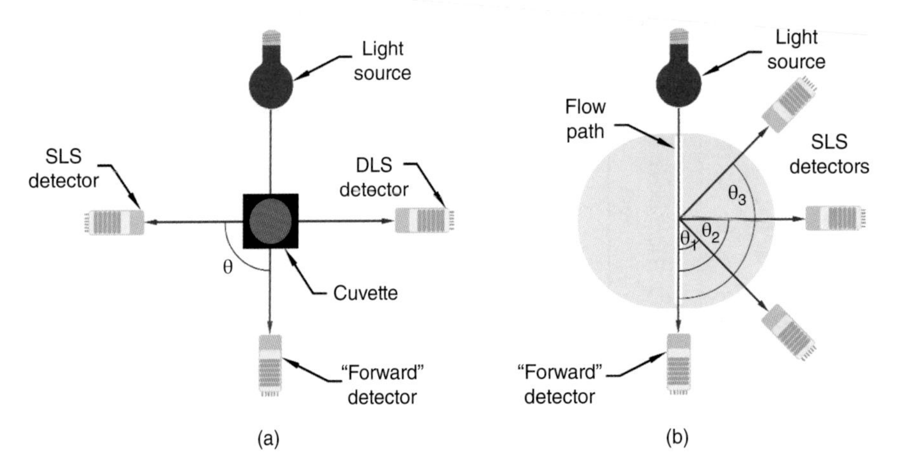

Figure 11.5 A simplified schematic view of the LS experiment (not to scale). In both parts, the detectors are in the same plane as the light source. (a) A cuvette-based system. Polarised laser light (red arrow) is emitted from the position of the light bulb. It impinges on a four-windowed or clear cuvette (centre). Light passing straight through the sample is detected by the 'forward' detector, which is primarily used for diagnostic purposes. Scattered light (blue arrows) is detected at right angles to the incident light. Different detectors are used for SLS and DLS measurements. Many different geometries are available, including placing the detector very near to the light source to detect 'backscatter'. (b) A flow-cell based system. A solution of the macromolecule flows through the cell (an inlet and outlet would be located below the cell in this view) and the incident laser light (red arrow) follows the same path. Scattered light is detected at multiple angles; three are shown here, but many more are possible.

(see Section 11.4.2). The necessary concentration depends on size; a large molecule will scatter light more efficiently, and thus can be used at a relatively lower concentration. For most moderately sized globular proteins, a concentration of between 1 and 2 mg/ml will result in good LS data. However, this recommendation may not hold for all proteins and will likely be inappropriate for highly asymmetric species; the reader is advised to perform experiments at multiple concentrations and perform extrapolations of the relevant quantities to zero concentration if there is any question of non-ideality.

11.3.3 LS Theory and Data Analysis

11.3.3.1 Static Light Scattering

Before commencing an overview of the theory of light scattering, it is worthwhile to first consider: 'Why do macromolecules scatter light?' To answer this question, we must examine light as an electromagnetic phenomenon. Light has both electric and magnetic fields associated with it. The magnetic field of light, even strong laser sources, is not sufficient to influence macromolecules in solution; to account for scattering, then, a description must be made of the interaction of the electric field component with matter. The electric field is capable of changing the distribution of electrical charge in macromolecules; in other words, it can *induce a dipole moment in them.* Because the electric field oscillates, the induced dipole moment oscillates as well, and this net movement of charge causes the macromolecule to act as a tiny transmitter, sending out light of the same energy, but in different directions. The transmitted light is the phenomenon measured in a light-scattering experiment.

The intensity of scattered light can be deduced by realising that the scattered electric field (E_s, which is proportional to the square root of the intensity) is due to the far-field component of Maxwell's field equations in the presence of an oscillating dipole [29]:

$$E_s(r,t) = \frac{-q}{c^2 r}\left(\frac{\mathrm{d}^2 p}{\mathrm{d}t^2}\right)\sin\phi. \tag{11.16}$$

Here, q is the net charge carried by the dipole, c is the speed of light, r is the distance from the scatterer to the detector, ϕ is the angle subtended by the arc connecting the electric-field vector and the detector, and p describes the light-induced oscillation of the dipole with respect to time. The second derivative of p, therefore, is describing the acceleration experienced by the dipole. With $p = \cos(2\pi ct/\lambda)$ (λ is the wavelength of the incident light) and $q = E_0\alpha$ (E_0 is the amplitude of the incident electric field and α is the particle's polarisability),

$$E_s(r,t) = \frac{4\pi^2 E_0\alpha}{r\lambda^2}\sin\phi\cos\left(\frac{2\pi ct}{\lambda}\right). \tag{11.17}$$

The two trigonometric terms on the right-hand side of Eq. (11.17) can be dropped; ϕ will be 90° in the experimental design of Figure 11.5a and the observed intensity is only concerned with the amplitude of the signal, not its time variation. With intensity being the square of the amplitude of the field, one arrives at an equation for so-called Rayleigh scattering:

$$\frac{I_s}{I_0} = \frac{16\pi^4\alpha^2}{r^2\lambda^4}. \tag{11.18}$$

Note that the intensity of the light will diminish as the square of the distance from the scatterer and as the fourth power of the wavelength. Importantly, Eq. (11.18) as presented here is for *isotropic scatterers*. Isotropic scattering occurs when the phase difference between light waves scattered from different parts of the macromolecule is small, i.e. when *the size of the particle is small relative to the wavelength of the incident light*. Here, 'small' is taken to mean that the largest dimension of the particle is less than 1/10th of the wavelength. Note that the diameter of the eukaryotic ribosome is about 30 nm and a typical wavelength for the incident radiation is about 650 nm; thus, most biological macromolecules meet this criterion. If the biomolecule is large based on the above definition, then there will be an angular dependence on scattering, and a correction factor called a *form factor* must be applied to the right-hand side of Eq. (11.18).

The only parameter that pertains to the scatterer in Eq. (11.18) is α, and thus this is the parameter that must have information on the molar mass. It can be shown that

$$\alpha = \frac{Mn_s}{2\pi N_A} \cdot \frac{\mathrm{d}n}{\mathrm{d}w}, \tag{11.19}$$

where n_s is the refractive index of the solvent and $\mathrm{d}n/\mathrm{d}w$ is the refractive-index increment of the solution (with w representing the weight concentration), a quantity that can be measured as a function of solute concentration. Substituting, one arrives at

$$\frac{I_s}{I_0} = \frac{4\pi^2 n_s^2 M^2 (\mathrm{d}n/\mathrm{d}w)^2}{r^2 \lambda^4 N_A^2}, \tag{11.20}$$

and, to characterise scattering from many entities, the right-hand side of Eq. (11.20) must be multiplied by the particle density $N = wN_A/M$:

$$\frac{I_s}{I_0} = \frac{4\pi^2 n_s^2 Mw (\mathrm{d}n/\mathrm{d}w)^2}{r^2 \lambda^4 N_A} \tag{11.21}$$

Terms in Eq. (11.21) can be gathered to give a functionally useful form:

$$R_r = KwM, \tag{11.22}$$

where R_r is the 'Rayleigh ratio' or $I_s r^2/I_0$ and K represents the gathered constant terms $4\pi^2 n_s^2(\mathrm{d}n/\mathrm{d}w)^2/\lambda^4 N_A$. In practice, a solvent with a known Rayleigh ratio is used to calibrate the system, which allows theoretical intensities to be related to the detector output, e.g. voltage. Then the intensity from the sample itself is taken, i.e. its Rayleigh ratio is measured. From Eq. (11.22), if we know the $\mathrm{d}n/\mathrm{d}w$ of the macromolecule (that for proteins can be calculated from the amino acid composition [18]) and w, the weight concentration, the molar mass is easily calculable.

11.3.3.2 Dynamic Light Scattering

As intimated by the name, DLS is concerned not with the intensity of the scattered light but with its fluctuations. Because the motions of molecules in solution is very rapid, a detector capable of microsecond resolution is required. This is easily achieved with modern electronic detectors. The theoretical treatment below is for small, isotropic scatterers, which scatter light without significant contributions from rotational diffusion or anisotropy. Analysis of these effects requires different analytic techniques and a different apparatus than that depicted in Figure 11.5a [30].

Noise, of course, is associated with the acquisition of practically all types of data and the electronic detectors used to observe LS are not immune to it. The key to DLS is the understanding that *the time-dependent noise detected from an LS experiment is not truly random*; large particles will cause long timescale fluctuations in scattered light intensity, while small particles will cause shorter timescale ones (Figure 11.6a). To interpret these fluctuations, the intensity signal is autocorrelated. Autocorrelation can be thought of as multiplying the signal (as a function of time) by the same signal that is offset by a small time increment (τ). At very small values of τ, there will be a high degree of correlation between the signal and its offset duplicate, but, if the noise is truly random, the correlation will decay at slightly higher τ values. The longer the timescale of the correlations, the higher the value of τ will be before there is observable decay in the correlation (Figure 11.6b). Mathematically, the reduced intensity autocorrelation function ($g_2(\tau)$) is defined as

$$g_2(\tau) = \frac{\langle I(t)I(t+\tau)\rangle}{\langle I(t)\rangle^2}, \tag{11.23}$$

where the angle-brackets indicate an average over all observed times t. Although it is possible to autocorrelate the scattering data with software, usually it is accomplished through hardware. Through autocorrelation,

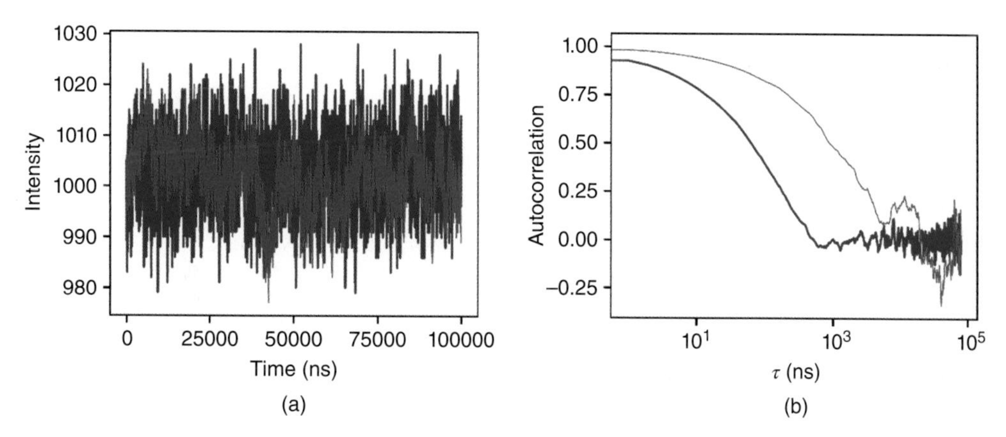

Figure 11.6 Scattered-light intensity variation and autocorrelation. (a) Simulated intensity variation from a fictional detector with ns time resolution. Two simulations were conducted on 200 particles in a 200 × 200 grid in which their random motions into and out of a central 100 × 100 'observation space' was monitored and converted to a pseudo-intensity. The blue trace monitors this property for particles that were allowed to move with 5× the velocity of those monitored in the red trace. This simulation mimics the study on small particles (blue trace) and larger ones (red trace). Note that this variation in number concentration is different from an actual DLS experiment, in which the intensity fluctuation is due to the differences in constructive and destructive interference engendered by the particles' relative positions at any given time point; phenomenologically, however, these variations can be treated similarly. Time steps of 1 ns were allowed. (b) Autocorrelation of the data in part (a). The colours are respective to the traces in part (a). The last 20% of this function was truncated because it becomes very noisy. The significant difference in the characteristic variation times in the data becomes very evident in this format. For ease and generality of implementation, a modified autocorrelation function was used to process these 'intensity' data: $1/(n-k)\sigma_I^2 \sum_{t=1}^{n-k}(I_t - \langle I \rangle)(I_{t+k} - \langle I \rangle)$, where n is the number of data points and the displacing index is k.

subtleties in the variation of the intensity data can be distilled into a simple function that captures the characteristic decay time in the data (see below and Figure 11.6b).

Another form of the autocorrelation function is very useful; it is called the reduced field autocorrelation function ($g_1(\tau)$) and can be calculated from $g_2(\tau)$:

$$g_1(\tau) = \sqrt{\frac{g_2(\tau) - 1}{\beta}}. \tag{11.24}$$

Here, β is an instrumental parameter. It is convenient to work with $g_1(\tau)$ when analysing DLS data because this function is directly related to the translational diffusion coefficient:

$$g_1(\tau) = e^{-\Gamma\tau}, \tag{11.25}$$

where $\Gamma = Dq^2$ (Γ is also known as the decay time) and $q = (4\pi n_s/\lambda)(\sin(\theta/2))$. Indeed, in the past, the natural log of $g_1(\tau)$ was plotted against τ, allowing Dq^2 to be determined using a linear regression [31]. With modern computers, DLS data from a sufficiently monodisperse solution can be directly fitted to arrive at D. Once D is known, it may be converted to a hydrodynamic radius (R_H) using the Stokes–Einstein relation:

$$R_H = \frac{kT}{6\pi\eta D}, \tag{11.26}$$

where η is the solution viscosity.

A powerful and commonly used method to analyse DLS data is the 'method of cumulants' [32]. This analysis strategy arises from probability theory; it must be first assumed that the field autocorrelation can be described as a superposition of a scaling distribution $G(\Gamma)$ and the corresponding decay terms:[5]

$$g_1(\tau) = \int G(\Gamma)e^{-\Gamma\tau}d\Gamma. \tag{11.27}$$

A further assumption is that $G(\Gamma)$ is monomodal and Gaussian-like. Taking the natural logarithm of both sides of Eq. (11.27) results in a 'cumulant-generating function',[6] i.e.

$$K(-\tau, \Gamma) = \ln(g_1(\tau)). \tag{11.28}$$

In this context, then, the first cumulant (κ_1) is the mean decay rate, $\overline{\Gamma}$. The next two cumulants (κ_2 and κ_3) are the variance and skewness of the distribution. In practice, it is rare to refine more than three cumulants, as this can lead to overfitting of the data. The cumulant method conveniently allows one to define the parameter of the 'polydispersity index' (or PDI) as

$$\text{PDI} = \frac{\kappa_2}{\kappa_1^2} = \frac{\sigma^2}{\overline{\Gamma}^2}. \tag{11.29}$$

PDI, in a single number, describes a normalised width of the distribution, and values below 0.15 are commonly considered to be 'monodisperse'.

When the assumption of monomodality is violated or undesirable, other methods are available to fit the autocorrelation function. Many of these fit distributions not unlike the $c(s)$ distribution to the data. The main difficulty with this approach is that many such distributions $G(\Gamma)$ fit the small amount of noisy data equally well; to choose between them, some external criterion must be imposed. A popular strategy in modern software is to choose the simplest, smoothest distribution [21, 23]. The means to achieve this goal vary, and several different (and sometimes proprietary) approaches are implemented in software that accompanies DLS instruments. However, many of these are based on ideas developed by Provencher and deployed in the software CONTIN [23]. The CONTIN approach is closely analogous to the Tikhonov–Phillips regularisation method implemented in SEDFIT (see Section 11.2.3 and Eq. (11.15)). CONTIN iteratively optimises γ based on Fisher statistics. SEDFIT also has routines that can fit DLS-derived autocorrelation functions with distributions using both Tikhonov–Phillips and maximum-entropy regularisations [33]. In the author's hands, the maximum-entropy method seems to afford a slightly higher resolution, though this has not been systematically investigated.

11.4 Protocols

11.4.1 DLS/SLS Protocol

The protocol detailed here is for a cuvette-based Wyatt DynaPro NanoStar apparatus using the Dynamics software package, though the basic principles will hold for any DLS instrument. It is assumed that the instrument and cuvette have been properly calibrated and that all software is already downloaded and installed. If using very high concentrations (>1 mg/ml), it may be useful to repeat this protocol at many different concentrations and extrapolate the resulting parameters of interest to zero concentration to correct for thermodynamic non-ideality.

1. Filter the sample through a compatible 0.1-μm syringe-tip filter if necessary. Centrifuge 50 μl of the protein preparation for 10 minutes at room temperature at 16100 × *g*.

---

5 Interestingly, the $G(\Gamma)$ distribution is analogous to the $c(s)$ distribution; in the first case, the distribution scales solutions to the Lamm equation and, in the latter, it scales respective exponential functions.

6 A cumulant-generating function $K(t)$ has the property that $K(t) = \kappa_1 t + \kappa_2 t^2/2 + \kappa_3 t^3/3! + \cdots + \kappa_n t^n/n! + \cdots$, where κ_n is the respective 'cumulant'.

2. Carefully remove 5 µl from the top of the supernatant and place it in a quartz cuvette that has been calibrated according to the manufacturer's instructions. Place the cuvette in the instrument.
3. Start the Dynamics software. Usually, a 'pre-set' protocol is available from the manufacturer during installation; select a pre-set that will record both DLS and SLS data. Supply the software with the correct dn/dw and concentration of the protein by defining a new sample under the 'Sample' section. Set the instrument to record data at 25 °C, taking three technical replicates of 10 averaged observations of 5 s each.
4. Press the Start Button in the software and acquire the data. Clean the cuvette according to the manufacturer's instructions when finished.
5. The molar mass of the protein will be automatically calculated and displayed in a table when the acquisition is complete.
6. Export the autocorrelation data to a file by choosing the autocorrelation view, expanding to select all data, turning off the display of the cumulant and regularisation fit information, right-clicking on the graph, choosing to export to a file, and supplying a filename.
7. Convert the files to SEDFIT format: within SEDFIT, select the command to convert Dynamics 7 DLS data to the SEDFIT format from the Options Menu. Follow the prompts, entering the necessary information. For these experiments, these data include η (0.0090635 Poise; this can be calculated for the experimental buffer using SEDNTERP [34]), solvent refractive index (1.333; this is the accepted value for PBS, which is close to most biological buffering systems for this parameter), λ (661 nm), θ (90°) and temperature (25 °C). Note that export to SEDFIT may not be possible for all data formats. The reader may need to use propriety software in these cases or to contact SEDFIT's author (P. Schuck) to implement a conversion protocol.
8. Load the converted DLS data into SEDFIT using the 'Load DLS data' command from the File menu.
9. Choose fitting limits by dragging the green lines to appropriate positions on the autocorrelation plot. Generally, the first few data points should be excluded, and including too much high-τ is not useful (the data can get noisy at long correlation times).
10. Choose the cumulant model from the Model menu under the DLS subcategory.
11. Choose the 'Parameters' command. Allow the refinement of the first and second cumulants by checking their respective boxes. Also allow 'Itot' and 'Baseline' to refine by checking their respective boxes. Click OK.
12. Choose Fit. The cumulant analysis will be completed. It is advisable to switch between the Simplex and Marquardt–Levenberg optimisation protocols (Options:Fitting Options) and repeating the fitting multiple times to ensure convergence (i.e. no decrease in the reported root-mean-square deviations (rmsd) upon repeated fittings).
13. Examine the cumulant values and *PDI*. Record them.
14. The autocorrelation data can be optionally plotted using GUSSI [35] by choosing 'Plot:GUSSI data-fit-residuals plot'.
15. If it is desirable, switch the model to the hydrodynamic radius distribution in the Model menu. Press Parameters. Input a resolution of 150, allow the refinement of the baseline, and choose appropriate minimum and maximum R_H values. Press OK. The maximum-entropy regularisation is selected by default, but this choice can be changed from the Options:Size Distribution Options menu.
16. Press 'Run'. The distribution will be calculated. The distribution can be copied (from the 'Copy' menu) and pasted into spreadsheet or graphing software, if desired.

11.4.2 SEC-MALS Protocol

The following protocol is for a system containing a Shimadzu high performance liquid chromatography (HPLC) pump and UV detector, a Wyatt TREOS II LS detector and a Wyatt Optilab T-rEX differential refractive-index (dRI) detector downstream of a size-exclusion chromatography (SEC) column, a GE Superdex 200 Increase 10/300 (see Figure 11.7). Wyatt's ASTRA 7 software was used to acquire and analyse

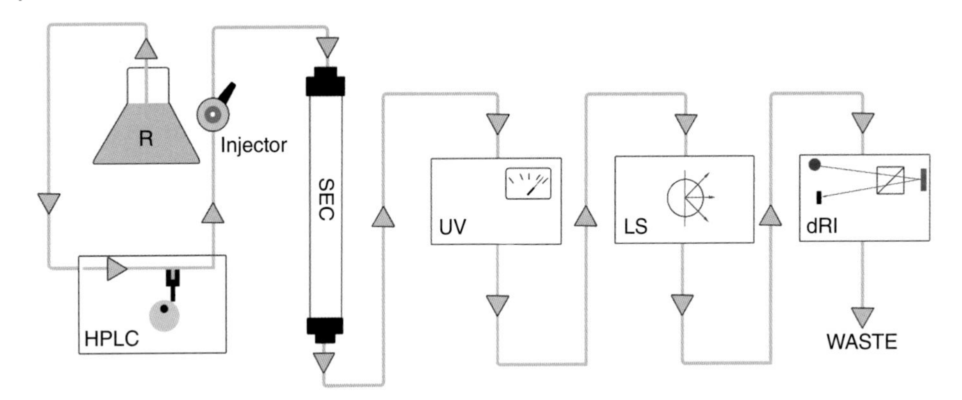

Figure 11.7 A schematic view of a SEC-MALS system. Flow direction is shown by the blue arrowheads. The buffer solution is drawn from the reservoir (R) by a high performance liquid chromatography (HPLC) pump; the sample is introduced at an in-line injector, then passes through the SEC column, a UV detector (UV), a light-scattering detector (LS) with a flow cell as shown in Figure 11.5b, a differential refractive-index instrument (dRI), and then finally to waste.

data. Again, it is assumed that all necessary instrumental calibrations have been accomplished according to the manufacturer's instructions.

1. Filter the protein through a compatible 0.1-μm syringe-tip filter.
2. Equilibrate the apparatus with at least two column-bed volumes of PBS at a flow rate of 0.5 ml/min (to be used throughout this protocol) with the purge valve on the dRI detector open; this instrument should be set to a temperature of 25 °C.
3. Select a protocol in the ASTRA software to collect molar mass data. This will have been set up when the LS instrument was installed. Supply the correct dn/dw in the 'Injector' part of the protocol (called 'dn/dc' in the software) and ensure that the dRI detector is the one being used for the determination of concentration (set 'concentration source' to RI in 'Configuration').
4. Close the dRI detector's purge valve to fix that instrument's reference solution; zero the UV and dRI detectors.
5. Inject 100 μl of the protein at a concentration of 1–2 mg/ml and collect data for 45 minutes.
6. When data acquisition is complete examine the baselines of the data traces in the 'Baselines' section. Adjust as necessary to establish appropriate baselines.
7. In the 'Peaks' section, define peaks as appropriate by clicking and dragging across them; it is usually best to select only the central portion of peaks. For each peak, make sure that the correct dn/dw ('dn/dc') is inputted in the table that appears below the peak graph.
8. The molar masses are available in the 'Molar Mass & Radius from LS' section of the protocol. A report should also be present at the end of the protocol recording the average molar masses in each peak and various mass moments. Record the necessary information and save the file.

11.4.3 SV Protocol

This protocol is for a Beckman XL-I centrifuge without a turbomolecular pump. To make it as general as possible, only the use of the absorbance optical system is described, as some centrifuges still in use do not have interference optics. However, the reader should note that the use of Rayleigh interferometry is more analogous to the LS experiments described above (Sections 11.4.1 and 11.4.2) because, like LS, interferometry detects all species in solution. If the experimenter has access to both AUC optical systems, a sometimes complex decision tree can be followed to decide on the best method for the experiment on hand [9]. In brief, if selectivity is required, absorbance is best, but speed and generality favour the interference method. It is common practice in the author's lab to use both simultaneously. Before the experiment, the experimenter should know the molar mass of the monomer, the $\bar{v}$ of the protein and the solution parameters ρ and η for the experimental temperature. All of these can be calculated using SEDNTERP [34].

1. Prepare at least 450 μl of the desired samples. For the examples below, there are three samples at different concentrations. Incubate at these concentrations overnight at 4 °C. Three concentrations are used to check for trends as a function of this parameter, which could indicate self-interactions. The long incubation period allows any slow dissociation process to come to equilibrium after dilution from an ostensibly high-concentration stock solution. Incidentally, this dilution/incubation technique may be used for LS as well, but those methods are typically performed at relatively higher concentrations (to achieve a good signal-to-noise ratio) and thus may not benefit as greatly as AUC.
2. Assemble, fill and insert into the rotor the necessary number of centrifugation cells (see [36]). For experiments in this chapter, 400 μl was always used.
3. Place the rotor into the centrifuge, insert the optical arm, and close the centrifuge. Engage the vacuum, set the desired temperature, and then set the speed to 0 and press 'Start'. Allow the rotor to equilibrate in this state for at least 2.5 hour.
4. Set the data-acquisition software to collect absorbance data at 280 nm from 6.0 to 7.25 cm; collect 999 scans and set them to be 1 minute apart (because the absorbance scans take ca. 1.5 minutes to collect, the 1 minute setting effectively causes continuous scanning).
5. Set the rotor speed (50 000 rpm in these instances) and start the rotation from the software. When the rotor has achieved the desired speed, click 'Start Method Scan'. Data acquisition will commence after some calibrations. Data acquisition is usually accomplished overnight. A typical run is ca. 14 hour, but one should allow data acquisition to proceed until all evidence of sedimenting boundaries is absent.
6. Stop data acquisition and rotation. Clean the centrifugation cells.
7. Use REDATE (http://biophysics.swmed.edu/MBR/software.html) to correct for time-stamp inaccuracies in the data, following that software's instructions [37].
8. Start SEDFIT and load the data using the Data menu. Select all data files in which there is non-zero data between the meniscus and about 7.1 cm (typically). This may take some trial and error. If there are more than 100 scans with such data, it may be advantageous to load only a subset, which SEDFIT allows during the loading process. A good target is between 50 and 100 scans.
9. Choose the fitting limits by dragging the green lines with the mouse. Only data within the fitting limits will be considered by the software. The left-fitting limit should be to the right of any meniscus optical artefacts and the right limits should be well away from the bottom, avoiding any strong signal gradients.
10. Choose the meniscus position by dragging the red line to the peak of the sharp, upward pointing optical data feature. Likewise, drag the blue line to the bottom position, which is usually where the data achieve a maximum on the right-hand side of the graph.
11. Choose the continuous $c(s)$ model from the Model menu.
12. Choose Parameters. In the Parameters window, choose a resolution; input $\bar{v}$, ρ and η where asked. Choose a resolution of 150 and a confidence level of 0.683; the latter allows the regularisation to modify the distribution so that the fit varies from the best unregularised fit by 1 σ. Allow the frictional ratio and meniscus to refine by checking their respective boxes and turn on time-independent noise fitting. The latter feature compensates for systematic noise features that are present in all data scans [24]. The default s limits (0–15 S) and frictional ratio (f_r) value (1.2) are acceptable for the analyses in this chapter. The f_r encapsulates diffusion information in SEDFIT. Press OK.
13. Press Run and examine the fit. If there is a strong upward feature at the resulting distribution's right-hand side, it is necessary to increase s_{max} in Parameters. A strong diagonal feature in the grey box in the middle of the program's window (the residuals bitmap) may indicate a poor initial choice for f_r, and that can also be adjusted in Parameters.
14. Press 'Fit'. The program will iterate until convergence. As before, it is prudent to alternate between the Simplex and Marquardt–Levenberg optimisation algorithms (see Step 12 in Section 11.4.1).

15. Press Control-M. Above each peak, a button will appear with a molar mass estimate. Clicking on the button will open a window with detailed information on M and the best s value for the peak. Record this information.
16. The data-fit-residuals and $c(s)$ plots can be exported to GUSSI for the production of publication-quality figures using the 'Plot' menu in SEDFIT. Also, conversion to $s_{20,w}$ was accomplished in that program by turning on Standardization (Axes:Standardization:Standardization On) and modifying the Standardization Parameters (Axes:Standardization:Modify Standardization Parameters).
17. Repeat Steps 7–15 for data from remaining centrifugation cells. To assemble several $c(s)$ distributions in one figure, one can copy the distribution from SEDFIT's Copy menu and paste it into an extant GUSSI $c(s)$ session.

11.5 Applications

In this section, two applications of AUC and LS are presented. Where possible and appropriate, the complementarity of the methods will be highlighted. Both applications pertain to proteins and both experiments ask the same question: what is the oligomeric status of the protein? In the first case (Tp0037), the protein is very nearly monodisperse and thus the characterisation is relatively simple. The second protein is bovine serum albumin (BSA), which is 'heterodisperse' (or 'paucidisperse'), i.e. there are several other species besides the main one. These other species complicate the analyses and result in slightly altered optimal experimental protocols.

11.5.1 Tp0037

Recombinant Tp0037, a protein named for its gene designation (*tp0037*), was a gift from the laboratory of M. Norgard. The protein was thought to be a D-lactate dehydrogenase, and its monomeric molar mass was known from its amino acid sequence: 39180 Da. These enzymes usually are dimeric in solution [38], and thus the question at hand was: 'Is Tp0037 also a dimer?'

11.5.1.1 DLS on Tp0037

To address this question, a series of DLS, SLS, and AUC experiments were performed. When undertaking a solution biophysical characterisation of a protein, it is advantageous to perform the DLS experiment first. This is because the DLS experiment will inform on the monodispersity (or lack thereof) in the sample, allowing subsequent experiments to be rationally designed. The concentration of the protein in the DLS experiment was 1.5 mg/ml. The fits were excellent, as evidenced by the low rmsds (0.0098, 0.0109 and 0.0115 autocorrelation units) and the small, random residuals (the residuals are the data minus the fit line; see Figure 11.8a). Because the first cumulant is equal to $\overline{\Gamma}$, the diffusion coefficient can be derived from it; in this case D is 6.13 ± 0.08 F (the symbol F stands for the non-SI unit of diffusion 'Fick', i.e. 1×10^{-7} cm^2/s). The PDI derived from this analysis is 0.01 ± 0.02 (in two of the replicates it was exactly 0.0), indicating a monodisperse sample. For an R_H distribution model (Figure 11.8b), the data are well-described by a single species with an intensity-averaged hydrodynamic radius of 3.94 ± 0.08 nm.

11.5.1.2 SLS on Tp0037

With its monodispersity confirmed, Tp0037 was deemed to be an excellent candidate for molar mass determination using the Nano star's on-board SLS detector. This approach may be considered a 'batch' method, as there is no attempt to separate impurities (indeed, according to the PDI, none are present). The average molar mass was $78\,100 \pm 100$ Da. Because this corresponds very well to the dimeric form of Tp0037 (theoretical molar mass of 78 380 Da), the hypothesis that the protein is a dimer in solution was formed.

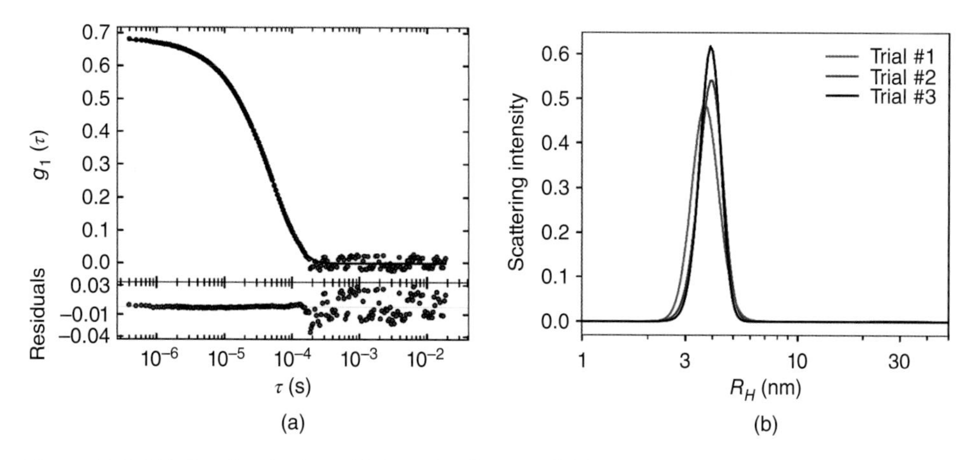

Figure 11.8 DLS data from Tp0037. (a) Cumulant fit to the field autocorrelation function. The $g_1(\tau)$ data are shown as red markers and a fit line (black) from the cumulant analysis is also drawn. (b) R_H distributions for the technical replicates.

11.5.1.3 SV on Tp0037

This result was buttressed by SV experiments. In these, three different concentrations (0.6, 0.2, and 0.06 mg/ml) of Tp0037 were placed in separate centrifugation cells. It is good practice to study several concentrations when studying the oligomeric state of a protein, as the concentration dependence on the s can offer clues regarding whether the dissociation constant of an oligomer is in the concentration range examined. All fits were judged to be excellent because of the low rmsds (0.009745, 0.007161 and 0.006860 AU, respectively) and random residuals (e.g. Figure 11.4a).

To determine the best s value of each experiment, two steps are taken. First, the s-values are data transformed to the 'hydrodynamic standard state', which is water at 20 °C (indicated by the subscript '20,w'). The conversion is calculated via the following equation:

$$s_{20,w} = s_e \frac{\eta_e}{\eta_{20,w}} \left(\frac{1 - \bar{v}_{20}\rho_{20,w}}{1 - \bar{v}_e \rho_e} \right), \tag{11.30}$$

where the subscript e is used to denote the experimental values. Although not absolutely necessary for the question at hand, the conversion has the benefit of allowing experimenters to compare s values directly; the analyst compensates for temperature and buffer composition rather than imposing that burden on other researchers. The second step is to integrate the $c(s)$ distributions to determine the signal-weighted sedimentation coefficients. The calculation, which is conveniently performed in SEDFIT or the graphing program GUSSI (the latter is designed to communicate with the former), is

$$s_{20,w}^{\text{best}} = \frac{\int_{s_{20,w}^{l}}^{s_{20,w}^{u}} s_{20,w} c(s_{20,w}) \mathrm{d}s_{20,w}}{\int_{s_{20,w}^{l}}^{s_{20,w}^{u}} c(s_{20,w}) \mathrm{d}s_{20,w}}, \tag{11.31}$$

where the superscripts 'u' and 'l' indicate the upper and lower limits of the integration, respectively. In this case, the three values were averaged. If the concentration range were wider and ventured into the multiple mg/ml range, this average would probably not be valid, as there are known non-ideality effects that cause an apparent diminution in s with increases in concentration [39]. The very small variance in the current measurements indicate that this is not a significant consideration in this case. The measured $s_{20,w}^{\text{best}}$ for Tp0037 is 4.501 ± 0.007 S.

Of course, the ultimate purpose of the experiment is to determine the molar mass of the protein. Simultaneously with the distribution, SEDFIT refines the frictional ratio, which can be converted to the diffusion coefficient as a function of s [28]:

$$D(s) = \frac{\sqrt{2}}{18\pi} kTs^{-1/2}(f_r\eta)^{-3/2}\left(\frac{1-\bar{v}\rho}{\bar{v}}\right)^{1/2} \tag{11.32}$$

With both s and D in hand, the program can calculate the molar mass using the Svedberg equation (Eq. (11.7); see Step 15 of Section 11.4.3). Again, the average of the reported molar masses was taken, arriving at 75800 ± 1600 Da. Thus the SV experiment is consistent with the contention that Tp0037 is a dimer.

With a seemingly unambiguous result arising from the excellent data derived from the DLS and SLS experiments (Sections 11.5.1.1 and 11.5.1.2; Figure 11.8), the reader may legitimately ask 'Why bother with the SV experiment?' The answer is bipartite. First, the light-scattering and centrifugal methods encompass differing physical principles (Sections 11.2.3 and 11.3.3) to derive similar molecular conclusions (see Section 11.5.3). Hence, rather than relying on a single physical method to arrive at a result, the experimenter has enabled stronger conclusions by including both. Second, the SV methodology has a higher resolution and almost invariably the researcher discovers information about the sample from SV analysis that was not evident in the LS experiment(s). Therefore, LS experiments may be thought of as quick means to get information on a protein sample, but SV reveals a more detailed picture of all detectable species.

11.5.2 BSA

Although the hydrodynamic properties of BSA are well-studied, here it is instructive to approach it as if they weren't, posing the same question as above: what is the oligomeric state of the protein? The same initial experimental approaches will be taken, but this time an alternative way to acquire the SLS data will be suggested. In all of these experiments, a single batch of BSA in 0.9% NaCl from Pierce (Prod. #23209) was passed through a 0.1-μm syringe-tip filter prior to experimentation. Where necessary, the protein was diluted in PBS.

11.5.2.1 DLS on BSA

Using exactly the same experimental and analytical protocol described for Tp0037, DLS measurements were taken on BSA. In this case, the sample is clearly not monodisperse and the example cumulant fit (Figure 11.9a) has a PDI of 0.2; the residuals from this fit have significantly systematic deviations in the 300-μs

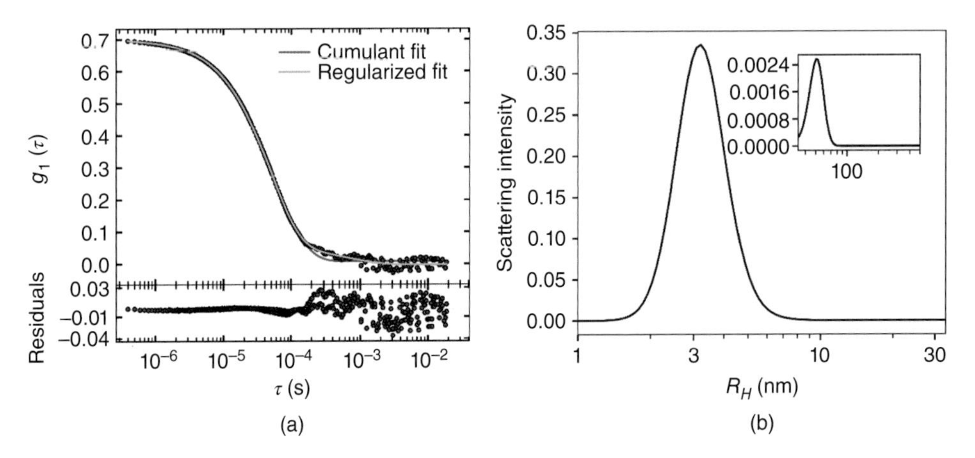

Figure 11.9 DLS data from BSA. (a) The cumulant and regularised distribution fits. The data are shown as blue circles; in the residuals, the cumulant differences are shown as red circles, whereas the regularised distribution differences are shown as blue circles. Only Trial 1 is shown. (b) The regularised distribution, again only for Trial 1. The contaminating 50-nm species is shown in the inset.

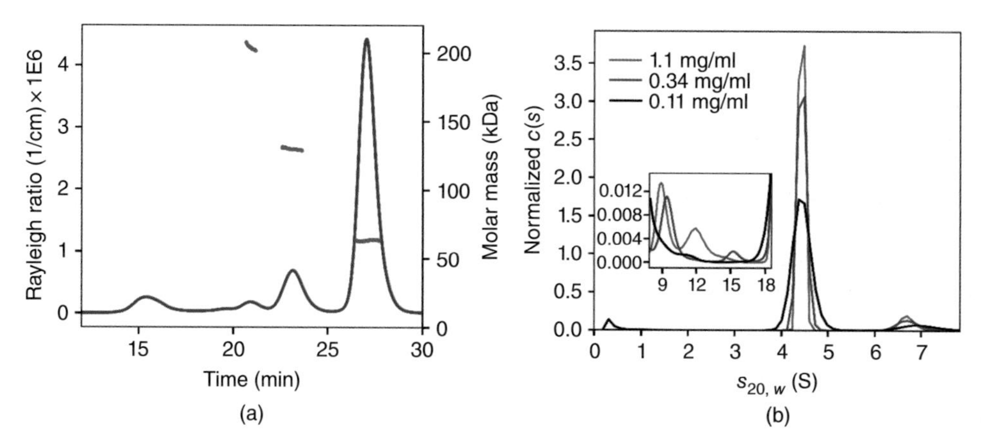

Figure 11.10 SEC-MALS and SV data for BSA. (a) SEC-MALS data. The blue trace shows the LS data from one detector. The red markers are the molar mass values for the three analysed peaks. (b) The *c*(*s*) distributions for BSA. The inset shows Peak 3 at about 9 S; the black distribution does not show this peak, probably because of the low signal-to-noise ratio.

range. A regularised fit eliminates this problem (Figure 11.9), but introduces a peak at ca. 50 nm. Therefore, there is no justification for proceeding with the SLS measurement in this 'batch' mode. To measure the molar mass, a method that can accommodate this level of heterogeneity must be utilised. Two such methods are detailed within this chapter: SEC-MALS (MALS stands for multiangle light scattering) and SV.

11.5.2.2 SEC-MALS on BSA

Size-exclusion chromatography–multiangle light scattering (SEC-MALS) is ideally suited to garner molar mass information on heterodisperse biological samples. Perhaps the most important aspect of this experimental strategy is the SEC. This technology separates the inputted macromolecules on the basis of size, with large particles eluting first from the SEC column. Hence, if several species are present and their respective sizes differ enough, their molar masses may be individually assessed in the SEC-MALS experiment. Downstream of the column is a UV detector, an LS detector, and a dRI detector (Figure 11.7). The first and third of these detectors can be used to calculate the concentration of the material eluting from the column (as long as the extinction coefficient or the dn/dw, respectively, of the material is known).

After the completion of the experiment, the data were analysed using Wyatt's ASTRA software. Three discernible peaks eluted from the column (Figure 11.10a). The first, smallest peak (Peak 3) crested at 20.9 minutes, the second (Peak 2) at 23.2 minute, while the third peak (Peak 1) at 27.1 minutes. Because each peak has many different data points, each with its own concentration and LS signal, the mass can be calculated many times for each peak (Figure 11.10a). When all of these are averaged, the overall molar mass reported from Peak 1 was 63 540 ± 80 Da, while those from Peaks 2 and 3 were 129 900 ± 100 and 204 000 ± 700 Da, respectively. Given that the theoretical molar mass of BSA is 66 433 Da, it is a reasonable hypothesis that Peaks 1, 2 and 3, respectively, represent a monomer, dimer and trimer of BSA.

11.5.2.3 SV on BSA

Another natural choice for studying a heterodisperse system like the BSA sample is SV. One of the strengths of the *c*(*s*) method is that it has a very high resolution and thus is able to discern species of different sizes. Determination of molar masses from this method may be a concern from a significantly heterodisperse or polydisperse sample, however. This is because of the assumption of uniform f_r for all species being analysed. For example, a typical purified protein will usually have a single species that represents >90% of the signal, with other species comprising the remainder. Thus, the fitted f_r, which will be a weighted average for all species, will be accurate for the dominant species, but there is no guarantee that the minority species share this value. This may be of particular concern where two species of dissimilar sizes have equal signal strength but drastically variant shapes. However, these cases are rare in purified protein samples and techniques have been introduced to minimise this problem [28].

Turning to BSA, the protein was subjected to SV. The three concentrations studied were 1.1, 0.34 and 0.11 mg/ml. The experimental temperature was 25 °C. All other experimental conditions were exactly as described for Tp0037. The same analysis parameters were also used, except that the highest s_e value considered was 20 S.

The $c(s)$ distributions resulting from the data analysis in SEDFIT show a dominant peak with $s_{20,w}^{\text{best}}$ of 4.448 ± 0.003 S (Figure 11.10b), with minor peaks occurring at 6.83 ± 0.13 S and 9.3 ± 0.3 S (the final species does not appear in the lowest concentration sample; this is likely due to its low abundance). The molar masses associated with these peaks are 63 800 ± 1200, 124 000 ± 6000 and 195 000 ± 6000 Da, respectively. As mentioned above, the value for the species in the first peak is likely to be well determined by the data, but those for minority species rely on the assumption that their respective f_r values are similar to that of the dominant one. This assumption, when adopted as above, results in molar masses that comport with those obtained in the SEC-MALS experiment. The SV data are therefore consistent with the SEC-MALS data. Certainly the assertion that the dominant species in this heterodisperse sample is the monomeric form of the protein is supported by the LS and SV data presented here.

11.5.3 Complementarity Between AUC and LS

The AUC and LS results presented in this chapter illustrate the strong compatibility of the methods. In the case of the SV and SLS data, the two approaches are *orthogonal*, i.e. they use different physical principles to measure similar quantities. To arrive at molar masses, SLS examines the intensity of scattered light, while AUC explores the movement (due to both diffusion and sedimentation) of species when subjected to a high centrifugal field. When a close agreement between the results of the two methods is achieved, the combined result is very well supported indeed. Besides the results presented above, another demonstration of these methods complementing one another for the characterisation of monoclonal antibodies is available [40].

DLS plays an extremely important complementary role in the experimental strategy presented herein. It can be used as a decision node; a low PDI and monomodality in a hydrodynamic-radius distribution (Figure 11.8b) demonstrates that the batch approach to molar-mass determination by SLS is appropriate. Higher PDI (e.g. >0.15) and a multi-modal hydrodynamic-radius distribution (Figure 11.9b) suggest that the batch mode of SLS is contraindicated. Instead, a separation technology like SEC should be used to conduct SLS on a truly monodisperse sample. The speed of the DLS measurement (minutes) and low sample requirements (typically ca. 10 pg) facilitate the experiment and, in the opinion of this author, all macromolecular samples destined for biophysical characterisation would benefit from this easy and powerfully informative experiment.

A final point of complementarity to mention in this context is the fact that the experiments measure identical quantities and thus there is the possibility of combining the analyses in a satisfying manner. For example, DLS is generally thought to be an excellent means of determining D, whereas this value is not as reliably determined from SV [28, 41]. However, the information on s in an SV experiment is extremely accurate. To obtain the molar mass, one could take the D obtained in a DLS experiment and the s from an SV experiment and combine[7] them with the Svedberg equation (Eq. (11.7)). In the data presented in this chapter for Tp0037, $s_{20,w}$ is 4.501 ± 0.007 S and $D_{20,w}$ from the DLS experiment is 5.45 ± 0.07 F. Inserting these values into Eq. (11.7) yields a molar mass of 76 000 Da, which is very close to the values obtained from the SLS and SV experiments (the propagated error in the calculation is 2000 Da). More sophisticated combinations of the data are possible;

7 If the experiments were carried out under slightly different conditions, effort must be made to correct these incompatibilities. A safe choice is to convert to $s_{20,w}$ and $D_{20,w}$; Eq. (11.30) can be used for the former, and the latter can be determined using the transformation $D_{20,w} = D_e(293/T_e)(\eta_e/\eta_{20,w})$. In the Svedberg equation, the density of water at 20 °C (0.99823 g/ml) must be used. This transformation was employed for the calculation performed later in this paragraph.

for example, the SV and DLS data can be combined into a single, global analysis using Schuck's SEDPHAT program [42].

11.5.4 Interacting Systems in LS and AUC

A huge literature exists for the use of LS and AUC for interacting systems. However, in this chapter, only the basic details are sketched, and the appropriate (but not comprehensive) literature cited for reference. SLS can be used in the SEC-MALS mode to determine the stoichiometry of a complex when the masses of the constituents are known (e.g. [43]). Also, Minton has devised a method by which the equilibrium dissociation constant for the interaction of non-identical partners (a 'hetero-association') can be derived using SLS [44, 45]. It relies on the fact that the observed molar mass is the *weight average* of all components present; therefore knowing the molar masses of the constituents and complexes can allow the deduction of species populations and thus of K_D. A very similar approach was pioneered by Reddick for DLS [46]. It uses the z-average D derived from DLS via information on the apparent R_H of the interacting system.

The use of SV for the study of interacting systems is widespread [47]. It can be employed to determine the stoichiometry of an associating complex using both mass and signal information [48, 49]. To obtain K_D, the concentration-profile data (e.g. Figure 11.4a) can be fitted directly using a modified kinetic Lamm equation [50–52]. This is best accomplished in a global analysis of several different loading concentrations of the constituents. However, these methods require extremely high sample purity and can be laborious and time-consuming to accomplish. An alternative approach is to summarise the sedimentation properties of the entire or part of the system [9, 26, 53, 54] by integrating $c(s)$ distributions using an approach similar to Eq. (11.31). When such s_w data are analysed, it is the signal-weighted average that is measured and the changes in s_w as a function of the inputted constituent concentrations can be analysed to yield K_D.

A special set of interactions that AUC and LS are well-suited to study are those of membrane proteins. There is a long history of utilising these methods to examine the interactions of such proteins with themselves and with the detergents necessary to solubilise them [55–57]. Interactions with and within nanodiscs have also recently been treated [58].

11.6 Conclusions

In this chapter, the theory and application of SV and LS for non-interacting solutes were covered. These methods offer a wealth of hydrodynamic information on biomolecules and they enjoy significant popularity among researchers studying their solution properties. One of the chief uses of the methods is the determination of the oligomeric state of a protein or other biomolecule; such was the case in the data presented above (Section 11.5). DLS is a useful technique to employ first, to gather data regarding the strategies to be used in further characterisations. It is also the method of choice to examine D and PDI. DLS contains some mass information in its own right (i.e. the hydrodynamic radius obtained therefrom may be related to a molar mass making assumptions of sphericity), but the determination of absolute molar mass is better carried out using SLS or SV. The batch mode of SLS can be used for monodisperse samples and a separation mode (SEC-MALS) can be used for heterodisperse samples. SV is useful in both circumstances. A researcher armed with these techniques is well positioned to provide useful information on the molecules under study that can inform biological models and also further biophysical characterisations. Finally, these methods can be and are employed to study interacting systems.

Acknowledgements

The author gratefully acknowledges Drs Michael Norgard and Ranjit Deka for the gift of Tp0037 and Dr Shih-Chia Tso for technical assistance.

References

1 Elzen, B. (1986). Two ultracentrifuges: a comparative study of the social construction of artefacts. *Soc. Stud. Sci.* 16: 621–662.
2 Timasheff, S.N. (1964). Light and small angle X-ray scattering and biological macromolecules. *J. Chem. Educ.* 41: 314–319.
3 Svedberg, T. (1937). The ultra-centrifuge and the study of high-molecular compounds. *Nature* 139: 1051–1062.
4 Svedberg, T. and Fåhraeus, R. (1926). A new method for the determination of the molecular weight of the proteins. *J. Am. Chem. Soc.* 48: 430–438.
5 Harding, S.E. (1997). Microbial laser light scattering. *Biotechnol. Genet. Eng. Rev.* 14: 145–164.
6 Richards, E.G. and Schachman, H.K. (1959). Ultracentrifuge studies with Rayleigh interference optics. I. General applications. *J. Phys. Chem.* 63: 1578–1591.
7 Kaszuba, M. et al. (2008). Measuring sub nanometre sizes using dynamic light scattering. *J. Nanopart. Res.* 10: 823–829.
8 LaBar, F.E. and Baldwin, R.L. (1963). The sedimentation coefficient of sucrose. *J. Am. Chem. Soc.* 85: 3105–3108.
9 Schuck, P. et al. (2016). *Basic Principles of Analytical Ultracentrifugation*. Boca Raton, FL: CRC Press.
10 Cole, J.L. et al. (2008). Analytical ultracentrifugation: sedimentation velocity and sedimentation equilibrium. In: *Biophysical Tools for Biologists. Volume One: In Vitro Techniques* (ed. J.J. Correia and H.W. Detrich III,), 143–179. Academic Press.
11 Ghirlando, R. (2011). The analysis of macromolecular interactions by sedimentation equilibrium. *Methods* 54: 145–156.
12 Laue, T.M. (1999). Analytical centrifugation: equilibrium approach. *Curr. Protoc. Protein Sci.* 18: 20.3.1–20.3.13.
13 Kar, S.R. et al. (2000). Analysis of transport experiments using pseudo-absorbance data. *Anal. Biochem.* 285: 135–142.
14 Lloyd, P.H. (1974). *Optical Methods in Ultracentrifugation, Electrophoresis and Diffusion: with a Guide to the Interpretation of Records*. Oxford: Clarendon Press.
15 Cölfen, H. et al. (2010). The open AUC project. *Eur. Biophys. J.* 39: 347–359.
16 Lamm, O. (1929). Die Differentialgleichung der Ultrazentrifugierung. *Ark. Mat. Astr. Fys.* 21B: 1–4.
17 Fujita, H. (1975). *Foundations of Ultracentrifugal Analysis*. New York: John Wiley & Sons.
18 Zhao, H., Brown, P., and Schuck, P. (2011). On the distribution of protein refractive index increments. *Biophys. J.* 100: 2309–2317.
19 Cao, W. and Demeler, B. (2008). Modeling analytical ultracentrifugation experiments with an adaptive space-time finite element solution for multicomponent reacting systems. *Biophys. J.* 95 (1): 54–65.
20 Schuck, P. (1998). Sedimentation analysis of noninteracting and self-associating solutes using numerical solutions to the Lamm Equation. *Biophys. J.* 75: 1503–1512.
21 Schuck, P. (2000). Size distribution analysis of macromolecules by sedimentation velocity ultracentrifugation and Lamm equation modeling. *Biophys. J.* 78: 1606–1619.
22 Schuck, P. et al. (2002). Size-distribution analysis of proteins by analytical ultracentrifugation: strategies and application to model systems. *Biophys. J.* 82: 1096–1111.
23 Provencher, S.W. (1982). A constrained regularization method for inverting data represented by linear algebraic or integral equations. *Comput. Phys. Commun.* 27: 213–227.
24 Schuck, P. and Demeler, B. (1999). Direct sedimentation analysis of interference optical data in analytical ultracentrifugation. *Biophys. J.* 76: 2288–2296.
25 Brown, P.H., Balbo, A., and Schuck, P. (2008). A Bayesian approach for quantifying trace amount of antibody aggregates by sedimentation velocity analytical ultracentrifugation. *AAPS J.* 10: 418–493.

26 Schuck, P. (2003). On the analysis of protein self-association by sedimentation velocity analytical ultracentrifugation. *Anal. Biochem.* 320: 104–124.
27 Brown, P.H., Balbo, A., and Schuck, P. (2007). Using prior knowledge in the determination of macromolecular size-distributions by analytical ultracentrifugation. *Biomacromolecules* 8: 2011–2024.
28 Brown, P.H. and Schuck, P. (2006). Macromolecular size-and-shape distributions by sedimentation velocity analytical ultracentrifugation. *Biophys. J.* 90: 4651–4661.
29 Feynman, R.P., Leighton, R.B., and Sands, M. (1964). *The Feynman Lectures on Physics: Volume II*. Reading, MA: Addison-Wesley Publishing Company.
30 Aragón, S.R. and Pecora, R. (1977). Theory of dynamic light scattering from large anisotropic particles. *J. Chem. Phys.* 66: 2506–2516.
31 Johnson, C.S. Jr., and Gabriel, D.A. (1981). *Laser Light Scattering*. New York, NY: Dover Publications.
32 Frisken, B.J. (2001). Revisiting the method of cumulants for the analysis of dynamic light-scattering data. *Appl. Opt.* 40: 4087–4091.
33 Schuck, P. et al. (2001). Rotavirus nonstructural protein NSP2 self-assembles into octamers that undergo ligand-induced conformational changes. *J. Biol. Chem.* 276: 9679–9687.
34 Laue, T.M. et al. (1992). Computer-aided interpretation of analytical sedimentation data for proteins. In: *Analytical Ultracentrifugation in Biochemistry and Polymer Science* (ed. S.E. Harding, A.J. Rowe and J.C. Horton), 90–125. Cambridge, UK: The Royal Society of Chemistry.
35 Brautigam, C.A. (2015). Calculations and publication-quality illustrations for analytical ultracentrifugation data. *Methods Enzymol.* 562: 109–134.
36 Balbo, A. et al. (2009). Assembly, loading, and alignment of an analytical ultracentrifuge sample cell. *J. Vis. Exp.* 33: e1530.
37 Zhao, H. et al. (2013). Recorded scan times can limit the accuracy of sedimentation coefficients in analytical ultracentrifugation. *Anal. Biochem.* 437 (1): 104–108.
38 Razeto, A. et al. (2002). Domain closure, substrate specificity and catalysis of D-lactate dehydrogenase from *Lactobacillus bulgaricus*. *J. Mol. Biol.* 318: 109–119.
39 Schachman, H.K. (1959). *Ultracentrifugation in Biochemistry*. New York, NY: Academic Press, Inc.
40 Nobbmann, U. et al. (2007). Dynamic light scattering as a relative tool for assessing the molecular integrity and stability of monoclonal antibodies. *Biotechnol. Genet. Eng. Rev.* 24: 117–128.
41 Schuck, P. (2010). On computational approaches for size-and-shape distributions from sedimentation velocity analytical ultracentrifugation. *Eur. Biophys. J.* 39: 1261–1275.
42 Zhao, H. and Schuck, P. (2015). Combining biophysical methods for the analysis of protein complex stoichiometry and affinity in SEDPHAT. *Acta Crystallogr.* D71: 3–14.
43 Padrick, S.B. et al. (2008). Hierarchical regulation of WASP/WAVE proteins. *Mol. Cell* 32 (3): 426–438.
44 Attri, A.K. and Minton, A.P. (2005). Composition gradient static light scattering: a new technique for rapid detection and quantitative characterization of reversible macromolecular hetero-associations in solution. *Anal. Biochem.* 346: 132–138.
45 Kameyama, K. and Minton, A.P. (2006). Rapid quantitative characterization of protein interactions by composition gradient static light scattering. *Biophys. J.* 90: 2164–2169.
46 Hanlon, A.D., Larkin, M.I., and Reddick, R.M. (2010). Free-solution, label-free protein-protein interactions characterized by dynamic light scattering. *Biophys. J.* 98: 297–304.
47 Schuck, P. and Zhao, H. (2017). *Sedimentation Velocity Analytical Ultracentrifugation: Interacting Systems*. Boca Raton, FL: CRC Press.
48 Balbo, A. et al. (2005). Studying multiprotein complexes by multisignal sedimentation velocity analytical ultracentrifugation. *Proc. Natl Acad. Sci. USA* 102: 81–86.
49 Padrick, S.B. and Brautigam, C.A. (2011). Evaluating the stoichiometry of macromolecular complexes using multisignal sedimentation velocity. *Methods* 54: 39–55.
50 Brautigam, C.A. (2011). Using Lamm-equation modeling of sedimentation velocity data to determine the kinetic and thermodynamic properties of macromolecular interactions. *Methods* 54 (1): 4–15.

51 Correia, J.J. and Stafford, W.F. (2009). Extracting equilibrium constants from kinetically limited reacting systems. *Methods Enzymol.* 455: 419–446.
52 Dam, J. et al. (2005). Sedimentation velocity analysis of heterogeneous protein-protein interactions: Lamm equation modeling and sedimentation coefficient distributions c(s). *Biophys. J.* 89: 619–634.
53 Dam, J. and Schuck, P. (2005). Sedimentation velocity analysis of heterogeneous protein–protein interactions: sedimentation coefficient distributions *c*(*s*) and asymptotic boundary profiles from Gilbert–Jenkins theory. *Biophys. J.* 89: 651–666.
54 Schuck, P. (2010). Sedimentation patterns of rapidly reversible protein interactions. *Biophys. J.* 98: 2005–2013.
55 Casassa, E.F. and Eisenberg, H. (1964). Thermodynamic analysis of multicomponent solutions. *Adv. Exp. Med. Biol.* 19: 287–395.
56 Salvay, A.G. et al. (2008). Analytical ultracentrifugation sedimentation velocity for the characterization of detergent-solubilized membrane proteins Ca^{++}-ATPase and ExbB. *J. Biol. Phys.* 33: 399–419.
57 Slotboom, D.J. et al. (2008). Static light scattering to characterize membrane proteins in detergent solutions. *Methods* 46: 73–82.
58 Inagaki, S., Ghirlando, R., and Grisshammer, R. (2013). Biophysical characterization of membrane proteins in nanodiscs. *Methods* 59: 287–300.

Further Reading

Berne, B.J. and Pecora, R. (2000). *Dynamic Light Scattering: With Applications to Chemistry, Biology, and Physics, Mineola*. NY: Dover Publications.
Cantor, C.R. and Schimmel, P.R. (1980). *Biophysical Chemistry Part II: Techniques for the Study of Biological Structure and Function*. New York, NY: W.H. Freeman and Company.
Schuck, P. (2016). *Sedimentation Velocity Analytical Ultracentrifugation: Discrete Species and Size-Distributions of Macromolecules and Particles*. Boca Raton, FL: CRC Press.

12

Application of Isothermal Titration Calorimetry (ITC) to Biomolecular Interactions

Graeme L. Conn

Department of Biochemistry, Emory University School of Medicine, Atlanta, GA, USA

12.1 Introduction

12.1.1 Why Measure Binding of Biological Molecules?

Proteins, nucleic acids (DNA and RNA) and other biomolecules do not work in isolation. Rather, the many biological functions of these molecules are defined by their often complex networks of intermolecular exchanges with a diverse array of other biomolecules and bioactive ligands. Enzymes and macromolecular machines like the ribosome, for example, must specifically assemble all required subunits, essential co-enzymes and/or co-substrates in order to efficiently catalyse turnover of their bound substrates. Further, enzyme activity may be controlled or localised through interaction with other regulatory binding partners. Similarly, biomolecules that play structural or transport roles in the cell must precisely associate to form their complex macromolecular assemblies. More broadly, regulation of cellular processes from genome replication and stability to each step of gene expression involves a complex and precisely coordinated collection of biomolecular interactions. As a result, understanding what drives biomolecular interactions is at the basis of understanding biology itself.

12.1.2 Approaches for Analysis of Biomolecular Interactions

Many approaches are available to study biomolecular interactions, each with its associated strengths and limitations. These approaches range from qualitative identification of potential binding partners to detailed, quantitative analyses of binding affinities and kinetics. The choice of an optimal approach is often dictated by the specific system under study, including availability or otherwise of purified biomolecules in a suitable form, as well as the ultimate goal of the experiment. A yeast two-hybrid screen, phage display or co-immunoprecipitation (e.g. from complex mixtures or cell lysate) may be a good starting point to identify potential interactions that can be further validated by other direct binding methods. Co-elution of purified binding partners from a size exclusion column might also serve as a simple qualitative assay for binding and the impact of changes (such as amino acid substitution or nucleic acid sequence changes) in one or both binding partners.

Classical biophysical approaches for quantitatively analysing binding often require a source of purified components, though a number of methods are equally applicable to complex mixtures, such as cell lysates, or even to interactions within whole cells. All methods also require a means of detecting one of the binding partners or the binding process itself. For protein–nucleic acid interactions, approaches such as filter binding and electromobility shift assay (EMSA) have a long-standing record of successful use. Such experiments typically use radioactive phosphorus (^{32}P)-labelling for its high sensitivity though fluorescent labelling of the nucleic acid is also possible. A limitation of these methods is that detection of binding requires separation of bound and free

Biomolecular and Bioanalytical Techniques: Theory, Methodology and Applications, First Edition. Edited by Vasudevan Ramesh.

ligand: this process can perturb the binding equilibrium, potentially leading to inaccuracies in the measured affinities and also making the approach less well suited for low affinity measurements.

Where fluorescently labelled proteins/polypeptides, nucleic acids or small molecule ligands are available, a variety of other methods becomes possible, including fluorescence anisotropy/polarisation (FP), Förster resonance energy transfer (FRET), fluorescence correlation spectroscopy (FCS), analytical ultracentrifugation (AU) and, more recently, microscale thermophoresis (MST). While each of these approaches has advantages and limitations, a key benefit of fluorescence-based approaches, and many other classical biophysical methods (including isothermal titration calorimetry, ITC), is that they are performed in free solution at thermodynamic equilibrium. It is also notable that many fluorescence-based approaches are equally applicable to studies of purified components as well as interactions within complex mixtures or whole cells.

For *in vitro* studies of purified binding partners, a key advantage of approaches like FP is that they do not require separation of bound and free ligand allowing binding to be quantified without perturbing the equilibrium and making them suitable for measurement of low affinity interactions. A further specific advantage of FP is the ability to set up the experiment in a straightforward manner as a competition assay in which a single fluorescently labelled ligand is pre-bound to the target molecule and this interaction is disrupted by titration of unlabelled competitor ligands. This approach can be used, for example, to relatively rapidly determine apparent affinities for libraries of protein/nucleic acid variants or small molecules for a given target. More recently, MST has also gained significantly in popularity as a versatile approach to study interactions of purified components and within complex mixtures. In MST, a solution or mixture containing the binding partners is placed with specialised thin capillaries and an infrared laser used to induce a microscopic temperature gradient that perturbs two parameters that both typically change upon binding: thermophoresis, or the movement of the molecule in the temperature gradient, and a temperature-related fluorescence intensity change that depends on the fluorophore's microenvironment. Key advantages of MST are that it is broadly applicable, performed in free solution, without the need for immobilisation, and requires only that one molecule is fluorescently labelled.

Finally, amongst the most widely used approaches for detailed, quantitative measurements of binding are surface plasmon resonance (SPR) and the conceptually similar 'dip and read' method, biolayer interferometry (BLI), though, as noted above, MST is also rapidly gaining in popularity. The power of these approaches lies in their capacity to directly measure kinetics of binding, i.e. rates of association (k_{on}) and dissociation (k_{off}), from which the equilibrium binding affinity (K_d) can be calculated. The main disadvantage of these approaches is the need to immobilise one of the binding partners on a sensor or tip surface (typically the smaller binding partner as the signal depends on changes in the immobilised mass on the surface). Nonetheless, the ability to define binding kinetics can provide important insights into the nature of binding interactions that are not usually possible with other methods, which only measure or approximate equilibrium binding affinities. However, as a widely applicable and quantitative method for analysis of biomolecular interactions, ITC has firmly established its place at the forefront of these approaches since its initial development several decades ago [1, 2]. Indeed, ITC is often referred to as the 'gold standard' for analysing intermolecular interactions.

12.1.3 Why Use ITC?

ITC has two key advantages: it is performed entirely in solution, meaning that there is no need to immobilise one of the interacting partners (as is the case for SPR or BLI), and it detects a universal signal, meaning no label is required (as is the case for EMSA, FP or MST). However, probably the foremost benefit of using ITC is that it allows the user to determine a complete thermodynamic profile of binding from a single experiment. Specifically, ITC can be used to *directly* determine binding affinity ($K_a = 1/K_d$), stoichiometry (n) and enthalpy (ΔH). From these, Gibbs free energy (ΔG) and entropy (ΔS) of binding can also be calculated (see Section 12.2). Measuring the enthalpy of binding provides insight into how favourable interactions contribute to binding, as it reflects the total contributions of hydrogen bonding as well as hydrophobic and electrostatic interactions. While generally more complex to interpret, entropy of binding, on the other hand, reports on the collective contributions of binding-induced changes in molecular freedom of motion and reorganisation of bound or free water molecules, ions, etc. Such information may not be readily discernible, even from high resolution structures, making ITC a powerful and often essential component of detailed structure–function studies of biomolecules. The insights ITC can provide on thermodynamics and kinetics of binding can also be particularly valuable in the process of drug development in identifying the best candidate(s) for further optimisation of their physicochemical properties from a group of leads that might otherwise appear equivalent on their binding affinities alone [3–5].

Following the introduction of the first commercial microcalorimeters over three decades ago, these instruments are now routinely found both in individual research labs as well as in shared equipment facilities in academic and industrial settings. More recent improvements in instrumentation have increased sensitivity while reducing the amount of sample required, opening the method to the study of harder to produce or purify biomolecules. In practical terms, while not every biomolecular interaction can be analysed by ITC, the method has without question found exceptionally wide application in the study of discrete biological systems and is applicable to measurement of many different types of binding interaction. The following sections will describe the basic principles of ITC theory, current instrumentation typically available at most research institutions and the application of ITC to the study of biological macromolecular interactions. Next, detailed protocols and a trouble-shooting guide will be provided, which are exemplified by two case studies with one example each of macromolecule–macromolecule and macromolecule–small molecule interaction.

Finally, before delving into the theory and practice of ITC, a brief note on nomenclature. For simplicity of the descriptions that follow, the molecule (titrand) that is placed in the ITC sample cell will be referred to as the 'macromolecule' and its binding partner (titrant) that is placed in the automated pipette used for injection (titration) during the experiment will be referred to as the 'ligand'. However, while often the most practical arrangement, in principle there is no requirement for a larger protein or nucleic acid macromolecule to be the binding partner placed in the sample cell, and the 'ligand' may equally be a second protein, DNA/RNA or small molecule. While this chapter will focus on the use of ITC for analysis of biomolecular binding interactions, the power of the approach and instrumentation extends to advanced methods for global ITC data analysis and measuring binding and measuring enzyme kinetics [6–8].

12.2 Principles and Theory of ITC

12.2.1 What Is ITC and What Does It Measure?

ITC is a non-destructive, label-free analytical technique that is performed entirely in solution. In principle, ITC can be applied to the study of any intermolecular interaction provided heat is either evolved (exothermic) or taken up (endothermic) upon binding. This 'universal signal' detected in ITC means that there is no requirement for labelling with radioisotope, chromophore or fluorophore, or for immobilisation of one binding partner to a surface. While the nature of the signal detected is one of ITC's principal advantages, it also results in an important limitation: all other processes occurring in the ITC cell upon titration of ligand will contribute to the signal. Although small background signals (such as heats arising from mechanical mixing of the ligand into the cell solution) can be readily accounted for, other more significant changes due to mismatches in solution conditions between macromolecule and ligand samples can cause significant problems. Complex binding interactions, e.g. where protein–ligand interaction leads to conformational change on a different timescale, or binding-induced aggregation, will also complicate ITC analysis and interpretation and may require other approaches for a complete resolution.

While some limitations exist, another key advantage of ITC is that experiments can be performed under a wide variety of different experimental conditions. Such flexibility may be useful to tailor solution conditions such that they are optimal for the specific biomolecules under study, which may require specific ionic strength or other additives like glycerol for increased solubility and stability. Flexibility in terms of solution and temperature parameters also allows the potential to adjust experimental conditions to maximise the signal detected.

Finally, variations in factors like pH, ionic strength or temperature can be used to directly learn more about the binding interaction. For example, experiments performed under different conditions of ionic strength can provide insight into the role of electrostatic interactions in binding, while variations in solution pH can provide information on the role of ionisation state and the mechanism of binding. In short, time invested in optimising the conditions used for a particular binding interaction is likely to be worthwhile and potentially highly informative on the nature of the interaction. With the speed and reduced sample requirement of the most recent versions of modern ITC instruments (see Section 12.2.2 below), as well as the ability in some cases to automate many aspects of data collection and initial analysis, such optimisations can be performed relatively quickly. Additionally, as ITC is performed in solution and is non-destructive, precious samples can in principle be recovered and used for additional ITC experiments (if the two binding partners are readily separable) or for other purposes.

Despite its flexibility and broad applicability, there are unfortunately still some instances where ITC may not be a useful approach. As noted above, the signal measured in ITC is proportional to binding enthalpy, meaning that interactions with very low ΔH may not be detectable. In such cases, the signal might be increased by altering the temperature at which titrations are performed or increasing the concentrations of both macromolecule and ligand to proportionally increase the signal generated. However, in practical terms (e.g. protein solubility) it may not always be possible to adjust the necessary parameters sufficiently. Additionally, due to the nature of the relationship between K_d and reactant concentrations in ITC, simply increasing concentrations may lead to other complications in data analysis. Nonetheless, with highly purified and carefully prepared samples, ITC is generally applicable to most biological interactions with binding affinities in the low nM to high μM range ($K_d \sim 10^{-8}$ to 10^{-4} M). Weaker binding affinities can also be measured reliably, though a full thermodynamic profile of binding may not be obtainable [9, 10]. Other approaches involving displacement of a weaker binding ligand can be used to extend this useful range to much tighter interactions ($K_d \sim 10^{-9}$ to 10^{-12} M) [11].

12.2.2 Overview of Current ITC Microcalorimeters

Commercial ITC microcalorimeters have been available for several decades and are produced today by a number of different manufacturers. Those made by MicroCal (currently part of Malvern Panalytical) and TA Instruments are among those most commonly found in most academic or industrial laboratories. The MicroCal/Malvern Panalytical VP-ITC and Auto-iTC$_{200}$ will be specifically referred to in descriptions later in this chapter, as these are the instruments with which the author is most familiar. However, the sample preparation and requirements, general principles of ITC experiment design and many practical aspects of performing a binding analysis will be more broadly applicable to newer instruments from MicroCal/Malvern Panalytical (such as the PEAQ-ITC range) or those from other manufacturers.

The basic principles of ITC instruments have not changed significantly in the last decade or so but two notable advancements in the technology have arisen during this time. First, the volumes of sample required to run an ITC titration have been substantially reduced compared to early instruments with the development of small cell (~200 μl) instruments such as the PEAQ-ITC and iTC$_{200}$ (MicroCal/Malvern Panalytical) and Affinity ITC and Nano ITC (TA Instruments). While these instruments have the same limited throughput and requirement for substantial user intervention in an experimental setup as older instruments, the reduced volumes of sample needed (both macromolecule and ligand) have opened the method to use greater effort to produce biomolecules. Importantly, despite the reduced sample requirements, these instruments can detect comparable or smaller heats of binding and operate with similar low levels of noise as instruments with larger cells. The second major recent advance has been the development of automation built around these same small-cell ITC instruments. Automation has the advantages of eliminating some of the steps that typically cause problems for novice users, such as proper sample loading, as well as allowing for unattended operation for multiple ITC experiments.

Manual instruments such as the iTC$_{200}$ or the older workhorse VP-ITC are most likely to be found in individual labs and may be shared by a relatively small number of users. Significant expertise in sample handling and instrument use, particularly sample loading, are critical for obtaining high quality data with these instruments. Automation brings additional cost for purchase and maintenance, meaning that instruments like the Auto-iTC$_{200}$ are more often found in managed shared facilities (e.g. [12]). Such ITC facilities can accommodate a much larger group of users with a wide range of expertise, supported by an instrument that is very easy to use. Whatever the specific type of instrument available, the principle of its operation and ultimate experimental output will be largely the same.

A typical ITC instrument (Figure 12.1a) consists of two identical coin-shaped or cylindrical cells made of a chemically inert and highly thermally conducting metal (such as gold or Hasteloy alloy). Each cell has a long, narrow neck through which it is filled with either the macromolecule sample or a reference solution (typically water). In the case of the sample cell, this long neck is also where the automated titration pipette is placed for injection of the ligand, the second component of the binding reaction. Within the instrument, the two cells are held under conditions of constant pressure (isobaric) and temperature (isothermal) within an adiabatic jacket, meaning that no heat is transferred into or out of this enclosed system. Under application of constant cooling, power is applied to heaters on each cell to maintain them at constant and equal temperature.

When a sample of the ligand is injected from the automated pipette, a binding reaction occurs in the sample cell and heat is either evolved (for an exothermic reaction) or taken up (for an endothermic reaction). In the more common case of an exothermic binding reaction, the heat increase in the cell due to binding is sensed by the instrument and a proportional reduction made to the power applied to the sample cell. In other words, the total heating power from the binding reaction and sample cell heater are maintained at a constant total level, with any heat input from binding compensated by a drop in the heat input from the cell heater. The effect of this power compensation is to maintain the sample and reference cells at an equal temperature ($\Delta T = 0$). Similarly, for an endothermic binding reaction, increased power will be applied to the cell heater to maintain the constant total heating power with the same final result.

The difference in power that needs to be applied to the sample cell heater is the direct output measurement in ITC. Further, the power difference is directly related to the molar enthalpy change associated with the binding process, allowing determination of binding thermodynamics. The specific measurement recorded by an ITC instrument is the time-dependent sample cell power input (i.e. power per unit of time, such as µcal/s or µJ/s) needed to keep the temperature of the sample and reference cells equal. This output is usually plotted in real-time during the course of the ITC experiment and appears as a series of sharp changes (peaks) in applied power, each corresponding to one injection of the ligand. In the case of an exothermic binding reaction, less power needs to be applied and the peaks therefore appear as negative deflections from the baseline. The example titration shown in Figure 12.1b corresponds to an exothermic binding reaction.

At the end of the ITC experiment the peaks corresponding to heater power are integrated with respect to time, from the beginning of the ligand injection to the point where the applied power returns to the pre-injection baseline value. This process of peak integration gives the total heat exchanged per injection and these heat effects can then be analysed as a function of the molar ratio of the binding partners to give the thermodynamic parameters of the interaction under study (Figure 12.1c). The entire ITC experiment and recording of sample cell power take place under computer control, and software associated with the instrument will usually generate an initial automated analysis of the raw data. As described further below, it is usually necessary to manually check and adjust the peak integrations and to perform additional analysis steps before the most accurate fit of the interaction thermodynamic parameters is obtained.

12.2.3 ITC and Binding Theory

The power of ITC comes from its ability to provide a complete thermodynamic analysis of a binding reaction from a single experiment. Specifically, the user can *directly* determine the affinity (K_a), stoichiometry (n) and enthalpy (ΔH) of binding. From these directly determined parameters, the Gibbs free energy (ΔG) and entropy (ΔS) of binding can also be calculated. The following descriptions present a simplified view of the theory behind ITC analysis.

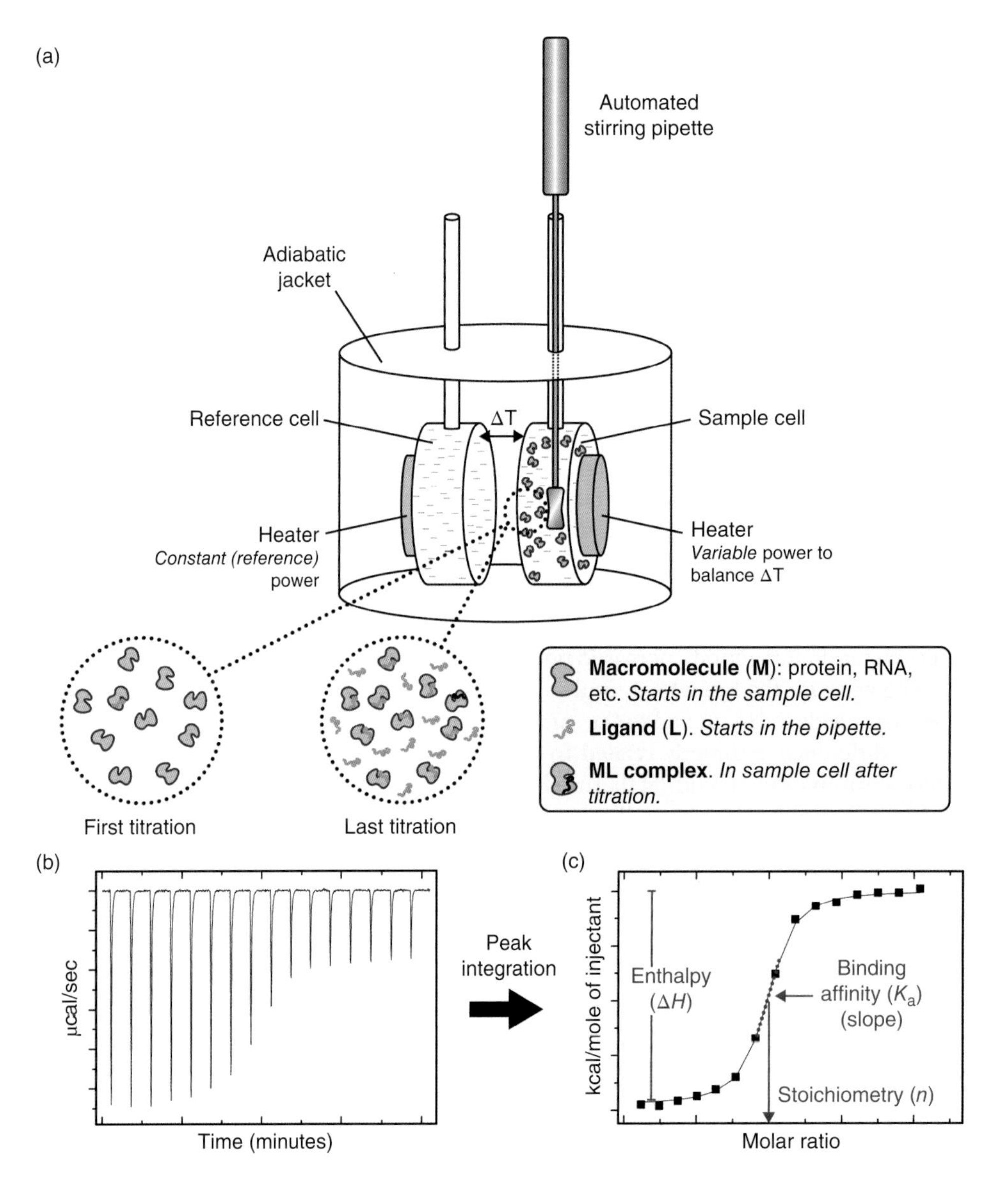

Figure 12.1 Overview of ITC instrument design and measurement. (a) Simplified schematic of a power compensation ITC instrument. Two coin-shaped cells within an adiabatic jacket hold the reference and macromolecule solutions while the ligand (titrant) is loaded in a specialised automated stirring pipette at the beginning of the experiment. Constant power is applied to the reference cell heater while the sample cell heater is controlled by a feedback loop that adjusts power to compensate for heat evolved or taken up during the binding reaction within the cell. The variation power supplied to the sample cell is calibrated to maintain the two cells at equal temperature (i.e. $\Delta T = 0$) and is the signal detected in ITC. (b) Ligand is injected into the cell in a series of injections of equal volume. In the example shown for a typical experiment on a MicroCal/Malvern Panalytical instrument, a negative deflection from baseline is due to an exothermic binding reaction: heat is evolved, meaning that power to the sample cell heater is reduced to compensate. (c) Values from integration of peaks in the raw ITC thermogram are plotted against the molar ratio of binding reaction components and fit to a specific binding model to yield molar enthalpy (ΔH), binding affinity (K_a) and molecular stoichiometry (n).

In an ITC experiment, the heat evolved (q) in the cell of volume V_o upon binding of the macromolecule (M) and titrated ligand (L) is proportional to the concentration of the macromolecule–ligand complex (ML) and related to ΔH of binding by Eq. (12.1). This can be expressed as the ratio of concentrations of complex and total macromolecule as in Eq. (12.2):

$$q = V_o\,[\mathrm{ML}]\Delta H, \tag{12.1}$$

$$\frac{[\mathrm{ML}]}{[\mathrm{M}]_T} = \frac{q}{V_o[\mathrm{M}]_T\Delta H}. \tag{12.2}$$

In the simplest case of a macromolecule with a single binding site for one ligand, the extent of reversible binding, i.e. concentration of the ML complex, is given by the equilibrium binding constant (Eq. (12.3)). Note that some ITC fitting software (e.g. Origin® supplied by MicroCal/Malvern Panalytical) generates the association equilibrium binding constant, K_a (units per molar, M^{-1}), whereas biologists typically discuss binding in terms of the dissociation binding constant, K_d (units of molar, M). Free energy of binding can be calculated from the K_a (or K_d) measured by ITC using Eq. (12.4), where R is the gas constant and T is temperature (in K):

$$K_a = \frac{[\mathrm{ML}]}{[\mathrm{M}][\mathrm{L}]} = \frac{1}{K_d}, \tag{12.3}$$

$$\Delta G = -RT\ln K_a = RT\ln K_d. \tag{12.4}$$

In terms of binding affinities and binding energies, it follows from these relationships that a larger value of K_a (or smaller value of K_d) refers to tighter binding between the macromolecule and ligand, as well as a more negative free energy of binding, ΔG. Finally, ΔG can be partitioned out into its enthalpic (directly determined by ITC) and entropic (ΔS) components in Eq. (12.5), allowing determination of each of these thermodynamic parameters:

$$\Delta G = \Delta H - T\Delta S \tag{12.5}$$

12.3 Protocols for Design, Implementation and Analysis of ITC Experiments

12.3.1 Where to Begin? Planning a First ITC Experiment

Having made the decision to study a binding interaction using ITC, the novice user (and even experienced ITC practitioner) is faced with a plethora of potential experimental variables when considering an analysis of a new system. For example, in sample preparation, which specific protein, nucleic acid or other biomolecule(s)/ligand(s) are the optimal ones to begin with? How much is needed and how should each component be purified? What experimental conditions, instrument settings and approaches for data analysis should be used?

Before beginning any new analysis, the most important first consideration is how much of each sample is needed and whether can these be feasibly produced at the level of purity needed for ITC. The answer to this question depends on both the instrument available and the system itself. For the MicroCal/Malvern Panalytical VP-ITC (~1.4 ml cell volume) or standard cell configurations of TA Instruments' ITC microcalorimeters (1 ml cell volume), up to 2 ml of macromolecule sample and 0.5 ml of ligand sample is needed for each titration experiment. For automated small-cell (~200 μl cell volume) instruments such as the Auto-iTC$_{200}$, these volumes are considerably less at 400 and 120 μl for macromolecule and ligand samples, respectively. For manually loaded small cell instruments, such as the MicroCal/Malvern iTC$_{200}$ or TA Instruments' small-cell configuration of the Nano ITC, there may up to a further 50% reduction in the volumes needed.

As discussed further below, macromolecule and ligand concentrations are important variables in optimising ITC experiments that also directly impact the quantity of material needed for each titration. Typically, the concentration of the macromolecule should be >10 times the expected K_d with the ligand sample at least ~10 times more concentrated, such that binding approaches saturation by the midpoint of the titration. However, for a new system with no prior knowledge of binding affinity, a first experiment with macromolecule and ligand concentrations of 10 and 100 μM, respectively, is a reasonable starting point (for a 1 : 1 binding stoichiometry). For a 50 kDa protein, this cell concentration corresponds to 0.5 mg/ml, or ~0.2 to 1 mg total protein depending on the sample cell size in the instrument to be used. For the same protein placed in the pipette, approximately two to three times more is required. For this reason, the harder to produce, more costly and/or most challenging to concentrate component is more often placed in the instrument sample cell. Of course, no matter which instrument is available or the optimal concentrations for the sample, significantly more material will be needed to complete a rigorous experiment complete with appropriate controls and titration replicates. However, if generating sufficient material for a pilot experiment is readily feasible, one should be able to quickly determine whether ITC will ultimately produce useful data for the system under study, thus making it worthwhile to further the investment in time and effort to produce the samples.

Most ITC instruments are supplied with software capable of simulating ITC experiments from a set of parameters (n, ΔH and K_d) provided by the user, allowing some optimisation of starting concentrations to be performed before even beginning a first experiment. Simulating ITC experiments can be particularly useful if some information is already known about the system in question, such as the approximate K_d from another binding measurement method or reasonably guessed from experiments with closely related molecules such as protein homologues. The simulation software is also a useful and simple way for the novice user to begin to get a feel for how sample concentration (both absolute and the ratio of macromolecule and ligand concentrations), K_d and ΔH influence the shape of the final binding isotherm obtained.

The preferred starting concentrations for an ITC experiment are commonly determined by consideration of a unitless parameter c, which is the product of the predicted affinity of the system and the total macromolecule concentration:

$$c = K_a.n[\mathrm{M}]_T = n[\mathrm{M}]_T/K_d. \quad (12.6)$$

Optimally, experiments should be performed in the range $5 < c < 100$. However, in most cases all three parameters measured by ITC (i.e. K_a, n and ΔH) can be determined with confidence for $1 < c < 1000$ in a single experiment. The macromolecule concentration should therefore be determined with these ranges of c values in mind. For example, for a binding reaction with a K_d of 10^{-6} M (or 1 μM), ideally a concentration of 5–500 μM should be used. For a given binding affinity, the curvature binding isotherm is dependent on the value of c (Figure 12.2). This makes it possible to manipulate the shape of the binding isotherm by adjusting the concentrations of the cell sample to optimise the quality of the data obtained. Of course, obtaining the fully optimal concentration may not always be possible: very low reactant concentrations might result in too small an experimental signal (which depends on ΔH of binding) whereas very high concentrations might result in heats that are too high or may simply be impractical or impossible for biological macromolecules (e.g. if a protein has a tendency to aggregate at a high concentration).

What happens outside of this optimal range of c values? For very tight binding reactions (where conditions of $c > 1000$ are likely) only ΔH and n can be accurately determined directly from an ITC experiment. Additional approaches, such as using competition experiments with a pre-bound competing ligand, may however still allow the user to determine a K_d [11]. In contrast, for weak binding events it may still be possible to obtain useful data working in the 'low c range', by using a very high concentration of ligand to drive binding to near saturation at the end of the titration [9, 10]. In this scenario, only the binding affinity can be measured with confidence. ΔH will not be well defined by the experiment and ΔH and n will be correlated in the fit of the data to a binding model. However, in cases where n is known with confidence and can be fixed in the fitting procedure if necessary, such experiments can still be very informative.

In summary, the availability of some prior knowledge (or reasonable guesses) of the system to be studied, use of ITC experiment simulation software and consideration of the c parameter can facilitate the early identification of suitable sample concentrations for an optimal ITC experiment. However, as noted above, in the absence of knowledge about a new system, a first guess of 10 and 100 μM concentrations in cell and

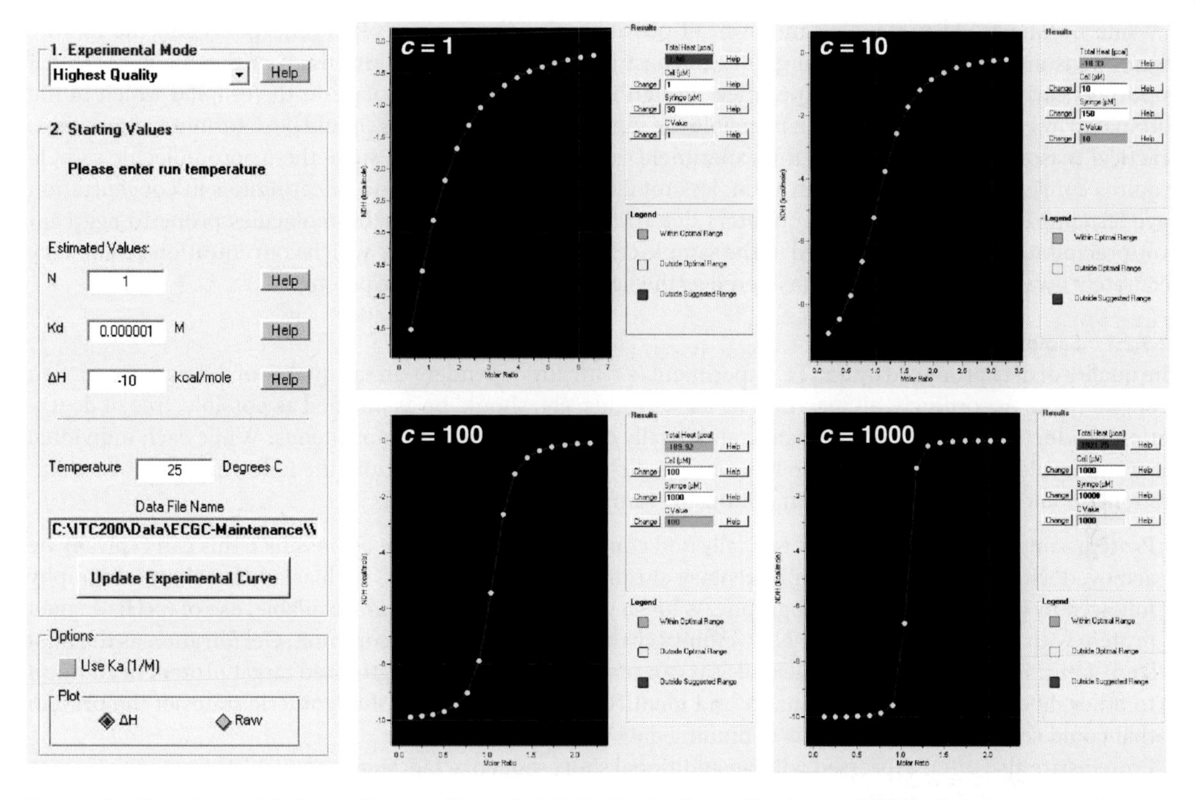

Figure 12.2 Experimental design and impact of *c* on the ITC binding isotherm. Simulation of ITC binding curves using the VP-ITC control software for a single set of user-defined parameters input through the panel shown on the left. In these examples the values of stoichiometry, *n*, K_d, and ΔH are fixed at 1, 1 μM and – 10 kcal/mol, respectively. Each plot was generated by editing the value of *c* and accepting the system generated values for macromolecule (cell) and ligand (pipette) sample. For the suboptimal extreme values of *c*, additional warnings are flagged (red box) for too low or too high heats for $c = 1$ and $c = 1000$, respectively. However, experimentally derived estimates of K_d and ΔH from a pilot ITC experiment could be used to more usefully optimise experimental conditions for subsequent experiments.

pipette, respectively, is a reasonable starting point for an anticipated 1 : 1 binding stoichiometry for a typical macromolecular interaction. From initial estimates of *n*, K_d and ΔH from this first pilot experiment, the same simulation tools can then be used to rapidly identify optimal concentrations for subsequent titrations.

12.3.2 Sample Requirements and Preparing for an ITC Experiment

The importance of sample preparation and planning for an ITC experiment cannot be overstated. The quality of data produced, and thus reliability of insights into binding, are only ever as good as the reagents used to generate them. The following sections delve deeper into many of the necessary practical considerations before embarking on an ITC experiment, but it is worth noting three particularly critical points from the outset:

- All samples – protein, nucleic acids, small molecules, etc. – should be as purified and homogeneous as is practical.
- Solutions containing the macromolecule and ligand must be as closely matched as possible to avoid excessive heats arising from simple mixing of the two different solutions.
- Ideally, the active concentrations of both components in the binding reaction should be very well defined. However, the accuracy of measurement of the ligand concentration is most critical as this directly impacts the accuracy of determining all three binding parameters (*n*, K_a and ΔH).

While no one set of protocols can cover all possible scenarios for a given ITC experiment, the following sections aim to serve as general guides for best practice in accomplishing each of the above goals. An important early consideration in experiment design is which molecule to place in the cell and which in the automated injection pipette. While in principle the same binding parameters should be determined regardless, practical reasons often dictate how the experiment is set up. For example, since the macromolecule sample requires considerably lower concentration, less total amount of material and uncertainties in concentration have less impact on the binding parameters determined, protein or other biomolecules prone to aggregation/precipitation may be best placed in the sample cell. Those molecules for which concentration can be very accurately measured may be best considered as the ligand and placed in the pipette.

12.3.2.1 Sample Preparation

The quality of data produced by an ITC experiment is critically dependent on the quality of the samples used to generate them. Accordingly, all macromolecules, ligands, etc. should be as purified as possible, free of degradation products and other contaminants, and ideally conformationally homogeneous. While each individual macromolecule or ligand will require its own specific protocol for purification, the following general considerations should be kept in mind when optimising sample preparation for ITC.

- Protein samples should be both chemically and conformationally as pure as possible. This can typically be achieved with two (or more) complementary chromatographic methods, such as affinity chromatography followed by gel filtration chromatography. While a vast range of options are available, use of 6×His-tagged protein with initial purification by Ni^{2+}-affinity chromatography is most common. Gel filtration as the final step of purification is particularly useful as it can remove misfolded or aggregated target protein in addition to other differently sized contaminants and identify potential mixtures of oligomeric states of the protein that could complicate later analysis of binding data.
- Proteins are also often expressed with an additional short sequence tag, such as hexahistidine, or as fusions with complete protein domains, like maltose binding protein (MBP), glutathione-*S*-transferase (GST) and many more. Such additions are routinely used to simplify and expedite purification, aid protein expression/solubility, or often both [13]. While appended tags or domains do not generally affect protein function, their possible impact on protein interactions or the potential to induce changes in protein oligomerisation should be kept in mind. Ideally, for example, for fusion proteins like GST, which can promote dimerisation, the GST tag should be removed during the purification protocol to eliminate potential confounding effects on the binding analyses.
- Chemically synthesised nucleic acids (DNA and RNA) and longer *in vitro* transcribed RNAs should be purified by high performance liquid chromatography (HPLC), fast protein liquid chromatography (FPLC) (e.g. gel filtration chromatography) or polyacrylamide gel electrophoresis (PAGE). The latter approach is typically done under denaturing conditions (e.g. 50% urea) but requires refolding, which may be challenging for some larger structured RNAs. Native gel purification is also possible.
- Small molecules and other ligands should also be as pure as possible. From commercial suppliers, purchase the highest purity grade available. For in-house syntheses, standard isolation procedures such as chromatographic methods, recrystallisation, etc., should be used with confirmation of identity and purity by mass spectrometry and nuclear magnetic resonance (NMR).
- Final samples for use in the ITC experiment should be free of aggregates and other particles. The final stage of sample preparation may involve concentration of protein or other macromolecules, e.g. in a centrifugal concentrator. Such devices can produce concentration gradients of the macromolecule, which may promote aggregation. The user should therefore be alert to signs of problems during this step. Before use, it may be desirable to filter or centrifuge the sample at high speed in a microfuge tube. Finally, the sample quality can be reassessed before use by re-running a small quantity on a gel filtration column or using a light scattering method like dynamic light scattering (DLS).
- Immediately before use, degassing of samples is often recommended to avoid formation of air bubbles during the ITC experiment. This may be particularly problematic for experiments performed at temperatures

higher than used for sample preparation or storage (e.g. within the Auto-iTC$_{200}$ autosampler tray). With the VP-ITC, degassing can be easily accomplished using the ThermoVac device typically supplied with this instrument. In our experience, however, sample degassing is not always necessary to obtain consistent high quality titrations but may be considered as an additional optimisation step if needed.

12.3.2.2 Buffer Considerations and Matching of Solutions

Many factors must be considered when deciding on the 'optimal' solution conditions for a given ITC experiment. Of course, a primary requirement is to use solution conditions – buffer/pH, ionic strength, reducing agent, organic solvent, or other additives, cofactors, etc. – that are best suited for the long-term (> several hours to days) solubility and stability of the macromolecule(s) or ligand to be studied. In the absence of homogeneous, well-behaved samples, no ITC experiment is going to succeed! Keeping this in mind, however, it is also important to note that some limitations on solution conditions do exist and, most critical for performing a successful ITC experiment, the ability is needed to precisely match the solution conditions for the macromolecule and ligand.

- Other than the two components of the binding interaction being studied, the two solutions containing the macromolecule and ligand should be as close to identical as possible. Even very small differences in composition or pH between the two solutions can cause significant artefactual signals in the ITC measurements due to large heats of dilution or ionisation. How this matching of macromolecule and ligand solutions is accomplished depends, in part, on the molecules under study.
- A commonly used approach for macromolecules is *exhaustive* dialysis against the experimental buffer solution to ensure exact matching of the sample(s). For example, two or three dialysis steps are required with >100- to 1000-fold volume difference between the sample and dialysis solution using dialysis tubing or disposable units (e.g. Slide-A-Lyzer™ or Spectra/Por® Float-A-Lyzer®) with an appropriate molecular weight cut-off (MWCO). A convenient approach is to perform two dialyses steps of three to four hours during the day prior to running the ITC experiment with a third overnight dialysis before the final steps of sample preparation the following morning. In cases where both binding partners are macromolecules, precise matching can be accomplished by placing both samples in separate dialysis tubing/units within the same large volume of dialysis buffer. Save the final dialysis buffer at the end of this process to preparing the ITC instrument and running control experiments, e.g. a ligand into buffer titration. As an alternative to dialysis, buffer exchange can also be accomplished using centrifugal concentration devices.
- With the shorter run times of columns now available, gel filtration chromatography can also be an excellent way to ensure buffers are well matched without the need for further dialysis. The reduced time in sample preparation may be particularly helpful in cases where proteins or other macromolecules are prone to aggregation or degradation. The final step of macromolecule purification can be performed directly in the buffer to be used for the ITC experiment and some of the buffer retained for dilution of other binding partners or for control titrations.
- Other ligands, such as small molecules, can also be dialysed along with macromolecules if membranes or dialysis units with a suitable MWCO are available. Alternatively, a commonly used approach for small molecules is to dissolve them in the buffer solution from the final step of dialysis used for the other binding partner (typically a macromolecule). Note, however, that care should be taken when using lyophilised samples (e.g. small molecules, peptides or nucleic acids) as salts present before lyophilisation will be retained in the final sample. These will be challenging to match in the other sample.
- While ITC experiments can be performed in most commonly used biological buffers, inadequate matching of pH for the macromolecule and ligand samples can cause significant issues. After preparing samples, it can therefore be useful to check that the final pH values are well matched (±0.05 pH units). The impact of pH discrepancies may also be exacerbated when using buffers with a large enthalpy of ionisation (ΔH_{ion}), such as Tris. Choosing a buffering system with a low (ΔH_{ion}), such as phosphate, may therefore be preferable so long as it is compatible with the samples under study. Alternatively, comparison of binding enthalpies and

affinities in multiple buffers with different ΔH_{ion} values can be informative on the nature of binding. For example, for binding reactions with a coupled protonation event, Tris buffer would yield a large enthalpic signal and measurements in multiple buffers can be used to quantify protons taken up or released upon binding [14, 15].

- If reducing agents are required for protein stability, the best choices are tris(2-carboxyethyl)phosphine (TCEP) or up to a few mM concentration of 2-mercaptoethanol (ß-mercaptoethanol). Use fresh stocks and, as noted above, make sure the concentration of reducing agent is closely matched between the macromolecule and ligand sample solutions. The commonly used reducing agent dithiothreitol (DTT) should usually be avoided if possible as it can cause serious problems with ITC baseline stability. However, specific compatibilities and tolerated reducing agent concentrations vary a little from instrument to instrument and should be checked before performing experiments.
- Small molecules may require a small amount of organic solvent (e.g. DMSO) in the final sample solution to maintain their solubility. In such cases, the same percentage of organic solvent should be added to the other binding partner (typically the macromolecule in the sample cell).

Even carefully following these suggested guidelines, slight differences in buffer composition between the ligand and macromolecule are possible due to co-solvents, salts or pH, particularly where samples are prepared independently (i.e. not co-dialysed), dissolved from lyophilised material or have components like organic solvents that need to be added immediately prior to experiments. In such cases, a test titration(s) of ligand into the macromolecule sample buffer and/or the ligand sample buffer into the macromolecule before performing the binding experiment may be advisable to check for large heats of dilution that could mask the binding signal. Finally, it is important to keep in mind that many of these steps used for buffer exchange and matching can result in potentially large changes in sample concentration. As a result, concentration measurements should be made following these steps, ideally on the final samples to be used in the ITC experiment.

12.3.2.3 Importance of Determining Accurate Sample Concentrations

The accuracy of data obtained from an ITC experiment is dependent on precise determination of sample concentrations. In particular, error in determining the concentration of the ligand in the pipette is directly proportional to the error of measurements of n, K_a and ΔH, and therefore the calculated free energy (ΔG) and entropic (ΔS) contribution to binding. In contrast, errors in cell component concentration only affect the accuracy of the determined stoichiometry. In cases where the concentration of one component is more reliably known, this may be an important consideration for experiment design: placing the component with the most accurate concentration in the pipette will result in the highest accuracy of the data produced in the ITC experiment.

Concentrations of each sample should be determined using an applicable method such as absorbance measurement, where a precise measured or calculated extinction coefficient is known, colorimetric assays such as BCA for proteins or HPLC with appropriate standards for small molecules. Ideally the concentrations of both binding partners should be accurately determined *after* buffer matching and final sample preparation. Two additional steps, described in the experimental protocols below, can also modestly impact the final sample concentrations. First, it is commonly recommended to degas the samples to avoid signal artefacts due to air bubbles or release of dissolved gases during the titration, particularly at higher temperatures. However, this also has the potential to reduce the sample volume via evaporation and thus increase concentration. Of course, since typically a little more sample is prepared than needed, the concentration can simply be rechecked to ensure there is no significant change. Second, rinsing of the sample cell will leave some residual buffer in the cell which will dilute the sample (by ~1% to 2%). While this is harder to accurately account for, as noted above, this change will only impact the determined binding stoichiometry and can likely be ignored in most cases.

12.3.3 Running the ITC Experiment

In each ITC experiment, a high concentration solution of ligand is injected (titrated) from an automated pipette (or syringe) in a series of equal injections into a more dilute sample of the macromolecule in the instrument cell. The final task, with final purified samples in hand, is therefore placing these two sample solutions in their respective components of the ITC instrument. Correctly loading the samples, for example without introducing any air bubbles, is essential for a successful experiment. How this is accomplished depends on the type of instrument available and is described below for both a manual VP-ITC and the fully automated Auto-iTC$_{200}$ MicroCal/Malvern Panalytical instruments. Newer manual instruments than the VP-ITC are likely to have partial automation that simplifies the sample loading process. Additionally, while the same general goals and principles apply, the practical procedures for setting up instruments from other manufacturers, such as the Nano ITC from TA Instruments, may also vary considerably (links to additional web resources for Nano ITC cell and syringe loading are provided under Website Resources at the end of this chapter).

Prior to beginning any experiment with precious samples, the user should make sure the ITC instrument is in good working order. Check any instrument logs for the date of last use and any notes from previous users to make sure the instrument was cleaned following manufacturer recommended or local instructions for cell and pipette cleaning. For automated instruments, visually inspect the fine tubing that carries samples and other reagents during the filling process for signs of precipitated materials or blockages. Running a test titration, which could simply be a small number of water-into-water injections, is a good idea in either type of instrument to confirm the instrument is operating as expected with a stable baseline close to the set reference power at the beginning of the test titration. Watching the pipette and cell loading processes on an automated instrument during this test titration is a good way to identify any potential issues before placing precious samples on the instrument.

12.3.3.1 Sample Loading and Experiment Set Up on a VP-ITC

Sample loading on the VP-ITC involves the use of a custom air-tight Hamilton syringe with a protective plastic sheath around the metal needle. Check that this sheath extends very slightly (~1 mm) beyond the tip of the metal needle to avoid scratching the surface of the ITC cell. Approximately 2 ml of the macromolecule and 0.5 ml of the ligand solutions are required for the filling processes for the cell and pipette, respectively. Also keep aside some of the final macromolecule dialysis buffer (10–20 ml) for control titrations and rinsing the cell during sample loading.

- The correct technique for sample cell filling is important to avoid introducing air bubbles that will impact the quality of the titration experiment. While the final addition of the macromolecule solution is most critical, using the following technique is advisable each time liquid is placed in the sample cell: (i) Fill the Hamilton syringe with liquid (~2 ml), then invert and tap to move air bubbles to the top of the barrel and expel them along with a very small amount of the liquid. (ii) Place the loaded Hamilton syringe carefully into the sample cell opening and lower until the tip gently touches the bottom of the cell. Raise the Hamilton syringe very slightly so that it is no longer touching the cell surface and then depress the plunger to fill the cell using a steady flowrate.
- Using the method described above, wash out the sample cell two to three times with water and then rinse two to three times with buffer (from the final dialysis). The sample cell is the more central of the two cell openings and is surrounded by a plastic overflow reservoir.
- Carefully load the macromolecule solution into the sample cell using the same technique. The final ~0.3 to 0.5 ml should be added in several short, rapid bursts that will dislodge any small air bubbles from the top of the cell. Lift the Hamilton syringe fully out of the sample cell and then slowly reinsert into the same opening at a sight angle until the tip touches a ledge located near the top of the opening (where the cell stem

contacts the plastic overflow reservoir). Draw up any liquid above this ledge by pulling up the Hamilton syringe plunger.

- Fill the reference cell with distilled water via the second opening adjacent to the sample cell. When filled with water, the reference cell does not need to be emptied and refilled between experiments; a newly refilled reference cell should be good for several days. However, if the experimental sample contains solvents such as DMSO, the reference cell should be filled with the same percentage of solvent in distilled water, or if the experimental buffer is of a particularly high ionic strength then the buffer may be used to fill the reference cell. This is necessary to maintain the heat capacities of the solutions in the reference and sample cells at a sufficiently close value. If reference solutions other than water are used, the reference cell should be emptied and thoroughly rinsed with distilled water immediately after the experiments are completed.
- Next, prepare the automated injection pipette by attaching a plastic syringe to the pipette fill port using clear tubing. Place the tip of the pipette needle into a microfuge tube filled with distilled water and use the attached plastic syringe to draw through the water after opening the fill port (note that opening and closing the pipette fill port is controlled by the instrument software). Repeat with the experimental buffer and then draw air through the pipette to remove all liquid. The plastic syringe can be removed and emptied between these steps if necessary.
- To fill the pipette with ligand solution, place the needle into a microfuge tube containing the ligand and draw the solution into the injection pipette until it is completely filled (i.e. when a small amount of ligand solution is visible in the tubing attached to the fill port). Close the pipette fill port, detach the tubing and plastic syringe and then purge/refill the pipette (using the instrument software) to remove any bubbles. Remove the ligand solution tube and gently dab the needle with a Kimwipe to remove any residual liquid from the outside of the pipette. Finally, place the injection pipette carefully into the sample cell.

Identifying the optimal experimental parameters for a given system is likely to take some trial and error. However, the following guidelines and suggestions should provide a suitable starting point for most experiments.

- The temperature at which the experiment will be performed is an important consideration (and as discussed elsewhere can be used in optimising experiments for challenging systems to study). While the VP-ITC can be run at temperatures between 2 and 80 °C, a first experiment at 25–37 °C is most common. Factors that might influence the final choice of experimental temperature include ensuring optimal protein activity, solubility and/or stability, best matching the biologically relevant temperature for the system under study or the need to correlate ITC analyses with data from other experiments performed at a specific temperature.
- The number and volume of injections are also important (and coupled) variables for the experimenter to consider. A typical first experiment on a new system the VP-ITC will likely use a relatively low number of injections of higher volume, e.g. 25–30 × 10 μl. The first injection is often inaccurate due to mixing of the macromolecule and ligand solutions at the pipette tip during the relatively long instrument baseline equilibration process before the titration begins. The first injection is therefore often set to a smaller volume (e.g. 2 μl) and this data point is subsequently ignored in the fitting process. Some experiments may also be more optimally performed with a larger number of small injections (e.g. 70 × 4 μl), including interactions with a strong heat signal or where binding is weak (low c range) as a larger number of injections will give more data points for accurate fitting. For systems with intrinsically low heat signals, fewer injections (minimally 10–15 for a reasonable binding isotherm) of larger volume may help, combined with increasing sample concentrations or a testing effect of altering temperature or solution conditions, e.g. pH or ionic strength.
- Most other instrument settings can be used at their default values, at least for a first experiment on a new macromolecule–ligand system. These include the reference power, stirring speed, and spacing and duration of each injection. In optimising ITC experiments the user can choose to increase spacing of injections if needed: it is critical that the system has sufficient time to equilibrate so that the heat signal returns to its baseline value before the next injection begins. Alternatively, for experiments with many small volume injections, where equilibration is rapid, the total time for the experiment can be significantly reduced by

decreasing the injection spacing. Stirring speed (default 310 rpm) can be reduced if there are any indications of denaturation (e.g. of protein samples) or increased for efficient mixing of more viscous samples (e.g. containing glycerol). The optimal reference power depends on the nature (exothermic versus endothermic) and scale of heats generated by binding while the duration of injection has only minimal impact on the peak profile and is usually left at its default value (0.5 μl/s).

At the end of the titration, the macromolecule–ligand mixture can be retrieved from the sample cell using the Hamilton syringe. Clean out the ITC cell with water and repeat the procedure for preparing the cell for sample loading to perform additional experiments. Ideally, at least one or two repeats of each macromolecule–ligand titration should be performed for reproducibility, in addition to any control experiments of ligand solution into buffer. At the end of a set of experiments (or between each experiment if the sample is heavily precipitated), clean the VP-ITC by manually filling and emptying the cell several times with detergent solution (e.g. 5% Contrad-70) and then rinsing using distilled water. Alternatively, the ThermoVac system, if available, can be used to flush larger volumes of these solutions through the cell. For more vigorous cleaning the cell can be filled with detergent of a higher concentration (up to 20% Contrad-70), for an extended time (60 minutes) and/or at an elevated temperature (65 °C).

12.3.3.2 Sample Loading and Experiment Set Up on an Auto-iTC$_{200}$

The Auto-iTC$_{200}$ instrument (and other semi- or fully automated systems) greatly improves the consistency of ITC experiments by automating sample loading, which is one of the most challenging aspects of performing an experiment for novice users. As with the VP-ITC, the software supplied with the instrument is used to define the experimental parameters and control the running of each ITC titration, but an additional benefit is the ability to queue many experiments for unattended operation. Samples are supplied in up to four 96-well deep well blocks and the cell can also be filled using a macromolecule solution from five 30 ml tubes. While this means that the Auto-iTC$_{200}$ could, in principle, run hundreds of experiments without user intervention, more commonly experiments run in smaller groups with periodic vigorous cleaning (which can also be automated) and manual checking of data quality.

- Each titration experiment on the Auto-iTC$_{200}$ requires 400 μl of macromolecule and 120 μl of ligand solutions. These volumes are needed for the automated liquid handling system to accurately fill the cell and automated pipette, respectively. They are significantly higher than the actual volumes used in the experiment (approximately 200 μl in the cell and 40 μl in the pipette) and also more than is required to manually load a standalone iTC$_{200}$ instrument.
- Configuring the system software with the planned experiment organisation (number of sample groups and number of experiments in each group) and selecting the desired run parameters *before* loading the sample block(s) is advisable. While sample loading on the Auto-iTC$_{200}$ is greatly simplified by using 96-well deep well blocks it is, of course, imperative than each sample solution or buffer is placed in its correct well. The final tab of the experiment setup shows a 'map' of the block and allows the user to select the site of the first sample (all subsequent sample positions are updated based on that selection).
- In the simplest setup, each titration uses only two samples, the macromolecule and ligand. However, up to four wells per titration will be needed if the experimental protocol includes a pre-rinse of the cell with buffer and the user selects to save the final macromolecule–ligand mixture to a clean empty well. Figure 12.3 shows examples of 96-well block maps for each of these scenarios with a small number of experiments.
- Once the complete set of samples is mapped on to the block, each solution is simply dispensed into its correct position, taking care to not introduce air bubbles, and the block is sealed with an adhesive film. Specialised films should be used that are designed for the Auto-iTC$_{200}$'s liquid handling cannulas to pierce to access the solution. The sealed block is then placed in the correct position in the temperature-controlled storage tray (4–25 °C).
- The experimental parameters that can be customised for each experiment are the same as those described for the VP-ITP: total number, volume, duration and spacing of injections, experiment temperature

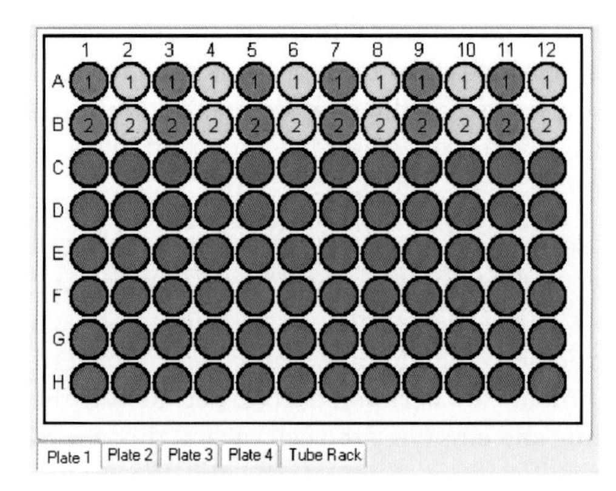

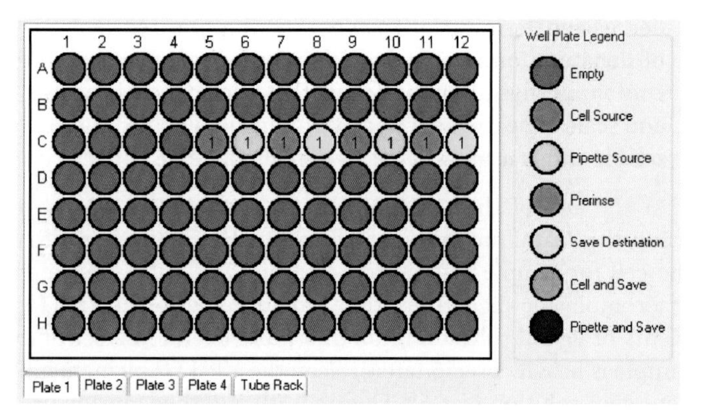

Figure 12.3 Example block layouts for an Auto-iTC$_{200}$ run. The final tab in the experiment setup of the Auto-iTC$_{200}$ software shows a map of the sample locations based on user-provided details of sample number and experiment type. This map should be carefully reviewed to ensure correct order of sample loading. The well for the macromolecule sample of the *first* titration in any run is user-selectable on this screen; all remaining sample locations are then mapped to the subsequent wells. Two examples are shown: (left) an experiment with 2 groups of 6 titrations (12 experiments in total) where the macromolecule and ligand in the cell at the end of the experiment will be discarded; (right) a more complex method involving a pre-rinse of the cell with the experiment buffer and saving the final cell solution to a new well at the end of each titration. It is important to note that regardless of the method selected and the order in which the reagents are used, the macromolecule and ligand sample are *always* placed in the first two positions for each titration. Once the samples are loaded and the run started, the complete set of experiments will be performed by the instrument without further input from the user.

(4–40 °C), reference power and stir speed. However, the defaults for these user optimisable settings are specific to the Auto-iTC$_{200}$ with its smaller capacity cell and pipette. A reasonable starting point for a new experiment on an Auto-iTC$_{200}$ is to use 16 injections of 2.5 μl (with default duration and spacing), with a reference power of 10 μCal/s and stir speed of 1000 rpm.

- A system cleaning procedure with water is performed automatically between each experiment. Protocols offering more rigorous cleaning of the cell and/or pipette can also be selected for some or all experiments and are recommended to be performed periodically even with well-behaved samples (e.g. every 5–10 titrations).

12.3.3.3 Dealing with Ligand 'Heats of Dilution'

Towards the end of the ITC experiment, the final few additions of ligand should result in relatively small peaks of equal height corresponding to the 'heat of dilution'. This heat arises from the mechanical mixing of the ligand into a sample containing the macromolecule already fully saturated by ligand binding. As discussed in more detail above, close matching of the macromolecule and ligand solutions is essential to avoid large heats of dilution that could mask the actual binding signal. These non-binding related heats need to be subtracted from each of the points in the macromolecule–ligand titration before further data analysis. This can be accomplished in several ways:

1) Perform a full titration control experiment (i.e. with the same number, volume and spacing of injections) of ligand solution into buffer and then subtract the integrated peak values from this control from those in the experiment during data analysis (see Section 12.3.4). This process can be automated within the Auto-iTC$_{200}$ software by designating control and experimental titrations so the reference run is subtracted from the macromolecule–ligand titration during the automated analysis.
2) Perform a separate, more limited titration of ligand into buffer to calculate a single average value for the corresponding heat of dilution and then manually subtract this value from each point in the experimental titration. This can be accomplished using the *Simple Math* option in the Origin software supplied with the VP-ITC or Auto-iTC$_{200}$.

3) A reasonable estimate of the heat of dilution can also often be obtained directly from the experimental (i.e. ligand into macromolecule) titration by averaging the heats associated with the final few (e.g. two or three) injections. Subtraction of this value from the experimental heats can then be accomplished as for approach 2 above. This method of estimating the heat of dilution makes the critical assumption that the system has essentially reached the point of saturation of macromolecule–ligand binding, which might not always be the case, e.g. where binding is weak or the concentrations of macromolecule and ligand used were not optimal. However, this method can be very useful if a ligand into buffer titration is not performed for some reason or if the heats of dilution obtained by the other approaches above are inconsistent with final heats in the experimental titration. (It is important to keep in mind that dilution of ligand into buffer is not precisely the same as dilution into solution containing the macromolecule, especially when relatively high concentrations are being used.)
4) Finally, more recent software supplied with some instruments (or developed independently by other researchers) allows heats of dilution to be fitted directly in the data analysis step. When available, this approach is likely to provide the simplest and most robust way to accurately account for these additional heats.

After heats of dilution have been accounted for by one of these approaches, the enthalpy of binding at saturation should approach zero and the binding isotherm can be fitted using an appropriate model to determine binding parameters (see Section 12.3.4 below).

12.3.3.4 Troubleshooting Some Common Issues

Several problems with ITC experiments are possible that can directly impact data quality. Some of these may be more common in one type of instrument than another, e.g. through the introduction of air bubbles into the sample cell due to a poor cell filling technique on a manual instrument that is less likely with an automated system (though these can have their own sample loading issues if not well maintained). Other problems with data quality may have their origin in problems with the samples themselves. Fortunately, many of these common problems give rise to identifiable signatures in the resulting titrations (Figure 12.4) that can point the user in the right direction to resolve them.

An ideal titration will begin a series of approximately equal, large peaks that become reduced as binding is saturated with further injections of ligand during the experiment. The final peaks, corresponding to the heats of dilution of ligand after binding is fully saturated, should be close to zero (Figure 12.4a). Underfilling of the cell or pipette will lead to erratic peaks and a shifting baseline that will likely make data unusable. A strongly downward stepping or shifting baseline may be indicative of an underfilled pipette, while upward changes may point to an underfilled or dirty cell. Such instrument issues are not likely to be common on a well-maintained and clean instrument but can happen on an Auto-iTC$_{200}$ instrument if some of the components or fine tubing needed to move the samples around become dirty or clogged. After ensuring that the instrument has sufficient supplies of water, detergent cleaning solution, methanol and N_2 gas, watching the processes of cell/pipette loading and cleaning should pin-point any problem areas.

Wildly oscillating or moving baselines, long times for equilibration of baseline or baseline values far from the set reference power may be resolved by thorough cleaning of the sample cell but could also be signs that the instrument has more serious issues that require expert service.

Other issues with sample quality or experiment design may also manifest in the following ways:

- Figure 12.4b shows *large heats of dilution at the end of the titration*. Large mismatches of the macromolecule and ligand buffer solutions will result in large heats of dilution that can potentially swamp the genuine binding heats. Check a buffer–buffer titration to confirm and improve buffer matching, ideally by exhaustive codialysis. A similar issue will arise on the Auto-iTC$_{200}$ if the methanol used to dry the tubing after system cleaning is not adequately removed before the next run. Check the tubing for blockages, ensure the system is being supplied with N_2 gas used to blow out residual methanol and/or watch the sample loading and system cleaning processes for signs of problems with liquid handling.

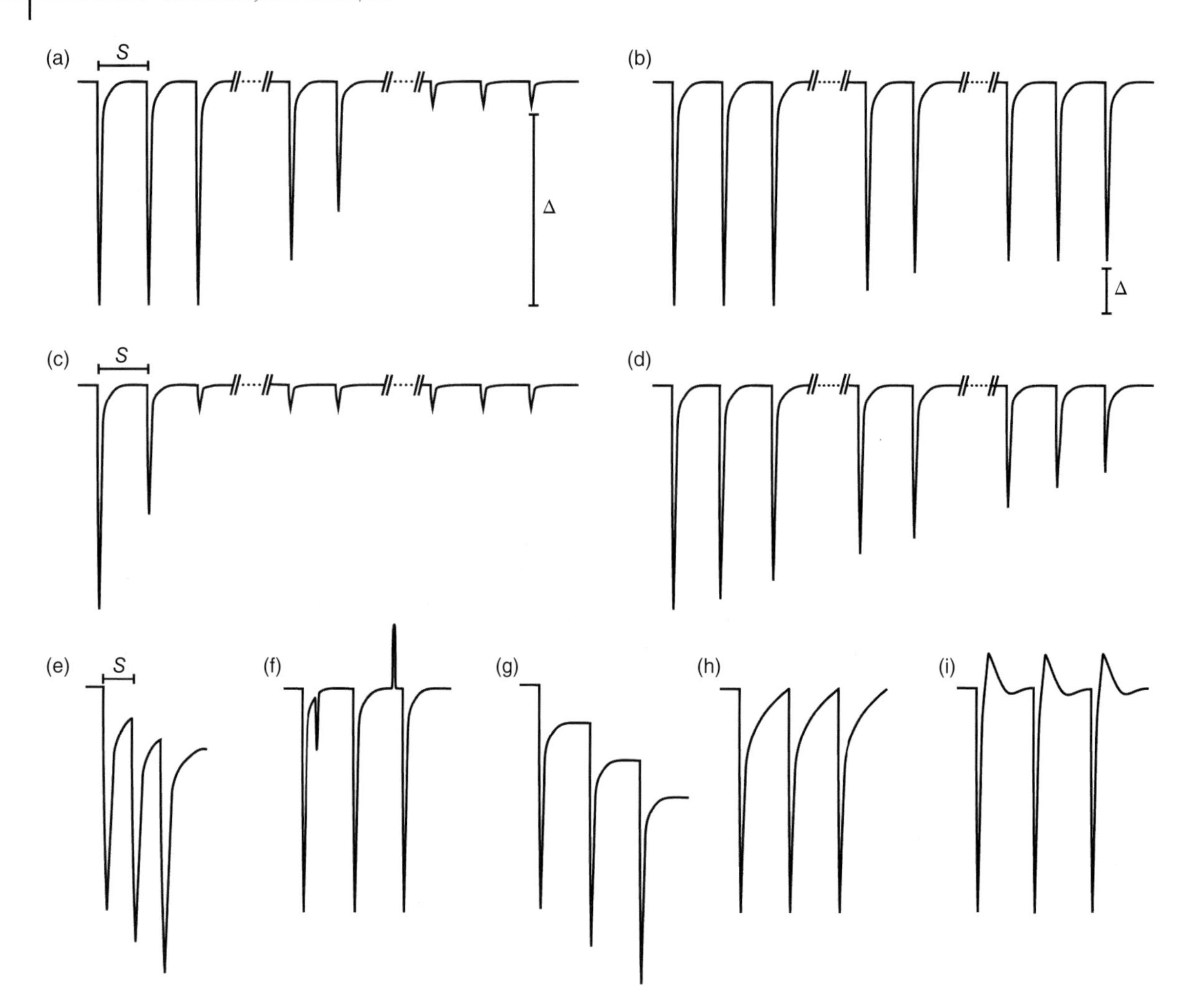

Figure 12.4 Signatures of some common problems in ITC experiments. (a) Schematic of an ideal titration in which initial large peaks of approximately equal size become smaller in the middle of the experiment (as binding sites become progressively more occupied) and finally ending with several peaks of equal value close to zero, corresponding to heats of dilution after saturation of binding. Poor-quality titrations may be indicative of (b) mismatch of macromolecule and ligand solutions resulting in large final peaks and a small difference (Δ) between peaks at the start and end of the experiment, (c) too high ligand concentration in the syringe relative to the macromolecule in the cell leading to rapid saturation of the binding signal, (d) too low ligand concentration in the syringe relative to the macromolecule in the cell meaning that binding is not saturated before the end of the titration, (e) slow binding events or inadequate spacing (s) of injections, (f) air bubbles introduced by a poor cell filling technique or spontaneous formation during the experiment, (g) a change in heat capacity of the system (e.g. enzymatic reaction or potentially a mismatch of buffers), (h) a slow equilibrium (e.g. degradation or aggregation) or other processes in addition to binding, or (i) two or more simultaneous processes occurring upon mixing with opposite heat effects.

- Figure 12.4c shows *binding* that *is too rapidly saturated – a very small number of large peaks are followed by many small peaks*. This may arise from inappropriate choice(s) of ligand and/or macromolecule concentration such that the relative of concentration in the pipette is too high (»10 times more than the cell concentration for an expected 1 : 1 binding). Check concentration measurements and either increase macromolecule concentration or decrease ligand concentration. Also consider the possibility that the binding stoichiometry may not be as expected, meaning that the relative concentrations of ligand and macromolecule should be adjusted (e.g. ligand concentration 20 times or more than the macromolecule for 2 : 1 stoichiometry).

- Figure 12.4d shows *binding that fails to saturate – small equal peaks corresponding to heats of dilution are not observed before the end of the titration.* This may arise for the same reasons as described above for too rapid saturation when relative concentration of the ligand is too low («10 times more than the macromolecule concentration for an expected 1 : 1 binding).
- Figure 12.4e shows *peaks that do not return to the baseline between experiments.* This will occur if binding results in broad peaks or there is inadequate spacing of injections used in the titration method (note the difference in spacing, *s*, in Figure 12.4a and c).
- Figure 12.4f shows *random spikes that appear during the titration.* Air bubbles can cause spurious peaks during the experiment and may impact data quality depending on their number and location: spikes in the baseline regions can simply be ignored during peak integration and discarding a small number of ligand injection peaks impacted by such artefacts should not greatly reduce data fitting quality. Air bubbles could arise due to the poor cell filling technique with a manual instrument (see Section 12.3.3.1) or from pipette filling issues on the Auto-iTC$_{200}$. In the latter case, inspect the pipette and watch the automated filling process for signs of problems during docking with the fill port adaptor, leaks from the pipette after filling, etc. Bubbles can also form spontaneously if samples are not adequately degassed, particularly when there is a significant difference in temperature between the sample storage and experiment.
- Figure 12.4g shows *an evenly stepping baseline.* This can occur when there is a change in heat capacity of the system, e.g. if an enzymatic reaction is occurring in the cell or potentially through a mismatch of the solutions containing the macromolecule and ligand. Some modification to the molecules or solutions containing them in the ITC experiment may be needed.
- Figure 12.4h shows *initially sharp peaks but with a long 'tail' before returning to the baseline.* Binding of the macromolecule and ligand may be accompanied by a second slow process (e.g. degradation or aggregation) in addition to binding. Alteration of the samples themselves (macromolecule or ligand) or the solution conditions may be needed; first, check for aggregation before and after the experiment (e.g. by visual inspection or using DLS or gel filtration chromatography).
- Figure 12.4i shows *peaks* that *are followed by an 'overshoot' in the opposite direction before returning to baseline.* One or more processes with opposite heat effects may be occurring upon mixing in addition to macromolecule–ligand binding. This situation could also arise due to a dirty sample cell, which is easily resolved by following a procedure for thorough cell cleaning with detergent.

12.3.4 ITC Data Analysis and Fitting to a Binding Model

Having performed a complete titration of ligand into macromolecule, along with any associated control experiments, the next steps are to perform initial adjustments of the raw data (e.g. baseline) and then to fit the measured heats to an appropriate model for the binding reaction, e.g. a single set of identical sites, two sets of independent sites, sequential binding, etc. A more detailed description of this process and the binding models for fitting ITC data can be found elsewhere (e.g. [16]). This fitting procedure allows the direct determination of binding affinity (K_d), stoichiometry (n) and enthalpy (ΔH) of binding with the goal of modelling the experimental data with the least error and using the simplest model that is in accordance with any prior information on the system under study. Free energy and entropy of binding (ΔG and ΔS, respectively) can then be calculated according to the equations of Section 12.2.3. Such data analysis and fitting of binding to one of several possible models is typically done in specially modified software supplied with the ITC instrument. With the VP-ITC and Auto-iTC$_{200}$ this analysis is done with a customised version of Origin software.

- First, open the analysis software, e.g. using the MicroCal Data Analysis Launcher, and load in the desired raw *.itc* titration file(s). For the VP-ITC, a single file is loaded with the *Read Data* button located in a menu on the top left of the screen; Auto-iTC$_{200}$ data can be loaded as single titrations using the iTC$_{200}$ option or as multiple files using the Auto-iTC$_{200}$ option from the launcher menu. The descriptions below cover the basic steps for analysing a single titration on either instrument.

- After opening the titration file, the Origin software will display the raw thermogram and perform an initial automated analysis of the data, including baseline fitting and peak integration, before presenting a view of the integrated ITC data in the *deltaH* window. Inspect the data points for signs of any potential problems such as significantly outlying values, which should be investigated in the raw titration data.
- Go to the raw titration data by switching to the *RawITC* window. Inspect the automatic fit baseline and integrated peaks for signs of any problems arising from air bubbles, abrupt baseline changes that are not well accounted for or other possible artefacts (see Section 12.3.3.4). A menu is displayed in the top left of the screen that allows the user to manually adjust the baseline points and the integration window around the peak. To do this, select *Adjust Integrations* and click on the first peak where adjustments are needed. A zoomed view of the peak base and fit baseline will be displayed along with new options to move the baseline points and reintegrate the peaks. The integration window can also be narrowed (e.g. to remove noise or artefacts outside of the peaks) by moving the blue vertical lines displayed on each side of the peak. Adjusting the baseline and integration window can often significantly improve the quality of the peak integrations and therefore quality of the final fit to the data. However, care should be taken in using these options as such decisions can be quite subjective and effort should be made to apply any changes consistently to each of the peaks in the thermogram.
- Return to the *deltaH* window and load the ligand–buffer control titration if this method is to be used to account for heats of dilution (see Section 12.3.3.3). Inspect the control titration and, if necessary, adjust as described above and then use the option in the *deltaH* window menu to *Subtract Reference Set*.
- Two other options on the *deltaH* window menu can be used at this point, if needed. *Modify Concentrations* can be used to adjust the concentration of the macromolecule or ligand, e.g. if analysis performed after the experiment was started revealed that the final concentrations differed from those entered in the instrument control software. This option can also be used as a starting point for further investigation if initial fitting of the data suggests the *active* concentration of the macromolecule and/or ligand was not as expected, e.g. a significantly non-integer value of n is obtained. Finally, use *Remove Bad Data Point* to remove the first data point in titrations where a smaller injection volume was used as well as any other point(s) where problems cannot be satisfactorily resolved using manual adjustment of the baseline and integration window. This step is performed last as further modifications of the baseline, integrations or concentrations will reset the plot to include the deleted points.
- If using *Simple Math* to account for heats of dilution (see Section 12.3.3.3), perform an initial correction at this point. Take careful note of the sign and value entered as this change is permanently applied to all integrated data points.
- Next, select the data fitting model: One Set of Sites, Two Sets of Sites, Sequential Binding Sites, Competitive Binding or Dissociation. In the absence of other information, begin with the simplest option for one binding site. Clicking on the desired binding model will open a new dialogue window in which initial values of the fit parameters will be displayed and can be adjusted if information from other methods is available. There is also an option to fix one or more of these values, if desired. For the one binding site model, the parameters displayed will be: stoichiometry (n), enthalpy (ΔH) and binding affinity (K_a). Click the button for single or multiple iterations of the fitting procedure and repeat until the fit has converged, i.e. when fit values and Chi^2 no longer change.
- Return to the *deltaH* window where the data, fit curve and binding parameters will be displayed. Inspect the quality of the fit to the integrated heats and make any further necessary adjustments, as described above. In particular, if using *Simple Math* to account for heats of dilution, check for systematic error in the fit to the final points, which would suggest too low or too high a value was entered. After any further adjustments, repeat the fit to the binding model. Finally, the MicroCal Origin software has a feature that allows the user to quickly generate a publication quality figure of the ITC experiment using the *Final Figure* option from the *Analysis* menu.

12.4 Example Applications of ITC to Analysis of Biomolecular Interactions

12.4.1 Case Study 1: Enthalpy versus Entropy Driven Transcription Factor (TF) Binding to Distinct DNA Sequences

Interaction of transcription factors (TFs) with their DNA binding sites is a key regulatory process in gene expression and an essential activity in living cells [17]. TFs directly control RNA transcription from the DNA genetic material by promoting (activator TF) or blocking (repressor TF) recruitment of RNA polymerase to the DNA [18]. Through these activities, TFs ensure both the correct level of expression in the correct cell type in multicellular organisms. TFs can act as either homodimeric complexes or as monomers, but always contain at least one DNA binding domain (DBD) that is responsible for directing specific interaction of the TF with its associated target DNA.

TF interaction with DNA can be driven by so-called 'direct read-out' through formation of direct hydrogen bonds and hydrophobic interactions between protein and DNA, as well as 'indirect read-out' of DNA shape and electrostatics. These contacts can also be made directly between the TF and DNA or mediated by water molecules, and each can therefore make distinct contributions to the enthalpy and entropy of binding. The ability of TFs to bind to multiple related sequences with biologically relevant affinities has been rationalised based on 'enthalpy–entropy compensation' arising from relatively subtle changes in the contributions of these different types of interaction. However, it is less clearly understood how some monomeric TFs are apparently able to bind more than one distinct DNA sequence, which would require more significant adjustment in the number and type of interactions necessary to maintain binding. To unravel the basis of this phenomenon in molecular detail, Morgunova et al. [19] performed a detailed mechanistic study of DNA binding by the posterior homeodomain protein HOXB13 and several other monomeric TFs, using X-ray crystallography, molecular dynamics (MD) simulations and ITC. These studies used two distinct high affinity DNA sequences CTCGTAAA (DNA^{TCG}) and CCAATAAA (DNA^{CAA}), which differ by the underlined nucleotides and to which HOXB13 binds with similar affinity despite their significant sequence difference.

X-ray crystal structures of HOXB13 bound to each DNA sequence revealed little difference in the HOXB13 protein structure in each complex, while the two DNAs differed in the bending of their backbone at the location of the distinguishing nucleotides (Figure 12.5a). Additionally, there were few differences observed in HOXB13's interactions with each DNA (Figure 12.5b). Together with extensive mutagenesis experiments, these structures suggested that the differences in protein–DNA interactions or commonalities in DNA structure cannot fully explain the dual sequence preferences of HOXB13 and other TFs. Rather, the major difference between the structures appeared to be the role of bridging water molecules in the interaction of HOXB13 with each DNA. Specifically, water molecules in the DNA^{CAA} complex appeared more stably bound and making interactions that were likely to correspond to stronger enthalpic contributions. In contrast, water molecules in the HOXB13-DNA^{TCG} complex appeared significantly more disordered, suggesting a high entropy state. These ideas were further supported by MD simulations and free energy perturbation calculations. However, to directly test their hypothesis of distinct enthalpic and entropic minima driving TF binding to distinct DNA sequences, the authors turned to ITC analysis with its power to provide a complete thermodynamic profile of HOXB13 binding.

For use in ITC experiments, HOXB13 was expressed in Rosetta 2(DE3)-pLysS *Escherichia coli* cells using an auto-induction protocol and purified by Ni^{2+}-affinity chromatography [19, 20]. DNAs were chemically synthesised by a commercial supplier as single strands and the complementary sequences mixed and annealed to form each double-stranded DNA. Both HOXB13 and each DNA were prepared in a solution containing 20 mM HEPES buffer pH 7.5, 300 mM NaCl, 10% glycerol and 2 mM TCEP, and titration experiments were performed on an iTC_{200} microcalorimeter at the Protein Science Facility at Karolinska Institute, Sweden. Each experiment was performed at 25 °C and involved 23 injections of HOXB13 'ligand' (150 μM) into DNA solution (12–16 μM) in the sample cell. All data were analysed using the customised Origin software supplied with the instrument and binding parameters (n, K_d, ΔH) determined by a non-linear least square fit of the

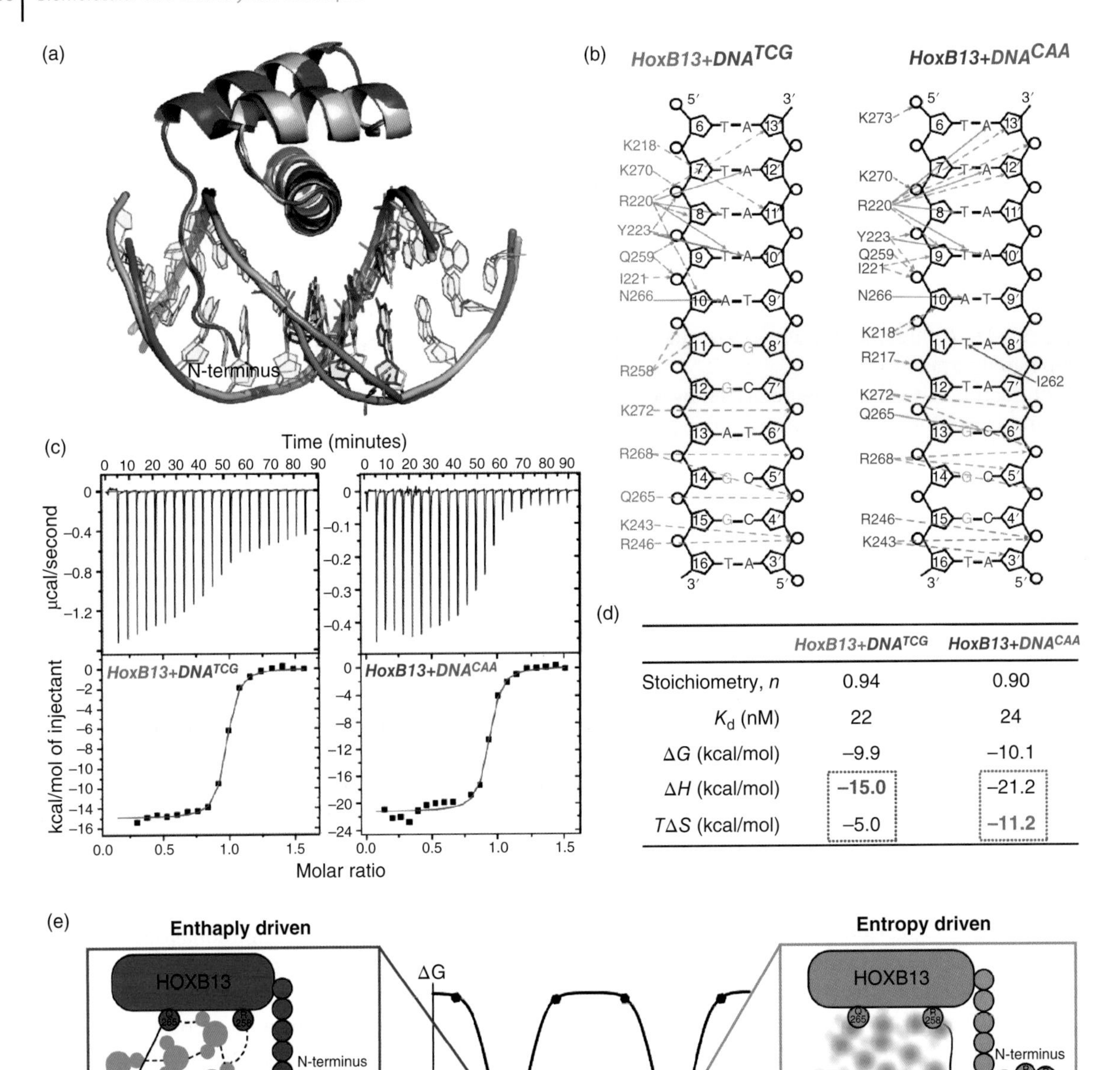

	HoxB13+DNA^TCG	HoxB13+DNA^CAA
Stoichiometry, *n*	0.94	0.90
K_d (nM)	22	24
ΔG (kcal/mol)	–9.9	–10.1
ΔH (kcal/mol)	**–15.0**	–21.2
$T\Delta S$ (kcal/mol)	–5.0	**–11.2**

Figure 12.5 Enthalpic and entropic minima control HOXB13 binding to two distinct DNA sequences. (a) Superposition of the HOXB13 DBD bound to DNA^{TCG} (tan protein/blue DNA) and DNA^{CAA} (red protein/green DNA). The two protein structures are essentially identical with a root mean square deviation of 0.81 Å for the 57 residues shown. DNA residues that differ between the DNAs are shown in orange and exhibit a distinct bending of the DNA backbone at this site. (b) Schematic of interactions formed between HOXB13 DBD and the two DNA sequences. Interactions with DNA backbone phosphate/deoxyribose and bases are indicated with dashed and solid lines, respectively. Shading denotes the regions with distinct DNA sequence. The TCG site lacks direct contacts to the DNA bases, whereas the CAA site is recognised by direct contacts by Gln-265 and Ile-262. Most other contacts are similar in both structures. (c) ITC analysis of HOXB13-DNA interactions for the two distinct sequences and (d) resulting thermodynamic parameters of binding. While both K_d and free energy are essentially identical for both, the DNA sequences differ strongly in the ΔH and $T\Delta S$ contributions to binding. (e) Model for the presence of two distinct optimal (low ΔG) DNA binding sequences, one driven by low enthalpy (left) and the other by high entropy (right), distinguished by differences in ordering of water molecules (blue spheres) and thus the interactions they mediate. Source: This figure is adapted from Morgunova E. et al., *eLife*, e32963, 2018 [19].

data to the one binding site model as implemented in the package. The $T\Delta S$ and ΔG of binding were obtained using Eq. (12.5). Each titration was performed in triplicate and average values for the determined binding parameters were reported and reproduced here.

ITC titrations of HOXB13 into each DNA revealed a very similar affinity and free energy of binding for both sequences, as anticipated (Figure 12.5c and d). The data also revealed, however, the strongly differing contributions of enthalpy and entropy in each case. For DNA^{CAA}, ΔH was significantly increased whereas ΔS was strongly reduced compared to DNA^{TCG} (Figure 12.5d). The authors were thus able to draw the conclusion from these direct experimental measurements that HOXB13 binding to one DNA sequence (DNA^{CAA}) represents an enthalpically driven minimum in binding free energy while the other (DNA^{TCG}) an entropically driven minimum (Figure 12.5e). This novel finding also demonstrates the power of ITC to provide a complete picture of binding thermodynamics to complement other biophysical approaches like X-ray crystallography in understanding the molecular basis of biomolecular function.

12.4.2 Case Study 2: Identification of a New Motif Essential for Co-substrate S-adenosyl-L-methionine (SAM) Binding by an rRNA Methyltransferase

RNA post-translational modifications such as nucleobase or ribose sugar methylations occur in many different types of RNA in both prokaryotes and eukaryotes. In ribosomal RNA (rRNA), for example, such modifications are critical for correct RNA folding and ribosome subunit assembly, as well as ribosome function in protein synthesis [21, 22]. Additionally, rRNA methylation can directly block the activity of ribosome-targeting antibiotics and is an important contributor to the escalating problem of bacterial resistance to antibiotics such as macrolides and aminoglycosides [23–25]. In some other rarer cases, specific rRNA modifications can also be required for proper antibiotic binding and therefore antimicrobial activity. One example of this scenario is the ribosome-targeting antibiotic capreomycin, which requires the ribose 2′-O-methylations incorporated by the 16S/23S rRNA methyltransferase TlyA for its antimicrobial activity [26]. Given the ubiquity of RNA modifications in biology and their important contributions to current biomedical problems, such as antibiotic resistance, identifying the enzymes responsible and defining their mechanisms of action, is an active area of investigation. An important aspect of such mechanistic understanding of rRNA methyltransferase enzyme activity is defining how they bind their obligatory co-substrate for the methylation reaction, S-adenosyl-L-methionine (SAM). In a recent study from our lab on the capreomycin resistance-related rRNA methyltransferase TlyA from *Mycobacterium tuberculosis*, we used ITC to define the role of a new protein motif critical for SAM binding by this enzyme [27].

At the outset of our study on TlyA, a key goal was to determine a high resolution crystal structure of the intact enzyme. However, when crystals of the full-length protein could not be obtained, we used a collection of proteases to identify stable TlyA fragments than might be more amenable to crystallisation. Limited proteolysis with the endopeptidase GluC, which selectively cleaves the polypeptide backbone to the C-terminal side of accessible glutamic acid residues, identified one such stable fragment. Further analyses suggested that GluC cleavage occurred primarily at glutamic acid 59 (E59), immediately adjacent to a short tetrapeptide linker between the TlyA N-terminal domains (NTD) and C-terminal domains (CTD), $\ldots{}^{59}\mathrm{E}{\downarrow}\underline{\mathrm{RAWVS}}^{64}\ldots$, where the site of cleavage is denoted by the arrow and the underlined sequence is the interdomain linker. A new construct was designed for expression of the isolated TlyA CTD (residues 64-268) beginning with the first residue of α-helix 1 (α1) in the CTD. Using this new construct, a crystal structure of the TlyA CTD was determined at 1.7 Å resolution. The TlyA CTD adopts a canonical Class I methyltransferase fold [28], including all the expected conserved motifs required for SAM binding. Despite this, we were unable to obtain a TlyA-CTD:SAM complex structure by co-crystallisation or soaking of preformed TlyA CTD crystals. We therefore used ITC to gain complementary insights into SAM binding by TlyA.

For these ITC analyses, 6× His-tagged wild-type TlyA and TlyA CTD proteins were expressed in *E. coli* BL21 (DE3) and purified using His-Trap HP and Superdex 75 16/60 columns on an ÄKTA Purifier FPLC for Ni^{2+}-affinity and gel filtration column chromatography, respectively. TlyA proteins were exhaustively dialysed

Table 12.1 SAM binding by TlyA proteins.[a)]

	Binding affinity	
Protein	K_d (μM)	**Fold-change**[b)]
TlyA	23.4 ± 2.9	—
CTD	No binding	—
^{RAWV}CTD	20 ± 1.1	—
TlyA-R60A	87.2 ± 8.5	3.6
TlyA-R60E	62.8 ± 23	2.7
TlyA-A61V	21.1 ± 4.3	1.0
TlyA-W62A	234 ± 56	10
TlyA-W62F	98.5 ± 27	4.3
TlyA-V63A	470 ± 19	20

a) Data are from Witek et al., *J. Biol. Chem.*, 2017 [27].
b) Relative to wild-type TlyA, indicated for RAWV linker substitutions only.

against 50 mM Tris buffer (pH 7.5) containing 120 mM NaCl and 10% glycerol and concentrated using a centrifugal concentration device to 60–100 μM. The final dialysis buffer solution was used to resuspend SAM at 1.5 mM final concentration. As binding was expected to be relatively weak, ITC experiments were performed at the highest practically achievable protein concentration and much higher ligand (SAM) concentration to drive binding towards saturation at the end of the titration (i.e. a 'low *c*' ITC experimental design). All ITC experiments were performed at 25 °C and used 16 injections of 2.4 μl of SAM into the cell containing the protein. After subtraction of residual heats derived from the average values of the final two or three injections, ITC data were fit to the model for one-binding site. The equilibrium dissociation constants (K_d) reproduced here (Table 12.1) are the averages of at least three experiments with their associated standard deviation.

Wild-type TlyA bound SAM with low micromolar affinity as expected (Figure 12.6a and Table 12.1), comparable to many other rRNA methyltransferase enzymes. In contrast, the isolated TlyA CTD protein failed to show any binding to SAM (Figure 12.6b), despite its crystal structure showing all expected conserved SAM binding motifs being present. While this result was surprising, it was consistent with our inability to obtain a TlyA CTD-SAM co-crystal structure. TlyA cleaved by GluC was examined next, both as the cleaved NTD/CTD mixture and the isolated CTD purified by gel filtration chromatography. Both samples were shown by ITC to retain the wild-type SAM binding affinity (data not shown). Together, these results suggested that some element outside the CTD is critical for SAM binding and a new protein expression construct was created to express the CTD protein including the four amino acids of the interdomain linker at its N-terminus (residues 60-268; ^{RAWV}CTD). Remarkably, addition of the RAWV tetrapeptide sequence fully restored SAM binding to the wild-type TlyA affinity (Figure 12.6c and Table 12.1).

These results suggested that the RAWV sequence of the TlyA interdomain linker might form a critical but previously unappreciated motif necessary for SAM binding by the TlyA family of rRNA methyltransferases. Site-directed mutagenesis was used to make one or two substitutions of each of the RAWV amino acids and the impact of these changes were tested by measuring SAM binding by ITC (Table 12.1). Consistent with sequence analysis of the RAWV tetrapeptide, which showed W62 and V63 to be most strongly conserved, changes at these two residues had the greatest impact on SAM binding affinity. X-ray crystallographic studies of the TlyA ^{RAWV}CTD protein revealed that the RAWV tetrapeptide is able to adopt two distinct conformations: a 'helix' form that extends $\alpha 1$ at the beginning of the CTD or a 'loop' structure that closely corresponded to

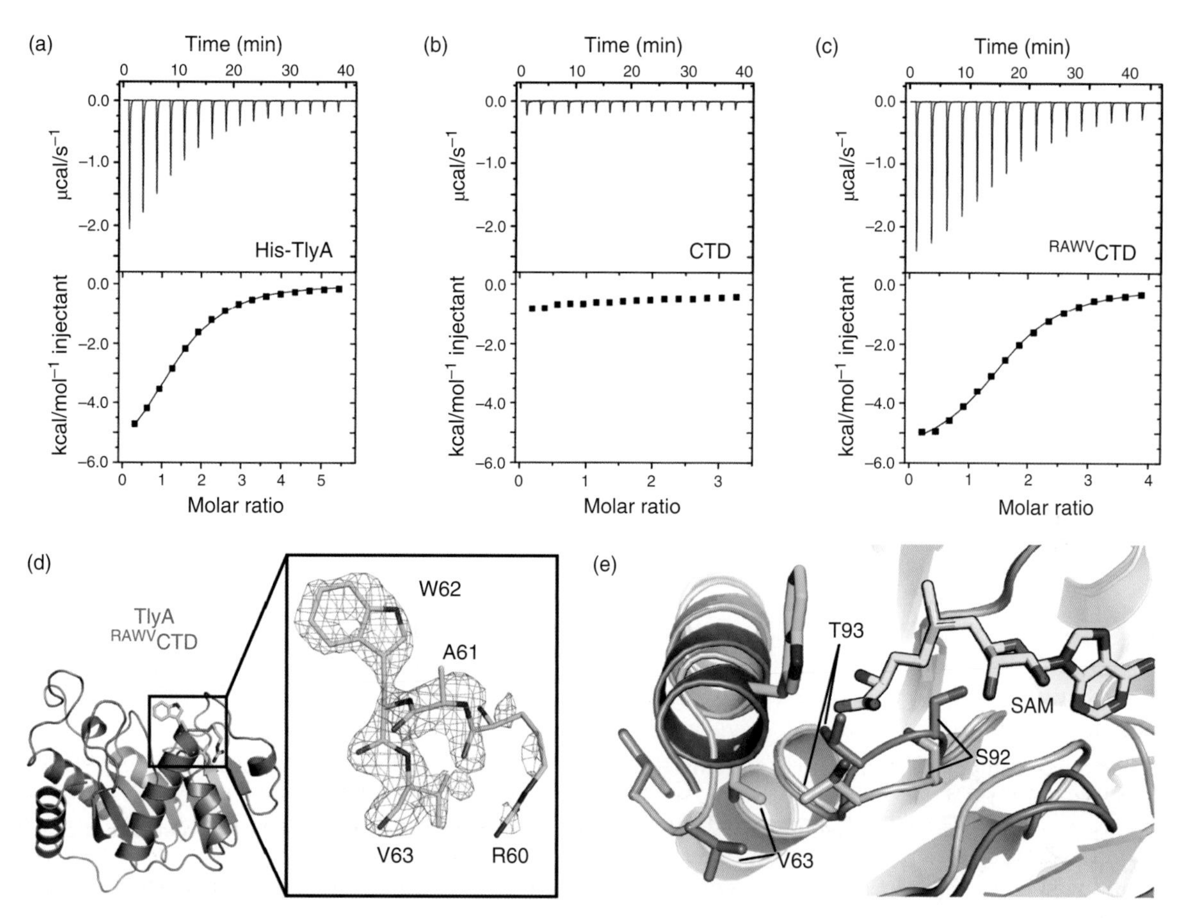

Figure 12.6 Identification of a novel motif critical for TlyA methyltransferase binding to SAM. ITC analysis of SAM binding to (a) full-length N-terminally 6× His-tagged TlyA (His-TlyA), (b) the isolated TlyA C-terminal Class I methyltransferase domain (CTD) and (c) the same isolated CTD but with the four-amino acid (RAWV) TlyA interdomain linker appended on its N-terminus (^{RAWV}CTD). (d) Crystal structure of the ^{RAWV}CTD protein with the RAWV interdomain linker in its 'helix' conformation. The zoomed region shows the RAWV linker in $2mF_o$-DF_c omit electron density contoured at 1σ. (e) Overlay of the ^{RAWV}CTD structures with RAWV interdomain linker in its 'helix' (cyan) and 'loop' (tan) conformations. Source: This figure is adapted from Witek et al., *J. Biol. Chem.*, 2017 [27].

the conformation of the equivalent sequence in several hemolysin proteins that had previously been used for homology modelling of TlyA [29].

Interactions made by V63 and other changes in the SAM binding pocket, which were only observed in the ^{RAWV}CTD 'helix' structure, also offered insight into a potential molecular basis for the role of the new RAWV motif in SAM binding (Figure 12.6d and e). The newly identified ability of the RAWV interdomain linker to adopt two distinct structural conformations, as well as its critical importance for SAM binding as revealed by ITC, suggested that this short sequence may play a critical functional role in substrate recognition by TlyA. Specifically, the question of how TlyA can specifically recognise and modify the 2′-OH of two cytosine residues within distinct structural contexts, 23S rRNA C1920 in the 50S subunit and 16S rRNA C1409 in the 30S subunit, may have its answer in a common substrate binding-induced interdomain linker conformational change that is linked to SAM binding affinity and thus TlyA activity. Fully delineating the molecular mechanisms of TlyA and other rRNA methyltransferase activities will require further investigations, of which ITC will likely be a critical component.

12.5 Concluding Remarks

ITC is firmly established as one of the best choices of approach to study many different types of biomolecular interaction. With its power to fully define the thermodynamics of binding in a single experiment, ITC has found broad application in studies of basic biological processes as well as in drug discovery. Further advances in instrument technology and the associated software have also simultaneously reduced the previously high sample quantity demands and simplified the processes of performing and analysing ITC data. As such, ITC is an approach that should be among the first considered for any study that aims to understand the nature and role of a biomolecular interaction.

Acknowledgements

I thank Dr Debayan Dey and Mr Zane Laughlin (Emory University) and the anonymous reviewers for their helpful comments and suggestions on this chapter. Our research is currently supported by NIH/NIAID grant R01-AI088025.

References

1 Velazquez-Campoy, A., Leavitt, S.A., and Freire, E. (2004). Characterization of protein-protein interactions by isothermal titration calorimetry. *Methods Mol. Biol.* 261: 35–54.

2 Velazquez-Campoy, A., Ohtaka, H., Nezami, A. et al. (2004). Isothermal titration calorimetry. *Curr. Protoc. Cell Biol.* 17–18.

3 Klebe, G. (2015). Applying thermodynamic profiling in lead finding and optimization. *Nat. Rev. Drug. Discovery* 14: 95–110.

4 Claveria-Gimeno, R., Vega, S., Abian, O., and Velazquez-Campoy, A. (2017). A look at ligand binding thermodynamics in drug discovery. *Expert Opin. Drug Discovery* 12: 363–377.

5 Di Trani, J.M., De Cesco, S., O'Leary, R. et al. (2018). Rapid measurement of inhibitor binding kinetics by isothermal titration calorimetry. *Nat. Commun.* 9: 893. PMCID: PMC5832847.

6 Zhao, H., Piszczek, G., and Schuck, P. (2015). SEDPHAT--a platform for global ITC analysis and global multi-method analysis of molecular interactions. *Methods* 76: 137–148. PMCID: PMC4380758.

7 Mazzei, L., Ciurli, S., and Zambelli, B. (2016). Isothermal titration Calorimetry to characterize enzymatic reactions. *Methods Enzymol.* 567: 215–236.

8 Falconer, R.J. (2016). Applications of isothermal titration calorimetry - the research and technical developments from 2011 to 2015. *J. Mol. Recognit.* 29: 504–515.

9 Turnbull, W.B. and Daranas, A.H. (2003). On the value of *c*: can low affinity systems be studied by isothermal titration calorimetry? *J. Am. Chem. Soc.* 125: 14859–14866.

10 Tellinghuisen, J. (2008). Isothermal titration calorimetry at very low *c*. *Anal. Biochem.* 373: 395–397.

11 Velazquez-Campoy, A. and Freire, E. (2006). Isothermal titration calorimetry to determine association constants for high-affinity ligands. *Nat. Protoc.* 1: 186–191.

12 Yennawar, N.H., Fecko, J.A., Showalter, S.A., and Bevilacqua, P.C. (2016). A high-throughput biological calorimetry core: steps to startup, run, and maintain a multiuser facility. *Methods Enzymol.* 567: 435–460.

13 Malhotra, A. (2009). Tagging for protein expression. *Methods Enzymol.* 463: 239–258.

14 Bradshaw, J.M. and Waksman, G. (1998). Calorimetric investigation of proton linkage by monitoring both the enthalpy and association constant of binding: application to the interaction of the Src SH2 domain with a high-affinity tyrosyl phosphopeptide. *Biochemistry* 37: 15400–15407.

15 Baker, B.M. and Murphy, K.P. (1996). Evaluation of linked protonation effects in protein binding reactions using isothermal titration calorimetry. *Biophys. J.* 71: 2049–2055. PMCID: PMC1233671.

16 Freyer, M.W. and Lewis, E.A. (2008). Isothermal titration calorimetry: experimental design, data analysis, and probing macromolecule/ligand binding and kinetic interactions. *Methods Cell Biol.* 84: 79–113.
17 Levine, M. and Tjian, R. (2003). Transcription regulation and animal diversity. *Nature* 424: 147–151.
18 Smith, N.C. and Matthews, J.M. (2016). Mechanisms of DNA-binding specificity and functional gene regulation by transcription factors. *Curr. Opin. Struct. Biol.* 38: 68–74.
19 Morgunova, E., Yin, Y., Das, P.K. et al. (2018). Two distinct DNA sequences recognized by transcription factors represent enthalpy and entropy optima. *eLife* 7: PMCID: PMC5896879.
20 Jolma, A., Yin, Y., Nitta, K.R. et al. (2015). DNA-dependent formation of transcription factor pairs alters their binding specificity. *Nature* 527: 384–388.
21 Polikanov, Y.S., Melnikov, S.V., Soll, D., and Steitz, T.A. (2015). Structural insights into the role of rRNA modifications in protein synthesis and ribosome assembly. *Nat. Struct. Mol. Biol.* 22: 342–344. PMCID: PMC4401423.
22 Decatur, W.A. and Fournier, M.J. (2002). rRNA modifications and ribosome function. *Trends Biochem. Sci.* 27: 344–351.
23 Wachino, J. and Arakawa, Y. (2012). Exogenously acquired 16S rRNA methyltransferases found in aminoglycoside-resistant pathogenic gram-negative bacteria: an update. *Drug Resist. Updat.* 15: 133–148.
24 Conn, G.L., Savic, M., and Macmaster, R. (2009). *DNA and RNA Modification Enzymes: Comparative Structure, Mechanism, Functions, Cellular Interactions and Evolution* (ed. H. Grosjean), 524–536. Austin, TX: Landes Bioscience.
25 Long, K.S. and Vester, B. (2009). *DNA and RNA Modification Enzymes: Comparative Structure, Mechanism, Functions, Cellular Interactions and Evolution* (ed. H. Grosjean), 537–549. Austin, TX: Landes Bioscience.
26 Monshupanee, T., Johansen, S.K., Dahlberg, A.E., and Douthwaite, S. (2012). Capreomycin susceptibility is increased by TlyA-directed 2'-O-methylation on both ribosomal subunits. *Mol. Microbiol.* 85: 1194–1203. PMCID: PMC3438285.
27 Witek, M.A., Kuiper, E.G., Minten, E. et al. (2017). A novel motif for S-adenosyl-l-methionine binding by the ribosomal RNA methyltransferase TlyA from *Mycobacterium tuberculosis*. *J. Biol. Chem.* 292: 1977–1987. PMCID: PMC5290967.
28 Schubert, H.L., Blumenthal, R.M., and Cheng, X.D. (2003). Many paths to methyltransfer: a chronicle of convergence. *Trends Biochem. Sci.* 28: 329–335.
29 Arenas, N.E., Salazar, L.M., Soto, C.Y. et al. (2011). Molecular modeling and in silico characterization of *Mycobacterium tuberculosis* TlyA: possible misannotation of this tubercle bacilli-hemolysin. *BMC Struct. Biol.* 11: 16. PMCID: PMC3072309.

Further Reading

Ladbury, J.E. and Doyle, M.L. (2005). *Biocalorimetry 2: Applications of Calorimetry in the Biological Sciences.* Wiley.
Privalov, P.L. (2012). *Microcalorimetry of Macromolecules: The Physical Basis of Biological Structures.* Wiley.

Website Resources

ITC experiment set up and analysis using a Microcal/Malvern Panalytical VP ITC instrument:
https://www.jove.com/video/2796/isothermal-titration-calorimetry-for-measuring-macromolecule-ligand.
Sample cell and syringe loading on a TA Instruments Nano ITC:
https://www.youtube.com/watch?v=VNf9ujlHrkc.
https://www.youtube.com/watch?v=EDgjtABjLc4.

13

An Introduction to Infra-red and Raman Spectroscopies for Pharmaceutical and Biomedical Studies

Ka Lung Andrew Chan

School of Cancers and Pharmaceutical Science, Institute of Pharmaceutical Science, King's College London, SE1 9NH, UK

13.1 Significance and Short Background

Infra-red and Raman spectroscopies are well-established analytical methods in pharmaceutical and biomedical sciences. Infra-red and Raman spectra often contain many sharp bands that are specific to the chemicals present in the sample. The acquired spectra can be used to characterise chemical composition, polymorphism, functional groups and molecular interactions within samples. These techniques are non-destructive and samples can be retrieved after measurement for further analysis. These analytical techniques are highly versatile; for example, they can be used for on-line process monitoring or the study of living samples, such as cell cultures. Measurements are highly reproducible and they have been used as gold standard methods in the quality control of pharmaceutical products. As the technology advances, large focal plane array (FPA) detectors are becoming routinely used for imaging, with a better, brighter light source and more sensitive detectors becoming available for fast data collection and development in chemometric techniques for data mining; novel applications are increasingly found in biomedical research including drug screening and disease diagnosis.

13.2 Theory

The theory of infra-red and Raman spectroscopies is extensive and it is not possible to cover all details in a single book chapter. This chapter is therefore intended to provide a snapshot of the essential understanding of the theory required by a non-spectroscopist to perform Fourier transform infra-red spectroscopy (FTIR) and Raman spectroscopic measurements, to understand the origin of spectral peaks and to interpret the meaning of spectral data.

13.2.1 Electromagnetic Radiation

Infra-red and Raman spectroscopies are often collectively referred to as vibrational spectroscopy, which is a study of the interactions between light and molecules where the interaction results in a change in the molecular vibrational energy state and the characteristics of light, such as intensity (I), wavelength (λ), propagation direction and polarisation. Light in the ultraviolet (UV), visible and infra-red ranges are the parts of the spectrum of electromagnetic radiation that are used in this technique. Electromagnetic radiation can be represented by the classical wave model or the particle model. The classical wave model describes the electromagnetic radiation as an oscillating electric and magnetic field that propagate through space at the speed of light ($\sim 3 \times 10^8$ m/s in vacuum). The classical wave model has helped scientists to understand some of the properties of electromagnetic radiation such as diffraction and refraction. However, the classical wave model does not explain the

Biomolecular and Bioanalytical Techniques: Theory, Methodology and Applications, First Edition. Edited by Vasudevan Ramesh.

particle behaviour of electromagnetic radiations such as the photoelectric effect. Albert Einstein had shown that there is a minimum frequency of radiation that is required for the ejection of electrons from metals when light is shone on the metal surface, and that it is independent of the radiation intensity. His observation can be explained by the particle model, where electromagnetic radiations are considered as particles called *photons*. Each photon contains a discrete amount of energy depending on the wavelength of the radiation, given by

$$E = \frac{hc}{\lambda} \tag{13.1a}$$

or

$$E = hc\omega \tag{13.1b}$$

or

$$E = h\nu. \tag{13.1c}$$

The symbol h is the Planck constant, which has a value of 6.626×10^{-34} J s, and that wavenumber, ω, and frequency, ν, are related to wavelength, λ, by

$$\omega = \frac{1}{\lambda}, \tag{13.2a}$$

$$\nu = \frac{c}{\lambda}. \tag{13.2b}$$

Note that the electromagnetic radiation of a specific wavelength carries a specific amount of energy and therefore energy is said to be *quantised*. The concept of quantisation of energy is important in many spectroscopic studies because, in analogy to the photoelectric effect, a molecule can only be excited to a higher electronic or vibrational energy state when the incident light carries enough energy. UV–visible absorption spectroscopy and fluorescent spectroscopy are concerned with the electronic energy state transitions while infra-red and Raman spectroscopies are concerned with the vibrational energy state transitions. The latter two spectroscopic methods produce spectra that are highly characteristic of the molecule.

13.2.2 Molecular Vibration

Atoms in a molecule are constantly oscillating even when they are in the non-excited ground state. The periodic movement of atoms in a vibrating molecule generates an oscillating local electric field in the molecule. These periodic fluctuations in the electric field enable the molecule to interact with light because light waves (or photons) are also oscillating electric fields. The frequency, ν, of molecular vibrations can be estimated using the simple harmonic oscillation model

$$\nu = \frac{1}{2\pi}\sqrt{\frac{k}{m}}. \tag{13.3}$$

From Eq. (13.3), where m is the reduced mass of the atoms involved in the covalent bond and k is the strength of the bond, we can see that a smaller mass of the system and a stronger chemical bond will result in a higher vibrational frequency. For example, stretching vibrations that involve hydrogen atoms are often found in the high frequency (or wavenumber) region of the spectrum because of the low atomic mass of hydrogen atom. C–H, N–H and O–H ('X'–H) stretching modes are found in the range of 2800–3800 cm^{-1}. Bands due to vibrations of single bonds (weaker), with the exception of X–H stretching modes, are often found below 1400 cm^{-1} while bands due to the stretching vibrations of double and triple bonds (stronger) are found in 1400–1800 and 2000–2400 cm^{-1} ranges, respectively. It is clear that the nature of the chemical bonding will determine the vibrational frequencies of the molecule. A list of vibrational frequencies of common function groups are listed in Table 13.1.

Table 13.1 A list of common molecular vibrational frequencies.

Wavenumber (cm^{-1})	Vibration mode
3600–2700	O—H stretching
3500–3100	N—H stretching
3050	C—H aromatic stretching
2900–2800	C—H aliphatic stretching
2390–2280	O=C=O asymmetric stretching, atmospheric CO_2
2260–2200	C≡N stretching
2250–2150	C≡C stretching
1800–1640	C=O carbonyl stretching
1690–1620	Amide I
1690–1590	C=C
1680–1580	O—H bending water (liquid)
1650–1500	N—H bending
1600	$(O—C=O)^-$ asymmetric stretching
1580–1500	Amide II
1450–1350	C—H bending
1400–1260	O—H bending
1400	$(O—C=O)^-$ symmetric stretching
1300–1050	C—O stretching
1250–1200	$(O—P=O)^-$ asymmetric stretching
1100–1000	$(O—P=O)^-$ symmetric stretching
900–700	C—H aromatic out of plan bending

13.2.3 Vibrational Energy

The energy required for vibrational energy transitions to occur can be schematically presented on an energy diagram as shown in Figure 13.1. Once the frequency of the molecular vibration is known, Eqs. (13.1a) to (13.3c) can be used to calculate the energy band gap (ΔE) between the excited and the ground vibrational states. One of the conditions for a molecule to absorb light is when the energy of the light matches the energy band gaps. Figure 13.1 schematically illustrates that the energy band gap for electronic transitions is much greater than vibrational transitions. For electronic transitions to occur, lights with wavelengths in the ultraviolet or visible range are required. However, the energy of light that is required to excite molecular vibration falls within the mid-IR range, which is between 2.5 and 40 μm (4000 and 250 cm^{-1}). The study of energy transitions within this wavelength range is therefore called vibrational spectroscopy, including infra-red (mid-infra-red) and Raman spectroscopies. Infra-red spectroscopy is the study of the absorption of infra-red light by the sample, which results in excitations between vibrational energy levels. Raman spectroscopy, on the other hand, is a study of the inelastic scattering of light, which results in a change in the vibrational energy level of the molecule through a temporal virtual energy state (Figure 13.1).

13.2.4 Modes of Vibration

A chemical bond can vibrate through a stretching and/or bending motion. Among others, there are two most important types of stretching mode (symmetric and asymmetric) and four types of bending mode (scissoring, rocking, wagging and twisting) depending on the structure of the molecule. For the same

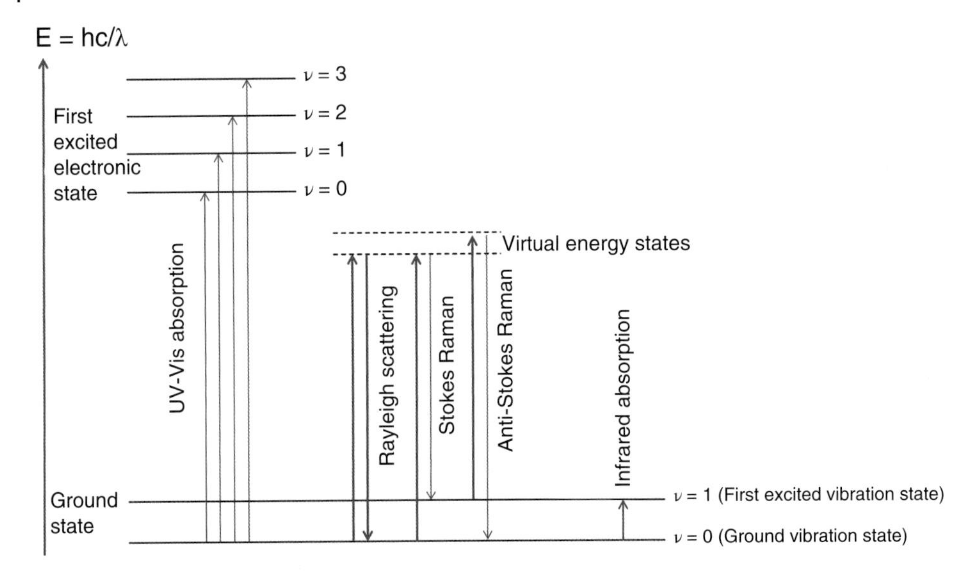

Figure 13.1 A schematic energy (not to scale) diagram showing the different excitations of a molecule by visible and infra-red light.

– Stretching (symmetric and asymmetric)

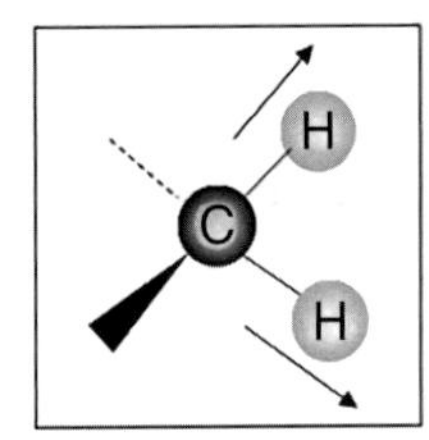

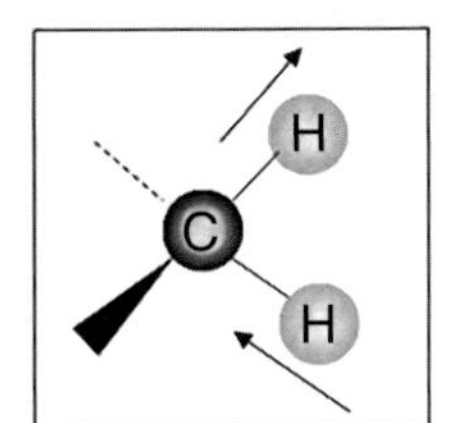

– Bending

(Scissoring, rocking, wagging and twisting)

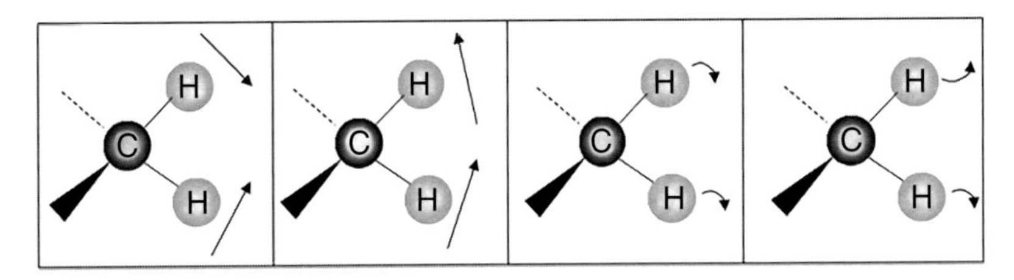

Figure 13.2 A schematic diagram showing the different modes of vibration for a CH_2 group (as an example).

functional group, stretching modes of vibration usually have higher energy (higher frequency/wavenumber) than bending modes of vibration. For example, the symmetric and asymmetric $\nu(CH_2)$ stretching modes of vibration are found in the 2800–3000 cm^{-1} region while the $\delta(CH_2)$ bending modes of vibration are found around ~1400 cm^{-1}. Note that stretching and bending modes of vibration are denoted by the symbols ν and δ, respectively. Examples of the various stretching and bending modes of vibration are illustrated in Figure 13.2.

13.2.5 Number of Vibrations

The total number of vibrations found in a molecule can be calculated by Eqs. (13.4a) and (13.4b) with

$$\text{Number of vibrations} = 3N - 5 \tag{13.4a}$$

for a linear molecule and

$$\text{Number of vibrations} = 3N - 6 \tag{13.4b}$$

for a non-linear molecule, where N is the number of atoms in the molecule. Diatomic molecules such as hydrogen, oxygen and nitrogen (H_2, O_2 and N_2) are linear molecules and therefore have just one vibration. Water (H_2O) is a non-linear molecule with three atoms and therefore has three modes of vibration (the symmetrical and asymmetrical stretching modes and a bending mode). In contrast, a small drug molecule such as ibuprofen ($C_{13}H_{18}O_2$) has 33 atoms, is a non-linear molecule, and will have 291 vibrations. It can be seen that with just a relatively small number of atoms, a small molecule can produce a large number of distinct vibrations, resulting in a unique set of vibrational energy band gaps. The vibrational spectrum of a molecule is often distinctive, which is the basis of the high chemical specificity offered by these vibrational spectroscopic techniques. However, not all vibrations will produce a spectral band because of the different *selection rules*, which we will discuss below.

13.2.6 Infra-red Spectroscopy

Infra-red spectroscopy is a study of the absorbance (or transmittance) of infra-red light through the sample. As a result of the absorption of infra-red light, the vibrational energy state rises from the ground state to the first excited state, as shown in Figure 13.1. Molecules can only absorb light that has the same frequencies as their molecular vibrations. This is one of the conditions for light absorption. The second condition is that *the molecular vibration must produce a change in dipole moment* so that the periodic fluctuation of the electric field generated from the vibrating molecule can interact with the light. These two conditions are the *selection rules* for infra-red absorption. Based on this principle, we can predict the types of covalent bond that will or will not produce infra-red bands. For example, diatomic molecules (e.g. H_2, O_2 and N_2) have no change in dipole moment from their vibrations and therefore they do not absorb infra-red light. These vibrations are considered as *infra-red inactive*. Covalent bonds with permanent dipoles that produce a larger change in dipole moment during molecular vibration will produce stronger infra-red absorption bands. For example, the carbonyl, C–O and N–O stretching modes of vibration can produce strong infra-red absorption bands with larger molar absorptivities (also referred to as the extinction coefficient, ε), while the C–H and C–C stretching modes of vibration have smaller molar absorptivities. However, the symmetric stretching mode of carbon dioxide, despite the large permanent dipole between the carbon and the oxygen atoms, is infra-red inactive because of the molecular symmetry. The changes in the electron environments from the symmetrical movement of the two oxygen atoms cancel each other out, leading to a zero overall change in dipole moment.

13.2.6.1 Infra-red Spectrum

An infra-red spectrum is a plot of the percentage transmittance, T (or $\%T$), or absorbance, A, against the wavenumber (cm^{-1}). Transmittance is defined as the intensity of the IR light measured with the sample, I, ratioed against the intensity of light measured without the sample, I_0:

$$T = \frac{I}{I_0}, \tag{13.5a}$$

$$\%T = T \times 100. \tag{13.5b}$$

The intensity measured without the sample is called the background spectrum. A suitable background spectrum is important because it can be used to remove spectral features that do not belong to the sample. This includes the intensity profile of the spectrometer and the absorption of IR light due to the molecules present in the atmosphere. Fluctuation in water vapour and carbon dioxide levels in the atmosphere are some of the most problematic issues in infra-red spectroscopic measurement. Inadequate control of these will result in compromises on the quality of the sample spectrum. Modern FTIR spectrometer software is often designed with automatic atmospheric correction to remove the unwanted water vapour and carbon dioxide contribution. However, the aim should be to ensure good control of atmospheric conditions during measurements rather than to rely on subtraction of the unwanted atmospheric vapour absorbances by algorithms.

13.2.6.2 Transmittance versus Absorbance

It is advantageous to present spectra as absorbances when quantitative analysis is important because the Beer–Lambert Law states that absorbance is equal to the multiple of the concentration of samples, c, molar absorptivity, ε, and path length, l (Eq. (13.6a)). A linear relationship between absorbance and concentration is expected when the molar absorptivity and the path length are kept constant for the absorbance range of between 0 and 1. Beyond absorbance of 1, the assumption of a linear response from the detector may not hold. Absorbance is defined as the negative log of transmittance and can be used to develop standard curves for the calculation of the concentration of samples. A transmittance spectrum can be converted to an absorbance spectrum using Eq. (13.6b). Spectral peaks in an absorbance spectrum point upward with a baseline of 0 because a stronger absorption of IR light by the sample results in a higher absorbance. Spectral 'peaks' in transmittance, however, point downward with a baseline of 1 (or 100% if $\%T$ is used) because absorption of IR light by the sample results in a lower transmittance of light. Thus,

$$A = \varepsilon c l, \tag{13.6a}$$

$$A = -log(T). \tag{13.6b}$$

13.2.7 Raman Spectroscopy

Raman spectroscopy is a study of inelastic scattering of UV, visible or near infra-red light. When the energy gap between the ground and the first excited electronic states is larger than the energy of the incident light, the light cannot cause an electronic excitation but it can polarise the electron cloud and induce a temporal dipole in the molecule. As the electric field of the light is oscillatory, the induced dipole in the molecule will also oscillate at the frequency of the incident light, which will then relax by emitting a photon. When the light photon is re-emitted immediately (on the order of 10^{-15} seconds) at the same energy, and hence frequency, as the incident light, the process is called *elastic scattering* or *Rayleigh scattering*. However, the induced oscillating dipole can interact with the vibrations of the molecule and produce an oscillation in the polarisability of the molecule. The relaxation of the induced dipole can therefore be affected by the molecular vibration resulting in the molecule re-emitting photons at a different energy and wavelength to the original incident light, as illustrated in Figure 13.1. The scattering process that results in a change in the wavelength of light due to inelastic scattering is called *Raman scattering* and the change in the wavelength in respect to the wavelength of the incident light is called *Raman shift*. If the light loses energy through the inelastic scattering process, it is called *Stokes Raman scattering* and if the light gains energy from the process, it is called *anti-Stokes Raman scattering*. The energy diagram in Figure 13.1 shows that the difference in energy from the incidence and emitted light is dependent on the vibrational energy band gaps. Raman spectroscopy is the study of the shift in the wavelength of light through the inelastic scattering process, which shows the vibrational energy levels of the molecule. Raman scattering can only occur when the *vibration induces a change in polarisability* of the molecule. The selection rules for Raman scattering are therefore different to infra-red absorption, which is one of the reasons why infra-red and Raman spectroscopies, although both providing information on the vibrational energy levels, are complementary techniques.

13.3 Technique/Methodology/Protocol

13.3.1 Infra-red Spectrometer

Infra-red spectrometers are designed to measure the absorbance or transmittance of light through a sample at different wavenumbers in the infra-red spectral region. The key components of an infra-red spectrometer include an infra-red source, a monochromator or interferometer and an infra-red detector. The infra-red source produces a constant flux of infra-red radiation that covers the desired measurement range. A Globar source is often used as the black body radiator to produce a broadband light in the mid-IR range that covers 4000–400 cm^{-1}. Some spectrometers are also equipped with a tungsten halogen lamp to produce near infra-red light for the measurement of the near infra-red spectrum. The infra-red light is detected by a semi-conductor infra-red detector. There are different types of detector available and the selection of detector depends on the requirements of the measurement. Generally room temperature operated infra-red detectors such as the deuterated tri-glycine sulphate (DTGS) and the lithium tantalate ($LiTaO_3$) detectors are less sensitive than the liquid nitrogen cooled mercury cadmium telluride (MCT) detectors, especially at low light conditions. Furthermore, MCT detectors are available in array format, enabling the measurement of multiple spectra from different areas of the sample simultaneously. These detectors include the FPA detector and the linear array detector. Using these detectors can significantly increase the speed of data acquisition for imaging applications [1–3]. However, MCT detectors are more expensive, especially those with the array format, more easily saturated and require cryogenic cooling, which increases the running cost.

A monochromator or an interferometer is used to separate the wavelengths of light so that the intensity of infra-red radiation at different wavelengths can be plotted individually. A dispersive infra-red spectrometer utilises a monochromator, which consists of a diffraction grating to disperse the light and a narrow slit for wavelength selection. This type of spectrometer has become less popular because the narrow slit gives a low throughput of light. Dispersive systems also require frequent wavenumber accuracy checks or calibration to ensure that the quality of the spectrum is acceptable. Pharmacopoeias have specific requirements for the standard of performance for instruments, which must be demonstrated to validate the measurement. For infra-red spectrometers, the performance is demonstrated through the measurement of a 35 μm thick polystyrene film. The sharpness of peaks and the wavenumber accuracy are both validated before the instrument can be used for pharmaceutical analysis.

To date, most infra-red spectrometers are based on the use of interferometers. The interferometer splits the broadband infra-red light beam from the source into two beams of equal intensity by means of a beam splitter. One of the beams is reflected off a fixed mirror, while the other is reflected off a scanning mirror before rejoining at the beam splitter, where the two beams interfere. The scanning mirror enables the distance travelled by the second beam to be precisely controlled so that the two beams can undergo different levels of interference depending on the wavelength of the light. A monochromatic HeNe laser beam is also shone into the interferometer where the interference pattern from the known-wavelength laser is used to measure the differences in distance between the two optical paths (the so-called optical path difference). The intensity measured at the position of the scanning mirror where the distances travelled by the two IR beams are equal is called the centre burst because, at this position, all wavelengths of light will be in constructive interference, resulting in the highest intensity. A plot of the scanning distance against the intensity of the IR light is called an interferogram. The intensity of light through the interferometer is the sum of all waves at different wavenumbers that created the interference pattern. Fourier transform (FT) is then applied to convert the interferogram into a single-beam spectrum where the intensity is plotted against wavenumber. Infra-red spectrometers that utilise an interferogram are called Fourier transform infra-red spectrometers (FTIR).

Advantages of FTIR over dispersive systems include the lower noise and faster measurement because a slit is no longer required for wavelength selection, leading to a much higher throughput of light. FTIR spectrometers also have the multiplex advantage, which means all wavelengths are measured at the same time in contrast to dispersive systems where different wavelengths of light are measured at different times. Finally, as the optical path difference is constantly measured against the interference pattern of the monochromatic laser, wavenumber accuracy is often much better and more consistent than a dispersive system.

13.3.2 Infra-red Sampling Protocol

There are three main measurement modes available for infra-red spectroscopic measurements. They are the reflection mode, transmission mode and attenuated total reflection (ATR) mode. Depending on the sample, different modes of measurement can be chosen, which determines the sample preparation method. Whichever mode of measurement is chosen, a suitable background spectrum is needed. A background spectrum is obtained from a measurement made without the sample. For many cases, it is the measurement of a plain substrate (without the sample) where the sample will be deposited. For examples, a clean IR transparent window or an empty sample chamber may be used for the transmission mode or a clean reflective slide or a gold mirror is used for the reflection mode and a clean ATR element is used for the ATR mode.

13.3.2.1 Reflection Measurements

In reflection measurements, the IR light is focused on the surface of the sample with the reflected light collected by an objective or mirrors followed by measuring the intensity at the detector. Path length through the sample is not always well defined in reflection modes and therefore quantification is not always possible. There are three main types of reflection, namely specular reflection, diffuse reflection, and transflection. Depending on the physical properties of the sample, the resultant spectra can be a combined effect of the three reflections. If the sample is thin and supported on a reflective substrate, the IR light can be transmitted through the sample, reflected off the substrate and then transmitted back through the sample before being measured. This results in the measurement being dominated by the transflection mode. If the sample is highly scattering, as typically with fine powders, the measurement will be dominated by the diffuse reflection mode. If the sample is thick, non-scattering and highly polished or the sample is a monolayer of molecule on a flat surface, the measurement will be dominated by the specular reflection mode.

Specular reflection occurs when the light interacts with the flat surface of the sample and is reflected but does not penetrate the sample (see Figure 13.3). In this case, the reflectance is dependent on the refractive index of the medium, n_1, and the sample, n_2, the angle of incidence and the polarisation of light. Equation (13.7) shows the reflectance, R, when the angle of incidence is 0°, which indicates that the reflectance measured is only dependent on the refractive index of the sample. A spectrum of refractive index of the sample against wavenumber is obtained. Thus,

$$R = \left(\frac{n_2 - n_1}{n_2 + n_1}\right)^2. \tag{13.7}$$

Specular reflection is useful in the measurement of the surface of a sample or monolayers as the light only interacts with the surface layer of the sample [4]. The refractive index spectrum can be used to calculate the absorption spectrum by the Kramers–Kronig transform, which is available in most advanced FTIR spectroscopic instrument software. Samples generally require some polishing (apart from monolayer samples) to ensure that a high specular reflectance is achieved and to remove the contribution from diffused scattering.

Transflection occurs when lights penetrate into a non-scattering sample on a highly reflective substrate (e.g. a metal surface) and then reflects off the substrate before emerging as reflected light through the sample

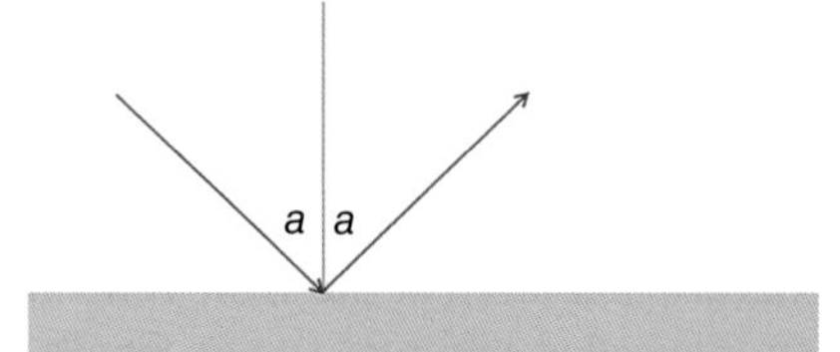

Figure 13.3 A schematic showing specular reflection of an infra-red beam shone on the surface of a sample at an angle of incidence *a*.

Figure 13.4 A schematic showing the transflection of an infra-red beam on a sample film deposited on a reflective substrate.

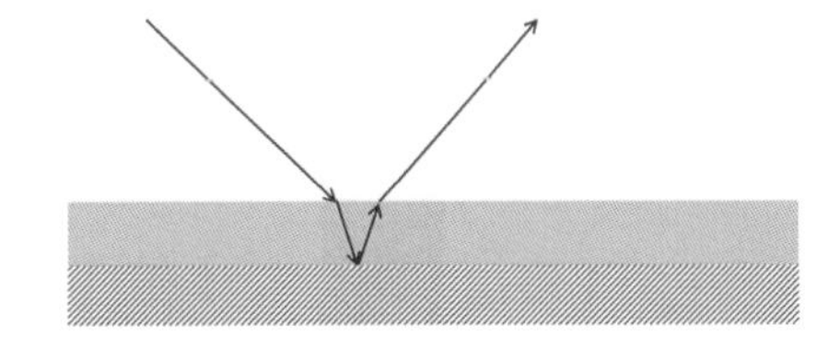

(Figure 13.4). The reflectance is dependent on the molar absorptivity of the sample with a path length of approximately twice the thickness of the sample. However, a recent study has shown that transflection measurements may not follow the Beer–Lambert law due to the electric field standing-wave effect [5]. Nevertheless, the method has gained popularity in the analysis of biological tissue sections due to the low cost of IR reflective low energy reflective glass slides [6]. As the IR light passes through the sample twice, the sample should be thin enough to avoid high absorbance. For tissue samples with strong protein amide bands, the ideal thickness of tissues should not be more than a few micrometres. Tissues can be embedded in an optimal cutting temperature (OCT) compound or wax block before being cryogenically microtomed to ~5 μm thick. Tissue sections are then directly deposited on the IR reflective substrates such as the low energy reflective glass side before being measured under a microscope in reflection mode. The OCT compound can be removed by washing the tissue in 0.9% saline followed by a quick rinse in distilled water and drying in air or nitrogen before taking FTIR measurements. Wax embedding medium can be removed by washing in hexane or measured together with the tissue and then mathematically de-waxed by spectral subtraction using spectroscopic software and algorithm [7].

In diffuse reflection, the light penetrates into the sample, is scattered and re-emerges from the surface as scattered light. Through this process, the direction of light is lost and the radiation is absorbed by the sample. The reflectance is dependent on the molar absorptivity of the sample and the scattering efficiency. Diffuse reflectance infra-red Fourier transform (DRIFT) is a popular technique that measures the diffused reflectance from powder samples. For example, DRIFT was used to detect melamine contamination in milk powder [8]. Powdered samples are often measured without preparation but would require an accessory specifically designed for DRIFT measurements. A DRIFT accessory typically contains a large parabolic mirror to focus the IR beam on to the powder sample and to collect the diffuse scattered light. For samples that have strong IR light absorption, samples in fine powder form can be diluted with KBr powder to lower the absorbance.

13.3.2.2 Transmission Measurements

Transmission mode measurements record the absorption of light through a sample, which can be a free-standing film or a film placed on an infra-red transparent substrate or a film sandwiched between two infra-red transparent windows. The thickness of the sample defines the path length of the measurement. Samples that are in powder form, such as many pharmaceutical ingredients, can also be measured in transmission mode. One of the classical techniques for powdered sample measurement is the nujol mull method. The sample is ground into a fine powder, mixed with a small amount of nujol (paraffin oil) to form a mull. The mull is spread between two NaCl plates (or other infra-red transparent plate) and then measured in transmission mode. Nujol, however, produces infra-red absorbance at ~3000–2800 and ~1460–1360 cm^{-1} for its CH stretching and bending vibrations, which can obscure the absorbance of the sample in these regions. For a measurement that is free from spectral interference, the powder sample can be made into a solid thin film by directly compacting between two diamond windows using a diamond cell or mixed with KBr powder (approximately 1 part of sample to 250 parts of KBr) before compacting into a KBr disc. The KBr disc can then be measured in transmission mode as a free-standing film or on an infra-red transparent substrate. The diamond press, Nujol mull and KBr disc methods are technically demanding and require practises for obtaining consistent, representative spectra.

FTIR imaging of biological tissues is often made in transmission mode and has shown to be a powerful method to obtain histological information without the need of staining [9]. These tissue samples can be

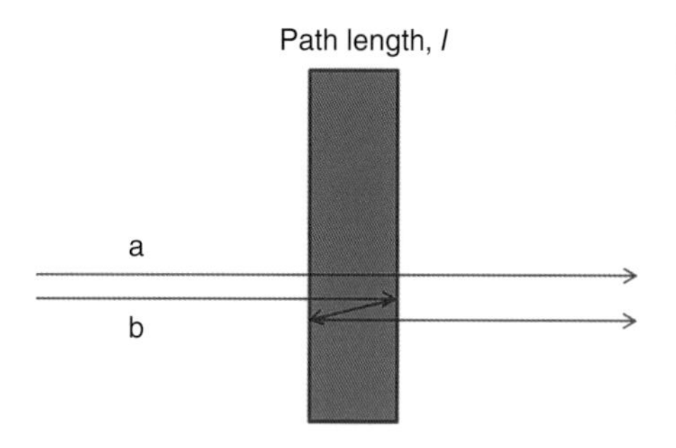

Figure 13.5 A schematic diagram showing beam a passing straight through the sample film and beam b internally reflected before exiting the sample film.

prepared by microtoming into thin sections, as is typical in histology. However, instead of depositing the tissue sections on glass slides, they are deposited on IR transparent substrates, such as CaF_2 or BaF_2 windows. Spectral contributions from the embedding medium can be removed by washing with 0.9% saline followed by washing in distilled water if an OCT compound was used or hexane if wax was used, or influence of these materials can be removed mathematically by spectral subtraction after the measurement taken during spectral pre-processing [7].

In transmission mode measurements, interference from the reflected light from the sample or the infra-red window can produce a fringe pattern on the baseline of the spectrum. As illustrated in Figure 13.5 using a free-standing film as an example, most of the IR light follows the path of beam 'a', which shows that the IR light passes straight through the free-standing sample film. However, a small percentage of the IR light will follow the path of beam 'b', which shows the light being reflected internally before emerging from the sample. When the wavelength of light is the same as the thickness of the film (path length l) the interference between 'a' and 'b' is constructive, but when the wavelength is double the thickness of the film, the interference is disruptive. Since the mid-IR spectrum wavelength span is from 2.5 to 25 μm, an alternation between the constructive and disruptive interference will result in a fringe pattern developed on the baseline of the measurement. The fringe pattern is weak compared to the absorbance spectrum when the refractive index of the sample, or the IR transparent window, is close to 1. However, when high refractive index materials are used, e.g. ZnSe windows, to sandwich a thin layer of sample, the fringe pattern can be significant. In this case, a wedge-shaped window may be used to reduce the unwanted fringe pattern.

While this is an artefact from the interference between the reflected and transmitted light, it can be a useful feature to determine the path length from the fringe pattern:

$$\text{Path length (mm)} = 10 \times \frac{\text{number of complete fringes}}{2 \times \text{wavenumber range (cm}^{-1}) \times \text{refractive index of film}}. \tag{13.8}$$

13.3.2.3 Attenuated Total Reflection (ATR)

The ATR mode is a highly versatile method that is suitable to measure solid and liquid samples with little sample preparation required. In this measurement mode, a high refractive infra-red transparent material (often referred as an ATR crystal if a crystal is used or more generally as an internal reflection element, IRE) is needed to generate an internal reflection at the sample–IRE interface. A common approach is to shine infra-red light to the IRE element from below towards the sampling surface at an incident angle that is greater than the critical angle, as illustrated in Figure 13.6. When the IR light is internally reflected, it penetrates into the sample by a few micrometres as an evanescence wave, as shown in Figure 13.6, which can be absorbed by the sample. The measurement of this absorption of light produces an ATR spectrum. The strength of the evanescent wave is strongest at the interface, which decays exponentially into the sample. The distance from the interface to where the evanescent wave decays to e^{-1} (~37%) is defined as the depth of penetration, d_p:

$$d_p = \frac{\lambda}{2\pi(n_1^2 \sin^2 \theta - n_2^2)}. \tag{13.9}$$

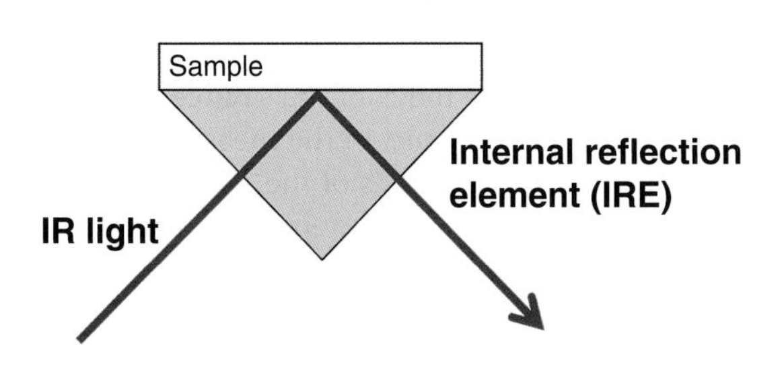

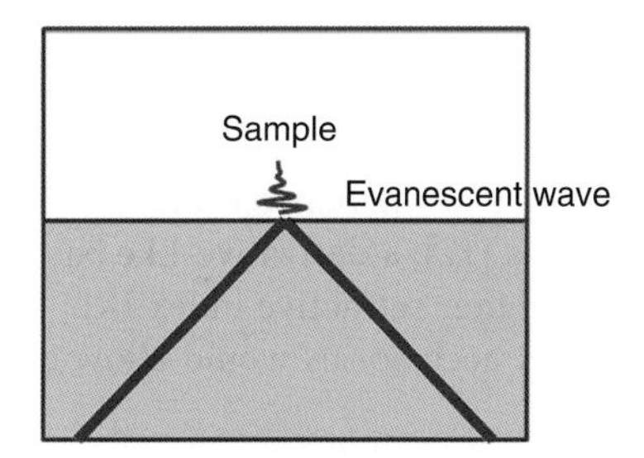

Figure 13.6 Schematic diagrams showing the ATR measurement mode. Left panel shows the internal reflection generated within the internal reflection element. Right panel is a close-up diagram showing the generation of the evanescent wave in the sample at the sample/IRE interface.

Equation (13.9) shows that the depth of penetration is a function of the wavelength of light, the refractive indices of the sample and the IRE and the angle of incidence, θ. These are the important parameters for the ATR measurement mode. For example, the different bands in a spectrum will have a different d_p depending on the wavelength of the band. Nevertheless, a typical ATR measurement produces a d_p of a few micrometres; therefore the ATR mode is often considered as a surface layer measurement technique. With the small value of d_p and the exponential decay in the strength of the evanescence wave, only the surface layer of the sample that is a few micrometres from the IRE is measured. A good contact between the IRE and the sample is important because an air gap as small as 1 μm between the sample and the IRE will result in a significant drop in the absorbance and invalidates the Beer–Lambert law. Many commercial ATR accessories have a pressure transducer sample press integrated above the measuring surface of the IRE so that a known reproducible pressure can be applied to compress the sample. Measurements of powder samples can be made without sample preparation by directly compressing a small amount of the powder on to the ATR element surface. FTIR measurements in the ATR mode were used in research to investigate the effect of composition in pharmaceutical formulation on the density of the compacted powder [10]. For the measurement of a liquid sample, intimate contact is achieved without the need of a press. However, a glass cover may be used to reduce evaporation of samples during measurement, especially when the liquid is volatile.

Although the Beer–Lambert law applies in ATR measurements, the path length is not as clear as in transmission measurements. The effective path length (also known as the equivalent path length) is the path length in transmission mode that would produce the same absorbance in an ATR mode measurement. It is also an important parameter to be considered for the design of an ATR experiment. The effective path length of an ATR measurement can be calculated using the following equations [11]:

$$\frac{de(S)}{\lambda} = \frac{\left({}^{n_2}/{}_{n_1}\right)\cos\theta}{\pi\left[1-\left({}^{n_2}/{}_{n_1}\right)^2\right]\left[\sin^2\theta-\left({}^{n_2}/{}_{n_1}\right)^2\right]^{0.5}}, \tag{13.10a}$$

$$\frac{de(P)}{\lambda} = \frac{\left({}^{n_2}/{}_{n_1}\right)\cos\theta\left[2\sin^2\theta-\left({}^{n_2}/{}_{n_1}\right)^2\right]}{\pi\left[1-\left({}^{n_2}/{}_{n_1}\right)^2\right]\left\{\left[1+\left({}^{n_2}/{}_{n_1}\right)^2\right]\sin^2\theta-\left({}^{n_2}/{}_{n_1}\right)^2\right\}\left[\sin^2\theta-\left({}^{n_2}/{}_{n_1}\right)^2\right]^{0.5}}, \tag{13.10b}$$

where $de(S)$ and $de(P)$ are the effective path lengths for the s- and p-polarised lights. For a non-polarised light, the effective path length will be the average of $de(S)$ and $de(P)$. It is important to note that the effective path length for ATR measurements does not have a physical reference of the distance the light travelled in the sample because light interacts with the sample through the evanescent wave. An important feature of an ATR spectrum is that the effective path length and the depth of penetration become smaller towards the higher wavenumber region. Another feature of an ATR spectrum is, from Eqs. (13.9), (13.10a) and (13.10b), that the effective path length and the depth of penetration are dependent on the refractive indices. This can lead to the spectral bands measured in the ATR mode being slightly red-shifted compared to the same spectrum measured in transmission mode. This is due to the dispersion of the refractive index of the sample near the absorption band. If the angle of incidence is close to the critical angle, for example, in the measurement of minerals enclosed in tissues [12], a derivative-like baseline near the absorption bands of the ATR spectrum can result. In this case, a higher refractive index IRE, such as germanium, which has a refractive index of 4, should be used. Some ATR accessories would allow the angle of incidence to be increased to avoid such a situation [13].

13.3.2.4 Multireflection ATR

The effective path length of a single reflection ATR measurement is typically below 2 μm, which is suitable for measuring highly absorbing substances such as water. However, for the measurement of weaker spectral bands or when the concentration of the solute of interest is low, increasing the path length would increase the absorbance to enable the detection and quantification of substances at lower concentrations [14]. In transmission mode, the path length can be adjusted by changing the thickness of the sample. In ATR mode, the path length can be increased by increasing the number of internal reflections on the sample using a multireflection (also called multibounce) ATR accessory. The schematic in Figure 13.7 shows a typical multireflection ATR accessory with a trapezoid ATR element. A recent study has shown that an optimised multireflection ATR measurement can improve the limit of detection from the mM level to approximately 20 μM in aqueous solution [14], significantly increasing the number of potential applications of attenuated total reflection Fourier transform infra-red (ATR-FTIR) studies, especially in biology where many compounds of interest are at low concentrations.

13.3.2.5 FTIR Microscope

Microscopes specifically designed for FTIR microscopy are commercially available. The optical components in these microscopes are specifically designed to be compatible with infra-red light because many of the materials used in a standard visible light microscope absorb in the mid-infra-red range. Aluminium or gold coated mirrors and all-reflective Schwarzschild objectives are used in an infra-red microscope to ensure the microscope can measure the full infra-red spectral range with minimal optical aberrations. The microscope objective is used to focus the IR beam to a relatively small spot. Optical apertures can then be applied to select the specific area or a feature of the sample to be measured. However, the minimum size of aperture that a microscope can operate to produce a spectrum of good signal-to-noise ratio is approximately ~300 μm^2 as the amount of light reaching the detector will reduce with the reducing size of the aperture. Using synchrotron infra-red radiation will increase the brightness of the IR light by a few orders of magnitude, allowing the aperture size to be reduced to 100 μm^2 or smaller. When an FPA detector is used in an IR microscope, diffraction limited imaging can be more easily achieved because no aperture is required. However, the size of the feature in a sample is ultimately limited by the diffraction of light (approximately 1.5λ with a standard objective of 0.4 numerical aperture, NA), which limits the size of the smallest feature that can be resolved by the infra-red imaging system.

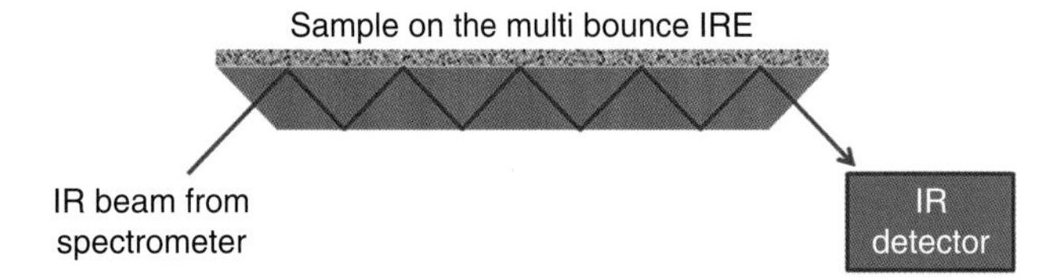

Figure 13.7 A schematic showing a nine reflections (five reflections on the sample side) IRE for ATR infra-red spectroscopy.

13.3.3 Raman Scattering

13.3.3.1 Spontaneous Raman Scattering

Spontaneous Raman scattering is a weak scattering process with signals approximately three orders of magnitude weaker than Rayleigh scattering. The intensity of the Raman scattering light, I_R, from a sample is related to the intensity, I_0, and the frequency, ν_0, of the illuminating light source through the following relationship:

$$I_R \propto I_0(\nu_0 \pm \nu_{vib})^4, \tag{13.11}$$

where ν_{vib} is the frequency of the molecular vibration. Using a high intensity light source can improve the Raman scattering signal. Most Raman spectrometers are therefore equipped with lasers as the monochromatic light source. However, increasing the intensity of illumination raises the risk of overheating of the sample. In some cases, samples can be burnt, vaporised or chemically degraded during measurement. Strategies to reduce the heat from the intense illumination include immersing the sample in water during measurement (using a water immersion objective) or spreading out the intensity of the laser from a spot to a line [15] or multiple spots by microlens array with fibre optics bundles [16]. The latter method also enables the collection of multiple spectra simultaneously, which can significantly increase the speed of data collection [17].

Some of the typical wavelengths used include 478, 514, 633, 785, 830 and 1064 nm. In the earlier studies, 785 and 830 nm lasers were the two most popular choices for biological specimen studies due to the lower risk of fluorescence and photo damage. For example, melanin can be measured with an 830 nm laser [18] but not with shorter wavelength lasers. However, longer wavelength lasers produce a weaker Raman signal when compared to shorter wavelength lasers and therefore require higher power and a longer acquisition time. To date, lasers of shorter wavelength are also used for biological applications when the sample does not fluoresce.

13.3.3.2 Resonance Raman Scattering

Another method to increase the Raman scattering intensity is by taking advantage of the resonance Raman effect. Resonance Raman scattering can be observed when the movement of an electron in an electronic transition is coupled to a vibrational transition. The Raman scattering signal for the associated vibration is greatly enhanced and the enhancement is greatest when the energy of the illuminating light source approaches the electronic transition energy. The signal can be enhanced by up to six to seven orders of magnitude higher than spontaneous Raman scattering, providing an increased sensitivity of the measurement. The requirement of the coupling between the vibrational states of the molecule to the electronic transition for the enhancement effect has also made resonance Raman scattering a highly selective method. It can be used to probe specific biomolecules such as nucleic acids and proteins and to deduce protein structure in a complex environment. For example, the spectral band of the haem moiety in a haemoglobin can be selectively enhanced using a 420 nm laser source while the Raman bands of tyrosine and tryptophan residues can be selectively enhanced using a 230 nm laser source [19]. However, the requirement of matching the laser frequency to the electronic transition energy limits the range of biological systems that can be studied by this method. Fluorescence at resonance excitation could overwhelm the resonance Raman signal. As the enhancement is increased when the energy of the light source approaches the energy of the electronic transition, tuneable lasers or broadband lasers (pulsed laser) can be used to avoid fluorescence by tuning the frequency of the laser to just below the resonance frequency. The flexibility offered by the tuneable laser also enables a greater range of biological systems to be studied by this method.

13.3.3.3 Non-linear Raman Scattering

Non-linear Raman scattering is a powerful approach that can significantly increase the Raman spectral signal to several orders of magnitude higher than a spontaneous Raman measurement. Because of the increased signal, a much higher speed of data acquisition can be achieved. These non-linear Raman scattering methods, including the stimulated Raman scattering (SRS) and coherent anti-Stokes Raman scattering (CARS), are multiphoton techniques based on a second- or third-order optical process [20, 21]. As a result, the SRS and CARS signal have, respectively, a second- or third-order relationship to the intensity of the illuminating light source and therefore a high intensity illumination at the sample must be achieved. For CARS, the photons must also reach the samples in a coherent manner. The non-linear relationship between the intensity of illumination and the signal ensures that CARS and SRS signals are only produced from the area of the tightly focused laser spot. The CARS and SRS signals decay rapidly outside the focus, which allows these techniques to scan in three dimensions to obtain three-dimensional images of the sample. The high intensity illumination can be provided by high power pico- or femtosecond pulse laser systems. Instrumentations for these types of measurement are more complex and expensive than a spontaneous Raman system. Further details of the theory of these techniques can be found elsewhere [20–24].

13.3.3.4 Surface Enhanced Raman Scattering

Surface enhanced Raman scattering (SERS) is a technique that has shown to significantly increase the Raman scattering signal of molecules that are in close proximity (a few nanometres) to metals with nanoscale structures. When the laser light strikes the surface of a metal with a nanoscale structure, the free surface electrons of the metal, called surface plasmons, can be excited if the frequency of the illuminating light matches the resonance frequency of the surface plasmons. These conditions can be met with nanoparticles of noble metals such as silver and gold, where the surface plasmons resonance frequencies are at approximately 400 and 550 nm, respectively. This will generate a large enhancement of the local electric field, leading to a six to seven orders of magnitude enhancement in the Raman signal from the molecules within that enhanced electric field. If the molecule in the enhanced electric field has an electronic excitation near the range of the surface plasmons resonance frequency, the SERS effect will be combined with the resonance Raman effect to produce the surface enhanced resonance Raman scattering with an overall 10^{14}-fold enhancement, allowing the detection of *single* molecules [25]. There are a large number of biomedical applications developed based on this technique, which are summarised in a recently published book [26].

13.3.4 Raman Spectrometer

A Raman spectrometer is designed to measure the intensity of the inelastic scattered light (i.e. the Raman signal) from a monochromatic laser. The measurement can be made using a dispersive system or an FT system. However, dispersive systems are more common than FT systems for Raman spectroscopy because dispersive Raman systems generally have a higher sensitivity than FT Raman systems. In a dispersive system, the scattered light collected from the sample is first passed through a filter (either a notch filter or an edge filter) that removes the laser line followed by shining the light on a diffraction grating to separate the wavelengths of the scatter light before being focused on the highly sensitive array detector (e.g. a Peltier cooled charge-coupled device [CCD] array detector). The different pixels on the array detector simultaneously measure the intensity of the Raman signal at different wavelengths from the dispersed light. The spectral resolution of the measurement is dependent on the type of grating used, the number of pixels the array detector has and the wavelength stability of the laser. A typical research Raman spectrometer would have a CCD camera containing a large number of pixels to measure the Raman signal at different wavelengths so that a full spectrum (e.g. from 3800 to 100 cm^{-1}) can be measured in one snapshot. However, changing the spectral resolution involves a change of hardware and alignments, which are mostly automated in modern systems. A Fourier transform Raman (FT-Raman) system utilises the same principles as the FTIR to separate the Raman intensity at different wavelengths. FT-Raman systems utilise a 1064 nm laser as the excitation light source such that the chance of sample fluorescence is minimised. However, the long excitation wavelength used in an FT-Raman system will produce a weaker Raman signal, according to Eq. (13.11), when compared to a dispersive system that uses a shorter excitation wavelength. To compensate for the weaker signal, a highly sensitive liquid nitrogen cooled germanium detector is used in an FT-Raman system and the measurements, generally, are slower than measurements made with a dispersive system.

As discussed above, the intensity from the Rayleigh scattering is much stronger than the weak Raman scattering and the laser line must be filtered before it reaches the detector. This can be done by either the use of an edge filter or a notch filter. The aim of these filters is to reduce the intensity of the Rayleigh line to a level that is similar or lower than the Raman lines so that the Rayleigh line does not overwhelm the signal of the Raman lines. Special filters have an extremely sharp cut-off transmittance profile so that Raman signals a few wavenumbers above the Rayleigh line can be measured are now available. This region of the spectrum, ~5–150 cm^{-1} is called the low frequency region (or far-IR spectrum if infra-red absorption is measured). The low frequency region can provide information on intermolecular forces. As a result, it is highly sensitive to the changes in the intermolecular structure and, therefore, is suitable to study transitions between amorphous and crystalline forms and changes in molecular mobilities in amorphous (glassy) materials. The study of pharmaceutical materials in the low frequency region has helped in the understanding of amorphous solid drug dispersion in a glassy matrix and the prediction of the stability of amorphous drug systems [27].

13.3.5 Raman Sampling Protocol

Most Raman measurements are acquired in backscattering mode, which means little sample preparation is required. A simple lens or objective can be used to focus the laser beam on to the sample, which can be a pharmaceutical tablet, drug powder, polymer blocks or films, tissue cross-sections or cells. The same objective or lens can be used to collect the backscattered light for detection. Raman spectra may be collected though containers or packages if the walls of the container or packages are transparent to the light in the range of the excitation laser used. Fibre optic probes are another well-developed technique for Raman spectral data collection. It is particularly useful for collecting spectra from a large object or remotely from a pipeline or reactor in a processing plant. In contrast to infra-red, fibre optics and probes in the visible range are widely available at a much lower cost. Some Raman instruments have fibre optics to transfer light between the spectrometer and the microscope with minimal loss in light intensity.

Raman spectra acquisitions generally do not require the collection of a background spectrum as it is a scattering technique rather than an absorption technique. However, some instruments offer the option of taking a background measurement, which is a measurement with the laser beam blocked or switched off. The subtraction of this background spectrum from the measurement can reduce the pixel to pixel variations along the detector array. More importantly, the detector used in a Raman spectrometer is highly sensitive to visible light. Stray light from the environment can produce spectral artefacts and increase background noise. The sample compartment and the collection optics of the instrument should be enclosed during data collection or the measurement should be collected in the dark.

As the scattered light is random in direction, a large NA objective is beneficial for maximising the amount of signal collected. A higher NA microscope objective with higher light collection efficiency will produce better Raman spectra than a lower NA objective. When a microscope with high magnification objectives is used to illuminate and collect the light, the measurement is called *Raman microspectroscopy* or simply *Raman microscopy*. If the microscope has a motorised stage or the mirror that directs the laser to the objective is motorised, Raman images can be obtained by rastering the sample point by point or line by line. The higher the magnification and the higher the NA the objective used, the smaller the laser spot will be generated and the better the lateral spatial resolution can be achieved. Since UV, visible or NIR lasers are used as the illuminating source, the spatial resolution obtained in *Raman mapping* is similar to UV, visible and NIR imaging. Note that Raman mapping may be referred as *Raman imaging* in the literature, especially when the speed of mapping is fast relative to the movement of the sample. For dynamic systems, such as living samples, the data acquisition rate will need to be sufficiently high so that there will be negligible movement of the sample within the imaged area for the duration of the imaging measurement. The lateral spatial resolution can be estimated using the Rayleigh criterion (Eq. (13.12)), where r is the distance between two adjacent points that are considered just resolved:

$$r = \frac{1.22\lambda}{2NA}. \tag{13.12}$$

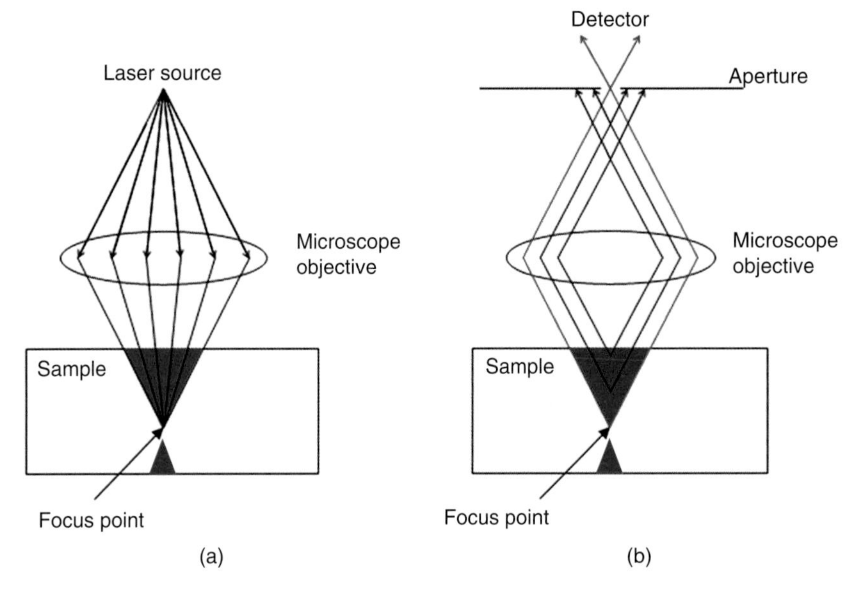

Figure 13.8 Schematic diagrams showing (a) the light is focused at a certain depth inside the sample. The red regions highlight the areas that are illuminated and can contribute to scattering. (b) shows that the backscattered light from the sample that is not from the focus point (black lines) are blocked by the pin hole (aperture) while the scattered light from the focus point is allowed to pass through.

In contrast to non-linear Raman such as SRS and CARS, the intensity of spontaneous Raman generally has a linear relationship with the intensity of the incident light so that the molecules in the volume above and below the focus can produce significant contributions to the overall Raman spectrum. This can result in a poor axial resolution for the measurement. To improve the axial resolution, a confocal microscope is used. A confocal microscope comprises a microscope objective with a pin-hole aperture. The pin-hole allows the scattered light from the focus of the illumination to pass thought to the detector while blocking light from other layers of the sample, as illustrated in Figure 13.8. Systems with fibre optics to transfer light between the microscope and the spectrometer are inherently confocal because the narrow entrance of the fibre optics produced the same effect as the pin-hole aperture. With the confocal setup, depth profiling (also called optical sectioning) of the sample can be achieved without cutting the sample, which is highly desirable for biological applications. One important point to note when performing depth profiling using the confocal mode is that an immersion objective should be used instead of a dry objective. A liquid medium (e.g. oil or water, depending on the type of immersion objective used) is added to the gap between the sample and the objective to reduce the refraction of light as it enters or exits the sample. Without the oil immersion objective, the actual sampling depth will be approximately twice the apparent depth and the axial resolution will be significantly degraded as a result of refraction of the laser at the sample–air interface [28].

When measuring a new sample with a Raman spectrometer or microscope, measurement parameters should be first optimised. The laser power should be first adjusted to a low setting (e.g. 1% power) to test if the sample will burn under this level of power. The laser power can then be slowly increased with the spectrum of the sample monitored over a period of exposure time to detect thermal degradation of the sample. Burnt organic samples can be detected by observing the appearance of the broad D and G bands of amorphous carbon at ~1350 and 1595 cm^{-1}, respectively. If a sample is sensitive to heat and is burnt by the laser, a lower laser power should be used. The highest laser power should only be used to maximise the Raman signal if the sample is stable. If the sample fluoresces, which produces an intense and broad baseline shift, the sample may be exposed in the laser while continuously acquiring spectra to observe if the fluorescence signal can be decreased by photo-bleaching. This procedure can be effective if the fluorescence signal originates from impurities or minor components in the sample that are not of interest in the study. However, if the fluorescence signal does not reduce by photo-bleaching, or if samples are sensitive to the strong illumination of light, as with many biological materials, a laser with a longer wavelength should be used instead. Sample heterogeneity can be an issue when high magnification objectives are used. In this case, measurements from several different areas of the sample or a hyperspectral image should be obtained to ensure that the spectra measured represent the bulk composition.

13.4 Applications

Infra-red and Raman spectroscopies are important techniques for pharmaceutical research and manufacturing quality control. For example, FTIR is often used as one of the key methods to confirm the identity of a pharmaceutical ingredient in a pharmacopoeia drug monograph. The identity confirmation test is much simplified when an FTIR measurement is included. In academic and industrial investigative studies, FTIR and Raman spectroscopic imaging are frequently applied to analyse the distribution, crystallinity and dissolution properties of pharmaceutical ingredients [29–33]. Increasingly they are also used in biological research, drug development and diagnosis. A summary of these applications can be found in a number of recent reviews [3, 34–38]. In this chapter, we will focus on two specific case studies of FTIR/Raman applications. One will be focused on the study of drug polymorph and amorphous systems, the other on the study of a living cell.

13.4.1 Case Study 1: Drug Polymorph Studies

Polymorphism in pharmaceutics is a term used to describe a drug solid that can exist in more than one crystalline form. A change in the crystalline structure can have pronounced effects on many physical properties of the drug that are important to the manufacturing process, such as particle shape and melting point, as well as the performance of the pharmaceutical product, such as the dissolution rate. The study of drug polymorphism is, therefore, an important area in pharmaceutical research and drug development. Vibrational spectroscopy has been an important analytical tool for this purpose. Although the chemical structure of the molecule remains the same in the different crystalline forms, changes in the molecular environment due to the different molecular arrangements could lead to observable differences in the vibration spectrum. Chemical bonds can be weakened or strengthened when intermolecular bonds are formed or broken. The changes in the bond strength, according to Eq. (13.3), will result in a shift in the wavenumber of the spectral band, allowing the different polymorphs of a drug to be distinguished and the functional group involved in the interaction to be identified. FTIR and Raman have become powerful tools in the development of pharmaceutical formulations and monitoring tools for process control [39, 40].

Polymorphic transition of crystalline materials, whether intentional or unintentional, can happen during manufacturing processing such as milling, granulation and extrusion. Accidental polymorphic conversion of a crystalline drug can affect the manufacturing process and results in out-of-specification products. Online monitoring using FTIR and Raman can be used to study changes in the polymorph in the manufacturing process to reduce problems associated with the changes in the physical properties of a drug. A recent study [41] has employed both in situ FTIR and in-line Raman to monitor the polymorphic transition of caffeine in a twin-screw melt granulator and in a rheometer. The results show that both Raman and FTIR have detected the conversion of caffeine from form II to I when heated in both the hot melt granulator and the rheometer, demonstrating that the rheometer can be used as a simulator for the hot melt granulator process studies.

While vibrational spectroscopy can often identify the different polymorphs, the limit of detection of these techniques is relatively poor compared to powder X-ray diffraction methods. For example, a study on the polymorphism of the drug imatinib has found that although the spectrum of α- and β-crystalline [42] could be easily distinguished by FTIR, the lack of a characteristic band of β-crystalline resulted in a rather poor limit of detection and, in comparison, powder X-ray diffraction was found to provide the best limit of detection for that particular system. Nevertheless, the advantages of vibrational spectroscopy over other analytical techniques, such as non-destructive measurements, non-invasiveness, low cost equipment and the technology can be adapted for online measurement, should be considered.

13.4.1.1 Amorphous Drug Studies

Molecules in amorphous solid drugs, in contrast to crystalline drugs, do not have long-range order. Instead, molecules have only short-range order, which can lead to profound differences in the physical properties of the drug. For example, due to the overall weaker molecular interactions for amorphous solids, the dissolution rate of an amorphous solid can be orders of magnitude higher than the crystalline form [43, 44]. This provides an opportunity to prepare drugs in amorphous form to harness the advantage of the improved dissolution rate. However, amorphous solids are thermodynamically unstable and they will eventually crystallise into the more stable forms. To date, difficulties in manufacturing process, poor amorphous stability and hard-to-predict crystallisation remain some of the key barriers for the wider use of amorphous drug in the pharmaceutical industry.

Since molecules in amorphous solids have only a short-range order, vibrational spectral bands of amorphous solids are significantly broader when compared to their crystalline counterparts. Raman and FTIR spectroscopy can be applied to distinguish amorphous form from their crystalline forms by the differences in the shape and position of the bands. For examples, amorphous and crystalline nifedipine were found to have distinct spectra (see Figure 13.9). An in situ Raman study of the crystallisation of amorphous nifedipine has shown that the drug first crystallised into the meta stable β-crystalline form before converting into the most stable α-crystalline form when exposed to high relative humidity and 40 °C (near the T_g) [45]. Raman mapping was also applied to detect trace crystalline griseofulvin and fenofibrate in solid dispersion prepared by hot-melt extrusion [46]. It was found that when combined with chemometric analysis, small drug crystals

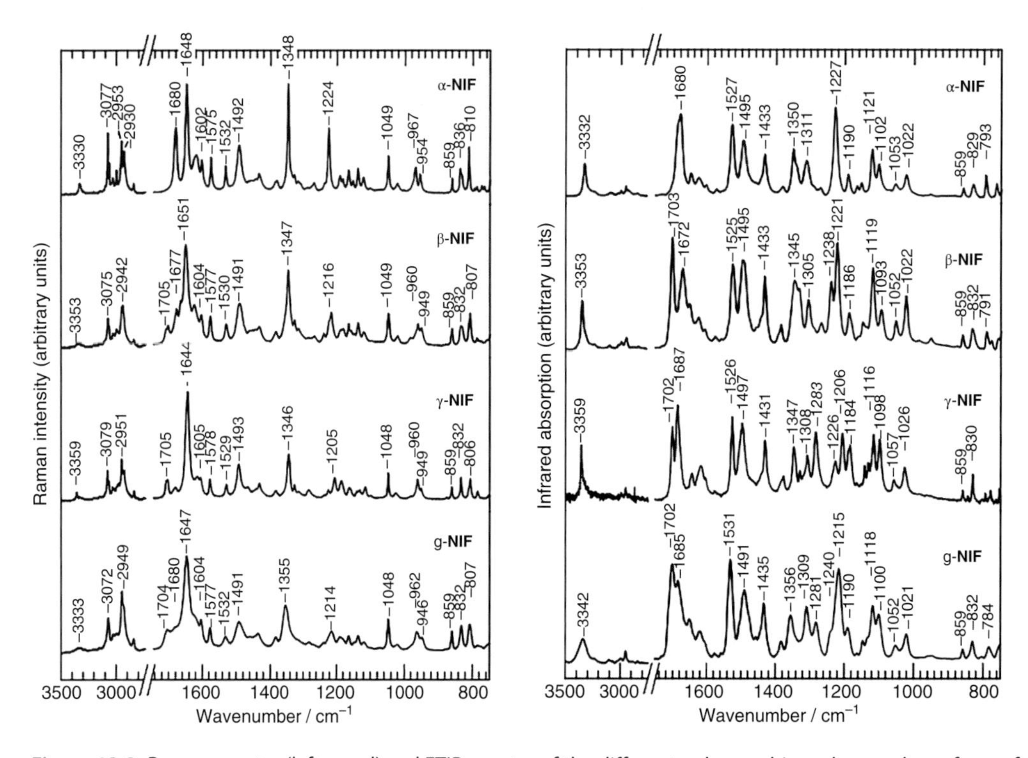

Figure 13.9 Raman spectra (left panel) and FTIR spectra of the different polymorphic and amorphous form of nifedipine. Source: Figure reproduced with permission from John Wiley & Sons reference [45].

a few micrometres in length can be detected within the amorphous formulation. In another study, the stabilities of co-amorphous nifedipine–nitrendipine systems at different drug ratios have been compared using FTIR imaging [47], where it has been shown that the 1 : 1 drug ratio has the best amorphous stability. Apart from detecting drug crystallisation in amorphous formulations, Rumondor et al. [48] have used FTIR to study the mechanism of amorphous drug recrystallisation in a polymer-drug amorphous solid solution system. Using the peak position of the N–H stretching mode vibration of nifedipine, they have deduced that the drug crystallised by first phase separating into amorphous drug-rich regions before crystallising. More recently, amorphous and crystalline nifedipine in different formulations of PEG and PVP were identified by FTIR spectroscopy, showing that vibrational spectroscopic methods are useful tools for studying recrystallisation of amorphous drugs [43, 49]. For further reading on the applications of infra-red and Raman spectroscopies on amorphous drugs, a comprehensive review on this topic can be found in [50].

13.4.2 Case Study 2: Live Cell Analysis

Cell-based assays form a large part of biological and medical research. However, bioanalyses of cells often require cell lysing or the addition of molecular probes that may disrupt the natural cellular environments or produce unwanted measurement artefacts. Fortunately, the majority of the cellular components, such as water, protein, lipids, carbohydrates, nucleic acids, are all directly detectable using vibrational spectroscopic techniques. The highly chemically specific, non-destructive and non-invasive nature of vibrational spectroscopic methods have become attractive analytical tools for live cell studies [34].

13.4.2.1 Live Cell Studies by Raman Spectroscopy

The desire to obtain high quality Raman spectra of biological samples, such as living cells, has motivated the development of highly efficient Raman spectrometers. In the early 1990s, most live cell studies employed long wavelength excitations, with, for example, 785 nm lasers, because they were considered less damaging to cells [51, 52]. Many early studies of live cells by Raman spectroscopy also involved customisation of the Raman microscope [15, 16, 53] with different substrates tested [54] to ensure the best results could be obtained [55]. Live cells are grown and measured on Raman-grade CaF_2 substrates, which are biocompatible [56] in order to minimise background signals that may interfere with the Raman signal from the living cells. To date, many modern research Raman microscopes are suitable for live cell measurements when a suitable live cell chamber is used without the restriction of the use of long-wavelength lasers. Measuring spectra of living cells with a good signal-to-noise ratio and short acquisition time allows the study of dynamic biochemical events. For example, a comparative Raman study of live and dead cells has shown a marked reduction in some of the Raman peaks associated with DNA (e.g. 782 cm^{-1}) and protein (e.g. 1004 cm^{-1}) [57], suggesting a breakdown of cellular components in the event of cell death. In 2004, Huang et al. discovered that a Raman spectral signature at ~1600 cm^{-1} reflects the mitochondrion metabolic activity [58], which has been called 'the Raman spectroscopic signature of life' [59]. In another study, an increase in lipid content was also observed in a time-course study of early cell death, showing that lipid condensation occurred within the cytosol [60]. Okada et al. have shown that Raman spectroscopy can be used to measure the oxidative state of cytochrome C from mitochondria in living cells in the study of the mechanism of apoptosis (controlled cell death) [61]. The work has shown that the intensity of the characteristic Raman bands of cytochrome C was decreased when oxidised. A reduction of these bands in cells as a result of apoptosis can be a result of cytochrome C oxidation, as reported in the previous work [61], or that cytochrome C was chemically degraded during apoptosis. While Okada's work was mainly based on the observation of individual peaks, Klein et al. have developed a barcode approach that can characterise the different compartments of a cell based on several peaks in the Raman spectra [62]. In that work, confocal Raman images were compared to immunofluorescence images of the same cell so that individual cellular components highlighted by the immunofluorescence images can be linked to a Raman spectrum collected from the same area. It was found that the Raman spectra collected from the individual cellular components have shown a set of characteristic peaks, from which a unique spectral barcode was generated to represent the individual cellular component [62].

Apart from cell death and cellular component characterisation, a drug-in-cell study was also an important area of research where Raman microscopy was shown to be a powerful tool. For example, in the development of anti-cancer agents, it is important to understand how the anti-cancer agent induces cell death so that a more targeted therapy may be developed. A Raman microscope can be used to image the location of the drug accumulated inside a cell, which can provide useful information to determine where the drug is active inside cells. For example, a study of paclitaxel (an anti-cancer drug that is thought to target tubulin) in cells using Raman mapping has shown that the drug was clustered in the cytoplasm region instead of microtubule bundles, which suggested that an alternative chemical pathway was taken, contrary to that commonly accepted [63]. More recently, Raman mapping was applied to study the cytotoxicity of nanoparticles [64, 65]. It was found that the nucleic acid band at 785 cm^{-1} and the RNA specific band at 810 cm^{-1} can be used as markers of oxidative stress for cells [65]. Another recent work has studied the effect of polyphenolic compounds in live cells. These compounds are thought to be beneficial to human health. However, in that study, when cells were exposed directly to these compounds in the cell culture, a change in the cytochrome C concentration and lipid condensation were observed, suggesting these compounds actually induce apoptosis [66].

To overcome the low signal from spontaneous Raman spectroscopic methods, surface enhanced Raman was found to be a powerful technique to increase the sensitivity and speed for live cell Raman imaging. For example, a study by Ock et al. employed surface-enhanced Raman to monitor anti-cancer drug release on gold nanoparticles with nanomolar range sensitivity [67]. Another study by Kang et al. [68] has shown that using gold nanoparticles with an intra-nanogap can significantly enhance the Raman signal, enabling the imaging of living cells at a rate of 10 Ms/pixel. Apart from surface-enhanced Raman methods, non-linear Raman spectroscopy is also increasingly applied to study living cells. For examples, CARS was used to study lipid hydrolysis in cells [69] and the effect of surfactants on the lipid components of living cells [70]. SRS is also highly suitable for live cell imaging because of the high sensitivity and imaging speed and the optical sectioning capability. For example, it has been applied to study protein degradation in live cells [71], choline metabolites [72] and uptake of tyrosine-kinase inhibitors in cells [73]. However, in contrast to spontaneous Raman, most of the stimulated Raman systems produce Raman images based on the measurement of a single wavenumber. Changing the spectral region of detection requires tuning of the laser. Nevertheless, it has been applied to detect polymeric nanoparticles in cells that are labelled with Raman active functional groups. These labelled nanoparticles produce Raman peaks at a spectral region with no or little background Raman signal from the cell, providing a useful tool in live cell labelling and cell sorting [74]. Further reading on the application of Raman spectroscopy on biological applications can be found in this recent critical review [38].

13.4.2.2 Live Cell Study by FTIR Spectroscopy

FTIR is a complementary analytical technique to Raman because functional groups that are not easily detected by Raman often produce strong IR absorption and vice versa. In contrast to Raman spectroscopy, FTIR does not suffer from fluorescence background issues, heating or photo-damage of the sample. However, there are two main challenges to overcome for FTIR measurement of living cells. First, water has a strong ν(O–H) band in the 3600–3000 cm^{-1} and δ(O–H) band at 1636 cm^{-1}. Unfortunately, the presence of water is essential if live cells are the sample to be measured because cells are cultured in aqueous medium and water is the main component (~70%) of a living cell. If water molecules are replaced by deuterated water, the position of the water bands can be shifted by approximately 30% (based on Eqs. (13.2a) and (13.2b)) because the atomic mass is doubled, allowing some of the spectral regions that were obscured by water bands to be analysed. However, replacement of water with deuterated water in cell cultures can disrupt the biology of the cell and often produce non-desirable results. To minimise the contribution of water from medium, special liquid cells designed to minimise the path length of the IR beam in the medium [75–77] or the ATR method, which can probe the attached cell without interference from the bulk medium, are used [78, 79]. When the ATR method is used, cells are directly grown on the measurement area of the IRE so that a good contact between cells and the IRE is naturally established. However, it was found that not all IREs are suitable for cell attachment. For example, germanium was found to be eroded by cells when they were grown for an extended period of time (>20 hours) [80]. The second challenge of measuring live cell with FTIR is that the relatively long wavelength of IR light limits the spatial resolution to approximately 10 μm. This prevents subcellular analysis because intracellular components are only a few micrometres in size or smaller. Thus many live cell studies were focused on the average signal from a population of cells [14, 77, 78, 81–83] or the whole single cell when an IR microscope was used [2, 76, 79, 84–86]. Recently, a number of different methods have been applied with the aim of improving the spatial resolution of FTIR images [79, 87–90]. It is expected that applications of high resolution FTIR imaging on live cell analysis will be more widely used in the near future.

Early studies of live cell with FTIR include the measurement of cell attachment [81] and cell death [79, 82, 91]. For example, in a study of starvation of leukemic monocytes using combined flow cytometry and FTIR microscopy [91], it was shown that after three days of incubation in a serum-deprived medium, the number of apoptotic cells increased compared to the control and there was good agreement between FTIR and flow cytometry data. Spectral changes between living cells and apoptotic cells were marked by the C–H stretching of vinyl moieties, carbonyl band of the phospholipid at 1745 cm^{-1} and the phosphate bands of the DNA and RNA at 1220 and 1241 cm^{-1}, highlighting the fact that detailed biochemical information can be extracted from the spectral data. Another study using multibounce ATR FTIR combined with MTT assay to analyse the effect of doxorubicin on three different cell lines has shown that FTIR can be used to assess drug resistance. FTIR data has also revealed major changes for the DNA bands at 1220, 1085, and 970 cm^{-1} of cells when treated with the drug, highlighting the DNA targeting mechanism of the drug [83].

13.5 Concluding Remarks

This chapter has discussed some of the basic theories, instrumentation and general considerations for FTIR and Raman spectroscopic measurements. The selected case studies revealed a small part of current applications of these technologies. It is not possible to cover all applications of these vibrational spectroscopic techniques in pharmaceutical, biological and medical science in a single chapter. Although the examples are not comprehensively described in this chapter, they demonstrate the potential and capabilities of these technologies. New applications, utilising powerful emerging technological advancements for FTIR and Raman spectroscopies, are continually being realised. It is expected that the development of these technologies will enable vibrational spectroscopic methods to be used across a wide range of industrial applications, medical applications and scientific research.

References

1 Salzer, R. and Siesler, H.W. (2014). *Infrared and Raman Spectroscopic Imaging, 2nd, Completely Revised and Updated Edition*, 2e. Wiley. 656 p.
2 Chan, K.L.A. and Kazarian, S.G. (2016). Attenuated total reflection Fourier-transform infrared (ATR-FTIR) imaging of tissues and live cells. *Chem. Soc. Rev.* 45 (7): 1850–1864.
3 Hermes, M., Morrish, R.B., Huot, L. et al. (2018). Mid-IR hyperspectral imaging for label-free histopathology and cytology. *J. Opt.* 20 (2): 023002.
4 Feliciano-Ramos, I., Caban-Acevedo, M., Scibioh, M.A., and Cabrera, C.R. (2010). Self-assembled monolayers of L-cysteine on palladium electrodes. *J. Electroanal. Chem.* 650 (1): 98–104.
5 Filik, J., Frogley, M.D., Pijanka, J.K. et al. (2012). Electric field standing wave artefacts in FTIR micro-spectroscopy of biological materials. *Analyst* 137 (4): 853–861.
6 Anderson, J., Dellomo, J., Sommer, A. et al. (2005). A concerted protocol for the analysis of mineral deposits in biopsied tissue using infrared microanalysis. *Urol. Res.* 33 (3): 213–219.

7 de Lima, F.A., Gobinet, C., Sockalingum, G. et al. (2017). Digital de-waxing on FTIR images. *Analyst* 142 (8): 1358–1370.

8 Mauer, L.J., Chernyshova, A.A., Hiatt, A. et al. (2009). Melamine detection in infant formula powder using near- and mid-infrared spectroscopy. *J. Agric. Food Chem.* 57 (10): 3974–3980.

9 Srinivasan, G. and Bhargava, R. (2007). Fourier transform-infrared spectroscopic imaging: the emerging evolution from a microscopy tool to a cancer imaging modality. *Spectroscopy* 22 (7): 30–43.

10 Elkhider, N., Chan, K.L.A., and Kazarian, S.G. (2007). Effect of moisture and pressure on tablet compaction studied with FTIR spectroscopic imaging. *J. Pharm. Sci.* 95 (2): 351–360.

11 Harrick, N.J. (1987). *Internal Reflection Spectroscopy*, 3e. New York: Harrick Scientific Corporation. 327 p.

12 Gulley-Stahl, H.J., Bledsoe, S.B., Evan, A.P., and Sommer, A.J. (2010). The advantages of an attenuated total internal reflection infrared microspectroscopic imaging approach for kidney biopsy analysis. *Appl. Spectrosc.* 64 (1): 15–22.

13 Frosch, T., Chan, K.L.A., Wong, H.C. et al. (2010). Nondestructive three-dimensional analysis of layered polymer structures with chemical imaging. *Langmuir* 26 (24): 19027–19032.

14 Chan, K.L.A. and Fale, P.L.V. (2014). Label-free in situ quantification of drug in living cells at micromolar levels using infrared spectroscopy. *Anal. Chem.* 86 (23): 11673–11679.

15 Hamada, K., Fujita, K., Smith, N.I. et al. (2008). Raman microscopy for dynamic molecular imaging of living cells. *J. Biomed. Opt.* 13 (4): 044027.

16 Okuno, M. and Hamaguchi, H. (2010). Multifocus confocal Raman microspectroscopy for fast multimode vibrational imaging of living cells. *Opt. Lett.* 35 (24): 4096–4098.

17 Qi, J., Li, J.T., and Shih, W.C. (2013). High-speed hyperspectral Raman imaging for label-free compositional microanalysis. *Biomed. Opt. Express* 4 (11): 2376–2382.

18 Feng, X., Moy, A.J., Nguyen, H.T.M. et al. (2017). Raman active components of skin cancer. *Biomed. Opt. Express* 8 (6): 2835–2850.

19 Ishita, M. (2012). *Resonance Raman Spectroscopy*. Chichester: Wiley.

20 Cheng, J.X. and Xie, X.S. (2004). Coherent anti-stokes Raman scattering microscopy: instrumentation, theory, and applications. *J. Phys. Chem. B* 108 (3): 827–840.

21 Freudiger, C.W., Min, W., Saar, B.G. et al. (2008). Label-free biomedical imaging with high sensitivity by stimulated Raman scattering microscopy. *Science* 322 (5909): 1857–1861.

22 Zumbusch, A., Langbein, W., and Borri, P. (2013). Nonlinear vibrational microscopy applied to lipid biology. *Prog. Lipid Res.* 52 (4): 615–632.

23 Muller, M. and Zumbusch, A. (2007). Coherent anti-Stokes Raman scattering microscopy. *ChemPhysChem* 8 (15): 2156–2170.

24 Ploetz, E., Marx, B., Klein, T. et al. (2009). A 75 MHz light source for femtosecond stimulated Raman microscopy. *Opt. Express* 17 (21): 18612–18620.

25 Kneipp, K., Wang, Y., Kneipp, H. et al. (1997). Single molecule detection using surface-enhanced Raman scattering (SERS). *Phys. Rev. Lett.* 78 (9): 1667–1670.

26 Schlucker, S. (ed.) (2011). *Surface Enhanced Raman Spectroscopy: Analytical, Biophysical and Life Science Applications*. Wiley-VCH Verlag GmbH & Co. KGaA.

27 Walker, G., Romann, P., Poller, B. et al. (2017). Probing pharmaceutical mixtures during milling: the potency of low-frequency Raman spectroscopy in identifying disorder. *Mol. Pharm.* 14 (12): 4675–4684.

28 Everall, N.J. (2009). Confocal Raman microscopy: performance, pitfalls, and best practice. *Appl. Spectrosc.* 63 (9): 245A–262A.

29 Smith, G.P.S., McGoverin, C.M., Fraser, S.J., and Gordon, K.C. (2015). Raman imaging of drug delivery systems. *Adv. Drug Deliv. Rev.* 89: 21–41.

30 Moriyama, K. (2016). Advanced applications of Raman imaging for deeper understanding and better quality control of formulations. *Curr. Pharm. Des.* 22 (32): 4912–4916.

31 Esmonde-White, K.A., Cuellar, M., Uerpmann, C. et al. (2017). Raman spectroscopy as a process analytical technology for pharmaceutical manufacturing and bioprocessing. *Anal. Bioanal. Chem.* 409 (3): 637–649.

32 Kazarian, S.G. and Ewing, A.V. (2013). Applications of Fourier transform infrared spectroscopic imaging to tablet dissolution and drug release. *Expert Opin. Drug Deliv.* 10 (9): 1207–1221.

33 Buckley, K. and Ryder, A.G. (2017). Applications of Raman spectroscopy in biopharmaceutical manufacturing: a short review. *Appl. Spectrosc.* 71 (6): 1085–1116.

34 Baker, M.J., Trevisan, J., Bassan, P. et al. (2014). Using Fourier transform IR spectroscopy to analyze biological materials. *Nat. Protoc.* 9 (8): 1771–1791.

35 Cialla-May, D., Zheng, X.S., Weber, K., and Popp, J. (2017). Recent progress in surface-enhanced Raman spectroscopy for biological and biomedical applications: from cells to clinics. *Chem. Soc. Rev.* 46 (13): 3945–3961.

36 Baker, M.J., Hussain, S.R., Lovergne, L. et al. (2016). Developing and understanding biofluid vibrational spectroscopy: a critical review. *Chem. Soc. Rev.* 45 (7): 1803–1818.

37 Kann, B., Offerhaus, H.L., Windbergs, M., and Otto, C. (2015). Raman microscopy for cellular investigations - from single cell imaging to drug carrier uptake visualization. *Adv. Drug Deliv. Rev.* 89: 71–90.

38 Krafft, C. and Popp, J. (2015). The many facets of Raman spectroscopy for biomedical analysis. *Anal. Bioanal. Chem.* 407 (3): 699–717.

39 Yu, L.X., Lionberger, R.A., Raw, A.S. et al. (2004). Applications of process analytical technology to crystallization processes. *Adv. Drug Deliv. Rev.* 56 (3): 349–369.

40 Gowen, A.A., O'Donnell, C.P., Cullen, P.J., and Bell, S.E.J. (2008). Recent applications of chemical imaging to pharmaceutical process monitoring and quality control. *Eur. J. Pharm. Biopharm.* 69 (1): 10–22.

41 Monteyne, T., Heeze, L., Oldorp, K. et al. (2016). Vibrational spectroscopy to support the link between rheology and continuous twin-screw melt granulation on molecular level: a case study. *Eur. J. Pharm. Biopharm.* 103: 127–135.

42 Atici, E.B. and Karliga, B. (2015). Quantitative determination of two polymorphic forms of imatinib mesylate in a drug substance and tablet formulation by X-ray powder diffraction, differential scanning calorimetry and attenuated total reflectance Fourier transform infrared spectroscopy. *J. Pharm. Biomed. Anal.* 114: 330–340.

43 Alqurshi, A., Chan, K.L.A., and Royall, P.G. (2017). In-situ freeze-drying - forming amorphous solids directly within capsules: an investigation of dissolution enhancement for a poorly soluble drug. *Sci. Rep.* 7: 2910.

44 Williams, H.D., Trevaskis, N.L., Charman, S.A. et al. (2013). Strategies to address low drug solubility in discovery and development. *Pharmacol. Rev.* 65 (1): 315–499.

45 Chan, K.L.A., Fleming, O.S., Kazarian, S.G. et al. (2004). Polymorphism and devitrification of nifedipine under controlled humidity: a combined FT-Raman, IR and Raman microscopic investigation. *J. Raman Spectrosc.* 35 (5): 353–359.

46 Widjaja, E., Kanaujia, P., Lau, G. et al. (2011). Detection of trace crystallinity in an amorphous system using Raman microscopy and chemometric analysis. *Eur. J. Pharm. Sci.* 42 (1–2): 45–54.

47 Chan, K.L.A., Kazarian, S.G., Vassou, D. et al. (2007). In situ high-throughput study of drug polymorphism under controlled temperature and humidity using FT-IR spectroscopic imaging. *Vib. Spectrosc.* 43 (1): 221–226.

48 Rumondor, A.C.F., Marsac, P.J., Stanford, L.A., and Taylor, L.S. (2009). Phase behavior of poly(vinylpyrrolidone) containing amorphous solid dispersions in the presence of moisture. *Mol. Pharm.* 6 (5): 1492–1505.

49 Iqbal, W.S. and Chan, K.L.A. (2015). FTIR spectroscopic study of poly(ethylene glycol)-nifedipine dispersion stability in different relative humidities. *J. Pharm. Sci.* 104 (1): 280–284.

50 Hedoux, A. (2016). Recent developments in the Raman and infrared investigations of amorphous pharmaceuticals and protein formulations: a review. *Adv. Drug Deliv. Rev.* 100: 133–146.

51 Puppels, G.J., Demul, F.F.M., Otto, C. et al. (1990). Studying single living cells and chromosomes by confocal Raman microspectroscopy. *Nature* 347 (6290): 301–303.

52 Notingher, I., Verrier, S., Romanska, H. et al. (2002). In situ characterisation of living cells by Raman spectroscopy. *J. Spectro.* 16 (2): 43–51.
53 Zoladek, A., Pascut, F., Patel, P., and Notingher, I. (2010). Development of Raman imaging system for time-course imaging of single living cells. *J. Spectro.* 24 (1–2): 131–136.
54 Draux, F., Jeannesson, P., Beljebbar, A. et al. (2009). Raman spectral imaging of single living cancer cells: a preliminary study. *Analyst* 134 (3): 542–548.
55 Chan, K.L.A. and Fale, P.L.V. (2015). Label-free optical imaging of live cells. In: *Biophotonics for Medical Applications* (ed. I. Meglinski), 215–241. Woodhead Publishing.
56 Wehbe, K., Filik, J., Frogley, M.D., and Cinque, G. (2013). The effect of optical substrates on micro-FTIR analysis of single mammalian cells. *Anal. Bioanal. Chem.* 405 (4): 1311–1324.
57 Notingher, I., Verrier, S., Haque, S. et al. (2003). Spectroscopic study of human lung epithelial cells (A549) in culture: living cells versus dead cells. *Biopolymers* 72 (4): 230–240.
58 Huang, Y.S., Karashima, T., Yamamoto, M. et al. (2004). Raman spectroscopic signature of life in a living yeast cell. *J. Raman Spectrosc.* 35 (7): 525–526.
59 Huang, Y.S., Nakatsuka, T., and Hamaguchi, H.O. (2007). Behaviors of the 'Raman spectroscopic signature of life' in single living fission yeast cells under different nutrient, stress, and atmospheric conditions. *Appl. Spectrosc.* 61 (12): 1290–1294.
60 Zoladek, A., Pascut, F.C., Patel, P., and Notingher, I. (2011). Non-invasive time-course imaging of apoptotic cells by confocal Raman micro-spectroscopy. *J. Raman Spectrosc.* 42 (3): 251–258.
61 Okada, M., Smith, N.I., Palonpon, A.F. et al. (2012). Label-free Raman observation of cytochrome c dynamics during apoptosis. *Proc. Natl Acad. Sci. USA* 109 (1): 28–32.
62 Klein, K., Gigler, A.M., Aschenbrenne, T. et al. (2012). Label-free live-cell imaging with confocal Raman microscopy. *Biophys. J.* 102 (2): 360–368.
63 Salehi, H., Derely, L., Vegh, A.G. et al. (2013). Label-free detection of anticancer drug paclitaxel in living cells by confocal Raman microscopy. *Appl. Phys. Lett.* 102 (11): 113701.
64 Efeoglu, E., Casey, A., and Byrne, H.J. (2017). Determination of spectral markers of cytotoxicity and genotoxicity using in vitro Raman microspectroscopy: cellular responses to polyamidoamine dendrimer exposure. *Analyst* 142 (20): 3848–3856.
65 Efeoglu, E., Maher, M.A., Casey, A., and Byrne, H.J. (2017). Label-free, high content screening using Raman microspectroscopy: the toxicological response of different cell lines to amine-modified polystyrene nanoparticles (PS-NH2). *Analyst* 142 (18): 3500–3513.
66 Mignolet, A., Wood, B.R., and Goormaghtigh, E. (2018). Intracellular investigation on the differential effects of 4 polyphenols on MCF-7 breast cancer cells by Raman imaging. *Analyst* 143 (1): 258–269.
67 Ock, K., Jeon, W.I., Ganbold, E.O. et al. (2012). Real-time monitoring of glutathione-triggered thiopurine anticancer drug release in live cells investigated by surface-enhanced Raman scattering. *Anal. Chem.* 84 (5): 2172–2178.
68 Kang, J.W., So, P.T.C., Dasari, R.R., and Lim, D.K. (2015). High resolution live cell Raman imaging using subcellular organelle-targeting SERS-sensitive gold nanoparticles with highly narrow intra-nanogap. *Nano Lett.* 15 (3): 1766–1772.
69 Chen, W.W., Chien, C.H., Wang, C.L. et al. (2013). Automated quantitative analysis of lipid accumulation and hydrolysis in living macrophages with label-free imaging. *Anal. Bioanal. Chem.* 405 (26): 8549–8559.
70 Okuno, M., Kano, H., Fujii, K. et al. (2014). Surfactant uptake dynamics in mammalian cells elucidated with quantitative coherent anti-stokes Raman scattering microspectroscopy. *PLoS One* 9 (4): e93401.
71 Shen, Y.H., Xu, F., Wei, L. et al. (2014). Live-cell quantitative imaging of proteome degradation by stimulated Raman scattering. *Angew. Chem. Int. Ed.* 53 (22): 5596–5599.
72 Hu, F.H., Wei, L., Zheng, C.G. et al. (2014). Live-cell vibrational imaging of choline metabolites by stimulated Raman scattering coupled with isotope-based metabolic labeling. *Analyst* 139 (10): 2312–2317.
73 Fu, D., Zhou, J., Zhu, W.S. et al. (2014). Imaging the intracellular distribution of tyrosine kinase inhibitors in living cells with quantitative hyperspectral stimulated Raman scattering. *Nat. Chem.* 6 (7): 615–623.

74 Hu, F.H., Brucks, S.D., Lambert, T.H. et al. (2017). Stimulated Raman scattering of polymer nanoparticles for multiplexed live-cell imaging. *Chem. Commun.* 53 (46): 6187–6190.
75 Marcsisin, E.J., Uttero, C.M., Miljkovic, M., and Diem, M. (2010). Infrared microspectroscopy of live cells in aqueous media. *Analyst* 135 (12): 3227–3232.
76 Birarda, G., Grenci, G., Businaro, L. et al. (2010). Infrared microspectroscopy of biochemical response of living cells in microfabricated devices. *Vib. Spectrosc.* 53 (1): 6–11.
77 Tobin, M.J., Puskar, L., Barber, R.L. et al. (2010). FTIR spectroscopy of single live cells in aqueous media by synchrotron IR microscopy using microfabricated sample holders. *Vib. Spectrosc.* 53 (1): 34–38.
78 Hutson, T.B., Mitchell, M.L., Keller, J.T. et al. (1988). A technique for monitoring mammalian-cell growth and inhibition insitu via Fourier-transform infrared-spectroscopy. *Anal. Biochem.* 174 (2): 415–422.
79 Kuimova, M.K., Chan, K.L.A., and Kazarian, S.G. (2009). Chemical imaging of live cancer cells in the natural aqueous environment. *Appl. Spectrosc.* 63 (2): 164–171.
80 Fale, P.L.V. and Chan, K.L.A. (2017). Preventing damage of germanium optical material in attenuated total reflection-Fourier transform infrared (ATR-FTIR) studies of living cells. *Vib. Spectrosc.* 91: 59–67.
81 Schmidt, M., Wolfram, T., Rumpler, M. et al. (2007). Live cell adhesion assay with attenuated total reflection infrared spectroscopy. *Biointerphases* 2 (1): 1–5.
82 Yamaguchi, R.T., Hirano-Iwata, A., Kimura, Y. et al. (2007). Real-time monitoring of cell death by surface infrared spectroscopy. *Appl. Phys. Lett.* 91 (20): 203902.
83 Fale, P.L.V., Altharawi, A., and Chan, K.L.A. (2015). In situ Fourier transform infrared analysis of live cells' response to doxorubicin. *BBA-Mol. Cell Res.* 1853 (10 Part A): 2640–2648.
84 Munro, K.L., Bambery, K.R., Carter, E.A. et al. (2010). Synchrotron radiation infrared microspectroscopy of arsenic-induced changes to intracellular biomolecules in live leukemia cells. *Vib. Spectrosc.* 53 (1): 39–44.
85 Quaroni, L. and Zlateva, T. (2011). Infrared spectromicroscopy of biochemistry in functional single cells. *Analyst* 136 (16): 3219–3232.
86 Doherty, J., Cinque, G., and Gardner, P. (2017). Single-cell analysis using Fourier transform infrared microspectroscopy. *Appl. Spectrosc. Rev.* 52 (6): 560–587.
87 Nasse, M.J., Walsh, M.J., Mattson, E.C. et al. (2011). High-resolution Fourier-transform infrared chemical imaging with multiple synchrotron beams. *Nat. Methods* 8 (5): 413–416.
88 Chan, K.L.A. and Kazarian, S.G. (2013). Aberration-free FTIR spectroscopic imaging of live cells in microfluidic devices. *Analyst* 138: 4040–4047.
89 Findlay, C.R., Wiens, R., Rak, M. et al. (2015). Rapid biodiagnostic ex vivo imaging at 1 mu m pixel resolution with thermal source FTIR FPA. *Analyst* 140 (7): 2493–2503.
90 Chan, K.L.A., Fale, P.L.V., Altharawi, A. et al. (2018). Subcellular mapping of living cells via synchrotron microFTIR and ZnS hemispheres. *Anal. Bioanal. Chem.* 410 (25): 6477–6487.
91 Birarda, G., Bedolla, D.E., Mitri, E. et al. (2014). Apoptotic pathways of U937 leukemic monocytes investigated by infrared microspectroscopy and flow cytometry. *Analyst* 139 (12): 3097–3106.

Further Reading

Salzer, R. and Siesler, H.W. (2014). *Infrared and Raman Spectroscopic Imaging, 2nd, Completely Revised and Updated Edition.* ISBN: 978-3-527-33652-4.
Sasic, S. and Ozaki, Y. (2011). *Raman, Infrared, and Near-Infrared Chemical Imaging.* Wiley.
Schlücker, S. (2011). *Surface Enhanced Raman Spectroscopy: Analytical, Biophysical and Life Science Applications.* Wiley-VCH Verlag GmbH & Co. KGaA. ISBN: 9783527325672, Online ISBN:9783527632756, https://doi.org/10.1002/9783527632756.
Siebert, F. and Hildebrandt, P. (2008). *Vibrational Spectroscopy in Life Science.* Wiley-VCH Verlag GmbH & Co. KGaA. ISBN: 9783527405060, Online ISBN: 9783527621347, https://doi.org/10.1002/9783527621347.

14

Fluorescence Spectroscopy and Its Applications in Analysing Biomolecular Processes

Nathan N. Alder

Department of Molecular and Cell Biology, University of Connecticut, Storrs, CT 06269, USA

14.1 Significance and Background

Luminescence is the process by which molecules emit light from electronically excited states that result from chemical reactions, mechanical forces or the absorption of *electromagnetic (EM) radiation*. *Photoluminescence* occurs when the excited state originates from the absorption of photons in the ultraviolet (UV) or visible region of the EM spectrum, a process called *photoexcitation*. The process of photoluminescence encompasses two phenomena, *fluorescence* and *phosphorescence*, which differ based on the electronic configuration of the excited state.

Fluorescence can be described as a series of photophysical events. First, a fluorescent molecule (*fluorophore*) in its ground state absorbs a photon and shifts to a high-energy excited electronic state on a picosecond (10^{-15} seconds) timescale. The fluorophore then occupies this excited state on the order of nanoseconds (10^{-9} seconds) before relaxing back to the ground state with the concurrent emission of a fluorescent photon. We can relativise this process to a more relatable timeframe: if the time for photon absorption and excitation were extended to one second, then the time required for the emission of fluorescent light would be several days. This comparison illustrates that during the fluorescence process, a fluorophore spends a significant amount of time in its excited state. During this time, the fluorophore has a chance to thoroughly interact with its microenvironment: its excited state energy can be modulated by the physicochemical features of the surrounding solvent, it can associate with neighbouring molecules by direct contact or through-space interactions and it can undergo Brownian tumbling based on factors that dictate its mobility. All of these processes in turn affect the fluorescence parameters that are measured experimentally. Hence, the excited state of the fluorophore serves as an experimental time window during which the fluorophore can 'collect' information that will be 'reported' to the observer in the form of fluorescent light.

Fluorescence spectroscopy, using both single-molecule and ensemble measurements, has many experimental applications, including the analysis of the interactions [1, 2] and conformational dynamics [3, 4] of macromolecules, as well as their locations when analysed in multiphase environments (e.g. containing both aqueous and membrane-bound compartments) [5]. By comparison with other experimental techniques, fluorescence-based approaches offer several practical benefits. The first is sensitivity. Depending on the efficiency with which a fluorescent probe absorbs and emits light, reproducible measurements can be made with fluorophores at concentrations in the nanomolar range or less. The second is that fluorescence measurements are made under equilibrium conditions. This allows for direct measurements of thermodynamic and binding parameters [6, 7]. Third, fluorescence measurements typically have a large linear dynamic range in which the measured signal is proportional to sample concentration. Fourth, fluorescence measurements can be made in real time, allowing the investigator to analyse the kinetics of a given process [8]. Fifth, fluorescence-based approaches are non-destructive and molecules can be analysed in their fully functional states. Sixth,

Biomolecular and Bioanalytical Techniques: Theory, Methodology and Applications, First Edition. Edited by Vasudevan Ramesh.

fluorescence measurements can be made in complex molecular environments, allowing fluorescently labelled molecules to reside in a native or native-like milieu during an experiment [9, 10]. A final related benefit is that of specificity. With the right choice of fluorescent probe, the behaviour of a single type of molecule can be investigated specifically, even in the presence of other molecules that may be fluorescent.

This chapter first covers the photophysical basis of fluorescence with a quantitative treatment of its associated molecular processes. It will then briefly cover practical aspects of instrumentation and fluorescent probes. Finally, it will present the theory and practice of some steady-state fluorescence-based techniques, each illustrated with sample protocols and simulated results. In each section, the reader will be referred to current studies from the recent literature to illustrate potential applications of the described approaches.

14.2 Theory and Principles

This section covers the fundamentals of the photophysical processes associated with fluorescence. For further details, the reader is referred to texts in the reference list [11–14].

14.2.1 Electromagnetic Radiation and Molecular Energy Levels

Fluorescence spectroscopy, like other forms of molecular spectroscopy, is based on the interaction of EM radiation with matter. EM radiation can be described both as a wave and as a stream of discrete particles called photons. A transverse EM wave consists of an oscillating electric ($\vec{E}$) and magnetic ($\vec{B}$) field, each in phase and oriented perpendicular to one another (Figure 14.1a). The sinusoidal $\vec{E}$ wave can be described by

$$c = \lambda \nu, \tag{14.1}$$

where c is the speed of light in a vacuum (3.00×10^8 m/s), λ is the *wavelength* (typically in nm), which defines the distance between successive peaks (alternatively reported as *wavenumber*, $\bar{\nu} = 1/\lambda$, in cm^{-1}), and ν is the *frequency* (in s^{-1} or Hz), which describes the number of wavefronts passing a fixed point in space per second. When considered as distinct particles, each photon of EM radiation contains a precise amount of energy termed a *quantum*. From Planck's law, this energy is

$$E_{photon} = h\nu = hc/\lambda = hc\bar{\nu}, \tag{14.2}$$

where h is the Planck constant (6.626×10^{-34} J s). The energy stored in EM radiation is therefore proportional to frequency (and wavenumber) and inversely proportional to wavelength. Fluorescence spectroscopy generally utilises EM radiation ranging from the UV through the visible regions of the EM spectrum (λ from ~250 to 700 nm, Figure 14.1b).

The phenomenon of fluorescence is based on transitions between quantised energy levels of fluorescent molecules (Figure 14.2a). Electronic energy levels (S_n) correspond to the potential energy of different electronic configurations, vibrational energy levels (V_n) pertain to oscillatory motions between bonds and rotational energy levels (R_n) correspond to molecular angular momentum. As shown in Figure 14.2a, electronic states can be depicted as plots of potential energy as a function of internuclear distance, following the formalism of the anharmonic Morse oscillator. Although such depictions strictly pertain to diatomic molecules, they can be used to schematically describe energy transitions of polyatomic molecules such as fluorophores as well. Each electronic state contains vibrational and rotational energy levels. As shown graphically, differences in energy between electronic states are large (on the order of 10^{-19} J), those between vibrational levels are smaller (about 10^{-20} J) and those between rotational states are even smaller (about 10^{-23} J). The fluorescence process is usually represented by a *Perrin–Jablonski diagram* (Figure 14.2b), which gives a simpler rendering of molecular energy levels. In such diagrams, electronic and vibrational levels are depicted horizontally (rotational levels are generally omitted), electronic transitions (light absorption or emission) are depicted as

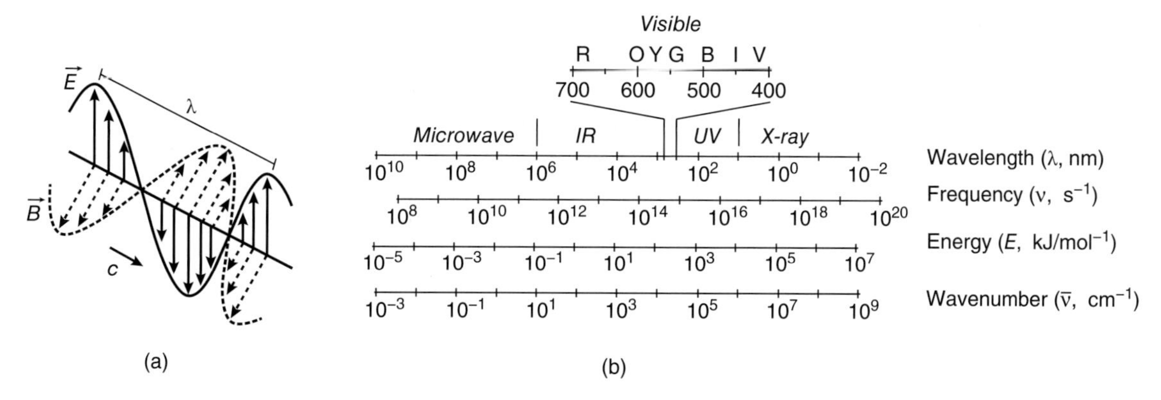

Figure 14.1 Light and the electromagnetic spectrum. (a) Light depicted as a transverse electromagnetic wave of perpendicular $\vec{E}$ and $\vec{B}$ fields characterised by wavelength (λ) and frequency (ν). (b) The electromagnetic spectrum characterised by four interrelated parameters.

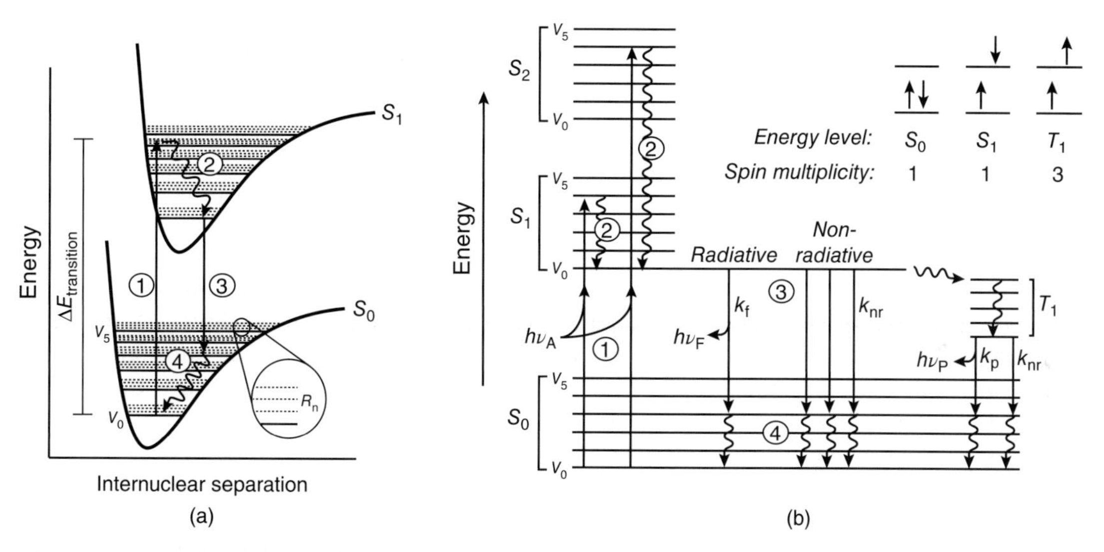

Figure 14.2 Energy level diagrams and the photophysical steps of fluorescence. Molecular energy level diagram (a) and Perrin–Jablonski diagram (b) depicting the radiative and non-radiative processes associated with fluorescence. See text for descriptions of steps 1 to 4.

vertical transitions due to their near-instantaneous nature and slower relaxation processes are depicted as squiggly lines. We will refer to such Perrin–Jablonski diagrams throughout this chapter.

When describing electronic transitions, one must also account for the *spin state* of the molecule (Figure 14.2b, inset). Electrons have two possible spin orientations ($s_i = +1/2$ or $-1/2$, represented by 'up' and 'down' arrows). Based on the Pauli Exclusion Principle, each orbital can contain two electrons with opposite spins, as depicted for valence electrons in the equilibrium (ground state). This results in a total spin ($S = \Sigma s_i$) of zero or a spin multiplicity ($M = 2S + 1$) of 1, resulting in a ground *singlet state* (S_0). Following excitation, an electron usually preserves its spin, resulting in an excited singlet state (S_1). If the electron undergoes spin conversion (e.g. from $s_i = -1/2$ to $+1/2$) following excitation, the total spin of the molecule will be 1 and the multiplicity will be 3, resulting in a *triplet state* (T_1). As discussed below, the conversion between S and T states has important implications for fluorescence spectroscopy.

Taken together, the molecular energy level diagrams of Figure 14.2 can be used to describe the photophysical processes of fluorescence, summarised as four steps: (1) the fluorophore absorbs a photon and becomes promoted from the ground (fundamental) electronic state (S_0) to a vibrational energy level of an excited electronic state ($S_{n>0}$); (2) the fluorphore relaxes to the lowest vibrational energy level of the first excited singlet state (S_1); (3) the fluorophore decays to a vibrational energy level of the ground electronic state, which may be associated with the emission of a fluorescent photon; and (4) the fluorophore relaxes back to a lower vibrational energy level of the ground state. The following section covers steps 1 to 4 in detail.

14.2.2 The Fluorescence Sequence

14.2.2.1 Step 1: Light Absorption

Any molecule that absorbs light is termed a *chromophore*. Under typical thermal conditions, a molecule at equilibrium will exist in its electronic ground state (S_0) in different rotational states of the lowest accessible vibrational level(s). Absorption of a photon ($h\nu_A$) by a ground state molecule occurs by the interaction of the electric field vector of the EM radiation with outermost (valence) electrons of specific molecular orbitals in the molecule, which promotes it to an excited electronic state. This entails the excitation of an electron from a ground state bonding or non-bonding orbital to a vacant orbital of higher energy (e.g. an anti-bonding orbital).

For absorption to occur, the energy of the photon (E_{photon}) must be equal to the energy difference between the ground state and an accessible excited state ($\Delta E_{\text{transition}}$). For instance, based on Eq. (14.2), absorbance of a photon from spectrally pure violet light with wavelength $\lambda = 420$ nm would promote a transition from the ground state to an excited state with an energy difference of 4.73×10^{-19} J (~280 kJ/mol), an energy consistent with electronic quantum transitions. For a given wavelength (and solvent), the efficiency with which a chromophore will absorb light is given by the Beer–Lambert law

$$A(\lambda) = \log\left(\frac{I_0}{I}\right) \varepsilon(\lambda)cl, \tag{14.3}$$

where $A(\lambda)$ is the *absorbance*, defined from the ratio of incident (I_0) to transmitted (I) light; $\varepsilon(\lambda)$ is the *molar absorption coefficient* (in M^{-1} cm^{-1}), which quantifies the probability of a chromophore absorbing light of a given wavelength under specified conditions; c is concentration (in M); and l is the light pathlength for measurement (in cm). Given the number of possible energetic transitions due to vibrational and rotational states among the electronic energy levels, molecules can absorb light over a range of wavelengths. This can give rise to absorbance spectra that are broadened, asymmetric and/or contain some fine structure.

Absorption of a photon ($h\nu_A$), a 'vertical' transition, occurs extraordinarily quickly, on a femtosecond (10^{-15} s) timescale. Attendant with the transition from the ground to an excited state is a redistribution of the cloud of electrons within the molecule that gives rise to an excited state dipole moment. However, as dictated by the Franck–Condon Principle, the molecule reaches the excited state before significant changes can occur in its own nuclear coordinates or in the orientations of surrounding solvent molecules. Hence, immediately upon absorption, the molecule is in the *Franck–Condon excited state*, having an altered electronic structure out of equilibrium with its surroundings.

14.2.2.2 Step 2: Relaxation to the Equilibrium Excited State

A molecule in the Franck–Condon excited state is inherently unstable. Due to instantaneous changes in electron distribution, properties such as the molecular dipole moment strongly alter the interaction of the fluorophore with its neighbouring molecules. Through the process of *vibrational relaxation*, the excited state molecule undergoes multiple vibration cycles, transferring excess energy as heat to surrounding molecules that collide with it. The fluorophore therefore decreases in energy through accessible vibrational modes. Moreover, because a fluorophore in the Franck–Condon excited state has an altered interaction with its solvent, the relaxation process often entails an adjustment of the interactions and orientations between the fluorophore and local solvent molecules. If the deactivation of excess vibrational energy involves transfer through different electronic energy states of the same spin multiplicity, this process is also referred to as *internal conversion*. The end result is that the fluorophore relaxes to the lowest vibrational energy state of the first excited electronic level (S_1) within a picosecond (10 ^{12} s) timescale. In this equilibrium excited state, the molecule has adjusted its interactions with neighbouring molecules to assume a more stable condition.

14.2.2.3 Step 3: De-excitation of the Excited State and Fluorescent Photon Emission

The excited molecule resides in the lowest excited singlet state S_1 on the order of nanoseconds (10^{-9} s), making this state the longest lived of all steps in the fluorescence process. The decay of the fluorophore back to one of the many vibrational levels of S_0 with the concurrent emission of a fluorescent photon is the origin of fluorescence. However, returning to the ground state can also proceed by a number of alternative non-radiative routes in which a photon is not emitted (Figure 14.3a). Each de-excitation pathway has an associated rate constant (k), which describes the rate at which the excited state is depopulated by that particular process. We consider the following de-excitation modes:

(i) *Fluorescence* (k_f): Relaxation from S_1 to S_0 with the emission of a photon ($h\nu$) is the formal definition of fluorescence. It should be noted that although the fluorophore resides in the equilibrium excited state for an extended time, the emission of a fluorescent photon itself occurs rapidly, on the same timescale as photon absorption, $\sim 10^{-15}$ s (emission is also a 'vertical' transition).

(ii) *Internal conversion* (k_{IC}): The excited fluorophore can return to the ground state by internal conversion, as described above. However, the efficiency with which this process occurs between electronic states is inversely related to the energy gap between the states. For most fluorophores, the energy difference between S_2 and S_1 is much smaller than the difference between S_1 and S_0; therefore, the rate at which internal conversion mediates the decay among higher electronic energy levels is generally much higher than the rate at which it depopulates the equilibrium excited state of S_1.

(iii) *Intermolecular processes* (k_{IM}): The excited state can be depopulated by chemical reactions or interactions with nearby molecules. We will later cover two of these processes, *fluorescence quenching* (Section 14.3.4) and *resonance energy transfer* (Section 14.3.5).

(iv) *Intersystem crossing* (k_{ISC}) *and phosphorescence* (k_p): The transition between electronic states of different multiplicities involves the spin reversal of the excited electron (see Section 14.2.1), resulting in *intersystem crossing* between the S and T states. Although this spin transition is in principle forbidden by the rules of quantum mechanics, it can be promoted by a process termed *spin–orbit coupling* (coupling between spin and orbital magnetic moments). Further, intersystem crossing is most probable between vibrational energy levels of the S and T states that are equal in energy (isoenergetic), or nearly

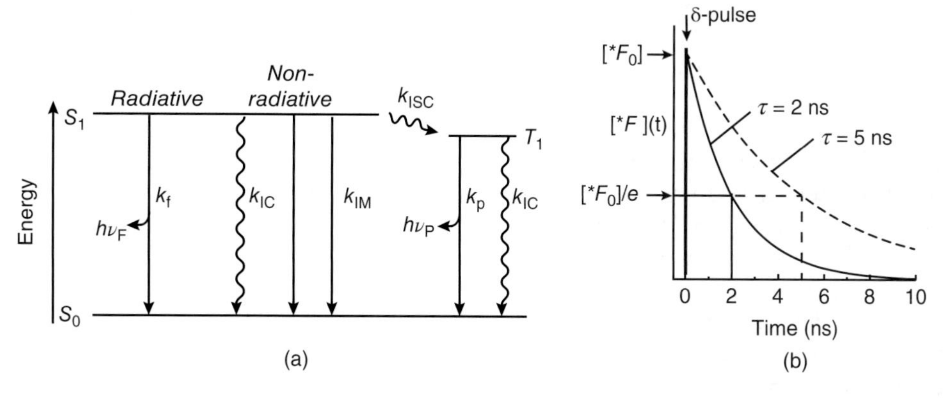

Figure 14.3 Pathways of excited state decay and fluorescence lifetime. (a) Simplified Perrin–Jablonski diagram showing potential de-excitation routes from the first excited singlet (S_1) and triplet (T_1) states. (b) Time-dependent exponential decay of fluorescence for a fluorophore with a shorter ($\tau = 2$ ns) and a longer ($\tau = 5$ ns) lifetime.

so. Following intersystem crossing, a molecule can relax to the lowest vibrational level of T_1, followed by the decay to S_0 (which, again, requires electron spin reversal). Return to the singlet ground state can occur via intersystem crossing or with the emission of a photon, which in the case of the T_1 to S_0 conversion is termed *phosphorescence*. Given the forbidden nature of these conversions, the k_p rate constant is usually low.

The process by which these many events can depopulate an excited state can be modelled within the framework of classical kinetics for branched pathways. To model the depopulation of singlet excited state S_1, we will consider the rate of fluorescence emission (k_f) along with a rate constant that includes all non-radiative pathways, $k_{nr} = k_{IC} + k_{IM} + k_{ISC}$. The process of de-excitation can be visualised as shown in Figure 14.3b. Imagine a population of fluorophore F subjected to an infinitely short (δ function) pulse of light. Some fraction of F will absorb the light and reach the excited state ($^*F$) so that the initial concentration of excited fluorophores is $[^*F_0]$. The rate law for the disappearance of $^*F$ can be modelled as an exponential decay:

$$-\frac{d[^*F]}{dt} = (k_r + k_{nr})[^*F]. \tag{14.4}$$

Integration of this differential equation gives the time evolution of the concentration of molecules in the excited state:

$$[^*F] = [^*F_0]e^{-(k_r+k_{nr})t}. \tag{14.5}$$

As an alternative measure of the amount of time a fluorophore remains in the excited state, we can define the *lifetime* (τ) of the excited state as the reciprocal of the rate constants:

$$\tau = \frac{1}{(k_r + k_{nr})}. \tag{14.6}$$

We can therefore recast Eq. (14.5) as

$$[^*F] = [^*F_0]e^{-(t/\tau)}. \tag{14.7}$$

From these definitions, we see that τ is the time required for the excited state to decay to $1/e$ of its original value (when $t = \tau$, $[^*F] = [^*F_0]/e$).

It is important to emphasise here that de-excitation is a stochastic process by which the S_1 state is spontaneously depopulated by a number of parallel pathways, both radiative and non-radiative. Fluorescence emission therefore 'competes' with other processes in a manner dependent on the relative rate constants of each pathway. Therefore, the amount of fluorescence that is measured in the time decay obtained in the experiment of Figure 14.3b would be proportional to our calculated value of $[^*F]$, but depend on the rates of other non-radiative de-excitation paths. Most fluorophores have lifetime values ranging from tens of picoseconds to tens of nanoseconds. In the absence of all non-radiative pathways, we define the radiative (or natural) lifetime as

$$\tau_N = \frac{1}{k_r}. \tag{14.8}$$

The fluorescence *quantum yield* (Φ) is the ratio of the number of photons emitted to the number of photons absorbed (i.e. the fraction of photons that return to S_0 from S_1 by emission of a fluorescent photon):

$$\Phi = \frac{k_r}{(k_r + k_{nr})} = \frac{\tau}{\tau_N} = k_r\tau. \tag{14.9}$$

Values of Φ can in principle range from 0 (a non-fluorescent chromophore) to a limiting value of 1 (a 'perfectly efficient' fluorophore).

14.2.2.4 Step 4: Vibrational Relaxation to the Ground State

Following the return to S_0, the molecule undergoes relaxation to the lowest vibrational energy levels to reach the equilibrium ground state.

14.2.3 Information Content of Fluorescence Excitation/Emission Spectra

In relation to the previous section, two fundamental types of information can be gleaned from spectral analysis of fluorophores. First, the wavelengths associated with absorbance/excitation and with emission (λ_{ex} and λ_{em}, respectively) are related to the energies of the $S_0 \rightarrow S_n$ and the $S_1 \rightarrow S_0$ transitions, respectively. From Figure 14.2 it is clear that the wavelengths of absorption will be lower (higher in energy) than those of emission (lower in energy). Second, the intensity of the emitted light will be a direct outcome of the probability of the fluorophore absorbing light of a particular wavelength (reflected in $\varepsilon(\lambda)$) and the relative efficiencies of radiative versus non-radiative paths for de-excitation (reflected in Φ). Fluorescence spectroscopy often involves measurements of emission intensity over a range of wavelengths of exciting and/or emitting light. Such measurements yield spectra that can provide a great deal of information about the fluorophore and its environment or can render spectral characteristics of an unknown sample (i.e. its fluorescence 'fingerprint').

Consider the spectra of a hypothetical fluorophore shown in Figure 14.4a, which includes an absorption/excitation spectrum (dashed line) and an emission spectrum (solid line) corresponding to transitions between ground (S_0) and excited (S_1) electronic states. An *absorbance spectrum*, measured in a spectrophotometer, shows the degree to which light at different wavelengths is absorbed by the chromophore (Eq. (14.3)). In this example, one might measure light absorbance from $\sim \lambda = 295$ to 450 nm. Four distinct transitions (upward arrows) are delineated as discrete *absorption bands* relating to the excitation of the molecule from the ground state (V_0 of S_0) to the different vibrational energy levels of the excited state (V_0 to V_3 of S_1). These transitions coincide with photon energies that match the $\Delta E_{transition}$ between ground and excited states

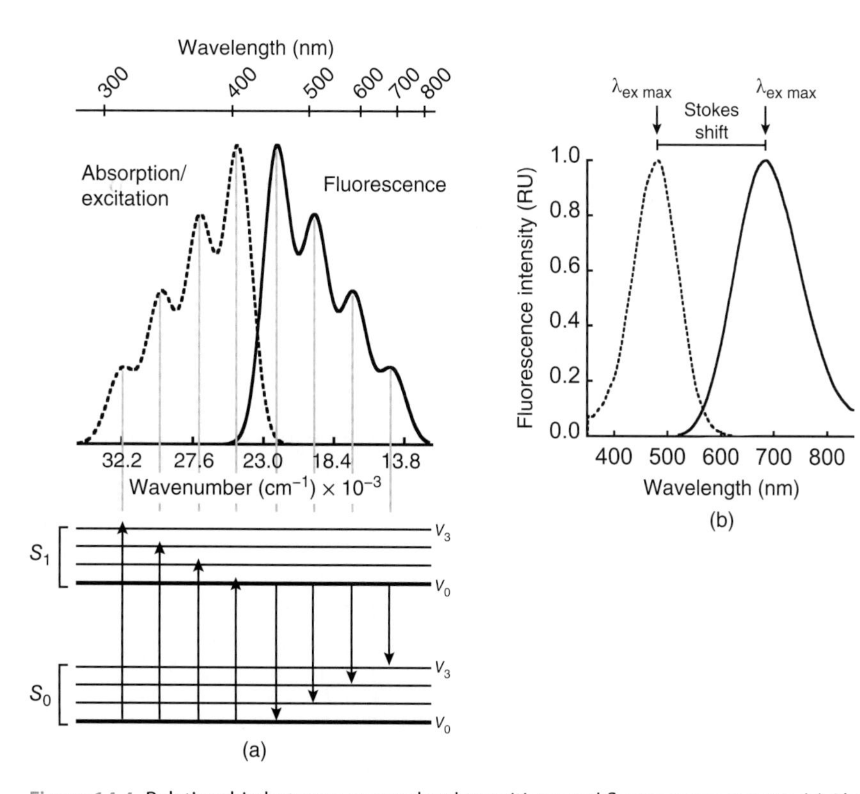

Figure 14.4 Relationship between energy level transitions and fluorescence spectra. (a) Absorption and fluorescence emission among the vibrational energy levels of S_0 and S_1 (below) are manifest as absorption/excitation and emission spectra (above). Note that the energy of transitions is a linear function of wavenumber ($\overline{\nu}$), not wavelength (λ). (b) Normalised excitation (dashed line) and emission (solid line) spectra for Di-4-ANEPPS.

(Eq. (14.2)). The density of rotational energy levels (not shown) within each vibrational energy level expands the number of energetically accessible transitions; hence, absorption spectra in reality appear as broadened distributions rather than discrete peaks. Moreover, the height of each peak is based on the relative efficiency of absorption, dictated by the value of $\varepsilon(\lambda)$ at each particular wavelength.

By comparison with an absorbance spectrum, which measures the amount of light absorbed, an *excitation spectrum* measures the amount of fluorescence that results from light absorption. Such spectra are measured in a fluorometer by exposing the sample to light with a range of excitation wavelengths (λ_{ex}) and reading the resulting total fluorescence intensity at a single wavelength (λ_{em}, generally the wavelength at maximal emission). In the example of Figure 14.4a, one might therefore excite the fluorophore over a range of wavelengths (say, λ_{ex} = 290–440 nm) and read the fluorescence emission at $\sim\lambda_{em}$ = 460 nm. The important point to note here is that for a given fluorophore, an excitation spectrum will generally have the same shape as its absorbance spectrum. To understand why this is so, consider the Perrin–Jablonski diagram (Figure 14.2b). Following excitation, the fluorophore will relax to the lowest vibrational energy level of S_1 prior to the emission of a fluorescent photon, regardless of the energy level to which it was initially excited (*Kasha's rule*). This is an outcome of the rapid rate of vibrational relaxation and internal conversion (picosecond timescale) relative to fluorescence emission (nanosecond timescale). Hence, in most cases, the fluorescence emission wavelength is independent of the wavelength of exciting light and the excitation spectrum can be considered an indirect readout of the absorption spectrum.

Now consider an *emission spectrum*, which is measured in a fluorometer by exposing the sample to light at a single λ_{ex} and reading the resulting fluorescence intensity over a range of λ_{em} values. Because fluorescence emission almost always occurs from the lowest vibrational level of S_1, the emission spectrum is largely independent of λ_{ex}. Moreover, because the fluorophore can relax to any vibrational level of S_0, and vibrational level spacing is largely similar between S_0 and S_1, excitation and emission spectra often appear as inverse images (*the mirror-image rule*). Finally, the difference in wavelengths between excitation and emission spectra is known as the *Stokes shift*, quantified as the difference in the wavelength of maximum excitation ($\lambda_{ex\ max}$) and that of maximum emission ($\lambda_{em\ max}$). Consider the excitation and emission spectra for an actual fluorophore, Di-4-ANEPPS, whose Stokes shift is quite large (~200 nm) (Figure 14.4b). Note that for actual fluorophores in solution at room temperature, vibrational structure is lost and spectra are broadened.

14.2.4 Solvent Effects

The polarity of solvent molecules (related to the solvent dielectric) in the vicinity of a fluorophore can significantly impact its spectral properties [16] (Figure 14.5a). In the ground state, polar solvating molecules around

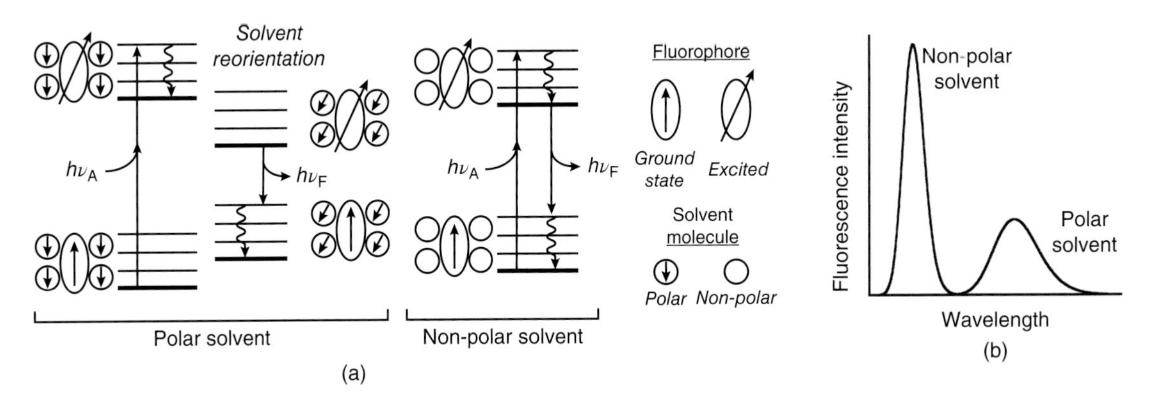

Figure 14.5 Solvent effects on fluorescence. (a) Energy level diagrams for solvent-dependent effects on relative energies of S_0 and S_1 states for a polar fluorophore. (b) Simulated emission spectra for a polarity-sensitive fluorophore in a polar and non-polar solvent.

a polar fluorophore orient optimally with the molecular dipole moment (Figure 14.5a, left). Upon excitation, fluorophores generally have an altered electronic distribution that increases the dipole moment and changes its orientation. In the Frank–Condon excited state, electronic redistribution occurs before solvent molecules have a chance to reorient and the solvent–fluorophore interaction is high energy and unstable. However, following vibronic relaxation through the energy levels of S_1, the polar solvent molecules can reorient for a more stabilising interaction that further lowers the energy of the equilibrium excited state. Both the permanent and induced dipoles of the solvent play a role in this process. This solvent relaxation has the effect of reducing the energy difference between S_1 and S_0, thereby causing an increase in emission wavelengths (also called a red shift). By contrast, when such a fluorophore is in a non-polar solvent, there is no solvent relaxation in the excited state and the energy gap between S_1 and S_0 is relatively higher, thereby causing a blue shift in emission (Figure 14.5a, right). Such solvent effects depend on the polarity of both the solvent and fluorophore molecules. Moreover, other features of the solvent can be important, such as specific interactions like the ability of solvent molecules to form hydrogen bonds with the fluorophore. As a result, the emission spectra of polarity-sensitive fluorophores are often blue-shifted and of higher intensity in a non-polar environment and red-shifted and of lower intensity in a polar environment (Figure 14.5b).

14.3 Techniques, Methodologies and Protocols

14.3.1 Instrumentation

The instrument used to perform analytical fluorescence spectroscopy is a *spectrofluorometer*. At its essence, the function of this instrument is twofold: (i) to generate the wavelength(s) of light (λ_{ex}) that excite the fluorescent analyte of interest and (ii) to measure the intensity of the resulting fluorescent light at the appropriate wavelength(s) (λ_{em}). The basic design of a research-grade spectrofluorometer is shown in Figure 14.6. The *light source* produces light for sample excitation. For steady-state measurements, the most common light source is the high pressure xenon arc lamp, which produces reasonably continuous spectral power distribution from 250 to 700 nm. Other potential sources include lasers, light-emitting diodes (LEDs) and laser diodes. Pulsed LEDs and laser diodes are commonly used for making time-domain lifetime measurements. *Monochromators* disperse the broadband source light into different wavelengths, typically with diffraction gratings. User-controlled settings for the excitation and emission monochromators define the wavelengths of incident light (λ_{ex}) and detected emission (λ_{em}). Light emerging from the monochromators has a Gaussian intensity distribution centred at λ_{ex} or λ_{em} and a spectral bandwidth (full width at half maximum) dictated by monochromator *slits*. Variable slit widths dictate light resolution and intensity passing through each monochromator (the smaller the slit, the higher the spectral resolution at the expense of lower intensity of passed light). *Detectors* convert emitted light into an electrical signal that is proportional to light intensity and interpreted by the system electronics. Most fluorometers use photomultiplier tubes, which greatly amplify the detected signal in a manner that can be controlled as the detector gain. *Polarisers* convert incident unpolarised light, which has electric vectors in equal distributions normal to the direction of light propagation, into polarised light with the electric vector amplitude greater at a particular angle. Fluorometers typically have linear birefringent polarisers (commonly Glan–Thompson polarisers) that can be selectively inserted in the excitation and emission light paths for anisotropy measurements. Finally, a *reference detector* measures emission from a reference fluorophore (a quantum counter) from a fraction of the excitation light to correct emission from the sample based on wavelength-dependent intensity of the exciting light.

14.3.2 Fluorophores

There are thousands of probes that can be used for different fluorescence-based approaches. Fluorophores are typically organic molecules consisting of planar or rigid ring structures with extended π systems: chemical

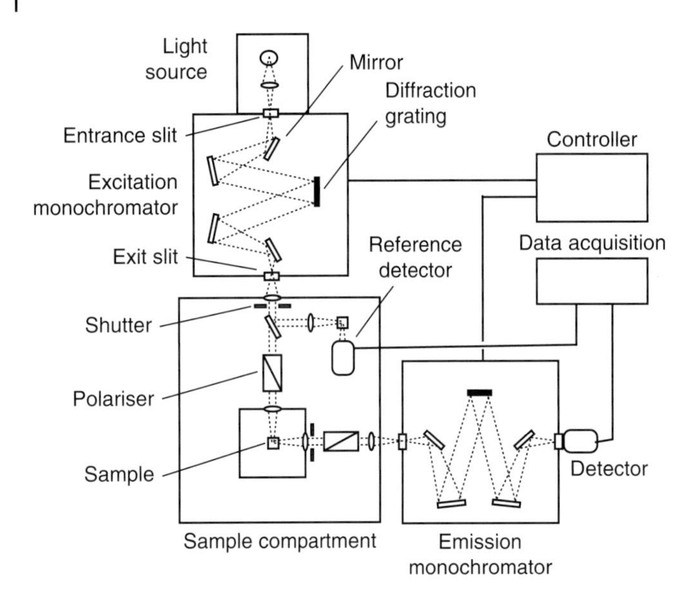

Figure 14.6 Schematic diagram of a standard L-format spectrofluorometer. The light path (dashed lines) is oriented such that the excitation light of a wavelength defined by the excitation monochromator (λ_{ex}) is focused on the sample and the emission light of a wavelength defined by the emission monochromator (λ_{em}) is detected at a 90° angle.

features that allow for light absorption and radiative emission. They can be broadly classified as *intrinsic*, or naturally occurring probes, and *extrinsic*, or synthetic probes that are added to a sample of biomolecules. Fluorophores are characterised by their spectral properties (ε (λ), $\lambda_{ex\ max}$, $\lambda_{em\ max}$, Φ, Stokes shift, lifetime, etc.) as well as other practical attributes such as photostability (ability to withstand continuous illumination without irreversible degradation). These features are all considered when selecting probes for fluorescence experiments. Comprehensive lists of commercially available fluorophores are available online, such on the Molecular Probes website (www.thermofisher.com/us/en/home/brands/molecular-probes), and several sites with interactive graphical viewers for fluorophore spectral data are also available (e.g. www.nightsea.com/sfa-sharing/fluorescence-spectra-viewers). A few example fluorophores are described below.

Intrinsic fluorophores of proteins include tryptophan, tyrosine and phenylalanine side chains, which contain optically active aromatic rings (Figure 14.7a). These side chains absorb light in the UV range; however, Trp and Tyr have larger $\varepsilon(\lambda)$ and Φ values, which accounts in part for the predominance of the Trp indole group (and, to a lesser extent, the Tyr phenol group) in the intrinsic fluorescence of natural proteins. Trp is a particularly useful probe because its fluorescence is highly sensitive to local polarity and can thus be used to monitor exposure to aqueous versus non-polar microenvironments that occur, for example, during protein unfolding. Other naturally occurring fluorophores include pyridine (NADH) and flavin (FAD^+) nucleotides as well as other aromatic enzyme cofactors. Fluorescent proteins including green fluorescent protein (GFP), yellow fluorescent protein (YFP) and cyan fluorescent protein (CFP) constitute a class of autofluorescent proteins that spontaneously form a fluorophore in their folded state and have been engineered to cover a broad spectral range [17].

Extrinsic fluorophores encompass a broad range of chemical structures [18, 19]. Such probes are designed for optimal absorbance, quantum yield and spectral range. Many extrinsic fluorophores can be classified based on the parent structures from which they are derived (Figure 14.7b). Fluoresceins and rhodamines, examples of xanthene-based fluorophores, have high extinction coefficients and small Stokes shifts, whereas fluorophores such as dansyl have large Stokes shifts (Figure 14.7c). In general, probes with larger π systems have absorbance and emission spectra at longer wavelengths. Many probes also have emission properties that are sensitive to the environment, which can be utilised for experimental purposes. For instance, the emission of fluorescein is sensitive to pH, whereas dansyl is highly sensitive to solvent polarity. Some families of fluorophores such as

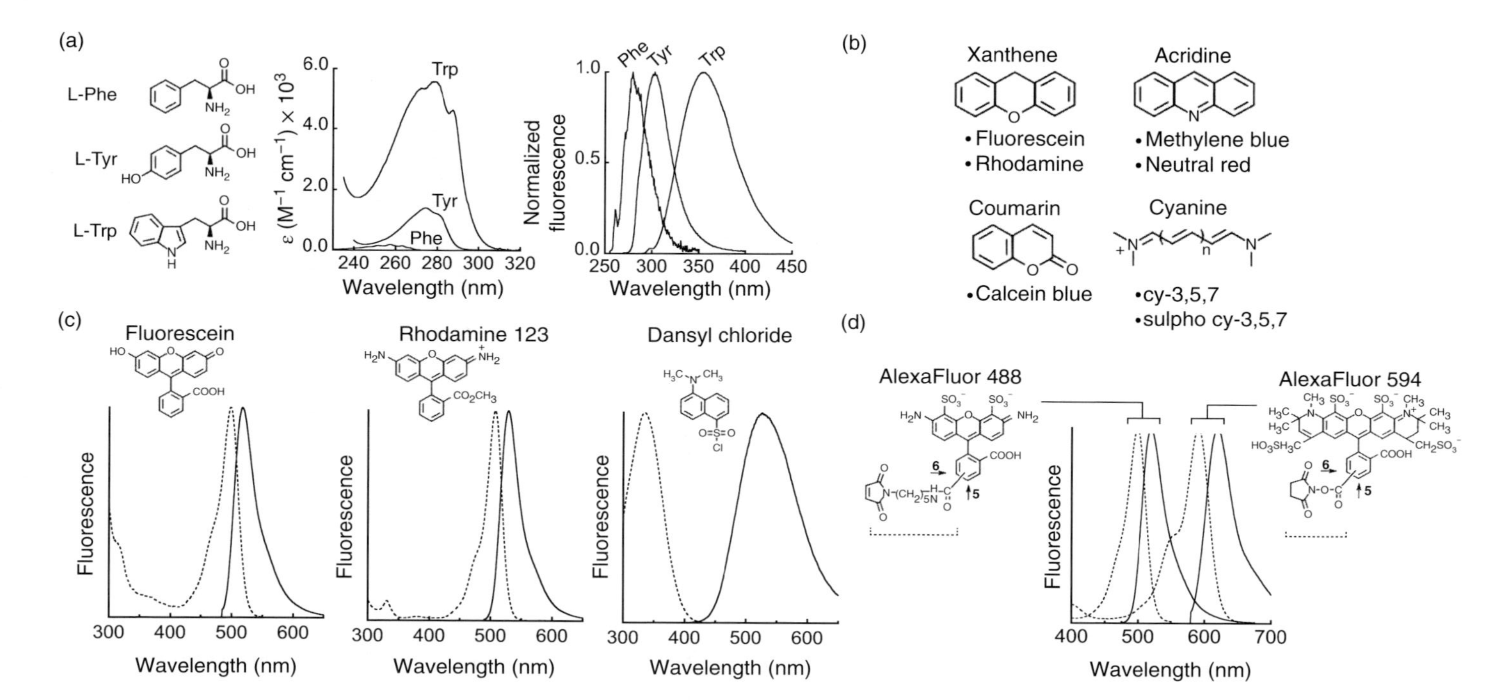

Figure 14.7 Chemical structures and spectral characteristics of common fluorophores. (a) Intrinsic fluorophores: absorbance and normalised fluorescence of aromatic protein side chains L-Phe, L-Tyr and L-Trp. (b) Examples of parent compounds of fluorescent dyes and representative fluorophores. (c) Normalised excitation (dashed lines) and emission (solid lines) spectra of fluorescein, rhodamine 123 and dansyl chloride. (d) Normalised excitation (dashed lines) and emission (solid lines) spectra of Alexa Fluor 488 maleimide and AlexaFluor 594 succinimidyl ester (mixed isomers). Chemical groups used for labelling are underscored by dashed lines.

Alexa dyes, two of which are shown in Figure 14.7d, are designed to absorb and emit across a wide spectral range. For labelling biomolecules, derivatives of extrinsic probes are available that react with naturally occurring functional groups. For example, maleimide and iodoacetamide moieties react with sulphhydryl groups on cysteine side chains, whereas sulfonyl chlorides and N-hydroxysuccinimide ester moieties react with primary amines, as illustrated in the structures for dansyl chloride and Alexa dyes (Figure 14.7c and d).

14.3.3 Measuring Fluorescence Spectra

Emission scans are one of the most common types of measurements in fluorescence spectroscopy. Some of the fundamental considerations in making such measurements can be illustrated by emission scans of dilute solutions of tryptophan in aqueous buffer (Figure 14.8). For these spectral measurements, two types of scans were taken: one for buffer only ('blank') and one with tryptophan in buffer ('sample'). Inspection of the blank traces (Figure 14.8, left) shows why it is critical to account for a signal that arises even in the absence of the sample being analysed. This background signal comes from two types of light scattering phenomena. First, *Rayleigh scattering* arises from the scattering of incident light on the solvent molecules without a change in wavelength. Hence, it is observed at emission wavelengths that are close to the selected λ_{ex} (depending on selected slit widths). Second, *Raman scattering* results from the inelastic scattering of incoming light by the solvent. For a given solvent, the Raman peak appears at a constant energy from the exciting light and will therefore shift depending on the λ_{ex} used. For water, the Raman peak is located at a wavenumber about 3600 cm^{-1} lower than that of the incident light; hence, when $\lambda_{ex} = 295$ nm, the Raman peak is at 330 nm $[= 10^7/(\overline{\nu}_{ex} - 3600\ cm^{-1})]$. Uncorrected emission scans with tryptophan present (Figure 14.8, centre) show a signal from both background scatter and from the fluorophore. To obtain the fluorescence originating only from the sample of interest, one simply subtracts the signal intensity of the blank from that of the uncorrected sample at each wavelength to obtain the blank-corrected spectra (Figure 14.8, right). Here we see a broad emission peak with a $\lambda_{em\ max}$

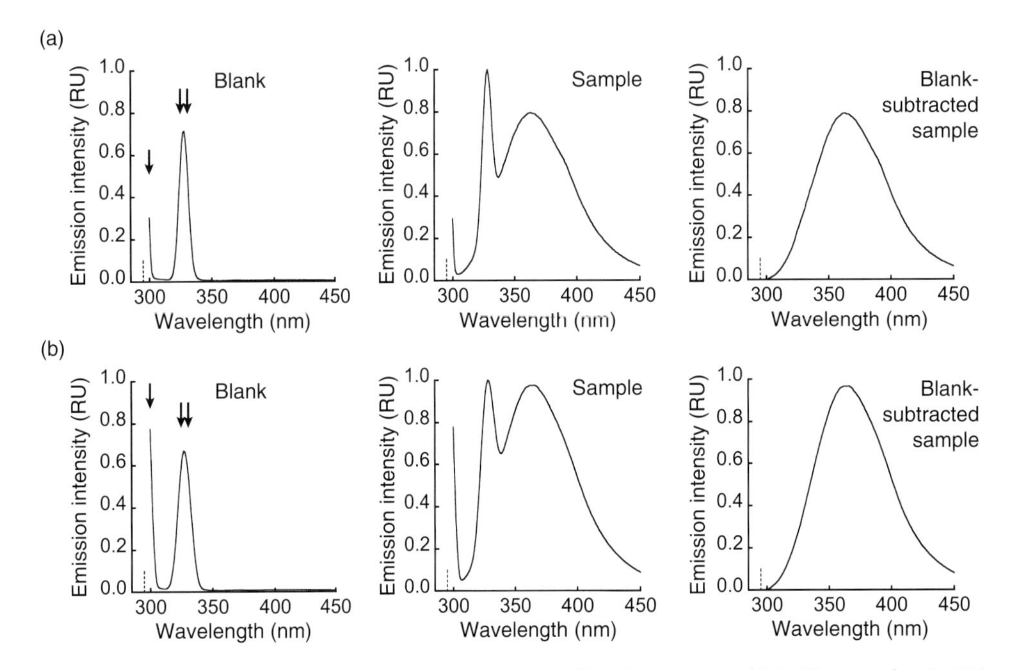

Figure 14.8 Fluorescence emission spectra measurements. Emission spectra of 0.8 μM tryptophan in 100 mM potassium phosphate buffer, pH 7.0 ($\lambda_{ex} = 295$ nm, denoted by dashed line; $\lambda_{em} = 300$–450 nm). Scans include blank (buffer only), sample (tryptophan in buffer) and blank-subtracted spectra. In blank spectra, the single arrow denotes Rayleigh scattering and the double arrow denotes the Raman peak. Spectra were collected with excitation/emission slits at 4/4 nm (a) and at 4/8 nm (b).

around 360 nm, typical of tryptophan in a polar solvent (compare with Figure 14.7a). Instrument settings are also important in the signal obtained from blanks and samples. For instance, when the emission slit widths are increased from 4 to 8 nm, Rayleigh scattering intensity increases significantly (Figure 14.8, compare a and b).

These relatively simple measurements highlight the importance of accounting for background signal, particularly when analysing a fluorophore in a complex background that may include large light-scattering particles and/or other endogenously fluorescent molecules whose signal is comparable to the fluorophore being analysed. Ideally, the blank will include everything that the sample has except for the fluorophore of interest. Moreover, blanks and samples must be measured with identical instrument settings (detector gain, slit widths, integration times, etc.) for accurate background subtraction.

14.3.4 Fluorescence Quenching

To this point, we have considered a number of intrinsic pathways that can depopulate the excited state of a fluorophore. We now turn to fluorescence quenching, an intermolecular process that alters fluorescence by interaction between the fluorophore and neighbouring solutes (ions or molecules). Here we consider two specific types of fluorescence quenching that entail direct contact between the fluorophore and quenching agents. *Dynamic quenching* involves diffusive collisions between a quencher and the fluorophore during its excited state lifetime (Figure 14.9a, left). This process effectively opens an additional de-excitation pathway that competes with emission, thereby decreasing both steady-state emission and the fluorescence lifetime. *Static quenching* involves the binding of a quencher to a fluorophore in the ground state, which renders it non-fluorescent (Figure 14.9a, right). This process therefore reduces emission intensity because fewer fluorophores can become excited, but it has no effect on lifetime because the de-excitation process for non-complexed fluorophores is not affected. Both dynamic and static quenching experiments can provide information regarding the accessibility of a fluorescent probe to quenching agents within a system because both phenomena require direct contact between fluorophore and quencher. We will now consider mathematical descriptions of each of these quenching processes in turn.

In *dynamic quenching*, quencher Q collides with $^*F$ at a rate defined by the bimolecular rate constant k_q (in units of M^{-1} s^{-1}). We can therefore consider $k_q[Q]$ as a pseudo-first order rate constant that describes a path for excited state depopulation in parallel with other non-radiative pathways. In this case, the rate of disappearance of $^*F$ following a short light pulse (Eq. (14.4)) can be rewritten as

$$-\frac{d[^*F]}{dt} = (k_r + k_{nr} + k_q[Q])[^*F] \tag{14.10}$$

and the fluorescence lifetime is

$$\tau = \frac{1}{(k_r + k_{nr} + k_q[Q])}. \tag{14.11}$$

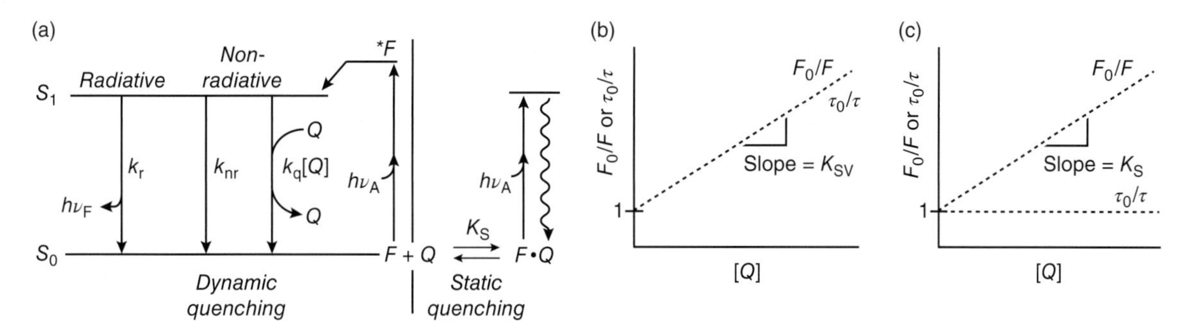

Figure 14.9 Fluorescence quenching processes. (a) During dynamic quenching, the collisional interaction between $^*F$ and Q creates a de-excitation pathway ($k_q[Q]$) in parallel with other non-radiative pathways (k_{nr}). During static quenching, F and Q form a non-fluorescent ground state complex. (b and c) Idealised Stern–Volmer plots for dynamic (b) and static (c) quenching.

Let us define F_0, Φ_0 and τ_0 as the fluorescence intensity, quantum yield and lifetime in the absence of quencher ($[Q] = 0$) and F, Φ and τ as the same parameters in the presence of quencher. We can then set up the following ratiometric relationships:

$$\frac{F_0}{F} = \frac{\Phi_0}{\Phi} = \left[\frac{\frac{k_r}{(k_r + k_{nr})}}{\frac{k_r}{(k_r + k_{nr} + k_q[Q])}}\right] = \frac{(k_r + k_{nr} + k_q[Q])}{(k_r + k_{nr})} = \frac{\left(\frac{1}{\tau}\right)}{\left(\frac{1}{\tau_0}\right)} = \frac{\tau_0}{\tau}. \tag{14.12}$$

Given that in the absence of Q, $\tau_0 = 1/(k_r + k_{nr})$, rearrangement of the fourth term in the above equation gives

$$\frac{F_0}{F} = 1 + \frac{k_q[Q]}{(k_r + k_{nr})} = 1 + k_q\tau_0[Q]. \tag{14.13}$$

For a single fluorophore and quencher under a given condition, k_q and τ_0 will be constants. We can therefore define the following *Stern–Volmer equation*:

$$\frac{F_0}{F} = 1 + K_{SV}[Q], \tag{14.14}$$

in which $K_{SV} = k_q\tau_0$ is the *Stern–Volmer constant*. Because $F_0/F = \tau_0/\tau$, this equation can also be written as

$$\frac{\tau_0}{\tau} = 1 + K_{SV}[Q]. \tag{14.15}$$

These relationships show that increasing $[Q]$, and therefore the number of possible encounters between Q and $^*F$, will result in a linear decrease in fluorescence intensity (and lifetime), resulting in an increase in F_0/F (and τ_0/τ). Dynamic quenching data are typically presented as *Stern–Volmer plots*, in which F_0/F (or τ_0/τ) are shown as a function of $[Q]$ at different concentrations, wherein the y-intercept is 1 and the slope is equal to K_{SV} (Figure 14.9b).

A wide variety of substances can act as collisional quenchers of different fluorophores. They can be free in solution (e.g. ions such as atomic halogen anions, transition metals or molecular oxygen) or covalently bound to larger molecules (e.g. paramagnetic moieties bound to lipid acyl chains or specific amino acid side chains in proteins). Moreover, the quenching mechanism will depend on the nature of the quencher and fluorophore. Dynamic quenching can occur by photophysical processes such as the formation of charge-transfer complexes. Among the most common quenching agents are heavy atoms (e.g. iodide and bromide), which promote spin–orbit coupling with $^*F$, resulting in intersystem crossing to the excited triplet state and non-radiative de-excitation due to the low efficiency of phosphorescent photon emission. For a given fluorophore and quencher, the k_q value reflects the frequency with which collisions occur, dependent not only on the accessibility of the Q to $^*F$ but also on factors such as the radii and diffusion coefficients of the two as well as quenching efficiency upon contact. Typical values of k_q are on the order of 10^{10} M^{-1} s^{-1}. Rapid collisional rates are required because in dynamic quenching, Q must make contact with $^*F$ during its excited state lifetime. For highly efficient quenching, k_q can approach rates near the limits of diffusion.

In most cases of *static quenching*, F in the ground state forms a non-fluorescent complex with a quencher Q. This 'dark complex', $F{\bullet}Q$, absorbs light, typically with altered absorption properties, but immediately returns to the ground state without emission of a fluorescent photon. Consider the equilibrium relationship of the 1 : 1 complex $F + Q \leftrightarrows F{\bullet}Q$ with association constant K_S:

$$K_S = \frac{[F \bullet Q]}{[F][Q]}. \tag{14.16}$$

Based on mass conservation, the total amount of fluorophore $[F]^{tot}$ is equal to the sum of free and complexed fluorophore; therefore, $[F]^{tot} = [F] + [F{\bullet}Q]$. Substituting this relationship into Eq. (14.16) and rearranging, we obtain

$$\frac{[F]^{tot}}{[F]} = 1 + K_S[Q]. \tag{14.17}$$

Noting that the relative fluorescence intensities in the absence and presence of Q will be proportional to $[F]^{tot}/[F]$, we obtain

$$\frac{F_0}{F} = 1 + K_S[Q]. \tag{14.18}$$

Thus Eq. (14.18) takes the same form as the Stern–Volmer relation for dynamic quenching (Eq. (14.14)); namely, there is a linear relationship between the proportional fluorescence decrease and $[Q]$. However, in this case the quenching constant is the complex association constant K_S. Moreover, because F•Q is non-fluorescent, the fraction of fluorophores in this complex are removed from observation but the emission properties of free F, including lifetime, remain unaffected. Thus, static quenching effectively just reduces the total number of fluorophores that are able to emit fluorescence within the system. As a result, with increasing $[Q]$ there will be a reduction in emission intensity (F_0/F) but τ_0/τ remains constant, as shown in the corresponding Stern–Volmer plot (Figure 14.9c). Hence, measurements of an excited state lifetime provide the most fundamental diagnostic feature to distinguish dynamic from static quenching.

14.3.4.1 Fluorescence Quenching Sample Experiment: Iodide Quenching of Tryptophan Fluorescence

One application of fluorescence quenching is to address the solvent accessibility of particular sites in macromolecules. Consider two proteins, each with a single environment-sensitive probe in either a polar environment (protein A) or a non-polar environment (protein B). We wish to directly test the hypothesis that the probe of protein A is solvent-accessible and the probe of protein B is not solvent-exposed, but likely to be buried in the non-polar core of the protein (Figure 14.10a). Many probes are quenched by water-soluble quenching agents such as acrylamide and halogens including iodide ions (I^-). We can therefore perform dynamic quenching experiments in which solutions of the two proteins are titrated with I^- and the resulting changes in steady-state emission and lifetime are analysed by Stern–Volmer plots.

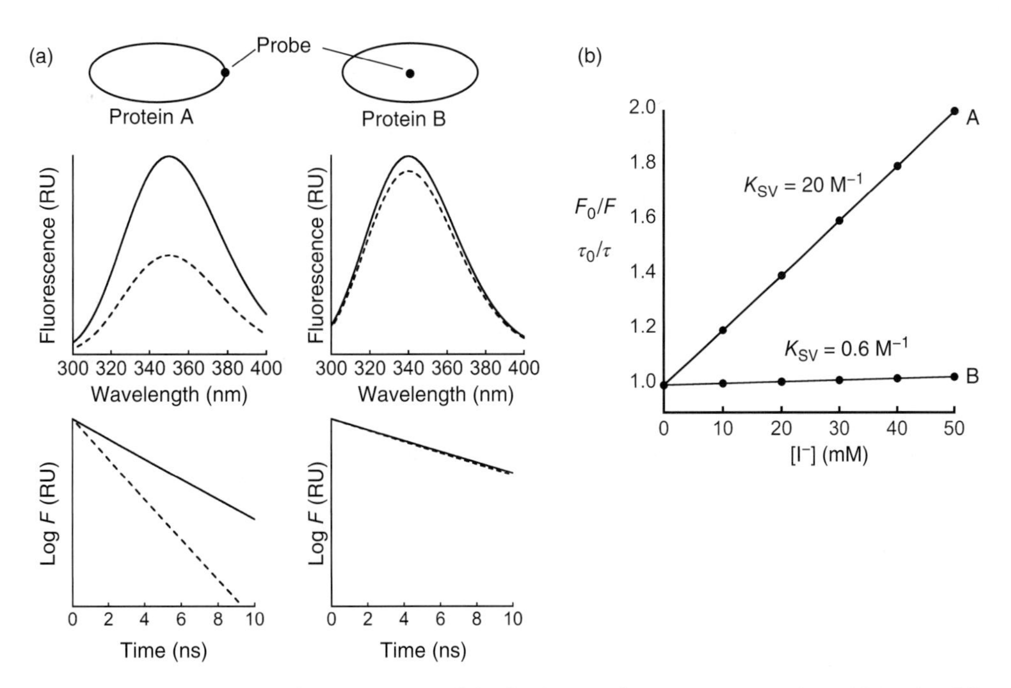

Figure 14.10 Monitoring probe solvent accessibility by dynamic fluorescence quenching. (a) Graphical illustration of experimental system with simulated emission scans and time-dependent lifetime decays in the presence (dashed lines) and absence (solid lines) of Q for the two proteins. (b) Simulated data for iodide quenching of proteins A and B represented as a Stern–Volmer plot.

Table 14.1 Sample preparation for dynamic quenching experiment.

Quencher stock solutions (100 μl each)				Sample preparation (500 μl each)			
Stock	$[KI]^{stock}$ (M)	Add (μl)		Sample	$[KI]^{final}$ (mM)	Add (μl)	
		1M KI $+Na_2S_2O_4$	1M KCl $+Na_2S_2O_4$			Protein stock	Quencher stock
Q1	0	0	100	1	0	475	25 (Q1)
Q2	0.2	20	80	2	10	475	25 (Q2)
Q3	0.4	40	60	3	20	475	25 (Q3)
Q4	0.6	60	40	4	30	475	25 (Q4)
Q5	0.8	80	20	5	40	475	25 (Q5)
Q6	1.0	100	0	6	50	475	25 (Q6)

14.3.4.2 Experimental Setup and Data Collection

Purified proteins A and B are suspended in a suitable buffer (e.g. 100 mM sodium phosphate, pH 7.0 with 25 mM NaCl) at a concentration sufficient for a good fluorescence signal. Stocks of 1M KI and 1M KCl are prepared, each containing 50 mM of sodium dithionite ($Na_2S_2O_4$), and mixed to prepare six 100 μl quenching stock solutions containing [KI] of 0, 0.2, 0.4, 0.6, 0.8 and 1.0 M. Samples for spectral analysis are then made by preparing six aliquots of 475 μl of protein stocks and mixing with 25 μl of the different quenching stocks, yielding 500 μl samples with [KI] of 0, 10, 20, 30, 40 and 50 mM of KI (Table 14.1). Steady-state emission spectra (e.g. λ_{ex} = 290 nm; λ_{em} = 300–400 nm) are taken for each sample and the emission intensity at the corresponding λ_{max} values are recorded. Measurements of fluorescence lifetime for each sample could be made in parallel.

14.3.4.3 Data Analysis

Simulated emission scans and lifetime measurements for this experiment are depicted in Figure 14.10a. For emission scans, the fluorescence intensity at $\lambda_{em\ max}$ in the absence of quencher (F_0) is calculated as a ratio of the intensity of each sample with increasing $[I^-]$ (F) to yield sample-specific F_0/F values, which are plotted as a function of the corresponding $[I^-]$ to generate Stern–Volmer plots. For lifetime measurements, the same calculations are made to calculate τ_0/τ for each sample and are graphically analysed in the same way. In these simulations, the unquenched probe lifetimes for each protein are τ_0 = 2.7 and 5.0 ns for proteins A and B, respectively, consistent with the different microenvironment polarity of each residue. From Eqs. (14.14) and (14.15), we see that linear regression of each F_0/F (or τ_0/τ) versus an $[I^-]$ dataset will give a slope corresponding to the K_{SV} value for each protein (Figure 14.10b). For protein A, a K_{SV} of 20 M^{-1} indicates that the Trp is highly accessible to added I^- ions, and we can calculate the biomolecular quenching constant as $k_q = K_{SV}/\tau_0 = 20\ M^{-1}/2.7 \times 10^{-9}\ s = 7.41 \times 10^9\ M^{-1}\ s^{-1}$, a high value consistent with a diffusion-limited collisional process. For protein B, a K_{SV} of 0.6 M^{-1} and $k_q = 0.6\ M^{-1}/5.0 \times 10^{-9}\ s = 1.20 \times 10^8\ M^{-1}\ s^{-1}$ confirms very low quencher access, consistent with burial of the probe in the non-polar protein core. The results for protein A confirm the nature of the quenching mechanism as dynamic, not static, because the fluorescence lifetime decreases with quencher concentration.

14.3.4.4 Experimental Considerations

1) When charged quenchers are used, it is important to maintain a constant ionic strength for all samples. In the above experiment, this is addressed by preparing quencher stock solutions as mixtures of KI and KCl.

2) I^- in solution can become oxidised to I_2, a non-polar compound that can partition into non-polar environments. In the above experiment, this is addressed by adding the reductant sodium dithionite to the quencher stocks.
3) Low accessibility to a soluble quencher often means that the fluorophore is physically located in a region that the quencher cannot reach (e.g. within a protein or in the non-polar core of a lipid bilayer). However, other factors such as the local electrostatic environment may be relevant. For example, if a Trp residue is surface-exposed but surrounded by a patch of acidic residues, this could cause electrostatic repulsion of anionic quenchers, resulting in a reduced K_{SV}. For such reasons, other independent experiments are often required to confirm fluorophore location.
4) Additional overviews and research on fluorescence quenching are provided in the reference list [20–22].

14.3.5 Förster Resonance Energy Transfer (FRET)

Förster Resonance Energy Transfer (FRET) is a process by which a fluorophore in the excited singlet state, termed the donor (D), transfers its excitation energy to a nearby fluorophore or chromophore, termed the acceptor (A), in a manner that does not require direct contact between the two molecules (Figure 14.11a). The transfer of energy occurs non-radiatively; that is, D does not emit a fluorescent photon as it relaxes to the ground state and A does not absorb a photon as it becomes excited. Rather, D and A behave like coupled oscillators with D transferring its excited state energy to A via long-range dipole–dipole interactions. During such resonant transitions, the energy available from D must match the energies to excite A to the various vibronic levels of its S_1 state. Hence, if we consider the excitation (or absorption) and emission spectra of D and A, it is clear that the emission spectrum of D must partially overlap with the excitation (absorption) spectrum of A (Figure 14.11b). If A is a non-fluorescent chromophore, then it acts as a long-distance quencher of fluorescence from D. If A is a fluorophore, then it can subsequently emit a fluorescent photon as it relaxes to the ground state.

Because FRET can occur over long distances (~10–100 Å, comparable to the sizes of macromolecules), it provides a powerful tool for analysing the proximity of specific sites within or between molecules. With the right experimental conditions, it can serve as a 'spectroscopic ruler' to estimate distances between D and A. Each D–A pair has a characteristic distance over which FRET will occur, determined by the parameter R_0, the *Förster distance*, described below. Steady-state experiments can measure FRET by a decrease in D emission or, alternatively, as an increase in A emission, provided that A is fluorescent. Time-resolved measurements can measure FRET as decreases in the D lifetime in the presence of A. As shown mathematically below, the

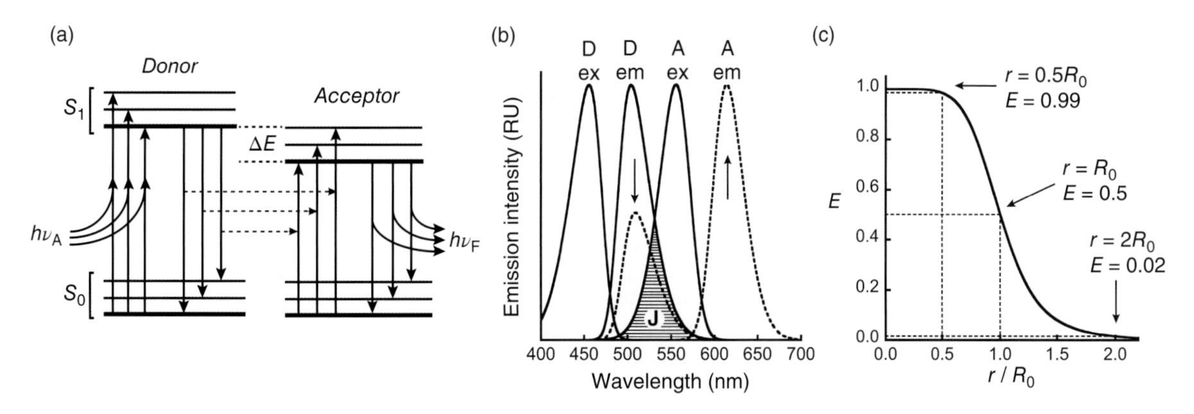

Figure 14.11 Förster resonance energy transfer. (a) Simplified Perrin–Jablonski diagram representing FRET. (b) Spectra of donor (D) and acceptor (A) fluorescence excitation (or absorption) and emission. The stippled area represents the integral overlap (*J*) between D emission and A excitation. Dashed lines show a FRET-associated decrease in D emission and an increase in A emission. (c) Variation in E as a function of r/R_0.

efficiency of FRET depends primarily on the distance between D and A, the extent of spectral overlap between D emission and A absorption and the relative orientations of D and A.

FRET creates a path for the de-excitation from the S_1 state of D in parallel with radiative and other non-radiative paths. The rate constant of energy transfer from D to A, k_{ET}, is defined as

$$k_{ET} = k_D \left(\frac{R_0}{r}\right)^6 = \frac{1}{\tau_D}\left(\frac{R_0}{r}\right)^6, \tag{14.19}$$

where k_D and τ_D are the emission rate constant and lifetime of D, respectively, in the absence of A; R_0 is the Förster distance and r is the distance between D and A. This equation shows that the rate of energy transfer is inversely proportional to the sixth power of r (i.e. $k_{ET} \propto r^{-6}$), underscoring the dominant role that the D–A distance has in energy transfer. It also shows that when $r = R_0$, the rate of energy transfer is equal to the intrinsic decay rate of D (i.e. $k_{ET} = k_D = 1/\tau_D$). Hence, R_0 defines the D–A distance at which FRET is 50% efficient. R_0, which ranges between 10 and 60 Å for typical D–A pairs, is calculated by the following equation:

$$R_0 = 0.211(\kappa^2 n^{-4} \Phi_D J)^{1/6}. \tag{14.20}$$

Equation (14.20) indicates that R_0 depends on the spectral characteristics of D and A as well as the environment through which energy transfer is occurring, as follows. First, the term J is the spectral overlap between D emission and A absorbance (Figure 14.11b). J is defined as

$$J = \frac{\int_0^\infty F_D(\lambda)\varepsilon_A(\lambda)\lambda^4 \, d\lambda}{\int_0^\infty F_D(\lambda)\, d\lambda}, \tag{14.21}$$

where $F_D(\lambda)$ represents the emission spectrum of D normalised so that $\int_0^\infty F_D(\lambda) d\lambda = 1$ and $\varepsilon_A(\lambda)$ is the wavelength-specific molar absorption coefficient of A. When $\varepsilon_A(\lambda)$ is in units of M^{-1} cm^{-1} and λ is in units of nm, the overlap integral J is in terms of M^{-1} cm^{-1} nm^4. Second, Φ_D is the quantum yield of D in the absence of A. Third, n is the average refractive index of the medium between D and A, typically taken to be 1.4 for biomolecules. Finally, the κ^2 term is an orientation factor that describes the spatial relationship between the emission dipole of D and the absorption dipole of A, where κ^2 can assume values between 0 (transition moments perpendicular) and 4 (transition moments collinear). However, when D and A are free to rotate at a rate faster than the lifetime of the excited state, one can assume isotropic dynamic averaging such that $\kappa^2 = 2/3$.

The efficiency of energy transfer, E, can be calculated as the fraction of photons absorbed by D that are resonantly transferred to A:

$$E = \frac{k_{ET}}{k_D + k_{ET}} = \frac{k_{ET}}{\tau_D^{-1} + k_{ET}}. \tag{14.22}$$

Combining Eqs. (14.19) and (14.22), we can recast E in terms of r and R_0 to obtain

$$E = \frac{R_0^6}{(r^6 + R_0^6)}. \tag{14.23}$$

Due to its inverse sixth power dependence on distance, E varies dynamically when r is between $0.5R_0$ and $2R_0$ (Figure 14.11c). Accurate distance measurements can only be made within this distance window because when $r < 0.5R_0$, energy transfer is essentially 100% efficient, and when $r > 2R_0$, essentially no energy transfer will occur. Rearrangement of Eq. (14.23) gives

$$r = R_0\left(\frac{1-E}{E}\right)^{1/6}, \tag{14.24}$$

which allows the calculation of r based on measured E and R_0 values.

The type of energy transfer described above refers to *hetero-FRET*, in which the donor and acceptor probes are distinct. A different kind of energy transfer termed *homo-FRET* can occur with a single type of fluorophore (typically containing a small Stokes shift), whereby excitation energy is reversibly transferred among the probes and FRET, is typically detected by decreases in anisotropy (see Section 14.3.6).

14.3.5.1 FRET Sample Experiment: Monitoring Molecular Interactions

The ability of FRET to detect changes in the distances between D and A moieties makes this an excellent technique for monitoring alterations in interatomic distances that occur by structural dynamics and associations of macromolecules. For example, by site-specific FRET pair labelling of two interacting proteins (incorporation of D on one protein and A on the other), one can monitor protein associations and ascertain the distance r between the probes in the interacting heterodimer. Consider a hypothetical case in which two proteins are individually labelled with extrinsic probes A and D whose R_0 is 50 Å (Figure 14.12a). Upon mixing the two proteins, FRET-based approaches can be used to monitor their binding kinetics in real time, their association constants, and distance between the probes in the interacting protein complex.

14.3.5.2 Experimental Setup and Data Collection

Most FRET experiments entail the preparation and measurement of four distinct samples: (i) DA contains both donor and acceptor probes, (ii) D contains donor only, (iii) A contains acceptor only and (iv) B is a blank without either probe (Figure 14.12b). FRET experiments are most straightforward if samples contain probes at equal concentrations (i.e. [D] in the D sample = [D] in the DA sample; [A] in the A sample = [A] in the DA sample; and [D] = [A] in the DA sample). In the present example, this would mean that the D and A sites on each protein are quantitatively labelled with the extrinsic probes, confirmed by biochemical analysis (protein concentration and absorbance scans to quantify D and A). If probe concentrations between samples are not equal, one must adjust the spectral data accordingly.

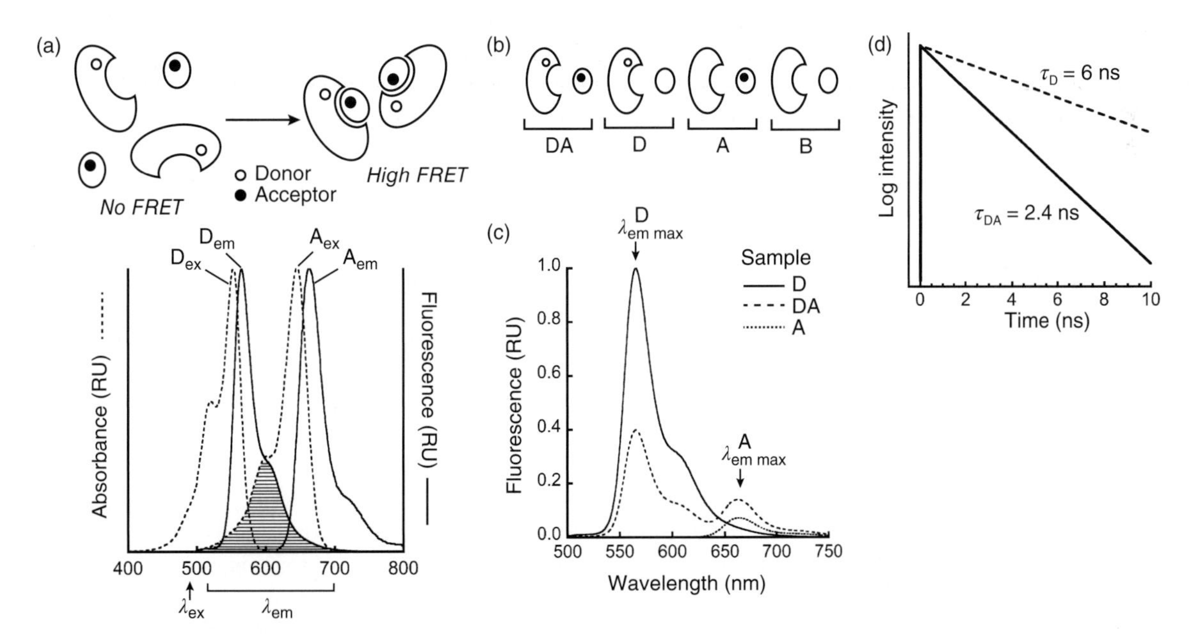

Figure 14.12 Monitoring protein interactions by FRET. (a) Experimental design. Above: proteins labelled with A or D do not show FRET individually but upon interaction, r is small enough for FRET to occur. Below: absorption and emission spectra of a hypothetical D–A pair with integral spectral overlap stippled. (b) Samples prepared for FRET analysis. (c) Simulated emission scans (λ_{ex} = 480 nm; λ_{em} = 500–700 nm) for D, D–A and A samples. (d) Simulated intensity decays for D and D–A samples with single exponential lifetimes.

We first describe FRET analysis using steady-state approaches. In the present case, samples for spectral analysis (the 'DA', 'D', 'A' and 'B' samples) are separately added to four cuvettes with the two proteins in equimolar amounts. Emission scans for each sample are then performed by exciting the D probe (here, $\lambda_{ex} = 480$ nm) and reading fluorescence in the wavelength range that includes emission spectra of both D and A probes (here, $\lambda_{em} = 500$–700). If the two proteins interact such that the D and A probes are proximal (within the range of R_0), the resulting spectra of each sample would appear similar to the simulated scans of Figure 14.12c. The 'D' sample shows the emission spectrum of the D probe only. The 'DA' sample reveals a decrease in D emission due to A-dependent energy transfer as well as the appearance of the emission spectrum of A, which is primarily due to FRET. The 'A' sample reveals slight A emission that originates from direct excitation of A at the exciting wavelength, a phenomenon termed 'crosstalk', which occurs when A has a non-zero extinction coefficient at λ_{ex} (i.e. D and A absorbance spectra partially overlap). Finally, the 'B' sample shows a signal from light scatter and background fluorescence, which can be appreciable, for instance, when the samples contain large particles or other molecules that are intrinsically fluorescent (but not likely to be sizeable in the present example).

14.3.5.3 Data Analysis

FRET is most easily and accurately quantified by measuring the A-dependent decrease in D fluorescence. Energy transfer causes a decrease in the donor quantum yield with the transfer efficiency quantified as

$$E = 1 - \left(\frac{\Phi_{DA}}{\Phi_D}\right), \tag{14.25}$$

where Φ_{DA} and Φ_D are the donor quantum yields in the presence and absence of acceptor, respectively. Direct measurements of Φ are not practical for most FRET experiments; therefore, assuming that D absorbance is identical between the two samples, this ratiometric term can be written in terms of D fluorescence intensities in the presence and absence of A (F_{DA} and F_D, respectively):

$$E = 1 - \left(\frac{F_{DA}}{F_D}\right). \tag{14.26}$$

In such cases, the donor emission wavelength(s) used for observation are chosen to minimise unwanted emissions from A.

Based on these relationships, the scans shown in Figure 14.12c can be used to calculate FRET efficiency. D fluorescence intensity can be measured as the integrated donor spectra for all samples; however, in cases where the D and A emission spectra overlap, this requires spectral deconvolution. For simplicity, we can take emission at a single wavelength ($\lambda_{em\ max} = 566$ nm) as our readout for fluorescence (F_{DA}, F_D, F_A and F_B). When the background signal is non-negligible, F_B must be subtracted from F_{DA}, F_D and F_A to obtain the net fluorescence signal that originates solely from the probes. In the present example, we assume that the scans are background-subtracted, which takes care of this term. Because there is no contaminating signal from A at this wavelength (no crosstalk), $F_A = 0$. We can then use Eq. (14.26) to calculate FRET efficiency, given relative intensity values $F_D = 1.0$ and $F_{DA} = 0.4$, which yields $E = 0.6$. From Eq. (14.24), we calculate $r = 47$ Å, which makes sense because at 60% FRET efficiency, r is expected to be lower than R_0.

FRET efficiency can also be calculated from the D-dependent increase in A fluorescence, termed 'sensitised emission'. Here the samples are again excited at the D excitation wavelength, but the signals in the acceptor emission spectral range are analysed. However, this approach is less straightforward than using D emission for two main reasons. First, due to overlap of the D and A emission spectra, there is often a contaminating signal from D in the A spectral range, termed 'bleed-through', which must be accounted for. Second, crosstalk from direct excitation of A can also contribute to the signal. In instances where D emission is negligible or can be deconvoluted from the signal, transfer efficiency is given by

$$E = \left(\frac{\varepsilon_A}{\varepsilon_D}\right)\left(\frac{F_{DA}}{F_A} - 1\right), \tag{14.27}$$

where ε_A and ε_D are the extinction coefficients of A and D, respectively, at the D excitation wavelength, and F_{DA} and F_A are the *acceptor* emission intensities, respectively.

Energy transfer efficiency can also be analysed by time-resolved techniques, for example by measuring the donor fluorescence lifetimes in the presence and absence of A (τ_{DA} and τ_D, respectively):

$$E = 1 - \left(\frac{\tau_{DA}}{\tau_D}\right). \tag{14.28}$$

Lifetime analysis of FRET would render something similar to the pattern shown in Figure 14.12d. For the experiment described above, assuming that τ_D is 6 ns and decays as a single exponential in the absence and presence of the acceptor, then τ_{DA} would be 2.4 ns if we assume 60% FRET efficiency. When measuring FRET, the use of time-resolved approaches has some advantages over using steady-state approaches, for example when samples are not 100% labelled with a single D or A probe. Based on our assumption that every 'D' sample has a D probe and every 'A' sample has an A probe, then an emission-based FRET efficiency of $E = 0.6$ can unambiguously indicate that the D–A distance $r = 0.94\,R_0$. However if, for instance, the 'A' samples were only 60% labelled with A probe and the *actual* D–A distance was much shorter ($r \ll R_0$, approaching $E = 1$), then the same steady-state efficiency could be obtained, significantly overestimating the D–A distance. By contrast, time-resolved FRET measurements can help overcome difficulties associated with incomplete D–A pairing. Provided that τ_D and τ_{DA} are sufficiently different and have relatively straightforward decay functions (e.g. one- or two-exponential decays), then one can use lifetime measurements to estimate E (Eq. (14.28)) and evaluate the fraction of D that are in D–A pairs.

14.3.5.4 Experimental Considerations

1) In designing FRET experiments, it is critical to choose D–A pairs with appropriate spectral properties (R_0, Φ, overlap integral, etc.) for the measurements being made. Whether examining inter- or intramolecular FRET, such experiments are most straightforward if there is a single D probe per 'D' site and a single A probe per 'A' site. If there are multiple copies of a single member in a cognate pair (e.g. two A probes for every D) then analysis of FRET data is more complex and distance measurements usually cannot be unambiguously calculated.
2) The κ^2 factor, reflecting the relative orientations of the D and A transition dipoles, is generally assumed to be 2/3 when the probes are freely rotating. This assumption can be supported experimentally by making anisotropy measurements, which reflect the rotational mobility of D and A, as described in the following section. If D and A are not sufficiently free to rotate, neglecting the κ^2 term can lead to erroneous FRET measurements.
3) Additional overviews and research on FRET are provided in the reference list [23–30].

14.3.6 Fluorescence Anisotropy

Chromophores contain absorption transition dipole moments ($\vec{\mu}_A$) related to the displacement of charges that occurs upon light absorption. They are represented by vectors within the coordinate system of atoms on the molecule (Figure 14.13a). Absorption occurs when the oscillating electric field vector ($\vec{E}$) of light interacts with the absorption dipole moment with a particular orientation. We can also consider the emission dipole moment of a fluorophore ($\vec{\mu}_E$) related to $\vec{\mu}_A$ by angle α. For many fluorophores, absorption and emission dipoles are collinear ($\alpha = 0°$); however, they may differ in cases where the fluorophore is excited to an electronic energy level higher than the singlet state from which emission originates. These factors dictate how fluorophores absorb and emit linearly polarised light, which is analysed by fluorescence anisotropy measurements.

Consider the instrument configuration shown in Figure 14.13b. Like all natural light sources, radiation from fluorometer lamps is unpolarised, meaning that its electric vectors are in all possible orientations. An excitation polariser in the light path along the x-axis produces plane-polarised light with $\vec{E}$ in a single orientation

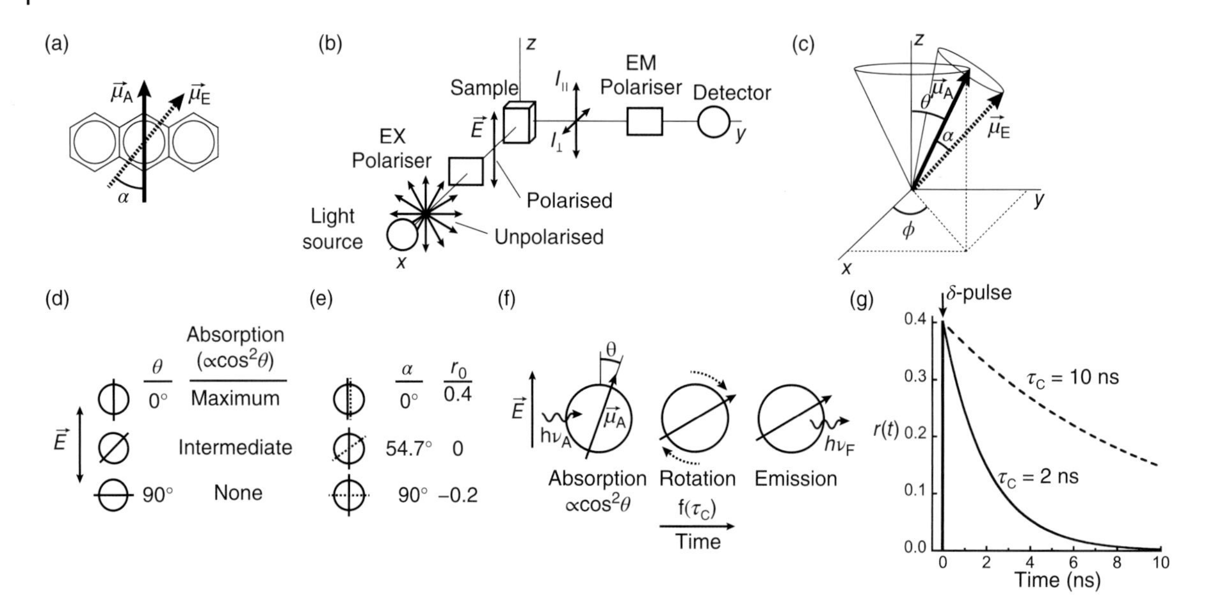

Figure 14.13 Principles of fluorescence anisotropy. (a) Hypothetical fluorophore with absorption ($\vec{\mu}_A$) and emission ($\vec{\mu}_E$) transition dipole moments related by angle α. (b) Instrument setup for anisotropy measurements with coordinates based on the laboratory axes. Light is incident along the x-axis and detected along the y-axis. Vertically and horizontally polarised light is parallel to the z- and x-axes, respectively. (c) Coordinate system for absorption and emission dipoles of a fluorophore. (d) Photoselection based on orientation between $\vec{E}$ and the absorption transition dipole. (e) Fundamental anisotropy based on orientation between absorption and emission dipoles. (f) Illustration of molecular rotation inducing time-dependent fluorescence depolarisation. (g) Time-dependent $r(t)$ decay following an instantaneous light pulse for $\tau_C = 2$ ns and $\tau_C = 10$ ns assuming absorption and emission dipoles are collinear ($\alpha = 0°$ and $r_0 = 0.4$).

(here, oriented vertically or parallel to the z-axis) prior to reaching the sample. The orientation of a single fluorophore in the sample chamber can be defined by our coordinate system such that θ and ϕ are the angles between the transition dipole and the z- and x-axes, respectively (Figure 14.13c). The probability of photon absorption by a fluorophore is directly proportional to $\cos^2\theta$, where θ is the angle between the transition dipole and $\vec{E}$, which is also the z-axis when light is vertically polarised. Hence, fluorophores with transition dipoles parallel to $\vec{E}$ ($\theta = 0°$) will have maximal absorption, those with perpendicular orientations ($\theta = 90°$) will not absorb light and those whose orientation is $0° < \theta < 90°$ will absorb with intermediate efficiency (Figure 14.13d). Therefore, while fluorophores in the sample are physically oriented in all possible directions (isotropically), plane-polarised light creates a subpopulation of excited fluorophores whose orientation is anisotropic and oriented along the z-axis, a phenomenon termed *photoselection.*

The polarisation state of the resulting fluorescence can be detected using an emission polariser oriented parallel to the $\vec{E}$ of polarised excitation (to detect intensity $I_{||}$) or perpendicular to the $\vec{E}$ vector (to detect intensity $I_\perp$). These intensity values are used to calculate the dimensionless parameter, *anisotropy* (r):

$$r = \frac{(I_{||} - I_\perp)}{(I_{||} + 2I_\perp)}. \tag{14.29}$$

These intensities can also be used to calculate a related parameter, *polarisation* (P):

$$P = \frac{(I_{||} - I_\perp)}{(I_{||} + I_\perp)}. \tag{14.30}$$

The denominators in the equations for r and P, respectively, represent total emission light intensity ($I_{||} + 2I_\perp$) and intensity in the direction of observation ($I_{||} + I_\perp$). Normalisation to total light intensity makes r more mathematically tractable than P; therefore, we will focus on anisotropy here. From Eq. (14.29), we see that if emitted light is completely polarised ($I_{\perp} = 0$) then $r = 1.0$, and if emitted light is completely depolarised ($I_{||} = I_{\perp}$) then $r = 0$. Note that because r is a ratiometric measure of emission intensity, it is independent of fluorophore concentration, at least for samples that are sufficiently dilute to avoid artefacts.

Several factors contribute to fluorescence anisotropy. One is of course photoselection; because the distribution of excited fluorophores is anisotropic, the emitted light will also be anisotropic, polarised in a plane defined by the emission moment. Assuming parallel transitions moments ($\alpha = 0°$), the relation between emission anisotropy and the angular distribution of transition moments is $r = (3 < \cos^2\theta > - 1)/2$. Due to photoselection, the probability distribution of excited molecules $<\cos^2 \theta>$ has a value of 3/5. This means that r has a maximal value of 0.4 ($\theta = 0°$) and a minimal value of -0.2 ($\theta = 90°$). A more general equation accounts for the orientation between $\vec{\mu}_A$ and $\vec{\mu}_E$:

$$r_0 = \frac{3 < \cos^2\alpha > -1}{5}. \tag{14.31}$$

In this equation, the term r_0 represents the *fundamental anisotropy* in the absence of other depolarising processes such as molecular tumbling or energy transfer (e.g. as occurs with homo-FRET); r_0 is measured under conditions that prevent molecular motions (e.g. by the immobilisation of molecules in clear glass or at low temperatures in glycerol solution). Values for r_0 range from 0.4 (the theoretical maximum when emission and transition dipoles are collinear) to -0.2 (when the dipoles are perpendicular) (Figure 14.13e).

To this point we have considered factors that influence anisotropy, which are based on the nature of incident and measured light and on the photophysical properties of the fluorophores. We now turn to ways in which fluorescence anisotropy can give information about the size, shape, flexibility and associations of macromolecules in solution. We will first consider the effects of molecular motions. In contrast to the rigid system used for the measurement of fundamental anisotropy, consider the other extreme in which photoselected fluorophores can tumble very rapidly during their excited state lifetimes. In this case, their orientations become randomised prior to photon emission and emission originates from a distribution of isotropically oriented molecules so that fluorescence is completely depolarised ($I_{||} = I_{\perp}$ and $r = 0$). Note that, in this case, it is the rate of rotational, not translational, diffusion that causes anisotropy reduction because molecular rotation alters the orientation of transition dipoles.

Now consider a macromolecule labelled with a fluorescent probe that is attached so that its dynamics are coupled to Brownian rotation of the larger molecule. When molecular rotation occurs on the same timescale as the excited state lifetime of the probe, then $0 < r < r_0$ and anisotropy measurements can reveal information about molecular motions. To illustrate this, imagine the time-resolved decay of anisotropy following an instantaneous pulse of vertically polarised light (Figure 14.13f). Upon excitation, the population of excited probes will have their absorption transition moments biased towards the z-axis due to photoselection. Following excitation, the molecules will randomly tumble, thereby reorienting their transition dipoles until their distributions are completely isotropic. Hence, emission anisotropy will be high immediately after excitation and will decrease over time. We can quantify the rate of molecular tumbling by the *rotational correlation time* (τ_C), defined as the average time it takes for a molecule to rotate $\cos^{-1}(1/e)$, or about 68.4°. Smaller molecules with τ_C in the picosecond range will show rapid anisotropy decay; larger molecules with τ_C in the nanosecond range will show slower anisotropy decay. This parameter also depends on molecular shape, as molecules that behave as spherical rotors will rotate faster than those that are elongated.

Time-resolved anisotropy decay can be calculated as

$$r(t) = r_0 \left[\frac{(3\cos^2\alpha - 1)}{2}\right] e^{-(t/\tau_C)}. \tag{14.32}$$

The τ_C-dependence of $r(t)$ decay is shown in Figure 14.13g. Steady-state anisotropy ($\bar{r}$) can be defined by the average of the anisotropy decay $r(t)$ normalised to the total intensity decay $I(t)$. In the case of a single exponential intensity decay:

$$\bar{r} = \frac{r_0}{\left(1 + {}^{\tau}/_{\tau_C}\right)}. \tag{14.33}$$

This gives a form of the *Perrin equation*, which allows the calculation of τ_C from anisotropy measurements. Thus, if $\tau \ll \tau_C$, there is little rotation during the fluorescence lifetime and measured r values will approach r_0; if $\tau \gg \tau_C$, then emission dipoles will randomly orient during the fluorescence lifetime and measured r values will approach zero. To gain insights into molecular motions, experimental conditions must be such that $\tau \approx \tau_C$.

14.3.6.1 Anisotropy Sample Experiment: Analysis of Molecular Rotational Dynamics

As a reporter of Brownian molecular rotation, fluorescence anisotropy is an excellent technique for analysing the sizes, flexibilities and interactions of macromolecules in solution. Here we describe an experimental approach for measuring anisotropy of a fluorescent molecule in solvents of different viscosities to obtain an estimate of its hydrated volume.

For spherical particles, τ_C can be described by the Stokes–Einstein relationship:

$$\tau_C = \frac{1}{6D_r} = \frac{\eta V_h}{k_B T}, \tag{14.34}$$

where D_r is the rotational diffusion coefficient (in s^{-1}), η is solvent viscosity (in Poise (P), where $1\,P = 0.1\,kg/m\,s$), k_B is the Boltzmann constant and V_h is the hydrated volume of the particle. This equation shows that rotational mobility is not only dependent on molecular size and shape but is also a function of solvent characteristics (greater viscosity causes higher τ_C and therefore higher r). Anisotropy measurements in solvents of known viscosity can be used to determine τ_C and therefore V_h.

14.3.6.2 Experimental Setup and Data Collection

A protein labelled with an extrinsic probe ($\tau = 15$ ns) is prepared in glycerol solutions of different viscosities to be used for steady-state anisotropy measurements. In this example, a series of samples (500 μl each) is prepared by mixing water, 50%(v/v) glycerol, 1M Tris-HCl, pH 7.5 ($[\text{Tris-HCl}]^{\text{final}} = 20$ mM) and 50 mM protein ($[\text{protein}]^{\text{final}} = 4$ μM) to produce solutions of known viscosities based on glycerol concentration (Table 14.2). Because viscosity is temperature-dependent, all preparatory and analytical steps must be performed at a controlled temperature (here, 25 °C).

A fluorescence anisotropy experiment for a fluorometer in the L-configuration equipped with polarisers is shown in Figure 14.14a. Anisotropy is determined by measuring emission intensities with P_{ex} and P_{em} (excitation and emission polarisers, respectively) in four different configurations: excitation with vertical polarisation

Table 14.2 Sample preparation for Perrin plot experiments.

Sample	% Glycerol (v/v)	η (cP)	Sample preparation (μl)			
			Water	50% glycerol	1M Tris pH 7.5	50 μM protein
1	0	0.893	450	0	10	40
2	5	1.039	400	50	10	40
3	10	1.219	350	100	10	40
4	15	1.444	300	150	10	40
5	20	1.729	250	200	10	40
6	25	2.094	200	250	10	40
7	30	2.569	150	300	10	40

Sample preparation and fluorescence experiments were performed at 25 °C (298.15 K).

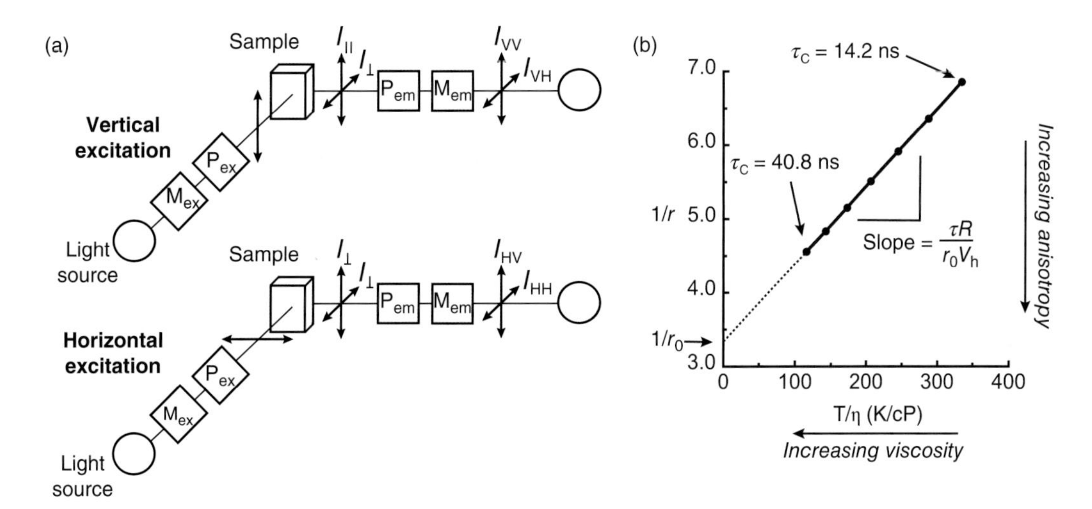

Figure 14.14 Monitoring protein dynamics by fluorescence anisotropy. (a) Diagram of L-format anisotropy measurements showing the measured intensity components with vertical excitation (I_{VV} and I_{VH}) and horizontal excitation (I_{HV}, I_{HH}). M_{ex} and M_{em} are excitation and emission monochromators, respectively; P_{ex} and P_{em} are excitation and emission polarisers, respectively. (b) Simulated data represented as a Perrin plot for a spherical fluorescent molecule ($\tau = 15$ ns). Graphical analysis yields the molecular volume (slope = $\tau k_B/r_0V_h$) or molar volume (slope = $\tau R/r_0V_h$).

(I_{VV} and I_{VH}) and with horizontal polarisation (I_{HV} and I_{HH}). Note that in the nomenclature for intensity readings, the first and second subscripts denote the orientations of the P_{ex} and P_{em}, respectively. Anisotropy is determined by $I_{||}$ and $I_{\perp}$ (Eq. (14.29)). However, because monochromators have wavelength-dependent transmission efficiencies for vertically and horizontally polarised light, it is important to adjust readings based on these efficiencies to obtain an unbiased measure of $I_{||}$ and $I_{\perp}$. To this end, a measured G factor accounts for these efficiencies and is usually determined as $G = I_{HV}/I_{HH}$. The G factor can be measured manually or automatically by the instrument. In either case, it must be known for the spectral settings of each experiment. The steady-state anisotropy is then calculated as

$$\bar{r} = \frac{(I_{VV} - GI_{VH})}{(I_{VV} + 2GI_{VH})}. \tag{14.35}$$

In the present example, the four intensity components are used in Eq. (14.35) to determine the anisotropy of each sample. Again, this can be performed manually by rotating the in-line polarisers for each of the four configurations or automatically by the fluorometer.

14.3.6.3 Data Analysis

Rearrangement of the Perrin equation (Eq. (14.33)) and substitution with Eq. (14.34) gives

$$\frac{1}{r} = \left(\frac{1}{r_0}\right) + \left(\frac{\tau}{r_0\tau_C}\right) = \left(\frac{1}{r_0}\right) + \left(\frac{\tau k_B T}{\eta V_h r_0}\right) \tag{14.36}$$

A plot of $1/r$ versus T/η, termed a *Perrin plot*, yields a straight line with a slope of $\tau k_B/V_h r_0$ and a y-intercept of $1/r_0$. One such plot from simulated data for this experiment is shown in Figure 14.14b. Extrapolation of the linear fit of the data to the y-axis (infinitely high viscosity) in this example shows an r_0 value of 0.3, and evaluation of the line gives the volume of the spherical particle ($V_h = \tau k_B/[r_0 \times \text{slope}] = 65.4\,\text{nm}^3$). From this information, based on Eq. (14.34), τ_C is seen to range from 14.2 to 40.8 ns, going from lowest to highest viscosity. Note that if absorption and emission dipoles are collinear, a measured value of r_0 less than the theoretical limit of 0.4 could indicate rapid rotational motions of the fluorophore that occur independently of the mobility of the macromolecule to which it is attached.

14.3.6.4 Experimental Considerations

1) For samples containing significant scattering or background fluorescence that contributes to anisotropy readings, blanks (containing everything except for the fluorophore) must be measured in parallel for each polariser configuration and subtracted from cognate measurements of fluorophore-containing samples prior to anisotropy calculation.
2) The presence of polarisers in the excitation and emission paths significantly attenuates light intensity. To obtain adequate signal to noise, it may be necessary to increase sample concentration or use probes with optimised spectral properties (e.g. higher Φ).
3) The use of 'magic angle' polariser conditions (e.g. vertically polarised excitation light and an emission polariser oriented 54.7° from the vertical) will select for $I_{\perp}$ emission twofold over $I_{||}$ emission. This enables measurements that are proportional to total light intensity ($I_T = I_{||} + 2I_{\perp}$).
4) Additional overviews and research on anisotropy are provided in the reference list [31–34].

14.4 Case Studies: Fluorescence Spectroscopy to Analyse Membrane Protein Structural Dynamics

Among the key advantages of fluorescence spectroscopy is that one can directly analyse structural features and physiologically relevant conformational changes of active macromolecules, even within complex molecular environments. These features are particularly useful when analysing structure–function relationships of membrane proteins. As an example, in the studies described below, multiple independent fluorescence-based techniques were used to give insights into the structural dynamics of Tim23, the central subunit of the mitochondrial TIM23 protein transport complex [9, 15] (Figure 14.15a).

14.4.1 Site-specific Labelling of Tim23 by Cotranslational Probe Incorporation

Purified proteins can be selectively labelled using extrinsic probes containing reactive moieties that recognise specific side chains (e.g. Figure 14.7d). An alternative to this chemical labelling approach is to incorporate the fluorescent probe as a non-natural amino acid into a target protein during its synthesis by including aminoacyl tRNA (aa-tRNA) analogues in the reaction [35]. To analyse Tim23, the fluorophore 7-nitrobenz-2-oxa-1,3-diazolyl (NBD) was site-specifically incorporated into the polypeptide during synthesis. Translation reactions were programmed with mRNA encoding monocysteine variants of Tim23 (each with unique in-frame 5′-UGC-3′ codons) and reactions contained NBD-Cys-tRNA^{Cys}, which incorporate NBD-labelled Cys at the cognate sites within Tim23 (Figure 14.15b). NBD-labelled Tim23 variants were then imported into isolated mitochondria by the endogenous protein biogenesis machinery. NBD fluorescence properties are dependent on the polarity and hydrogen bonding capacity of its microenvironment [36, 37]; hence, it is an excellent reporter of the environment of local regions within membrane proteins.

14.4.2 Fluorescence Analysis of Tim23 Structural Features

In these studies, mitochondria containing NBD-Tim23 served as samples for spectral analysis. Due to the high scatter and endogenous fluorescence of mitochondria, it was crucial to account for signal originating from background (i.e. from sources other than NBD-Tim23; see Sections 14.3.1 and 14.3.3). The background-corrected emission scans of Figure 14.15c revealed: (i) that the measured emission was specific for NBD incorporated into Tim23, because Cys-less (ΔCys) constructs gave a negligible signal and (ii) that the environment-sensitive NBD served as a good reporter for local polarity, because sites in the N-terminal domain (e.g. site 30) displayed red-shifted emission with reduced intensity and sites within the membrane (e.g. site 161) displayed blue-shifted emission with higher intensity. When NBD was incorporated at sequential

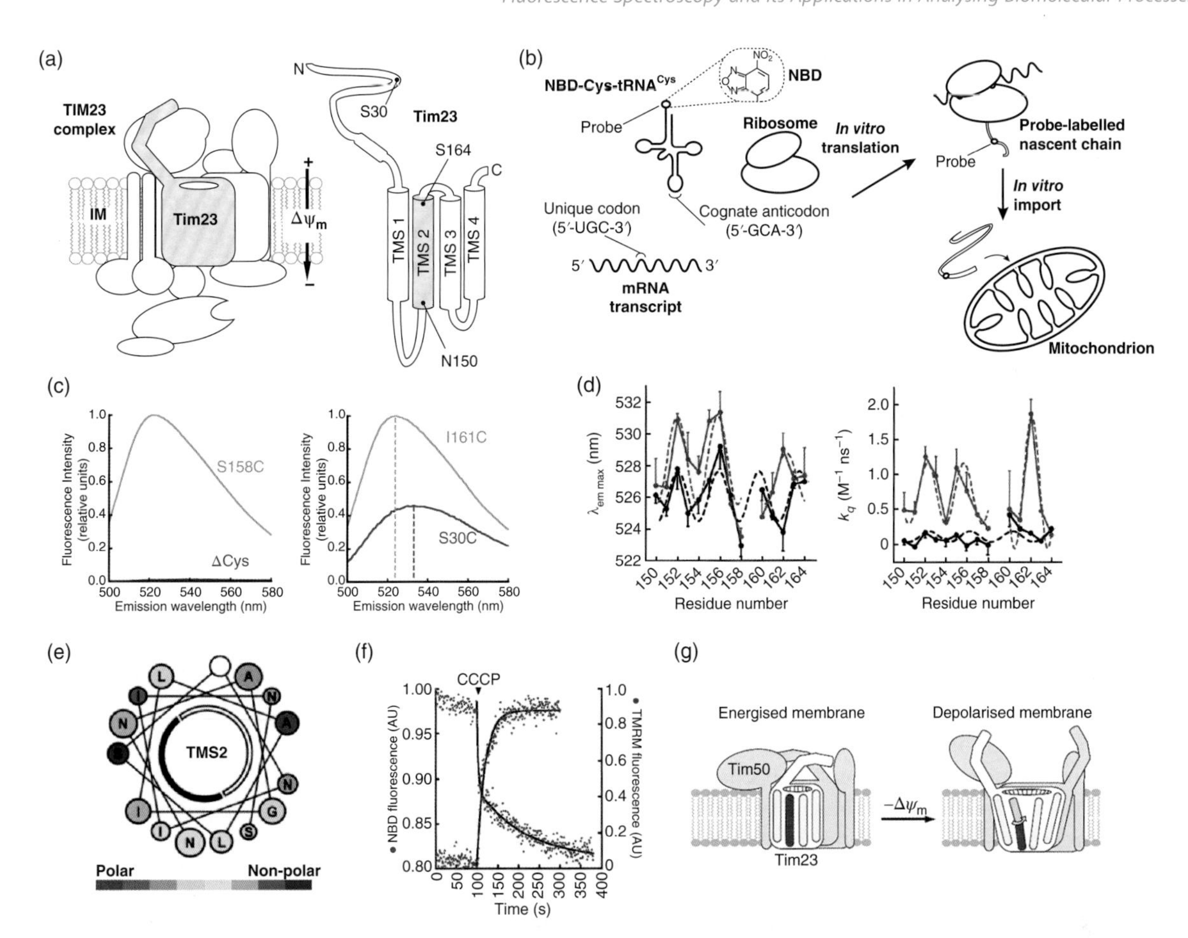

Figure 14.15 Fluorescence-based analysis of Tim23 structural dynamics. (a) The TIM23 complex. Left: the mitochondrial TIM23 protein translocation complex is a multisubunit assembly within the inner membrane (IM). Right: Tim23 has a bipartite domain organisation with a C-terminal membrane bound region of four predicted transmembrane segments (TMSs) and an intrinsically disordered N-terminal region. Shown are approximate locations of key amino acids. (b) Cotranslational site-specific incorporation of non-natural (fluorophore-containing) amino acids during cell-free translation and import into isolated mitochondria. (c) Emission spectra of NBD-Tim23. Left: comparison of emission scans for Tim23 constructs lacking (ΔCys) and including (S158C), an NBD incorporation site. Right: comparison of emission scans for Tim23 with NBD probe in a polar (S30C) or non-polar (S161C) microenvironment. (d) Fluorescence properties of NBD probes along the TMS2 helical axis, including $\lambda_{em\,max}$ (left) and k_q (right), in mitochondria that were fully energised (black) or depolarised with the protonophore CCCP (red). Data points are means ($n > 3$ independent measurements) with standard deviations and dashed traces show non-linear least-squares fits to a harmonic wave function. (e) Helical wheel projection of TMS2 showing polarity of side chain microenvironment in energised mitochondria based on steady-state and time-resolved measurements of NBD fluorescence. (f) Time course measurements of NBD-detected structural changes (blue) and membrane potential detected by the potentiometric dye TMRM (red). Fits of the data (black traces) are based on monoexponential increase (TMRM) or biexponential decay (NBD). (g) Working model for membrane potential-coupled Tim23 structural dynamics. Source: panels a, c, d, e, f, g reproduced with permission from Elsevier, taken from reference [15]. Panel b reproduced with permission from Springer Nature.

sites along Tim23 TMS2, a striking periodic pattern emerged in fluorescence properties such as $\lambda_{em\,max}$ (Figure 14.15d, left, black symbols). Fits to a harmonic wave function (dashed lines) revealed a periodicity near 3.6 residues per turn, consistent with TMS2 residing in an amphipathic environment with one helical face towards a non-polar environment and the opposing face towards a polar region (potentially facing an aqueous channel) (Figure 14.15e). Subsequent analyses confirmed that probes on the putative channel-facing side were indeed sensitive to the presence of substrates engaged with the complex [15].

14.4.3 Fluorescence Analysis of Tim23 Conformational Dynamics

The inner membrane of active mitochondria contains a transmembrane electric potential ($\Delta\psi_m$). The TIM23 complex requires the $\Delta\psi_m$ to drive protein import and Tim23 is known to be a voltage-gated channel. To address possible $\Delta\psi_m$-coupled structural changes within Tim23, NBD probes along TMS2 were again analysed following depolarisation by the protonophore CCCP, which caused two observable structural alterations (Figure 14.15d, left, compare red and black symbols) [9]. First, there was an increase in polarity of channel-facing probes near the N-terminus of TMS2, indicating greater exposure of the channel to the aqueous space. These results were confirmed by Stern–Volmer analyses of accessibility of dynamic quenching agents added externally, which showed much greater exposure following depolarisation (Figure 14.15d, right). Second, there was an observed disruption of the helical pattern near the C-terminus of TMS2 following depolarisation, suggesting partial unfolding of the helix. These structural changes could also be monitored in real time by measuring time courses of NBD fluorescence following depolarisation (Figure 14.15f). Taken together, these results formed the foundation of a working model for the structural dynamics of the Tim23 channel that are coupled to the energised state of the membrane (Figure 14.15g).

14.5 Concluding Remarks

Fluorescence spectroscopy provides a range of technical approaches for the study of biomolecules under equilibrium conditions, with high sensitivity, and in real time. This chapter has presented a range of possible steady-state approaches for addressing specific experimental questions. It should be emphasised that multiple independent fluorescence techniques could be employed to address a given biological question. In fact, the use of independent experimental approaches is encouraged to bolster confidence in the results of each individual technique. Consider the analysis of binding between macromolecules X and Y (Figure 14.16). Different fluorescence-based strategies can be used to measure a saturation binding curve or time course data to obtain equilibrium binding information. (i) Using an environment-sensitive probe on X with titration of

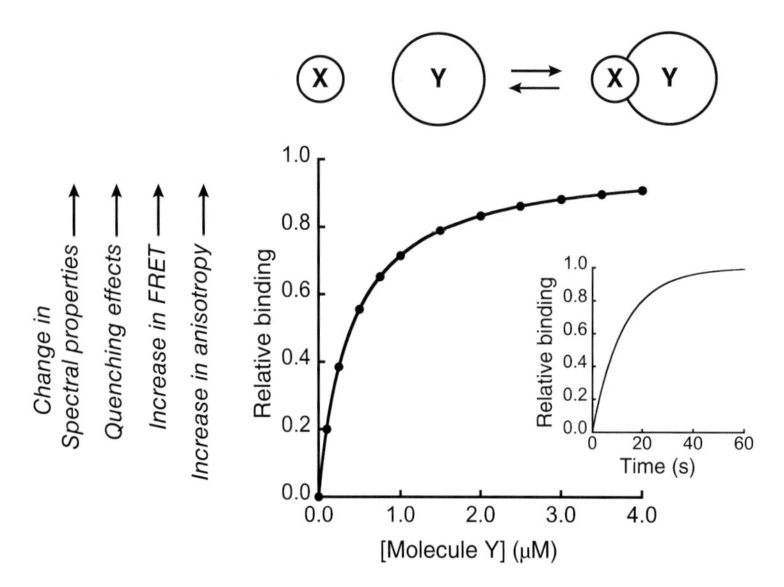

Figure 14.16 Multiple independent fluorescence techniques for equilibrium binding analysis. Hypothetical binding isotherm for molecules X and Y fit as a saturable Langmuir curve ($K_D = 0.4\ \mu M$). Inset: simulated time course for the addition of 1 nM X and 1 nM Y at time = 0 s ($k_{on} = 2 \times 10^5\ M^{-1}\ s^{-1}$; $k_{off} = 0.08\ s^{-1}$). Both types of measurements can be made by several fluorescence-based approaches.

unlabelled Y could be used as a measure of binding, provided that the binding of Y alters the spectral features (emission intensity, $\lambda_{em\ max}$) of the probe on X. (ii) Fluorescence quenching could be used if the binding of Y to X shields the probe from soluble quenching agents or, conversely, if Y contains a quenching agent near the probe at the binding interface. (iii) FRET could be used if X contains a donor probe and Y contains an acceptor probe by monitoring the progressive decrease in X emission with added Y, provided that the distance of the probes in the bound state is in the range of R_0. (iv) Fluorescence anisotropy could also be used if the binding of Y measurably changes the rotational motion of the probe on X.

Of the many experimental considerations noted in this chapter, one bears particular emphasis. The choice of fluorophore and the site to which it is bound, particularly if an extrinsic probe is used, must be carefully considered when optimising the experimental design. The probe must be located on a macromolecule such that it acts as a good reporter for the process being analysed, but must not itself have any influence on the process or alter the function of the molecule(s) being interrogated. The spectral features of the probe (absorbance, emission, lifetime) must also be compatible with the experiment. As examples, in FRET experiments, the donor and acceptor probes must have the correct spectral properties for efficient energy transfer; in anisotropy experiments, the lifetime of the probe must be a good match for its rotational mobility; and when dealing with complex molecular environments, the probe should ideally have emission in a range that can be distinguished from background fluorescence. As with any quantitative analytical approach, careful planning of fluorescence-based experiments facilitates data acquisition and interpretation.

Acknowledgements

I wish to thank Professor Arthur Johnson and all members of the Alder Research Group for their critical reading of this chapter and insightful comments. Current research in the lab of N.N.A. is supported by National Institutes of Health Grant 1R01GM113092.

References

1 Strianese, M., Staiano, M., Capo, A. et al. (2017). Modern fluorescence-based concepts and methods to study biomolecular interactions. *Mol. Syst. Des. Eng.* 2: 123–132.
2 Mocz, G. and Ross, J.A. (2013). Fluorescence techniques in analysis of protein-ligand interactions. *Methods Mol. Biol.* 1008: 169–210.
3 Weiss, S. (2000). Measuring conformational dynamics of biomolecules by single molecule fluorescence spectroscopy. *Nat. Struct. Biol.* 7: 724–729.
4 Czar, M.F. and Jockusch, R.A. (2015). Sensitive probes of protein structure and dynamics in well-controlled environments: combining mass spectrometry with fluorescence spectroscopy. *Curr. Opin. Struct. Biol.* 34: 123–134.
5 Johnson, A.E. (2005). Fluorescence approaches for determining protein conformations, interactions and mechanisms at membranes. *Traffic* 6: 1078–1092.
6 Suryawanshi, V.D., Walekar, L.S., Gore, A.H. et al. (2016). Spectroscopic analysis on the binding interaction of biologically active pyrimidine derivative with bovine serum albumin. *J. Pharm. Anal.* 6: 56–63.
7 Engstrom, H.A., Andersson, P.O., and Ohlson, S. (2005). Analysis of the specificity and thermodynamics of the interaction between low affinity antibodies and carbohydrate antigens using fluorescence spectroscopy. *J. Immunol. Methods* 297: 203–211.
8 Meyer-Almes, F.J. (2015). Kinetic binding assays for the analysis of protein-ligand interactions. *Drug Discovery Today Technol.* 17: 1–8.
9 Malhotra, K., Sathappa, M., Landin, J.S. et al. (2013). Structural changes in the mitochondrial Tim23 channel are coupled to the proton-motive force. *Nat. Struct. Mol. Biol.* 20: 965–972.

10 Woolhead, C.A., McCormick, P.J., and Johnson, A.E. (2004). Nascent membrane and secretory proteins differ in FRET-detected folding far inside the ribosome and in their exposure to ribosomal proteins. *Cell* 116: 725–736.
11 Lakowicz, J. (2006). *Principles of Fluorescence Spectroscopy*, 3e. New York, NY: Springer Science + Business Media, LLC.
12 Valeur, B. (2012). *Molecular Fluorescence Principles and Applications*, 2e. Weinheim: Wiley-VCH.
13 Goldys, E.A. (ed.) (2009). *Fluorescence Applications in Biotechnology and the Life Sciences*. Hoboken, NJ: Wiley Blackwell.
14 Gore, M.G. (ed.) (2000). *Spectrophotometry and Spectrofluorimetry: A Practical Approach*. Oxford: Oxford University Press.
15 Alder, N.N., Jensen, R.E., and Johnson, A.E. (2008). Fluorescence mapping of mitochondrial TIM23 complex reveals a water-facing, substrate-interacting helix surface. *Cell* 134: 439–450.
16 Gemeda, F.T. (2017). A review on effect of solvents on fluorescent spectra. *Chem. Sci. Int. J.* 18: 1–12.
17 Day, R.N. and Davidson, M.W. (2009). The fluorescent protein palette: tools for cellular imaging. *Chem. Soc. Rev.* 38: 2887–2921.
18 Hawe, A., Sutter, M., and Jiskoot, W. (2008). Extrinsic fluorescent dyes as tools for protein characterization. *Pharm. Res.* 25: 1487–1499.
19 Ptaszek, M. (2013). Rational design of fluorophores for *in vivo* applications. *Prog. Mol. Biol. Transl. Sci.* 113: 59–108.
20 Vos, E.P., Bokhove, M., Hesp, B.H., and Broos, J. (2009). Structure of the cytoplasmic loop between putative helices II and III of the mannitol permease of *Escherichia coli*: a tryptophan and 5-fluorotryptophan spectroscopy study. *Biochemistry* 48: 5284–5290.
21 Wang, Z., Wang, N., Han, X. et al. (2017). Interaction of two flavonols with fat mass and obesity-associated protein investigated by fluorescence quenching and molecular docking. *J. Biomol. Struct. Dyn.* 1–10.
22 van de Weert, M. and Stella, L. (2011). Fluorescence quenching and ligand binding: a critical discussion of a popular methodology. *J. Mol. Struct.* 998: 144–150.
23 Bhatia, S., Krishnamoorthy, G., and Udgaonkar, J.B. (2018). Site-specific time-resolved FRET reveals local variations in the unfolding mechanism in an apparently two-state protein unfolding transition. *Phys. Chem. Chem. Phys.* 20: 3216–3232.
24 Dimura, M., Peulen, T.O., Hanke, C.A. et al. (2016). Quantitative FRET studies and integrative modeling unravel the structure and dynamics of biomolecular systems. *Curr. Opin. Struct. Biol.* 40: 163–185.
25 Dyla, M., Terry, D.S., Kjaergaard, M. et al. (2017). Dynamics of P-type ATPase transport revealed by single-molecule FRET. *Nature* 551: 346–351.
26 Liao, J.Y., Song, Y., and Liu, Y. (2015). A new trend to determine biochemical parameters by quantitative FRET assays. *Acta Pharmacol. Sin.* 36: 1408–1415.
27 Liu, Y., Chen, L.Y., Zeng, H. et al. (2018). Assessing the real-time activation of the cannabinoid CB1 receptor and the associated structural changes using a FRET biosensor. *Int. J. Biochem. Cell Biol.* 99: 114–124.
28 Ma, L., Yang, F., and Zheng, J. (2014). Application of fluorescence resonance energy transfer in protein studies. *J. Mol. Struct.* 1077: 87–100.
29 Preus, S. and Wilhelmsson, L.M. (2012). Advances in quantitative FRET-based methods for studying nucleic acids. *ChemBioChem* 13: 1990–2001.
30 Voith von Voithenberg, L. and Lamb, D.C. (2018). Single pair Forster resonance energy transfer: a versatile tool to investigate protein conformational dynamics. *Bioessays* 40.
31 Gradinaru, C.C., Marushchak, D.O., Samim, M., and Krull, U.J. (2010). Fluorescence anisotropy: from single molecules to live cells. *Analyst* 135: 452–459.
32 James, N.G. and Jameson, D.M. (2014). Steady-state fluorescence polarization/anisotropy for the study of protein interactions. *Methods Mol. Biol.* 1076: 29–42.

33 Kuznetsova, I.M., Sulatskaya, A.I., Maskevich, A.A. et al. (2016). High fluorescence anisotropy of thioflavin T in aqueous solution resulting from its molecular rotor nature. *Anal. Chem.* 88: 718–724.
34 Mishra, S., Meher, G., and Chakraborty, H. (2017). Conformational transition of kappa-casein in micellar environment: insight from the tryptophan fluorescence. *Spectrochim. Acta, Part A* 186: 99–104.
35 Schwall, C.T. and Alder, N.N. (2013). Site-specific fluorescent probe labeling of mitochondrial membrane proteins. *Methods Mol. Biol.* 1033: 103–120.
36 Lancet, D. and Pecht, I. (1977). Spectroscopic and immunochemical studies with nitrobenzoxadiazolealanine, a fluorescent dinitrophenyl analogue. *Biochemistry* 16: 5150–5157.
37 Lin, S. and Struve, W.S. (1991). Time-resolved fluorescence of nitrobenzoxadiazole-aminohexanoic acid: effect of intermolecular hydrogen-bonding on non-radiative decay. *Photochem. Photobiol.* 54: 361–365.

Further Reading

Albani, J. (2007). *Principles and Applications of Fluorescence Spectroscopy*. Wiley Blackwell.
Lakowicz, J. (2006). *Principles of Fluorescence Spectroscopy*, 3e. New York, NY: Springer Science+Business Media, LLC.
Valeur, B. (2012). *Molecular Fluorescence Principles and Applications*, 2e. Weinheim: Wiley-VCH.
Goldys, E.A. (ed.) (2009). *Fluorescence Applications in Biotechnology and the Life Sciences*. Hoboken, NJ: Wiley Blackwell.
Gore, M.G. (ed.) (2000). *Spectrophotometry and Spectrofluorimetry: A Practical Approach*. Oxford: Oxford University Press.

15

Circular Dichroism and Related Spectroscopic Techniques

Sophia C. Goodchild, Krishanthi Jayasundera and Alison Rodger

Department of Molecular Sciences, Macquarie University, NSW, 2109, Australia

15.1 Significance and Background

Humans navigate the world in which we live largely by vision. Our photoreceptors are able to detect light mainly between 390 and 720 nm. Along with the use of infra-red (IR) light (which we refer to as heat), sound waves and contact, this means our spatial resolution can range from hundreds of micrometres to of the order of a kilometre. Our spatial resolution is thus about seven orders of magnitude, but we use only a small part of the electromagnetic spectrum and despite our dynamic range we do not approach the molecular level. If we could see the molecular level, we would be overwhelmed by data but would be able intuitively to grasp how biological and other molecular systems work.

We can use different microscopy techniques to improve the dynamic range of our vision. Classical optical microscopy takes our resolution to about half the wavelength of the light used. When one uses electron rather than light microscopy, has ideal samples, low temperature, and quite a lot of computing power, it is now possible to 'see' subnanometre structures. The various forms of probe microscopy are also in the same regime, with scanning tunnelling microscopy having lateral resolution of down to ~0.1 nm and atomic force microscopy a bit higher. These techniques give us the impression of being able to see the molecular world and gives rise to the question of whether spectroscopic techniques have seen their day.

Spectroscopic techniques measure the interaction of radiation with matter and are loosely separated from microscopy techniques in that they usually involve scanning over a wavelength, frequency or energy range, and typically average over many molecules at one time with little spatial resolution. The spectral response depends on the electrons, protons or neutrons of the molecules in the sample interacting with the radiation, and can often be interpreted to give information about molecular electronics and structure. Microscopy and spectroscopy are gradually converging with spectral resolution microscopy [1] and spatial resolution spectroscopy [2]. However, there is a long way to go before microscopy-collected datasets approach the information content of spectroscopic experiments. With spectroscopy, we need to deal with the fact that a 1 mM sample in a 1 cm path length cell contains 10^{15} molecules in a 1 mm^2 light beam. As spectroscopic measurements average over all species through which the light beam passes, we may need to change variables such as concentration, temperature, solvent or sample preparation to determine information for a single species.

In this chapter the use of some different spectroscopy techniques, including some that use polarised light, is outlined. Application to biomolecules is the focus and a case study is presented where we use spectroscopy to determine when a protein molecule has changed its structure. The case study is representative of a current challenge, which is to determine whether or not different biopharmaceutical formulations of a drug are the same. This challenge has come into focus as the patents of biopharmaceuticals are expiring, creating the opportunity to develop so-called biosimilar drugs. We desperately need analytical methodologies to determine how 'similar' a proposed product is to the original innovator product. In contrast to small molecule drugs, the

Biomolecular and Bioanalytical Techniques: Theory, Methodology and Applications, First Edition. Edited by Vasudevan Ramesh.

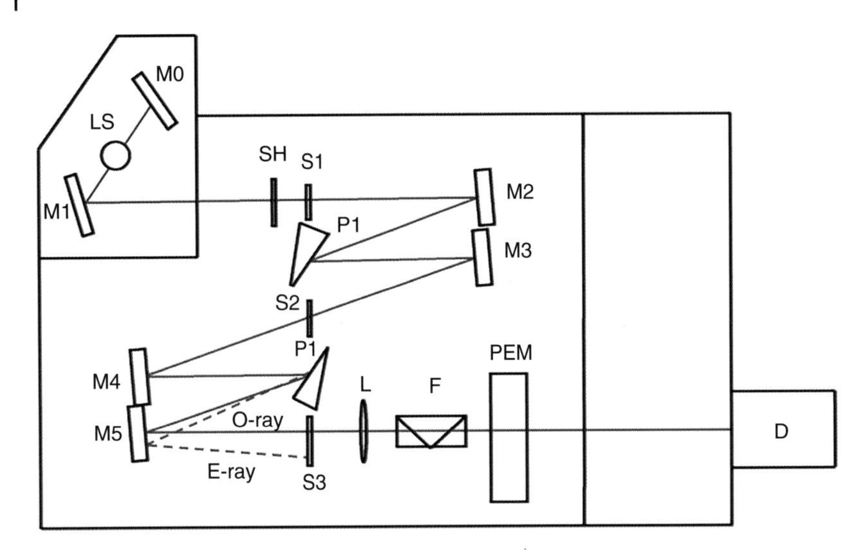

Figure 15.1 A schematic of the Jasco J-815 CD spectrometer, redrawn from the user manual. LS stands for light source, M for mirror, S for slit, P for prism, L for lens, F for polariser, PEM for photoelastic modulator and D for detector. These components are discussed in the main text. E-ray and O-ray denote the extraordinary ray and ordinary ray, respectively.

activity of a protein biopharmaceutical is dependent not only on its primary structure (what atom is bonded to what) but also on its secondary and tertiary structures. Within any solution-phase sample there will be a distribution of geometries, either as an equilibrium between different structures, so on average the same, or actually different isolatable structures. We will outline how circular dichroism (CD) spectroscopy can be used to estimate the secondary structure of unknown proteins. IR absorbance spectroscopy of proteins also contains secondary structure information and we consider how it might be used as an alternative to CD. Another spectroscopy technique that can be implemented on a CD instrument, flow linear dichroism (LD), provides complementary structural information about molecules with a high aspect ratio that can be flow oriented.

All spectroscopic instruments have the same basic components: a light source, a sample and a detector. Most ultraviolet (UV)–visible spectrometers have a monochromator to control the wavelength of light incident on the sample. CD and LD experiments in addition require the light to be polarised, usually before it is incident on the sample. A schematic of a CD spectropolarimeter is shown in Figure 15.1.

15.2 Theory/Principles

Spectroscopy involves measuring how molecules interact with light. Absorbance spectroscopy is dominated by the transitions between the ground state energy level and an excited state that follow from the interaction of the electric field of the light with the molecule. If a molecule absorbs a photon of frequency ν, it increases its energy by

$$\Delta E = h\nu = hc/\lambda, \tag{15.1}$$

where h is Planck's constant, λ is wavelength and c is the speed of light. UV and visible light cause transitions between electronic energy levels and IR radiation causes transitions between vibrational energy levels. An absorbance transition is largest when the electric field, $\boldsymbol{E}$, of the light is parallel to the electric dipole transition moment, $\boldsymbol{\mu}$, that links the ground and excited state. The absorbance intensity of a transition is proportional to [3]

$$(\boldsymbol{E}\cdot\boldsymbol{\mu})^2 \tag{15.2}$$

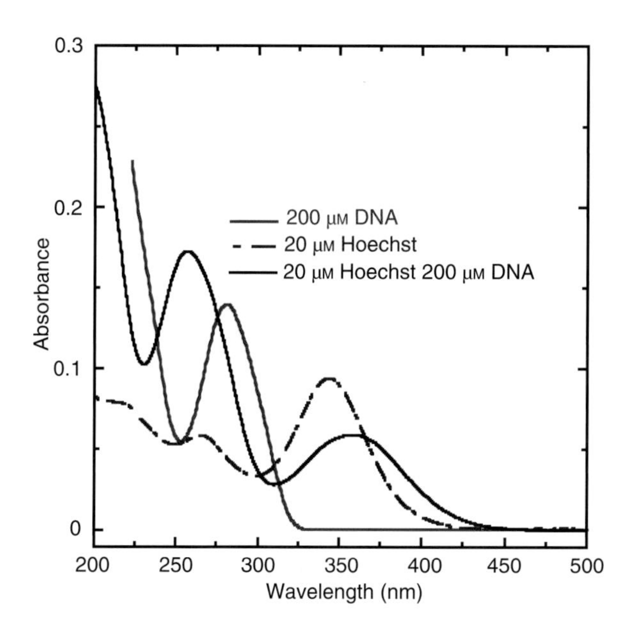

Figure 15.2 Absorbance spectrum of calf thymus DNA (200 μM base), Hoechst (20 μM) and DNA (200 μM) plus Hoechst (20 μM) in water in a 1 mm pathlength cuvette.

and the direction of $\boldsymbol{\mu}$ is called the transition polarisation. The magnetic field can also induce a transition, but the effect is much weaker than the result of the electric coupling. Hence, in practice, even for transitions where Eq. (15.2) is zero for the average structure, the small contributions from when it vibrates away from equilibrium dominate the magnetic component. In a collection of molecules, the photons absorbed by different molecules will be of slightly different energies so what we measure is a curve such as the one in Figure 15.2, where the signal that is plotted is a measure of the probability that a transition will occur at that energy (or wavelength). Such a plot of the absorbance of light verses λ or ν is known as an absorption spectrum. The Beer–Lambert law relates the absorbance of light to properties of the samples:

$$A = \varepsilon C \ell \tag{15.3}$$

for A being absorbance, ε the wavelength dependent extinction coefficient (which hides all the electronic structural information), C the concentration and ℓ the pathlength of the sample. Absorbance is dimensionless so the unit and value of the extinction coefficient depends on the unit of concentration and pathlength used. For proteins, in the case of molar concentrations, the unit of the extinction coefficient is usually $\text{mol}^{-1}\ \text{dm}^3\ \text{cm}^{-1}$ and the value can be tens of thousands, while in the case of mg/ml concentration, it is $\text{mg}^{-1}\ \text{cm}^2$ and its value is in the range of 0.3–3.

Figure 15.2 illustrates the kind of information that can be deduced from a normal absorption spectrum. The dashed blue line is the spectrum of Hoechst, a bis-benzimide compound that binds DNA. Given the concentration of the sample and the pathlength are known, the extinction coefficient at each wavelength can be determined. When DNA is added (the solid black line) we see the DNA signal (solid red line) added to that of the Hoechst below 300 nm. The Hoechst absorbance intensity changes significantly when it binds to the DNA, in this case because its geometry changes towards a more planar structure.

15.2.1 Circular Dichroism Spectroscopy

If we use polarised light in a spectroscopy experiment, we can enhance the information available for systems that interact differently with different polarisations. The most commonly used polarisation type is circularly polarised light whose electric and magnetic fields trace helices in space and time. For solutions where the molecules are randomly oriented, the only kind of molecule that interacts differently with left and with right circularly polarised light are chiral molecules – that is, molecules whose mirror images are not superposable on each other, e.g. [3–6]. The difference in absorbance of the two polarisations gives rise to the spectroscopy called CD, which is the difference in absorption of left and right circularly polarised light:

$$CD = A_l - A_r = \Delta\varepsilon C\ell. \tag{15.4}$$

One can use CD quite effectively without understanding what it is, but it is helpful to have at least a pictorial understanding of the origin of the signal before attempting to interpret data. The key feature that gives rise to a non-zero CD spectrum is that the electron redistribution that happens during the transition is helical. This is mathematically expressed by two vectors, one of which describes the linear direction of the electron density change, $\boldsymbol{\mu}$, which arises from coupling with the electric field of the light, and one of which describes the circling of charge about that direction, which we denote $\boldsymbol{m}$, which arises from coupling with the magnetic field as discussed above. Simultaneous linear and circular rearrangements of the electrons gives us a helix. The Rosenfeld equation for CD magnitude is [3, 7, 8]

$$R = \mathrm{Im}(\boldsymbol{\mu} \cdot \boldsymbol{m}) \tag{15.5}$$

where Im denotes 'imaginary part of', since the magnetic dipole operation is imaginary. The sign of the CD depends on the handedness of the helix of electron redistribution – which has a 50% chance of being the same as the helix of the molecular bonding framework. The challenging task is to relate the helical motions of the electrons to the arrangement of the atoms and bonds in space.

CD is now a routine tool in many laboratories. The most common applications include proving that a chiral molecule has indeed been synthesised or resolved into pure enantiomers and probing the structure of biological macromolecules, in particular determining the secondary structure content of proteins. It is also useful for probing the binding of molecules to a chiral molecule, as the CD spectrum is perturbed by the interaction. In particular, if an achiral molecule binds to a chiral molecule, an induced CD signal will appear in the absorption bands of the achiral molecule. Most CD experiments involve randomly oriented samples. However, if the samples are oriented, even unintentionally, great care must be taken to ensure that one is not measuring LD rather than oriented CD due to instrumentation imperfections. If a sample is oriented with the unique axis along the light path, its CD will be independent of its rotation about that axis. However, any LD signal will invert when the sample is rotated 90°.

Most chiral molecules can be divided up into smaller subunits in which electronic transitions are largely located, called chromophores. For nucleic acids and proteins, the chromophores are approximately achiral. Thus, we can often understand the CD of a molecule as arising from the coupling of electronic transitions in different parts of the molecule. If we consider two monomers, the dimer system has two new transitions corresponding to collective charge displacements, described as the sum or difference of the two monomeric electric dipole transition moments to give two new ones [3]:

$$\boldsymbol{\mu}(\text{dimer}) = \frac{1}{\sqrt{2}}[\boldsymbol{\mu}(\text{monomer a}) \pm \boldsymbol{\mu}(\text{monomer b})] = \frac{1}{\sqrt{2}}[\boldsymbol{\mu}^{\text{a}} \pm \boldsymbol{\mu}^{\text{b}}], \tag{15.6}$$

where the factor of $1/\sqrt{2}$ ensures that the total absorption intensity is conserved. However, as a result of the + and − signs in Eq. (15.6), the absorption intensities of the two transitions may differ from each other. For example, if the monomeric transition moments are parallel, the plus combination will be a transition that contains all the absorption intensity of the two monomers while the minus combination will have zero absorption (the two anti-parallel monomeric dipoles exactly cancelling each other). In the spectrum, this extreme case will look like a shift of the monomer absorption band to either shorter (H-aggregate) or longer (J-aggregate) wavelengths, depending on the changes to the transition energies. The coupling is called exciton coupling and the direction of shift depends on whether the geometry of the system means that dipole combination is, respectively, repulsive (destabilising) or attractive (stabilising) when one puts + and − signs on each end of the dipole.

Figure 15.3 CD spectrum of Λ-[Ru(1,10-phenanthroline)$_3$]$^{2+}$ (black solid line) and Δ-[Ru(1,10-phenanthroline)$_3$]$^{2+}$ (pink dashed line). $\Delta\varepsilon$ in units of mol^{-1} dm^3 cm^{-1}.

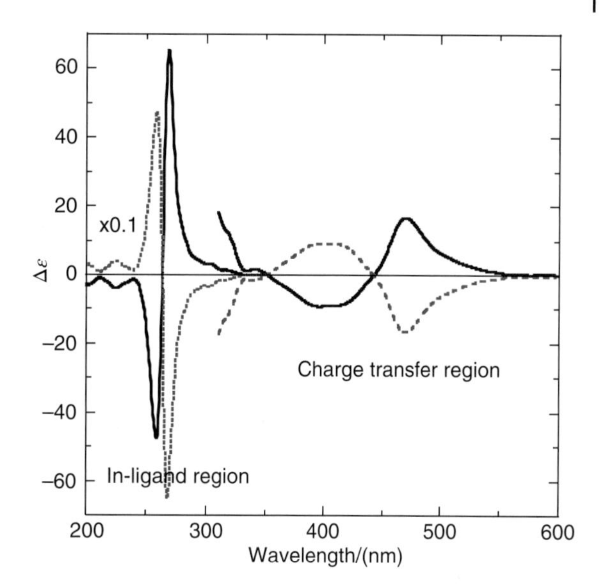

If the monomer transitions in Eq. (15.6) are oriented parallel, anti-parallel or orthogonal to each other, then they make an achiral system as the mirror image is the same as the original and we get no CD. However, any other angle gives a system where there is a net circulation of charge between the ground and excited states, which interacts with both the electric and magnetic fields of the light (Eq. (15.5)). The mirror image system of the one we start with has the opposite handed helices and so for each transition will have a CD spectrum that is −1 times the original, as illustrated in Figure 15.3. If the two monomers are identical so their transitions are degenerate, the energies of the transitions of the coupled system is

$$\varepsilon^{\pm} = \varepsilon \pm V, \tag{15.7}$$

where

$$V = \frac{\boldsymbol{\mu}^{a}\cdot\boldsymbol{\mu}^{b} - 3\boldsymbol{R}_{AB}\cdot\boldsymbol{\mu}^{a}\boldsymbol{\mu}^{b}\cdot\boldsymbol{R}_{AB}}{R_{AB}^{3}} \tag{15.8}$$

for $\boldsymbol{R}_{AB}$ the vector from the origins of monomer A to that of B and we get an obvious +/− couplet in the CD spectrum due to overlap and cancellation. If the monomers are different the transition energies are very close to the uncoupled ones and we do not see the overlapping bands, but the CD signals at the two energies are opposite in sign. A real molecule has more than one transition and so what we observe is always a combination of many different interactions.

Figure 15.4 shows the absorbance (lower dashed line) and CD (upper solid line) spectrum for a chiral molecule where two anthracenes are held in a skewed arrangement. In the 254 nm region of the spectrum we see essentially a single exciton couplet from that intense transition. The CD is zero at the absorbance maximum where there is cancellation of the positive and negative components. In the 380 nm region of the spectrum we see evidence of the exciton coupling the vibronic components, but the multiple couplings of neighbouring components result in a less symmetric spectrum.

15.2.2 Linear Dichroism

When the electric field of the light oscillates in a plane, the result is linearly polarised light. The polarisation of the light is parallel to its electric field. From Eq. (15.2) we can see that oriented systems may interact differently with different polarisations of linearly polarised light and so may give an LD spectrum: the

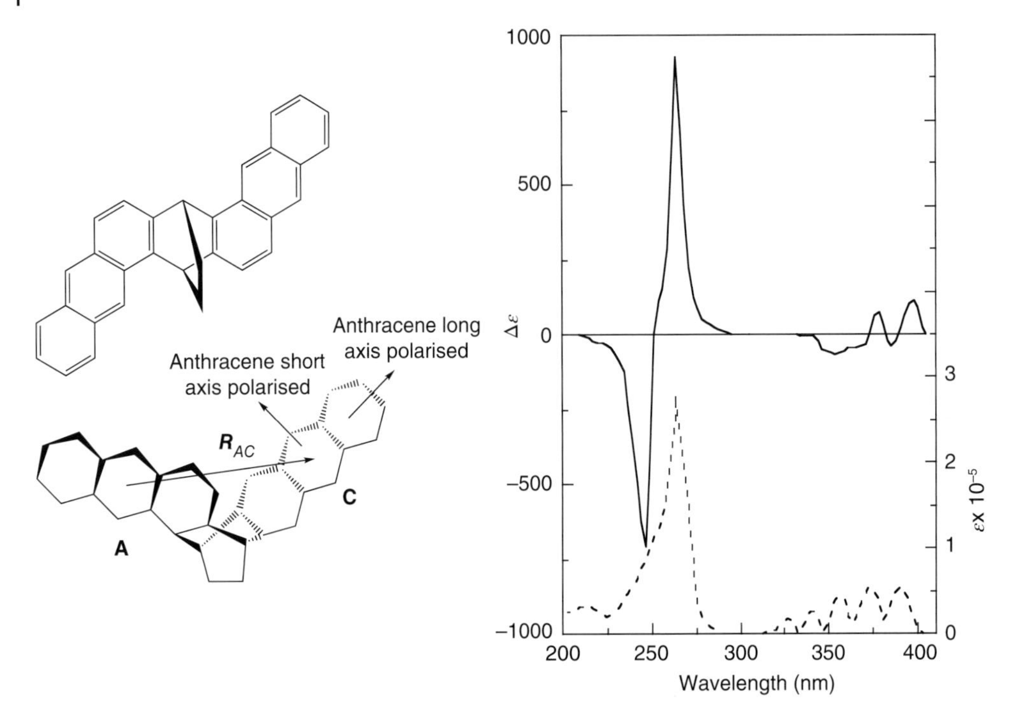

Figure 15.4 Molecular geometry (left), absorption (lower dashed line) and CD (upper solid line) of 6*R*,15*R*-(+)-dihydro-6,15-ethanonaphthol[2,3-*c*]pentaphene. Source: Spectra sketched from reference [9].

difference in absorption of light linearly polarised parallel (//) and perpendicular (⊥) to an orientation axis is, e.g. see [3, 10–12],

$$LD = A_{//} - A_{\perp}. \tag{15.9}$$

Figure 15.5 illustrates many of the possibilities of LD spectroscopy. In this case anthracene has been deposited on a stretched oxidised polyethylene film (a high quality plastic bag held in a film stretcher). When the sample was prepared to ensure monomers dominated (0.1 mg/ml stock solution, black solid lines), we simply see a spectrum that tells us the polarisation of the transitions relative to the stretch direction. The 254 nm region is parallel to the stretch direction in accord with its known polarisation being along the long axis of anthracene. It follows the shape of the normal absorption spectrum. The 380 nm region is negative in accord with it being short-axis polarised and, thus, perpendicular to the 254 nm region. The vibronic components follow the negative of the absorbance spectrum. At first sight the positive 325 nm signals are perplexing since they are part of the 380 nm vibronic progression. However, the closer the transitions get to the large 254 nm band the more it couples into the transition contributing long-axis polarised intensity, so the high energy part of the progression has the same sign as the 254 nm band.

When the sample is deposited from an 8 mg/ml solution (pink dashed lines), it is no longer monomeric and we see exciton shifts of bands and the appearance of exciton couplets in the spectrum. Since the two new transitions from any exciton coupling are always polarised perpendicular to each other, in the simple film orientation system used for anthracene in Figure 15.5 they will almost certainly be of opposite sign. Anthracene forms a range of different oligomeric structures and different experiments can yield quite different spectra depending on how the sample dries on to the film [13].

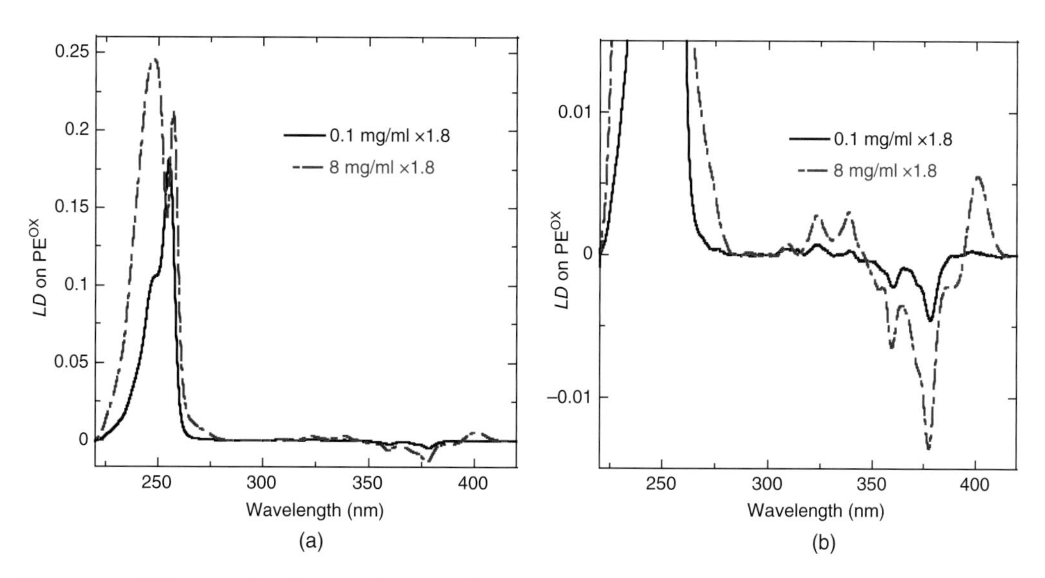

Figure 15.5 (a) Spectra of anthracene deposited from 0.1 mg/ml and 8 mg/ml stock solutions in chloroform on to oxidised polyethylene (PE^{OX}) and stretched by a factor of 1.8. (b) A matching baseline without anthracene was subtracted in each case. Source: Data are from reference [13].

In more complex orienting systems, such as dyes binding to flow-oriented DNA, more complex options are possible. Quantitative analysis using the ratio of LD and absorbance, the reduced LD, is often helpful:

$$LD^r = \frac{LD}{A_{iso}} = \frac{3}{2}S(3\cos^2\alpha - 1), \tag{15.10}$$

where α is the angle between the orientation direction and the transition dipole moment, S is the orientation parameter and the *iso* subscript is to remind us that this is strictly the absorbance of an isotropic sample, not an oriented sample. S is usually the 'Achilles' heel' of quantitative LD analysis as it typically varies from experiment to experiment and unless there is an internal standard (such as a DNA base signal when we are doing a ligand binding experiment, as discussed below) it can be challenging to estimate.

15.3 Technique/Methodology/Protocol

15.3.1 The Sample: Concentration, Pathlength and Matrix

Any absorption spectroscopy technique has a dynamic range limited by the need to have enough photons reaching the detector for us to be able to count and a signal above the background noise level. What this means varies from instrument to instrument and as a function of wavelength. It also depends not only on the sample but also its matrix.

For biomolecules one cannot design a spectroscopy experiment without considering the buffer and any associated salts. For example, for a backbone protein CD experiment on a bench top (i.e. not synchrotron) instrument we usually aim to collect data from 260 nm to at least 190 nm and ideally a bit lower in wavelength. Many buffer components, including chloride ions, sugars and amino acids as well as the more obvious buffer components such as the sulphite of HEPES and MOPS, absorb light in this region, thus reducing the effective lamp intensity for the sample. Likewise, sugars and amino acids are chiral themselves, so also contribute to the CD spectrum. Chloride ions in 'standard' PBS (phosphate buffered saline) vary from laboratory to laboratory, but in all cases are more than one wants to use in a CD experiment as most of the photons will be absorbed by chloride by 215 nm. So step (i) for any spectroscopy experiment is to measure an absorbance spectrum of

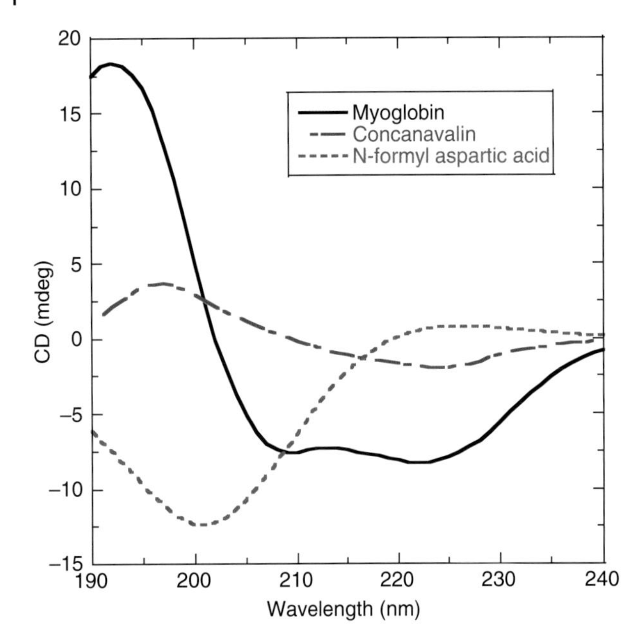

Figure 15.6 CD spectra of myoglobin (77% α-helix, 0% β-sheet), concanavalin (0% α-helix, 40% β-sheet) and N-formyl aspartic acid (100% random coil) all at 0.1 mg/ml in water in a 1 mm cuvette.

the buffer, solvent or other matrix. This can usually be done in the CD instrument if that option is checked in the parameter menu.

Let us assume that we have found a largely invisible buffer (phosphate and TRIS without extra chloride are a good start for backbone protein CD), we then must consider how much sample is needed. A rough rule-of-thumb is that 0.1 mg/ml in a 1 mm pathlength cuvette for backbone protein spectroscopy is ideal. Typical spectra for predominantly helical myoglobin (solid black line), β-sheet rich concanavalin (dashed pink line) and a random coil polymer (N-formyl aspartic acid, green dotted line) measured at 0.1 mg/ml in water are illustrated in Figure 15.6.

So what is the significance of 0.1 mg/ml and a 1 mm pathlength? The average amino acid weight in a protein is between 105 and 110 Da or g/mol, so 0.1 mg/ml is approximately 1 mM residue concentration. Below 250 nm, most of the absorbance is due to the amino bonds of the protein backbone, with only a little coming from any side chains. Since any protein is a mixture of amino acids, it follows that the concentration of the amides is the key. The protein molar concentration is not relevant as different sized proteins will have different numbers of amide bonds. Following Eq. (15.3), a 1 mM residue concentration in a 1 mm pathlength gives an absorbance of approximately 1, which is optimal for instrument performance. If one wishes to use a higher concentration then, following Eq. (15.3), one must reduce the pathlength such that the total absorbance of the sample does not exceed approximately 2. In practice, the extent to which one can do this depends on the operator's ability to assemble demountable cells reproducibly below 0.1 mm (the smallest available fixed pathlength quartz cuvette). It should also be remembered that any non-protein species that absorbs reduces the photons available to the protein – hence the discussion about buffers above. If buffer components are chiral, they not only absorb photons but also give a CD signal and so baseline correction must be very carefully done.

It should be noted that we tend not to use the backbone region of the protein absorbance spectrum to determine concentration. This is partly because of the potential contribution from buffer components and light scattering (LS) but also because there is no maximum just gradually increasing signal as one decreases the wavelength below about 250 nm. However, in some cases the protein backbone absorbance spectrum can be useful for estimating protein concentration and typical extinction coefficients are available in reference [6].

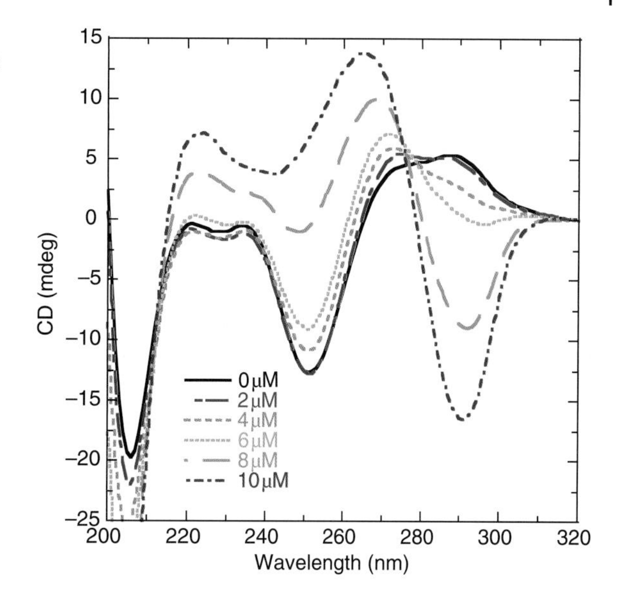

Figure 15.7 Poly[d(G-C)$_2$] DNA CD spectrum (50 μM in water) with increasing concentrations of spermine converting it from right-handed B-DNA to left-handed Z-DNA. Source: Data replotted from reference [14].

By way of contrast, the so-called aromatic region, which is dominated by tyrosines and tryptophans along with a small contribution from disulphide bonds, can be used to estimate protein concentration (typically at 280 nm). The CD signal in the aromatic region can also provide a fingerprint to compare proteins. However, in this case care must be taken to choose an appropriate combination of protein concentration and pathlength based on the protein being analysed. Unless the percentage of aromatic residues and disulphide bonds is known, then it is really just a sophisticated guess to say a 1 mg/ml sample in a 1 cm cuvette has an absorbance of 1. For proteins with known sequences, the ExPASy ProtParam website provides a useful tool to estimate the 280 nm extinction coefficient from approximate extinction coefficients for residues. These 280 nm protein extinction coefficients typically range from approximately 10 000–200 000 mol^{-1} dm^3 cm^{-1}. Strictly speaking, unfolded proteins should be used to determine the protein concentration using these extinction coefficients.

Similar to the situation with the backbone region of proteins, we can identify ideal concentrations for DNA spectroscopy since DNAs are composed of the same four components. For DNA we usually expressed concentration in terms of the concentration of bases (or base pairs). A random sequence genomic DNA has an extinction coefficient of about 6600 mol^{-1} dm^3 cm^{-1} per DNA base at 258 nm so an ideal concentration for a UV absorbance or CD experiment is approximately: 150 μM in a 1 cm path length cuvette for DNA spectroscopy as this gives a 258 nm absorbance of about one, which optimises instrument performance. While this could be expressed as 0.05 mg/ml, for DNA sample preparation it is better to avoid assuming one's sample only contains DNA and to use the Beer–Lambert law to determine the concentration. The choice of a 1 cm pathlength for DNA samples is due to the empirical observation that the Beer–Lambert law (Eq. (15.3)) tends to break down when the base concentration goes above 200 μM. The DNA CD spectrum for 100% alternating G-C DNA with increasing concentrations of spermine is given in Figure 15.7. As the concentration of the highly cationic spermine is increased, the DNA converts from standard B-form to Z-form with approximate inversion of the CD spectrum. The inversion is only approximate as, although the helix changes handedness, the final structure is not the mirror image of the original.

For other samples, some idea of the extinction coefficient is needed to determine the optimal concentration and pathlength combination for an experiment. As a general rule, UV–visible spectroscopy extinction coefficients range up to 20 000 mol^{-1} dm^3 cm^{-1}. By way of contrast, IR samples are seldom greater than 100 mol^{-1} dm^3 cm^{-1}. Unfortunately for IR spectroscopy of proteins and DNAs, water has a significant absorbance in the same region and omitting it from the sample cannot be assumed to have no effect on the biomacromolecule structure. We return to this issue below.

15.3.2 The Instrument

15.3.2.1 UV/Visible Instruments

The key to collecting spectroscopic data, as noted above, is to ensure sufficient photons reach the detector and also that the instrument correctly records intensity at the indicated wavelength/wavenumber on the output file. To have a non-zero signal the photon count needs to be measurably different from the corresponding background and noise. To obtain the true spectrum of the analyte(s), it is essential to subtract any buffer or solvent signal (as discussed above), as well as the signal of the instrument and the cuvette. For this reason, many users take their buffer or solvent as the background. However, for biomolecules and particularly for CD and IR absorbance experiments, it is generally advisable to use air as the background and measure the buffer/solvent as a 'sample' that can subsequently be subtracted from the spectra of samples containing the analyte(s) of interest. This aids in seeing any issues relating to the buffer/solvent or the instrument. Some sample issues were discussed above. It is also usually advisable to use a quartz cuvette for CD experiments, even if one is working in the visible region, as they tend to have a smaller intrinsic CD signal. In addition, it is always essential to ensure that the light beam is passing through the sample and not hitting the side of the cuvette or the meniscus of the sample. The size of the light beam can depend on the choice of parameters so it must be checked when these are changed. Most absorbance spectroscopies are performed in the transmission mode with the light beam passing through the sample.

A CD instrument is required for a CD experiment. It produces alternatively left and right circularly polarised light, usually at 50 kHz in the UV–visible (electronic transitions) region of the spectrum. Different instruments use different parameters and mean different things by their labelling. For example, is a 1 nm bandwidth an indication of the spread of the wavelength of the beam at half-height or somewhere else? Some parameter sets are clearly inappropriate for some applications, for example, if one is undertaking a CD experiment with a scan speed of 100 nm/min and 10 seconds response time, then the resulting data will be averaged over nearly 20 nm whatever bandwidth one chooses, thus giving a distorted spectrum when plotted as a function of wavelength. However, for some experiments, such as kinetics, the noise reduction of a larger bandwidth is very attractive – assuming the beam size is not also distorted (and it should be noted that on many UV instruments the beam size gets larger with decreasing wavelength – and you cannot easily see its size in the UV).

As CD is the measurement of a small difference between large absorbances we would want to push the parameters without compromising the data. A good starting set of parameters for proteins and nucleic acids is typically 100 nm/min, 1 nm bandwidth and 1 second resolution. Because these molecules all have broad bands, one can usually compromise somewhat and, for example, increase the bandwidth to 2 nm. Testing whether the spectrum overlays with a safe (but more time consuming) and a less safe (but less time consuming) parameter set is an essential part of experiment design. We have found that increasing the scan speed can compromise data quality with some instruments – presumably due to moving parts in the instrument not quite catching up. Therefore we usually compromise with larger bandwidths and/or resolutions. Averaging over more than one spectrum or increasing the resolution improves the signal-to-noise ratio by the square root of the factor.

15.3.2.2 Infra-red Absorbance

As discussed in Chapter 13, IR spectra give the energies and intensities of the vibrational modes of the molecules in the sample, which for proteins and other macromolecules contain information about secondary structure. As H_2O has an absorbance maximum at 1644 cm^{-1} with $\varepsilon \sim 21.7$ mol^{-1} dm^3 cm^{-1}, given that water concentration is 55.5 M, then we need an 8 μM pathlength to have an absorbance in the region of 1. In this experiment a 20 mg/ml protein will contribute about 5% of the total intensity, which makes baseline correction challenging. In our experience, effective baseline correction in transmission IR spectroscopy requires the sample and baseline to have the same path length, which we find extremely hard to deliver as the viscosity of buffer and protein-in-buffer samples are very different. Limited scaling (less than 0.05) of the baseline is possible to try to deliver a flat sample spectrum at 2125 cm^{-1} (which corresponds to a liquid water libration that is not present for the protein or DNA so is indicative of good correction). An experimentally simpler option is to use attenuated total reflectance (ATR) where the absorbance of the sample defines the penetration depth of the evanescent wave. As the water absorbance dominates the protein or nucleic acid, this can be assumed to be approximately the same and subtraction with minimal scaling delivers the required flat region near 2125 cm^{-1}. Further correction to remove vapour contributions to the spectrum may be required, as illustrated below.

In an ATR experiment the sample is dropped on to the surface of a dense internally reflecting crystal. The light beam interacts with the sample via the electric field of its evanescent way, which does not transmit through the sample, as in a transmission experiment, but has intensity which decays exponentially from the surface. The rate of decay depends on the refractive index of the sample, which is turn varies with sample absorbance. Thus, unfortunately the ATR spectrum has a different shape from the corresponding transmission one. Given that water absorbance dominates the protein or nucleic acid, the refractive index of a protein or nucleic acid sample is approximately that of water and we have developed a method for transforming an ATR spectrum into what would be obtained in a transmission experiment [15]. As we approximate the refractive index in this method, it is not perfect but it enables structure fitting to be performed on ATR data [15].

15.3.2.3 Linear Dichroism

LD experiments require data to be collected with two perpendicular linear polarisations. Although one could simply rotate a polariser or rotate the sample with respect to a fixed polariser, adapting a CD instrument to produce linearly polarised light invariably produces better-quality spectra. The simplest way to do this is to insert a quarter-wave plate to convert alternating circularly to alternating linearly polarised light. However, most electronic CD instruments now come with a software option to increase the voltage on the photoelastic modulator and the user needs to do nothing except remember to change the software setting.

Another essential feature of an LD experiment is sample orientation. Although there are many ways to achieve this, the most common way to orientate small molecules is to absorb them on to, or into, a polymer film either before or after the film is stretched. For biomacromolecular systems of high aspect ratio, Couette flow is more commonly used to orientate the long axis of the sample with the direction of flow. A film stretcher and a microvolume Couette flow cell are illustrated in Figure 15.8. A recent innovation has simplified LD for polar small molecules. Previously the best data were collected with polyvinyl alcohol films, which required the analyte to be included in a polymerising film and the sample to be left for a few days to dry before stretching under gentle heat [3]. By way of contrast, polyethylene (in the form of commercial plastic bags) could be used for hydrophobic analytes. We recently discovered that placing polyethylene in an oxidising plasma asher for a

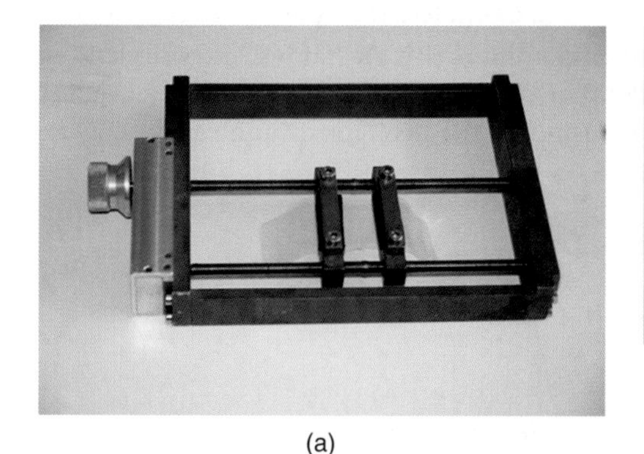

(a)

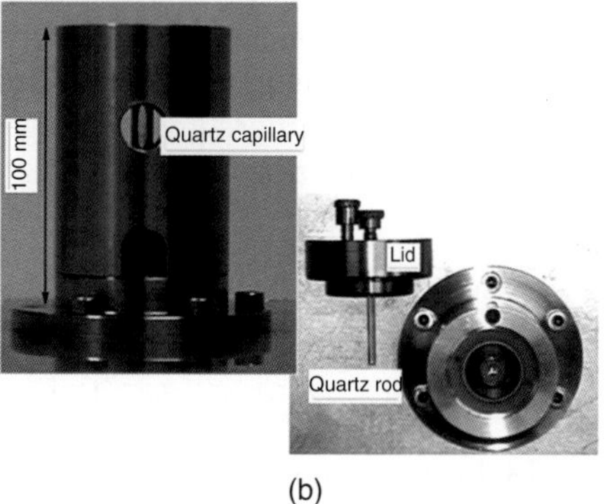

(b)

Figure 15.8 (a) A film stretcher, (b) microvolume Couette flow cell.

few seconds made the surface sufficiently hydrophilic for charged and polar molecules to bind and align with their long axes along the polymer stretch direction, but had no effect on the polymer absorption spectrum in the UV–visible region. Most modern LD instrument define // (Eq. (15.9)) to be horizontal.

15.3.3 Structure Fitting of Proteins from Spectral Data

Sometimes one uses spectroscopy simply to determine whether two samples are structurally the same. This can be done by visually comparing the shape of the two spectra. However, for many purposes it is helpful to reduce the complexity of visual comparison of spectra by extracting an alternate simpler information set. Using protein backbone CD spectra (including data from 250 to 190 nm or lower wavelength) is a well-established method to estimate the average secondary structure of a protein, based on the CD spectra of a dataset of proteins with a known secondary structure content. However, as the CD signal is proportional to protein concentration, the accuracy of these secondary structure fitting methods is most affected by the accuracy of the knowledge of the concentration of the sample. If the methods [16–23] that are collated on, for example, the Dichroweb platform [24] give similar answers then one can believe them. However, if they differ and particularly if Selcon [17] refuses to give an answer, then one needs to question the data quality that is being used. We developed a neural network approach, SSNN, that with good data gives similar answers to the other methods with the same reference set, has an accompanying scaling option that can be used to estimate the real concentration of the sample and then give a secondary structure estimate [22, 23, 25]. Of course, each stage of data manipulation or approximation introduces the potential for error.

Baseline-corrected protein IR transmission spectra also depend on the secondary structure of proteins. The protein amide I band occurring between 1700 and 1600 cm^{-1} is fairly independent of the ions in the buffer and is widely accepted to reflect the secondary structure [15, 26]. Extracting secondary structure content of proteins from IR data is typically done by band fitting with the relative areas of fitted bands at 1645–1660 cm^{-1} attributed to α-helix, at 1620–1640 and 1670–1695 cm^{-1} to the β-sheet, at 1620–1640 and 1650–1695 cm^{-1} to turns and at 1640–1657 and 1660–1670 cm^{-1} to other structures [27]. In our hands, this approach can work reasonably well if the spectrum has a good signal-to-noise ratio, the baseline subtraction is perfect and the spectrum has no unusual features. However, our fits do vary noticeably. We have therefore been working to apply our SSNN [22, 23] neural network approach for CD to IR data and have generalised the method, now calling it SOMSpec (self-organising map spectroscopy). It generally seems to be more effective than the band-fitting approach, though it depends on the quality of the reference set used. The spectral NRMSD (normalised root mean squared deviation), defined as

$$NRMSD = \sqrt{\left(\sum_i (x_{i,experiment} - x_{i,\text{model}})^2/N\right)/(M-m)} \tag{15.11}$$

where x_i is the value at each wavenumber, N is the number of data points, M is the largest intensity and m is the smallest, so $(M–m)$ is the range that gives a numerical measure of the goodness of spectral fit that is equally weighted across the spectrum. A more limited range focusing on the peak maximum region may be more useful. SOMSpec is available from reference [28].

15.4 Applications

15.4.1 Comparison of Different α-Lactalbumin Protein Structures

Historically, it was assumed that the amino acid sequence of a protein defines a single, unique native structure, with at most small local rearrangements to accompany function. However, we now understand that the process of protein folding is far more complex – many proteins exist in a dynamic ensemble of different structures and/or adopt intermediate folding states or form different structures under different environmental conditions. Hence, both the sequence and structure of a protein are important considerations in comparing the biosimilarity of different protein preparations for applications such as biopharmaceuticals. In contrast to traditional protein structure determination methods, such as X-ray crystallography, spectroscopic techniques provide several distinct advantages for characterising and comparing different protein structures, including the ability to work in different solvent environments and look at multiple different structures simultaneously and in real time. Different spectroscopic methods can also be useful in providing complementary structural information.

An illustrative example of the application of CD and related spectroscopy to look at protein structure is α-lactalbumin (α-LA). α-LA is a relatively small protein from milk that has been extensively studied as a model of protein folding. α-LA has a single strong Ca^{2+} binding site. The crystal structures of the Ca^{2+} bound holo α-LA reveals a compact globular structure consisting of an α-helical domain and a small β-sheet domain that flank the Ca^{2+} binding cleft (Figure 15.9a) [29]. Several other α-LA structures have also been identified. These include: the Ca^{2+} minus apo-α-LA form [30], molten globules formed at low pH (pH 3–5) or elevated temperature, and partially folded intermediates formed in moderate guanidine hydrochloride (GuHCl) concentrations

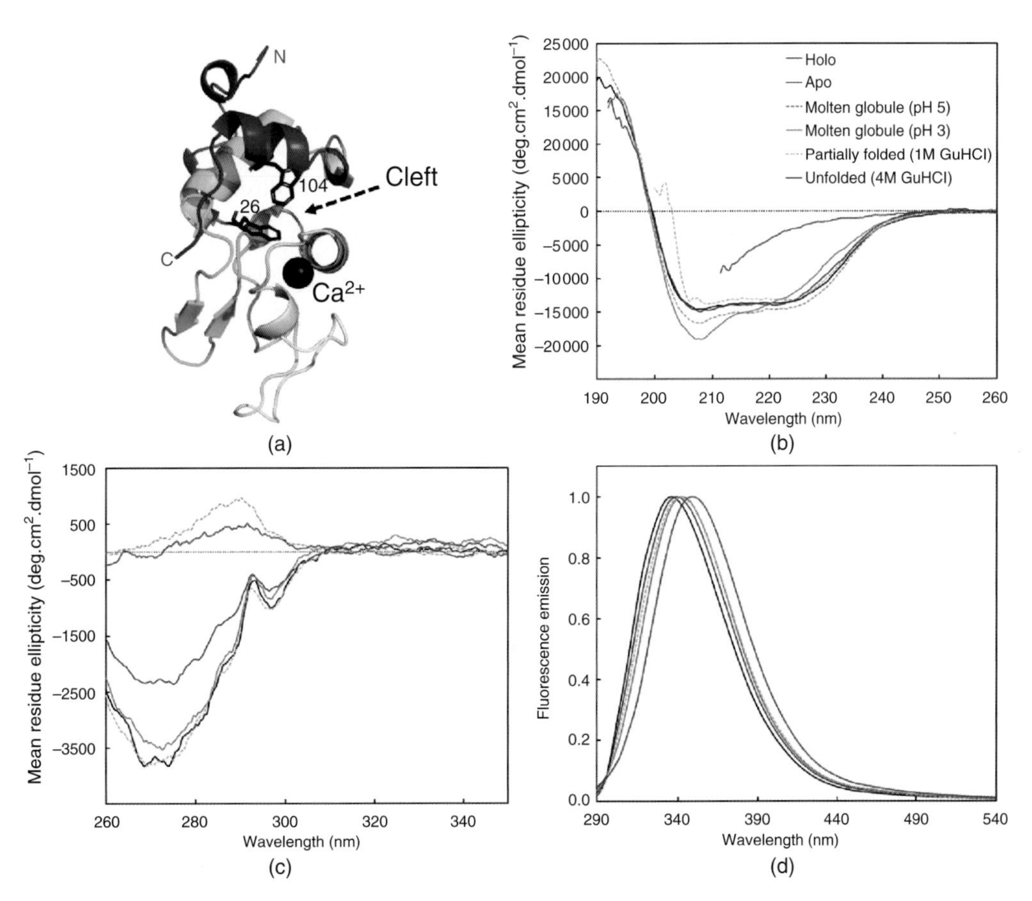

Figure 15.9 (a) Crystal structure of holo bovine α-LA (PDB 1HFZ rendered using Pymol). (b) Protein backbone CD, (c) aromatic region CD and (d) fluorescence emission from 280 nm excitation (i.e. from both tryptophan and tyrosine) for the *holo*, *apo*, molten globules, partially folded and unfolded α-LA states. Note: (i) Molar ellipticity differs from $\Delta\varepsilon$ by a factor of 3298. (ii) The maximum fluorescence intensity of all samples has been normalised to 1 to highlight the shift in maximum emission wavelength.

[31]. The therapeutic potential of a partially unfolded α-LA form stabilised by integration of an oleic acid co-factor (also known as a HAMLET complex) is also currently being explored as a tumour suppressor [32].

As discussed above, CD in the protein backbone region (260–190 nm and lower) is commonly used to determine the secondary structure content of proteins. For example, the CD spectra of holo bovine α-LA (in 50 mM sodium phosphate pH 7 + 2 mM $CaCl_2$, solid black line, Figure 15.8b) shows the negative bands at 222 and 208 nm and the positive band at 193 nm, characteristic of primarily α-helical proteins.

In this case, the CD spectra were also collected in a series of buffer conditions designed to favour alternate structures of α-LA – including the *apo* (50 mM sodium phosphate pH 7 + 1 mM ethylenediaminetetraacetic acid [EDTA]), molten globule pH 5 (50 mM sodium phosphate pH 5 + 2 mM $CaCl_2$), molten globule pH 3 (50 mM sodium phosphate pH 3 + 2 mM $CaCl_2$), partially folded (50 mM sodium phosphate pH 7 + 1 M GuHCl) and unfolded (50 mM sodium phosphate pH 7 + 4 M GuHCl) states. Each sample was prepared at ~0.1 mg/ml and the CD spectra were measured in a 1 mm pathlength cuvette. To account for any error in concentration, the absorbance of each sample was measured at 280 nm and the concentration was estimated using the Beer–Lambert Law (Eq. (15.3)) and an extinction coefficient of 28 460 M^{-1} cm^{-1}, obtained by providing amino acid sequence of the bovine α-LA to the ProtParam tool available on EXPASY [33]. These concentrations (c) were then used to convert the CD in units of mdeg (as output from the instrument) to molar ellipticity (θ):

$$\theta = \frac{(CD/\text{mdeg}) \times MW}{10\ell Cr}, \tag{15.12}$$

where MW is the molecular weight (14 154 g/mol for α-LA), ℓ is the pathlength in cm (here 0.1 cm), C is the molar concentration and r is the number of amino acids (123 for α-LA), to enable direct visual comparison of the different spectra (Figure 15.9b). Molar ellipticity, which is widely used in the literature, differs from the Beer–Lambert extinction coefficient $\Delta\varepsilon$ (Eq. (15.4)) by a factor of 3298. The unit of θ is deg cm^2/dmol, which is a historical legacy. Using the above-mentioned units in Eq. (15.12), a factor of 10 is needed in the denominator for the correct conversion.

Comparison of the α-LA protein backbone CD spectra reveals some clear differences in the secondary structure of the different α-LA structures. The spectra of the holo and apo structures almost completely overlay, while the molten globule pH 5 and partially folded spectra follow a very similar shape, suggesting that the native secondary structure is also maintained in these states. The shape of the molten globule pH 3 spectra is significantly different from that of the apo spectra, with a decrease in the CD signal in the region of 222 nm and a shift of the 208 nm negative maximum towards 200 nm, corresponds to a loss of secondary structure, while the unfolded spectrum (4 M GuHCl) more closely resembles that of a random coil presumably with a negative maximum at 200 nm. Qualitative comparison of the spectral shapes and amplitudes can point out the fact that there are structural differences between α-LA samples in different buffers. However, the real question, what the structural changes are and their extent, can be answered reliably only by analysing the spectra for the secondary structure contents with the methods presented above.

In this case, a 50 mM sodium phosphate buffer was chosen to buffer the samples as it is largely invisible in CD. However, to form the different α-LA structures, inclusion of some highly absorbent species was required, including GuHCl (in the case of the partially unfolded and unfolded samples), HCl (in the case of the pH 3 sample) and EDTA (in the case of the apo samples). Thus, these spectra have been cropped at lower wavelengths. Where absorbance is high (i.e. transmission is low) the high tension (HT) voltage that is applied to amplify the detector's sensitivity is also high. Since different wavelengths have different light energies, the HT accommodates for fluctuations in the light levels by changing the gain. When the HT increases above a threshold of ~600 V there are not enough photons being sampled to measure a reliable or valid CD signal and these data points should not be used. While this typically negates performing secondary structure fitting for such samples, useful structural information can still be obtained, especially when comparing the CD spectra of different protein sample as in this example.

CD in the aromatic region can also provide useful information about protein structure. Figure 15.9c shows the CD spectra for the same α-LA samples as in Figure 15.9b, but this time measured at concentrations of ~0.4 mg/ml in a 1 cm pathlength cuvette. Again, the holo and apo spectra overlay. The loss of the CD signal around 280 nm in the partially unfolded and, to a greater extent, in the molten globule pH 5 spectra are consistent with a loss of tertiary structure, which removes the well-defined chiral environment from around the aromatic residues. In accord with this explanation, the pH 3 spectra in the aromatic region more closely resembles that of the unfolded state than of a folded one. However, from the backbone CD we know the secondary structure of this sample is not unfolded. Hence, taken together, the backbone and aromatic CD of the pH 7 holo sample and pH 5 and pH 3 samples are consistent with a classic molten globule transition in which a native-like secondary structure is maintained but a defined tertiary structure is progressively lost as the pH is lowered.

One may then ask: how can we distinguish between the holo and the apo α-LA structures as there is no discernible difference in either the protein backbone or aromatic region CD for these two samples? Absorbance in the aromatic region also results in fluorescence emission predominantly derived from the tryptophan (Trp) residues (for further discussion, see [34]). The wavelength of maximum emission is highly sensitive to the solvation and rotational restriction of the Trp residues and can thus be used as a site-specific probe of protein folding. In folded proteins, where the Trp residues are buried in the protein structure, fluorescence emission occurs at a lower wavelength (e.g. the α-LA holo sample, 336 nm), while for unfolded proteins, in which the Trp residues are maximally solvated, maximum fluorescence emission occurs at longer wavelengths (e.g. the α-LA unfolded state, 349 nm, see Figure 15.9d). The maximum emission of the partially unfolded and molten globule α-LA structures are all intermediate between those of the folded and unfolded states. Of particular note, the maximum fluorescence emission of the apo state (342 nm) is significantly higher than that of the holo state. This is the result of a change in the local environment of two Trp residues (Trp 26 and Trp 104) located in an aromatic cluster in the binding cleft of bovine α-LA, as absence of the stabilising Ca^{2+} ion in the apo form allows the binding cleft to adopt a more 'open' conformation [30]. It should be noted that we have chosen to plot the fluorescence data normalised to 1 to emphasise the wavelength shifts. Fluorescence magnitudes do vary with environment, but the influences are much more complex and can be misleading.

The ability to monitor protein structural changes in real time, for example during thermal unfolding, can also be particularly informative when multiple spectroscopic methods are used in combination. Figure 15.10 shows the melting curves obtained for holo α-LA using three different spectroscopic methods: (i) the change

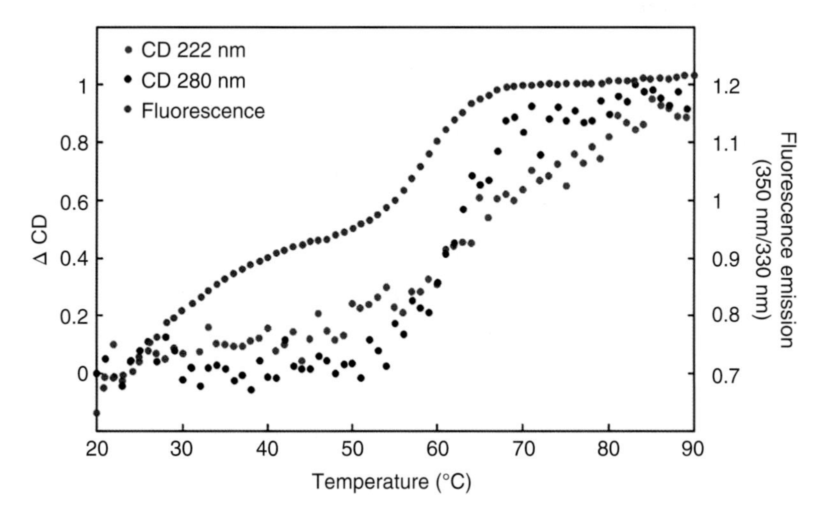

Figure 15.10 Holo α-LA melting curves measured using CD at 222 and 280 nm and shift in fluorescence emission from 280 nm excitation. Note the change in the CD signal has been normalised to between 0 and 1 (refer to Figure 15.9 for absolute magnitudes).

in protein backbone CD signal monitored at 222 nm; (ii) the change in aromatic CD signal monitored at 280 nm; and (iii) the shift in maximum Trp fluorescence emission wavelength (plotted as the ratio of emission at 350 nm to that at 330 nm). Initial destabilisation of the α-LA binding site cleft at slightly elevated temperatures (20–45 °C) does not result in significant change to the overall secondary or tertiary structure and is therefore not discernible in the CD spectra. However, this initial destabilisation is visible in the fluorescence melting curve. At higher temperatures (above approximately 55 °C), the 280 nm CD melting curve mirrors that of the fluorescence curve due to loss of tertiary structure as α-LA transitions through the molten globule state, while loss of secondary structure with increasing temperature is far more gradual, as can be seen in the 222 nm CD melting curve.

Fourier transform infra-red (FTIR) can also be a useful spectroscopic method to characterise and compare different protein structures. Original spectra and baseline corrected spectra are shown in Figure 15.11. The Amide I band is centred at about 1650 cm^{-1} and is generally deemed to correlate with secondary structure. The 1550 cm^{-1} Amide II band seems to be influenced by the salts in the solution. As discussed above, a significant drawback of FTIR is that the protein signal overlaps with that of water (for example, see Figure 15.11a, where the black protein in water spectrum overlays almost exactly the dotted red buffer spectrum). While subtraction of appropriate baseline spectra can be used to eliminate the water FTIR signal, at the protein concentration chosen for these experiments (which is lower than what is usually used) this can make these data hard to interpret. We can see the effect of water vapour most obviously in the Apo spectrum of Figure 15.11 (blue solid line). The FTIR-ATR data are consistent with little secondary structure change between the three samples considered (the wavenumber maxima of the Amide I bands are similar). However, data quality is poor at such a low concentrations where water dominates the original spectra.

High protein concentrations (50–200 mg/ml) are typically used for protein FTIR spectra. However, these concentrations are above the solubility limit of a lot of proteins and thus not always a feasible option. One possible solution is to work in D_2O, which removes the issue of the water background. However, biopharmaceutical products are not formulated in D_2O. An alternative approach would be to measure the FTIR-ATR of solid protein samples. This is an attractive solution for comparing protein therapeutic preparations, which are often produced as lyophilised proteins. However, methods to extrapolate secondary structure or compare similarities between different protein structures have not yet been developed. We are currently working on establishing methods to use FTIR-ATR spectroscopy on aqueous or solid protein samples to generate secondary structure estimates accounting for refractive index and light intensity variations with wavenumber. This requires first transforming ATR data into the equivalent transmission form [15] and using a self-organising structure fitting approach. [22].

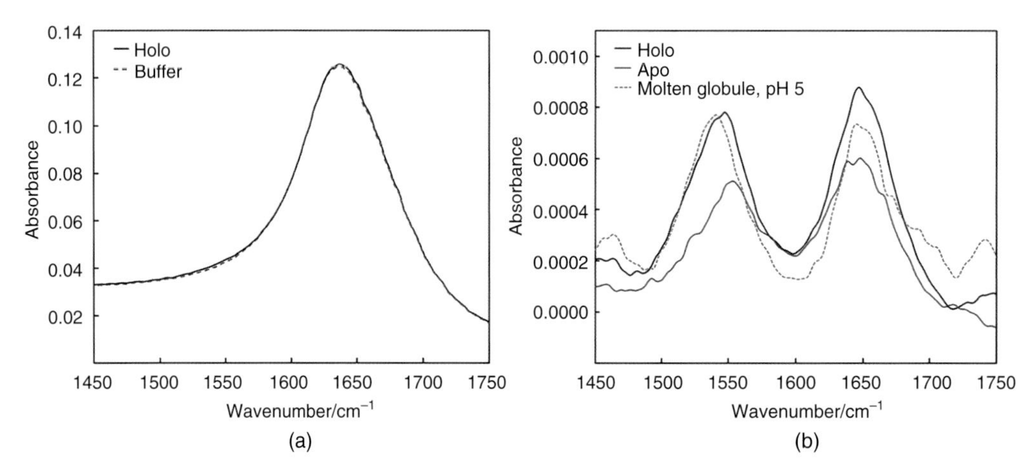

Figure 15.11 (a) Holo α-LA FTIR-ATR spectra and corresponding buffer baseline. (b) Buffer subtracted FTIR-ATR spectra of α-LA (2 mg/ml) in holo, apo and molten globule pH 5 forms. Note that chemical incompatibility with the ZnSe ATR crystal meant that the FTIR-ATR spectra of the molten globule pH 3 and GuHCl samples could not be performed.

15.4.2 Linear Dichroism Examples

15.4.2.1 Flow Oriented DNA and Ligand Binding

When long pieces of DNA are flow oriented in a Couette flow cell there is a preferential alignment of the sample. At 1000 rpm in a microvolume cell, the orientation parameter $S = 0.1$. Figure 15.12 illustrates the type of spectrum one expects: zero LD until about 300 nm, where the DNA bases start absorbing, and then a negative LD signal that mirrors the shape of the DNA absorbance spectrum. Its negative sign follows from the fact that the transitions are all in the plane of the aromatic bases of DNA and these are oriented approximately perpendicular to the helix axis at an average angle of about 86°. When the groove binder Hoechst 33258 is added to the solution, the spectrum gains additional signals from the Hoechst molecules that bind in the minor groove of the DNA. The 370 nm transition of Hoechst is polarised along its long axis [35], which is therefore oriented at about 45° from the helix axis, so less than 54.7° makes the LD (Eqs. (15.9) and (15.10)) positive. Groove binding ligands do not significantly stiffen or lengthen DNA, so the fact that the 250 nm LD in the presence of Hoechst is more negative than the DNA signal indicates that the Hoechst transition in this region is approximately short-axis polarised so is more than 54.7°.

15.4.2.2 M-13 Bacteriophage

Bacteriophage are well-defined assemblies of proteins and nucleic acids. In the case of M13, it is a filamentous structure with DNA in the middle and a very limited number of proteins forming the capsid structure. M13 bacteriophage are about 800 nm in length and have a very a large persistence length (>1 μM) and so orient extremely well, giving an enormous protein LD signal in the backbone region below 250 nm [36–38]. The spectra shown in Figure 15.13 have a negative signal at 280 nm, which faintly shows the three bands of tryptophans sometimes seen in the aromatic region of protein CD spectra. The 250 nm region is due to the DNA whose bases are oriented perpendicular to the long axis of the phage. The backbone region of the spectrum has a 188 nm negative signal and 205 nm positive signal, which are dominated by the two components of the α-helix π–π^* transition. The 222 nm n–π^* region has a similar but much less intense couplet (positive maximum and a minimum that would be negative were it not overlaid on the tail of the large positive 205 nm

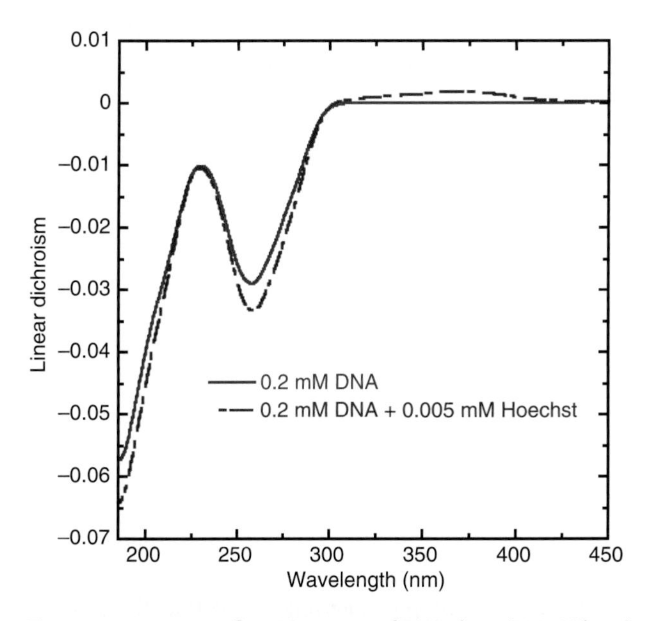

Figure 15.12 Couette flow LD spectra of DNA alone (200 μM base) and calf thymus DNA (200 μM base) in the presence of Hoechst (5 μM) in water.

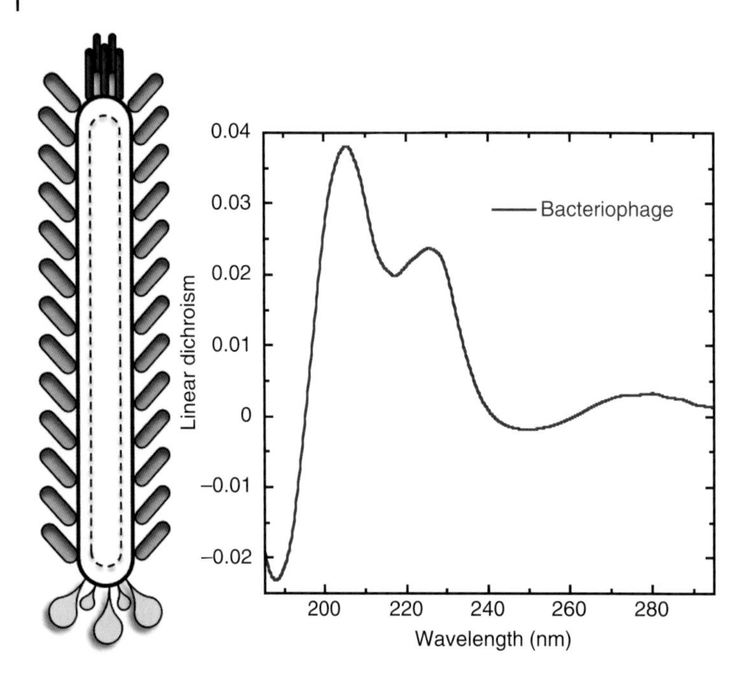

Figure 15.13 Schematic of M-13 bacteriophage indicating the different coat proteins and the central DNA molecule. LD of 0.1 mg/ml M13 bacteriophage in water.

signal), which we speculate arises from the intensity that the electronically forbidden transition 'borrows' from the intense allowed neighbouring π–π^* transition.

15.5 Concluding Remarks

The goal of this chapter has been to illustrate how spectroscopic techniques can be used to provide structural data about biomacromolecules. Compared with many other techniques, spectroscopy gives qualitative and quantitative data relatively quickly, which means answers to more questions can be established using these methods. However, spectroscopy is usually most powerful when used in combination with other methods so clear questions and experiment design is an essential part of successful spectroscopy.

References

1 Zimmermann, T., Rietdorf, J., and Pepperkok, R. (2003). Spectral imaging and its applications in live cell microscopy. *FEBS Lett.* 546 (1): 87–92.
2 Lasch, P. and Naumann, D. (2006). Spatial resolution in infrared microspectroscopic imaging of tissues. *Biochim. Biophys. Acta Biomembr.* 1758 (7): 814–829.
3 Nordén, B., Rodger, A., and Dafforn, T.R. (2010). *Linear Dichroism and Circular Dichroism: A Textbook on Polarized Spectroscopy*, 304. Cambridge: Royal Society of Chemistry.
4 Berova, N., Nakanishi, K., and Woody, R.W. (eds.) (2000). *Circular Dichroism Principles and Applications.* New York: Wiley-VCH.
5 Woody, R.W. (2009). Circular dichroism spectrum of peptides in the poly(Pro)II conformation. *J. Am. Chem. Soc.* 131: 8234–8245.

6 Kelly, S.M., Jess, T.J., and Price, N.C. (2005). How to study proteins by circular dichroism. *Biochim. Biophys. Acta* 1751: 119–139.
7 Rosenfeld, L. (1928). Quantenmechanische Theorie der natürlichen optischen Aktivität von Flüssigkeiten und Gasen. *Z. Phys.* 52: 161–174.
8 Schipper, P.E. and Rodger, A. (1983). Symmetry rules for the determination of the intercalation geometry of host/guest systems using circular dichroism: a symmetry adapted coupled-oscillator model. *J. Am. Chem. Soc.* 105: 4541–4550.
9 Harada, N., arada, N., Takuma, Y., and Uda, H. (1976). The absolute stereochemistries of 6,15-dihydro-6,15-ethanonaphtho[2.3-c]pentaphene and related homologs as determined by both exciton chirality and X-ray Bijvoet methods. *J. Am. Chem. Soc.* 98: 5408–5409.
10 Nordén, B. (1977). General aspects on linear dichroism spectroscopy and its application. *Spectrosc. Lett.* 10: 381–400.
11 Halsall, D.J., Dafforn, T.R., Marrington, R. et al. (2004). A linear dichroism technique for quantitative PCR applications. *IVD Technol.* 6: 51–60.
12 Marrington, R., Dafforn, T.R., Halsall, D.J. et al. (2005). Validation of new microvolume Couette flow linear dichroism cells. *Analyst* 130: 1608–1616.
13 Razmkhah, K., Gibson, M.I., Chmel, N.P., and Rodger, A. (2014). Oxidized polyethylene films for orienting polar molecules for linear dichroism spectroscopy. *Analyst* 139: 1372–1382.
14 Rodger, A., Sanders, K.J., Hannon, M.J. et al. (2000). DNA structure control by polycationic species: polyamines, cobalt ammines, and di-metallo transition metal chelates. *Chirality* 12: 221–236.
15 Pinto-Corujo, M., Sklepari, M., Ang, D. et al. (2018). Infra-red absorbance spectroscopy of aqueous proteins: comparison of transmission and ATR data collection and analysis for secondary structure fitting. *Chirality* .
16 Woody, R.W. (1994). Circular dichroism of peptides and proteins. In: *Circular Dichroism Principles and Applications* (ed. K. Nakanishi, N. Berova and R.W. Woody). New York: VCH.
17 Sreerama, N. and Woody, R.W. (2000). Estimation of protein secondary structure from circular dichroism spectra: comparison of CONTIN, SELCON, and CDSSTR methods with an expanded reference set. *Anal. Biochem.* 287: 252–260.
18 Sreerama, N. and Woody, R.W. (1993). A self-consistent method for the analysis of protein secondary structure from circular dichroism. *Anal. Biochem.* 209: 32–44.
19 Nakanishi, K., Berova, N., and Woody, R.W. (1994). *Circular Dichroism: Principles and Applications.* New York: VCH.
20 Johnson, W.C.J. (1988). Secondary structure of proteins through circular dichroism spectroscopy. *Ann. Rev. Biophys. Biophys. Chem.* 17: 145–166.
21 Johnson, W.C. (1999). Analyzing protein circular dichroism spectra for accurate secondary structures. *Proteins Struct. Funct. Genet.* 35: 307–312.
22 Hall V, Sklepari M, Rodger A (2014) Protein secondary structure prediction from circular dichroism spectra using a self-organizing map with concentration correction. *Chirality* 26: 471–482. doi:10.1002/chir.22338.
23 Hall, V., Nash, A., and Rodger, A. (2014). SSNN, a method for neural network protein secondary structure fitting using circular dichroism data. *Anal. Methods* 6 (17): 6721–6726.
24 Whitmore, L. and Wallace, B.A. (2004). DICHROWEB: an online server for protein secondary structure analyses from circular dichroism specroscopic data. *Nucleic Acids Res.* 32: W668–W673.
25 Hall, V., Nash, A., Hines, E., and Rodger, A. (2013). Elucidating protein secondary structure with circular dichroism and a neural network. *J. Comput. Chem.* 34: 2774–2786.
26 Corujo MP, Praveen A, Steel MJ, Ang D, Chmel N, Rodger A. *Attenuated Total Reflectance Infra Red Absorbance Spectroscopy for Proteins in H_2O: Electromagnetic Fields at Boundaries.* Submitted 2018.
27 Haris, P.I. (2013). Infrared spectroscopy of protein structure. In: *Encyclopedia of Biophysics: European Biophysical Societies' Association* (ed. G.K. Roberts).

28 Ang D, Dukor R, Pinto-Corujo M, Reason A, Rodger A. SOMSpec: a general purpose neural network-based tool for rapid protein secondary structure prediction from circular dichroism and infra-red absorbance spectra. In preparation 2019.
29 Acharya, K.R., Stuart, D.I., Walker, N.P.C. et al. (1989). Refined structure of baboon α-lactalbumin at 1.7 Å resolution: comparison with C-type lysozyme. *J. Mol. Biol.* 208 (1): 99–127.
30 Chrysina, E.D., Brew, K., and Acharya, K.R. (2000). Crystal structures of apo- and holo-bovine α-lactalbumin at 2.2-Å resolution reveal an effect of calcium on inter-lobe interactions. *J. Biol. Chem.* 275 (47): 37021–37029.
31 Dolgikh, D.A., Gilmanshin, R.I., Brazhnikov, E.V. et al. (1981). α-Lactalbumin: compact state with fluctuating tertiary structure? *FEBS Lett.* 136 (2): 311–315.
32 Ho, J.C.S., Nadeem, A., and Svanborg, C. (2017). HAMLET – a protein-lipid complex with broad tumoricidal activity. *Biochem. Biophys. Res. Commun.* 482 (3): 454–458.
33 Gasteiger, E., Hoogland, C., Gattiker, A. et al. (2005). Protein identification and analysis tools on the ExPASy server. In: *The Proteomics Protocols Handbook* (ed. J.M. Walker). Humana Press.
34 Lakowicz, J.R. (2006). *Principles of Fluorescence Spectroscopy.* Springer-Verlag.
35 Moon, J.-H., Kim, S.K., Sehlstedt, U. et al. (1996). DNA structural features responsible for sequence dependent binding geometries of Hoechst 33258. *Biopolymers* 38: 593–606.
36 Clack, B.A. and Gray, D.M. (1992). Flow linear dichroism spectra of four filamentous bacteriophages: DNA and coat protein contributions. *Biopolymers* 32: 795–810.
37 Pacheco-Gomez, R., Roper, D.I., Dafforn, T.R., and Rodger, A. (2011). The pH dependence of polymerization and bundling by the essential bacterial cytoskeltal protein FtsZ. *PLoS One* 6 (6): e19369. https://doi.org/10.1371/journal.pone.0019369.
38 Pacheco-Gomez, R., Kraemer, J., Stokoe, S. et al. (2012). Detection of pathogenic bacteria using a homogeneous immunoassay based on shear alignment of virus particles and linear dichroism. *Anal. Chem.* 84 (1): 91–97.

Further Reading

Berova, N., Nakanishi, K., and Woody, R.W. (eds.) (2000). *Circular Dichroism Principles and Applications.* New York: Wiley-VCH.
Gasteiger, E., Hoogland, C., Gattiker, A. et al. (2005). Protein identification and analysis tools on the ExPASy server. In: *The Proteomics Protocols Handbook* (ed. J.M. Walker). Humana Press.
Lakowicz, J.R. (2006). *Principles of Fluorescence Spectroscopy.* Springer-Verlag.
Nakanishi, K., Berova, N., and Woody, R. (eds.) (1994). *Circular Dichroism: Principles and Applications.* New York: VCH.
Nordén, B., Rodger, A., and Dafforn, T.R. (2010). *Linear Dichroism and Circular Dichroism: A Textbook on Polarized Spectroscopy.* Cambridge: Royal Society of Chemistry.
Wallace, B.A. and Janes, R. (eds.) (2009). *Modern Techniques for Circular Dichroism Spectroscopy.* Amsterdam: IOS Press.

16

Principles and Practice in Macromolecular X-Ray Crystallography

Arnaud Baslé and Richard J. Lewis

Institute for Cell and Molecular Biosciences, Newcastle University, Newcastle upon Tyne, NE2 4HH, UK

16.1 Significance and Short Background

To understand the function of proteins we need to observe them on the atomic scale in 3D. Proteins can be as small as 5 nm and even large complexes, like ribosomes, are only 30 nm in diameter. The length of covalent bonds in proteins, ~0.15 nm, is simply too short to be observed using the visible portion, 400–700 nm, of the electromagnetic spectrum. A form of illumination has to be used that is better matched to the dimensions of the object under study, and that means X-rays. It was first established over a century ago that X-rays could be used to determine molecular structures and X-rays have since been applied to molecules ranging in size and complexity from table salt to the ribosome, the molecular machine powering protein synthesis in all cells. The first protein structures, myoglobin and haemoglobin, were solved ~60 years ago. There were just 13 structures in 1976 and 1994 was the first year that over 1000 new structures were deposited in a calendar year. At the time of writing there are over 132 000 PDB entries based on X-ray data. The explosion in crystallographic analysis can be traced to the development of molecular biology in the 1980s, personal computing and third generation synchrotron light sources in the 1990s.

Our molecular understanding of the fundamentals of life, DNA replication and transcription; RNA translation and protein synthesis; trans-membrane trafficking and transport; ATP generation and metabolism, have all depended upon the application of crystallography. X-ray crystallography not only provides critical information on the structure, function and mechanism of proteins but the molecular understanding of disease, and drug discovery and development are also underpinned by crystallography. Of course, there are other methods that can be applied to the problem of protein structure determination; cryo-electron microscopy (cryoEM) is currently in vogue for its applicability to large proteins and macromolecular complexes, and X-ray free electron laser (XFEL) facilities are likely to revolutionise structural biology because of their time-resolved illumination sources six orders of magnitude brighter than synchrotrons. Here we concern ourselves only with X-ray crystallography, with a focus on practical application.

16.2 Theory and Principles: Overview

Protein crystallography is underwritten by physics and mathematics, but nowadays most practitioners are life scientists who seek to understand biological phenomena. X-rays are used as their wavelengths are better matched to the lengths of covalent bonds in proteins, which are simply too short to be seen by white light. However, there is no microscope that can focus X-rays to provide the resolution required for determining atomic positions in proteins. The scattered X-rays from the electron orbitals of atoms in an individual protein molecule are too weak to be measured but the scattering signal can be amplified by measuring diffraction

Biomolecular and Bioanalytical Techniques: Theory, Methodology and Applications, First Edition. Edited by Vasudevan Ramesh.

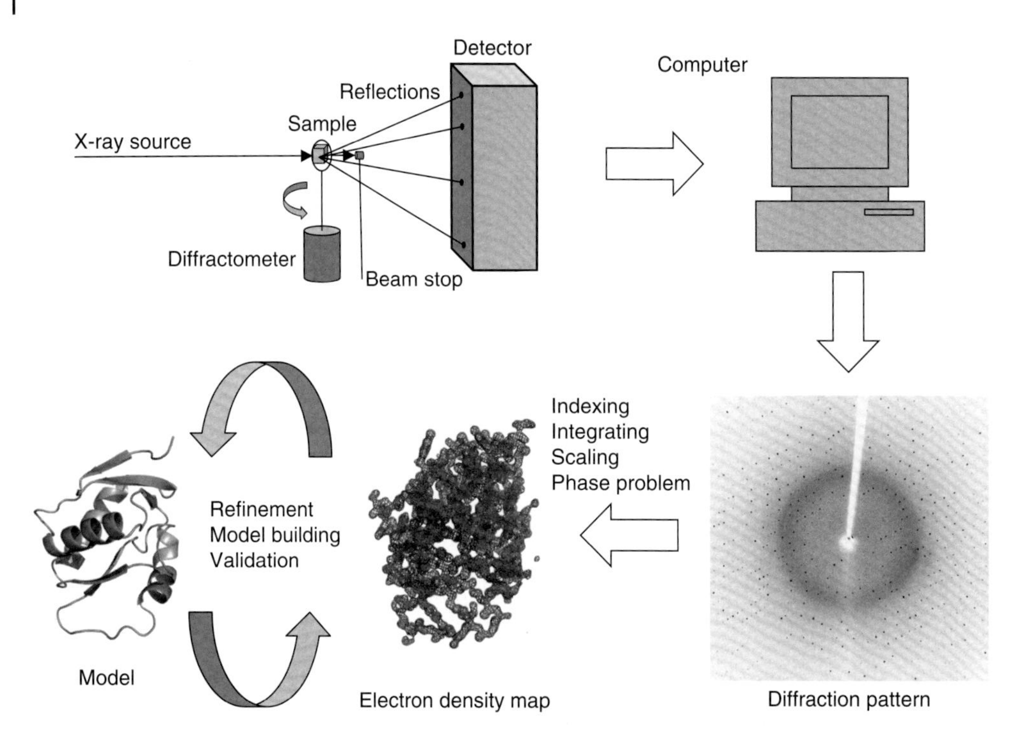

Figure 16.1 Schematic of experimental setup. The typical setup comprises an X-ray source, a diffractometer, a beam stop and a detector, all under computer control. The reflections are indexed, integrated and scaled, and once the phase problem is solved, a model can be built and refined against the collected data.

from protein crystals. The recorded diffraction results from the interaction of the incident X-ray beam with electrons from all the protein's atoms, averaged across all the molecules in the path of the X-ray beam. For instance, tens of thousands of these measurements, called reflections, are required to solve the structure of a 43 kDa protein, which typically contains ~3000 non-hydrogen atoms. The very high data-to-parameter ratio of X-ray crystallography is one of the reasons why it is a very powerful and robust technique. The diffracted reflections are X-ray waves and thus contain three essential elements for calculating an electron density map: (i) the wavelength of the incident beam; (ii) the amplitude of the wave, which is calculated directly from the measured intensities; (iii) the phase of each wave. However, the phase information cannot be recorded directly and has to be determined experimentally or inferred from related structure determinations. The loss of the phase information is problematic because the phase is an exponential term in the mathematical equation that defines the electron density map and as such it dominates the equation product. Only once the phase of each reflection is determined or inferred can an electron density map be calculated from which an atomic model can be built (Figure 16.1). Therefore, each model deposited at the PDB is an interpretation of its electron density. The model may contain errors, some of them substantial, but the data collected are a physical phenomenon resulting from the exposure of the crystal to X-rays. Modern tools allow careful inspection and validation of both model and data to avoid misinterpretation.

16.2.1 Basics of Crystallisation

Protein crystallisation was first developed in the late nineteenth century as a method to demonstrate protein purity, and crystallisation is still used during the chemical purification of organic compounds. However, most proteins do not tolerate the extremes of temperature, pH and solvents used for crystallising small molecules and predicting how any protein will crystallise is effectively impossible.

A crystal is an ordered repetition of a building block called the asymmetric unit, which is the smallest packing arrangement of the protein before crystallographic symmetry operators are applied to generate the effectively infinite crystal lattice. The packing of protein to build a three-dimensional crystal is achieved by a series of specific salt bridges, hydrogen bonds and van der Waals' contacts between protein molecules. It is generally accepted that proteins destined for crystallisation should be as chemically pure as possible, which is trivial to assess by SDS polyacrylamide gel electrophoresis and mass spectrometry. The chances of growing crystals are improved by reducing conformational heterogeneity, though this is harder to monitor and achieve. Occasionally the presence of serendipitous impurities in the preparation or deliberately included ligands will aid crystal lattice formation but, on the whole, impurities decrease crystal quality by intercalating between asymmetric units, which may hinder crystal growth completely. It is important to have a monodisperse (a particle of uniform size), stable sample as crystallisation can take up to a few months. Proteins that fold poorly, have low thermal stability or are prone to degradation are less likely to crystallise. Time invested in protein quality control (e.g. circular dichroism for protein folding, thermal melts for protein stability) can save time overall. Protein flexibility, instability, denaturation and polydispersity are most likely to hinder crystallisation.

16.2.1.1 Crystallisation

Crystallisation is a two-step process. First, a supersaturated solution of the molecule must form and be induced to form critical nuclei, microscopic clusters of the molecule in solution. Second, the nuclei must grow into a three-dimensional crystal by the deposition of additional molecules on to the growing surface. The formation of nuclei is a thermodynamic process [1] that occurs once the protein reaches supersaturation, but exceeding the supersaturation limit results in protein precipitation. Supersaturation can be achieved by mixing a highly concentrated solution of the protein with crystallisation solutions, usually at a 1 : 1 ratio, and this mixture is left to equilibrate. In the case of vapour diffusion, the most popular protein crystallisation technique, the concentration of the components of the crystallisation drop reach equilibrium with those in the crystallisation well solution. Water is drawn from the drop to the well solution and eventually the protein reaches supersaturation (Figure 16.2).

There are three variants of vapour diffusion, hanging, sitting and sandwich, where water equilibrates between the drop and reservoir (Figure 16.3). Crystals can be obtained by dialysis where water and small molecule components of the crystallisation solution equilibrate between two compartments separated by a semi-permeable membrane. In the batch method, the crystallisation reagent and protein mixture is covered by oil and water molecules evaporate slowly through the oil. Proteins can also be crystallised in capillaries where the protein and crystallisation solution are not mixed but are just in contact and the two solutions counterdiffuse; somewhere in this dynamic arrangement a supersaturated zone may form. For a review of other common methods, see McPherson and Gavira [2]. Finally, crystallisation of membrane proteins forms a distinct challenge as substantial parts of these protein surfaces are highly hydrophobic as they would ordinarily be embedded in a lipid membrane. Membrane proteins therefore need to be solubilised and crystallised in the presence of detergents, or lipidic cubic phases, introducing further important variables to crystallisation. Membrane protein crystallisation has been reviewed recently by Parker and Newstead [3].

16.2.2 Basics of X-Ray Crystallography

A single protein molecule in solution will diffract X-rays but will generate a scattering signal below the detection capabilities of any detector; even if a suitably sensitive detector did exist, the protein would likely be destroyed by absorbing the applied radiation before a useful signal was recorded. Protein crystals thus amplify the scattering signal and mitigate the damaging effects of the radiation. Whilst crystals of small molecules (M_r $\sim < 500$ Da) typically do not contain solvent, proteins are highly solvated molecules and consequently macromolecular crystals have solvent contents in the 20–80% range, which can be exploited in soaking experiments of inhibitors, effectors and other ligands, or compounds for experimental phasing. High solvent content is

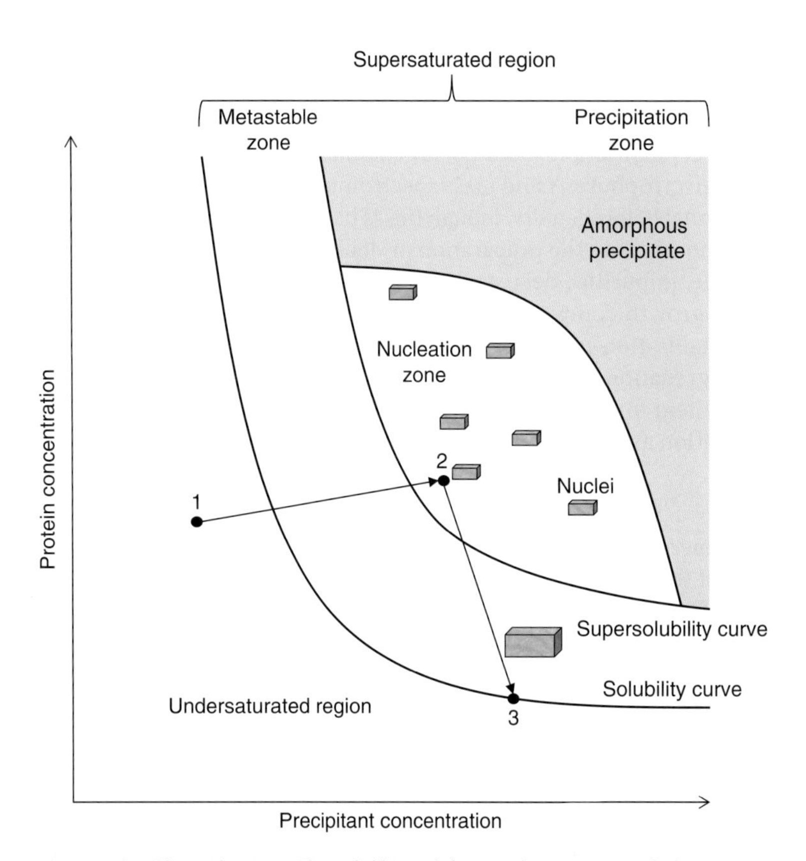

Figure 16.2 Phase diagram. The solid lines delineate the protein and precipitant concentration required to reach the metastable zone and the concentration of either should be increased to migrate from the undersaturated region (1) to the nucleation zone (2) in order to obtain crystals. The grey, shaded region delineates the precipitation zone to avoid where protein will form amorphous precipitates. Crystals grow in size in the path from (2) to (3) and when this point is reached crystal growth will stop (3).

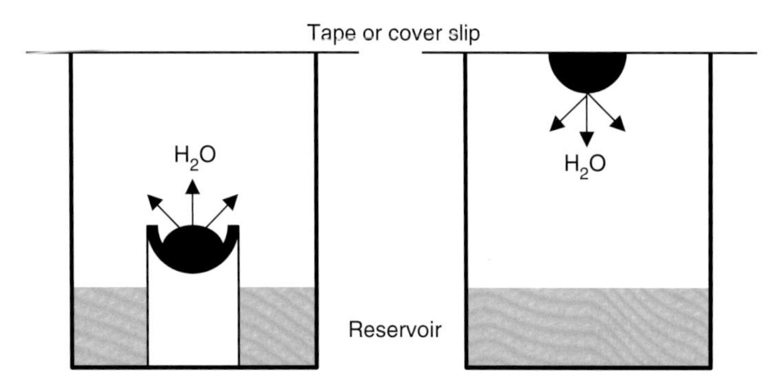

Figure 16.3 Sitting and hanging drop vapour diffusion. Protein and reservoir are mixed and are represented here as a dark drop. The mixture is left to either sit on a pedestal or hang by surface tension on a cover slip and the experiment is sealed with either tape or a cover slip. Over time water evaporates from the drop to equilibrate with the reservoir, represented in light grey, and the drop shrinks in volume.

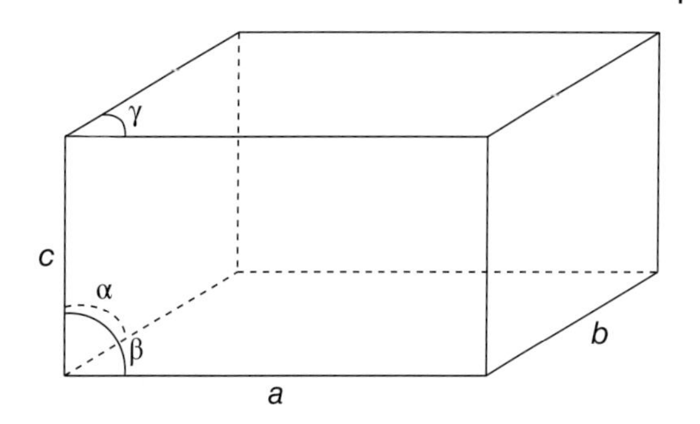

Figure 16.4 Unit cell parameters. The lengths (*a*, *b*, *c*) and angles (α, β, γ) of a typical unit cell are represented.

advantageous in phase improvement strategies but creates problems in cryocrystallography because of potential ice formation.

In order to interpret the X-ray data, we must first understand the principles by which molecules assemble into crystalline three-dimensional repeating arrays. The unit cell (Figure 16.4), the smallest volume to describe a crystal lattice, repeats identically in three dimensions to form the crystal and is characterised by three distances (measured in Å) a, b, c and three angles α, β, γ (measured in °). There are 230 unique ways that molecules can be arranged symmetrically within a unit cell and still obey the translational symmetry in all three directions. Hence there are 230 possible individual space groups that describe the symmetry within the unit cell. If the space group of the crystal cannot be determined correctly, the crystal structure cannot be solved. Each space group is formed by a unique combination of 7 lattice systems, 14 Bravais lattices and 32 crystallographic point groups. Each of these terms is described briefly in the next two paragraphs, but for a comprehensive description the reader should consult the International Tables for Crystallography [4].

Symmetry rules define the building block of each unit cell, called the asymmetric unit, though the asymmetric unit may contain more than one copy of the molecule. The seven lattice systems impose symmetry rules on the crystal in defining its space group. These lattice systems include triclinic, which imposes no symmetry restraints on the unit cell parameters, and tetragonal, with one fourfold rotation axis, requiring that unit cell dimensions $a = b$ and all three angles are 90°. The most complex lattice system is cubic, in which $a = b = c$ and all three angles are 90°. The 14 Bravais lattices are formed by combining a lattice system with fixed lattice points, or origins, found at the corners of each unit cell only, called primitive (or *P* for short). Additional lattice points can be accommodated at the centre of each face of the unit cell, called face-centred (*F*), at the centre of the unit cell, called body-centred (*I*) or on one parallel face, called *C*-centred (*C*). Crystallographic point groups are a set of symmetry operators applied to the asymmetric unit to generate the unit cell and leave a lattice point fixed while moving other parts of the crystal to symmetry-equivalent positions. Other than in the triclinic space group *P*1, each unit cell comprises more than one asymmetric unit. In the tetragonal space group *P*422 there are 8 asymmetric units and in one of the most extreme cases, the cubic space group *F*432, there are 96.

Four symmetry operators define point groups: rotation and screw axes, mirror and glide planes. However, because amino acids (except glycine) contain chiral centres, mirror and glide planes are incompatible with chiral protein crystals and consequently there are just 65 macromolecular space groups. Rotation axes are n-fold where n is 2, 3, 4 or 6, because only these rotations can be applied to generate infinite repeats with no gaps, called the crystallographic restriction. Crystallographic and macromolecular symmetry axes can coincide: OmpF crystallises with one molecule in the asymmetric unit of the trigonal space group *P*321 (PDBid 2OMF) and the biological, trimeric structure of OmpF is reassembled by application of the crystallographic threefold axis (Figure 16.5). Macromolecules with symmetries distinct from two-, three-, four-, and sixfold can still crystallise: GroEL contains sevenfold symmetry [5] and the RNA-binding protein TRAP has 11-fold symmetry [6]. The incompatibility of these symmetries with the crystallographic restriction means that entire assemblies are present within crystallographic asymmetric units. Screw axes combine n-fold/360° rotations with translations along the crystallographic axis and are demarked with a subscript number related to the fraction of the unit

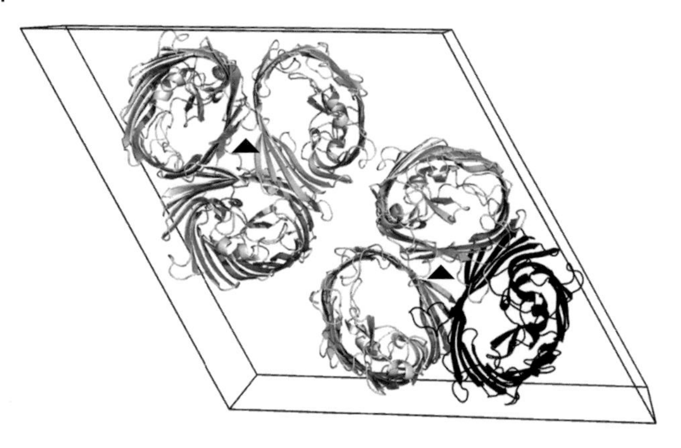

Figure 16.5 OmpF (PDBid 2OMF) unit cell construction. In this example OmpF has crystalized in space group *P*321. The unit cell represented in this figure contains six asymmetric units. Only one molecule (black) is built in the asymmetric unit and five symmetry operations permit the building of the unit cell. The threefold molecular axes are marked with a solid triangle and are parallel to the crystallographic threefold axis, going into the face of the page.

cell dimension of the translation. For instance, the space group $P2_1$ means that there is one twofold axis of rotation combined with a translation of half a unit cell length along the same crystallographic axis.

16.2.2.1 Bragg's Law

Father and son, William Henry and William Lawrence Bragg, are generally viewed as the Godfathers of crystallography and shared the Nobel Prize for Physics in 1915 – when Lawrence was just 25 years old. The Braggs observed that crystalline materials diffracted X-rays in an ordered manner and William Lawrence proposed that crystals could be viewed as a set of parallel planes separated by the distance d. The electrons interfering with incident X-rays would produce constructive interference only if the phase shift of waves reflected by different planes was a multiple of 2π. William Henry built a diffractometer to record each reflection on a photographic plate enabling atomic distances in simple salt crystals, such as NaCl, to be determined. Bragg's law is normally expressed as

$$2d \sin \theta = n\lambda,$$

where d is the interplanar distance, θ is the incident or scattering angle, n is a positive integer and λ is the wavelength of the incident beam (Figure 16.6). Note that Bragg's law also applies to electron and neutron diffraction, such is its applicability to the fundamentals of diffraction.

Waves that traverse larger interplanar spacings are diffracted by smaller θ angles and are usually observed towards the centre of the detector. These waves convey low resolution information on larger features such as the overall solvent distribution and the protein envelope. These low resolution reflections are essential for the structure solution process and have higher than average intensities. Reflections with high θ angles are observed towards the perimeter of the detector due to the small interplanar spacing d (Figure 16.7). These reflections carry information about spatially close features and are the high resolution data essential for building an atomic model of the protein. However, the high resolution reflections can be difficult to measure accurately over background and are lost first if the crystal suffers from radiation damage.

The pattern of reflections recorded is derived from the diffraction of X-rays from points in the reciprocal lattice, a coordinate system related to real space by the following operations:

$$a^* = 1/a,$$
$$b^* = 1/b,$$
$$c^* = 1/c.$$

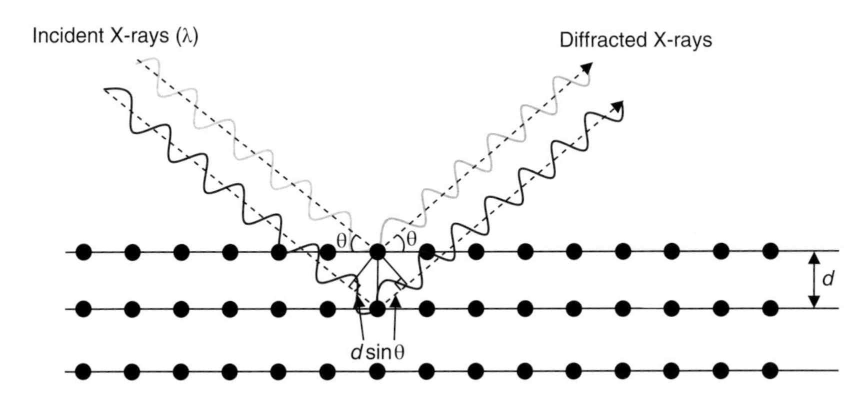

Figure 16.6 Bragg's law ($n\lambda = 2d \sin \theta$). Three lattice planes are represented with the interplanar distance d. X-rays of wavelength λ are represented as a dotted line with an incident angle θ. For the diffracted X-rays to arrive in phase on the detector, the distance travelled by the lower of the two reflections must be an integral number of wavelength, $n\lambda$, which must equal twice $d \sin \theta$.

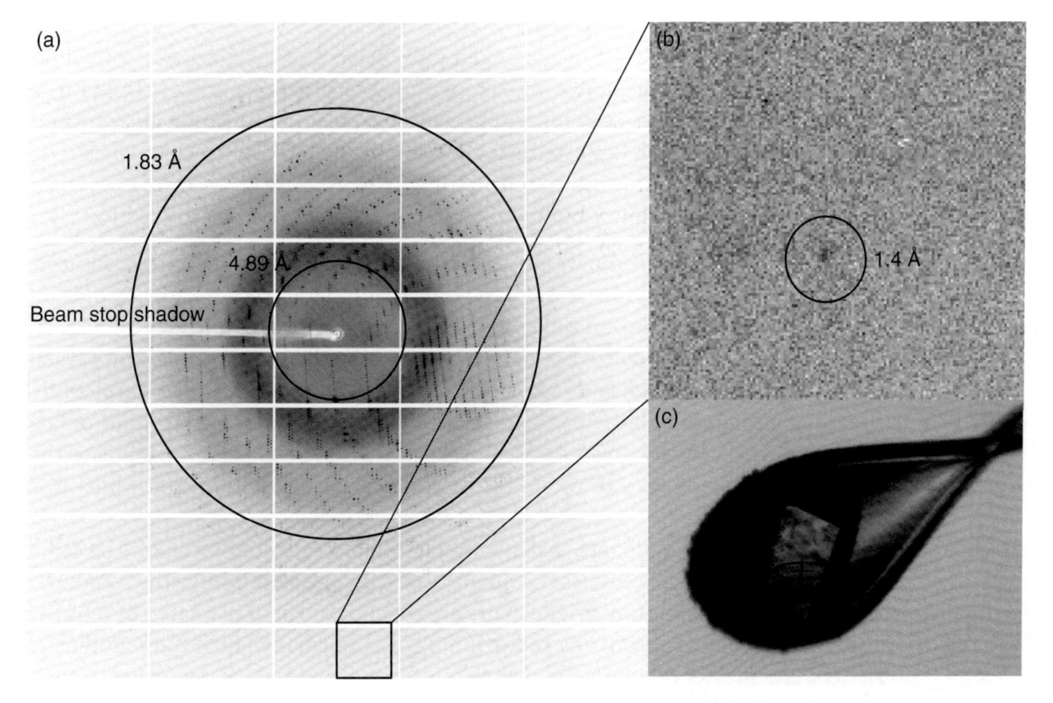

Figure 16.7 Diffraction pattern. (a) Macromolecular diffraction recorded on a Pilatus 6M detector (0.5° oscillation). Two rings have been displayed with the corresponding resolution. The slightly darker background outside of the 4.89 Å ring is due to the solvent scatter between 3.7 and 2.5 Å resolution. (b) Magnification of panel A with contrast adjustment; a barely visible reflection at 1.4 Å is circled. (c) The sample mounted in a nylon loop with the beam shape (80 μm × 20 μm) and its position marked with an ellipse. Please note that this dataset was integrated successfully to 1.18 Å, $I/[\sigma]I$ of 1.4 and a $CC_{1/2}$ of 0.6 for the highest resolution shell (1.2–1.18 Å).

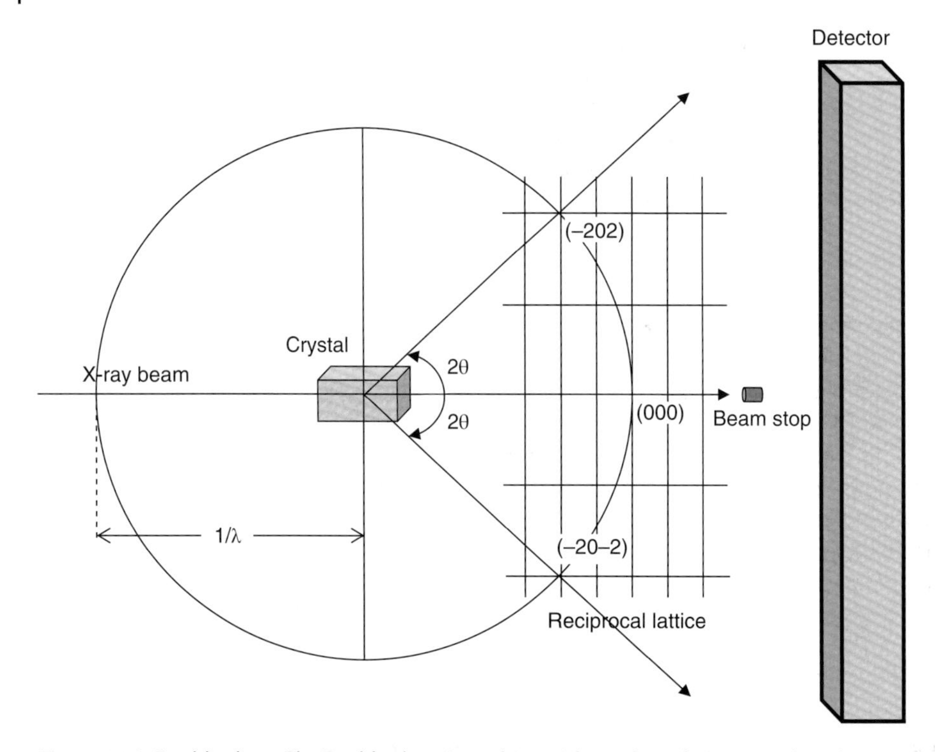

Figure 16.8 Ewald sphere. The Ewald sphere is a sphere with a radius of $1/\lambda$ centred on the crystal. The reciprocal lattice is represented with its origin (0, 0, 0) where the incident beam crosses the Ewald sphere. Only the points where the surface of the Ewald sphere coincide with a reciprocal lattice point (e.g. (−2, 0, 2) and (−2, 0, −2)) will result in a reflection recorded on the detector.

We record the intensity of each reflection on the detector in real terms (i.e. a detector coordinate in x and y, and an intensity for each pixel) from which their coordinates in reciprocal space (h, k, l), called Miller indices, can be determined during the indexing and initial processing of the diffraction data. In short, low resolution reflections have low Miller indices and higher resolution reflections have one or more high Miller indices.

The Ewald sphere (Figure 16.8) is a geometrical construction that helps to explain diffraction. The crystal is located at the centre of the Ewald sphere, which has a radius of $1/\lambda$, where λ is the radiation wavelength. Diffraction can only be recorded when reciprocal lattice points coincide with the sphere surface, but only a small number of points fulfil this condition at any given time. To record the data required to calculate the structure of the crystallised molecule it is necessary to rotate the crystal around a rotation axis whilst it is illuminated by the X-ray beam.

The symmetry relationships within the unit cell dictate that symmetry-related reflections exhibit the same intensities. Therefore, symmetry within the crystal reduces the quantity of data needed to solve the structure. Some special relationships require further explanation. Friedel pairs are reflections that are related by a transformation through the origin such that the intensity of any reflection, $I_{h,k,l}$, is normally equal to the intensity of its Friedel pair $I_{-h,-k,-l}$. Friedel's law breaks down in the presence of heavier atoms (S, Se or any transition metal) that are anomalous scatterers when the energy of the incident radiation is at or above an atomic absorption edge of the heavy atom. Other reflections may be related to one of the Friedel pairs, for example in the presence of a twofold symmetry axis, and such reflections are called Bijovet pairs and can be used to strategically orient the crystal for data collection.

Finally, all X-ray detectors are capable of recording the intensity of the diffracted X-rays, which can be considered as wave forms. A simple wave is a sinusoidal function (Figure 16.9) and though diffracted X-rays are complex waves, they are repetitive. The mathematical function that relates the electron density of the

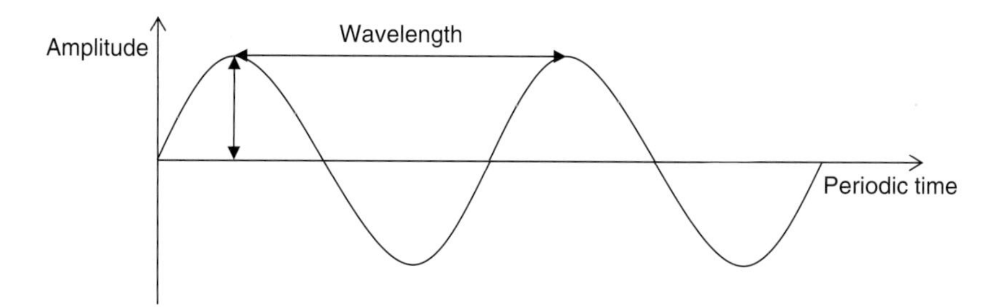

Figure 16.9 Sine wave. The amplitude and the wavelength are represented on a typical sine wave.

crystallised molecule to the diffracted waves is the Fourier transform, named after the eighteenth century French mathematician Joseph Fourier, and the operation from reciprocal space to real space is an inverse Fourier transform. In brief, two parameters are needed for the Fourier transform. The first is the amplitude of the wave, which is proportional to the root mean square of the intensity, and the second is the phase, which is the angle at which the wave peaks.

16.2.3 Scaling

Symmetry-equivalent reflections may not experience the same environment – for instance, variations in crystal quality, shape and size, and the path of the X-ray beam through the loop and cryoprotectant can affect the signal-to-noise ratio of the recorded reflections. The crystal may have been centred poorly, and the X-ray dose received may vary during the experiment. Many modern detectors are mosaic assemblies and each tile in the detector can have slight variations in sensitivity. The beam flux may also reduce during the time of the data collection. Therefore, each reflection should be recorded multiple times before merging, averaging and outputting a single mean intensity, a process called scaling. Data for multidataset techniques also need to be on the same relative scale and cross-dataset scaling must be done after internal scaling has been completed. The amplitude of each reflection needs to be calculated, which is proportional to the root mean square of its intensity, for calculating the structure factors necessary for Fourier syntheses. Typically, 5% of the measured data are randomly marked at the end of scaling for cross-validation, the R_{free}, which are used to provide an unbiased metric of a quality model.

16.2.4 Radiation Damage

All biological materials suffer from radiation damage when exposed to ionising radiation and radiation damage in protein crystallography is a significant challenge, especially considering third generation synchrotron beamlines where fluxes of 10^{12} photons per second and beam cross-sections of 20–50 μm are not uncommon. One of the first properties of the crystal that is lost during radiation damage is diffraction strength. The higher resolution reflections fade quickly and disappear, a phenomenon first described in the 1960s [7]. The diffraction lifetime is related to the X-ray dose and Henderson has proposed a generalised limit for protein crystallography of 2×10^7 Gy (J/kg), the X-ray dose that can be absorbed by a cryocooled protein crystal before the mean reflection intensity is halved [8]. The dose constraint has been refined subsequently to $3–4 \times 10^7$ Gy [9]. The radiation absorbed by the sample leads to crystal decay, increasing disorder within the sample, and a consequent reduction in reflection intensity through two phenomena. Primary damage is the ionisation of an atom due to photoelectric absorption or Compton scattering and only a reduction in the total applied dose can reduce the effects of primary damage. Secondary damage occurs from the formation of secondary electrons that diffuse and induce further damage [10] but its accrual rate can be reduced by maintaining the sample at cryogenic temperatures. A constant stream of dry nitrogen gas at 100 K is directed over the crystal during data collection to slow secondary radiation damage. Covalent bonds break on absorbing sufficient energy, resulting in the destruction of the protein. The overall loss of diffraction intensity affects the whole experiment but localised specific damage can also occur, such as disulphide bond destruction, leading to artefacts in the crystallographic analysis. Radiation damage should be avoided as much as possible as even mild damage can complicate structure solution and analysis unnecessarily.

16.2.5 Matthews' Coefficient

Matthews' coefficient (V_M) is a measure of the solvent content of protein crystals and typically ranges between 27 and 65%, corresponding to V_Ms of 1.62 and 3.53 $Å^3$/Da, respectively [11]. Crystals with low V_Ms tend to diffract well and those with high V_Ms tend to diffract modestly. The most likely number of copies of the crystallised molecule in the crystallographic asymmetric unit can be calculated in several ways (e.g. http://csb.wfu.edu/tools/vmcalc/vm.html, http://www.ruppweb.org/mattprob/default.html and the CCP4 suite), knowledge of which is imperative for molecular replacement and density modification routines.

16.2.6 Non-crystallographic Symmetry

Proteins, especially symmetric ones, can pack with multiple copies within the asymmetric unit. The spatial relationships between each molecule do not obey crystallographic symmetry rules; instead they follow non-crystallographic symmetry (NCS) operators. Detecting and defining NCS operators can be critical to successful structure solution and in the special case of icosahedral virus crystallography, NCS can be exploited to solve new structures without recourse to experimental phasing [12].

There are two forms of NCS. First, protein molecules orientated in the same way within the asymmetric unit can be separated by a simple translation, and the translational symmetry is detected by a native Patterson function. Alternatively, and more commonly, molecules within the asymmetric unit are related by set of rotations alone, as first detected for haemoglobin [13] in a calculation called a self-rotation function. Both types of NCS are detected by different applications of the Patterson function, which is calculated directly from the intensities of the diffraction data and is thus independent of phase information. The resultant Patterson map containing peaks corresponding to the vector between every atom in the asymmetric unit and every other atom. A simple structure containing three atoms will yield a Patterson map containing just six peaks, but a protein of 43 kDa will yield an uninterpretable Patterson map containing 9 million peaks. Subsets of these peaks will superimpose by correctly applying NCS to determine the spatial relationships between molecules in the asymmetric unit.

16.2.7 Structure Factors

While the amplitude of each reflection is calculated from its measured intensity, this information is insufficient to calculate an electron density map. The Fourier synthesis to calculate the electron density map depends upon the structure factor, $F_{h,k,l}$, of each measured reflection. Structure factors encompass both amplitude and phase information of each reflection and each $F_{h,k,l}$ is thus related to the corresponding reflection intensity, $I_{h,k,l}$. Once phase information has been obtained, each $F_{h,k,l}$, can be derived and an electron density map calculated.

16.2.8 Phase Problem

It is impossible to measure reflection phases directly and therefore structure factors cannot be calculated for the Fourier transform to generate an electron density map. This is the crystallographic phase problem that can be solved either by molecular replacement, and modern software pipelines perform initial searches automatically, or by experimental phasing (see [14, 15]). We introduce both in the next sections but do not discuss direct methods that are used mostly to solve small molecule structures and sometimes to solve atomic resolution structures of small proteins.

16.2.8.1 Molecular Replacement

Molecular replacement relies on similar protein sequences having similar 3D folds. The amino acid sequence of the target is used to find a potential molecular replacement search model from the PDB based on sequence homology only. If the target and search model share at least 50% sequence identity, molecular replacement will likely succeed. In exceptional cases molecular replacement can still succeed even if the identity drops to 13% [16]. In molecular replacement the search model is repositioned so that it matches the packing arrangement of the target protein in its crystal form. The procedure has two components, a rotation search to orient the search model correctly and a translation search that slides the rotated search model in three orthogonal directions until it occupies the same space and orientation as the target structure. If molecular replacement has worked, the calculated phases will be a reasonable approximation, enabling meaningful electron density maps to be calculated, and model building and refinement can commence.

16.2.8.2 Experimental Phasing: SAD

The current method of choice for experimental phasing is single-wavelength anomalous dispersion (SAD). It relies upon collecting data on an anomalous scattering-containing sample from incident radiation at or above the scatterer's atomic absorption energy. Selenomethionine is most commonly used for anomalous scattering measurements, with data collected above the Se K edge, 12 657 eV (0.9795 Å). Reflections no longer scatter in phase with the incident beam with radiation at or above the electron transition energy, the diffracted X-rays no longer scatter in phase with the incident beam, $I_{h,k,l}$ no longer equals $I_{-h,-k,-l}$ and Friedel's law breaks down. Phase information can be obtained by careful measurement of the intensity differences between Friedel pairs, called anomalous differences (Figure 16.10).

The heavy atom positions are found by calculating a Patterson function from the anomalous differences. The Patterson peaks represent vectors between anomalous scatterers within and between asymmetric units. As anomalous differences are miniscule in comparison to normal intensities, the corresponding intensities of $I_{h,k,l}$ and $I_{-h,-k,-l}$ must be measured accurately; highly redundant data collection is one solution to this problem. The Patterson map is complicated by the presence of multiple heavy atom positions (there are N^2–N peaks in the Patterson function, where N is the number of heavy atoms) and partial occupancy and/or mobile atoms reduce Patterson map peak heights. If the position(s) of the heavy atom(s) can be deduced, phase estimates can be calculated. The phase problem is represented geometrically with the elegant Harker diagram (Figure 16.10): here the structure factor F_{PH+} and its Bijvoet pair F_{PH-} result from the real contributions of the protein (F_P) and heavy atom (F_H) and imaginary contributions F_{H+} and F_{H-}, which are oriented 90° away from the normal contribution F_H but in opposite directions. Two circles of radius F_{PH+} and F_{PH-} can be drawn with the enantiomer shifted from the origin by the enantiomer imaginary contribution (Figure 16.10).

SAD is very effective at finding heavy atom positions when data are collected carefully but initial phases are dependent solely upon the anomalous differences and tend to be poorer than phases from MAD experiments. Single and multiple isomorphous replacement can be combined with anomalous scattering (SIRAS and MIRAS, see section 16.3.5.4), but unlike SAD and MAD these experiments rely upon having multiple isomorphous crystals available and the hit-and-miss approach of heavy atom derivatisation.

16.2.8.3 Experimental Phasing: MAD

The normal and anomalous scattering changes, which depend on the wavelength of the incident beam, can also be used to determine phases. This method is called multiple-wavelength anomalous dispersion (MAD) and unlike SAD addresses phase ambiguity directly [17]. However, multiple datasets are required from ideally a single crystal and only heavy atoms with a transition edge in the range of a tunable synchrotron beamline can be used. Three datasets are typically collected: (i) the peak, where the anomalous scattering is at its highest; (ii) the high energy remote, where the normal scattering is close to its native value; (iii) the inflexion, where the normal scattering is at its lowest (Figure 16.11). Two-wavelength experiments can mitigate radiation damage when collecting multiple datasets from a single crystal [18]. Heavy atom positions are found as in SAD and the third dataset resolves phase ambiguity because three circles in a Harker can only intersect at one point.

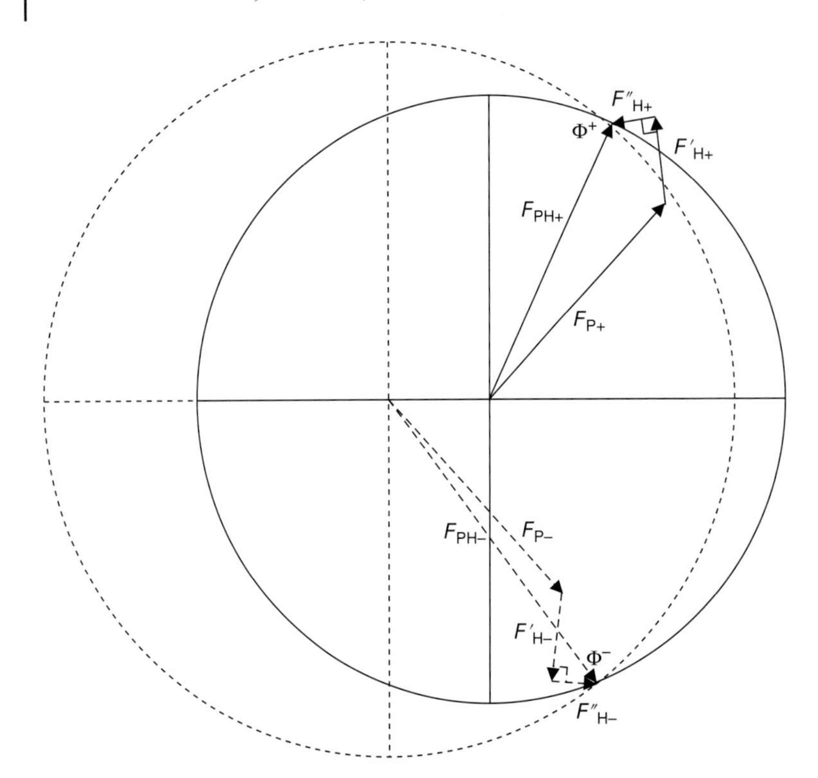

Figure 16.10 Harker diagram. Reflections are presented as vectors. A given reflection F_{PH+} results from the normal contribution of the protein (F_{P+}) as well as the normal (F'_{H+}) and imaginary contribution (F''_{H+}) of the heavy atom. Two circles can be drawn with a radius of F_{PH+} (solid line) and for the opposite Friedel pair a radius of F_{PH-} (dashed line) having their origins shifted with the anomalous contributions F''_{H+} and F''_{H-}. The positions at which the circles intersect give two possible phase angles (φ^+ and φ^-) for this reflection. Only one of the two is correct and the phase ambiguity will have to be solved by a subsequent step.

16.2.8.4 Experimental Phasing: SIR and MIR

Single isomorphous replacement (SIR) and multiple isomorphous replacement (MIR) were the methods of choice to solve the phase problem before the advent of tunable third generation synchrotrons beamlines. Both require multiple datasets, a native and at least one heavy atom derivative. Heavy atom derivatives arise when protein crystals are soaked in solutions of gold, mercury, platinum, etc. salts if the heavy atom is bound at specific points and if the crystal remains isomorphous with the native. The Patterson function is again used to identify the heavy atom positions, which are used to estimate initial phases. All MIR datasets have to be in the same space group and highly isomorphous with cell parameters that differ by less than 0.5%. Phase ambiguity will be resolved in MIR with two or more derivatives.

16.2.9 Model Building and Refinement

Atomic models can be built after successful molecular replacement or the calculation of initial experimental phases. The protein model is manipulated using a computer graphics programme (e.g. FRODO and O [19], but nowadays Coot [20]) by reference to the electron density map, followed by rounds of model refinement. The changes to the model are restrained to known stereochemical parameters, such as the covalent bond lengths and angles, and non-covalent interactions such as van der Waals' contacts. A restraint is an ideal target value to which a function may converge. Refinement is an iterative process with the ambition of interpreting correctly every feature in the electron density map including protein atoms, bound ligands, ions and solvent. During refinement the observed data will be compared to calculated data from the model (atomic coordinates in x, y and z, atomic occupancies and a model for the positional error of each atom [or group of atoms] called the B factor). The two most important reliability factors, R_{work} and R_{free}, are calculated; the R_{work}

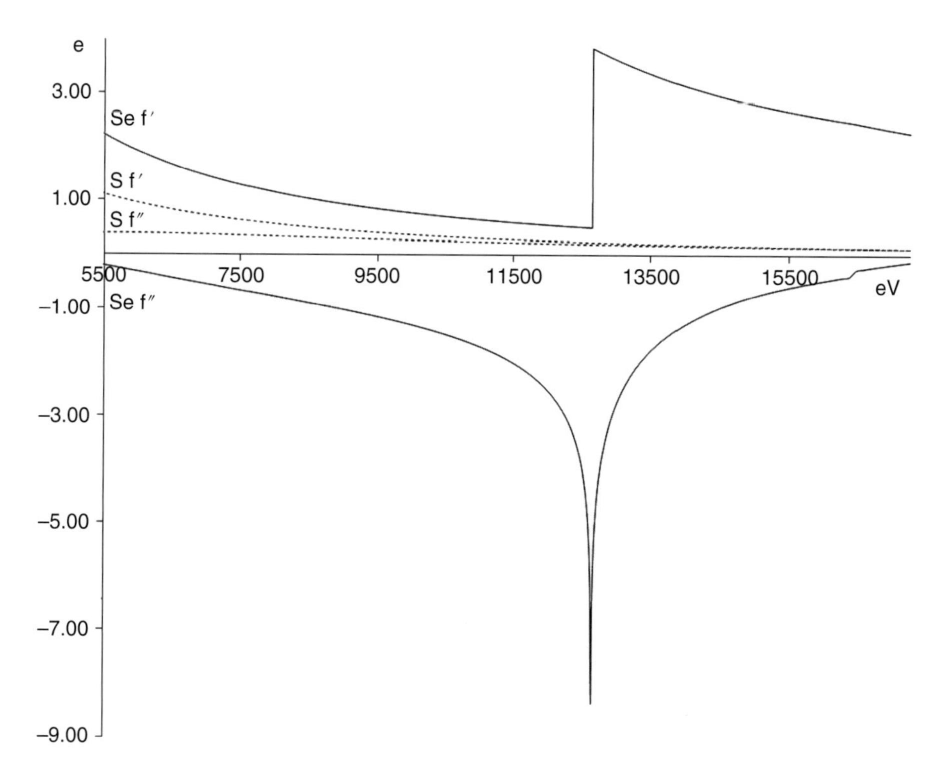

Figure 16.11 Edge plot. Selenium normal (f') and anomalous (f'') edge plots are represented as solid lines. For comparison sulphur is represented with dashed lines but does not have an edge between 5500 and 17 000 eV. Most standard synchrotron beamlines operate around the Se edge atomic absorption. Source: The scattering factor data were downloaded from http://skuld.bmsc.washington.edu/scatter/AS_form.html.

is calculated using 95% of the data used in refinement and R_{free} is calculated on the residual 5% that are set aside for cross-validation [21].

Some side chains or loops in the model may occupy multiple conformations because they are in different orientations in the asymmetric units through which the X-ray beam passes during the experiment. At low resolution these differences may be invisible, but at resolutions better than 1.8 Å each conformation and the relative occupancy of each can be modelled. Ligands may bind only partially and the occupancy can be adjusted to reflect the electron density.

The atom positional error is represented by the B factor (Å^2) as a spherical distribution and is normally modelled isotropically for each atom. Residue groups or overall B factor models may be more appropriate at very low resolution. Anisotropic B factor models can be modelled with ultrahigh resolution datasets. An additional thermal description of the model can be achieved with the TLS (translation libration screw) method (reviewed in [22]) with isotropic B factors.

16.3 Methodology

In the following section we cover current practice in most aspects of the entire structure solution pipeline.

16.3.1 Construct Design

Potential disordered regions, especially at the N- and C-termini, should be removed from the recombinant construct and XtalPred [23] and PONDR [24] used to predict disorder at the amino acid level. Secondary structure predictions from JPred [25] can check that terminal helices are not cut in half when designing constructs.

The crystallisation of thermostable orthologues or engineered variants [26] when the mesophilic equivalent protein has poor solution and/or thermostable characteristics should be attempted rather than focusing on just one family member. Alternate expression systems should be considered and the Oxford Protein Production Facility (OPPF at https://www.oppf.rc-harwell.ac.uk/OPPF/public/services) offers high throughput cloning and expression services.

16.3.2 Crystallisation Screening

Successful chemical conditions to crystallise groups of proteins have been identified and made commercially available as crystallisation kits for soluble proteins (e.g. PACT, JCSG+), monoclonal antibodies (e.g. GRAS), nucleic acids (e.g. HELIX) and their binding proteins (e.g. Natrix), protein complexes (e.g. Wizard), kinases (e.g. Kinase) and membrane proteins (e.g. MembFac, MemGold, Lipidic-Sponge Phase). The screens are biased towards what has worked previously, but new screens have emerged (MORPHEUS II and MIDAS) that explore different chemical space. Screens based on one fixed chemical family are also available (MORPHEUS, MPD and ammonium sulphate). Since optimisation may still be necessary to convert poor crystals into well-diffracting ones, additive screens that contain low concentrations of ions, detergents and solvents may improve crystal quality. The websites of screen suppliers (www.moleculardimensions.com, www.jenabioscience.com, www.qiagen.com and www.hamptonresearch.com) are useful resources.

Most screens will work with protein concentrations of 5–50 mg/ml. The protein should be kept in a simple buffer, and metal ions and phosphates should be avoided to stop salt crystal formation during screening. Small proteins (<30 kDa) might need higher concentrations to crystallise: 190 mg/ml was required for the 15 kDa CBM62 [27]. Larger or less soluble proteins can be crystallised at concentrations as low as 1 mg/ml. Crystallisation should start with a sparse matrix screen and the experiment inspected immediately with a stereo microscope; many heavy precipitates suggest that the protein concentration is too high. If the drops are mostly clear the protein concentration is likely to be too low.

Crystallisation robots are commonplace and permit screening hundreds of conditions using small amounts of protein, 50–300 nl, per drop, and 70–100 μl of crystallisation screen per condition. Most crystallisation robots use almost exclusively multiwell plate format labware. There are many manufacturers and distributors of multiwell plates that can be used, with variations in reservoir volumes, number of wells and shapes (flat bottom or round bottom wells) for the experiment. Initial screening is usually performed in 96-well format plates with small drops. Subsequently, initial hits are optimised on a larger scale. Anaerobic chambers can be used to store crystallisation experiments if the protein is redox sensitive.

When crystallising protein–ligand complexes, the ligand will generally need to be at a concentration of ~20-fold higher than K_d for the protein in order to saturate the binding site. Fundamental biochemistry thus ought to be performed before commencing potentially lengthy crystallisation screenings. Protein–protein complexes ought to be co-expressed and only those fractions that contain both (or all) proteins purified. Alternatively, pre-purified proteins should be mixed and the complex purified away from uncomplexed components by size exclusion chromatography.

Sitting drop vapour diffusion is the most common crystallisation technique because it matches best the experimental setups of crystallisation robots. Protein and crystallisation conditions are dispensed on to a plastic shelf. This is a simple and rapid procedure with most robots. For instance, it takes five minutes to dispense screen solutions from a deep well block into a 96-well plate using a hand-held multichannel pipette and two minutes for a TTP Labtech Mosquito to dispense protein and crystallisation solution for the entire plate at two drop ratios. Therefore, 192 experiments take less than 10 minutes to set up using only 28.8 μl of protein. Once the tray is finished the user seals all the drops using a clear plastic adhesive sheet and the tray is left at the chosen temperature for the precipitant concentration in the drop to equilibrate with that in the well.

Hanging drop experiments can also be established using robotics but this technique is normally favoured for manual crystallisation optimisation experiments. Here the optimisation reagents are pipetted directly into a 24-well tissue culture tray and high vacuum grease around the rim of each well is used to provide an airtight seal. The protein and mother liquor are mixed on plastic or silanised glass coverslips, which are inverted over the greased well rim and an airtight seal made by gentle pressing.

In sandwich drop experiments, the protein is mixed with the crystallisation solution and 'sandwiched' between two flat glass or plastic surfaces. The sandwich is placed on to a support above the well solution in a tissue culture plate and sealed with vacuum grease and coverslips. Though less trivial to set up, the sandwich method can be advantageous because the sandwiched drop surface area is greatly reduced, which slows the equilibration rate substantially. Slowing the crystallisation process can lead to fewer, larger, better quality crystals, but the sandwich process does not lend itself easily to the economies of scale and throughput of sitting drop crystallisation. In capillary dialysis crystallisation a narrow bore capillary is filled with protein by capillary action and sealed at one end with low melting point wax. The open end of the capillary is placed into a plug of agarose gel into which the crystallisation reagent has been impregnated. Protein and crystallisation reagent diffuse to set up a range of unique crystallisation conditions within one capillary. The batch method has also been miniaturised; here small volumes of protein and crystallisation screen are mixed in a multiwell batch plate and maintained under a layer of paraffin oil to prevent the drop from drying out.

In the absence of crystals or poorly diffracting ones, the construct could be modified, or limited proteolysis in situ could be used to remove disordered regions [28]. Lysines can be methylated [29] and lysines and other residues mutated to alanine [30] to reduce surface entropy. Cysteines can be modified [31] to reduce redox sensitivity. Differential scanning fluorimetry and other biophysical techniques [32] may be used to improve the protein stability as a function of buffer composition (i.e. buffer type, pH, salt).

16.3.2.1 Crystallisation Optimisation

Diffraction data of sufficient quality to solve structures can be obtained for most 'normal' projects using third generation synchrotron beamlines and crystals harvested direct from screening experiments that may be no bigger than 10 μm × 10 μm × 10 μm in size. However, if screen yields crystals that are too small (larger proteins or protein complexes often yield small, poorly diffracting crystals) or will be the subject of ligand soaks or experimental phasing, the crystallisation condition must be reproduced and optimised. Optimisation involves varying the concentration of the components of the crystallisation condition one at a time, and is performed normally in a 6 × 4 Linbro tissue culture tray. Typically, the precipitant concentration or type is varied in the longest dimension and the salt concentration or type in the shortest. Hanging drop vapour diffusion in Linbro plates is simple to achieve and is convenient for opening and closing the drop for harvesting. The cover slip is suitable for sample manipulation without changing the microscope focus: the less time spent trying to keep visual track of the crystal, the easier it is to harvest. The pH can be screened by repeating the optimisation tray at a different constant pH or by changing the buffer molecule itself, as in rare occasions it can contribute to the crystal lattice (e.g. [33]).

16.3.2.2 Crystallisation Seeding

If the crystallisation experiment reaches the metastable zone but without forming nuclei spontaneously (Figure 16.2), the introduction of seeds can produce controlled and reliable crystallisation of the protein. Seeds can be obtained from low quality crystals, crystalline precipitate and even inert PTFE or ceramic shards. There are two common seeding approaches. In macroseeding, a crystal (or crystals) is broken into several pieces and each macroseed is transferred into single crystallisation drops that have already equilibrated with their mother liquor, but in which no crystals have grown. The macroseed can be washed with mother liquor with a lower precipitant concentration to partially dissolve the outer layers of the macroseed. The seed encourages protein molecules to arrange themselves on to the introduced crystal lattice and thus a larger crystal can grow.

There are several ways to initiate crystal growth by microseeding, including in microfluidics platforms [34], and several approaches may be tried to find the best outcome. First, a microseed stock is generated by harvesting a small number of poor-quality crystals into an Eppendorf tube containing a few μl of crystallisation solution. The crystals are ground with the blunt end of a pipette tip and a $\sim 10^{-1}$–$\sim 10^{-6}$serial dilution of the microseeds is made. A dilution series is required so that just a few good-quality crystals grow instead of a shower of thousands of small crystals. The protein is doped with aliquots of the microseed stocks and robotic crystallisation trays are set up normally. Alternatively, equal volumes of protein, microseed and crystallisation solution are mixed for setting up a manual crystallisation plate by optimising the original condition. Cross-seeding can be used to re-screen conditions where nucleation did not occur with the protein alone. Finally, microseeds can be introduced to crystallisation drops that have reached equilibrium but without producing crystals with the use of a thin, flexible fibre such as a cat's whisker. Here the whisker is dipped into a seed stock and drawn over the top of 1–3 drops in succession, so a few microseeds are transferred from the whisker to the drop to nucleate crystal growth.

16.3.3 Cryoprotection

The natural form of crystalline ice on earth is sometimes referred to as hexagonal ice. It was predicted by Hass and Rossmann ~50 years ago that crystalline ice formation in the solvent channels of protein crystals will disrupt or break the crystal lattice [35]. Cryoprotectants are small molecules such as organic solvents, salts or small organic compounds that inhibit the formation of hexagonal ice in protein crystals [36]. Alcohols have low melting points but are sometimes incompatible with proteins. Salts are widely used on the roads in winter to prevent ice formation and saturated salts can also be used to cryoprotect protein crystals [37]. Polyethylene glycols (PEGs) are widely used as precipitants in protein crystallography, and hence PEG 400 is a very popular cryoprotectant. Other than undiluted, saturated salts, most cryoprotectants are used in the 20–30% range. Ideally the cryoprotectant replaces water and maintains the reservoir composition, assuming that stock solution concentrations can be made high enough to supplement the crystallisation condition with an appropriate cryoprotectant. It is tedious and time consuming to make each mother liquor and cryoprotectant combination from scratch for a large number of crystallisation conditions and often it is possible to simply dilute the reservoir directly with the cryoprotectant (e.g. 8 μl of reservoir and 2 μl of PEG 400 to achieve a 20% PEG 400 cryoprotectant). Some cryoprotectants are immiscible with the reservoir (for instance ammonium sulphate and ethylene glycol) and inspection of the cryoprotectant under the microscope will identify unwanted phase separation.

If the sample diffracts badly, its diffraction at room temperature should be tested to determine if the cryoregime is the root cause of the poor diffraction. The crystal should be mounted in a loop with a viscous oil, such as paratone N, which will protect it from dehydration at room temperature for long enough to establish the sample's base diffraction properties. Some crystallisation conditions are already cryoprotected because of their composition (e.g. the entire MORPHEUS screen) and no additional cryoprotection steps are required.

Glycerol has been used extensively as a cryoprotectant and the concentration required to cryoprotect common crystallisation conditions has been published [38]. Glycerol increases protein solubility and can dissolve protein crystals [39]. Glycerol can also be found in the binding pockets of carbohydrate binding and processing proteins instead of the native ligand when glycerol is used as a cryoprotectant because 20% glycerol, 2.1 M, outcompetes the binding of the desired ligand that is normally often present at mM levels (e.g. [40]).

There are many methods and cryoprotectants available and we direct the reader to Ciccone et al. [39] and Parkin et al. [41].

16.3.3.1 Mounting

Increased automation, especially at synchrotrons, has necessitated mount type and length standardisation. Several crystal mounts, pins and loops are available from commercial suppliers (e.g. www.mitegen.com, www.hamptonresearch.com and www.moleculardimensions.com). The mounts come in a variety of shapes, sizes and materials, including nylon, polyimide and other polymers (Figure 16.12). To reduce background on the diffraction image and to aid sample visualisation, the size of the loop should be matched to that of the crystal.

If no crystal manipulation is necessary, samples can be harvested and cryocooled directly. Otherwise 0.5–1 μl of the cryoprotectant should be placed next to the sample-containing drop, ideally in the same focal

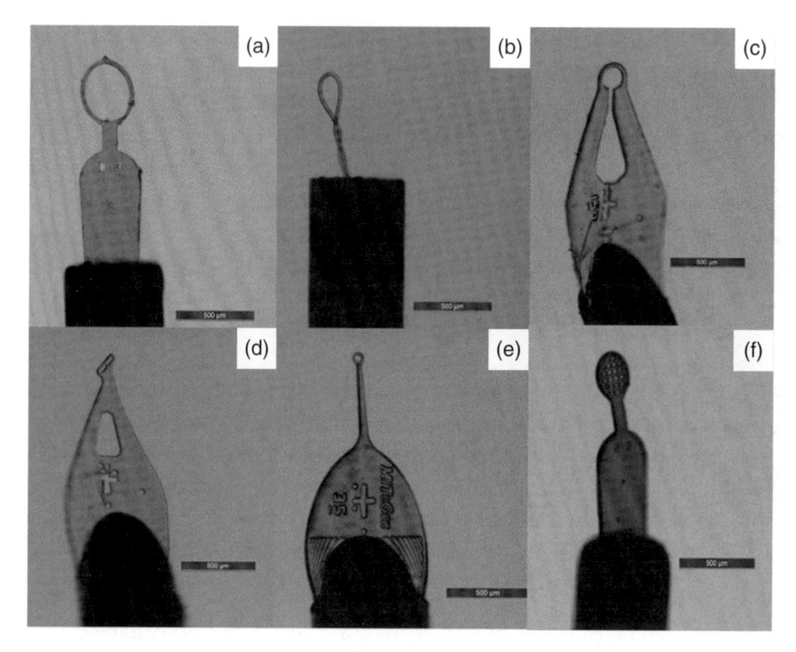

Figure 16.12 Typical and specialised mounts. (a) Litholoop (Molecular Dimensions). (b) Nylon loop (Hampton Research). (c) Micromount (Mitigen). (d) Inclined mount that can be used for rods or needles (Mitigen). (e) Dual thickness microloop useful with viscous material such as LCP (Mitigen). (f) Mesh loop that can be used to support plates or harvest microcrystals (Molecular Dimensions).

plane (Figure 16.13). The crystal is picked up with the loop and transferred to the adjacent cryoprotectant drop. The loop is rinsed in the cryoprotectant, away from the crystal, before the crystal is removed from the cryoprotectant and cryocooled. This procedure should be completed within a few seconds to avoid dehydrating the sample, so all necessary tools should be to hand. If longer times (minutes to overnight) are required for ligand soaking and/or cryoprotection, the drop must be kept sealed within the experimental chamber.

Acupuncture needles to remove skin (made of unfolded proteins) from the surface of the drop are useful and can also be used to break crystals free from the well or separate multiple, intergrown crystals. A cat's whisker glued to the end of a laboratory pipette tip, mounted on a pencil for handling, is useful as it is just stiff enough to gently push a crystal around without damaging it.

16.3.3.2 Cryocooling

The sample must be cryocooled rapidly once cryoprotected. Since speed is essential to vitrify water without crystalline ice formation, cryocooling by plunging the sample directly into a liquid nitrogen bath is best. Slow cryocooling can result in hexagonal ice formation, even if the sample is cryoprotected adequately. Hexagonal ice has a translucent, white appearance and should be avoided. The cryocooled samples are placed into a pre-labelled cryovial while being retained under liquid nitrogen by careful juxtaposition of the mount and cryovial (Figure 16.14); such samples can be stored indefinitely in liquid nitrogen dewars.

When liquid nitrogen is exposed to ambient air, moisture condenses on the cold nitrogen gas forming flocculent ice, which should not build up in the liquid nitrogen dewar. Flocculent ice can be removed from liquid nitrogen by filtering through 'blue roll' in a metal funnel. The cryomount components can build up static and attract flocculent ice, and this should be avoided. Filling the liquid nitrogen bath and dewars used in cryocooling (Figure 16.15) completely will dilute the potential effect of condensing moisture. Replacing liquid nitrogen frequently or recycling liquid nitrogen by filtering through blue roll into a clean dry dewar every five samples will reduce the risk of flocculent ice. A tall dewar should be used because flocculent ice tends to sink in liquid nitrogen.

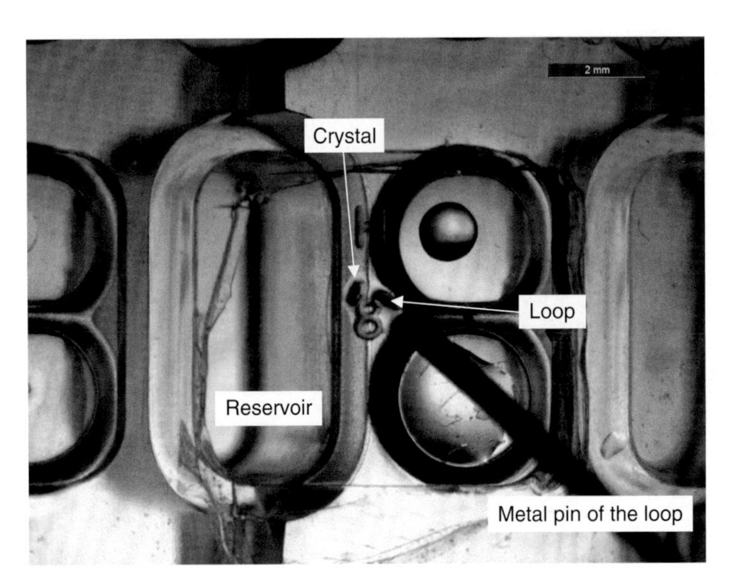

Figure 16.13 Cryoprotecting on a 96 well plate format. The condition H8 was opened using a scalpel. A drop of cryoprotectant has been placed on the centre ledge between the reservoir and wells. A crystal and a nylon loop are in the solution and the metal pin of the loop is visible in the bottom right hand corner of the image, but slightly out of focus.

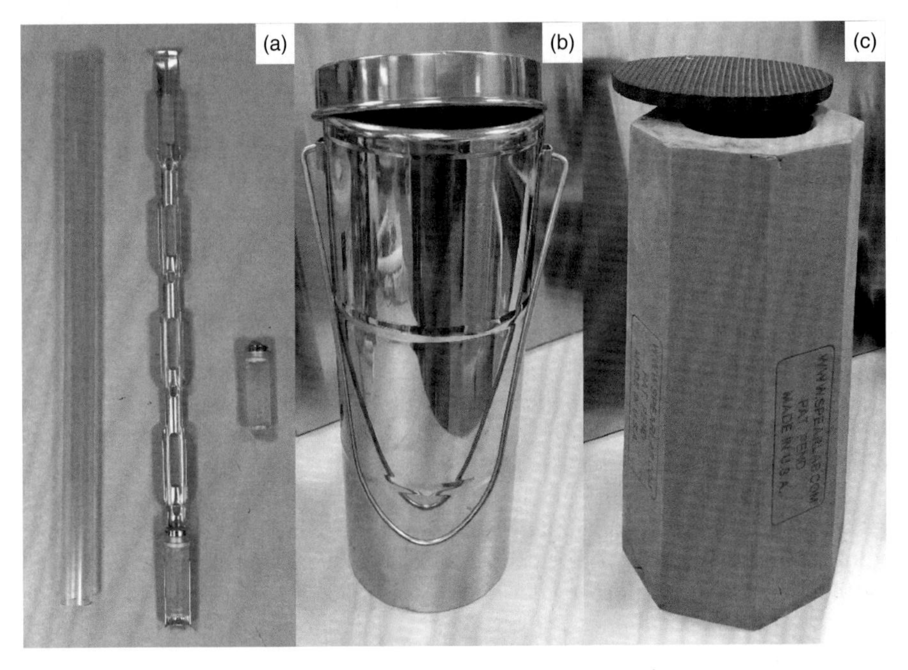

Figure 16.14 Sample handling material. (a): Metal cane with one vial and its mount. The cane can hold five vials. A single vial is on the right side of the cane. On the left there is a plastic sleeve in which the cane is sheathed prior to storage. (b): Tall metal dewar used to carry canes. (c) Tall foam dewar. Both are high enough to take a cane under liquid nitrogen.

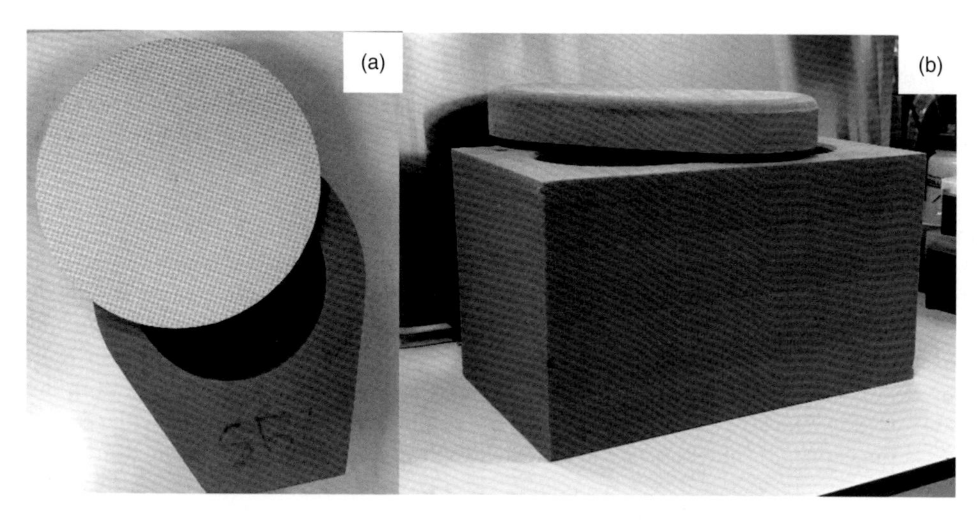

Figure 16.15 Foam dewars. (a) A typical foam dewar used for cryocooling samples. (b) A specialised dewar used for carrying sample holders called pucks, which hold up to 16 samples at once.

16.3.3.3 Storage and Transportation

Cryocooled samples can be kept indefinitely in storage dewars that are topped-up regularly with liquid nitrogen. 'Dry shippers', dewars containing an inner, sponge-like compartment that can be saturated with liquid nitrogen, are used in transportation. Saturated dry shippers can maintain sub-100 K temperatures for more than one week [42]. Dry shippers should be dried out completely and routinely to remove water. A dry shipper at room temperature can be chilled to operational in less than 30 minutes if refilled with liquid nitrogen as soon as it evaporates. The shipper is ready for use once the liquid nitrogen within stops boiling, but prior to transportation the excess liquid nitrogen must be decanted.

16.3.4 Data Collection Strategies

Data collection strategies depend on the type of available X-ray source and goniometry [43–45]. Many laboratories have stopped investing in in-house X-ray facilities, because of the current excellent access to synchrotrons and the high recurrent cost of maintaining in-house facilities; therefore, much of the following concerns synchrotron data collection strategies. The primary goal is normally to obtain a complete diffraction dataset to the highest resolution possible while avoiding data corruption through radiation damage. The more brilliant the incident X-ray beam, the stronger is the signal. A bigger crystal increases the signal but may have more internal defects, such as higher mosaic spreads, and a better sample may be required. Each crystal has an inherent diffraction limit and increasing the X-ray dose will not result in higher resolution; instead poorer quality data will be collected because of radiation damage.

The following discussion focuses on 'normal' crystallographic problems. The challenges posed by very large unit cells and crystals exquisitely sensitive to radiation damage (e.g. membrane proteins, viruses and the ribosome) are not covered here, and the reader should consult an excellent recent review that addresses these problems and how they can be overcome [46].

16.3.4.1 Mounting and Centring the Sample

The crystal is placed on the goniometer either manually or through a robotic sample changer and maintained at a cryogenic temperature throughout. The sample is centred in the intersection of the X-ray beam and the rotation axis of the goniometer and while historically the centring was done manually it is now commonly software-driven.

Any surface ice detected ought to be removed by manual washing of the sample with a 25 ml plastic pipette that has been filled with dry liquid nitrogen. Some synchrotron beamlines provide a crystal wash device. If a wash device is not available and/or the diffractometer environment has insufficient space to permit manual washing of the sample, simple unmounting of the sample into liquid nitrogen before returning it to the goniometer is often sufficient to remove at least some of the surface ice.

16.3.4.2 Collecting the First Few Images and Indexing

The inherent diffraction properties of the crystal are assessed and the crystal symmetry, which impacts on the data collection strategy, determined. Then two, three or even four initial images are collected at relative phi/omega positions of 0°, 90°, and if necessary 45° or 30° and 60°. The oscillation angle depends on the unit cell parameters, the strength and quality of the diffraction and the detector characteristics, but ordinarily an oscillation angle of 0.5° is sufficient. The exposure time and transmission have to be determined and in the absence of previous data on the crystal type, or experience with the equipment, it is best to start with conservative values and increase the exposure time and/or the transmission until the crystal's intrinsic diffraction limit is reached.

These initial images are used to determine the basic symmetry of the crystal, its unit cell parameters and the orientation of the crystallographic symmetry axes in relation to the axes of the experimental setup. Indexing software calculates the 3×3 orientation matrix that describes the relationship between the reflections to the position of the lattice plane. The nine values of the matrix can be considered as a combination of the six unit cell dimensions and three orientation values. The likely Bravais lattice is chosen as the actual space group might not be discernible until the data have been processed or even until the structure is solved. More than one data collection strategy might be suggested but additional considerations will apply in experimental phasing problems, which are described in sections 16.3.5.1 to 16.3.5.4. For more detailed information on data collection strategies the reader is referred to these excellent articles [44, 46].

Some detector types, such as silicon pixel area detectors (PADs), yield optimal signal-to-noise ratios by 'fine slicing' with 0.1° oscillation angles per image because of the fast, noise-less readout of PADs. It is almost impossible to see the weak high resolution reflections under these conditions, but current data processing packages handle these data easily. Finally, the exposure time and transmission should be considered. For example, if the best resolution is attained using 0.5° oscillations for a 0.5 s exposure at 50% transmission, then in fine slicing mode 0.1° oscillations at 0.1 s exposure and 50% transmission should be selected. If the data acquisition rate of the detector is sufficiently fast the exposure time can be halved and the transmission doubled to keep the same overall dose; in this scenario the parameters will be 0.1°oscillations at 0.05 s and at 100% transmission. The crystal system and an exposure time per degree at a given transmission are input factors to programmes such as RADDOSE [47] that supplement the data collection strategy to avoid exceeding the Henderson limit [8]. The weak data – those with low background-to-noise ratios – in the highest resolution zones of the image contain the crucial atomic detail of the structure; if the data processing reveals that they have been recorded poorly they can be discarded later.

The diffraction pattern should contain well-resolved spots, without overlaps or elongated, 'streaky' profiles. Ideally, the diffraction should have the same resolution limits in all directions and similar resolution limits on any image. If this is not the case the sample is anisotropic, a phenomenon illustrated well in plate-like crystals where the diffraction may be reasonable when the X-ray beam traverses the long dimensions of the crystal but very weak when traversing the shortest. The incident beam can be made smaller on some synchrotron beamlines and the entire crystal scanned in a grid to find a region with improved diffraction. The incident beam should not exceed the sample size to reduce background scatter. Diffraction from crystalline ice forms rings at resolutions of 3.90, 3.70, 3.44, 2.67 and 2.25 Å if the cryoprotection regime has been unsuccessful and the diffraction from the macromolecule is likely to be badly affected. Poor inherent diffraction can be improved by annealing; the poor mosaic spread of a badly diffracting crystal suffering from internal disorder can be improved by a warming and cooling cycle. In the absence of additional crystals or crystal forms, annealing is worth trying but its success rate is somewhat limited.

Finally, multiple crystal lattices can be detected during the initial stages of data collection. Multiple lattices result in additional reflections that cannot be predicted by any indexing solution. Some reflections overlap and their deconvolution is not straightforward as the relative contribution from each lattice is not trivial to determine. Therefore, only one crystal per loop should be mounted. At synchrotrons it may be possible to position the loop or minimise the beam cross-section so that only one crystal is illuminated by the X-ray beam and hence only a single diffraction pattern is recorded.

16.3.4.3 Sacrificial Sample

If more than one similar sample is available, one crystal can be 'sacrificed' by assessing its radiation sensitivity on the synchrotron beamline so that the user can use exposure and transmission settings that are close to what the next sample is likely to tolerate. In addition to the potential loss of a sample through mishandling, a sacrificial crystal means that several samples of the same crystal should be sent to each synchrotron trip. Fortunately, the universal puck system can hold up to 112 samples in a dry shipper (Figure 16.16) and higher density systems are becoming available.

16.3.4.4 Indexing and Integration

Data indexing and integrating programmes are now highly automated, but it is important to ensure the right Bravais lattice has been selected so the correct data collection strategy is decided. iMosflm [48], XDS [49] and DIALS [50] are excellent for indexing and integrating and complement the Xia2 [51] and AutoPROC [52] automated pipelines.

16.3.4.5 Scaling

The Phenix [53] module Xtriage and the CCP4 [54] packages Aimless [55] and Pointless [56] are under constant development. Scaling of diffraction data is straightforward as many tasks are automated and various diagnostic subroutines direct decision making. The diagnostics include space group determination and ice ring and anisotropy detection. The output reports should be considered and action taken to mitigate data defects. A key decision is the high resolution cut-off [55]. While it is desirable to have as much atomic detail as possible, it is detrimental to subsequent processes to include data that are simply not observed. Many metrics are computed to determine the resolution limit and the following are merely a guide. Ideally, the high resolution shell should have a completeness above 95%, a ½ correlation coefficient (CC½) above 50% and a signal-to-noise ($I/[\sigma]I$) ratio above 1.5. However, some compromises may be necessary and lower data completeness levels can still be useful as every reflection carries some information about the entire structure. In SAD experiments, the data collection strategy is the most critical step and the anomalous signal (often 1 Å less than native resolution) is much more important than the high resolution cut-off. Highly redundant data are critical for experimental phasing, especially for weak anomalous scattering signals, and should be as high as possible without incurring radiation damage. If radiation damage is detected during scaling, the affected images should be removed.

16.3.5 Native Data Collection, Molecular Replacement and Ligand Soaks

The simplest structures to solve have the simplest data collection requirements. The data collection requirements are low when the structure is already known and the experimental question is 'How does this small molecule bind to my protein?' or 'How does this mutation affect the protein structure?' The aim is to collect complete, as high resolution data as possible without compromising data quality. The signal-to-noise ratio is set by the exposure time and transmission per image and the overall dose should be pushed close to the Henderson limit. The resolution should be below 2 Å to observe most of the ordered water molecules that play important roles in mediating protein–ligand complexes. The data completeness ought to exceed 90–95% and the data redundancy, the number of times single reflections have been recorded, can be as low as 2–3. In molecular replacement, the data should be complete and well-recorded and it would be better to double the

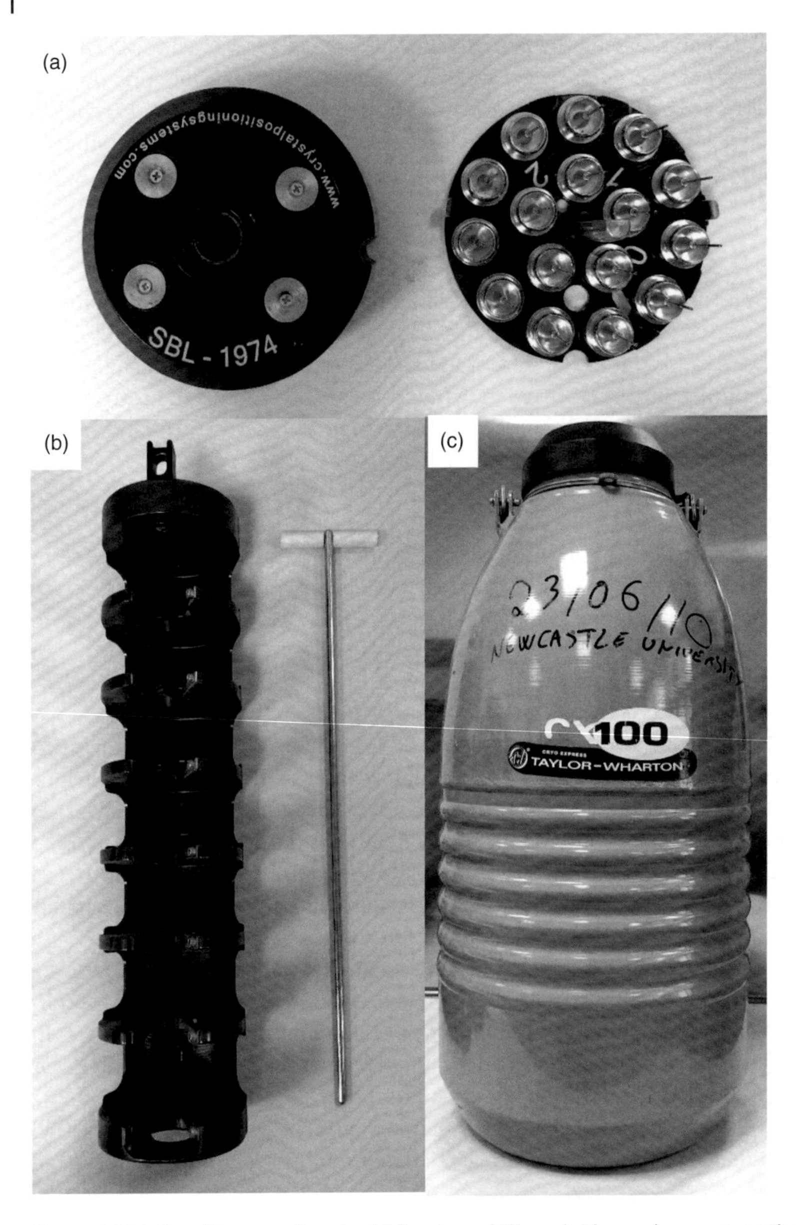

Figure 16.16 Dry shipper and pucks. (a) A universal V1-puck. The puck cover is on the left and the puck plate on the right. 16 mounts are on the plate. (b) A puck shelf for transportation of up to 7 pucks. (c) A dry shipper where the puck shelf can be loaded and kept cold for >2 weeks under regular conditions.

redundancy and ensure 100% completeness for the loss of 0.2 Å in resolution. These important parameters are all defined in the data collection strategy.

16.3.5.1 Experimental Phasing

If molecular replacement cannot or fails to solve the structure, phase angles for each reflection must be determined de novo. The most commonly used method to this end is anomalous scattering from selenomethionine-labelled proteins. Briefly, the recombinant protein is expressed in a medium in which methionine is replaced by selenomethionine, where selenium replaces the naturally occurring sulphur

[17, 57, 58]. Methionine-auxotrophic *E. coli* strains ensure that natural methionine is not biosynthesised and bacterial growth thus depends upon selenomethionine in the medium. Selenomethionine-labelled proteins should purify and crystallise similarly to the wildtype, though additional care (and reducing agents) might be necessary to avoid selenium oxidation. Inherent transition metals, such as zinc, can also be used for phasing and before a selenomethione-substituted sample is made the potential metal content of native crystals should be tested by a fluorescence emission scan on a tunable synchrotron beamline. The two approaches to structure solution based on anomalous scattering are covered next.

16.3.5.2 Single Wavelength Anomalous Scattering – SAD

SAD experiments, like all others, depend upon a well-cryoprotected sample that diffracts sufficiently well to be indexed. However, additional factors have to be considered. An incident radiation energy (or wavelength) must be selected for recording sufficient anomalous differences to solve the structure. On a tunable, synchrotron X-ray beamline, the user can collect data at the optimal wavelength and webservers (e.g. www.bmsc.washington.edu/scatter) used to estimate the likely SAD signal at any given wavelength for any atom (Figure 16.11).

On tunable synchrotron beamlines the sample can be scanned for element-specific fluorescence emission, allowing fast identification of potential scatterers and the precise absorption edge that is related to each scatterer's local environment. In the absence of a scan the theoretical energy should guide the SAD data collection strategy, for example, above the Se *K* atomic absorption edge (12 657.8 eV; 0.9795 Å). The anomalous scatterer does not require an accessible edge for successful SAD. For example, SAD phasing can be done on light atoms such as sulphur or iodine at low energies (high wavelength) at a synchrotron (e.g. [59]) or on a home source (e.g. [60–62]). The Diamond synchrotron beamline I23 is dedicated to low energy data collection and the beamline operates under vacuum to minimise air absorption of scattered X-rays. It is critical in SAD experiments to measure the intensities of Friedel pairs accurately to determine their anomalous differences. In comparison to the absolute intensity of each reflection, the intensity differences between Friedel pairs are very weak and their accurate measurement is challenging. Each Friedel pair is measured multiple times to achieve the necessary accuracy. Since solving the anomalous scatterer substructure is heavily dependent upon highly redundant data (perhaps greater than 50-fold for weak anomalous scatterers such as sulphur [61, 63]) very conservative data collection strategies are required for the sample to survive such large X-ray doses. The anomalous scatterer is prone to suffering the greatest radiation damage exposed to a wavelength close to its atomic absorption edge [64]; therefore, the transmission and/or exposure time per image should be reduced. Under such conditions the crystal's inherent diffraction limit might not be reached and higher resolution data for refinement should be collected from another sample. The weak anomalous signal is the important measurement in SAD experiments and diffraction data in a SAD experiment are typically collected 0.5–1 Å lower in resolution than the sample's diffraction potential. NucB, for instance, was solved by sulphur SAD using data to 2.26 Å, but refined against data to 1.35 Å [59].

It can be important to measure Friedel pairs in close temporal proximity to minimise the problem of reduced diffraction intensity as a function of absorbed X-ray dose. The small intensity differences between Friedel pairs are lost if the crystal suffers radiation damage. This problem is addressed by the 'inverse beam', where a few degrees of data are measured followed by a 180° rotation to measure the equivalent wedge in order to collect the Friedel pairs as close together as possible. This process is repeated throughout the entire data collection strategy. The beamline goniometry may permit Bijvoet pairs (Friedel pair symmetry equivalents) to be collected on the same image if crystallographic symmetry axes can be aligned appropriately on the goniometer. The synchrotron beamline data collection strategy programmes should be used to determine how the data are best collected [65–67].

16.3.5.3 Multiple Wavelength Anomalous Scattering – MAD

MAD data collection is dependent upon the accessibility of an atomic absorption edge for the anomalous scatterer. Classically, MAD is performed by collecting diffraction data at wavelengths corresponding to the peak, high energy remote and inflection of the atomic absorption edge, and the intensity differences between reflections at the different wavelengths underpin the phasing problem [17, 68]. A fluorescence emission scan is recorded to determine the sample's inflection and peak, which may be separated by only 4 eV, a tiny difference in comparison to the selenium *K* atomic absorption edge of 12 657 eV. Peak and inflection energies can vary between samples [69] and it is important to conduct the fluorescence scan on the sample to be used for data collection.

The maximum anomalous differences measured at the atomic absorption peak are supported in MAD by the maximum dispersive differences between the high energy remote and the inflection point. Both the dispersive and the anomalous differences therefore contribute to the phasing potential in MAD. Redundant data, not as high as in SAD, are required for the peak data set in three-wavelength MAD. The potential for radiation damage means the transmission and/or exposure time must be considered so that complete, redundant data can be collected at each wavelength before the crystal suffers from the absorbed radiation dose. SAD and MAD each require well-measured data to yield a successful structure solution. The choice of method adopted may be influenced by the solvent content of the crystal, the type of anomalous scatterer, prior knowledge on the sensitivity of the sample to X-rays and the experience of the beamline staff.

16.3.5.4 Other Experimental Phasing Techniques

In isomorphous replacement 'native' crystals are soaked, or co-crystallised, with heavy atoms to form a 'derivative' [70]. Reflection intensity differences between the native and derivative(s) are used in the structure determination process. The change in reflection intensity is a function of the number of heavy atoms introduced, their atomic number and the overall mass of the target protein. For a 100 kDa protein, the introduction of a single copper, zinc or iron ion will change the average reflection amplitude (the square root of the intensity) by ~5%, whereas a single uranium, platinum, gold or mercury ion will result in a much bigger change, 10–20%. The heavy atom(s) may have occupancies of less than unity, reducing the mean amplitude difference in comparison to the native and increasing the difficulty in solving the heavy atom substructure.

Under certain circumstances a single heavy atom derivative – SIR – can be sufficient for structure solution if the phase ambiguity can be solved by anomalous scattering or density modification, but usually multiple different derivatives – MIR – need to be identified. The native and derivative crystals must be isomorphous, i.e. the crystal symmetry, unit cell dimensions, and crystal packing must be unaffected by the addition of the heavy atom as unit cell dimension changes of 0.5% result in intensity changes of ~20% [71] and the failure of the structure solution. The crystallisation conditions and the buffers and cryoprotectants used in heavy atom soaking should therefore not differ from those used to handle the native crystal. Isomorphous replacement and anomalous scattering can be combined in methods called SIRAS and MIRAS.

In the final part of this section we consider slightly more 'exotic' means by which structures can be phased experimentally. It is possible to exploit radiation damage in a process called radiation-damaged induced phasing (RIP). The process is predicated on deliberate, partial destruction of sample integrity, but it is usually difficult to judge how much radiation damage is sufficient to produce a substantial enough change to solve the structure [72]. UV radiation can also be used instead of X-rays to induce damage to the protein crystal [72]. However, RIP is not universally popular because a well-conducted SAD or MAD experiment will yield superior phases. Inert gases have been used in structural studies of globins for over 50 years [73]. Myoglobin crystals in capillaries filled with nitrogen at 140 atm bind non-reactive gas molecules in hydrophobic cavities [74]. Pressure cells have been developed so that protein crystals mounted in loops can be 'soaked' in xenon gas, which acts as the 'derivative' at 0.5–5 MPa before flash-freezing and collecting anomalous scattering data from the bound xenon(s). Cavities can also be targeted by halide salts; here native crystals can be derivatised by cryoprotecting with solutions containing halide ions (up to 1 M concentration) for less than one minute [75]. Water molecules in surface features are displaced by the halide ions and their binding, albeit at low occupancy, can provide sufficient anomalous scattering signal. Several different soaking regimes, where time of soak and/or concentration of halide are varied empirically, may be necessary for successful structure solution. Large polymetal clusters, including clusters of 6 tantalums and 18 tungstens, have phased large macromolecules like ribosomal subunits [76]. The 'magic triangle', 5-amino-2,4,6-triiodoisophthalic acid bromide (aka I3C), which contains three iodine atoms in an triangular shape, has also found utility in experimental phasing [77]. The polymetals and I3C are giant features in initial low resolution electron density maps [78], but because of their large size they tend to be used for solving the structures of megalithic macromolecules where space might be available for their binding. Finally, probabilistic relationships and phase set reliability determination can be used to generate a model derived solely from the measured reflection intensities without the need for phasing by molecular replacement or experimental procedures, a process called ab initio phasing. The number of independent atoms (normally >1000 atoms) and the relatively poor diffraction of most protein crystals (worse than ~1.2 Å) has limited the impact of ab initio methods on macromolecular crystallography. However, two recent software developments are lowering the requirements for ab initio phasing – Arcimboldo and Ample – for data as low as 2 Å. Both programmes use fragments such as small α-helices combined with density modification routines to obtain initial phase information and this approach has proven successful for data as low as 2 Å [79–81].

Many heavy atom salts are seriously deleterious to health on contact with skin or by inhalation. Some accumulate as a neurotoxin, others are carcinogenic and/or teratogenic and uranium is radioactive. Therefore, critical attention to safety considerations must be paid at all times when working with heavy atoms.

16.3.6 Data Processing

After the data have been collected the intensities of all reflections measured are averaged. Friedel pairs for anomalous scattering have to be separated. The occasional rogue reflection, which might have impinged upon the detector at the same place as an ice ring or at the junction between panels in a mosaic detector, has to be recognised and discarded. Synchrotron beamlines usually have semi-automatic, background-running routines for data processing. It is probably unnecessary to repeat the data processing manually for simple structure solutions unless data need to be trimmed because of radiation damage, for instance. However, for more difficult structure solutions, indexing, integration and scaling routines should be re-run and optimised. At this juncture, the structure will be solved by difference Fourier methods, molecular replacement, experimental phasing or potentially by direct methods (which is deliberately not discussed further as it is a procedure beyond the scope of this chapter) and each of the other methods is described briefly in sections 16.3.6.1 to 16.3.6.3.

16.3.6.1 Difference Fouriers

The simplest scenario is when the target macromolecule has crystallised in the same crystal form from which the structure has already been solved and the question is, for instance, 'Where does my ligand bind?' The differences in amplitude between reflections from the unbound protein and the ligand-bound form are used with phases calculated from the protein model to generate electron density maps (aka difference Fouriers) identify the location of the bound ligand.

16.3.6.2 Molecular Replacement

The first step is to identify a PDB search model from the PDBe (www.ebi.ac.uk/pdbe/entry/search/index/?advancedSearch:true=). While a high resolution structure of the same protein, or protein family, might theoretically be the best search model, the PDB validation metrics will confirm its overall quality. The PDB search model should be edited to remove ligands and solvent, additional copies if there is more than one molecule per asymmetric unit and all macromolecules that are not relevant for the search. Search models can be edited extensively to match best the target molecule. For instance, the target sequence can be modelled by homology on to the search model [82] or converted to a polyalanine chain if the sequences are too divergent. Proteins with multiple domains can be split into their components in case the domain arrangements in the target differ from the search model. If the target is multimeric a search model can be constructed as a single chain to search for multiple copies of the protein at once. Molecular replacement pipelines facilitate the process and automate search model preparation. For example, Balbes [83] uses an internal database of known domains to solve molecular replacement structures using only the target protein sequence and the diffraction data set provided. Mr Bump [84] also only requires the protein sequence and the observed data and locates and prepares search models automatically with its own algorithms. Molecular replacement programmes with more user control include Molrep [85] and Phaser [86], for which the edited search model has to be provided. When several structures with similar folds are available, a structural ensemble, built by careful superposition of the individual models, is an effective approach in Phaser. The ensemble should be edited to remove structurally distinct regions such as flexible surface loops. Arcimboldo [87] and Ample [88] pipelines solve the phase problem using in silico generated fragments based solely on sequence homology. Some rounds of refinement are typically run automatically to assess the solution quality. If the R_{work} is below 45% the phase problem for that structure is probably solved. Solutions above 50% need to be inspected closely, as they are either wrong or require significant and iterative adjustments to the model. The reliance on previously determined structures is not normally a problem in molecular replacement, but if the structures of search and target molecules are too distinct or too incomplete, then molecular replacement will fail.

16.3.6.3 Experimental Phasing

Here the correct location of the heavy atom(s) (from 2 sulphurs in a disulphide to 18 tungstens in a polymetal cluster) is used to calculate experimental phases. If the heavy atom location(s) can be determined by the isomorphous differences between native and derivate datasets, or from the anomalous differences in anomalous scattering, an electron density map can be calculated. These map calculations use initial phase estimates that are sometimes far from the correct value and consequently initial electron density maps can be poor, limited to contrast between solvent and protein. Secondary structure features may be apparent, but it may not be possible to map the 1D protein sequence to the 3D electron density. Breaks in the peptide chain are common, rendering map interpretation difficult to impossible. Thankfully, electron density maps from well-conducted experiments are usually much better than this gloomy description. Density modification routines generally enhance the quality of the initial electron density map greatly. There are two main density modification procedures, solvent flattening and averaging. Noise in the electron density map is present randomly whereas the electron density in the solvent regions between molecules should be flat and close to zero. 'Flattening' the electron density in solvent regions will increase the signal-to-noise ratio of the rest of the map and improve the electron density map quality [89]. This process is more effective the greater the solvent content and can lead to substantial improvements in map quality and the ease by which it is interpreted.

In averaging, a mask is defined around the protein (or even a domain) and the spatial relationship of this part of the structure to equivalent parts is defined; these are NCS operators and knowing them is essential for averaging of the electron density corresponding to equivalent parts of the structure. Averaging can be performed between different crystal forms of the same protein, called multicrystal averaging, if multiple datasets with some phase information for at least one of the datasets are available and if the symmetry relationships between equivalent parts of the structure are known. Averaging is a very effective density modification tool once the masks and symmetry operators are defined correctly, and its power is proportional to the square root of the number of copies being averaged.

16.3.7 Model Building and Refinement

The model is subjected to alternating cycles of refinement (e.g. Refmac [90], Phenix.refine [53], Buster [91], SHELXL [92]) and rebuilding (e.g. Coot [20]) no matter how the structure was solved. Automated routines found in ArpWarp [93], Phenix autobuild [53] and Buccaneer [94] can expedite refinement by autobuilding the protein chain, thus removing the subjectivity of the human eye, and can add ordered solvent molecules. Though these routines are robust and reliable for building the protein chain, nothing beats an experienced eye when solving sporadic crystallographic molecular puzzles. Unexpected electron density 'blobs' that correlate to molecules from the crystallisation solution, cryoprotectant, or purification conditions that have serendipitously co-crystallised with the protein often inform on the biochemistry of the target [e.g. 95]. There are substantial differences in the level of detail provided by the model dependent upon the resolution and the quality of the data from which it is built and many refinement protocols can be followed, but there is insufficient space here to go into details. Briefly, a model refined with low resolution data (>3 Å) will benefit from conservative protocols such as rigid body refinement and NCS restraints if possible. Less conservative approaches can be used at high resolution (<1.5 Å), including approaches more typical of chemical crystallography [92]. B factors can be modelled anisotropically for which there should be at least six times the number of non-H atoms in the asymmetric unit as there are unique reflections in the dataset. For multidomain proteins the TLS protocol can be advantageous and there are on-line resources (http://skuld.bmsc.washington.edu/~tlsmd) to generate TLS groups. Ordered water molecules and bound ligands should be built as late on in the refinement process as possible in order to avoid biasing the model.

Broad consensus on how best to model disorder has yet to be reached. Alternative conformations can generally be modelled at high resolution and limited information on the position of long side chains such as lysines and arginines is common because of their disorder. The disordered atoms can be removed from the model, but users of the PDB who are not crystallographers may not notice and may even mistake affected residues for alanine. Alternatively, the occupancy of disordered atoms can be set to zero and so will be ignored during refinement, but the naïve PDB user will not easily notice this. Finally, the B factor of the disordered atoms can be allowed to rise to meaningless values (e.g. above 100 $Å^2$); the PDB user must display B factors while viewing the protein to identify disordered regions that have been treated in this way, which is trivial for the popular graphics programmes PyMol (https://pymol.org) and CCP4MG [96].

Refinement is completed when every feature in the electron density map is accounted for rather than fixating on a target final R_{work}/R_{free}. These values report on the global fit of the model to the data and the ambition is to lower them throughout refinement. If the data are being interpreted correctly, both R_{work} and R_{free} will diminish. If the R_{free} rises, the model is probably being overfitted or fitted incorrectly to the data. However, it is impossible to account for everything in every structure and those parts of the electron density that cannot be explained rationally should not be modelled. Towards the end of the refinement, strong peaks in the difference map should have been accounted for and the electron density contour level displayed should be adjusted to avoid looking only at noise. Throughout refinement a variety of geometrical parameters are reported, including the root mean square deviation for bond angles and bond length, and it is generally accepted that these should be no more than 2° and 0.02 Å, respectively. Weighting schemes within refinement packages may be adjusted to ensure that these values remain close to the accepted target. Model building is completed when the electron density is interpreted to the best of the user's knowledge. The deviance of the geometry and other reliability indicators of the structure from well-refined models in the PDB of the same resolution should give rise to concern, a comparison that can be achieved with the polygon tool in Phenix [97].

16.3.7.1 Validation

It is likely during model validation that a return to model building and refinement will be necessary to correct any anomalies. Validation of the model is vital to ensure that it makes chemical, physical and biological sense so users have confidence in it. Validation of the model starts during its building – hydrophobic amino acids dominate protein cores and charged and polar residues are found mostly on the protein surface – and refinement. Aberrant stereochemistry should be addressed during model building and refinement because outliers to accepted targets are flagged by refinement programmes and some molecular graphics packages. Validation is formed of two parts, one concerning the quality of the model and a second that assesses the fit of the model to the data. Cross-validation methods (the R_{free}) that have been adopted from statistics should be applied during refinement and validation [21].

As a prelude to deposition at the PDB, the structure should be checked with Molprobity, distributed with Phenix and also available from its webserver (http://molprobity.biochem.duke.edu). Molprobity performs a series of analyses including optimising the intramolecular hydrogen bond network by 'flipping' asparagine, glutamine and histidine side chains where necessary. The validation menu in Coot is extensive and should be used to identify and fix problems. The PDBe validation service (https://validate-rcsb-1.wwpdb.org) runs many deposition checks before the model is deposited and freely available. Finally, the PDB and its mirrors (RCSB, PDBe, PDBj) display graphical assessments of model quality in comparison to all other structures.

16.3.7.2 Deposition and Publication

It is a condition of publication in journals that both model and the data from which it is derived are deposited at the PDB with their release coincident with publication of the accompanying paper. PDB deposition ensures that model and data are archived together for others to use, and potentially reanalyse. Deposition also ensures that publically funded researchers' work is publically available. Depositing the structure entails validation and sanity checks on both structure and data, and provides an opportunity to record additional information about the structure solution that data mining routines cannot extract from the submitted files, including the crystallisation conditions. The PDB deposition generates a validation report required for the reviewing stage in many journals.

16.4 Applications

Next we summarise two applications of the methods and procedures that we have highlighted in this chapter; one concerns molecular replacement and throughput and the second describes exploiting anomalous scattering from sulphur.

16.4.1 Molecular Replacement, the Latest Automated Software Approaches and Drug Discovery

A drug discovery project, either at the fragment screening stage or during medicinal chemistry compound optimisation, requires a steady source of good quality crystals that diffract better than 2 Å to infer useful information about the binding mode of the ligand, the involvement of ordered water molecules and to track conformational changes in the protein. The crystal's solvent channels must allow molecules to access the ligand-binding site if soaking experiments are used instead of co-crystallisation. The crystal form should be robust to withstand organic solvents, such as DMSO, which may be required to solubilise the ligand. Whether co-crystallisation or soaking approaches are used will depend upon the project and the equipment available. For example, ligands can be dispensed using ultrasound with the Labcyte Echo instruments, which dispense very small volumes of potentially rare or expensive ligands in a high-throughput, highly automated fashion [98]. Hundreds of diffraction datasets might be collected over the course of a typical drug discovery project and to support such projects in academia the Structural Genomics Consortium and Diamond Light Source have developed the LabXChem pipeline, which allows hundreds of samples to be mounted in a day using robotics. The crystal mounts can be handled on any synchrotron beamline and the dedicated XChem beamline, I04-1, has been designed specifically to handle high throughput projects. More details can be found on the beamline's home page (www.diamond.ac.uk/Beamlines/Mx/I04-1.html). With automated sample centring, the whole experiment, from crystal mounting to structure solution, can be performed with little user intervention. Over the past decade many synchrotron sites have developed software pipelines to process data automatically. Space limitations do not permit a complete description of these pipelines, but the user uploads essential information (protein sequence, PDB search model, cell dimensions, space group) for each project to the IspyB database [99]. Xia2 [51] and Dimple [100] routines can be invoked for an automatic structure solution. The presence of bound ligands can be detected automatically using PanDDa [101], obviating the need to inspect manually hundreds of solutions that, in the case of fragment-based screening, contain no bound molecule.

16.4.2 Use of Sulphur as an Anomalous Scatterer for Solving the Phase Problem

The first protein structure solved by SAD measurements was crambin [62], but the methodology has moved on considerably in the intervening 30 years, so to be current we summarise here the recent structure determination by sulphur-SAD of a biofilm-degrading endonuclease, NucB [59]. NucB is a 12 kDa protein with four

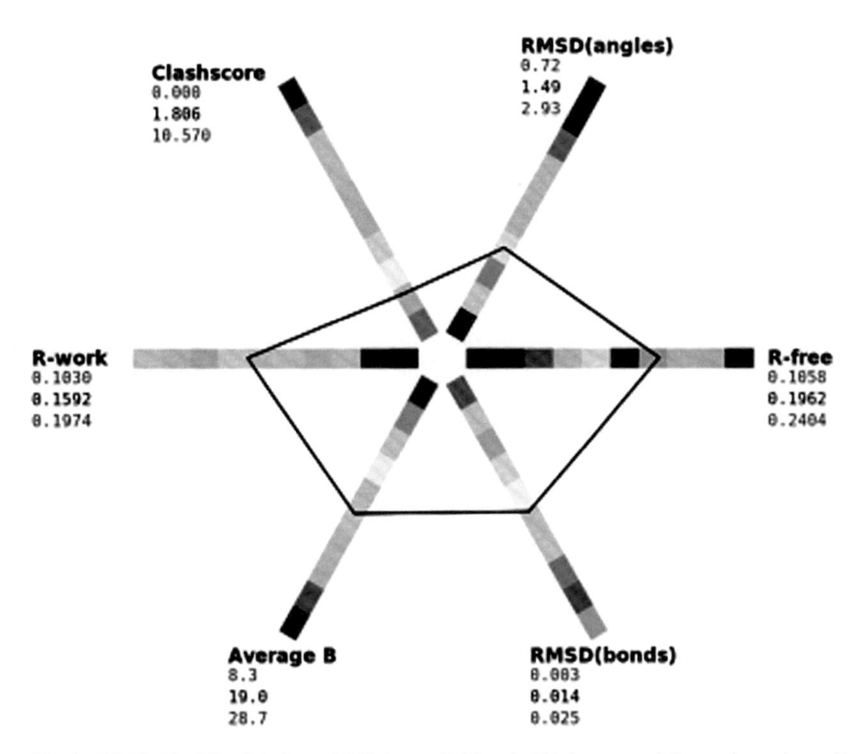

Figure 16.17 Model validation with Polygon in Phenix. We have used the endonuclease NucB model (PDBid 5OMT) to generate the Polygon statistics plot. Polygon has compared the submitted model with 734 models at a similar resolution range. For each statistic the lower and higher limits and the mean is given. Each statistic is binned and a dark grey square equates to a low number of models, while the lighter grey squares represent a bin with many models. The submitted model statistics are plotted and link together with a solid line.

sulphur atoms from two cysteines in a disulphide and two methionines. The crystals used for phasing and high resolution refinement were obtained from different conditions, but both were used directly from initial robotic screens without subsequent optimisation. Both conditions contained PEGs, so reservoir solutions were supplemented with 20% PEG 400 for cryoprotection. The anomalous scattering data for phasing were collected on the Diamond microfocus beamline, I24. The X-ray energy was set to 6500 eV (1.907 Å), which represents a compromise between a low energy for maximising the anomalous differences and the absorption of the crystal's scattering by air. With improvements in both hardware and software since these data were collected, 7500 eV should now be used. Though the data were collected on a microfocus beamline, the beam dimensions were 50 μm × 50 μm and the microfocus capabilities of the beamline were not used. The photon flux of I24 is one order of magnitude above the standard macromolecular beamlines at Diamond and a 9% beam transmission, 0.1° oscillations with 100 ms exposure times were used. The beamline did not have a multiaxis goniometer at the time of data collection so 999° of data were collected to increase data redundancy to 32-fold for accurate measurement of weak anomalous differences. The crystals diffract much further than the 2.26 Å limit imposed in this phasing experiment in which data quality was the primary driver and a useful anomalous signal was maintained to the resolution cut-off. The positions of the four sulphur atoms were found using ShelXD [102] from which initial phases were obtained. High resolution data, to 1.35 Å, were subsequently collected on the same $P4_12_12$ crystal form, but from a crystal grown under different crystallisation conditions. The experimental phases were extended to the high resolution and the model was built automatically with ArpWarp [93]. The data collection and refinement statistics for the NucB structure solution can be found in Basle et al. [59] and we have plotted some of the validation results from this structure as an example of the tools that are available to assess PDB model quality (Figure 16.17).

16.5 Concluding Remarks

Here we have provided basic theoretical background to X-ray crystallography and a deeper overview of the practicalities as used today in our academic laboratory. During our careers there have been substantial and significant changes to crystallography. For instance, there is far more processing power in today's average smartphone than there was in the desktop computers on which our PhD work was completed and one of us remembers his first synchrotron trip and the data that were collected on to film. No doubt further significant advances are ahead, excluding the impact of X-FELS and cryo-EM in the immediate future. Users should remain abreast of technical advances that drive crystallography and subscribing to the CCP4 bulletin board is vital for this, for open answers to subscribers' questions and a good place to search for jobs. The books given in Further Reading should be consulted for a more in-depth explanation of the topics we have merely introduced.

Acknowledgements

The authors would like to thank Ehmke Pohl, Peter Moody, Simon Booth, Vincent Rao, Daniel Wood, Sema Ejder and Jon Marles-Wright for suggestions on how to improve this chapter; Juan Sanchez-Weatherby for the tip on removing flocculent ice from liquid nitrogen; Simon Booth for help with photography; Chloe, Rupert and Humphrey for supplying our lab with cats' whiskers; and the following funding agencies for their support of the Newcastle Structural Biology Laboratory at various points since its inception in 2003: the Wellcome Trust, the Royal Society, the BBSRC, the MRC and last, but by no means least, the European Union.

References

1 Van Driessche, A.E.S., Van Gerven, N., Bomans, P.H.H. et al. (2018). Molecular nucleation mechanisms and control strategies for crystal polymorph selection. *Nature* 556: 89–94.
2 McPherson, A. and Gavira, J.A. (2014). Introduction to protein crystallization. *Acta Crystallogr. F Struct. Biol. Commun.* 70(Pt 1): 2–20.
3 Parker, J.L. and Newstead, S. (2016). Membrane protein crystallisation: current trends and future perspectives. *Adv. Exp. Med. Biol.* 922: 61–72.
4 Hanh, T. (2002). *International Tables for Crystallography, Volume A: Space Group Symmetry*. New York: Springer.
5 Svensson, L.A., Surin, B.P., Dixon, N.E., and Spangfort, M.D. (1994). The symmetry of *Escherichia coli* cpn60 (GroEL) determined by X-ray crystallography. *J. Mol. Biol.* 235 (1): 47–52.
6 Antson, A.A., Brzozowski, A.M., Dodson, E.J. et al. (1994). 11-fold symmetry of the trp RNA-binding attenuation protein (TRAP) from *Bacillus subtilis* determined by X-ray analysis. *J. Mol. Biol.* 244 (1): 1–5.
7 Blake, C. and Phillips, D.C. (1962). Effects of X-irradiation on single crystals of myoglobin. In Proceedings of the Symposium on the Biological Effects of Ionizing Radiation at the Molecular Level. *Int. Atomic Energy Agency* 183–191.
8 Henderson, R. (1990). Cryo-protection of protein crystals against radiation damage in electron and X-ray diffraction. *Proc. R. Soc. Lond. Ser. B Biol. Sci.* 241 (1300): 6–8.
9 Owen, R.L., Rudino-Pinera, E., and Garman, E.F. (2006). Experimental determination of the radiation dose limit for cryocooled protein crystals. *Proc. Natl Acad. Sci. USA* 103 (13): 4912–4917.
10 Garman, E. (2010). Radiation damage in macromolecular crystallography: what is it and why should we care? *Acta Crystallogr. D* 66 (4): 339–351.
11 Matthews, B.W. (1968). Solvent content of protein crystals. *J. Mol. Biol.* 33 (2): 491–497.
12 Acharya, R., Fry, E., Stuart, D. et al. (1989). The three-dimensional structure of foot-and-mouth disease virus at 2.9 Å resolution. *Nature* 337 (6209): 709–716.

13 Rossmann, M.G. and Blow, D.M. (1962). The detection of sub-units within the crystallographic asymmetric unit. *Acta Crystallogr.* 15 (1): 24–31.
14 Isaacs, N. (2016). A history of experimental phasing in macromolecular crystallography. *Acta Crystallogr. Sect. D Biol. Crystallogr.* 72 (3): 293–295.
15 Taylor, G.L. (2010). Introduction to phasing. *Acta Crystallogr. D Biol. Crystallogr.* 66(4): 325–338.
16 Scapin, G. (2013). Molecular replacement then and now. *Acta Crystallogr. D Biol. Crystallogr.* 69(11): 2266–2275.
17 Hendrickson, W. (1991). Determination of macromolecular structures from anomalous diffraction of synchrotron radiation. *Science* 254 (5028): 51–58.
18 Gonzalez, A. (2003). Faster data-collection strategies for structure determination using anomalous dispersion. *Acta Crystallogr. D Biol. Crystallogr.* 59 (2): 315–322.
19 Jones, T.A., Zou, J.-Y., Cowan, S.W., and Kjeldgaard, M. (1991). Improved methods for building protein models in electron density maps and the location of errors in these models. *Acta Crystallogr. A* 47 (2): 110–119.
20 Emsley, P., Lohkamp, B., Scott, W.G., and Cowtan, K. (2010). Features and development of Coot. *Acta Crystallogr. D Biol. Crystallogr.* 66(4): 486–501.
21 Brünger, A.T. (1992). Free R value: a novel statistical quantity for assessing the accuracy of crystal structures. *Nature* 355: 472–475.
22 Urzhumtsev, A., Afonine, P.V., and Adams, P.D. (2013). TLS from fundamentals to practice. *Crystallogr. Rev.* 19 (4): 230–270.
23 Slabinski, L., Jaroszewski, L., Rychlewski, L. et al. (2007). XtalPred: a web server for prediction of protein crystallizability. *Bioinformatics* 23 (24): 3403–3405.
24 Romero, P., Obradovic, Z., and Dunker, A.K. (2004). Natively disordered proteins: functions and predictions. *Appl. Bioinforma.* 3 (2-3): 105–113.
25 Cuff, J.A., Clamp, M.E., Siddiqui, A.S. et al. (1998). JPred: a consensus secondary structure prediction server. *Bioinformatics* 14 (10): 892–893.
26 Magnani, F., Serrano-Vega, M.J., Shibata, Y. et al. (2016). A mutagenesis and screening strategy to generate optimally thermostabilized membrane proteins for structural studies. *Nat. Protoc.* 11 (8): 1554–1571.
27 Montanier, C.Y., Correia, M.A., Flint, J.E. et al. (2011). A novel, noncatalytic carbohydrate-binding module displays specificity for galactose-containing polysaccharides through calcium-mediated oligomerization. *J. Biol. Chem.* 286 (25): 22499–22509.
28 Dong, A., Xu, X., Edwards, A.M., Midwest Center for Structural Genomics et al. (2007). In situ proteolysis for protein crystallization and structure determination. *Nat. Methods* 4 (12): 1019–1021.
29 Walter, T.S., Meier, C., Assenberg, R. et al. (2006). Lysine methylation as a routine rescue strategy for protein crystallization. *Structure* 14 (11): 1617–1622.
30 Derewenda, Z.S. and Vekilov, P.G. (2006). Entropy and surface engineering in protein crystallization. *Acta Crystallogr. D Biol. Crystallogr.* 62 (1): 116–124.
31 Eiler, S., Gangloff, M., Duclaud, S. et al. (2001). Overexpression, purification, and crystal structure of native ERα LBD. *Protein Expr. Purif.* 22 (2): 165–173.
32 Groftehauge, M.K., Hajizadeh, N.R., Swann, M.J., and Pohl, E. (2015). Protein-ligand interactions investigated by thermal shift assays (TSA) and dual polarization interferometry (DPI). *Acta Crystallogr. D Biol. Crystallogr.* 71 (1): 36–44.
33 Rismondo, J., Cleverley, R.M., Lane, H.V. et al. (2016). Structure of the bacterial cell division determinant GpsB and its interaction with penicillin-binding proteins. *Mol. Microbiol.* 99 (5): 978–998.
34 Schieferstein, J.M., Pawate, A.S., Varel, M.J. et al. (2018). X-ray transparent microfluidic platforms for membrane protein crystallization with microseeds. *Lab Chip* 18 (6): 944–954.
35 Haas, D.J. and Rossmann, M.G. (1970). Crystallographic studies on lactate dehydrogenase at −75 degrees C. *Acta Crystallogr. B* 26 (7): 998–1004.

36 Petsko, G.A. (1975). Protein crystallography at sub-zero temperatures: cryo-protective mother liquors for protein crystals. *J. Mol. Biol.* 96 (3): 381–392.

37 Rubinson, K.A., Ladner, J.E., Tordova, M., and Gilliland, G.L. (2000). Cryosalts: suppression of ice formation in macromolecular crystallography. *Acta Crystallogr. D Biol. Crystallogr.* 56 (8): 996–1001.

38 Garman, E.F. and Mitchell, E.P. (1996). Glycerol concentrations required for cryoprotection of 50 typical protein crystallization solutions. *J. Appl. Crystallogr.* 29 (5): 584–587.

39 Ciccone, L., Vera, L., Tepshi, L. et al. (2015). Multicomponent mixtures for cryoprotection and ligand solubilization. *Biotechnol. Rep.* 7: 120–127.

40 Tailford, L.E., Ducros, V.M., Flint, J.E. et al. (2009). Understanding how diverse beta-mannanases recognize heterogeneous substrates. *Biochemistry* 48 (29): 7009–7018.

41 Parkin, S. and Hope, H. (1998). Macromolecular cryocrystallography: cooling, mounting, storage and transportation of crystals. *J. Appl. Crystallogr.* 31 (6): 945–953.

42 Owen, R.L., Pritchard, M., and Garman, E. (2004). Temperature characteristics of crystal storage devices in a CP100 dry shipping dewar. *J. Appl. Crystallogr.* 37 (6): 1000–1003.

43 Bourenkov, G.P. and Popov, A.N. (2006). A quantitative approach to data-collection strategies. *Acta Crystallogr. D Biol. Crystallogr.* 62 (1): 58–64.

44 Dauter, Z. (1999). Data-collection strategies. *Acta Crystallogr. D Biol. Crystallogr.* 55 (10): 1703–1717.

45 Krojer, T., Pike, A.C.W., and von Delft, F. (2013). Squeezing the most from every crystal: the fine details of data collection. *Acta Crystallogr. D Biol. Crystallogr.* 69 (7): 1303–1313.

46 Aller, P., Geng, T., Evans, G., and Foadi, J. (2016). Applications of the BLEND software to crystallographic data from membrane proteins. *Adv. Exp. Med. Biol.* 922: 119–135.

47 Bury, C.S., Brooks-Bartlett, J.C., Walsh, S.P., and Garman, E.F. (2018). Estimate your dose: RADDOSE-3D. *Protein Sci.* 27 (1): 217–228.

48 Powell, H.R., Johnson, O., and Leslie, A.G. (2013). Autoindexing diffraction images with iMosflm. *Acta Crystallogr. D Biol. Crystallogr.* 69(7): 1195–1203.

49 Kabsch, W. (2010). XDS. *Acta Crystallogr. D Biol. Crystallogr.* 66 (2): 125–132.

50 Winter, G., Waterman, D.G., Parkhurst, J.M. et al. (2018). DIALS: implementation and evaluation of a new integration package. *Acta Crystallogr. D Struct. Biol.* 74(2): 85–97.

51 Winter, G., Lobley, C.M., and Prince, S.M. (2013). Decision making in xia2. *Acta Crystallogr. D Biol. Crystallogr.* 69 (7): 1260–1273.

52 Vonrhein, C., Flensburg, C., Keller, P. et al. (2011). Data processing and analysis with the autoPROC toolbox. *Acta Crystallogr. D Biol. Crystallogr.* 67(4): 293–302.

53 Adams, P.D., Afonine, P.V., Bunkoczi, G. et al. (2010). PHENIX: a comprehensive python-based system for macromolecular structure solution. *Acta Crystallogr. D Biol. Crystallogr.* 66(2): 213–221.

54 Winn, M.D., Ballard, C.C., Cowtan, K.D. et al. (2011). Overview of the CCP4 suite and current developments. *Acta Crystallogr. D Biol. Crystallogr.* 67(4): 235–242.

55 Evans, P.R. and Murshudov, G.N. (2013). How good are my data and what is the resolution? *Acta Crystallogr. D Biol. Crystallogr.* 69(7): 1204–1214.

56 Evans, P.R. (2011). An introduction to data reduction: space-group determination, scaling and intensity statistics. *Acta Crystallogr. D Biol. Crystallogr.* 67(4): 282–292.

57 Nettleship, J.E., Assenberg, R., Diprose, J.M. et al. (2010). Recent advances in the production of proteins in insect and mammalian cells for structural biology. *J. Struct. Biol.* 172 (1): 55–65.

58 Walden, H. (2010). Selenium incorporation using recombinant techniques. *Acta Crystallogr. D Biol. Crystallogr.* 66 (4): 352–357.

59 Basle, A., Hewitt, L., Koh, A. et al. (2018). Crystal structure of NucB, a biofilm-degrading endonuclease. *Nucleic Acids Res.* 46 (1): 473–484.

60 Abendroth, J., Gardberg, A.S., Robinson, J.I. et al. (2011). SAD phasing using iodide ions in a high-throughput structural genomics environment. *J. Struct. Funct. Genom.* 12 (2): 83–95.

61 Dauter, Z., Dauter, M., de La Fortelle, E. et al. (1999). Can anomalous signal of sulfur become a tool for solving protein crystal structures? *J. Mol. Biol.* 289 (1): 83–92.

62 Hendrickson, W.A. and Teeter, M.M. (1981). Structure of the hydrophobic protein crambin determined directly from the anomalous scattering of sulphur. *Nature* 290: 107–113.

63 Liu, Q., Dahmane, T., Zhang, Z. et al. (2012). Structures from anomalous diffraction of native biological macromolecules. *Science* 336 (6084): 1033–1037.

64 Murray, J.W., Rudino-Pinera, E., Owen, R.L. et al. (2005). Parameters affecting the X-ray dose absorbed by macromolecular crystals. *J. Synchrotron Radiat.* 12(3): 268–275.

65 de Sanctis, D., Oscarsson, M., Popov, A. et al. (2016). Facilitating best practices in collecting anomalous scattering data for de novo structure solution at the ESRF structural biology beamlines. *Acta Crystallogr. D Struct. Biol. Crystallogr.* 72(3): 413–420.

66 El Omari, K., Iourin, O., Kadlec, J. et al. (2014). Pushing the limits of sulfur SAD phasing: de novo structure solution of the N-terminal domain of the ectodomain of HCV E1. *Acta Crystallogr. D Biol. Crystallogr.* 70(8): 2197–2203.

67 Finke, A.D., Panepucci, E., Vonrhein, C. et al. (2016). Advanced crystallographic data collection protocols for experimental phasing. *Methods Mol. Biol.* 1320: 175–191.

68 Dauter, Z. (2013). SAD/MAD Phasing. *Adv. Methods Biomol. Crystallogr.* 135–149.

69 Ducros, V.M.A., Lewis, R.J., Verma, C.S. et al. (2001). Crystal structure of GerE, the ultimate transcriptional regulator of spore formation in *Bacillus subtilis*. *J. Mol. Biol.* 306 (4): 759–771.

70 Pike, A.C., Garman, E.F., Krojer, T. et al. (2016). An overview of heavy-atom derivatization of protein crystals. *Acta Crystallogr. D Struct. Biol.* 72(3): 303–318.

71 Crick, F.H.C. and Magdoff, B.S. (1956). The theory of the method of isomorphous replacement for protein crystals. I. *Acta Crystallogr.* 9 (11): 901–908.

72 de Sanctis, D. and Nanao, M.H. (2012). Segmenting data sets for RIP. *Acta Crystallogr. D Biol. Crystallogr.* 68(9): 1152–1162.

73 Schoenborn, B.P., Watson, H.C., and Kendrew, J.C. (1965). Binding of xenon to sperm whale myoglobin. *Nature* 207: 28–30.

74 Tilton, R.F., Kuntz, I.D., and Petsko, G.A. (1984). Cavities in proteins: structure of a metmyoglobin xenon complex solved to 1.9 Å. *Biochemistry* 23 (13): 2849–2857.

75 Dauter, Z., Dauter, M., and Rajashankar, K.R. (2000). Novel approach to phasing proteins: derivatization by short cryo-soaking with halides. *Acta Crystallogr. D Biol. Crystallogr.* 56 (2): 232–237.

76 Schluenzen, F., Tocilj, A., Zarivach, R. et al. (2000). Structure of functionally activated small ribosomal subunit at 3.3 Å resolution. *Cell* 102 (5): 615–623.

77 Beck, T., Krasauskas, A., Gruene, T., and Sheldrick, G.M. (2008). A magic triangle for experimental phasing of macromolecules. *Acta Crystallogr. D Biol. Crystallogr.* 64(11): 1179–1182.

78 Szczepanowski, R.H., Filipek, R., and Bochtler, M. (2005). Crystal structure of a fragment of mouse ubiquitin-activating enzyme. *J. Biol. Chem.* 280 (23): 22006–22011.

79 Bibby, J., Keegan, R.M., Mayans, O. et al. (2012). AMPLE: a cluster-and-truncate approach to solve the crystal structures of small proteins using rapidly computed ab initio models. *Acta Crystallogr. D Biol. Crystallogr.* 68(12): 1622–1631.

80 Caballero, I., Sammito, M., Millán, C. et al. (2018). ARCIMBOLDO on coiled coils. *Acta Crystallogr. D Biol. Crystallogr.* 74(3): 194–204.

81 Rodríguez, D.D., Grosse, C., Himmel, S. et al. (2009). Crystallographic ab initio protein structure solution below atomic resolution. *Nat. Methods* 6: 651–653.

82 Kelley, L.A., Mezulis, S., Yates, C.M. et al. (2015). The Phyre2 web portal for protein modeling, prediction and analysis. *Nat. Protoc.* 10: 845–858.

83 Long, F., Vagin, A.A., Young, P., and Murshudov, G.N. (2008). BALBES: a molecular-replacement pipeline. *Acta Crystallogr. D Biol. Crystallogr.* 64(1): 125–132.

84 Keegan, R.M. and Winn, M.D. (2008). MrBUMP: an automated pipeline for molecular replacement. *Acta Crystallogr. D Biol. Crystallogr.* 64(1): 119–124.
85 Vagin, A. and Teplyakov, A. (1997). MOLREP: an automated program for molecular replacement. *J. Appl. Crystallogr.* 30 (6): 1022–1025.
86 McCoy, A.J., Grosse-Kunstleve, R.W., Adams, P.D. et al. (2007). Phaser crystallographic software. *J. Appl. Crystallogr.* 40(4): 658–674.
87 Sammito, M., Meindl, K., de Ilarduya, I.M. et al. (2014). Structure solution with ARCIMBOLDO using fragments derived from distant homology models. *FEBS J.* 281 (18): 4029–4045.
88 Bibby, J., Keegan, R.M., Mayans, O. et al. (2013). Application of the AMPLE cluster-and-truncate approach to NMR structures for molecular replacement. *Acta Crystallogr. D Biol. Crystallogr.* 69(11): 2194–2201.
89 Wang, B.-C. (1985). Resolution of phase ambiguity in macromolecular crystallography. *Methods Enzymol.*, Academic Press. 115: 90–112.
90 Murshudov, G.N., Skubak, P., Lebedev, A.A. et al. (2011). REFMAC5 for the refinement of macromolecular crystal structures. *Acta Crystallogr. D Biol. Crystallogr.* 67(4): 355–367.
91 Blanc, E., Roversi, P., Vonrhein, C. et al. (2004). Refinement of severely incomplete structures with maximum likelihood in BUSTER-TNT. *Acta Crystallogr. D Biol. Crystallogr.* 60(12): 2210–2221.
92 Sheldrick, G. (2015). Crystal structure refinement with SHELXL. *Acta Crystallogr. C* 71 (1): 3–8.
93 Langer, G., Cohen, S.X., Lamzin, V.S., and Perrakis, A. (2008). Automated macromolecular model building for X-ray crystallography using ARP/wARP version 7. *Nat. Protoc.* 3 (7): 1171–1179.
94 Cowtan, K. (2006). The buccaneer software for automated model building. 1. Tracing protein chains. *Acta Crystallogr. D Biol. Crystallogr.* 62(9): 1002–1011.
95 Kawai, Y., Marles-Wright, J., Cleverley, R.M. et al. (2011). A widespread family of bacterial cell wall assembly proteins. *EMBO J.* 30 (24): 4931–4941.
96 McNicholas, S., Potterton, E., Wilson, K.S., and Noble, M.E. (2011). Presenting your structures: the CCP4mg molecular-graphics software. *Acta Crystallogr. D Biol. Crystallogr.* 67(4): 386–394.
97 Urzhumtseva, L., Afonine, P.V., Adams, P.D., and Urzhumtsev, A. (2009). Crystallographic model quality at a glance. *Acta Crystallogr. D Biol. Crystallogr.* 65(3): 297–300.
98 Yin, X., Scalia, A., Leroy, L. et al. (2014). Hitting the target: fragment screening with acoustic in situ co-crystallization of proteins plus fragment libraries on pin-mounted data-collection micromeshes. *Acta Crystallogr. D Biol. Crystallogr.* 70 (5): 1177–1189.
99 Delageniere, S., Brenchereau, P., Launer, L. et al. (2011). ISPyB: an information management system for synchrotron macromolecular crystallography. *Bioinformatics* 27 (22): 3186–3192.
100 Wojdyr, M., Keegan, R., Winter, G. et al. (2014). DIMPLE - a pipeline for the rapid generation of difference maps from protein crystals with putatively bound ligands. *Acta Crystallogr. A* 69: s299–s299.
101 Pearce, N.M., Krojer, T., Bradley, A.R. et al. (2017). A multi-crystal method for extracting obscured crystallographic states from conventionally uninterpretable electron density. *Nat. Commun.* 8: 15123.
102 Sheldrick, G.M. (2008). A short history of SHELX. *Acta Crystallogr. A* 64(1): 112–122.

Further Reading

Blow, D. (2002). *Outline of Crystallography for Biologists.* Oxford University Press.
Chayen, N., Helliwell, J., and Snell, E. (2010). *Macromolecular Crystallization and Crystal Perfection.* Oxford University Press.
Methods in Enzymology Volume 276. Carter, C. Jr., and Sweet, R. (eds.) (1997). *Macromolecular Crystallography, Part A.* Academic Press.
Methods in Enzymology Volume 277. Carter, C. Jr., and Sweet, R. (eds.) (1997). *Macromolecular Crystallography, Part B.* Academic Press.

Methods in Enzymology Volume 358. Carter, C. Jr., and Sweet, R. (eds.) (2003). *Macromolecular Crystallography, Part C*. Academic Press.

Doublie, S. (2007). *Macromolecular Crystallography Protocols, Volume 2: Structure Determination*. Humana Press.

Rupp, B. (2010). *Biomolecular Crystallography: Principles, Practice and Application to Structural Biology*. Garland Science.

Sherwood, D. and Cooper, J. (2010). *Crystals, X-rays and Proteins*. Oxford University Press.

17

Biomolecular NMR Spectroscopy and Structure Determination of DNA

Tony Cheung and Vasudevan Ramesh

School of Chemistry, University of Manchester, Manchester, M13 9PL, UK

17.1 Significance and Background

17.1.1 Analytical NMR: Discovery and Evolution

It may be fair to say that there is no other analytical technique that has had as dramatic an impact on several fields of science (Physics, Chemistry, Biology and Medicine) as nuclear magnetic resonance (NMR) spectroscopy since its discovery in 1945 [1]. The justification for this observation is clearly manifest in the award of four Nobel Prizes to scientists who have demonstrated the role of NMR in each of these sciences by their pioneering contributions [1].

The experimental evidence for the phenomenon of NMR in bulk matter was first provided in 1945, independently, by physicists Felix Bloch and Edward Purcell at Stanford and Harvard, respectively [1]. This was followed by intense activities in the 1960s and 1970s when the NMR phenomenon was successfully developed as a pulsed Fourier transform experimental technique with wide applications in Chemistry [2], culminating in the experimental demonstration of two-dimensional (2D) NMR in 1976 by Richard Ernst at ETH, Zurich, that won him a Nobel Prize in 1991 [3]. Soon the exciting applications of NMR in chemistry were further advanced into biology and medicine (magnetic resonance imaging) in the 1980s resulting in the Nobel Prize to Paul Lauterbur (Illinois/Chicago) and Sir Peter Mansfield (Nottingham) in 2003 [4].

One of the defining moments for biomolecular NMR came when Kurt Wüthrich (Nobel Prize, 2002) at ETH, Zurich, showed unambiguously that the three-dimensional NMR structure of a protein in the solution state could be determined to a high resolution and accuracy [5, 6]. This NMR solution structure matched well with the structure of the same protein determined separately in the crystalline state by X-ray crystallography [7].

NMR has a number of unique features that make it a better analytical tool to solve structural problems in chemistry and biology. The power and range of NMR is further enhanced when it is applied in combination with other techniques such as mass spectrometry, with or without chromatographic separation (LC-MS-NMR) [8]. During the last five years, there have been rapid developments in the application of NMR in almost all sectors of industry, globally. Areas such as nuclear, polymers, agriculture, pharmaceuticals, dyes and chemicals, food analysis, water, solar dye cells, etc., have been witness to exciting new NMR applications [9]. NMR has now made very significant progress in metabolomics and is used as a comprehensive profiling tool to identify qualitatively and quantitatively a large number of metabolites present in a biological system [10].

17.1.2 The Aim of the Chapter

The principal aim of this chapter is to provide advanced undergraduates (Years 3 and 4) and early postgraduate (MSc, PhD) students studying chemistry, biology and physics, who are new to research, an opportunity

Biomolecular and Bioanalytical Techniques: Theory, Methodology and Applications, First Edition. Edited by Vasudevan Ramesh.

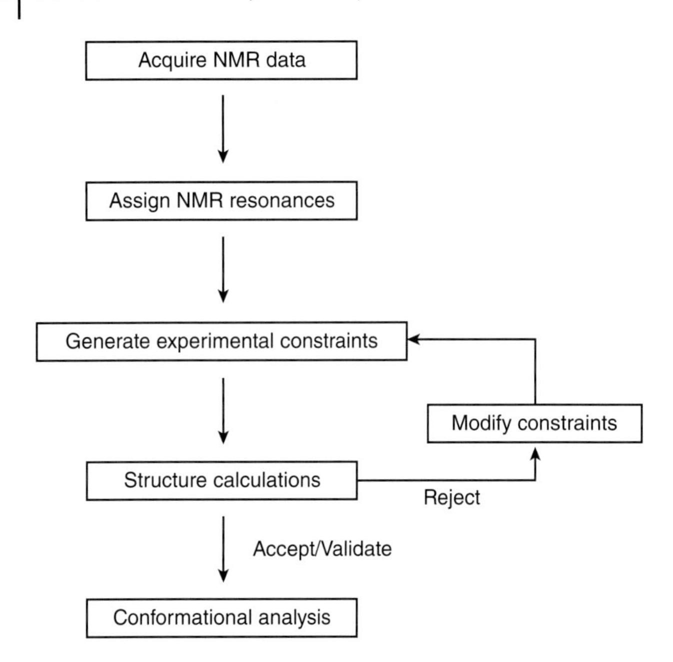

Figure 17.1 Flow diagram outlining the various stages of NMR structure determination.

to learn about the application of NMR spectroscopy to determine the structure of biologically significant macromolecules [5]. Understanding the structure at the molecular level is essential in order to elucidate the mechanism of action of small and large biomolecules, which play important roles in biological functions.

It is expected that the new researchers would already have had some exposure to the basic principles of NMR and interpretation of one-dimensional (1D) proton and carbon NMR spectra of small organic molecules as part of their core undergraduate taught curriculum [11]. Consequently, this chapter will be devoted to the application of two-dimensional (2D) NMR to solve the structure of large biomolecules with emphasis on experimental methodology involving NMR data acquisition, analysis, interpretation and structure calculation, as shown in the scheme in Figure 17.1. Due to this focus and space limitation, the quantum mechanical description of the theory behind 2D NMR will not be provided and the readers are referred to the excellent books recommended at the end of the chapter.

The NMR structure determination of proteins using multidimensional NMR is already well established and the methodology is well-documented in books and excellent reviews published in the literature [12–14]. In comparison, structure determination of nucleic acids (DNA, RNA) is relatively rare but increasing [15–18]. In this chapter we will demonstrate the application of NMR structure determination methodology with two case studies: (i) a canonical 17mer GC DNA duplex and (ii) a selectively modified S^6-GC 13mer thioDNA.

17.2 Basic NMR Theory

A short description of the basic principles and key experimental parameters of NMR is described in the following sections.

17.2.1 Basic Principles of NMR [19–21]

The property of nuclear spin (I) is fundamental to the NMR phenomenon and I may have values that are multiples of ½, for example, 1/2, 1, 3/2 and 2, etc. Values of the nuclear spin are dependent on the precise

Table 17.1 List of common nuclei studied in NMR spectroscopy.

Nuclei	Spin (*I*)	Natural abundance (%)	γ (rad/gauss s)
^{1}H	½	99.99	26.75
^{2}H	1	0.01	4.107
^{13}C	½	1.11	6.728
^{15}N	½	0.37	−2.713
^{19}F	½	100	25.181
^{31}P	½	100	10.839

nuclear configuration of protons and neutrons. Nuclei with spin $I = 0$ are termed NMR inactive or NMR silent as they have no magnetic moment (μ); consequently, they cannot exhibit NMR. A list of common spin active nuclei used in NMR is shown in Table 17.1.

The interaction of the nuclear magnetic moment (μ) with an external magnetic field (B_0) causes a splitting in the energy of the spin states. When no magnetic field is present there are a number of spin states but these appear degenerate, i.e. all of the possible nuclear spin states have the same energy. When a magnetic field is introduced, the degeneracy state is split into its individual energy states; the number of energy states is governed by the $2I + 1$ rule. In the case of spin ½ nuclei, there exists two energy states, a low energy state (α) and a high energy state (β), whose spin populations are governed by the Boltzmann distribution (Figure 17.2).

At thermal equilibrium, the Boltzmann distribution shows that there is a slight excess of spins in the lower energy α state. The population excess in this α state determines the transition of spins into the upper energy β state. Transitions occur upon irradiation, normally achieved using a radiofrequency (RF) pulse, B_1, which exactly matches the energy gap ($h\nu$). This is known as the resonance condition and the frequency at which the RF pulse is applied is termed the Larmor frequency (ν). The Larmor frequency, at which the states absorb energy, is directly dependent on the gyromagnetic ratio (γ) and the strength of the external magnetic field (B_0):

$$\nu = \frac{\gamma}{2\pi} B_0. \tag{17.1}$$

This absorption of energy is detected as a signal whose intensity is proportional to the population difference, $N_\alpha - N_\beta$. The energy gap between energy states is different for each type of nucleus and therefore each one requires a different resonance frequency.

The gyromagnetic ratio, γ, is a proportionality constant, which differs for each nucleus and is a measure of the strength of the nuclear magnetic moment. Consequently, the detection sensitivity during an NMR experiment depends on the γ value of the nuclide. The larger the value of γ, the more sensitive a nucleus is and,

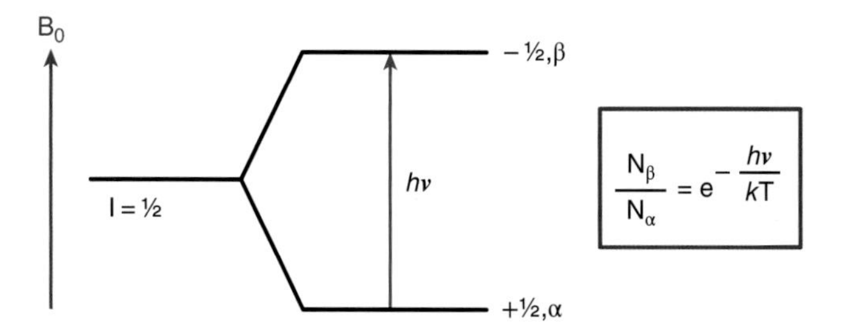

$$\frac{N_\beta}{N_\alpha} = e^{-\frac{h\nu}{kT}}$$

Figure 17.2 Energy level diagram for spin ½ nuclei showing splitting of the degenerate spin states into lower (+1/2, α) and higher (− 1/2, β) energy states upon the application of an external magnetic field B_0. The distribution of spins between the two states (N_β/N_α) is governed by the Boltzmann equation shown alongside.

consequently, the easier it is to observe. The strength of the external magnetic field (B_0) also influences the sensitivity as it increases the population difference between the spin states in accordance with the Boltzmann distribution. Consequently, there will be a greater number of spins in the lower energy state, thus increasing the probability of absorption.

17.2.2 NMR Experimental Parameters [19–21]

When one carries out an NMR experiment, there are a number of important parameters that must be considered. These are chemical shift, line integral and spin–spin coupling, which are discussed below.

17.2.2.1 Chemical Shift

If one considers a molecule with a range of functional groups and proton connectivities, the chemical shift (δ) defines the position of each NMR signal corresponding to these protons. The NMR signals are spread over a range of frequencies as each proton has a different local electronic environment. Consequently, each one is shielded to a different extent when placed in an external magnetic field (B_0).

A relative spectral scale is used where one measures the difference in frequency ($\Delta\nu$) between the sample (ν_{sample}) and that of a reference compound (ν_{ref}), usually tetramethyl silane. Because $\Delta\nu$ is dependent on B_0, one defines a dimensionless quantity δ, the chemical shift, as shown in Eq. (17.2), where ν_0 in the denominator is the observing or spectrometer frequency. The factor 10^6 in the numerator is used to simplify the numerical values and hence the δ values are always expressed in parts per million:

$$\delta\ (\text{ppm}) = \frac{\nu_{sample} - \nu_{ref}}{\nu_0} \times 10^6. \tag{17.2}$$

17.2.2.2 Line Integral

Resonance signals have different intensities and their peak areas can be integrated to provide the relative number of protons giving rise to each signal. This is very useful in interpreting one-dimensional proton NMR spectra of small organic compounds. They are also useful in quantitative analysis of mixtures. In two-dimensional NMR, the nuclear Overhauser effect spectroscopy (NOESY) cross-peaks are integrated to extract experimental distance constraints (Section 17.6.3), which form an important input in structure determination (Figure 17.1).

17.2.2.3 Spin–Spin Coupling

Spin–spin coupling, also known as scalar coupling, arises from the interaction between nuclear magnetic moments and the bonded electron spins in a spin system, e.g. A–B–C–D. When two nuclei are coupled, signal splitting is observed. The distance between the peaks in a splitting pattern is given by the coupling constant, J, which has units of Hertz (Hz). The most common and useful coupling encountered in NMR is the vicinal (three-bond) proton–proton coupling, $^3J_{H\text{-}H}$.

Vicinal coupling constants depend on the dihedral angle between the coupled protons and this is described by the *Karplus* equation (Eq. (17.3)), where J is the coupling constant, ϑ is the dihedral angle and X, Y and Z are values that depend on the bond type and on substituents:

$$^3J = X\cos^2\vartheta + Y\cos\vartheta + Z. \tag{17.3}$$

17.2.3 Longitudinal (T_1) and Transverse Relaxation (T_2) [19–21]

The population difference between the α and β spin states (Figure 17.2) corresponds to bulk magnetisation. At thermal equilibrium, the precession of the nuclei lies along the direction of the external magnetic field, B_0, and the bulk magnetisation in this direction is denoted M_0. Nuclei that align with the external magnetic field are by convention said to be in the 'Z-axis'. Therefore the magnetisation of the Z-axis (M_z), is in fact equal to M_0.

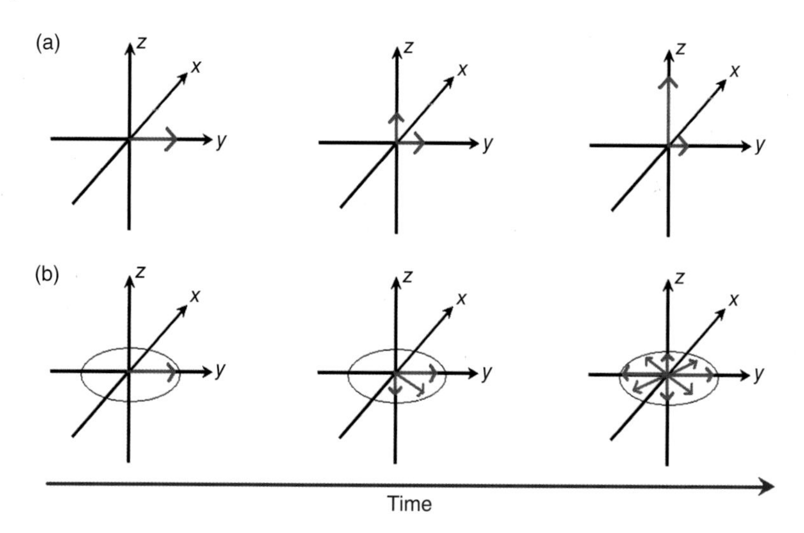

Figure 17.3 Vector diagram showing (a) longitudinal relaxation and (b) transverse relaxation.

In the pulsed NMR method, a RF pulse consisting of a band of frequencies and typically of a few μs in duration is applied to excite all the nuclei simultaneously. Applying a $90°_x$ pulse along the X-axis rotates the magnetisation to lie along the Y-axis with more nuclei in the higher energy state. This energy absorbed from the pulse needs to be lost to the lattice (surrounding molecules) and the time it takes for the magnetisation to return to equilibrium, i.e. magnetisation along the Z-axis, is known as the longitudinal (or spin-lattice) relaxation time, T_1 (Figure 17.3).

In addition to longitudinal relaxation, transverse (spin–spin) relaxation also exists and describes the process by which magnetisation decays in the XY plane. In the transverse relaxation process, the bulk magnetisation is rotated into the XY plane where some spins will experience a larger magnetic field and these spins start to process faster than the Larmor frequency. This causes the spins to fan out along the XY plane until no net magnetisation is present. The time taken for the magnetisation to return to equilibrium in this instance is known as the transverse (or spin–spin) relaxation time T_2 (Figure 17.3).

Equilibrium cannot be achieved along the Z-axis unless magnetisation on the Y-axis is completely removed; for this reason, one would expect T_1 to always be greater or equal to T_2.

The practical significance of T_2 values lies in their relationship to the resonance line width; the shorter the relaxation time T_2, the greater the line width and vice versa. The line width of resonance signals is important in NMR spectroscopy as the narrower the line width the greater the resolution, which in turn yields a more complete and accurate assignment of resonances. The line width is normally measured at half the peak height ($\Delta\nu_{1/2}$) and in high-resolution NMR spectroscopy this should be as small as possible to produce well-resolved peaks.

17.2.4 Nuclear Overhauser Effect (NOE) [22]

The nuclear Overhauser effect (NOE) is a dipolar effect that results from the 'cross-relaxation' between nuclear spins that are in close proximity in three-dimensional space. The effect results in an increase in the intensity of one proton, say H_x, after the irradiation of a neighbouring proton H_A (Figure 17.4).

To briefly explain the NOE effect, a homonuclear two-spin system (AX) will be considered. In this instance, it will be assumed for the sake of simplicity that the two spins are not scalar coupled with each other; in other words, both A and X will each produce a singlet peak. Figure 17.4 shows the three stages leading to the NOE effect: equilibrium, saturation and relaxation. At equilibrium (Figure 17.4, top left), the dipolar coupling between the two nuclear spins A and X, each with spin states α and β, results in four energy states with spin

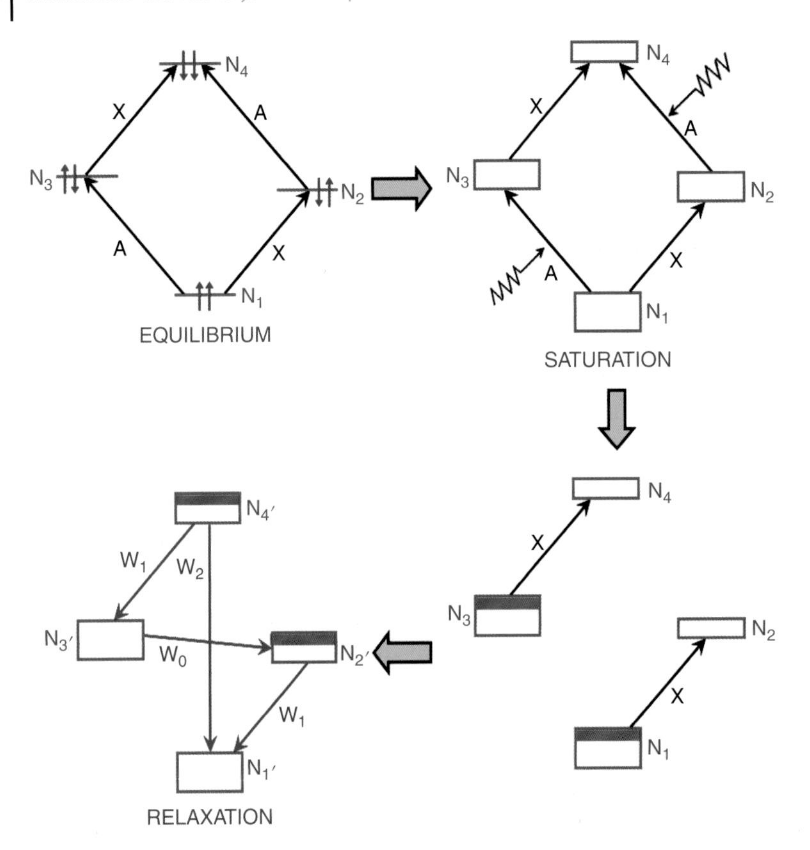

Figure 17.4 The NOE effect as observed in a two spin system, AX, energy level diagram. At equilibrium the population difference between the four energy levels is the same and intensities of A and X are equal. Saturation of A causes the population to be equalised between N_1 and N_3 and N_2 and N_4. W_0, W_1 and W_2 are the possible relaxation processes that occur to restore equilibrium, which affect the intensity of the X-transition. Relaxation along W_2 creates a positive NOE and that along W_0 creates a negative NOE.

populations N_1 to N_4. N_1 corresponds to the lowest energy state αα (↑↑) with the largest number of spins and N_4 the highest energy state ββ (↓↓) with the smallest number of spins. N_2 and N_3 correspond to the intermediate energy states αβ (↑↓) and βα (↓↑), respectively, with spin population N_2 being slightly higher than N_3 due to different Larmor frequencies of A and X. The energy of transition of the spins of A nuclei from N_1 (αα) to N_3 (βα) and N_2 (αβ) to N_4 (ββ) are both equal and similarly it is equal for spins of X nuclei from N_1 (αα) to N_2 (αβ) and N_3 (βα) to N_4 (ββ), resulting in the two signals A and X.

To observe an NOE, one of the signals, e.g. the signal due to A, is irradiated continuously (saturation) (Figure 17.4, top right), which makes the spin population equal between N_1 and N_3 and also between N_2 and N_4 (bottom right). Consequently, the transition of spins due to A is no longer possible and hence there is no signal due to A. The effect of saturation does not affect the transitions N_1 to N_2 or N_3 to N_4 due to the spins of X (Figure 17.4, bottom right). Following the perturbation to spin populations a new population distribution is established amongst the four states, $N_{1'}$ to $N_{4'}$ (Figure 17.4, bottom left). By comparing the population difference before and after saturation, it can be noted that $N_{1'} < N_1$ and $N_{4'} > N_4$, which means that that ratio $N_{1'}/N_{4'}$ is smaller than the equilibrium ratio N_1/N_4.

The system tries to restore equilibrium through the relaxation processes W_1 (single quantum), W_2 (double quantum) and W_0 (zero quantum) (Figure 17.4, bottom left). Of these processes, the W_2 and W_0 are the dominant dipole-dipole relaxation pathways and are closely connected to the NOE effect. The W_2 relaxation increases the population difference between $N_{1'}$ and $N_{2'}$ and $N_{3'}$ and $N_{4'}$, causing an increase in the signal intensity of the X-transitions. However, the W_0 relaxation reduces the population difference between $N_{1'}$ and $N_{2'}$ and $N_{3'}$ and $N_{4'}$, causing a reduction in the signal intensity of the same X-transitions. Thus, depending on the magnitude of W_2 and W_0 relaxation, the signal intensity can either increase (positive NOE) or decrease (negative NOE), respectively.

Since dipolar coupling is observed as 'through-space' interactions, it is distance dependent and gives information about the three-dimensional structure of the molecule. Normally, dipolar couplings are observed if the distance between the two interacting nuclei is less than 5 Å. If the distance is greater than 5 Å, the NOE signal is too small to be observed.

The distance is not the only factor that affects the NOE signal; the correlation time, τ_c, is also very important. The correlation time is defined as the time it takes a molecule to rotate 1 radian and since the rate of rotation is affected by molecular size it directly influences the intensity and sign of NOE signals observed.

17.2.5 NMR Rate Processes [5]

Chemical exchange processes are important in NMR spectroscopy as they affect the linewidths and peak intensity of resonances. Consequently, this phenomenon can be used to investigate the dynamics in a given system. If, at equilibrium, a molecule adopts two conformations, A and B, a different set of resonances will be observed for each, ν_A and ν_B (Figure 17.5):

$$\mathbf{A} \underset{k_B}{\overset{k_A}{\rightleftharpoons}} \mathbf{B}$$

As previously noted, the chemical environment has a direct effect on the chemical shift of a nucleus. However, the peak positions of resonances are also affected by the rate of interconversion between A and B, k_A and k_B. Both k_A and k_B are concentration independent since the conformational changes at equilibrium occur within the same molecule.

In systems where the interconversion rate is slow on the NMR timescale, a separate peak is observed for each conformation (Figure 17.5). The relative intensity of each peak depends on the molar concentrations of conformations A and B at equilibrium. In slow exchange processes, k_A and k_B are much less than $\Delta\nu$, where $\Delta\nu$ is the difference in chemical shifts measured in Hz between the conformations.

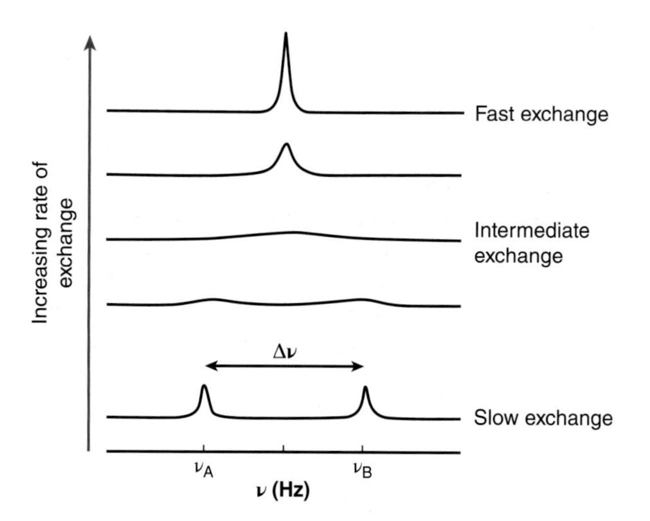

Figure 17.5 An illustration of the differences in linewidth and peak intensity between slow, intermediate and fast NMR exchange rate processes for two conformations, A and B, at equilibrium.

As the interconversion rate increases, it will eventually reach a point where k_A and k_B will be approximately equal to Δv; this is known as intermediate exchange. During intermediate exchange, there is a broadening of the two peaks for each conformation brought about by a partial averaging of the two chemical shifts (Figure 17.5).

In fast-exchanging systems, k_A and k_B are significantly greater than Δv. The effects on the different chemical shifts are averaged, resulting in one peak observed corresponding to both conformations (Figure 17.5). The position of the single peak will depend on the ratio between the concentrations of the two conformations at equilibrium.

17.3 Multidimensional NMR Spectroscopy

17.3.1 One-Dimensional NMR (1D NMR) [20, 21]

The one-dimensional (1D) NMR experiment consists of two parts, the first part being the preparation and the second being detection. During preparation, the bulk magnetisation of the sample is perturbed by an RF pulse or set of pulses dependent on the type of experiment undertaken before detection. During detection, the decaying magnetisation called the free induction decay (FID) is recorded as a function of a single time variable, t_2. The FID is subsequently Fourier transformed from time domain (t_2) to frequency domain (F_2) data to produce the one-dimensional spectrum

17.3.2 Two-Dimensional NMR (2D NMR) [20, 21]

In the case of two-dimensional NMR, the FID is detected and recorded as a function of two time variables, t_1 and t_2 (Figure 17.6). This creates two frequency/chemical shift dimensions (axes) instead of one. The advantage of 2D experiments is that it resolves overlapped signals into a second dimension. Furthermore, additional information is gained about the coupling within the molecule being analysed, which would not be evident in the equivalent 1D spectrum.

All 2D NMR experiments have the same basic format, which includes four parts: preparation, evolution, mixing and detection (Figure 17.6). Depending on the type of experiment undertaken, the preparation step involves exciting the spins with a single or set of RF pulses; this is followed by the evolution step, which is the key step in generating the second dimension. During the evolution period the magnetisation evolves for an incrementable time. Following evolution is the mixing step, which consists of either a single pulse or a series of pulses or just a delay of fixed duration for cross-relaxation, diffusion, etc. The mixing step is important as it is the stage at which magnetisation is transferred from one spin or coherence to another, i.e. it defines the type of correlation that will give rise to cross-peaks.

The data are recorded initially at $t_1 = 0$ and the pulse sequence is allowed to proceed; this generates an FID, which is recorded. The system is then allowed to return to equilibrium before the process is repeated with $t_1 = t_1 + \Delta t$, where Δt is known as the incremental time. This process is repeated many times over until enough data are collected, with Δt increasing incrementally with every additional run, i.e. $2\Delta t$, $3\Delta t$, $4\Delta t$, …, $n\Delta t$. It is very common for the number of Δt increments to be 300–500 in the case of homonuclear ^1H 2D experiments. The detected FID signal, which is a function of two time variables t_1 and t_2, is Fourier transformed to produce a spectrum that has two frequency dimensions (F_1 and F_2).

Figure 17.6 Generic pulse sequence scheme for a 2D NMR experiment.

17.3.3 Three-Dimensional NMR (3D NMR) [21]

As molecular size increases one is faced with three problems in NMR spectroscopy: the number of protons, the length of correlation time (rate of rotation of molecule) and the resonance line width all increase proportionally with size. This leads to overlapping of resonances and lower sensitivity in the case of experiments that rely on scalar coupling.

If resonances are unresolved in 2D spectra, it is possible to expand the data acquisition into a third dimension (Figure 17.7). This is important as it increases the dispersion of peaks thus simplifying the spectrum, as each plane in the 3D data only contains a few peaks. A basic 3D pulse sequence is similar to a 2D pulse sequence but with an additional evolution (t_2) and mixing period before the detection step (t_3) (Figure 17.8).

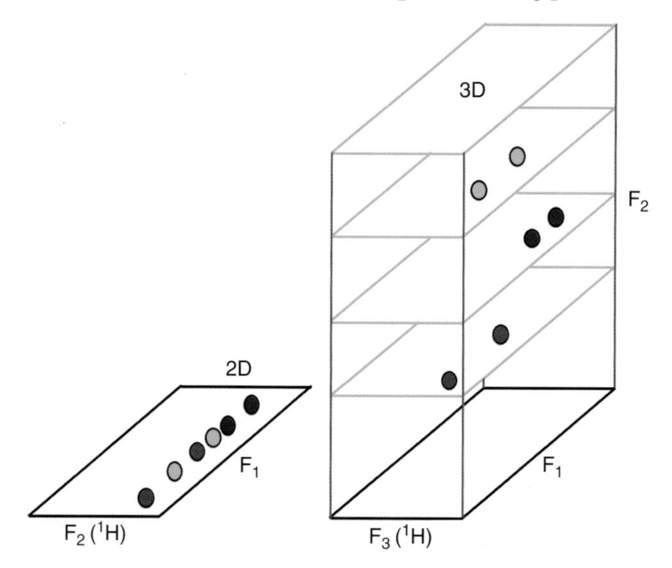

Figure 17.7 Scheme showing the resolution of degenerate resonances in 2D (left) by expanding into a third dimension (right) in NMR spectroscopy.

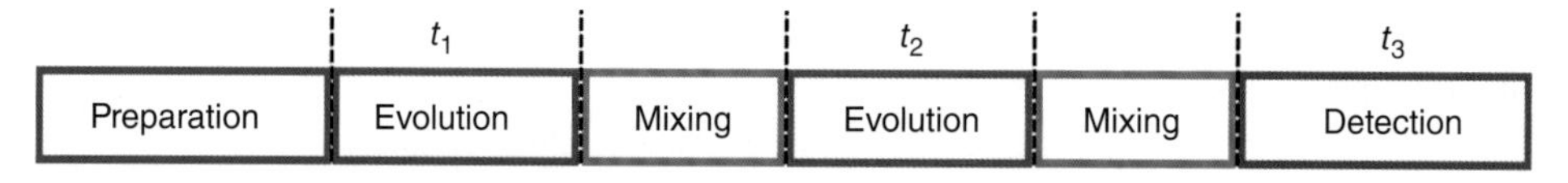

Figure 17.8 Generic pulse sequence scheme for a 3D NMR experiment.

Most 3D NMR techniques transfer the magnetisation through one bond (^1J) or two bond (^2J) couplings. This is more efficient than ^3J coupling as the transfer times between spins are shorter during relaxation and the sensitivity of the experiment is also increased.

17.4 NMR Instrumentation and Experiments [23]

17.4.1 NMR Instrumentation

NMR instrumentation is expensive to purchase, run and maintain. High field machines (800 MHz for protons or greater) are preferred for structure determination of large biomolecules (>10 kDa) due to the higher sensitivity and larger signal dispersion they provide. However, most universities and research institutes will have medium field 500 or 600 MHz NMR spectrometers, which can be used for structure determination of small biomolecules <10 kDa. An introduction to the workings of an NMR spectrometer and data acquisition by various pulse sequences is given in reference [23].

The NMR data in our case studies, described in Sections 17.7 and 17.8, were obtained using Bruker Avance 600, 700 and Varian Inova 800 MHz spectrometers equipped with 5 mm $^{1}H/^{13}C/^{15}N$ triple resonance cryogenically cooled pulsed field gradient (PFG) probes. Application of PFGs [24, 25] allows one to effectively suppress any signals in the spectrum in which one is not interested, thereby allowing the other interesting signals to be recorded with improved sensitivity. These probes can detect ^{1}H, ^{13}C and ^{15}N and being cryogenically cooled reduces noise in the spectrum. The PFG probe in the Bruker Avance spectrometers could also be changed for one that could be used to carry experiments detecting ^{31}P and ^{19}F.

17.4.2 NMR Experiments

Further to one-dimensional data acquisition, two-dimensional NMR experiments are used to aid assignment of the DNA spectra. Here, a brief description of the main NMR experiments carried out is given.

17.4.2.1 Solvent Water Suppression by Pre-saturation [26]

Molecules such as proteins and nucleic acids are of biological or biomedical interest and in order to observe all the possible proton environments in these molecules, they must be studied in water. This ensures that labile protons such as imino and amino protons can be observed. Typically, experiments are carried out in 90% $^{1}H_2O$ and 10% $^{2}H_2O$. The added $^{2}H_2O$ provides a ^{2}H signal, commonly called the lock signal, to enable the NMR spectrometer to continuously monitor and correct any field frequency changes during data acquisition.

Solvent suppression methods are used to minimise the solvent proton signal. This is required as the concentration of protons in the sample is very small (~mM), when compared to that in the water solvent (111 M). Consequently, the goal is to reduce the magnitude of the solvent signal before the NMR signal reaches the receiver so that the sample resonances can be observed with greater sensitivity. In the case studies work covered in this chapter, the main water suppression methods used were pre-saturation [26] and water suppression by gradient-tailored excitation (WATERGATE) [24, 25].

Pre-saturation suppresses the water signal by application of a weak, continuous RF pulse prior to the excitation and acquisition steps in an NMR experiment (Figure 17.9). This results in saturation of the water resonance, rendering it nearly unobservable. The main disadvantage is that exchangeable resonances may also be suppressed due to chemical exchange with saturated water protons but non-exchangeable resonances remain less affected. It is possible to minimise this drawback by changing conditions such as pH and temperature, which reduce the exchange of labile protons with water.

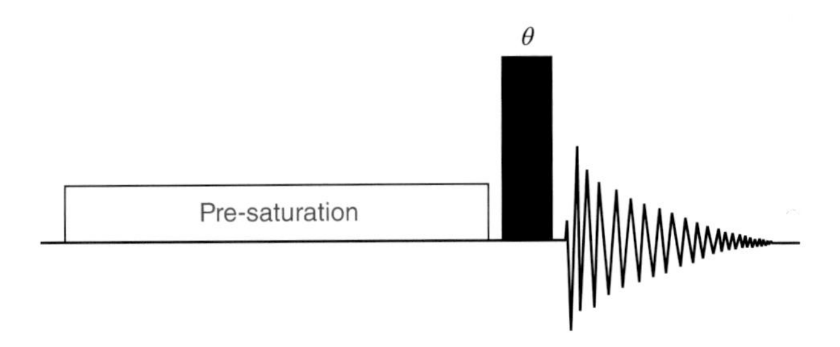

Figure 17.9 Pulse sequence to suppress solvent water proton resonance by the pre-saturation method [26].

17.4.2.2 Solvent Water Suppression by Gradient Method [24, 25]

WATERGATE (Figure 17.10) differs from pre-saturation as it involves the destruction of solvent magnetisation by means of PFGs [24, 25]. Briefly, in NMR experiments involving PFG, a large *z*-gradient is deliberately introduced along the sample's vertical axis, which results in the defocusing of transverse magnetisation that is

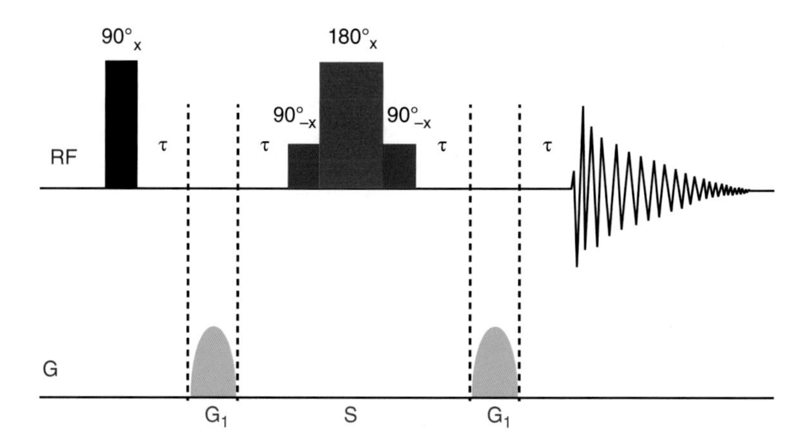

Figure 17.10 Pulse sequence for the WATERGATE water suppression method [24, 25]. Two selective $90°_{-x}$ pulses are used to rotate the water magnetisation to prevent its rephasing during the second gradient and are placed symmetrically around the non-selective $180°_x$ pulse. Four delay periods (τ) are used to allow for gradient recovery.

responsible for the NMR signal. However, by applying a second gradient the magnetisation can be refocused with a selection of coherence components or signals of interest and removing those that are undesired. Thus PFGs ensure that no magnetisation due to the solvent is observable prior to acquisition. The WATERGATE technique does not carry the disadvantages of pre-saturation and both exchangeable and non-exchangeable resonances can be observed.

The WATERGATE pulse sequence begins with an initial $90°_x$ pulse and is followed by a G_1–S–G_1 spin echo scheme, which causes the destruction of the solvent magnetisation. The two gradients (G_1) dephase the solvent magnetisation, leading to its destruction whereas all other resonances are refocused by the spin-echo (S), thus making them observable. Consequently, at the point of acquisition, there is no water signal left.

17.4.2.3 X {^{1}H} Decoupled 1D Experiment [27]

With a majority of biological molecules, NMR data recorded for different spin active nuclei can provide important structural and functional information. For example, in DNA, data collected on ^{31}P chemical shifts will give information about the phosphate backbone. In contrast, data collected on ^{15}N chemical shifts can provide information on base-pairing.

Proton decoupling is a non-selective (broadband) method which involves the decoupling of ^{1}H nuclei in order to enhance the signal of X nuclei and simplify the spectrum [27]. It does this by removing all the splittings due to ^{1}H–X couplings, which focuses all the X nuclei into a single sharp and intense line. Experimentally, proton decoupling is achieved by repetition of composite pulses during t_2 detection of the X nuclei (Figure 17.11).

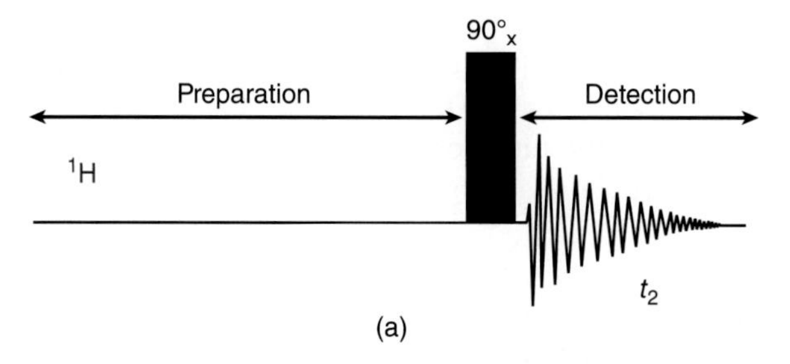

Figure 17.11 Comparison between (a) normal 1D ^{1}H-NMR and (b) X {^{1}H} decoupled 1D pulse sequence where X represents any spin active heteronucleus [27].

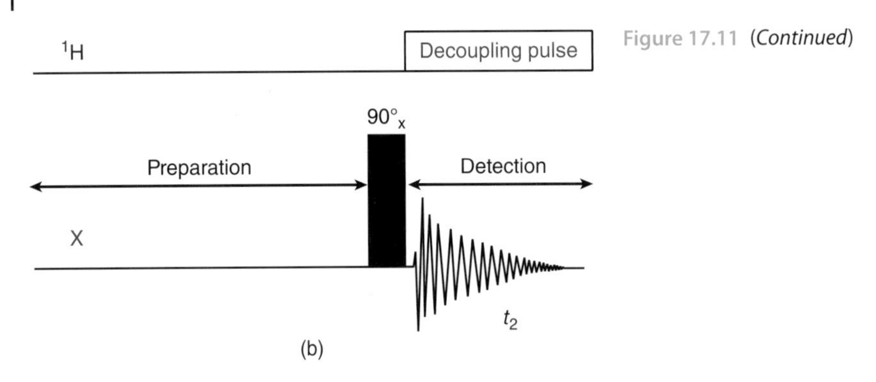

Figure 17.11 (*Continued*)

For 2D experiments requiring decoupling, composite pulse decoupling (CPD) is normally used. These are decoupling pulse techniques including sequences such as MLEV-16 [28] and WALTZ-16 [29]. The first CPD sequence consists of a continuous series of 180° pulses applied to the nucleus that requires decoupling. For spin 1/2 nuclei, the 180° pulse inverts the α and β states causing rates of precession to be switched. The switch in precession rates causes refocusing of the spins and if this is done at a fast rate relative to the relevant coupling constant, the coupling is suppressed. Improvements and modifications have been made to this initial method to remove defects and/or to tailor them for specific purposes.

17.4.2.4 Correlated Spectroscopy (COSY) [30, 31]

The correlated spectroscopy (COSY) [30, 31] experiment is used to identify protons that are scalar coupled to each other, usually within three bonds (^3J) (Figure 17.12). Two 90° pulses are required, which lie on either side of the evolution period. The first 90° pulse excites the spin population and creates coherence on each spin, which then evolve during t_1. If spins are J coupled to each other, coherence transfer occurs between them during the second 90° pulse, which is the mixing period.

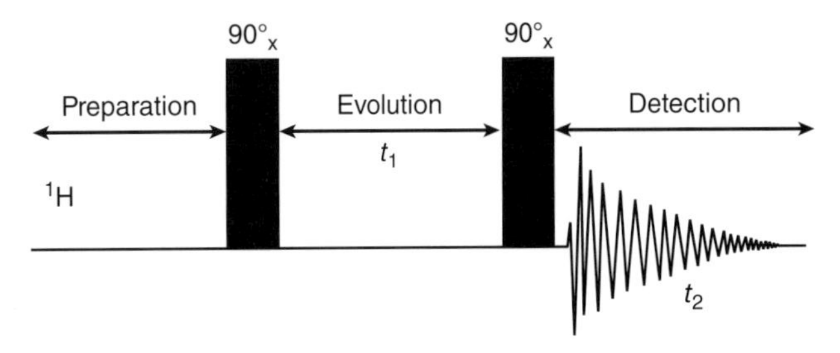

Figure 17.12 Pulse sequence for COSY experiment.

17.4.2.5 Double Quantum Filtered Correlated Spectroscopy (DQF-COSY) [32]

The double quantum filtered correlated spectroscopy (DQF-COSY) [32] experiment (Figure 17.13) is a phase-sensitive COSY experiment where all the observed resonances are filtered using double quantum coherence and appear in an anti-phase manner. By performing the filtration step, which occurs between the second and third 90° pulses, only signals arising from double quantum coherence are detected, thus increasing spectral resolution.

A DQF-COSY experiment can be used for identifying scalar (J) couplings between protons. A crucial advantage of this experiment is that it has absorption-mode diagonal peaks, not dispersion-mode ones. That greatly narrows the diagonal and is advantageous for all multiplicities. This is particularly advantageous for nucleic acids as correlations between the sugar proton resonances appear quite close to the diagonal and cannot be readily identified.

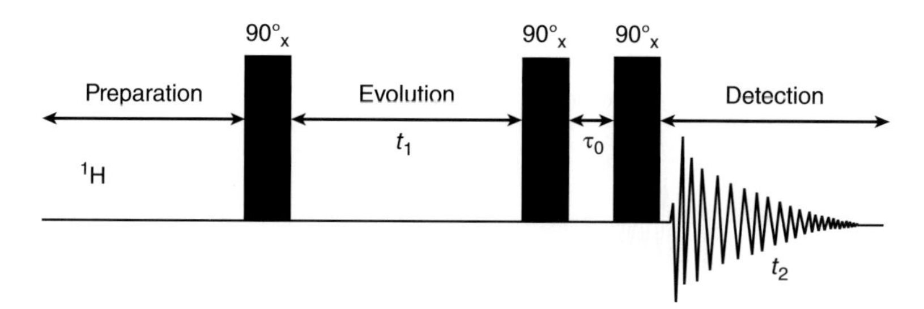

Figure 17.13 Pulse sequence for DQF-COSY experiment [32]. The delay time, τ_0, is of fixed duration and is typically a few microseconds.

17.4.2.6 Total Correlation Spectroscopy (TOCSY) [33, 34]

The total correlation spectroscopy (TOCSY) [33, 34] experiment (Figure 17.14) is performed to gain information on scalar coupled spins within a spin system. Once magnetisation is transferred into the first nucleus the magnetisation can then be passed on to subsequent nuclei within the same spin system. A main advantage of TOCSY is that (i) the peak multiplet structure is all in-phase, so there is no cancellation as in COSY when couplings are small relative to line widths, and (ii) cross-peaks are observed due to relayed transfer.

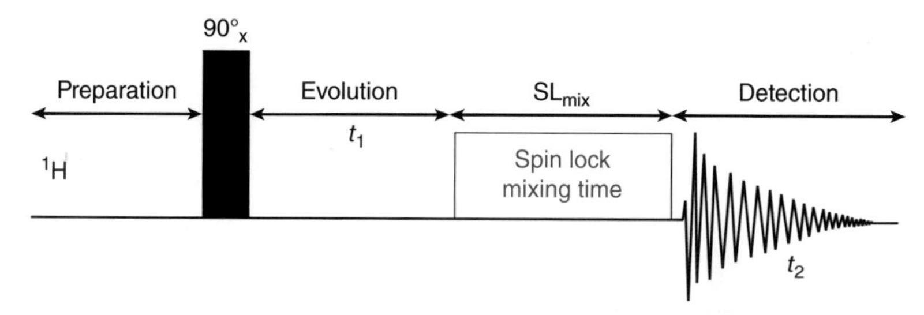

Figure 17.14 Pulse sequence for the TOCSY experiment. The long spin-lock mixing pulse executes the magnetization transfer and replaces the single mixing pulse in the COSY experiment.

A spin lock pulse (SL_{mix}) is added after the evolution period and during this time the chemical shifts are averaged out by applying a continuous sequence of pulses. This distributes the magnetisation throughout the entire spin system. The most common spin lock pulse scheme is known as MLEV-17, which is a series of closely spaced 180° pulses that allows homonuclear coupling to evolve while refocusing chemical shift evolution.

One can increase the number of correlations observed within a spin system by adjusting the spin lock mixing time and this is one of the advantages of the TOCSY experiment. Shorter mixing times mean that only correlations between adjacent nuclei may be seen, whereas longer mixing times lead to more signals being observed as the time allowed for magnetisation transfer is longer.

17.4.2.7 Nuclear Overhauser Effect Spectroscopy (NOESY) [35, 36]

The NOESY [35, 36] experiment (Figure 17.15) is used to determine dipolar coupling between spins, which are close together in three-dimensional space, experiencing cross-relaxation. Cross-peaks will be observed typically if two nuclei are ≤5 Å apart in space. The magnitude of the cross-relaxation will depend not only on the distance between the spins in space but also on the correlation time of the molecules, τ_c.

The NOESY pulse sequence has three 90° pulses and transfer of magnetisation occurs via population transfer (Z-magnetisation) as opposed to coherence (XY-magnetisation). To achieve this, a 90° pulse at the end of the evolution period creates Z-magnetisation and one at the end of the mixing period recreates transverse magnetisation for detection. A final 90° pulse is applied before the FID signal is detected and stored during t_2.

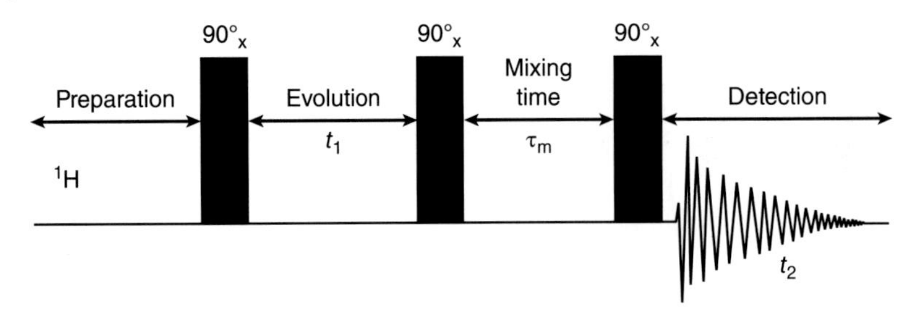

Figure 17.15 Pulse sequence for the NOESY experiment.

17.4.2.8 Heteronuclear Single-Quantum Correlation Spectroscopy (HSQC) [37, 38]

The heteronuclear single-quantum correlation spectroscopy (HSQC) [37, 38] experiment (Figure 17.16) is an NMR technique that correlates protons to a directly attached heteroatom, which is most commonly ^{13}C or ^{15}N. In very basic terms, the HSQC is a double insensitive nuclei enhanced by polarisation transfer (INEPT) [39] experiment, as magnetisation is transferred from proton to ^{13}C/^{15}N and then back to the proton again.

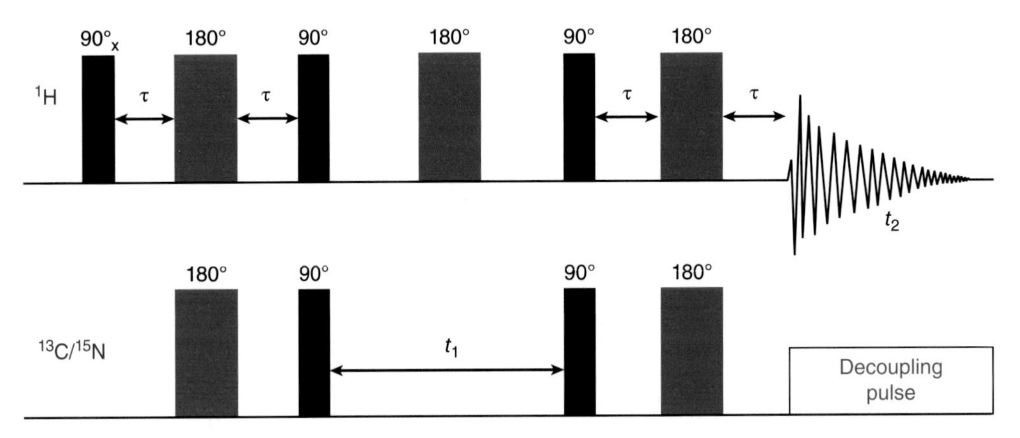

Figure 17.16 Pulse sequence for the HSQC experiment, where τ represent delay times, which are dependent on the value of J_{HX}; typically a value of 1/(4 J) is used.

The experiment first begins with proton magnetisation, which is followed by the first INEPT step, where proton chemical shift evolution is refocused by the double 180° pulses. Following refocusing of the magnetisation, it is transferred to the heteroatom (^{13}C, ^{15}N) where it is allowed to evolve with the chemical shift of the heteroatom. During this period, proton coupling is removed by use of a 180° pulse halfway into the t_1 period. Following the evolution step, a reverse INEPT step is carried out, which transfers the magnetisation back to the proton. When the magnetisation is back on the proton, it is refocused by the two 180° pulses. Magnetisation of the proton is detected during the detection time (t_2).

17.4.2.9 CPMG–HSQC–NOESY [40]

^{31}P resonances can provide important information about the phosphate backbone of DNA such as dynamics. However, carrying out ^{1}H-^{31}P NMR experiments prove to be uninformative as phosphorus resonances give very broad linewidths. One way to tackle this is to apply a Carr–Purcell–Meiboom–Gill (CPMG) pulse train, which is a series of closely linked 180° pulses applied during periods of coherence transfer between the phosphorus and the scalar coupled H3′ and H5′/5″ protons (see later in Figure 17.40). This results in enhanced sensitivity due to optimum coupling refocusing between nuclei in the phosphate backbone. The NOESY element of this experiment allows the identification of other sugar and aromatic base protons, which are close in three-dimensional space (see later in Figure 17.40). The CPMG–HSQC–NOESY pulse sequence is shown in Figure 17.17.

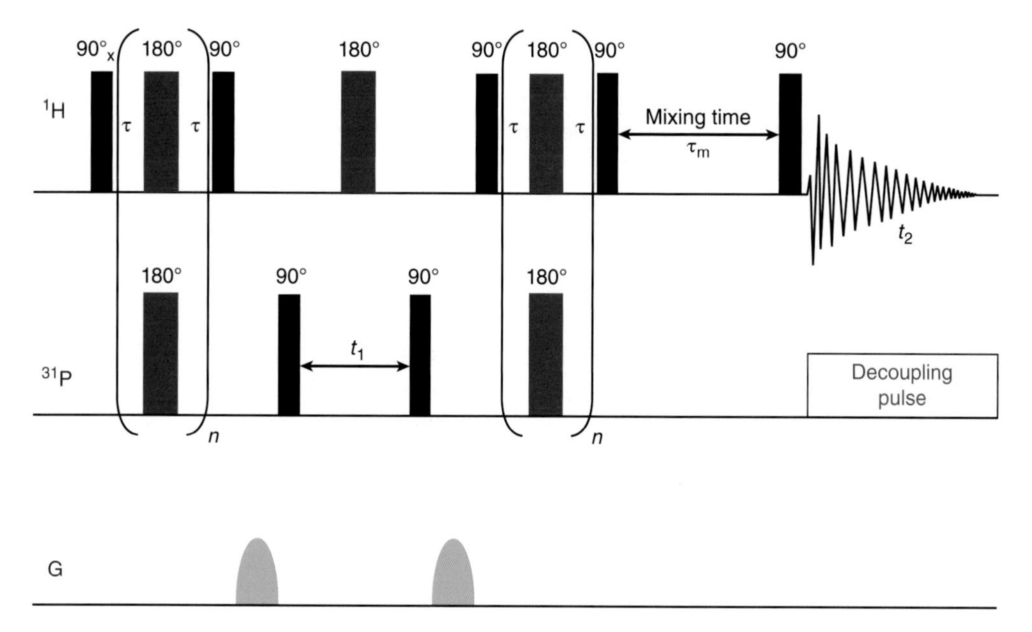

Figure 17.17 Pulse sequence for ^{1}H-^{31}P CPMG-HSQC-NOESY experiment, where τ represent delay times, which are set to 100 μs and the mixing period (τ_m) was set to 500 ms.

17.5 Structure and Conformational Parameters of DNA

17.5.1 Basic Building Blocks of DNA [41, 42]

In nature there are two types of nucleic acids, which are deoxyribonucleic acid (DNA) and ribonucleic acid (RNA). DNA plays a primary role as it is the carrier of genetic information in all cellular organisms. The double helix structure of DNA was elucidated by James Watson and Francis Crick in 1953 [43] and their discovery is often regarded as one of the greatest scientific achievements in history. The determination of the DNA structure meant significant advances were possible in the field of genetics.

The basic building blocks of a nucleic acid molecule are known as nucleotides [41, 42], which contain three components to their structure: a heterocyclic aromatic base, a pentose sugar and a phosphate group. In DNA, there are four aromatic +bases and these can be substituted monocyclic pyrimidines or bicyclic purines. The pyrimidine bases are cytosine (C) and thymine (T) and the purine bases are adenine (A) and guanine (G) (Figure 17.18).

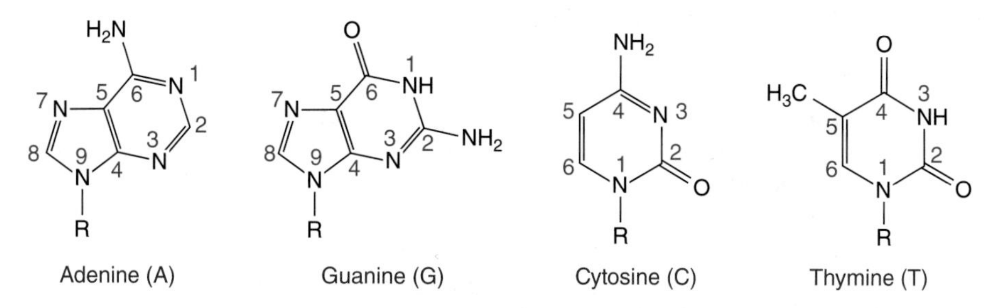

Figure 17.18 Structure of purine (A, G) and pyrimidine (C, T) bases present in DNA molecules. The common numbering scheme of the atoms is shown for each base. The R group represents bond connectivity to the sugar–phosphate group, which is discussed later.

The sugar–phosphate group has equal importance as it forms the backbone of DNA [41, 42] molecules; in DNA, the sugar is deoxyribose, which is shown in Figure 17.19.

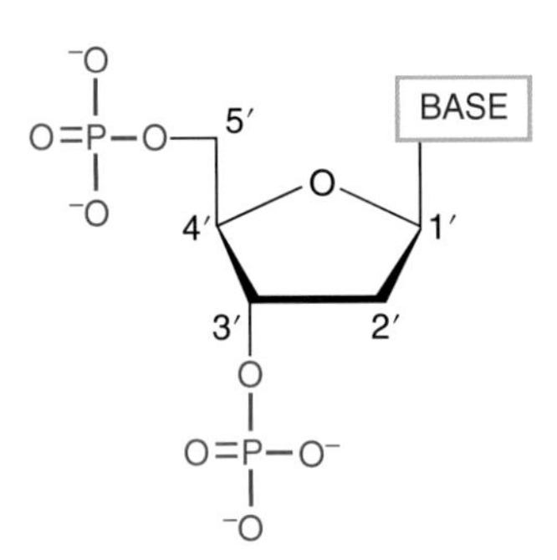

Figure 17.19 Structure of the deoxyribose sugar–phosphate group.

17.5.2 Phosphodiester Linkage [41, 42]

Nucleotides can covalently link together to form long chains, which are known as polynucleotides; covalent bonding occurs between the phosphate and sugar groups (Figure 17.20). For any one nucleotide, the phosphate group attached at the 5′ position of the sugar ring is linked to the hydroxyl group at the 3′ position on the

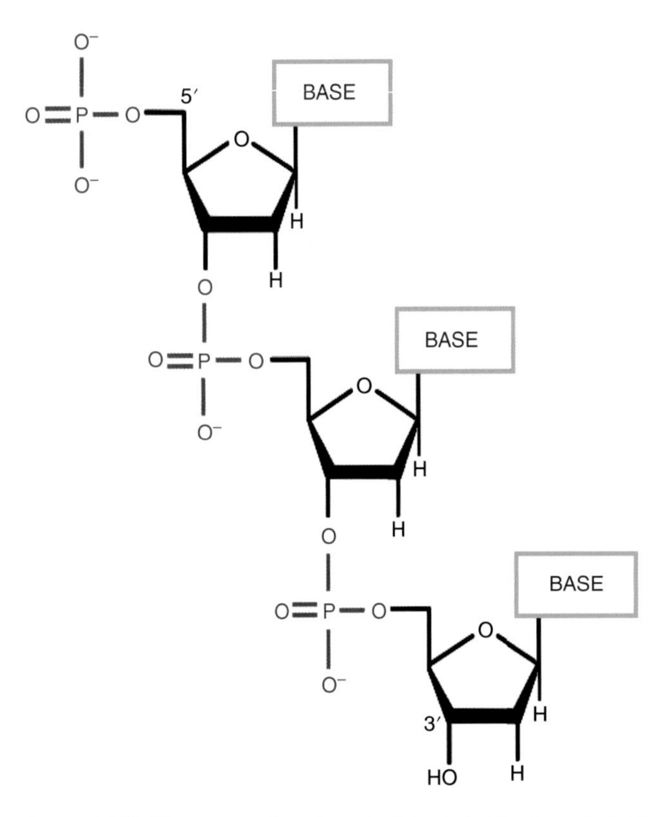

Figure 17.20 Illustration of a polynucleotide chain. The three individual nucleotides are joined by a series of phosphodiester linkages shown in red. The other components of nucleotide structure, deoxyribose sugar and heterocyclic aromatic base, are shown in black and green respectively [1].

preceding nucleotide. Each phosphate–hydroxyl bond is known as a phosphodiester linkage [41, 42]. With the exception of the first and last nucleotides in a polynucleotide chain, which have free 5′ phosphate and 3′ hydroxyl groups, respectively, all of the 5′ and 3′ groups are involved in phosphodiester linkages.

17.5.3 Watson–Crick Hydrogen Bonded Base Pairs [42, 43]

Base pairs form according to complementary pairing rules [41, 43]. These state certain associations are preferred between purine and pyrimidine bases. Each base pair is formed through hydrogen bonds between bases, which align on opposing polynucleotide strands. Complementary base-pairing only allows G to pair with C and A to pair with T. GC and AT base-pairing involves three and two hydrogen bonds, respectively, and these base pairs are often known as canonical or Watson–Crick base pairs (Figure 17.21). Canonical base pairs not only confer stability to the double helical structure of DNA but also align the functional groups in order to interact with other molecules such as proteins [41, 42].

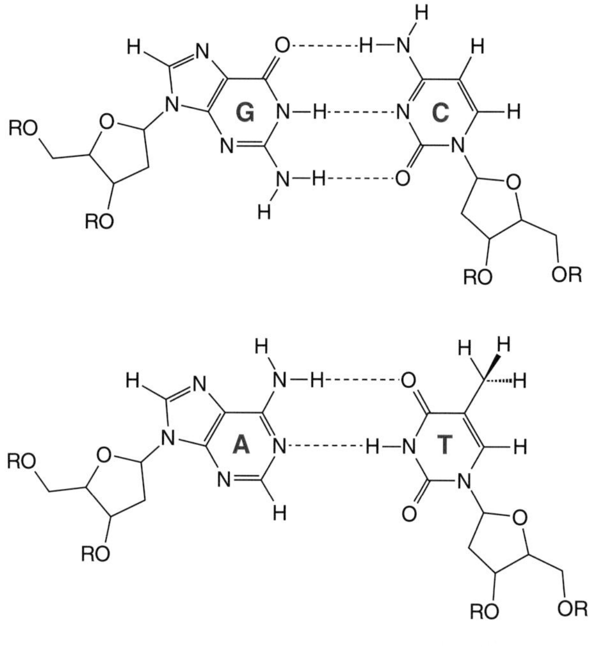

Figure 17.21 Illustration of GC (top) and AT (bottom) canonical Watson–Crick base pairs. The R groups denote the attachment sites for the phosphorus atoms at 5′ and 3′ hydroxyl groups of the sugar ring.

Stacking between adjacent bases is caused by interaction of π-orbitals between the aromatic rings of bases [44, 45]. This stacking interaction contributes to the overall stability of nucleic acid structure. Base stacking between purine–purine bases are commonly found to be the most stable, followed by pyrimidine–purine and pyrimidine–pyrimidine interactions, which has implications in mismatch DNA repair [46].

17.5.4 Models of DNA

17.5.4.1 B-DNA [47, 48]

The B-form (Figure 17.22, left) is the most common conformation adopted by DNA in aqueous solution and its main helical parameters are given in Table 17.2. Each molecule consists of two polynucleotide chains (blue and red), which twist to form a right-handed helix with a diameter of approximately 20 Å. The duplex winds

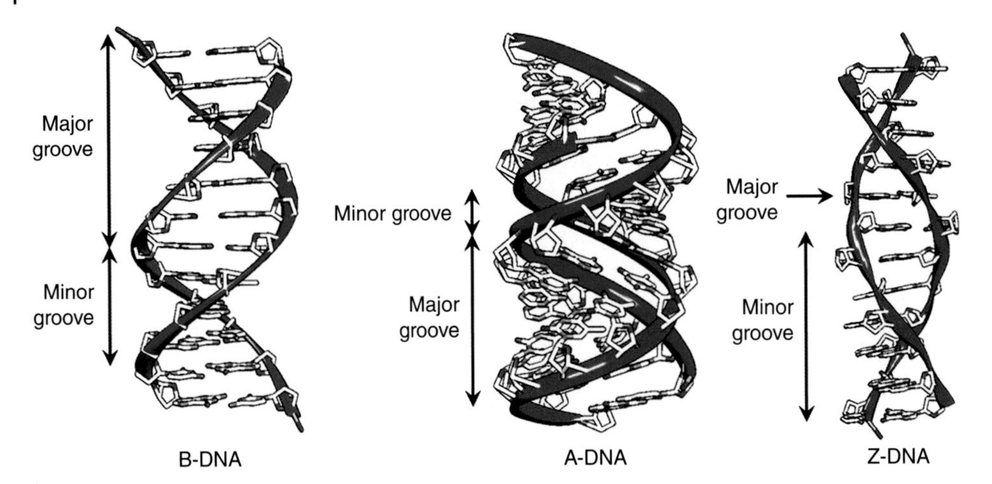

Figure 17.22 Molecular models illustrating the three main conformations of DNA: B-DNA, A-DNA and Z-DNA.

Table 17.2 Summary of main helical parameters showing the comparison between B-, A- and Z-DNA conformations.

Helix parameter	B-DNA	A-DNA	Z-DNA
Helix sense	R	R	L
Sugar pucker	$C_{2'-\text{endo}}$	$C_{3'-\text{endo}}$	$C_{3'-\text{endo}\ (syn)}$
Residues per turn	10	11	12
Helical twist per base pair (°)	36.0	32.7	−9.0, −51.0
Rise per base pair (Å)	3.3–3.4	2.9	3.7
Base tilt (°)	−6.0	20.0	−7.0
Major groove width (Å)	11.7	2.7	8.8
Minor groove width (Å)	5.7	11.0	2.0
Major groove depth (Å)	8.8	13.5	3.7
Minor groove depth (Å)	7.5	2.8	13.8

about a common axis and in such a way that it is impossible to separate the polynucleotide chains without first unwinding the helix.

The polynucleotide chains have both 5′ and 3′ ends and when a double helix forms they combine in an anti-parallel manner. Thus, the 5′ end only joins with a 3′ end and, similarly, a 3′ end only joins with a 5′ end during such that the two strands run in opposite directions (see later in Figure 17.44c).

If a helix were viewed down its central axis, one would see the sugar–phosphate backbone on the periphery of the DNA molecule while the bases occupy the central core of the helix. In this conformation, extra stability is gained because the repulsion caused by opposing sugar–phosphate groups is minimised. The base pairs are placed so that they lie in a plane, almost perpendicular to the helix itself.

In an ideal B-DNA molecule, the helical axis approximately passes through the centre of each base pair. This causes the formation of two exterior grooves, major and minor grooves, in the helix structure. The way that the base pairs are aligned also allows base stacking to occur, stabilising the double-helix structure even further.

17.5.4.2 A-DNA [47, 48]

A-form DNA (Figure 17.22, middle) initially appears to be very similar to B-DNA but in fact the bases are tilted 20° more towards the helix. This gives A-form DNA a much wider and flatter appearance with a deep major groove and a very shallow minor groove (Table 17.2). In contrast to DNA, double-stranded RNA tends to adopt the A form in solution.

17.5.4.3 Z-DNA [49]

Z-form DNA (Figure 17.22, right) is not known to naturally occur in living cells as a major conformation of DNA. However, Z-DNA does form in sequences that have alternating purines and pyrimidines, e.g. d(CG). Z-DNA is different from conventional B- or A-DNA as it is a left-handed helix. This means that unlike B-DNA, which winds in a clockwise manner, Z-DNA winds in an anti-clockwise manner (Table 17.2).

17.5.5 Conformation of DNA [44, 45]

The five-membered deoxyribose's sugar pucker is determined by the conformation in which the DNA exists. B-DNA, the most common structure of DNA, has a $C_{2'}$-*endo* pucker, A-DNA has a $C_{3'}$-*endo* and Z-DNA has both sugar puckers present [2]. Both sugar puckers are shown in Figure 17.23 and the various structures of DNA will be discussed later in this section.

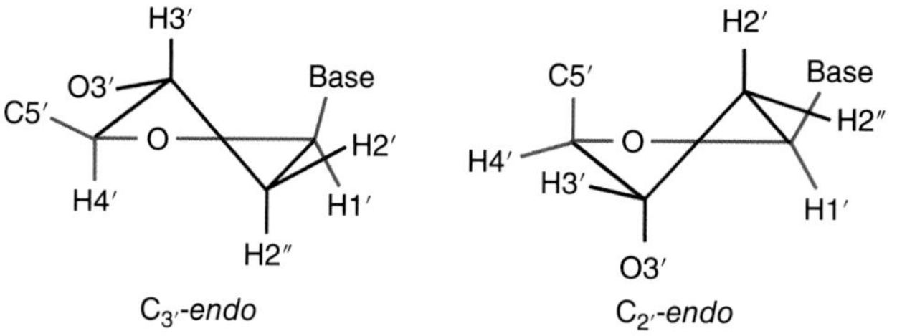

Figure 17.23 Illustration of sugar pucker conformations in nucleic acids: $C_{3'}$-*endo* (A-DNA) and $C_{2'}$-*endo* (B-DNA).

Further to the sugar pucker, nucleotide conformation is also dependent on a number of dihedral angles, which arise from the sugar–phosphate group (Table 17.3 and Figure 17.24). Here α, β, γ, ε and ζ define the conformation of the phosphate backbone only, whereas δ describes the phase amplitude and is both deoxyribose sugar and phosphate backbone dependent; χ is known as the glycosidic angle and defines the position of the base itself with respect to the deoxyribose sugar ring.

Table 17.3 Backbone and dihedral angles in nucleic acids listed with the atoms by which they are defined.

Dihedral angle	Defined by atoms
α	$O3'_{i-1}$-P_i-$O5'_i$-$C5'_i$
β	P_i-$O5'_i$-$C5'_i$ – $C4'_i$
γ	$O5'_i$-$C5'_i$-$C4'_i$-$C3'_i$
δ	$C5'_i$-$C4'_i$-$C3'_i$-$O3'_i$
ε	$C4'_i$-$C3'_i$-$O3'_i$-P_{i+1}
ζ	$C3'_i$-$O3'_i$-P_{i+1}-$O5'_{i+1}$
χ (Pu)	$O4'_i$ – $C1'_i$ – $N9_i$ – $C4'_i$
χ (Py)	$O4'_i$ – $C1'_i$ – $N1_i$ – $C2'_i$
ν_0	C4′ - O4′ - C1′ - C2′
ν_1	O4′ - C1′ - C2′ - C3′
ν_2	C1′ - C2′ - C3′ - C4′
ν_3	C2′ - C3′ - C4′ - O4′
ν_4	C3′ - C4′ - O4′ - C1′

i, *i* – 1 and *i* + 1 correspond to the atoms on the same, previous and following nucleotide, respectively.

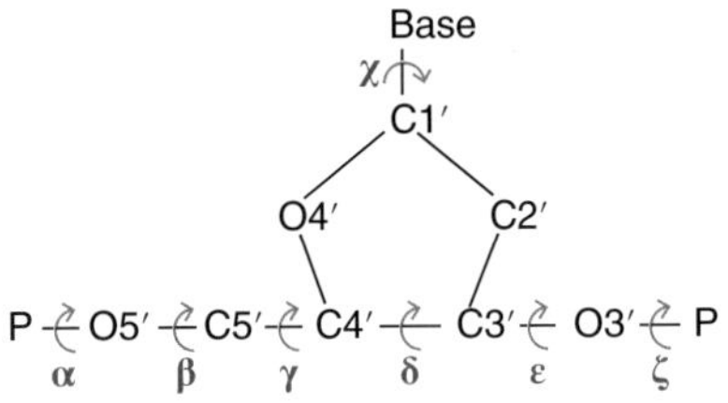

Base
χ
C1′
O4′ C2′
P – O5′ – C5′ – C4′ – C3′ – O3′ – P
α β γ δ ε ζ

Figure 17.24 Backbone dihedral angles, which define the conformation of the nucleotide and orientation of the phosphate backbone in nucleic acids.

Five dihedral angles exist that characterise the sugar pucker, ν_0 to ν_4 (Table 17.3 and Figure 17.25), and, furthermore, the sugar conformation can also be defined by the pseudorotation phase angle (P) and amplitude (φ) of δ and χ.

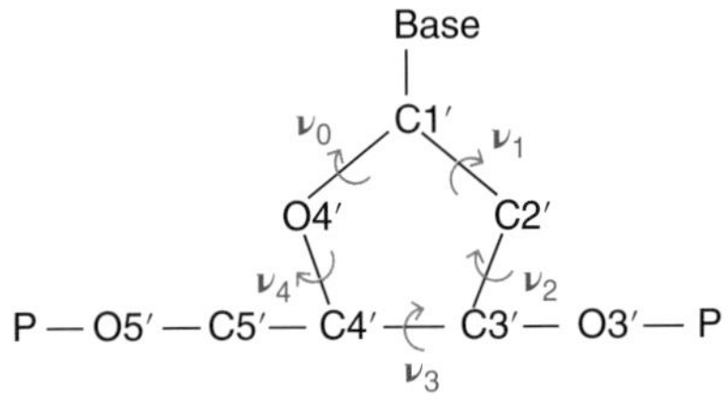

Base
C1′
ν_0 ν_1
O4′ C2′
ν_4 ν_2
P — O5′ — C5′ — C4′ – C3′ — O3′ — P
ν_3

Figure 17.25 Deoxyribose sugar ring dihedral angles, which define the sugar pucker in nucleic acids.

17.6 NMR Structure Determination [50–55]

Structure determination is the process of producing a three-dimensional refined and validated structure from experimental data by computational simulation using molecular mechanics and molecular dynamics (MD).

To successfully generate an NMR structure, a sequence of steps has to be undertaken (Figure 17.1). Experimental constraints are derived from the analysed NMR data before they are submitted to a structure determination program. Following structure calculations, the resulting structures are analysed, checked, and the constraints are modified until a reliable structure is produced for detailed conformational analysis.

17.6.1 NMR Assignment Strategy for DNA

The methodology for the assignment of NMR resonances in nucleic acids is quite well established [5, 56, 57] and is outlined in Figure 17.26.

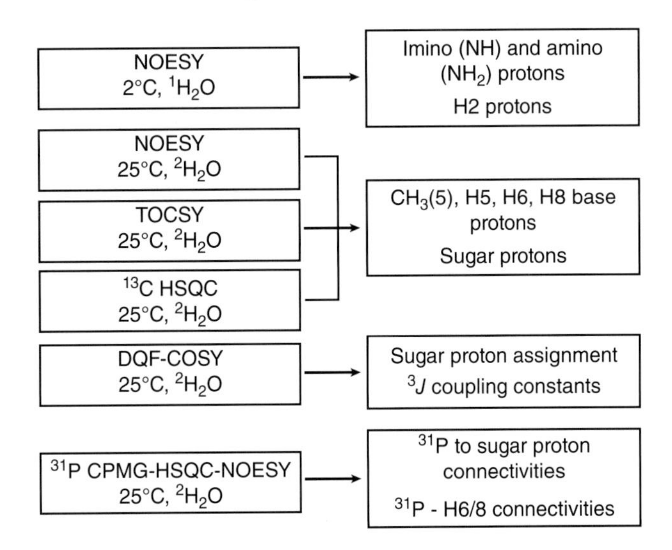

Figure 17.26 NMR experimental strategy towards assignment of ^1H, ^{13}C and ^{31}P resonances in nucleic acids.

Spectra measured in $^{1}H_2O$ at 2 °C are used to identify exchangeable imino (NH) and amino (NH_2) protons whereas spectra measured in $^{2}H_2O$ at 25 °C are used to identify non-exchangeable deoxyribose sugar and aromatic protons. ^{13}C and ^{31}P resonances are identified in natural abundance heteronuclear spectra measured in $^{2}H_2O$ at 25 °C.

17.6.2 Molecular Modelling

Molecular modelling can be used to generate three-dimensional representations of molecules in order to understand their structure and behaviour under given conditions [50–55]. Positions of atoms or molecules in the modelled system are represented by Cartesian coordinates (x, y and z) of the atoms or by internal coordinates, in which all the atoms in the system are represented relative to each other.

Molecular dynamics (MD) is a type of computer simulation in which atoms and molecules are allowed to interact for a period of time. This gives an indication of the internal motion of atoms in a molecule or system. The geometry optimisation (GO) of the molecule is normally carried out prior to MD simulations. GO simulations locate local minima on the potential energy surface rather than the global minimum.

17.6.3 Distance Constraints

In NMR structure determination, the most important and useful parameter is the NOE [22]. The theory of NOE states that the dipolar cross-relaxation rate is proportional to the inverse sixth power of the distance (r_{ij}) between two interacting ^{1}H spins, i and j:

$$\mathrm{NOE}_{ij} \sim 1/r_{ij}{}^{6}. \tag{17.4}$$

The distance constraints are generated from the very large number of cross-peaks observed in NOESY spectra using programs such as CcpNMR analysis [58]. Such programs use built-in protocols to automatically generate distance constraints from $^{1}H_2O$ and $^{2}H_2O$ NOESY spectra that have been assigned.

A reference NOE distance (r_{ref}) is set from which all other distances (r_i) are calculated. This is normally set to a known covalent distance, which is unlikely to change in the structure. The default distance function 'standard intensity' (S_{ref}) is selected and the peak intensities (S_{i}) are measured based on integrated peak volumes:

$$r_i = r_{\mathrm{ref}}\,(S_{\mathrm{ref}}/S_i)^{1/6}. \tag{17.5}$$

In DNA, the reference distance selected is the intraresidue guanine NH–NH_2(1) proton distance (2.5 Å) for experiments in $^{1}H_2O$ and cytosine H5–H6 proton distance (2.4 Å), for experiments in $^{2}H_2O$ [6].

The lower limits for distance constraints is set to 1.7 Å, which is the van der Waals radius in atoms. Even though NOE connectivities are usually for distances up to 5 Å, the upper distance limit for distance constraints is often set to 6.5 Å to account for factors such as spin diffusion, which is the NOE generated between two nuclei spreading or diffusing to other nuclei of the molecule at longer mixing times. Error bounds given to these limits depended on the degree of NOE overlap seen in the spectrum, but typically range between 10 and 20%. Once the distance constraints are calculated they are exported in a format that can be recognised by the structure calculation program.

17.6.4 Dihedral Angle Constraints

Dihedral angles give information about the sugar pucker, glycosidic dihedral angle and sugar orientation with respect to the backbone.

The conformation adopted by the sugar ring can be determined by correlating with the coupling constants ($^{3}J_{\mathrm{H1'\text{-}H2'}}$) from the DQF-COSY experiment (Figure 17.27). Large values of 7–11 Hz for this coupling typically demonstrate that the sugar pucker is $C_{2'}$-*endo* and hence the B-form DNA in conformation, and, conversely, small values of 2–3 Hz indicate a $C_{3'}$-*endo* sugar pucker and hence the A-form conformation DNA. Depending

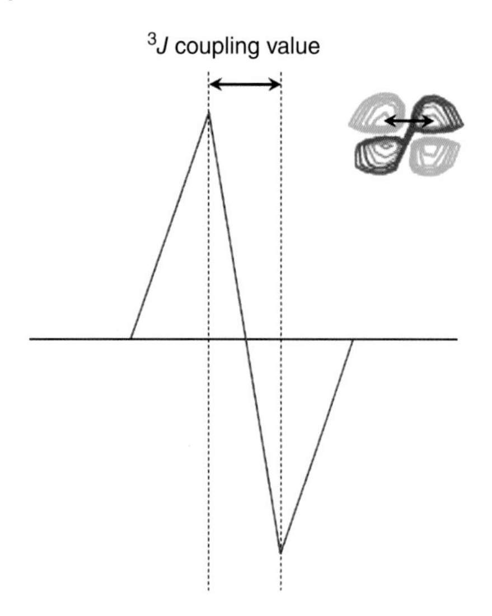

Figure 17.27 Visualisation of ^3J coupling and extraction of coupling constant from DQF-COSY spectra.

on the $^3J_{H1'-H2'}$ values obtained from the DQF-COSY spectrum, dihedral angles are assigned with reference to the standard values for B- or A-form DNA given by Markley et al. [59].

If no experimental data are available to derive the backbone dihedral angles (α, β, γ, ε and ζ), the values for these angles can be assigned, using the above approach, once the sugar pucker is identified for each nucleotide.

The assigned values and error limits for each dihedral angle can be modified and narrowed further based on the structures produced from initial calculations. Larger limits around the standard values were used initially to allow for the greater dynamic movement of the DNA molecule. However, in such cases, one will find that structures will converge towards a common set of dihedral angles in line with a preferred conformation of the molecule. This allows a more appropriate value to be assigned to each dihedral angle with narrower error limits.

17.6.5 Hydrogen Bond/Planarity Constraints [54, 55]

Exchangeable imino protons are involved in base-pairing (see later in Figure 17.36) and they appear low field in a spectrum due to stacking of the aromatic base pairs in the DNA structure. Hydrogen bond constraints are included upon observation of internucleotide imino to imino (10–15 ppm) and imino to amino (6–9 ppm) exchangeable NOEs in NOESY spectra, which give evidence for base-pairing. Planarity constraints ensure that each base pair remains planar in the DNA molecule. These are included during initial sets of calculations, to aid the helical shape of the DNA, but are removed during the refinement step of the structure calculation once planarity is observed in the structures, even in the absence of these constraints.

17.6.6 Structure Calculations [54, 55]

With any structure determination calculation, a starting point must be created before experimental constraints can be added to the calculation. Requirements for the calculation include a topology file and a structure file. The topology file gives information on the molecule including atom types, masses and charges and bonding, while the structure file gives information on atom coordinates. These are combined to generate a protein structure file (psf), which can be entered into a structure determination program (Figure 17.28).

Once a psf file has been generated, experimental constraints can be added before structure determination calculations are carried out to constrain the structures to the distances and angles calculated from the NMR data. If experimental constraints are excluded from the calculation, assumptions must be made to restrain the structure and dynamics to avoid the production of very random starting structures. For example, for nucleic

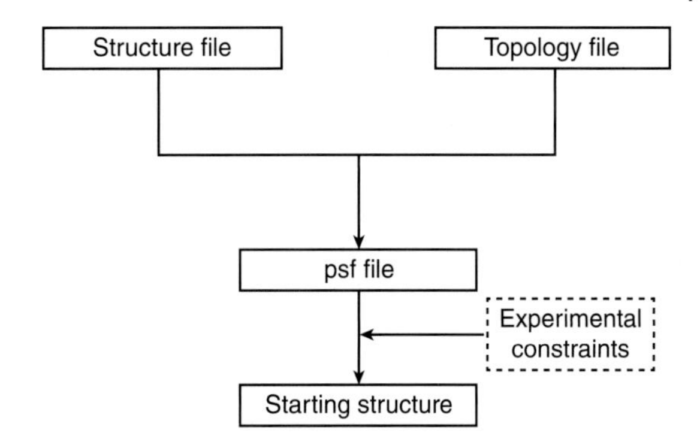

Figure 17.28 Scheme illustrating requirements for structure determination.

acids, if canonical base pairs are present, the base pairs can be set to have standard hydrogen bonding constraints.

The determination of three-dimensional solution structures from NMR data, using the program such as Xplor NIH [54, 55], typically follow the scheme shown in Figure 17.29.

A hundred structures are generated initially using restrained (NOE distance, hydrogen bond and dihedral angle restraints) simulated annealing calculations. During this step, the aim is to find optimum structures by freely exploring the conformational space, starting from the input structure file.

Randomisation is then carried out on these 100 structures with the aim to find the lowest energy conformations. A long period of restrained molecular dynamics is carried out with the starting temperature set at 3500 K. This ensures that all atoms are free of any interactions and thus allows them to move randomly through conformational space. Following this, the temperature of the system is reduced slowly by 12.5 K per 0.2 ps to find the lowest energy conformations.

The lowest energy structures generated during simulated annealing are put through a refinement step, which reduces the number of restraint violations and aims to generate improved structures with a lower root mean square deviation (RMSD) when compared to simulated annealing. Having a low RMSD value is important as it provides a measure of how well structures overlap each other and thus how similar the structures are from the structure calculations. Structures are rejected at this point based on high energy and RMSD when compared to the average of an ensemble of structures.

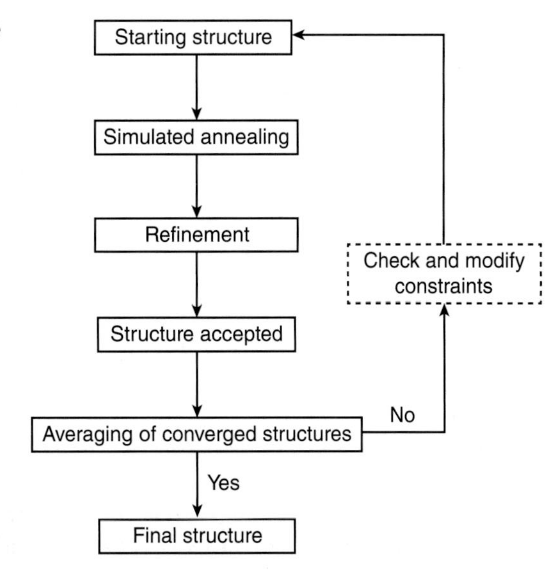

Figure 17.29 Flowchart showing the scheme followed for structure determination.

The 10 lowest RMSD structures at this stage are put through a final refinement calculation and the final structure is accepted on the basis of low RMSD, low energy and the smallest number of restraint violations.

17.6.7 Conformational Analysis [15, 60]

The final structure obtained from the structure calculation can be further analysed to verify if the overall conformation is correct, which can be carried out using software such as w3DNA software [60]. Such software analyses the three-dimensional structure of nucleic acids by characterising all base–base interactions and the helical character of the base pairs. The calculated values are based on comparison with a standard nucleic acid reference frame [61, 62] (Figure 17.30) and can be classified into three groups.

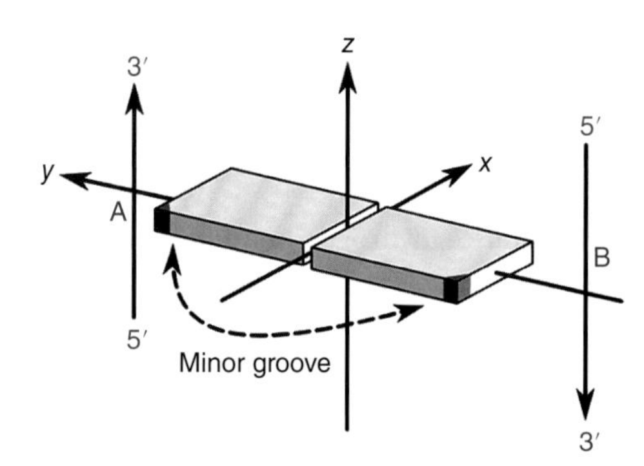

Figure 17.30 Standard reference frame used for determination of parameters in the conformational analysis of nucleic acid structures. A base pair, between polynucleotides A and B, is illustrated by the two rectangular cuboids. The Cartesian axes are shown with respect to the base pair.

Local base pair parameters (Figure 17.31) include values of shear (S_x), stretch (S_y), stagger (S_z), buckle (κ), propeller twist (π) and opening (σ). These define the relative position and orientation of the complementary bases in a base pair.

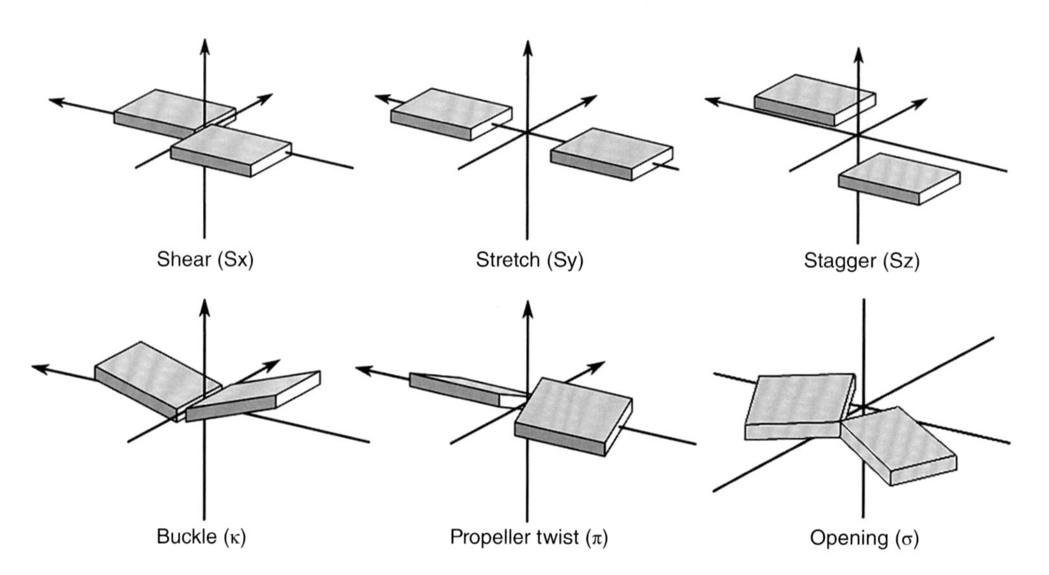

Figure 17.31 Illustration of local base pair relationships in conformational analysis of nucleic acid structures.

Local base pair step parameters (Figure 17.32) define the stacking geometry from a local point of reference and include shift (D_x), slide (D_y), rise (D_z), tilt (τ), roll (ρ) and twist (Ω).

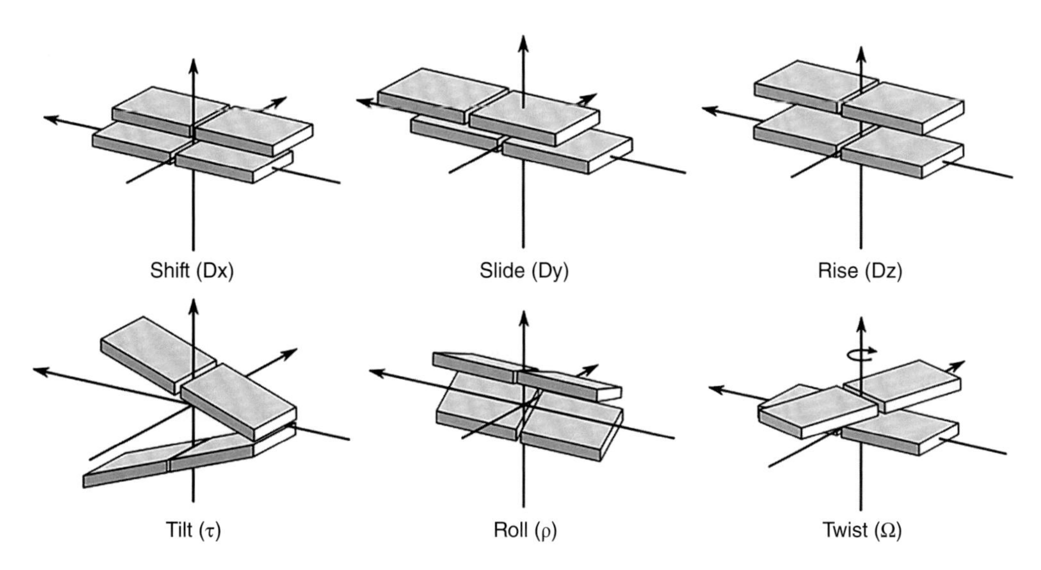

Figure 17.32 Illustration of local base pair step parameters in conformational analysis of nucleic acid structures.

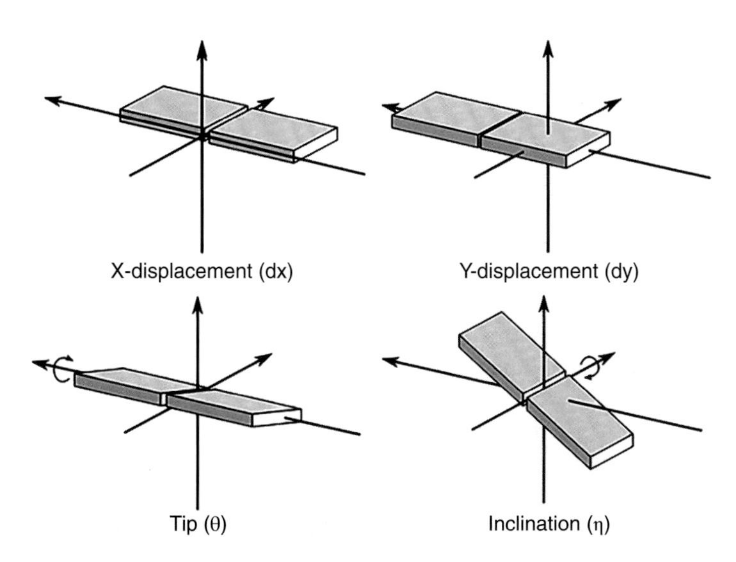

Figure 17.33 Illustration of local base pair helical parameters in conformational analysis of nucleic acid structures.

Local base pair helical parameters (Figure 17.33) define the regularity of the helix based on the calculated conformation and include x-displacement (d_x), y-displacement (d_y), tip (ϑ) and inclination (η).

It is important to note that local base pair helical parameters also include helical rise (D_z) and helical twist (Ω). These function in the same manner as rise and twist in local base pair step parameters (Figure 17.32) but describe the conformation of the double-helix DNA structure as a whole. In an ideal, B-form DNA the values of helical rise and helical twist should be very similar to the values of rise and twist.

17.6.8 Deposition of Structures in Databases

Structures elucidated by NMR spectroscopy are deposited in three main databases: nucleic acid database (NDB) [63], protein data bank (PDB) [64] and biological magnetic resonance bank (BMRB) [65].

When submitting an NMR structure, the Cartesian coordinate and constraint data files are processed and released by the PDB and the structure is assigned a unique ID number. NMR spectral data including chemical shifts and coupling constants are processed and archived by the BMRB [65].

17.7 Case Study 1: NMR Structure Determination and Conformational Analysis of 17mer Canonical GC DNA

As a case study in the application of NMR for structure determination, we will study a 17mer GC DNA duplex whose sequence is given in Figure 17.34, from sample preparation to final structure determination. The sequence chosen corresponds to the DNA used for the X-ray structure determination of *E. coli* MutS protein complexed to a mismatch GT DNA [66]. For the purpose of this NMR case study, the mismatch GT base pair in the middle of the sequence has been substituted by a canonical GC base pair G_9C_{26}, shown enclosed by a box (Figure 17.34).

5′- A_1 G_2 C_3 T_4 G_5 C_6 C_7 A_8 G_9 G_{10} C_{11} A_{12} C_{13} C_{14} A_{15} G_{16} T_{17} -3′
3′- T_{34} C_{33} G_{32} A_{31} C_{30} G_{29} G_{28} T_{27} C_{26} C_{25} G_{24} T_{23} G_{22} G_{21} T_{20} C_{19} A_{18} -5′

Figure 17.34 17mer G_9C_{26} DNA duplex investigated by NMR.

17.7.1 Sample Preparation of 17mer GC DNA

The two strands of the 17mer GC DNA homoduplex were chemically synthesised, high-performance liquid chromatography (HPLC) purified and obtained commercially. The oligonucleotides were obtained in a lyophilised salt form containing sodium phosphate ready for use in the NMR sample.

In order to prepare DNA duplexes from the synthesised oligonucleotides, they were mixed in equimolar amounts. Concentration was determined by ultraviolet (UV) absorbance at 260 nm and both oligonucleotides were made up into a sample of equal volume containing approximately 100 absorbance units each in 50 mM PO_4^{3-} and 50 mM NaCl at pH of 6.2. The sample was then annealed by heating to 90 °C for three minutes followed by slow cooling to room temperature. The sample was then stored at 4 °C.

NMR samples of the DNA duplexes were prepared in either 90% ^1H_2O and 10% ^2H_2O, or 100% ^2H_2O. Samples were prepared in ^1H_2O to enable the observation of water exchangeable imino and amino protons. Samples containing just ^2H_2O were used to obtain NMR spectra with fewer signals to assist the assignment of non-exchangeable aromatic and sugar proton resonances. Samples in ^2H_2O were lyophilised in order to remove any traces of residual water from the sample, before the addition of high purity ^2H_2O (≤99.9%).

17.7.2 Acquisition, Processing and Analysis of 17mer GC DNA NMR

17.7.2.1 Data Acquisition

1D ^{1}H-NMR spectra were usually recorded with 128–512 transients, with the proton transmitter offset (carrier frequency) set to be the same as the ^1H_2O proton frequency. Experiments in ^1H_2O solvent were measured at 1–5 °C to retard the exchange of labile NH and NH_2 protons of DNA and thus enhance peak intensity and definition. Those in ^2H_2O were measured at 25 °C to observe fewer resonances and with higher resolution.

For samples in ^2H_2O, the pre-saturation technique [26] was favoured and for samples in ^1H_2O, WATERGATE was favoured [25]. The typical proton spectral width for the ^2H_2O samples was 12 ppm; for ^1H_2O samples this was increased to 24 ppm, to allow low field exchangeable protons to be observed. The typical ^{13}C spectral width for the ^2H_2O was 120 ppm. The typical ^{31}P spectral width for the ^2H_2O samples was around 8 ppm.

2D NMR experiments were set up with the number of transients set to be 256 or less. TD1 and TD2 indicate the number of data points for each time domain and the set values were optimised depending on the experiment type. For example, for a NOESY experiment TD1 and TD2 were typically set with optimum values of 512 and 4096 respectively. The number of transients and data points was carefully chosen to give data with the best sensitivity and spectral resolution within the time available.

As with 1D, in 2D experiments, the carrier position for ^{1}H was set on the $^{1}H_2O$ proton frequency. Experiments involving the detection of ^{13}C and ^{31}P were carried out in $^{2}H_2O$ at 25 °C and the carrier positions for each were set to 100 and −3.52 ppm, respectively.

Mixing times for the NOESY spectra varied from 150 to 250 ms, for the TOCSY, it was set to 75 ms and in the case of ^{1}H-^{31}P CPMG-HSQC-NOESY, the mixing time was 500 ms.

17.7.2.2 NMR Data Processing and Analysis

NMR data, as acquired by the spectrometer, needs to be processed into a format that can be used further for analysis. There are a number of software programs that are able to achieve this in addition to those that will be given in this section.

1D NMR data was processed using Spinworks [67] and 2D NMR data was processed using NMRPipe [68]. Data processed by NMRpipe can be checked using NMRdraw [68], which is a visual interface for multidimensional NMR data. The processed data was converted into UCSF format in order to be viewed so that assignments can be made in software such as SPARKY [69].

SPARKY [69] was used to view 2D NMR spectra and assign resonances. The completed assignment file was transferred into the program CcpNMR analysis [58]. This program, similar to SPARKY, allows the viewing and assignment of 2D and higher dimension NMR spectra but also generates NOE distance constraints in an automated manner.

17.7.3 NMR Assignment of 17mer-GC DNA

Under aqueous conditions, it can be assumed that a double helical DNA molecule adopts a right-handed B-DNA conformation. Identification and assignment of exchangeable and non-exchangeable proton resonances in DNA is relatively easier to carry out due to a wider chemical shift range of resonances between 0.0 and 15.0 ppm (Figure 17.35).

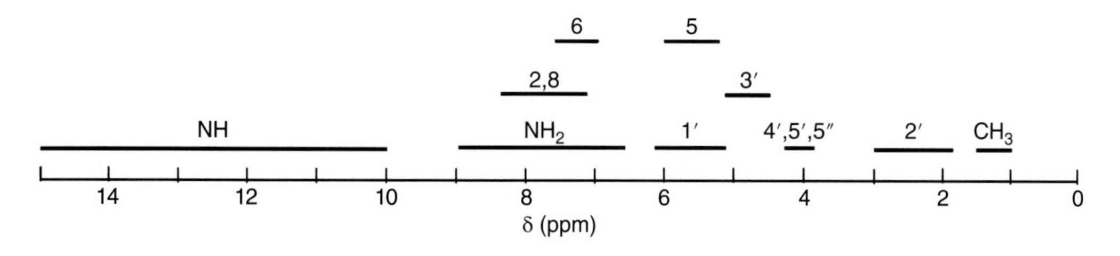

Figure 17.35 The typical chemical shift range of the various protons of DNA.

17.7.3.1 Identification and Assignment of Exchangeable Proton Resonances

NMR spectra recorded in $^{1}H_2O$ were used to identify exchangeable imino (NH) and amino (NH_2) protons as well as confirm base-pairing and the nucleotide sequence in the 17mer GC DNA sample.

Assignment of Imino NH Resonances The imino protons of the 17mer GC DNA can all be observed between 12 and 15 ppm using the 1D ^{1}H-NMR spectra recorded in $^{1}H_2O$. Sixteen imino protons can be assigned through imino–imino proton sequential NOEs (Figure 17.36) starting at T27 ($\delta = 13.89$); T27 NH proton can be assigned from an NOE to its base-paired A8 NH_2(1) proton. Figure 17.37 shows the sequential connectivity from T27 to G16.

Guanine
Cytosine
Thymine
Adenine

Figure 17.36 NOE connectivities between exchangeable protons in DNA. The intranucleotide imino-amino and internucleotide sequential imino–imino proton connectivities are shown by solid arrows while internucleotide imino–amino and imino H2 proton connectivities are shown by dashed arrows.

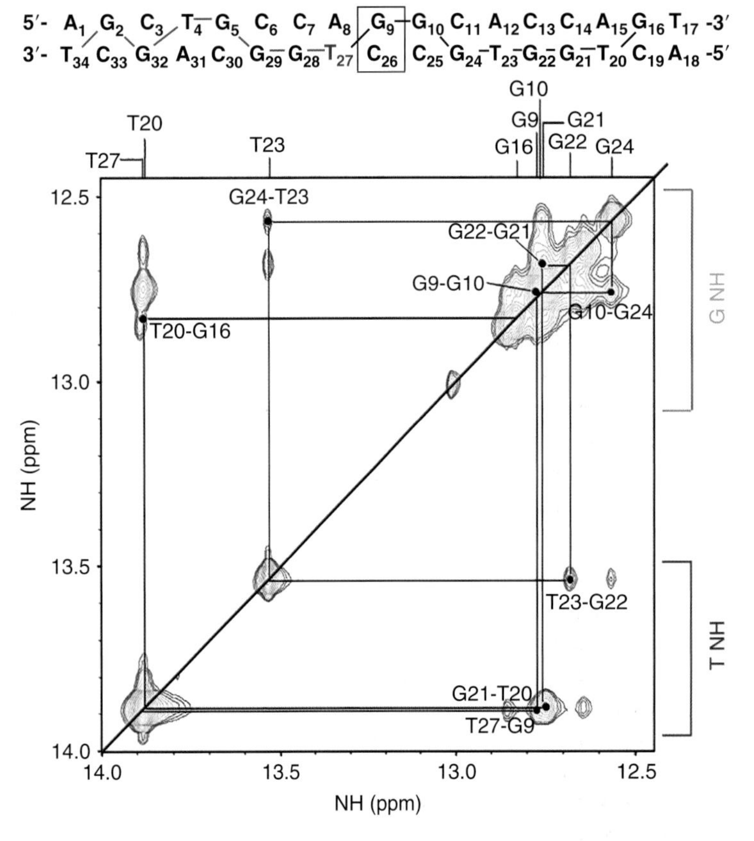

Figure 17.37 700 MHz NOESY ($\tau_m = 250$ ms) spectrum of 17mer canonical GC DNA (0.8 mM, 90% ^1H_2O + 10% ^2H_2O, 50 mM PO_4^{3-} and 50 mM NaCl, pH 6.2) at 2 °C. The sequential connectivity stretching from T27 to G16 is indicated by the black line.

Assignment of Amino NH_2 Resonances Although adenine, cytosine and guanine have amino protons, they can be easily distinguished in the NOESY spectrum. Hydrogen-bonded adenine and cytosine amino proton [NH_2(1)] resonances are typically observed between 7.6 and 8.5 ppm. Non-hydrogen bonded amino protons [NH_2(2)] resonate further up-field from this region. For adenine, they typically resonate between 6.0 and 6.5 ppm and for cytosine between 5.5 and 6.0 ppm. For guanine residues, both hydrogen and non-hydrogen bonded amino protons resonate between 6.0 and 7.0 ppm.

NOE crosspeaks were observed between imino protons and the amino protons in canonical GC and AT base pairs within the duplex and inter/intrastrand NOE cross-peaks aided in the assignment (not shown). The hydrogen bonded amino protons were easily assigned as they gave strong NOE cross-peaks to the imino protons.

17.7.3.2 Identification and Assignment of Non-exchangeable Proton and Carbon Resonances

Assignment of Adenine H2 Resonances Non-exchangeable adenine H2 resonances give strong NOEs to their base-paired thymine imino proton resonances (Figure 17.36) and were assigned by using the NOESY spectrum measured in ^1H_2O (Figure 17.38, left panel). Corresponding C2 resonances were well resolved from other regions in the ^{1}H-^{13}C HSQC spectrum (150–160 ppm) measured separately with ^2H_2O and were assigned by correlating with the H2 assignment from the NOESY spectrum (Figure 17.38, right panel).

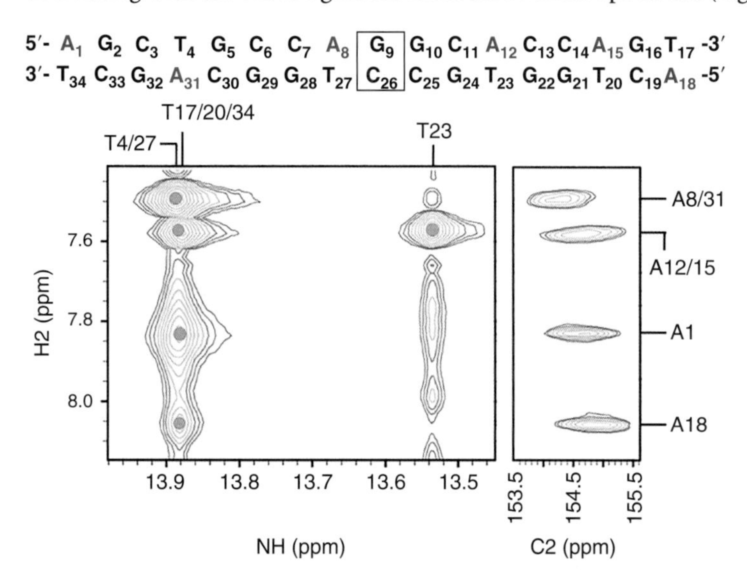

Figure 17.38 (Left panel) 700 MHz NOESY (t_m = 250 ms) spectrum of 17mer canonical GC DNA (0.8 mM, 90% ^1H_2O + 10% ^2H_2O, 50 mM PO_4^{3-} and 50 mM NaCl, pH 6.2) at 2 °C; (right panel) 600 MHz ^{1}H-^{13}C HSQC spectrum of 17mer canonical GC DNA (1 mM, 100% ^2H_2O, 50 mM PO_4^{3-} and 50 mM NaCl, pH 6.2) measured at 25 °C. The assignment of adenine H2 resonances is based on the NOE connectivities to thymine H3 imino resonances. Correlation to the ^{1}H-^{13}C HSQC spectrum identified the chemical shifts of C2 resonances.

Assignment of Cytosine H5-H6 and Thymine CH_3-H6 Correlations These correlations were identified using DQF-COSY, TOCSY. As they are scalar coupled, these give strong peaks in both spectra. The chemical shifts of cytosine and thymine H6 (7.0–7.5 ppm) are distinct from the H8 resonances of adenine (8.0–8.5 ppm) and guanine (7.5–8.0 ppm), which aids in their assignment.

Eleven cytosine H5-H6 correlations are expected for 17mer canonical GC DNA and all were observed to be well resolved in the DQF-COSY and TOCSY spectra (Figure 17.39), with the corresponding C5 and C6 resonances being assigned in the ^{1}H-^{13}C HSQC spectrum (not shown). The thymine CH_3-H6 correlations were assigned in an analogous manner (not shown).

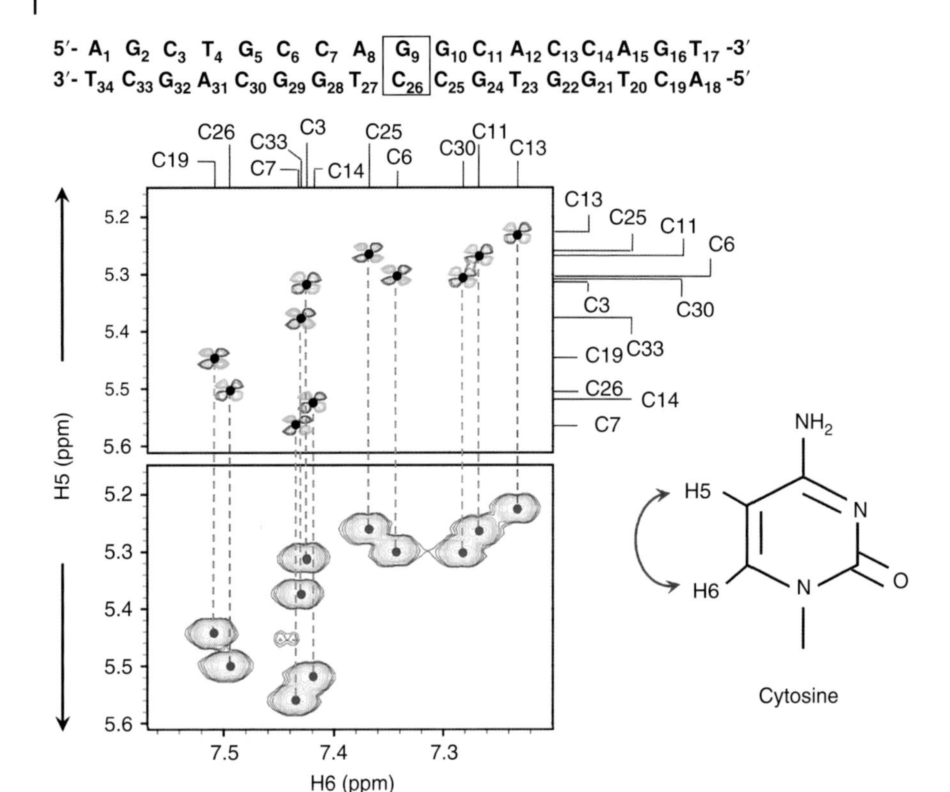

Figure 17.39 Stack plot of spectra showing the assignment of 11 cytosine H5-H6 connectivities in 17mer canonical GC DNA (1 mM, 100% $^{2}H_2O$, 50 mM PO_4^{3-} and 50 mM NaCl, pH 6.2) at 25 °C. The top panel shows the DQF-COSY spectrum (800 MHz) and the bottom panel shows the TOCSY spectrum (600 MHz, SL_{mix} = 75 ms).

Assignment of Aromatic H6/8 – Sugar H1′ Sequential Connectivity In B-DNA, the sugar pucker is $C_{2'}$-*endo*, resulting in the H1′ being in close proximity to H6/H8 (3.6–3.9 Å); these connectivities can be clearly observed in the NOESY spectrum. The chemical shifts of cytosine and thymine H6 (7.0–7.5 ppm) are distinct from the H8 resonances of adenine (8.0–8.5 ppm) and guanine (7.5–8.0 ppm). In addition, H1′ protons resonate between 5.0 and 6.5 ppm and this wide dispersion in chemical shifts allows cross-peaks to be clearly observed since overlapping of resonances is reduced.

Sequential assignment of non-exchangeable protons was achieved by identifying intra- and internucleotide NOEs between aromatic H6/8 and sugar H1′ protons, as highlighted in the scheme in Figure 17.40.

The H6/8-H1′ sequential connectivity was identified for both strands in the NOESY spectrum measured in $^{2}H_2O$ at 25 °C, confirming the sequence of the DNA. The sequential connectivity for A18-T34 is shown in Figure 17.41 (Panel B).

The assignment was carried out in the 3′ direction for both strands, A1 to T17 and A18 to T34, and all intranucleotide and internucleotide connectivities were successfully identified, leading to the full assignment of all 34 H6, H8 and H1′ resonances. The C6, C8 and C1′ resonances were assigned by correlation with the ^{1}H-^{13}C HSQC spectrum (Figure 17.41, Panels A and C) in support of the sequential assignment (Figure 17.41, Panel B).

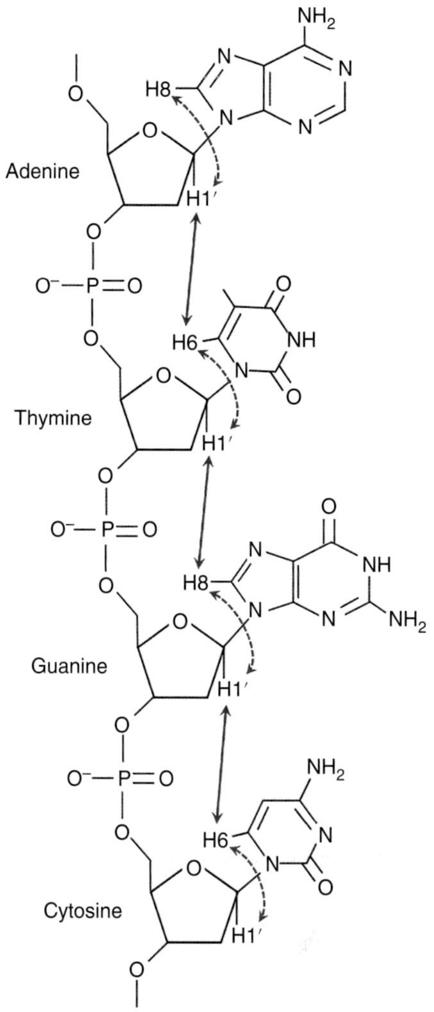

Figure 17.40 Scheme showing the H6/8 – H1′ sequential assignment in DNA. Dashed and solid arrows show the intranucleotide and internucleotide H6/H8-H1′ connectivities, respectively.

Assignment of H2′ and H2″ Resonances There are two protons H2′ and H2″ at the C2′ position and these resonate in the high field of all other sugar protons (δ = 1.8–3.0 ppm). H2′ and H2″ can be differentiated in the DQF-COSY spectrum as the coupling constant of $^3J_{\mathrm{H1'\text{-}H2'}}$ correlations, which are larger than $^3J_{\mathrm{H1'\text{-}H2''}}$ correlations (Figure 17.42)

Interestingly, it was found during assignment that for both pyrimidines and purines, H2′ protons resonated at a higher field than H2″ protons. All H1′-H2′ correlations were identified in the DQF-COSY spectrum on the basis of their coupling constants (7–11 Hz, Figure 17.42). However, since the coupling constant of H1′-H2″ correlations are very small (2–3 Hz), the peaks were difficult to observe in the DQF-COSY spectrum. This problem was not encountered in the TOCSY spectrum, where each H1′-H2″ correlation was seen as an intense peak because they are in phase (not shown).

The chemical shifts of H2′/2″ protons were confirmed in the NOESY spectrum as both exhibit NOEs to the same H1′ resonance for each nucleotide (not shown). H1′-H2″ NOESY cross-peaks are more intense as the H1′-H2″ distance is shorter than H1′-H2′ for all sugar conformations [5].

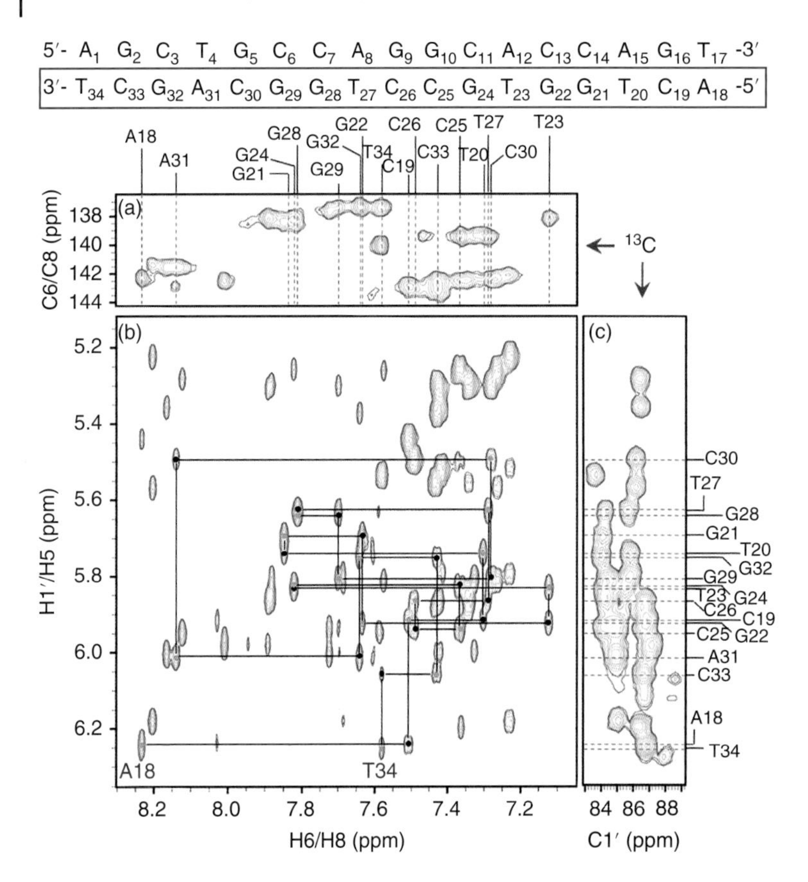

Figure 17.41 600 MHz NOESY (τ_m = 250 ms) spectrum (b) showing H6/8-H1′ sequential NOE connectivities between A18-T34 in 17mer canonical GC DNA (1 mM, 100% ^2H_2O, 50 mM PO_4^{3-} and 50 mM NaCl, pH 6.2) at 25 °C. Intranucleotide connectivities are shown by blue circles and internucleotide connectivities by purple circles. 600 MHz ^{1}H-^{13}C HSQC spectra (a and c) show the corresponding assignment for C6/C8 (a) and C1′ (c) resonances, respectively.

By distinguishing the different chemical shifts of the H2′ and H2″ resonances, H6/H8-H2′/H2″ NOE connectivities could be assigned, which give important distance constraints in structure calculations. NOEs between aromatic H6/8 and sugar H2′/2″ protons can be distinguished by the difference in their intramolecular distances where H6/8-H2′ and H6/8-H2″ have values of 2.0–3.6 Å and 3.4–4.5 Å, respectively. The corresponding C2′ resonances were identified and assigned in the ^{1}H-^{13}C HSQC spectrum by correlating with the chemical shifts of the H2′ and H2″ resonances (not shown).

Assignment of H3′, H4′ and H5′/H5″ Resonances To assign the H3′, H4′ and H5′/H5″ sugar resonances, the direct method is to identify ^3J scalar-coupled correlations in the DQF-COSY or TOCSY spectrum as these give intense cross-peaks. However, this proved difficult as these regions were highly overlapped over a narrow chemical shift range, so distinguishing one resonance from another was not possible (not shown).

The H3′ resonances were assigned by identifying cross-peaks to H1′ resonances in the NOESY spectrum where H3′ protons resonate between 4.5 and 5.1 ppm (not shown). This was aided by the fact that there was a clear divide in the chemical shift depending on the base type (pyrimidine or purine). H3′ protons of pyrimidines typically resonate a high field of 4.9 ppm and those of purines a low field of 4.9 ppm

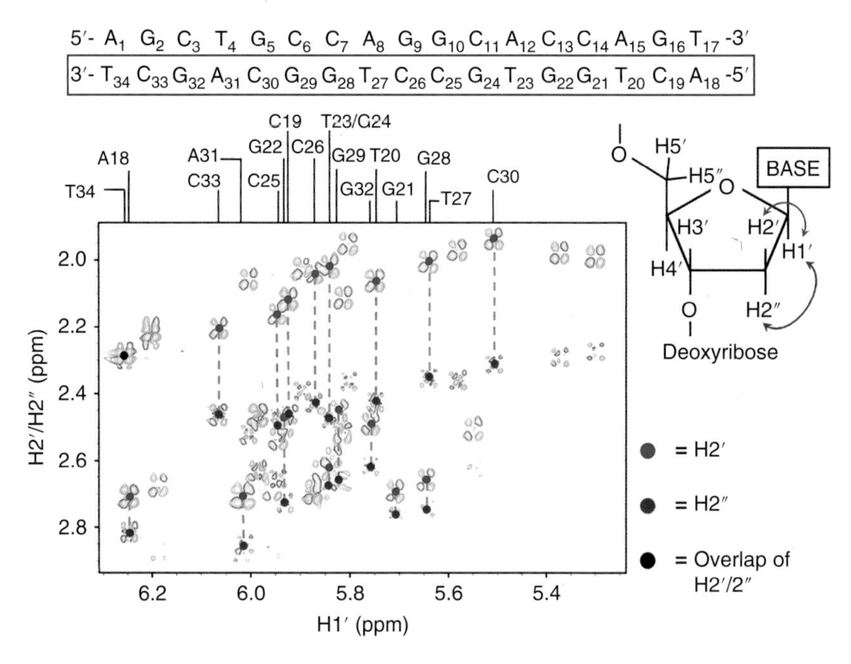

Figure 17.42 800 MHz DQF-COSY spectrum showing identification and assignment of H2′/H2″ resonances for A18-T34 in 17mer GC DNA (1 mM, 100% ^2H_2O, 50 mM PO_4^{3-} and 50 mM NaCl, pH 6.2) recorded at 25 °C.

Strong NOEs were also observed between aromatic H6/H8 and H3′ protons, which also allowed the H6/H8-H3′ sequential connectivity to be identified for each strand (not shown). This confirmed the assignment of H3′ resonances from the H1′-H3′ region.

The assignment of H4′ and H5′/H5″ resonances proved more difficult as they overlapped in the same region of the NOESY spectrum. Fortunately, the H4′ resonances for purines (4.3–4.5 ppm) appear at a different chemical shift range to pyrimidines, which overlap with H5′/5″ resonances (4.0–4.3 ppm), which allowed the H4′ resonances to be more easily assigned. All C3′, C4′ resonances were also assigned in the ^1H-^{13}C HSQC spectrum by correlating with the chemical shifts of the H3′ and H4′ resonances (not shown).

H5′ and H5″ resonances were assigned similarly to H2′/2″ resonances with H1′-H5″ NOEs being more intense than those for H1′-H5′ correlations (not shown). The same observation applied for NOEs between aromatic H6/8 and sugar H5′/5″ protons as those for H6/8-H5″ appeared more intense (not shown). The H5′/5″ resonances of the A1 and A18 residues were readily identified as they appeared clearly separate from other resonances, which allowed the subsequent assignment of their respective C5′ resonances in the ^1H-^{13}C HSQC spectrum (not shown).

Assignment of Phosphorus Resonances The ^{31}P resonances were assigned based on H3′-^{31}P cross-peaks observed in the ^1H-^{31}P CPMG-HSQC-NOESY spectrum (Figure 17.43, top panel). Using this approach all ^{31}P resonances were assigned with the exception of those for C6, C7, G10 and C11 residues. ^{31}P chemical shifts for residues C7 and G10 were assigned by identifying the NOEs to aromatic H6/8 and sugar H1′ protons (not shown). Cross-peaks were not observed for C6 in all regions of the ^1H-^{31}P CPMG-HSQC-NOESY spectrum and thus its ^{31}P resonance could not be assigned. The assignment of ^{31}P resonances using H3′-^{31}P cross-peaks for residues A1-T17 is shown in Figure 17.43 (top panel).

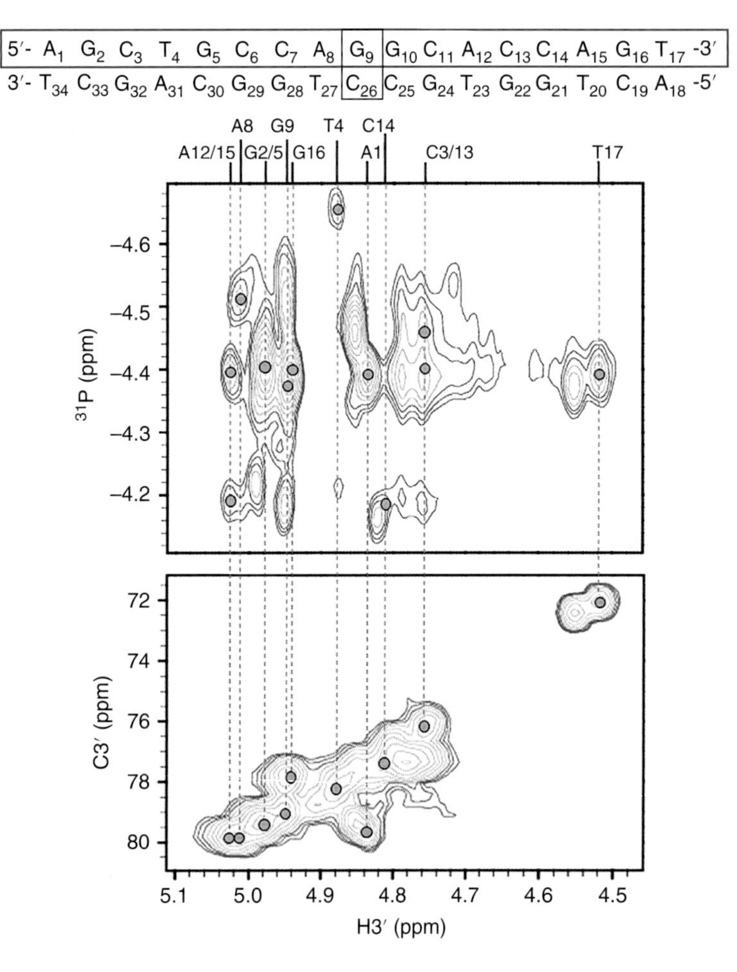

Figure 17.43 (Top panel) 600 MHz ^{1}H-^{31}P CPMG-HSQC-NOESY (τ_{mix} = 500 ms) spectrum and (bottom panel) 600 MHz ^{1}H-^{13}C HSQC spectrum. The assignment of ^{31}P resonances is shown by means of H3′-^{31}P HSQC correlations for residues A1-T17 in 17mer canonical GC DNA (1 mM, 100% $^{2}H_2O$, 50 mM PO_4^{3-} and 50 mM NaCl, pH 6.2) measured at 25 °C. Unlabelled peaks correspond to assigned NOESY and HSQC peaks for residues A18-T34.

17.7.4 Table of Assignments

Based on the identification and assignment procedures described above, the chemical shifts of exchangeable and non-exchangeable proton, carbon and phosphorus resonances of 17mer GC DNA are summarised in Tables 17.4 and 17.5.

17.7.5 Determination of Experimental Constraints for Structure Calculation

NMR distance constraints were extracted from the $^{1}H_2O$ and $^{2}H_2O$ NOESY experiments (τ_m = 250 ms) measured at 2 and 25 °C, respectively, using the CcpNMR analysis program [58]. Distances were grouped as strong (1.5–2.5 Å), medium (2.6–4.0 Å) and weak (4.1–6.0 Å), which are determined by CcpNMR [58] analysis with respect to the reference distances stated in Section 17.4.2.2.

Dihedral angle constraints were constrained according to the sugar pucker type determined from $^{3}J_{H1'\text{-}H2'}$ coupling constants in the DQF-COSY spectrum and were assigned based on standard B-DNA values.

Table 17.4 Summary of chemical shifts for ^{1}H proton resonances in 17mer canonical GC DNA.

Nucleotide number	Imino H1	Amino $NH_2(1)$	Amino $NH_2(2)$	H2	$CH_3(5)$	H5	H6	H8	H1′	H2′	H2″	H3′	H4′	H5′	H5″
A1		8.14	6.24	8.03				8.01	5.98	2.46	2.63	4.84	4.21	3.69	3.69
G2	12.83	6.84						7.89	5.87	2.70	2.70	4.98	4.41	4.26	4.11
C3		8.14	5.99			5.31	7.43		6.00	2.06	2.52	4.75	4.27	4.10	4.18
T4	13.97				1.60		7.33		5.81	2.11	2.49	4.88	4.15		4.09
G5	12.83	6.40						7.88	5.85	2.64	2.68	4.98	4.38	4.08	4.13
C6		8.14	5.83			5.30	7.34		5.89	2.02	2.39	4.77	4.19	4.01	4.12
C7		8.53				5.56	7.43		5.29	1.99	2.27	4.79	4.03	3.97	
A8		7.91	6.38	7.58				8.13	5.95	2.69	2.86	5.01	4.35	4.14	3.96
G9	12.85	6.39						7.59	5.54	2.50	2.63	4.95	4.33		4.15
G10	12.84	6.39						7.58	5.81	2.46	2.66	4.91	4.34		4.15
C11		8.24	5.79			5.27	7.28		5.57	1.97	2.36	4.80	4.12		4.04
A12		7.91	6.07	7.66				8.21	6.19	2.67	2.87	5.00	4.40	4.25	4.13
C13		8.07	5.80			5.23	7.23		5.79	1.96	2.36	4.75	4.14	4.26	4.04
C14		8.51	5.89			5.52	7.42		5.35	1.98	2.29	4.81	4.06	3.99	
A15		7.92	6.35	7.66				8.17	6.01	2.71	2.85	5.03	4.38	3.98	4.11
G16	12.91	6.67	6.24					7.73	5.94	2.53	2.66	4.94	4.37	4.07	4.22
T17	13.97				1.55		7.37		6.20	2.21	2.21	4.52	4.07		4.23
A18				7.92				8.23	6.25	2.71	2.81	4.85	4.27	3.75	3.75
C19		8.25	5.98			5.44	7.51		5.92	2.12	2.46	4.78	4.24	4.19	4.10
T20	13.96				1.51		7.31		5.74	2.06	2.43	4.86	4.15	4.07	4.09
G21	12.84	6.39						7.85	5.69	2.70	2.74	4.99	4.36	4.20	4.13
G22	12.77	6.51	6.80					7.65	5.93	2.50	2.73	4.89	4.40	4.10	4.22
T23	13.62				1.31		7.13		5.83	2.02	2.47	4.85	4.16	4.22	4.11
G24	12.65	6.41						7.82	5.82	2.59	2.66	4.95	4.34	4.10	4.15
C25		8.07	5.82			5.26	7.37		5.94	2.16	2.47	4.78	4.23		4.13
C26		8.37	5.83			5.50	7.49		5.87	2.03	2.43	4.76	4.12	4.23	4.07
T27	13.97				1.59		7.29		5.63	2.01	2.35	4.82	4.09	4.01	
G28	12.94	6.84						7.82	5.64	2.66	2.74	4.98	4.32	4.15	4.09
G29	12.86	6.90						7.71	5.80	2.51	2.66	4.95	4.36		4.15
C30		8.29	5.85			5.30	7.29		5.50	1.95	2.31	4.81	4.10	3.99	
A31		7.90	6.25	7.58				8.14	6.02	2.71	2.87	5.02	4.37	4.19	4.11
G32	12.83	6.56						7.64	5.75	2.48	2.62	4.95	4.35		4.19
C33		8.25	5.69			5.38	7.43		6.06	2.20	2.46	4.75	4.21	4.06	4.13
T34	13.36				1.71		7.59		6.26	2.29	2.29	4.55	4.08		

Boxes shaded in grey indicate the absence of the proton in the specified residue and those shaded in blue indicate unassigned protons.

Finally, hydrogen bond constraints were included in the structure calculation to loosely constrain base pairs, which could be identified through imino–imino proton NOE connectivities in the $^{1}H_2O$ NOESY spectrum.

17.7.6 Structure Calculation

Experimental constraints used for structure calculations using the Xplor-NIH program are given in Table 17.6.

Table 17.5 Summary of chemical shifts for ^{13}C and ^{31}P resonances in 17mer canonical GC DNA.

Nucleotide number	C2	C5	C6	C8	C1′	C2′	C3′	C4′	C5′	^{31}P
A1	154.7			142.4	84.9	39.6	79.7	89.6	64.4	−4.41
G2				138.0	84.4	40.2	79.3	87.4		−4.43
C3		98.5	143.3		87.2	40.0	76.1	85.8		−4.43
T4		14.7	139.2		85.5	39.2	78.1	85.8		−4.68
G5				138.0	84.4	40.4	79.3	87.3		−4.43
C6		98.3	142.4		86.5	40.0	77.4	85.7		
C7		98.5	142.5		86.5	39.4	76.3	85.80		−4.18
A8	154.2			141.5	84.3	40.3	79.8	86.5		−4.54
G9				137.3	83.7	40.8	79.1	86.9		−4.38
G10				137.3	84.4	39.6	78.7	86.9		−4.38
C11		98.1	142.4		86.1	39.8	76.7	85.5		
A12	154.4			141.4	85.1	40.4	79.8	87.4		−4.42
C13		98.2	142.1		85.8	39.8	76.1	85.8		−4.48
C14		98.5	142.5		86.5	39.8	77.4	85.80		−4.21
A15	154.8			141.5	84.9	40.3	79.8	87.3		−4.22
G16				137.7	84.3	40.8	77.7	87.2		−4.42
T17		14.0	139.4		86.5	41.3	72.1	87.2		−4.41
A18	154.8			142.4	86.8	40.2	79.5	89.3	64.3	−4.48
C19		98.5	142.8		86.9	39.2	77.4	85.8		−4.49
T20		14.7	139.3		85.8	39.0	77.9	85.8		−4.36
G21				138.4	84.0	40.2	79.3	86.9		−4.24
G22				137.2	85.1	41.2	78.4	87.4		−4.50
T23		14.3	138.2		85.4	39.7	77.5	85.8		−4.58
G24				138.2	84.4	40.8	79.1	86.9		−4.38
C25		98.1	142.5		86.7	40.2	77.4	85.8		−4.41
C26		98.7	142.8		86.7	39.9	77.4	85.6		−4.48
T27		14.7	139.3		85.8	39.7	77.5	85.8		−4.41
G28				138.2	84.2	40.1	79.3	86.9		−4.43
G29				137.2	84.4	40.8	79.1	87.0		−4.40
C30		98.2	142.4		86.2	39.7	77.4	85.5		−4.40
A31	154.2			141.5	84.9	40.4	79.8	87.2		−4.42
G32				137.2	84.0	40.8	77.7	86.9		−4.21
C33		98.4	142.5		86.7	40.5	76.1	85.5		−4.38
T34		14.4	140.0		86.8	41.5	72.5	87.2		−4.40

Boxes shaded in grey indicate the absence of the resonance in the specified residue and those shaded in blue indicate unassigned resonances. Boxes shaded in red indicate that the chemical shifts of C5′ resonances are highly overlapped and between 66.2 and 69.0 ppm.

From the 100 structures generated by Xplor-NIH, 10 were selected based on their lowest energy and lowest root mean square deviation (RMSD = 0.50), from which the average structure was generated. A good planarity of base pairs can be clearly identified and a canonical B-DNA type conformation is adopted (Figure 17.44).

Table 17.6 Summary of experimental constraints used in the structure calculation of 17mer canonical GC DNA using the Xplor-NIH program [54, 55].

Constraint type	No. of constraints
Total NOEs	486
Strong (1.5–2.5 Å)	48
Medium (2.6–4.0 Å)	382
Weak (4.1–6.0 Å)	56
Hydrogen bond	102
Total dihedral angle	306
Helix (α, β, γ, ε, ζ)	204
Glycosidic (χ)	34
Ribose pucker (υ_1, υ_2, δ)	68
NOE constraints per residue	14.3
NOE and dihedral constraints per residue	23.3

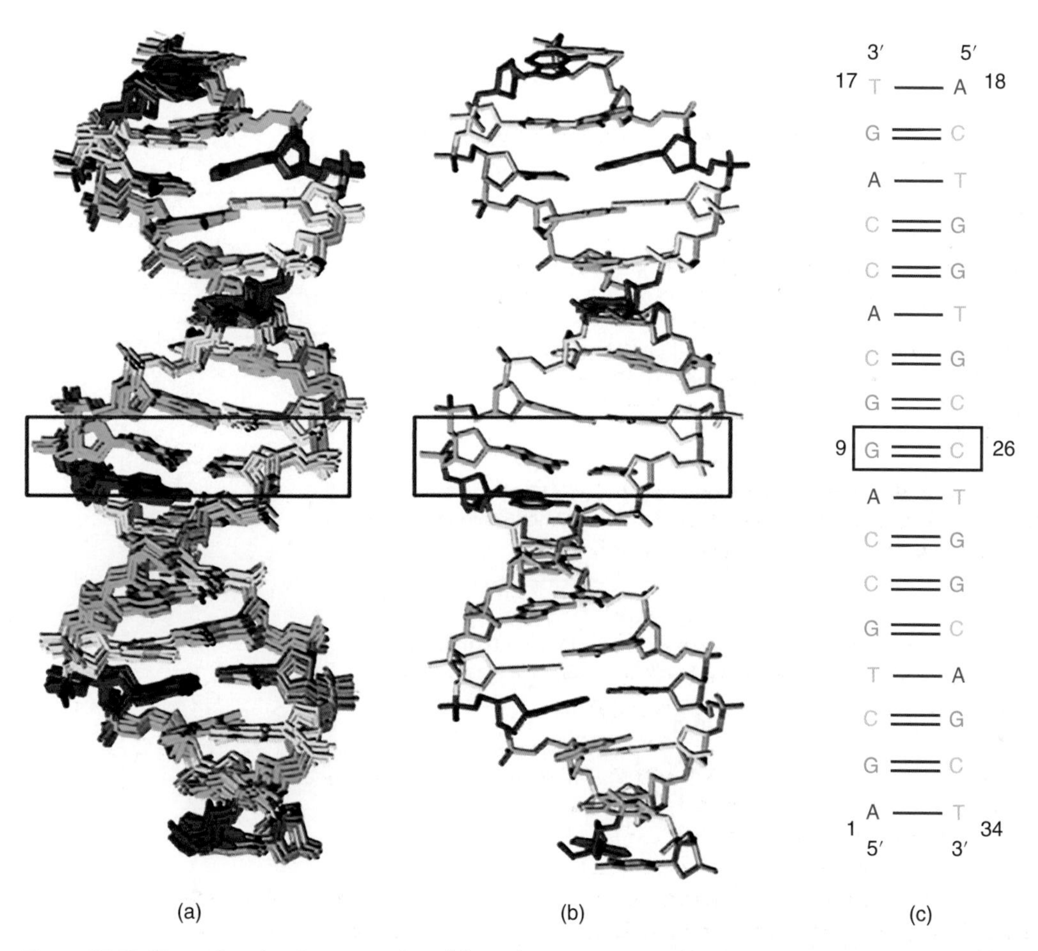

Figure 17.44 Illustration showing an overlay of the 10 lowest energy and lowest RMSD NMR structures (a) and the final average structure for 17mer GC DNA (b) calculated from Xplor-NIH. The sequence is shown on the right (c) and the position of the G9-C26 base pair is outlined by the purple box.

17.7.7 Conformational Analysis

To investigate the quality of the canonical 17mer GC DNA NMR structure, the helical parameters and dihedral angle values obtained for the final NMR structure were compared with those for a canonical 12mer DNA solution structure (PDB: 1DUF). Both structures were run through a w3DNA program and the average values are compared in Tables 17.7 to 17.11.

Overall, the values for the 17mer GC DNA show good similarity when compared to the 12mer DNA. Ideally, the larger 17mer GC DNA structure will be more susceptible to dynamic movement and thus greater flexibility. This will affect not only the backbone dihedral angles but also the position of adjacent base pairs. This accounts for the large differences for the roll, x-displacement and inclination values between the two DNA structures.

Differences of 14.9 and 20.5 for the dihedral angles, δ and ε, respectively, show clear evidence of this as they lie directly along the helical backbone in DNA. A large difference was also observed in the value of ν_4, which contributes to the sugar pucker itself. However, w3DNA analysis showed that despite the large difference of

Table 17.7 Comparison of local base pair parameters between a 12mer canonical GC DNA (PDB: 1DUF), 17mer canonical GC DNA (Case Study 1) and 13mer 6-thioguanine modified GC DNA (Case Study 2).

	Average value		
Helical parameter	**12mer DNA**	**17mer canonical GC DNA (Case Study 1)**	**13mer 6-TG modified GC DNA (Case Study 2)**
Shear (S_x)	0.00	−0.02	0.02
Stretch (S_y)	−0.39	−0.28	−0.31
Stagger (S_z)	−0.28	0.00	0.35
Buckle (κ)	−0.02	1.27	−2.26
Propeller twist (π)	−11.76	0.41	5.65
Opening (σ)	0.99	−1.30	−4.39

Values were calculated using w3DNA and for 17mer canonical GC DNA are based on the average structure shown in Figure 17.44b in Case Study 1. For 13mer 6-TG modified GC DNA, the values are based on the average structure shown in Figure 17.49b in Case study 2.

Table 17.8 Comparison of local base pair step parameters between a 12mer canonical GC DNA (PDB: 1DUF), 17mer canonical GC DNA (Case Study 1) and 13mer 6-thioguanine modified GC DNA (Case Study 2).

	Average value		
Helical parameter	**12mer DNA**	**17mer canonical GC DNA (Case Study 1)**	**13mer 6-TG modified GC DNA (Case Study 2)**
Shift (D_x)	0.01	−0.02	−0.01
Slide (D_y)	−0.46	−0.63	−0.98
Rise (D_z)	3.41	3.58	3.65
Tilt (T_τ)	0.02	0.03	−0.76
Roll (P_ρ)	2.32	5.72	−1.33
Twist (Ω)	34.45	34.29	34.11

Values were calculated using w3DNA and for 17mer canonical GC DNA are based on the average structure shown in Figure 17.44b in Case Study 1. For 13mer 6-TG modified GC DNA, the values are based on the average structure shown in Figure 17.49b in Case Study 2.

Table 17.9 Comparison of local base pair helical parameters between a 12mer canonical GC DNA (PDB: 1DUF), 17mer canonical GC DNA (Case Study 1) and 13mer 6-thioguanine modified GC DNA (Case Study 2).

	Average value		
Helical parameter	**12mer DNA**	**17mer canonical GC DNA (Case Study 1)**	**13mer 6-TG modified GC DNA (Case Study 2)**
x-displacement (d_x)	−1.28	−2.04	−1.56
y-displacement (d_y)	−0.01	0.03	−0.05
Inclination (η)	4.41	9.53	−2.20
Tip (θ)	−0.04	−0.06	1.38
Helical rise	3.32	3.40	3.57
Helical twist	35.01	35.04	35.00

Values were calculated using w3DNA and for 17mer canonical GC DNA are based on the average structure shown in Figure 17.44b in Case Study 1. For 13mer 6-TG modified GC DNA, the values are based on the average structure shown in Figure 17.49b in Case Study 2.

Table 17.10 Comparison of backbone dihedral angles between a 12mer canonical GC DNA (PDB: 1BNA), 17mer canonical GC DNA (Case Study 1) and 13mer 6-thioguanine modified GC DNA (Case Study 2).

	Average value		
Dihedral angle	**12mer DNA**	**17mer canonical GC DNA (Case Study 1)**	**13mer 6-TG modified GC DNA (Case Study 2)**
α	−62.8	−64.2	−70.7
β	−169.3	−169.1	−155.2
γ	59.5	48.3	58.9
δ	122.7	137.6	136.0
ε	164.2	143.7	131.4
ζ	−105.0	−99.8	−102.2
χ	−117.3	−110.1	−116.5

Values were calculated using w3DNA and for 17mer canonical GC DNA are based on the average structure shown in Figure 17.44b in Case Study 1. For 13mer 6-TG modified GC DNA, the values are based on the average structure shown in Figure 17.49b in Case Study 2.

Table 17.11 Comparison of sugar ring dihedral angles between a 12mer canonical GC DNA (PDB: 1BNA), 17mer canonical GC DNA (Case Study 1) and 13mer 6-thioguanine modified GC DNA (Case Study 2).

	Average value		
Dihedral angle	**12mer DNA**	**17mer canonical GC DNA (Case Study 1)**	**13mer 6-TG modified GC DNA (Case Study 2)**
ν_0	−38.7	−22.4	−27.2
ν_1	40.8	32.3	35.8
ν_2	−30.5	−29.6	−30.3
ν_3	17.3	17.6	15.4
ν_4	21.7	2.8	7.1
Amplitude	46.4	32.3	35.2
Phase	127.6	156.5	150.0

Values were calculated using 3DNA and for 17mer canonical GC DNA are based on the average structure shown in Figure 17.44b in Case Study 1. For 13mer 6-TG modified GC DNA, the values are based on the average structure shown in Figure 17.49b in Case Study 2.

18.9, all the sugar groups in the 17mer canonical GC DNA had a $C_{2'}$-endo conformation, which is expected in B-conformation DNA.

In terms of the helical parameters, greater dynamic movement would be supported primarily by large differences in the local base pair step and helical parameters as these describe the relationships between adjacent base pairs rather than look at each base pair individually. As the data shows, a noticeable difference was observed for x-displacement (0.76), but the differences observed for roll (3.4) and inclination (5.12) are far more significant and give further confirmation of the greater dynamic nature of the 17mer GC DNA structure when compared to the 12mer DNA.

17.8 Case Study 2: NMR Structure Determination and Conformational Analysis of 13mer 6-Thioguanine Modified GC DNA [70–74]

17.8.1 Structure of 6-Thioguanine DNA [70, 71]

6-thioguanine (6-TG) differs from guanine as the oxygen atom on the carbonyl group is replaced by a sulphur atom (Figure 17.45). 6-TG is well known in medical research as an anti-leukaemia agent, but prolonged exposure to UVA light causes oxidation of the sulphur atom, which abolishes its anticancer properties [70, 71].

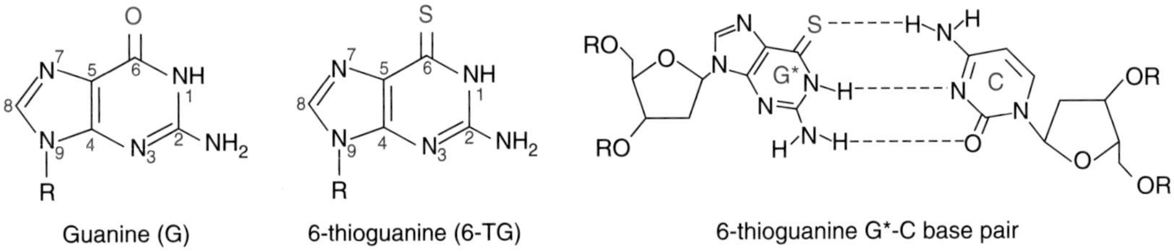

Guanine (G) 6-thioguanine (6-TG) 6-thioguanine G*-C base pair

Figure 17.45 (Left and centre) Comparison between guanine (G) and 6-thioguanine (6-TG) base structures. (Right) This shows that canonical base-pairing is conserved even when guanine is substituted with 6-TG.

When compared to oxygen, Sulphur is a poor hydrogen bond acceptor and this is observed in a 6-TG substituted GC base pair [72]. The weakening of the hydrogen bond between S6 on 6-TG and the hydrogen bonded amino proton on the opposite cytosine results in approximately a 10° opening of the base pair towards the major groove. The extent of the weakening of the hydrogen bond was investigated by Guerra et al. [73], where ab initio theory was used to calculate the bonding energies and charge distribution of AT and GC base pair mimics in which the O· · ·H—N hydrogen bonds were modified to S· · ·H—N hydrogen bonds. A number of mimics were used where the number of sulphur substitutions varied.

The study showed that the substitution of sulphur in AT base pairs caused a decrease in the interaction energy of the S4· · ·H61—N hydrogen bond (ΔE_{int} = −11.1 kcal/mol) when compared to a Watson-Crick O4· · ·H61—N hydrogen bond (ΔE_{int} = −13.0 kcal/mol) [74]. This change was also accompanied by an increase in the bond length from 2.85 to 3.35 Å, most likely caused by the sulphur atom being larger than the oxygen atom. The increased S4· · ·H61—N bond length also caused the second hydrogen bond (N1· · ·H3) to increase by 0.05 Å.

17.8.2 NMR Assignment of 13mer 6-Thioguanine Modified GC DNA

In this case study, the truncated sequence of the 13mer thioDNA is derived from the 17mer GC DNA (Case Study 1) by deleting the initial two and last two bases of the latter DNA and selectively modifying the middle guanine base G7 to a thio G7 ($O^6G \rightarrow S^6G$), as shown in Figure 17.46. The thio modified base G is shown in

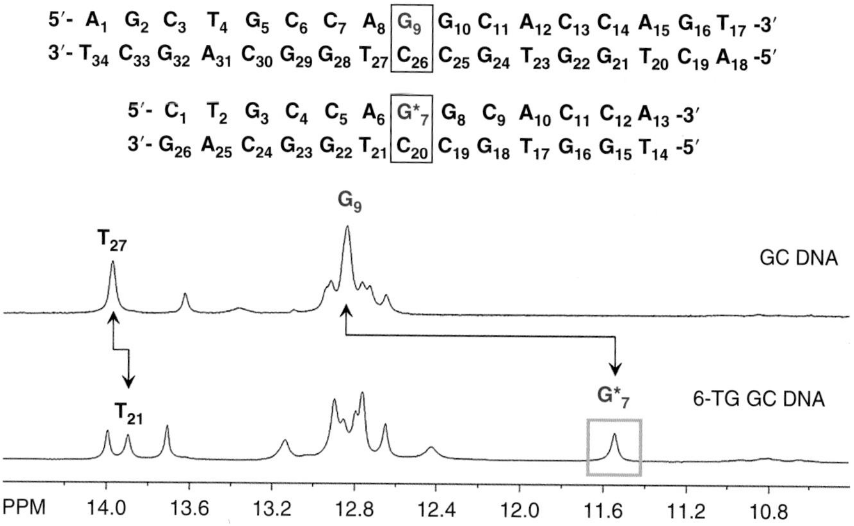

5′- A_1 G_2 C_3 T_4 G_5 C_6 C_7 A_8 G_9 G_{10} C_{11} A_{12} C_{13} C_{14} A_{15} G_{16} T_{17} -3′
3′- T_{34} C_{33} G_{32} A_{31} C_{30} G_{29} G_{28} T_{27} C_{26} C_{25} G_{24} T_{23} G_{22} G_{21} T_{20} C_{19} A_{18} -5′

5′- C_1 T_2 G_3 C_4 C_5 A_6 G^*_7 G_8 C_9 A_{10} C_{11} C_{12} A_{13} -3′
3′- G_{26} A_{25} C_{24} G_{23} G_{22} T_{21} C_{20} C_{19} G_{18} T_{17} G_{16} G_{15} T_{14} -5′

Figure 17.46 Stack of 800 MHz 1D ^{1}H-NMR spectra showing a comparison of canonical 17mer GC DNA (1 mM, 90% ^1H_2O + 10% ^2H_2O, 50 mM PO_4^{3-} and 50 mM NaCl, pH 6.2) and 13mer 6-thioguanine (6-TG) modified GC DNA (0.8 mM, 90% ^1H_2O + 10% ^2H_2O, 50 mM PO_4^{3-} and 50 mM NaCl, pH 6.2) measured at 2 °C. The 6-thioguanine residue is highlighted and assigned in red in both the sequence and bottom spectrum.

red with an asterisk. Except for these two changes, the remainder of the sequence is identical between the two DNAs.

The NMR sample preparation, data acquisition, processing and analysis of the 13mer 6-TG modified GC DNA were carried out in a similar manner to that of 17mer GC DNA in Case Study 1.

17.8.2.1 Identification and Assignment of Exchangeable Proton Resonances

Hydrogen bonding in a 6-TG modified GC base pair is similar to a canonical GC base pair so a B-form DNA structure was still expected to be observed.

All imino proton resonances were observed in the expected chemical shift range with the exception of the 6-TG modified G7 residue. The chemical shift of the 6-TG imino proton was expected to change due to a change in the base pairing and this difference ($\Delta\delta = 1.32$ ppm) is clearly observed in its 1D ^{1}H-NMR spectrum when compared to that of canonical GC DNA (Figure 17.46).

A key observation in the assignment of imino protons was the broad peak at 12.43 ppm in Figure 17.46, which could not be identified through imino–imino NOE connectivities. The peak was most probably due to a minor secondary conformation of the 6-TG modified base (G7), as evidenced by its disappearance after the addition of *E. coli* MutS protein [17].

Although the imino proton corresponding to the G7 base is clearly shifted to the high field, its pattern of NOE cross-peaks in the NOESY spectrum was similar to that observed for a canonical GC base pair (Figure 17.47). This provides convincing evidence that the 6-TG modified residue base pair interacts in a similar way to a standard guanine base.

The imino proton resonances were assigned following the imino–imino proton sequential connectivity starting from G7 ($\delta = 11.54$ ppm), which appeared clearly separate from other resonances. Assignment was unambiguous as sequential connectivity to G8 and T21 were clearly resolved and observed. Twelve imino proton resonances were assigned in the NOESY spectrum; only T14 remained unassigned. The sequential connectivity from G7 to G26 is shown in Figure 17.48.

Imino proton assignments were confirmed by assigning NOEs to proximal base-paired amino protons, which helped identify the previously unassigned T14 residue. Another distinct feature of this region was the

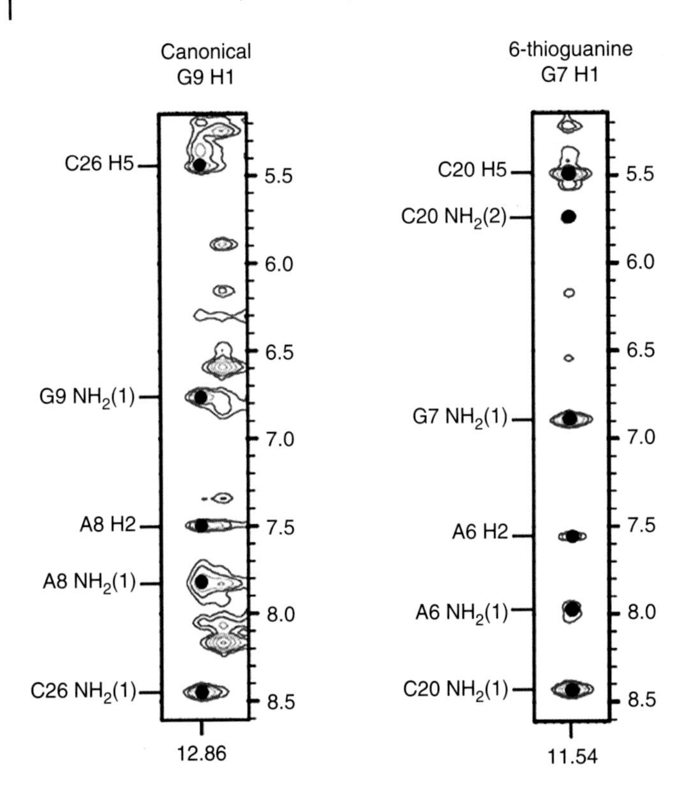

Figure 17.47 (Left panel) 700 MHz NOESY (τ_{mix} = 250 ms) panel for residue G9 in 17mer canonical GC DNA measured at 2 °C. (Right panel) 800 MHz NOESY (τ_{mix} = 250 ms) panel for residue G7 in 13mer 6-TG modified GC DNA measured at 2 °C. A comparison illustrating the similarities in base pair NOE cross-peaks observed from the respective imino protons.

strong NOE between the imino proton of the 6-TG modified imino G7 residue and the hydrogen bonded cytosine NH_2 of C20 to which it is base-paired (Figure 17.47).

The identification and assignment of the non-exchangeable proton, carbon and phosphorus resonances of 13mer 6-TG modified GC DNA were carried out using the same methodology as discussed before for 17mer GC DNA (Section 17.7.3.2)

17.8.3 Table of Assignments

Based on the identification and assignment procedures described above, the chemical shifts of exchangeable and non-exchangeable proton, carbon and phosphorus resonances of 13mer 6-TG modified GC DNA are summarised in Tables 17.12 and 17.13.

17.8.4 Structure Calculation

Experimental distance and dihedral angle constraints were obtained using the CcpNMR [58] analysis program as outlined in Sections 17.6.3 and 17.6.4. The number of NMR experimental constraints used is summarised in Table 17.14 for the structure calculation of 6-TG modified 13mer GC DNA using the Xplor-NIH program.

No constraints were included for H4′, H5′ and H5″ NOEs as they were too overlapped in the spectrum. For the same reason, NOE correlations between sugar protons were not included in the structure calculation with the exception of H1′-H2′/2″, H1′-H3′ and H2′/2″-H3′ whose NOEs were well-resolved.

The best 10 structures show a good overlay (RMSD = 0.46) of B-form DNA duplexes (Figure 17.49a). The average structure (Figure 17.49b) was then generated from these 10 structures and the results of these two stages are shown in Figure 17.49.

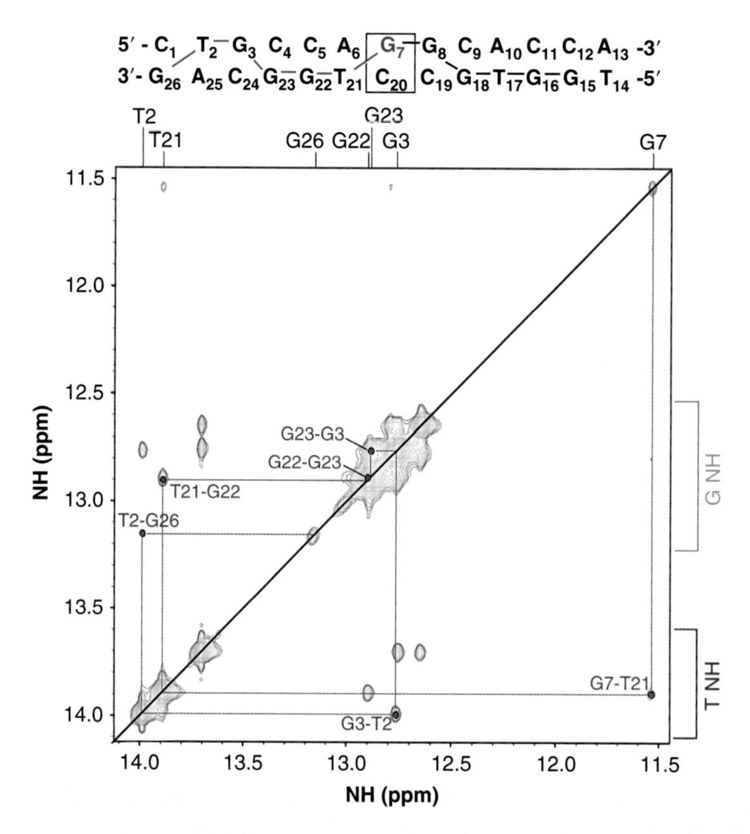

Figure 17.48 800 MHz NOESY (τ_m = 250 ms) spectrum of 13mer 6-TG modified GC DNA (0.8 mM, 90% ^1H_2O + 10% ^2H_2O, 50 mM PO_4^{3-} and 50 mM NaCl, pH 6.2) at 2 °C. The respective imino protons were identified and assigned based on the sequential connectivity stretching from G7 to G26, which is indicated by the pink trace.

17.8.5 Conformational Analysis

Conformational analysis was carried out using w3DNA on the final structure obtained from Xplor-NIH (Figure 17.49b) and the helical parameters were analysed to monitor and compare both DNA structures analysed in Case Study 1 (Tables 17.7 to 17.11).

Since the inclusion of a modified 6-TG base was the only selective change in the structure compared to canonical GC DNA, the helical parameters around this base pair are of great interest. There were large changes to the values of buckle and opening when compared to both DNA structures in Case Study 1. The data comparing each individual base pair (not shown) indicated a significant change in the buckle value from the modified G*7-C20 base pair to the succeeding base pairs G8-C19 and C9-G18 (total difference 39.49). In support of this, the modified G*7-C20 base pair had an opening value (−13.46) that is much larger than the average value across the whole DNA helix of −4.39. These differences show that the effect of the substitution of the sulphur atom in guanine is not only local but spreads further across the structure. Large differences observed for the tilt and tip parameters give further evidence that the modified base pair does affect the conformation of the DNA, as reported in earlier literature [72–74].

The backbone dihedral angles given by w3DNA for the 6-TG modified 13mer GC DNA showed good consistency with the DNA structures in Case Study 1. The values for ε varied the most but all the nucleotides were $C_{2'}$-*endo* resulting in a B-form DNA conformation.

Table 17.12 Summary of chemical shifts for ^{1}H proton resonances in 13mer 6-TG modified GC DNA.

Nucleotide number	Imino H1	Amino NH_2(1)	Amino NH_2(2)	H2	CH_3 (5)	H5	H6	H8	H1′	H2′	H2″	H3′	H4′	H5′	H5″
C1		7.86	5.82			5.97	7.86		5.92	2.18	2.57	4.68	4.12	3.78	3.82
T2	13.98				1.71		7.52		5.82	2.20	2.53	4.90	4.20	4.07	
G3	12.76	6.43						7.95	5.90	2.69	2.71	5.00	4.40	4.09	4.16
C4		8.13	5.86			5.35	7.41		5.95	2.08	2.44	4.80	4.23		
C5		8.44	5.73			5.58	7.44		5.34	1.96	2.27	4.80	4.05		
A6		8.01	6.18	7.56				8.12	5.83	2.69	2.83	5.00	4.35	4.04	
G7	11.54	6.90						7.85	5.66	2.57	2.78	4.97	4.37	4.16	
G8	12.79	6.52						7.58	5.83	2.44	2.66	4.91	4.36	4.17	
C9		8.20	5.76			5.28	7.31		5.56	2.01	2.37	4.82	4.14		
A10		7.84	6.11	7.71				8.24	6.19	2.70	2.86	5.01	4.41	4.12	4.26
C11		8.11	5.74			5.30	7.27		5.84	1.92	2.32	4.75	4.12		
C12		8.12	5.75			5.64	7.45		5.76	2.02	2.30	4.79	4.07		
A13		7.91	6.08	7.73				8.25	6.34	2.48	2.67	4.69	4.19	4.08	
T14	13.71				1.58		7.34		5.87	1.79	2.24	4.62	4.03	3.65	3.65
G15	12.84	7.03						7.98	5.65	2.76	2.79	4.96	4.34	3.97	4.05
G16	12.76	6.60						7.78	6.01	2.59	2.78	4.94	4.42	4.19	
T17	13.71				1.37		7.17		5.86	2.03	2.46	4.88	4.19	4.15	4.22
G18	12.63	6.30						7.86	5.80	2.63	2.63	4.97	4.35		
C19		7.96	5.72			5.29	7.39		5.97	2.26	2.47	4.79	4.26		
C20		8.43	5.73			5.50	7.52		5.82	2.00	2.43	4.72	4.10		
T21	13.89				1.62		7.38		5.68	2.08	2.41	4.85	4.14	4.06	4.04
G22	12.90	6.82						7.73	5.67	2.68	2.77	4.99	4.36	4.04	4.11
G23	12.89	6.35						7.73	5.82	2.53	2.67	4.96	4.37		
C24		8.21	5.86			5.38	7.32		5.47	1.90	2.25	4.79	4.09		
A25		7.97	6.41	7.70				8.17	6.04	2.69	2.86	5.01	4.37	3.99	4.10
G26	13.15	7.11						7.71	6.03	2.27	2.44	4.63	4.17	4.12	4.24

Boxes shaded in grey indicate the absence of the proton in the specified residue and those shaded in blue indicate unassigned protons.

17.9 Conclusion

We have demonstrated that homo- and heteronuclear 2D NMR methods can be successfully applied to determine the three-dimensional structure and conformational analysis of DNA duplexes in the solution state. The results of the two case studies highlight the importance of prior knowledge of the structure of uncomplexed DNA before embarking on the structure determination of DNA–protein or DNA–drug complexes.

Equipped with the NMR structure of DNA, one can characterise the base pair opening and closing kinetics by measuring the rates of exchange of the individual imino proton resonances of unmodified and selectively modified DNAs [75]. A combined knowledge of the structure and kinetics of DNA can provide a valuable insight into the mode of recognition of base pairs by DNA repair proteins such as *E. coli* MutS.

For NMR studies of larger DNA molecules (>25mers), isotope (^{13}C, ^{15}N) enrichment [76] of DNA is required for an unambiguous resonance assignment by heteronuclear 3D NMR experiments [21] as being done for isotopically labelled proteins [12–14] and RNAs [77, 78].

Table 17.13 Summary of chemical shifts for ^{13}C and ^{31}P resonances in 13mer 6-TG modified GC DNA.

Nucleotide number	C2	C5	C6	C8	C1′	C2′	C3′	C4′	C5′	^{31}P
C1		99.3	143.3		88.4	40.7	76.7	88.2	63.0	−4.59
T2		14.6	139.6		87.0	39.3	78.7	86.4		−4.46
G3				138.4	84.3	40.3	79.6	87.2		−4.45
C4		98.5	142.6		86.6	40.0	77.3	85.8		−4.58
C5		98.7	143.3		86.4	39.6	77.2	85.7		−4.45
A6	154.4			141.4	84.8	40.2	79.6	87.6		−4.45
G7				138.0	83.9	40.8	78.8	87.6		−4.57
G8				137.2	84.8	41.5	78.7	87.2		−4.59
C9		98.1	142.5		86.2	39.9	76.8	85.6		−4.46
A10	154.9			141.5	85.2	40.7	79.6	87.4		−4.59
C11		98.5	142.2		86.0	39.8	75.7	85.1		−4.39
C12		98.8	143.5		88.7	39.4	77.9	85.8		−4.24
A13	154.8			142.2	85.7	42.0	73.0	88.0		−4.25
T14		14.1	139.2		87.5	39.4	78.8	88.8	64.0	−4.52
G15				138.6	84.7	39.6	79.1	87.6		−4.59
G16				137.8	84.9	40.9	79.0	87.2		−4.33
T17		14.7	138.4		85.5	39.4	77.6	85.7		−4.64
G18				138.1	84.3	40.5	79.1	87.6		−4.57
C19		98.2	142.5		86.7	39.6	76.8	86.0		−4.39
C20		98.7	142.5		86.1	40.2	76.1	85.4		−4.49
T21		14.1	139.6		85.9	39.2	78.0	86.0		−4.22
G22				137.6	84.3	40.3	79.6	87.4		−4.54
G23				137.6	84.4	40.9	79.2	87.0		−4.32
C24		98.3	142.6		86.2	39.8	77.2	85.4		−4.42
A25	154.8			141.7	85.1	40.2	79.6	87.0		−4.45
G26				138.5	84.3	42.4	72.4	87.4		−4.47

Boxes shaded in grey indicate the absence of the resonance in the specified residue. Boxes shaded in red indicate that the chemical shifts of C5′ resonances are highly overlapped and appear between 65.9 and 69.1 ppm.

Table 17.14 Summary of NMR experimental constraints used in the structure calculation of 13mer 6-TG modified GC DNA carried out with the Xplor-NIH program.

Constraint type	No. of constraints
Total NOEs	368
Strong (1.5–2.5 Å)	30
Medium (2.6–4.0 Å)	311
Weak (4.1–6.0 Å)	27
Hydrogen bond	78
Total dihedral angle	252
Helix ($\alpha, \beta, \gamma, \varepsilon, \zeta$)	174
Glycosidic (χ)	26
Ribose pucker ($\upsilon_1, \upsilon_2, \delta$)	52
NOE constraints per residue	14.1
NOE and dihedral constraints per residue	23.8

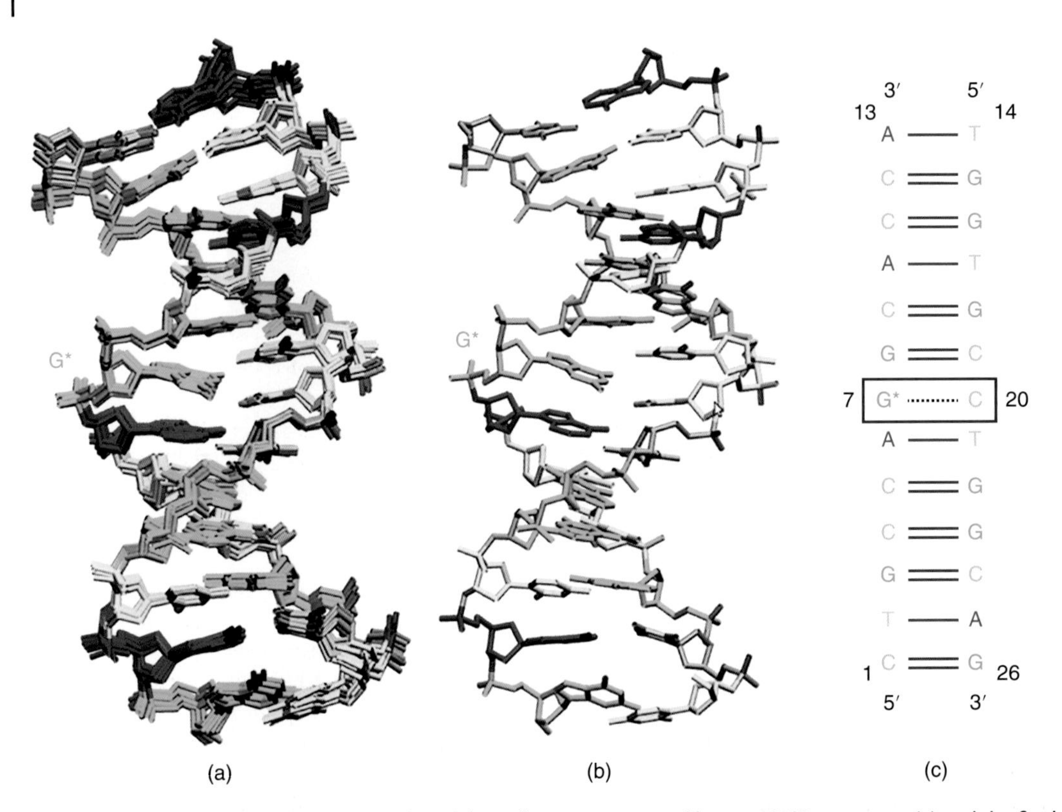

Figure 17.49 Illustration showing an overlay of the 10 lowest energy and lowest RMSD structures (a) and the final structure (b) for 13mer 6-TG modified GC DNA calculated from Xplor-NIH. The sequence of the thio-modified DNA is shown on the right (c) and the position of the 6-TG (G*) base is highlighted on the structures.

References

1 Emsley, J.W. and Feeney, J. (2007). Forty years of progress in nuclear magnetic resonance spectroscopy. *Prog. Nucl. Magn. Reson. Spectrosc.* 50 (4): 179–198.

2 Shoolery, J.N. (1995). The development of experimental and high resolution NMR. *Prog. Nucl. Magn. Reson. Spectrosc.* 28: 37–52.

3 Ernst, R.R., Bodenhausen, G., and Wokaun, A. (1987). *Principles of Nuclear Magnetic Resonance in One and Two Dimensions*. Oxford: Clarendon Press.

4 P. Lauterbur and P. Mansfield, The Nobel Prize in Physiology or Medicine 2003, http://www.nobelprize .org/nobel_prizes/medicine/laureates/2003.

5 Wüthrich, K. (1986). *NMR of Proteins and Nucleic Acids*. New York: Wiley.

6 Wüthrich, K. (1995). *NMR in Structural Biology, World Scientific Series in 20th Century Chemistry*, vol. 5. World Scientific.

7 Billeter, M., Kline, A.D., Braun, W. et al. (1989). Comparison of the high-resolution structures of the alpha-amylase inhibitor tendamistat determined by nuclear magnetic resonance in solution and by X-ray diffraction in single crystals. *J. Mol. Biol.* 206: 677–687.

8 Lin, Y., Schiavo, S., Orjala, J. et al. (2008). Microscale LC-MS-NMR platform applied to the identification of active cyanobacterial metabolites. *Anal. Chem.* 80: 8045–8054.

9 Ramesh, V. (ed.) (2016). *Nuclear Magnetic Resonance*, vol. 45. Cambridge: Specialist Periodical Reports, RSC.

10 Markley, J.L., Brüschweiler, R., Edison, A.S. et al. (2017). The future of NMR-based metabolomics. *Curr. Opin. Biotechnol.* 43: 34–40.

11 Claridge, T.D.W. (2009). *High-Resolution NMR Techniques in Organic Chemistry*, 2e. London, UK: Elsevier Academic Press.

12 Cavanagh, J., Fairbrother, W.J., Palmer, A.G. III et al. (2007). *Protein NMR Spectroscopy: Principles and Practice*, 2e. San Diego, USA: Elsevier Academic Press.

13 Wider, G. (ed.) (2006). NMR of proteins in solution (a review). *Magn. Reson. Chem.* 44: S1.

14 Sugiki, T., Kobayashi, N., and Fujiwara, T. (2017). Modern technologies of solution nuclear magnetic resonance spectroscopy for three-dimensional structure determination of proteins open avenues for life scientists. *Comput. Struct. Biotechnol. J.* 15: 328–339.

15 Wijmenga, S.S. and van Buuren, B.N.M. (1998). The use of NMR methods for conformational studies of nucleic acids. *Prog. Nucl. Magn. Reson. Spectrosc.* 32: 287–387.

16 Conte, M.R., Bauer, C.J., and Lane, A.N. (1996). Determination of sugar conformations by NMR in larger DNA duplexes using both dipolar and scalar data: application to d(CATGTGACGTCACATG)2. *J. Biomol. NMR* 7 (3): 190–206.

17 T. Cheung, NMR structural studies of mismatched DNA base pairs and their interaction with E. coli MutS protein, PhD thesis (2010), University of Manchester, Manchester, UK.

18 Mohammed, S., Phelan, M., Rasul, U., and Ramesh, V. (2014). NMR elucidation of the role of Mg^{2+} in the structure and stability of the conserved RNA motifs of the EMCV IRES element. *Org. Biomol. Chem.* 12: 1495–1509.

19 Hore, P.J. (2015). *Nuclear Magnetic Resonance (Oxford Chemistry Primers)*, 2e. Oxford: Oxford University Press.

20 Friebolin, H. (2010). *Basic One- and Two-Dimensional NMR Spectroscopy*, 5e. Wiley & VCH.

21 Keeler, J. (2010). *Understanding NMR Spectroscopy*, 2e. Wiley Blackwell.

22 Neuhaus, D. and Williamson, M.P. (2000). *The Nuclear Overhauser Effect in Structural and Conformational Analysis*, 2e. Wiley.

23 Frenkiel, T. (1993). Instrumentation and pulse sequences, Chapter 3. In: *NMR of Macromolecules: A Practical Approach* (ed. G.C.K. Roberts), 35–70. IRL Press (OUP).

24 Hurd, R.H. (1990). Gradient-enhanced spectroscopy. *J. Magn. Reson.* 87: 442–428.

25 Hwang, T.L. and Shaka, A.J. (1995). Water suppression that works. Excitation sculpting using arbitrary wave-forms and pulsed-field gradients. *J. Magn. Reson.* 112: 275–279.

26 Hoult, D.I. (1976). Solvent peak saturation with single phase and quadrature Fourier transformation. *J. Magn. Reson.* 21: 337–347.

27 Freeman, R., Hill, H.D.W., and Kaptein, R. (1972). Proton-decoupled NMR spectra of carbon-13 with the nuclear Overhauser effect suppressed. *J. Magn. Reson.* 7: 327–329.

28 Levitt, M.H., Freeman, R., and Frenkiel, T. (1982). Broadband heteronuclear decoupling. *J. Magn. Reson.* 47: 320–330.

29 Shaka, A.J., Keeler, J., Frenkiel, T., and Freeman, R. (1983). An improved sequence for broadband decoupling: WALTZ-16. *J. Magn. Reson.* 52: 335–338.

30 Aue, W.P., Bartholdi, E., and Ernst, R.R. (1976). Two-dimensional spectroscopy: application to nuclear magnetic resonance. *J. Chem. Phys.* 64: 2229–2246.

31 Bax, A. and Freeman, R. (1981). Investigation of complex networks of spin-spin coupling by two-dimensional NMR. *J. Magn. Reson.* 44: 542–561.

32 Rance, M., Sørensen, O.W., Bodenhausen, G. et al. (1983). Improved spectral resolution in COSY ^{1}H NMR spectra of proteins via double quantum filtering. *Biochem. Biophys. Res. Commun.* 117: 479–485.

33 Braunschweiler, L. and Ernst, R.R. (1983). Coherence transfer by isotropic mixing: application to proton correlation spectroscopy. *J. Magn. Reson.* 53: 521–528.

34 Bax, A. and Davis, D.G. (1985). MLEV-17-based two-dimensional homonuclear magnetization transfer spectroscopy. *J. Magn. Reson.* 65: 355–360.

35 Macura, S. and Ernst, R.R. (1980). Elucidation of cross relaxation in liquids by two-dimensional NMR spectroscopy. *Mol. Phys.* 41: 95–117.
36 Macura, S., Huang, Y., Suter, D., and Ernst, R.R. (1981). Two-dimensional chemical exchange and cross-relaxation spectroscopy of coupled nuclear spins. *J. Magn. Reson.* 43: 259–281.
37 Bodenhausen, G. and Ruben, D.J. (1980). Natural abundance nitrogen-15 NMR by enhanced heteronuclear spectroscopy. *Chem. Phys. Lett.* 69: 185–189.
38 Norwood, T.J., Boyd, J., Heritage, J.E. et al. (1990). Comparison of techniques for 1H-detected heteronuclear 1H-15N spectroscopy. *J. Magn. Reson.* 87: 488–501.
39 Morris, G.A. and Freeman, R. (1979). Enhancement of nuclear magnetic resonance signals by polarization transfer. *J. Am. Chem. Soc.* 101: 760–762.
40 Luy, B. and Marino, J.P. (2001). ^{1}H-^{31}P CPMG-correlated experiments for the assignment of nucleic acids. *J. Am. Chem. Soc.* 123: 11306–11307.
41 Blackburn, G.M., Gait, M.J., Loakes, D., and Williams, D.M. (eds.) (2005). *Nucleic Acids in Chemistry and Biology*, 3e. UK: Royal Society of Chemistry Publishing.
42 Neidle, S. (2008). *Principles of Nucleic Acid Structure*. Elsevier.
43 Watson, J.D. and Crick, F.H.C. (1953). Molecular structure of nucleic acids: a structure for deoxyribonucleic acid. *Nature* 171: 737–738.
44 Bloomfield, V.A., Crothers, D.M., and Tinoco, I. Jr. (2000). *Nucleic Acids: Structures, Properties and Functions*. Sausalito, California, USA: University Science Books.
45 Saenger, W. (1984). *Principles of Nucleic Acid Structure*. USA: Springer-Verlag.
46 Modrich, P. (1997). Strand-specific mismatch repair in mammalian cells. *J. Biol. Chem.* 272: 24727–24730.
47 Dickerson, R.E., Drew, H.R., Conner, B.N. et al. (1982). The anatomy of A-, B-, and Z-DNA. *Science* 216: 475–485.
48 Conner, B.N., Takano, T., Tanaka, S. et al. (1982). The molecular structure of d(^{I}CpCpGpG), a fragment of right-handed double helical A-DNA. *Nature* 295: 294–299.
49 Wang, A.H.J., Quigley, G.J., Kolpak, F.J. et al. (1979). Molecular structure of a left-handed double helical DNA fragment at atomic resolution. *Nature* 282: 680–686.
50 Pearlman, D.A., Case, D.A., Caldwell, J.W. et al. (1995). AMBER, a package of computer programs for applying molecular mechanics, normal mode analysis, molecular dynamics and free energy calculations to simulate the structural and energetic properties of molecules. *Comput. Phys. Commun.* 91: 1–41.
51 Brooks, B.R., Bruccoleri, R.E., Olafson, B.D. et al. (1983). CHARMM: a program for macromolecular energy, minimization, and dynamics calculations. *J. Comput. Chem.* 4: 187–217.
52 Christen, M., Hünenberger, P.H., Bakowies, D. et al. (2005). The GROMOS software for biomolecular simulation: GROMOS05. *J. Comput. Chem.* 26: 1719–1751.
53 Leach, A. (2001). *Molecular Modeling: Principles and Practice*, 2e. Prentice Hall.
54 Schwieters, C.D., Kuszewski, J.J., Tjandra, N., and Clore, G.M. (2003). The Xplor-NIH NMR molecular structure determination package. *J. Magn. Reson.* 160: 65–73.
55 Schwieters, C.D., Kuszewski, J.J., and Clore, G.M. (2006). Using Xplor-NIH for NMR molecular structure determination. *Prog. Nucl. Magn. Reson. Spectrosc.* 48: 47–62.
56 Wijmenga, S.S., Mooren, M., M.W., and Hilbers, C.W. (1993). NMR of nucleic acids: from spectrum to structure, Chapter 7. In: *NMR of Macromolecules: A Practical Approach* (ed. G.C.K. Roberts), 217–288. IRL Press (OUP).
57 Liu, W., Vu, H.M., and Kearns, D.R. (2002). 1H NMR studies of a 17-mer DNA duplex. *Biochim. Biophys. Acta* 1574: 93–99.
58 Vranken, W.F., Boucher, W., Stevens, T.J. et al. (2005). The CCPN data model for NMR spectroscopy: development of a software pipeline. *Proteins* 59: 687–696.
59 Markley, J.L., Bax, A., Arata, Y. et al. (1998). Recommendations for the presentation of NMR structures of proteins and nucleic acids. *Pure Appl. Chem.* 70: 117–142.

60 Zheng, G., Lu, X.-J., and Olson, W.K. (2009). Web 3DNA – a web server for the analysis, reconstruction, and visualization of three-dimensional nucleic-acid structures. *Nucleic Acids Res.* 37: W240–W246.
61 Dickerson, R.E. (1989). Definitions and nomenclature of nucleic acid structure components. *Nucleic Acids Res.* 17: 1797–1803.
62 Olson, W.K., Bansal, M., Burley, S.K. et al. (2001). A standard reference frame for the description of nucleic acid base-pair geometry. *J. Mol. Biol.* 313: 229–237.
63 Berman, H.M., Olson, W.K., Beveridge, D.L. et al. (1992). The nucleic acid database. A comprehensive relational database of three-dimensional structures of nucleic acids. *Biophys. J.* 63: 751–759.
64 Berman, H.M., Westbrook, J., Feng, Z. et al. (2000). The protein data bank. *Nucleic Acids Res.* 28: 235–242.
65 Ulrich, E.L., Akutsu, H., Doreleijers, J.F. et al. (2007). *Nucleic Acids Res.* 36: D402–D408.
66 Obmolova, G., Ban, C., Hsieh, P., and Yang, W. (2000). Crystal structures of mismatch repair protein MutS and its complex with a substrate DNA. *Nature* 407: 703–710.
67 K. Marat, SpinWorks 3.1.8.1, University of Manitoba, Winnipeg.
68 Delaglio, F., Grzesiek, S., Vuister, G.W. et al. (1995). NMRPipe: a multidimensional spectral processing system based on UNIX pipes. *J. Biomol. NMR* 6: 277–293.
69 T. D. Goddard and D. G. Kneller, SPARKY 3, University of California, San Francisco.
70 Ren, X., Li, F., Jeffs, G. et al. (2010). Guanine sulphinate is a major stable product of photochemical oxidation of DNA 6-thioguanine by UVA irradiation. *Nucleic Acids Res.* 38: 1832–1840.
71 Swann, P.F., Waters, T.R., Moulton, D.C. et al. (1996). Role of postreplicative DNA mismatch repair in the cytotoxic action of thioguanine. *Science* 273: 1109–1111.
72 Somerville, L., Krynetski, E.Y., Krynetskaia, N.F. et al. (2003). Structure and dynamics of thioguanine-modified duplex DNA. *J. Biol. Chem.* 278: 1005–1011.
73 Guerra, C.F., Baerends, E.J., and Bickelhaupt, F.M. (2008). Watson–Crick base pairs with thiocarbonyl groups: how sulfur changes the hydrogen bonds in DNA. *Cent. Eur. J. Chem.* 6: 15–21.
74 Vilani, G. (2009). Properties of the thiobase pairs hydrogen bridges: a theoretical study. *J. Phys. Chem. B* 113: 2128–2134.
75 Szulik, M.W., Voehler, M.V., and Stone, M.P. (2014). NMR analysis of base-pair opening kinetics in DNA. *Curr. Protoc. Nucleic Acid Chem.* 59: 7.20.1–7.20.18.
76 Nelissen, F.H.T., Tessari, M., Wijmenga, S.,.S., and Heus, H.A. (2016). Stable isotope labeling methods for DNA. *Prog. Nucl. Magn. Reson. Spectrosc.* 96: 89–108.
77 Fürtig, B., Richter, C., Wöhnert, J., and Schwalbe, H. (2003). NMR spectroscopy of RNA. *ChemBioChem* 4: 936–962.
78 Sakamoto, T., Otsu, M., and Kawai, G. (2018). NMR studies on RNA. In: *Experimental Approaches of NMR Spectroscopy-Methodology and Application to Life Science and Materials Science* (ed. The Nuclear Magnetic Resonance Society of Japan), 439–459. Springer.

Further Reading

Claridge, T.D.W. (2016). *High Resolution NMR Techniques in Organic Chemistry*, 3e. Elsevier Science.
Evans, J.N.S. (1995). *Biomolecular N.M.R. Spectroscopy*. USA: Oxford University Press.
Hinchcliffe, A. (2008). *Molecular Modelling for Beginners*, 2e. Wiley.
Hore, P., Jones, J., and Wimperis, S. (2015). *NMR: THE TOOLKIT How Pulse Sequences Work (Oxford Chemistry Primers)*, 2e. Oxford: Oxford University Press.

18

Cryo-TEM and Biological Structure Determination

Szymon W. Manka and Carolyn A. Moores

Institute of Structural and Molecular Biology, Birkbeck College, Malet Street, London, WC1E 7HX, UK

18.1 Significance and Background

The electron microscope (EM) has been a key tool for biologists since its invention by Ernst Ruska in the 1930s (Table 18.1). Use of EM to visualise cell samples led to the discovery of many key aspects of cell ultrastructure and function, while EM imaging of molecular samples was first used for 3D structure determination of macromolecular complexes in the late 1960s [1].

EM of biological samples brings challenges. The first is the fundamentally destructive nature of the electron radiation used in the imaging experiment. The second is that the EM must be operated under vacuum – an environment at odds with the aqueous milieu of life – because electrons are strongly scattered by air molecules. Furthermore, while EM can be used to scan the surface of a sample (scanning electron microscopy, SEM), this chapter is about transmission electron microscopy (TEM), in which the electrons are transmitted by the sample (i.e. pass through it). Therefore, the sample must be thin (<0.5 μm). Finally, image contrast depends on scattering of electrons by the sample, but the low mass atoms from which most biological samples are built only scatter electrons very weakly. As a result of all these factors, a set of compromises during sample preparation and imaging must be negotiated to ensure that the sample under study is both sufficiently stable to observe and in a physiologically relevant state.

Early progress was made by visualising dehydrated samples – both macromolecules and cells – stained with heavy metals that could be inserted into the microscope vacuum and produce high contrast images. Tissue specimens were additionally chemically fixed with crosslinking agents to stabilise them, prior to embedding in plastic resin to enable thin sections to be cut. Despite these seemingly harsh treatments, many significant insights into biological mechanisms were (and continue to be) made using these preparations. However, such treatments present limitations, including that the resulting images are of the stain distribution rather than the biological sample itself and that the ultimate resolution of the resulting images is limited by the grain size of the stain molecules (>1 nm).

Thus, it was a truly significant breakthrough when methods for cryo-electron microscopy (cryo-EM) were developed in the 1980s. Here, ultrarapid sample freezing allowed preservation of the sample's hydrated state in a thin layer of so-called amorphous (non-crystalline) or vitreous ice, which is itself more or less transparent to electrons (Figure 18.1a). These specimens, maintained at cold temperatures, are solid and can therefore be inserted into the microscope vacuum. In this frozen state, the biological sample – both molecular and cellular – can be visualised directly without intermediary stain. It thus provides a direct window into the atoms of molecules and molecules of cells. In the absence of stain, the contrast of the resulting images is very low.

Biomolecular and Bioanalytical Techniques: Theory, Methodology and Applications, First Edition. Edited by Vasudevan Ramesh.

Table 18.1 Nobel Prizes awarded in EM (www.nobelprize.org).

Name	Year	Nobel citation
Joseph John Thompson	Physics, 1906	'In recognition of the great merits of his theoretical and experimental investigations on the conduction of electricity by gases'
Aaron Klug	Chemistry, 1982	'For his development of crystallographic electron microscopy and his structural elucidation of biologically important nucleic acid–protein complexes'
Ernst Ruska	Physics, 1986	'For his fundamental work in electron optics, and for the design of the first electron microscope'
Jacques Dubochet, Joachim Frank, Richard Henderson	Chemistry, 2017	'For developing cryo-electron microscopy for the high-resolution structure determination of biomolecules in solution'

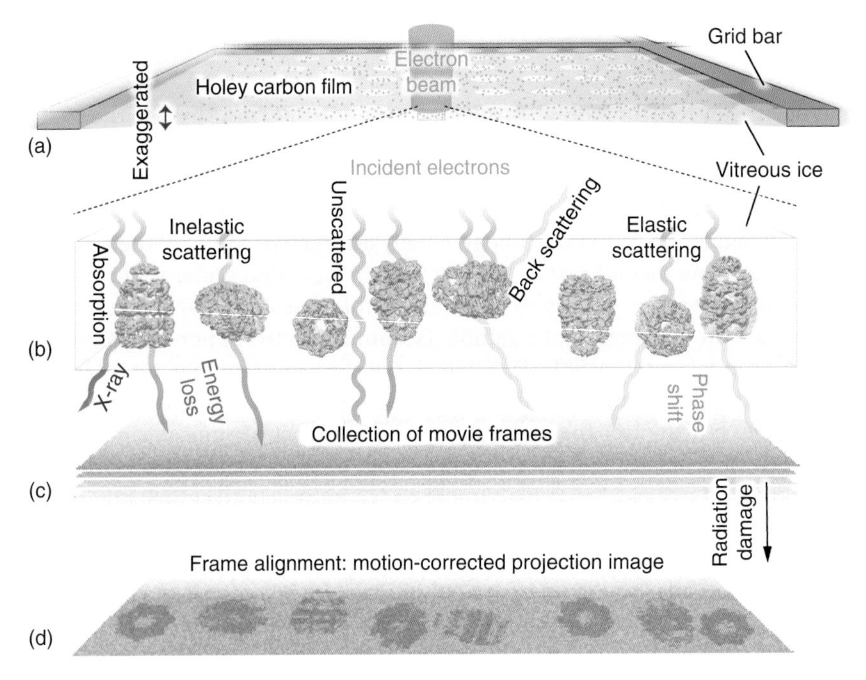

Figure 18.1 Single-particle cryo-TEM data collection. (a) Frozen-hydrated (vitrified) sample exemplified by GroEL/ES chaperonin protein (Electron Microscopy Data Bank [EMDB] accession code: EMD-2325) [2] is illuminated over holes in the carbon film, where it is suspended in a thin film of vitreous ice. (b) The protein in random orientations scatters electrons elastically or inelastically. Wavy lines represent the dual nature of electrons. (c) Noisy projection images are acquired as movie frames. The later the frame, the more the accumulated radiation damage to the sample. (d) The frames are aligned and summed, producing a composite micrograph with increased signal-to-noise ratio (SNR) and reduced sample motion.

Furthermore, the electron dose must be limited during data collection because the samples are also very fragile in the electron beam. These experiments bring their own challenges. Nevertheless, cryo-TEM enables unprecedented mechanistic insight into the operation of the molecular machinery that drives the life of the cell.

The optical arrangement of a TEM is such that images closely approximate 2D projections of the 3D sample. Such 2D visualisation can be informative, but 3D information provides the most complete account of spatial organisation. TEM can be used for 3D structure determination; to do this, many different 2D views of a particular sample must be collected and combined computationally to yield its 3D structure. According to the type of sample, these multiple views are acquired in two distinct data collection regimes:

1) *Single particle/single particle-like averaging (SPA)*: Data collection involves acquisition of many thousands of images of a population of molecules or complexes that are randomly oriented and assumed to be biochemically and structurally homogenous; structure determination involves calculation of how the individual 2D projection images relate to each other; the images are then combined to reconstruct the sample 3D structure (Figure 18.1).
   - *Sample type*: Macromolecules or macromolecular complexes (Case Study 1).
2) *Electron tomography (ET)*: Data collection involves repeated tilting and imaging of a particular feature inside the microscope; this imaging modality provides a defined range of views although they are limited due to the physical layout of the TEM, with the maximum tilting angle being 70°; the set of 2D views are subsequently combined computationally to reconstruct the 3D structure of the sample (see later in Figure 18.6).
   - *Sample type*: Large, complex and heterogeneous macromolecular complexes, organelles, cells, cell or tissue sections (Case Study 2).

Thus, cryo-TEM allows the study of molecular machinery at a range of sizes, complexities and resolutions. With recent developments in hardware and software, cryo-TEM is now a central technique in structural and cell biology, and provides an essential complement to other approaches at both high and low resolutions. This chapter will first introduce basic theory and principles of biological cryo-TEM, before describing practical aspects of the method in more detail, ending with two case studies.

18.2 Theoretical Principles of Biological Cryo-TEM

It is useful to think about electron radiation by analogy with other, more commonly known types of electromagnetic radiation, such as visible light or X-rays. Like photons, electrons have a dual (particle and wave) nature. For some aspects of their behaviour it is useful to describe them as streams of particles and for others, as moving wavefronts with a certain amplitude and relative phase. In terms of optical properties electrons resemble light, while in terms of their interaction with biological molecules, they resemble X-rays (see below and also [3]).

18.2.1 Use of Electron Radiation for Imaging of Biological Machinery

To use any kind of radiation for imaging at a certain resolution, it must: (i) be focusable and (ii) have a wavelength that matches the resolution sought. In other words, for a wave to *experience* an object as an obstacle it needs to be of a similar size. For example, raindrops cannot inflict any meaningful perturbation on ocean waves. Light microscopy is a powerful imaging methodology, because photons can be easily focused (with a precisely shaped piece of glass, known as a lens) and observed – eyes are detectors of unparalleled performance. However, atoms in molecules are roughly 1 Ångstrom (0.1 nm) apart. Therefore, visible light with the shortest wavelength of ~400 nm cannot provide sufficient resolution for molecular structure determination. X-rays have a short wavelength (~Ångstrom) and excellent penetration of solid matter (fracture clinics, airport security), but are hard to focus. Neutrons, currently used in diffraction experiments like X-rays, can be emitted with sub-Ångstrom wavelengths. However, at present it is not known how to make a neutron lens.

Electron radiation has the two crucial properties (ability to be focused and the short wavelength) that led to its use in EMs for high resolution imaging. First, the wavelength of the electron radiation is in the picometre range and is thus suitable for investigating biological structure, theoretically to atomic level resolution. Second, the charge carried by electrons enables their deflection by an electromagnetic field and hence electron beams can be focused with magnetic lenses – as light is focused by glass lenses – to produce images.

The imaging ability of cryo-TEM has a great advantage for structure determination over X-ray or neutron diffraction experiments. Diffraction patterns (also produced in microscopes at the back focal plane of the objective lens, Figure 18.2, 18.3) contain only amplitude information, leading to the classical phasing problem of X-ray and neutron diffraction techniques. Crucially, information about both amplitudes and phases of the scattered/diffracted waves are captured in TEM images.

18.2.2 Electron Beam Interaction with a Biological Sample

Unlike X-rays, electrons have a resting mass that imposes limits on their ability to penetrate samples. The so-called mean free path of an electron is in the range of ~200–300 nm depending on microscope voltage. Thus, the ideal cryo-TEM sample would only be marginally thicker than the longest dimension of the molecule or region of interest, such that the sample is as thin as possible, while not being squashed (Figure 18.1b).

Atoms of the specimen interact with incident electrons in various ways, depending on their electrostatic (Coulomb) potential (proportional to atomic number) and the beam energy [4]. Considering interactions with heavy atoms, it is sufficient to interpret incident electrons (also called primary electrons) as a stream of particles with defined velocities and trajectories, being scattered at high angles – including backscattering – by the large electrostatic fields of those heavy atoms. Such scattering is negligible in biological TEM, since biological matter is mainly built of light elements, such as hydrogen (H), carbon (C), oxygen (O), nitrogen (N) and phosphorus (P). These atoms mostly only weakly scatter primary electrons.

To understand weak scattering and its consequences, the particle approximation of an electron is no longer helpful. It is best to think of each primary electron as a single plane wave, traversing space in a certain direction and with a certain energy (wavelength). This interpretation is overall much more broadly applicable, but also more abstract. Every such electron passes through the specimen individually and interacts with the whole sample (Figure 18.2). Weak scattering means that a portion of this incident wave passes through the sample without effect, i.e. unscattered, but the electrostatic potentials of sample atoms – multiple scattering centres – will also cause perturbations that can be envisaged as ripples in this single plane wave. These ripples interfere, giving rise to new wave fronts (Figure 18.2). Upon such interaction with the sample, the primary electron wave can lose some of its energy, exciting the sample electrons to higher orbitals. This is immediately followed by a decay to a lower energy state, accompanied by X-ray emission or by an ejection of a so-called secondary electron in any direction from an outer orbital. This is called inelastic scattering, resulting in a longer wavelength of the transmitted wave (Figure 18.1b). However, when the primary electron wave does not lose energy after impinging on sample atoms, the information about them is transmitted by the scattered component waves, retaining the original wavelength, but with a phase altered by ~ ¼ of a wavelength (~90°) relative to the unscattered component wave. This is called elastic scattering. At voltages used in typical cryo-TEM experiments, inelastic scattering is around three times more likely than elastic scattering [5].

Elastic scattering contributes useful structural information to the final image, while inelastic scattering lowers the signal-to-noise ratio (SNR) in the image. This is because the inelastically scattered wave: (i) is focused more strongly (out of the image plane) due to its lower energy (longer wavelength) compared to both the unscattered and the elastically scattered component waves and causes blurring of fine image features; (ii) causes deposition of energy in the sample, resulting in damage through ionisation, induction of free radicals and emission of X-rays. Due to this radiation damage, the maximum dose in a typical cryo-TEM experiment must be limited (~50 e-/$Å^2$ on the sample). However, the finest structural details are immediately compromised as soon as the sample is exposed to the electron beam. In particular, negatively charged side chains of acidic amino acid residues are exceptionally susceptible and are almost immediately destroyed in imaging experiments. A final consideration for imaging is that in order to accurately capture structural information about the sample, an elastically scattered electron wave should only be scattered once before transmission through the sample. In samples that are thicker than ~200 nm, multiple scattering events can also affect the SNR.

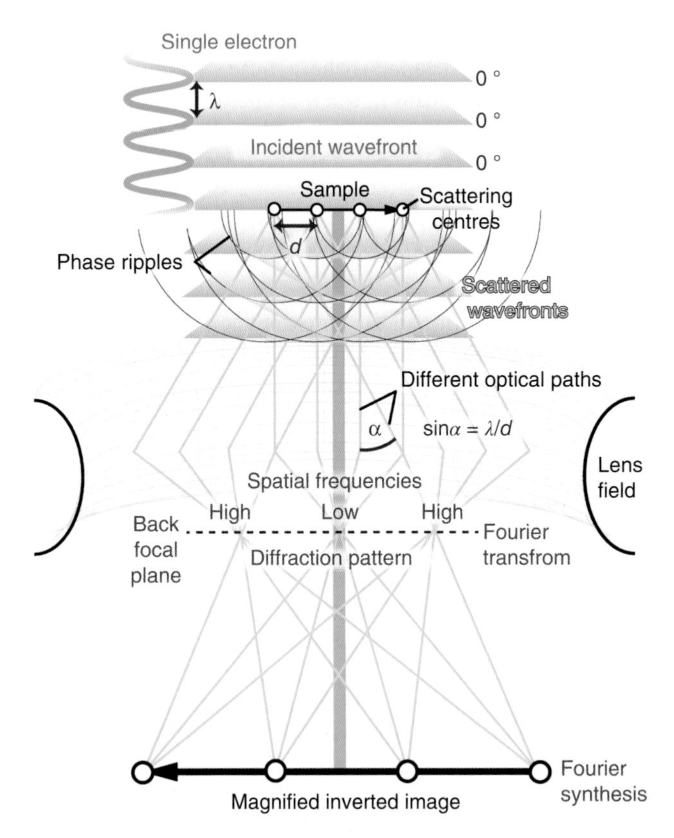

Figure 18.2 TEM image formation. Every electron travels through the microscope column as a plane wave, here depicted with green gradients (moving wavefront) at 0° phase values. This single electron interacts with the whole sample – all of its multiple scattering centres. The diagram assumes only elastic scattering events, which lead to multiple disturbances within the electron wave. These disturbances are depicted as concentric circles, or phase ripples within the wave, since each scattering centre causes an additional oscillation within that single plane wave. These ripples constructively interfere in multiple directions, forming new wavefronts (yellow gradients), which can be presented as rays (yellow beams). The unscattered component wave travels in the direction of the optical axis of the lens (green beam). The scattering angle α is the angle between that axis and the direction of the scattered component wave or ray. This angle is inversely proportional to the distance (*d*) between the scattering centres producing the scattered wave (Bragg's law: equation). The scattered component waves are focused together with the unscattered (primary) wave to form a diffraction image or pattern (Fourier transform) at the back focal plane of the lens (where *d* corresponds to spatial frequency) and to a virtual image at the image plane (Fourier synthesis). The figure was inspired by lectures from Prof. Grant Jensen, Caltech, CA, USA.

Constructive interference of the elastically scattered waves will occur when Bragg's law is satisfied (equation in Figure 18.2) as determined by the spacing of the scattering centres (sample atoms). These scattered component waves can be depicted as rays perpendicular to their fronts in order to follow their fate in an optical system (Figure 18.2). This is further explored in the next section, which focuses on image formation.

18.2.3 Image Formation: The Weak Phase Object Challenge

Regardless of the type of cryo-TEM imaging experiment – single particle data collection or successive tomographic tilt views – the primary image is always a 2D projection of a 3D specimen (Figure 18.1d, single particle example). Density variation in the direction perpendicular to the electron beam is projected and recorded as intensity variation in the resultant image. The image contrast is the difference between the intensities of the brightest and the darkest point in the image, divided by the average intensity of the image. Modulation in these intensities – defined as amplitude contrast – can arise directly from strong scattering of incident electrons by the sample, i.e. scattering at high angles, including backscattering mentioned earlier. These strongly scattered waves can be blocked by diaphragms called apertures accompanying lenses in microscopes (see later). Therefore, a simplistic explanation for the emergence of the amplitude contrast in a microscope is that rays travelling around a dense object pass through and rays colliding with it are removed by the optical system. However, since cells, viruses and biomolecules are composed mainly of H, C, O, N and P, as noted above, they produce very low intensity modulation (amplitude contrast) against the background of similar atoms from the surrounding vitrified solvent. A thin, unstained biological specimen generates amplitude contrast estimated to be ~7% at 120 kV and ~4% at 300 kV [6]. Therefore, biological samples are generally considered as weak phase objects; the projection of their electron potential is encoded in the phase variation of the emergent composite wave [7].

The direction (angle) and strength (amplitude) of each scattered wave depends on the distance (d in Figure 18.2) between the scattering centres (sample atoms) and their number, respectively. Thus, fine features in the sample (e.g. distance between atoms in the adjacent amino acid side chains in a protein) produce scattering at relatively high angles (scattering angle α in Figure 18.2), whereas larger features (atoms spaced between particular secondary structures or domains of a protein) produce scattering at lower angles. All of the oscillations within the single electron wave, including all of the scattered components and the unscattered primary component, can be considered as separate rays of electron radiation, as mentioned above (Figure 18.2). An electron lens collects all those rays and focuses them, giving rise to positive (constructive) and negative (destructive) interference of the scattered waves with the unscattered primary wave. The resultant waves manifest in a so-called diffraction pattern at the back focal plane of the lens (Figure 18.2), where each instance of productive interference contributes a discrete signal – a single spot in the diffraction pattern. Such signal decomposition is known as Fourier transformation and each constituent wave is a separate Fourier component that represents specific distance between scattering centres (d in Figure 18.2). In Fourier space (also called reciprocal space) this distance is called spatial frequency. The fine features (small d), generating high scattering angles, correspond to high spatial frequencies (edges of the diffraction pattern), while the coarser features, generating low scattering angles, correspond to low spatial frequencies (middle of the diffraction pattern). Further on the optical path of the lens is the image plane, where the rays are recombined to form a projection image approximation of the sample (Fourier synthesis) (Figure 18.2).

Crucially, phase contrast produced in biological TEM must be expressed as (or converted to) amplitude contrast to be detected by an imaging device that can only measure signal intensity modulation. Therefore, the phase shift produced by the sample needs to be such that it produces detectable amplitude contrast in the image: maximum contrast is derived from either ~180° (negative interference) or 360° (identical to 0°, positive interference) phase shifts. Overall visibility of any object in the image is critically dependent on the phase difference between the scattered waves at low spatial frequencies corresponding to object shape in the image and the unscattered wave. Since low spatial frequencies are represented by rays nearly parallel to the optical axis of the lens (Figure 18.2), their optical path length is similar to that of the unscattered wave component. This means that the two waves remain only ~90° shifted in phase, are unproductive in terms of image formation and thus these critical low frequencies remain invisible at focus under perfect lens conditions (assumed in Figure 18.2). In practice, adequate amplitude contrast such that the sample can be seen is obtained by imposing additional shifts in the phase of the scattered component waves. A later section will explain how this is achieved in a TEM.

18.2.4 Basic Anatomy of a TEM

At the top of a TEM column is an electron source or gun (Figure 18.3a). Electrons are extracted by high voltage (100–300 kV) from so-called filaments. The extraction voltage defines the wavelength of the electron radiation. There are three types of filaments: (i) bent tungsten wire heated up to 3000 °C in low end microscopes, which thermally emit electrons like a light bulb; (ii) lanthanum hexaboride (LaB6) crystal, with emission from a smaller area (~5 μm) and with lower temperature; and (iii) a field emission gun (FEG) in high performance

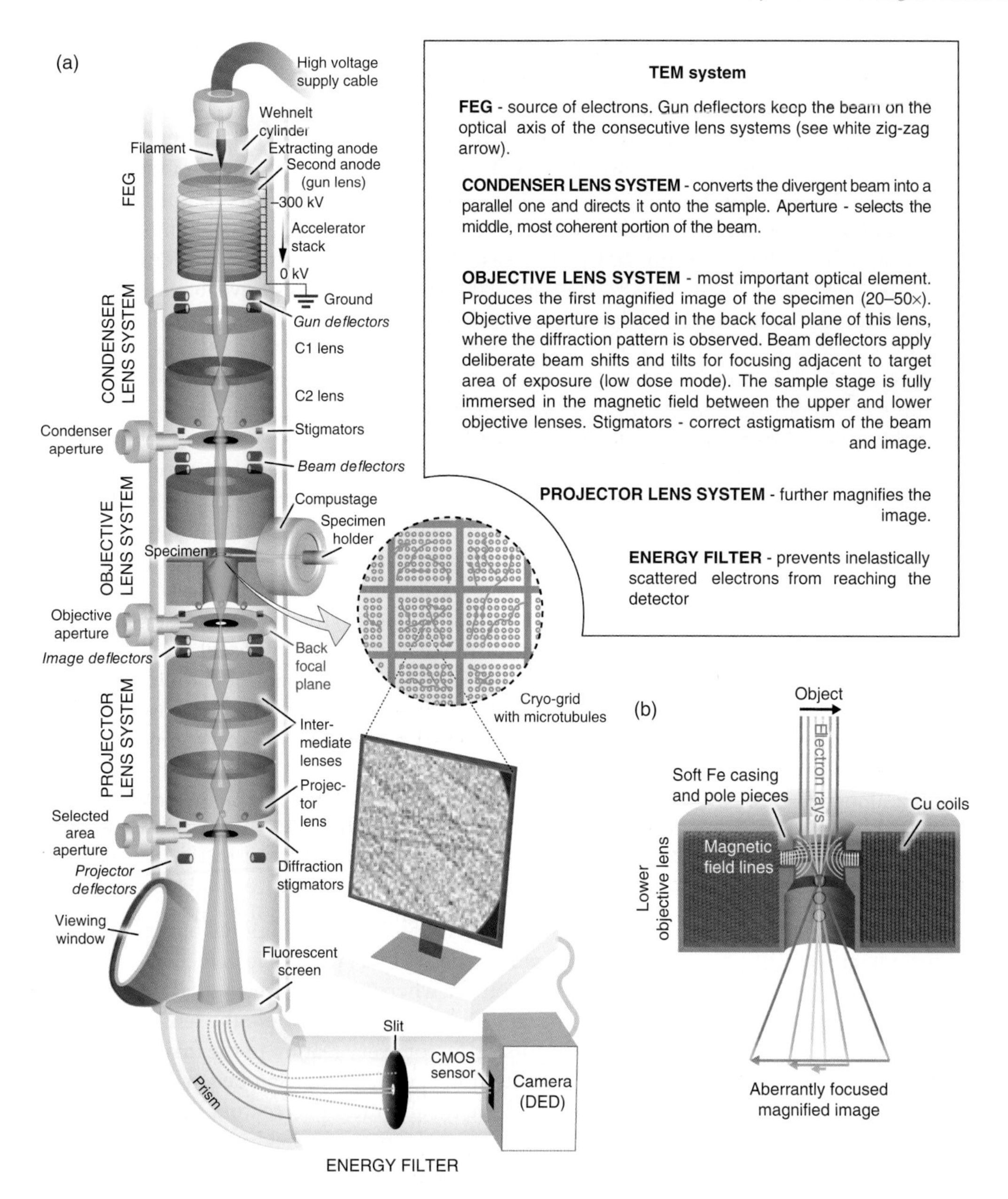

Figure 18.3 Schematic representation of a high resolution TEM and the spherical aberration of the objective lens. (a) Simplified schematic showing one of many possible TEM systems and its major parts. Roles of the particular segments of the system are outlined in the frame. The first image is formed by the objective lens; it is magnified by the projector lens system, can be visualised on the fluorescent screen at the bottom of the microscope column and is ultimately captured on the camera. The post-column energy filter bends radiation with a prism, letting only the elastically scattered electron waves pass through the slit (solid line) and blocking the inelastically scattered waves (dotted line). All lenses are presented as complete discs, except the lower objective lens, which is shown as a cross-section in panel (b). The cooling and vacuum systems and lead shielding protection from emitted X-rays are not included. (b) Illustration of the spherical aberration (Cs) of the objective lens. The points of particular ray crossings are indicated with circles.

microscopes, where electrons escape from the very sharp tip (~10 nm) of a tungsten crystal coated with ZrO_2 to lower the work function of tungsten for electrons. This allows them to be pulled off the tip by a strong and uniform electric field with uniform energies, even at room temperature. Each of these sources emit electron beams with different homogeneity or coherence. There are two types of coherence: (i) spatial coherence, where electrons emerge in the same direction, and (ii) temporal coherence, where electrons emerge with the same speed or energy, allowing uniform focusing of all the rays. Tungsten wire filaments produce relatively incoherent electrons, LaB6 filaments produce beams with intermediate coherence and field emission from a small area produces the most coherent beam. In this last case, the electrons are then accelerated in the accelerator stack, e.g. for 300 kV settings through voltages of −300 to 0 kV (electrons travel towards positive potential – here ground) to ~3/4 of the speed of light, before entering the optical system of the microscope column (Figure 18.3a).

As previously mentioned, the column of a TEM instrument needs to be under ultrahigh vacuum to avoid performance-dampening contamination and scattering from gas molecules that generate image noise. EM lenses are magnetic fields induced by the passing of current through copper coils (Figure 18.3b, white lines). First, the beam passes through the condenser lens system, which converts the divergent beam into a parallel one and directs it on to the specimen (Figure 18.3a). The specimen is located inside the objective lens system, producing the first magnified image of the specimen. The projector lens system further magnifies the image that can be displayed on a fluorescent viewing screen (not included in most modern TEM designs) or recorded by an imaging device (Figure 18.3a). All lens systems have four basic components (starting from top): (i) deflectors, moving the incoming beam to the optical axis of the microscope; (ii) the lenses themselves, focusing the beam at the specified distance (magnification is image distance from the centre of the lenses, divided by the distance from the same point to the object); (iii) stigmators, correcting for imperfection in the shape of the magnetic field of the lenses (see below); and (iv) apertures, which limit the most incoherent rays at the periphery of the beam, thus increasing contrast (Figure 18.3a). Paradoxically, it is thanks to a common imperfection of lenses (Figure 18.3b) that we can see anything in classical biological cryo-TEM. The next section will discuss lens aberrations alongside other systematic defects of a TEM and their influence on image formation.

18.2.5 Systematic TEM Defects and Their Influence on Image Formation

Electromagnetic lenses come with typical performance limitations familiar from traditional optical systems, of which the most prominent are: (i) spherical aberration, where the periphery of the lens focuses the beam more strongly than its centre (Figure 18.3b); (ii) chromatic aberration, where longer wavelength (lower energy) waves are focused more strongly than shorter wavelengths; and (iii) astigmatism, where the lens focuses the beam more strongly in one direction, producing ellipsoidal distortion of every feature in the image. All these defects have obviously adverse consequences for the transmitted image, but some can be partly offset with available or emerging technologies, such as spherical and chromatic aberration correctors (Cs and Cc correctors, respectively). Because astigmatism results from an asymmetric magnetic field in the lens, it can be largely compensated for by additional stigmator coils. Chromatic aberration is detrimental because the temporal incoherence (significant energy spread) of the beam – due to, for example, voltage variation at the source, as well as the inelastic (energy-loss) events – causes degradation of the image by blurring its fine details. This can be partly remedied with a pre-column monochromator and/or an in-column (omega) or post-column (GIF) energy filter set to a desired electron energy range (Figure 18.3a) to sharpen the image.

Crucially, spherical aberration of the objective lens, aside from its obvious resolution-degrading effect, plays a key role in imaging. Such distortion facilitates increases in the phase shift between the scattered component waves and the unscattered component wave, which converts the undetectable phase contrast into detectable amplitude contrast in the image. The Cs is a fixed parameter specific to a microscope, but the overall effect is oscillatory due to the wave nature of the signal (phase contrast), and is described by the phase contrast transfer

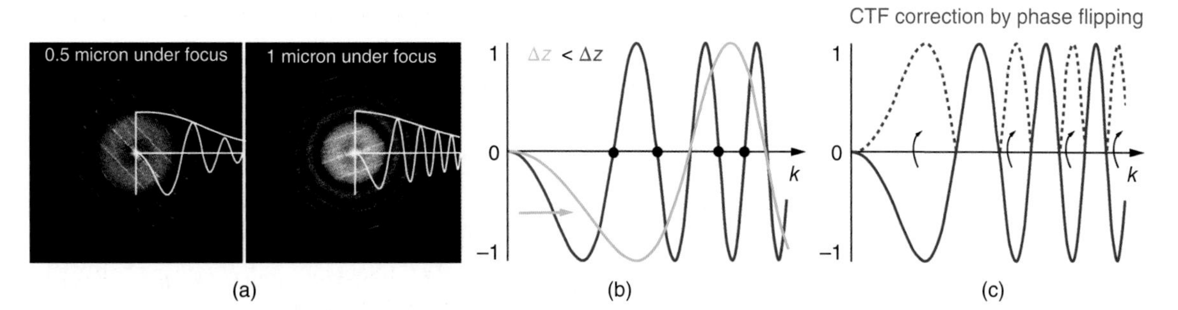

Figure 18.4 Defocus effects and CTF correction. (a) Two example Fourier transforms of images taken at two different defoci. Contrast transfer function (CTF) behaviour in response to the different amount of defocus is represented by Thon rings. The rate of oscillation is proportional to the defocus level. Every other Thon ring has a negative contrast value. The signal falloff (envelope function) is indicated by the line connecting positive CTF amplitudes. The images from which these Fourier transforms were calculated were of microtubules (see Case Study 1), and therefore diffraction from the regular microtubule array (called layer lines) are also visible; (b) Superposition of idealised CTF plots (without signal dampening) for two different defocus values (Δz); k is spatial frequency. Spatial frequencies absent from the image represented by the dark blue curve (zero crossings) are partly compensated in the image represented by the light blue curve (nodes indicated with black circles). (c) Simple CTF correction by phase flipping.

function (CTF) of the microscope:

$$CTF = \sin\left[-\pi \Delta z \lambda k^2 + \frac{\pi C_s \lambda^3 k^4}{2}\right]'$$

where Δz = defocus, λ = wavelength and k = spatial frequency. It was found that image contrast increases with increasing defocus, revealing that defocusing combined with spherical aberration increases the observed phase shift [8]. In practice, micrographs are taken at up to several micrometres under focus (negative defocus).

Imaging out of focus is a hallmark of the classical biological cryo-TEM and produces specific effects in the projection images and their Fourier transforms. The primary consequence of defocusing is the modulation of the CTF oscillation, giving rise to the so-called Thon rings in the Fourier transform of micrographs (Figure 18.4a). The rings represent the decomposition of the signal from the amorphous material (specimen support film and/or specimen) in the image into discrete spatial frequencies (Fourier components), equivalent to the diffraction pattern (image formed at the back focal plane of the objective lens). As previously discussed, in such a spectrum, the low spatial frequencies (corresponding to coarse features of the image) are located near the origin (middle) of the spectrum and the higher spatial frequencies (representing the finer features) progress towards its outer edge. The phases of the Thon rings alternately switch between positive and negative contrast with the overall rate dependent on the amount of defocus: the further away from focus, the more frequent the oscillation, with oscillations increasing towards higher spatial frequencies (the effect of Cs) (Figure 18.4).

At every CTF node, where the function crosses the 0-amplitude axis, there is no contrast transfer, so the relevant spatial frequency information is absent. The loss of information, strongly skewed towards high frequencies, increases with the number of nodes, resulting in a characteristic image coarsening with increasing defocus. The specimen outline becomes more visible but at the expense of structural detail (loss of high resolution information). This is illustrated in Figure 18.4b, where the first contrast lobe of the curve representing lower defocus (light blue) has a maximum shifted towards higher frequencies and with fewer zero crossings in the higher spatial frequency region compared to the curve representing higher defocus (dark blue). Furthermore, the oscillations on either side of the horizontal axis indicate that certain Fourier components are inverted.

18.2.6 CTF Detection and Correction

Since any amount of defocus inevitably causes loss of information, it is necessary to compensate for the loss by collecting data at a range of defoci to *fill in the zeros* of the CTF, especially for high resolution information. The CTF also inverts Fourier components, which corrupts the image. Therefore, to obtain a more faithful projection image of the sample, and thus correctly interpret the data, the CTF lobes with negative amplitudes need to be inverted. This can be achieved, for example, by dividing the Fourier transform of the micrograph by the CTF of the microscope, after which the image is recovered by inverse Fourier transform. An alternative strategy, called phase flipping (as illustrated in Figure 18.4c), involves simply reversing the sign of the CTF lobes with negative amplitudes. Variations on these approaches are described in the literature [9]. These are the most basic version of the procedure known as CTF correction, which first requires detection of the exact defocus value of the micrograph. This can be done in Fourier space by fitting a simulated CTF curve to Thon rings derived from the image. CTF correction parameters are usually established for each micrograph in the dataset before it is incorporated into any 3D reconstruction pipeline, where the correction is applied.

Thus, with the contrast gain through spherical aberration and defocusing comes the necessity of CTF correction. This becomes much more challenging in cryo-ET tilt series. An alternative way to introduce phase shift to the scattered wave than defocus imaging is to use a phase plate inserted at the diffraction (back focal) plane of the objective lens. The concept of a phase plate comes from light microscopy; its early adaptations for EM – such as a Zernike phase plate (ZPP) – despite promising results, had practical limitations, including lifetime and alignment [10]. The more recently introduced Volta phase plate (VPP) is a continuous piece of carbon film, constantly heated to ~250 °C, which, by some mechanism not fully understood, produces a nearly ~90° phase shift in the scattered electron waves. In principle, the VPP works in focus, but in practice it is not easy to find the exact point of focus. Thus, data collected with a VPP is usually minimally defocused and CTF corrected [10, 11].

18.2.7 Resolution Limiting Factors

Besides stable aberrations discussed earlier, there are a number of irregular disturbances that limit the information transfer of a TEM, including: mechanical vibrations in the column and fluctuations in lens currents (as resistivity is sensitive to temperature instability of the lens cooling system). Distortion and dampening of the signal due to both stable and unstable defects is therefore intrinsic to TEM imaging; it manifests in decay of the CTF amplitude towards high frequencies (Figure 18.4a), because the finer the detail, the more it suffers from those imperfections (dampening factors). The stable dampening factors combine to generate a net envelope function of a given microscope that describes the total systematic signal falloff. The CTF is always multiplied by this envelope function, so these two functions always appear together and can be further modulated by irregular (incidental) disturbances, superposing their own envelope function on the CTF. The gradual attenuation of high spatial frequencies is best illustrated by the fading of Thon rings (Figure 18.4a). The extent of Thon ring visualisation gives an idea about the resolution of the data collected in each image and may be used as a guide for selection of micrographs to be included in the 3D reconstruction. More sophisticated CTF correction calculations involve compensation for the amplitude decay towards high spatial frequencies.

In real space, the same concept is described by the point spread function (PSF) of the microscope, which is an inverse Fourier transform of the CTF. PSF can therefore be thought of as being like the brush with which the microscope paints the image, meaning that every point of the object is convoluted in its projection image with the unique PSF of a given TEM system. This distortion is partially removed by CTF correction, but the TEM projection image always suffers non-recoverable loss of information at CTF zeroes.

A major challenge in data collection for biological TEM comes from the damage to the sample incurred during imaging and from the beam-induced sample movement. To consider this more completely, we now describe the properties of cryo-TEM samples and the procedures for preparing them.

18.3 Experimental Approaches in Biological Cryo-TEM

18.3.1 Sample Preparation for Cryo-TEM

Sample quality is widely recognised as a major bottleneck in cryo-TEM structure determination experiments [12]. While each sample requires bespoke optimization, general considerations are summarised here. Central to any cryo-TEM experiment is the preservation and maintenance of the sample in a layer of so-called vitreous ice [13]. Vitreous ice is a non-crystalline, glass-like form of solid water, in which the biological sample remains hydrated. Vitreous ice is metastable, readily transforming into other crystalline forms of ice if warmed above −150 °C. Therefore, once vitrification has been achieved, the sample must be maintained at liquid nitrogen temperatures during transfer into the microscope and for data collection.

A TEM sample must be prepared such that it is sufficiently thin for imaging. Typically, the sample is applied to a support layer, often made of a thin (5–10 nm) electron-transparent film of carbon; this support layer is itself supported by a copper mesh disc, 3 mm in diameter and together with a mesh and support layer constitute 'the EM grid' (Figure 18.1a). Variations in the material of both the mesh and the support have been implemented, but it is important that both: (i) are somewhat physically tough (given their dimensions) in order to withstand sample preparation steps, (ii) are non-ferromagnetic to avoid distortion of the electron beam, (iii) can rapidly conduct heat away from the sample and (iv) are electrically conductive to minimise charge accumulation arising from incident electrons. Prior to sample application, the normally hydrophobic support surface is typically treated (for example by exposure to low energy plasma) to make it hydrophilic and thereby enable sample spreading and adherence.

The goal in typical molecular cryo-TEM experiments is to determine an atomic/near-atomic resolution structure or set of structures. Thorough biochemical and biophysical analysis of the sample to establish its compositional homogeneity is vital prior to initiation of structural experiments. This should include >98% purity by sodium dodecyl sulphate-polyacrylamide gel electrophoresis (SDS-PAGE), unambiguous chromatograms from size exclusion chromatography, ideally size exclusion chromatography-multiangle light scattering (SEC-MALS) analysis, yielding approximate molecular dimensions and validation of biologically relevant activity. Cryo-TEM is an inefficient and very expensive tool for screening biochemical sample quality. The best image contrast will be achieved if the sample is embedded in a film of pure frozen water. While the molecular stability of molecular complexes in water is unlikely and their physiological relevance uncertain, EM-optimised buffers should also be established.

With a biochemically high quality sample in hand, it is very important to first check its stability, behaviour and appearance by negative stain TEM [14]. In a typical negative stain experiment, a relatively low concentration (~0.01–0.05 mg/ml) of sample is applied to the carbon surface of a TEM grid. Excess liquid is blotted away and a thin layer of heavy metal stain – such as 1% (w/v) uranyl acetate (pH 4.2) or 2% methylamine tungstate (pH 6.8) – is applied on top of the sample and allowed to dry (~1 minute on the lab bench). The stain dries around the sample of interest, forming a cast of heavy metal that strongly scatters electrons in the TEM, thereby offering a high (amplitude) contrast view of the stained sample. Commonly used stains have high density and stability, small grain size and are chemically inert, giving confidence that the resulting images provide reliable information about the sample. However, it will be clear from this outline that the limitations of negative stain experiments include a requirement that the samples are dehydrated, potentially distorted by the stain and that the images provide a view of the cast formed by the stain rather than of the sample itself. Nevertheless, negative stain TEM experiments have the distinct advantage of being quick and straightforward to perform, allowing multiple samples/conditions to be efficiently screened.

Once sample quality and parameters for stability have been established, preparation of cryo-TEM samples can proceed. As a rule of thumb, higher sample concentrations (~0.1–2 mg/ml) will be required for cryo-TEM experiments than for negative stain studies of the same sample. However, there are a very large number of parameters that can affect sample distribution on cryo-TEM grids – e.g. buffer contents, grid type, grid surface treatment – such that protein concentration will be just one of multiple experimental variables to be

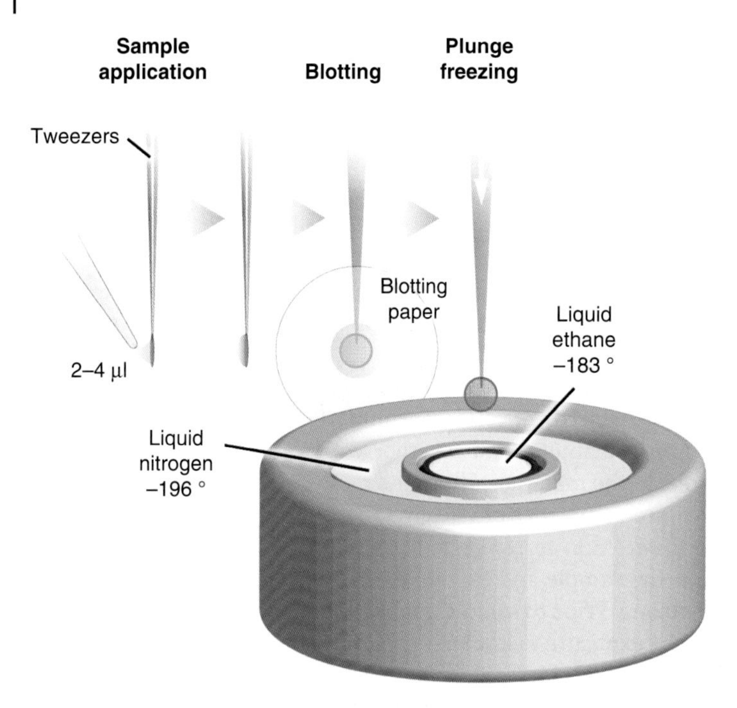

Figure 18.5 Cryo-TEM sample preparation. A cryo-TEM grid, the surface of which has been treated to facilitate sample spreading, is held in tweezers and a small volume of sample is applied. The majority is removed by on-face blotting and the thin layer of water is immediately plunged into liquid ethane, cooled by liquid nitrogen. The vitrified grid must be kept at liquid nitrogen temperatures thereafter. These steps are usually performed in an enclosed chamber with high humidity to prevent water evaporation from the sample.

optimised during cryo-TEM grid preparation. As for negative stain experiments, cryo-TEM sample preparation typically involves application of a small volume of sample to the treated surface of a carbon-coated copper grid. However, unlike negative stain experiments, most cryo-TEM samples will be applied to a layer of carbon that has holes in it, typically with a diameter of between 1 and 4 μm. During sample preparation, excess liquid is blotted away from the surface of the grid, leaving a very thin (~100 nm) layer of sample suspended within the holes in the carbon support. The grid is then plunged immediately into liquid cryogen maintained at ~−183 °C, at which temperature the sample is vitrified (Figure 18.5). The rate of freezing is critical to the success of this step, requiring freezing speeds of 10^5–10^6 deg/s [13], thereby preventing reordering of water molecules into ice crystals. The choice of cryogen is clearly vital, since it must be liquid below −180 °C. Typically this is liquid ethane, liquid propane or a mixture of the two [15]. Although liquid nitrogen is a much more commonly available cryogen, it cannot be used for vitrification because when it contacts warmer objects – such as the EM grid – it boils, forming an insulating layer of nitrogen gas. This so-called Leidenfrost effect means that the rapid rates of freezing necessary for vitrification cannot be achieved. Liquid nitrogen (at −196 °C) is, however, routinely used for preservation of cryo-TEM samples, both for storage and during imaging in the microscope.

Water is a poor thermal conductor, so the samples must be sufficiently thin (<1 μm) to achieve the rapid freezing required for vitrification. However, with these thin samples, fast freezing can be achieved using homemade devices that effectively drop the grid (held tightly in a pair of tweezers) a distance of ~30 cm into the cryogen under gravity. Hydraulically controlled devices are also available commercially. The rate of water evaporation from these thin samples is very high and, if uncontrolled, can result in increases in sample ionic strength, which can in turn affect sample stability and appearance. This can be readily controlled by enclosing the grid preparation steps in a small chamber with high humidity [16].

A successful freezing experiment will result in multiple well-preserved cryo-TEM grids, typically stored in small plastic grid boxes under liquid nitrogen. In theory, the sample will be evenly distributed across the entire grid surface in a uniform layer of ice and present with all possible orientations. On a single such cryo-TEM grid with 80 000 holes (~800 grid squares and ~100 holes per square) and a 'typical' sample distribution yielding ~1500 particles perhole, a dataset arising from such an idealised grid would be more than adequate to yield an atomic resolution reconstruction [5]. In practice, this rarely happens. For example, there is often a gradient of ice across the grid, giving areas where the ice is too thick to visualise particles and/or too thin, thereby excluding larger particles [17]. Crystalline ice contamination can adhere to the surface of the vitreous ice layer occluding embedded particles; this contamination often occurs during storage or transfer into the microscope or less often during data collection. A topic of great current interest for the cryo-TEM field is the effect of the air–water interface in the thin sample film immediately prior to freezing. Although the precise nature of these effects is not well understood, the way in which a given sample interacts with this interface may profoundly affect the dispersion that is ultimately visualised in the TEM and may include denaturation of sensitive samples previously visualised by negative stain. Experimental conditions for optimal sample preparation must be empirically determined for each new sample [12].

18.3.1.1 Sample Preparation of Thicker Specimens

Cryo-Focused Ion Beam Milling (FIB) The plunge freezing approach described above can be applied equally to molecular complexes, small organelles, bacteria and thinner regions of eukaryotic cells. However, not all of these 'vitrifiable' samples are sufficiently thin for cryo-TEM imaging to be feasible. For these thicker (>500 nm) samples, the relatively new approach of cryo-focused ion beam (cryo-FIB) milling can be used [18]. In this treatment, the plunge-frozen sample is thinned in situ by a beam of ions that cuts a thin section (~10 μm in x, y; 200–300 nm in z) in its middle. Aided by SEM imaging within the cryo-FIB system, a layer of protective organometallic platinum is first deposited on the sample surface. The gallium ion beam is then directed at a glancing angle at the sample surface, leaving a thin plank of undistorted vitreous section in the middle (the so-called lamella). This is then transferred into a cryo-TEM for tilt series data collection. The details from the middle of the cell that emerge from such treatment are truly spectacular. Hwever, this methodology is still very much at the cutting edge of cryo-TEM cell biology methods. For example, it is currently not possible to target particular regions for milling. The milling itself is also slow (~1 hour per lamella) and devitrification of the sample by the FIB is common, rendering the throughput of these experiments painfully slow – less than 10 lamella a day from an expert user. Numerous technical developments, including the ability to correlate fluorescence light microscopy with FIB/SEM imaging, would allow targeting of particular regions of interest [19].

Cryo-EM of Vitreous Sections (CEMOVIS) There are a large number of biological samples – larger cells, tissue – that are too thick to be vitrified by plunge freezing at atmospheric pressure. However, vitrification of thicker samples can be achieved with high pressure freezing (HPF). Since the crystallisation of ice is temperature- and pressure-dependent, HPF devices apply high pressure (210 MPa) as samples are frozen, thereby preventing ice crystal formation [20, 21] and allowing vitrification of samples up to 200 μm thickness. Samples are placed in small, disc-shaped carriers (for example with internal dimensions ~100 μm depth, ~200 μm diameter), the exact design of which depends on the HPF manufacturer. Good heat conduction throughout the sample is essential, so air-filled spaces must be avoided because the air acts as an insulator and disrupts freezing. Furthermore, air-filled spaces collapse under high pressure, potentially deforming the sample. Therefore, samples are surrounded by inert cryoprotectant filler solutions such as biological safety advisor (BSA) or dextran.

Following freezing, samples are removed from sample holders and cut into thin (~50 nm) slices with a cryo-microtome at liquid nitrogen temperatures using specialised diamond knives. The resulting ribbon of frozen sections is laid across an EM grid and transferred into the TEM for imaging or cryo-ET data collection. This overall workflow – called a cryo-electron microscope of vitreous section (CEMOVIS) [22] – is technically difficult and has not been widely adopted. Sectioning of the frozen sample block can cause compression in the direction of cutting and cutting itself can be uneven, causing distorting crevasses on the surface of the section. CEMOVIS currently remains the main method for preparing thick biological samples for cryo-TEM data collection. However, use of cryo-FIB milling is also being actively investigated for processing of these thick samples [23].

18.3.2 Cryo-TEM Data Collection

Once a cryo-TEM sample has been prepared, it must be transferred into the microscope and maintained there without devitrification. The grid is held within the sample holder by a metal ring around its perimeter, with the entire holder maintained under liquid nitrogen. According to the holder/microscope combination, the grid may be inserted into the microscope in a so-called side entry holder. In this configuration, the grid is kept cold inside the microscope via a liquid nitrogen-containing external dewar, which conducts heat away from it. This dewar requires topping up every few hours and causes instability because, for example, vibrations from the bubbling liquid nitrogen cause vibrations of the sample itself, blurring features of interest in the images. More modern microscopes use individual cartridges for holding grids, which are inserted via a top entry stage and maintained by internal cooling with liquid nitrogen, allowing for much improved sample stability [24]. Automated sample transfer systems that allow loading of multiple grids have brought further improvements in sample stability and data collection efficiency.

An important parameter to consider in advance of cryo-TEM data collection is the desired/realistically achievable resolution of the structural experiment. In data collection terms, this will depend on the magnification at which data are collected and the physical pixel size of the image detector (see below); together these parameters define the digital pixel size in the images. As stated by the Nyquist theorem, the best attainable resolution in any imaging experiment is twice the pixel size. Whether the properties of a given specimen will allow this theoretical resolution to be achieved cannot be determined a priori. However, it is important that data collection parameters do not impose limits on this unnecessarily.

18.3.2.1 Image Acquisition for Cryo-TEM: The Resolution Revolution with Direct Electron Detectors (DEDs)

Historically, photographic film plates (6.5 × 9 cm) covered by a thin (0.17 mm) electron-sensitive emulsion were used for image capture in cryo-TEM. This allowed very large fields of view to be captured and with the small grain size of the emulsion allowing very fine sampling. However, the use of film routinely introduced water contamination in the TEM column and the films needed to be developed, dried and scanned (digitised) prior to image processing. The introduction of charge-coupled devices (CCDs) allowed immediate digital image capture and was thus much more practical, especially for collection of ET tilt series and in the development of automated data collection. However, a scintillator layer that converts incident electrons to photons is required, together with a thick fibre optic layer that transmits the image information converted to photons to the actual CCD chip. This complex signal transmission causes spreading (blurring) of the signal from each electron across many CCD pixels with variable intensities. The practical consequence of this is that, while high resolution structures are achievable with CCDs, data must be collected at very high magnification – with a concomitant reduction in the total field of view – to compensate for information blurring. Thus, very large datasets are required to achieve reconstructions of equivalent overall quality to those collected on film [25].

Development and commercial availability of direct electron detectors (DEDs) (also known as direct detector devices (DDDs)) was a major technological breakthrough for the field [26]. DEDs have the ability to capture information from incident electrons directly. This direct electron readout is enabled by radiation-hardened, back-thinned monolithic active-pixel sensors (MAPSs) combined with a complementary metal-oxide-semiconductor (CMOS) chip with amplifiers built into each pixel that converts information from the incident electron into voltage [27]. As a consequence, the information content of the data obtained using DEDs per unit of electron dose is much better compared to CCDs [28]. This is expressed using the detective quantum efficiency (DQE) parameter, which is a measure of the preservation of the SNR in the image (output SNR) in relation to the true image (input SNR). Furthermore, data readout from DEDs is sufficiently fast that instead of recording a snapshot of the sample over ~1 second exposure, movies with multiple frames are collected. DEDs can operate in two modes: (i) the integrating mode involves summing of the charges of incoming imaging electrons and (ii) the counting mode involves monitoring the arrival of each individual imaging electron on the detector, allowing a more precise record of their position and a more precise record of the structural information they carry as a result. Collecting data in the counting mode requires a higher DED frame rate compared to the integration mode (hundreds of frames per second) and a very low electron dose rate to ensure that electrons do not coincide as they arrive at the detector (called coincidence loss). This in turn means that longer exposure times are required.

Collecting movies of cryo-TEM samples confirmed a vital facet of their behaviour: the samples move during electron beam illumination as a result of the interaction of electrons with the grid material and the ice [29]. Such movement causes significant information blurring in single, long snapshots. However, with individual movie frames now available, some of this movement can be corrected for by frame realignment. Some modifications of sample preparation procedures can further minimise the physical sources of sample instability [30]. With all the benefits of the 'resolution revolution', an ever-increasing number of structures of a wide variety of macromolecular complexes have been determined at crystallographic resolutions [31]. With such improvements have come the plausibility of using cryo-EM in drug discovery [32].

18.3.2.2 Data Collection Strategies for SPA

Once a cryo-TEM sample is in the microscope, data collection can begin. For cryo-TEM SPA data collection, the goal is to collect as many 2D projections of the 'single particle' in as many orientations within the ice as possible to ultimately achieve the highest possible resolution. So-called low dose imaging is used, which takes into account that the region of interest – for example the ice in one of the holes in the arrays on a cryo-TEM grid – can only be imaged once before it is damaged or destroyed by the imaging electrons. Therefore, after collecting an overview of the grid at very low magnification (~2000×) and very low electron dose, optimisation of the imaging conditions occurs immediately adjacent to the region of interest. Conducted at high (50 000–100 000×) nominal magnification, this will involve checking for sample movement, correcting image distortions such as objective lens astigmatism and setting the image defocus. It is only after this is completed that the highest quality data are collected on the region of interest itself. The quality of the sample in the hole – particle distribution, ice quality and thickness – can only be fully evaluated after the image/movie has been collected. The exact imaging conditions depend on the particular sample, detector and target resolution, but the magnification used will be the same as for imaging optimisation (i.e. 50 000–100 000×).These well-defined imaging routines can be performed manually, in particular during sample optimization. However, automated data collection programmes can follow these steps iteratively, allowing thousands of images or movies to be collected from a single good grid over the course of several days [33]. Efficient data collection relies on even and consistent distribution of a sample across the grid to avoid accumulation of useless (no particles, too thick ice) data.

18.3.2.3 Data Collection Strategies for Cryo-Electron Tomography (Cryo-ET)

All the considerations of the sensitivity of the specimen to electron dose and the use of low dose imaging also apply to cryo-ET samples. However, the goals of cryo-ET data collection are different to SPA. Tomography is particularly useful for gaining structural insight into irregular objects, such as cellular compartments, isolated organelles or viruses with flexible capsids, and for in situ studies of cellular components, such as cytoskeletal filamentous networks or membrane protein complexes [[34], pp. 7–12]. In such samples, averaging of different copies of the object of interest is rarely possible, so 3D information must be acquired using a different approach. Different views of the individual object are obtained by physically tilting the sample holder – and thus the sample of interest – within the TEM, enabling a series of 2D projections to be recorded over a range of angles (the 'tilt series'; see Figure 18.6). At least 30 projections of the same region are collected through a tilt series.

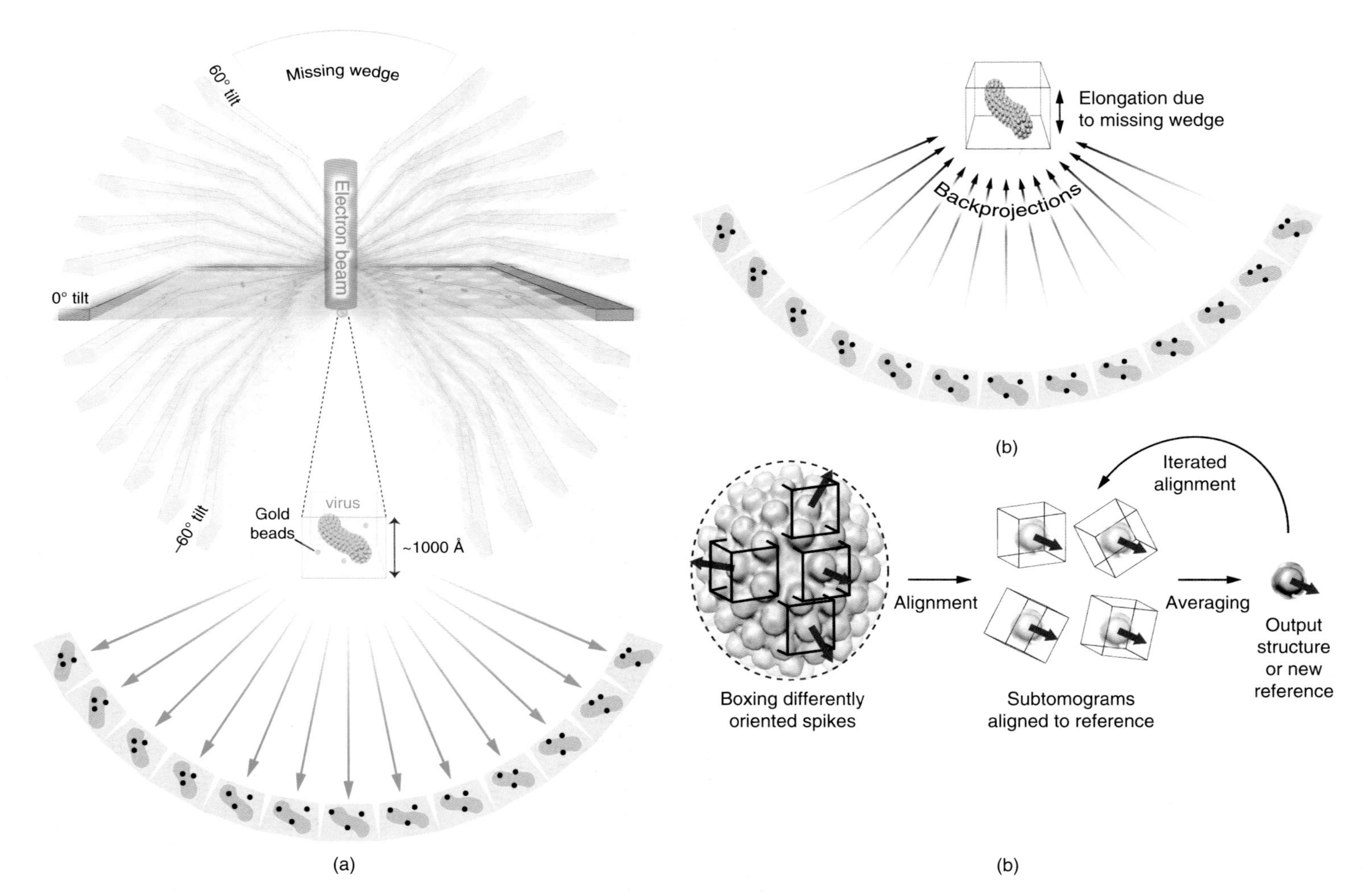

Figure 18.6 Tomographic data collection and subtomogram averaging. (a) Tilt series of an example, irregular Ebola-like virus. Gold beads serve as fiducials for alignment of the projections. (b) Slightly longitudinally distorted (elongated) 3D reconstruction through back-projections with missing views due to tilt limitations. (c) Schematic procedure of subtomogram averaging for the refinement of the viral capsid spike particle in 3D.

The extent of sample tilting is physically limited by the way the grid sits within the microscope, yielding maximum tilts of $\sim \pm 70°$. Thus, a considerable portion of spatial information (in Fourier space) is missing from the tilt series (the 'missing wedge'), ultimately causing distortion in the 3D reconstruction. In addition, as sample tilt increases, so does its thickness, causing a higher number of inelastic scattering events and increasing the noise further in already noisy data. The use of energy filters – either within the TEM column or after it – that exclude inelastically scattered electrons can improve cryo-ET data SNR (Figure 18.3a). Even in modern microscopes, tilted samples are fundamentally less stable within the microscope. However, because cryo-ET experiments do not impose any requirements on the type of sample to be studied, they are very widely applicable. As well as complex cell samples, SPA samples can also be studied in cryo-ET experiments, which may be useful, for example, to investigate the variability of a particle population or its distribution in the ice layer.

18.3.2.4 Special Cases of Cryo-TEM Samples and Data Collection Experiments

Not all cryo-TEM experiments can be strictly classified as SPA or cryo-ET. For example, the first near-atomic resolution TEM structure of bacteriorhodopsin was determined using electron crystallography [35]. In such experiments, both image (amplitude and phase) and diffraction (phase) data are collected from sets of 2D crystalline arrays, with data collected from both untilted and tilted specimens. 2D crystalline datasets may also suffer from the missing wedge issue associated with ET data, leading to non-isotropic resolution in the final reconstruction. There are also relatively few samples that form the crystalline arrays required for this type of experiment. However, early work on electron diffraction (ED) was important in the evolution of structural EM. Inclusion of data collection and image processing details specific to this sample type is beyond the scope of the current chapter but have been reviewed elsewhere [9, 36–38].

3D Micro-ED is another type of cryo-TEM data collection experiment and offers an alternative to X-ray crystallography for very small (<1 μm) 3D crystals [39]. The crystals are subjected to plunge freezing for cryo-TEM but only diffraction data are collected from continuously tilted crystals, a data collection scheme that requires adjustment of the imaging setup in the TEM. These data yield multiple views of the reciprocal lattice (diffraction patterns). Since only amplitude information is collected, the same phase problem that exists for X-ray diffraction data must also be solved in micro-ED. However, molecular replacement methods can be applied or, in the best case, high resolution structures (1.0 Å) can be solved ab initio [40].

Specialised data processing software exists to deal with data arising from these experiments. Although many of the general principles from more common approaches apply to such algorithms, the specifics will not be described further here. Next, we provide an overview of the image processing pipelines for standard SPA and cryo-ET experiments.

18.3.3 Image Processing for Cryo-TEM Samples

18.3.3.1 Cryo-TEM Pre-processing

As described above, modern image acquisition experiments using DEDs involve collection of movies. Prior to any structure determination experiment, the individual frames from these movies are aligned, thereby minimising blurring effects from beam-induced sample movement. These series of frames also capture the gradual accumulation of radiation-induced damage during imaging. Graduated weighting of these frames – known as dose-weighting, in which structural information from early frames is incorporated into the structures with greater weight than that in later frames – is also iteratively incorporated in subsequent processing steps.

In a given cryo-TEM experiment, each image will be collected at a unique location on the grid and local variations in, for example, ice and support film thickness will cause variations in contrast. Therefore, another important pre-processing step, called normalisation, is performed in which the mean density of all images in a dataset is set to zero and the standard deviations of their densities are scaled against a single (arbitrary) value. This is very important for ensuring that density variations relating only to sample structure drive all subsequent steps. CTF determination is also usually undertaken early in the image processing pipeline, but correction and iterative refinement of the relevant parameters can be incorporated into later refinement steps.

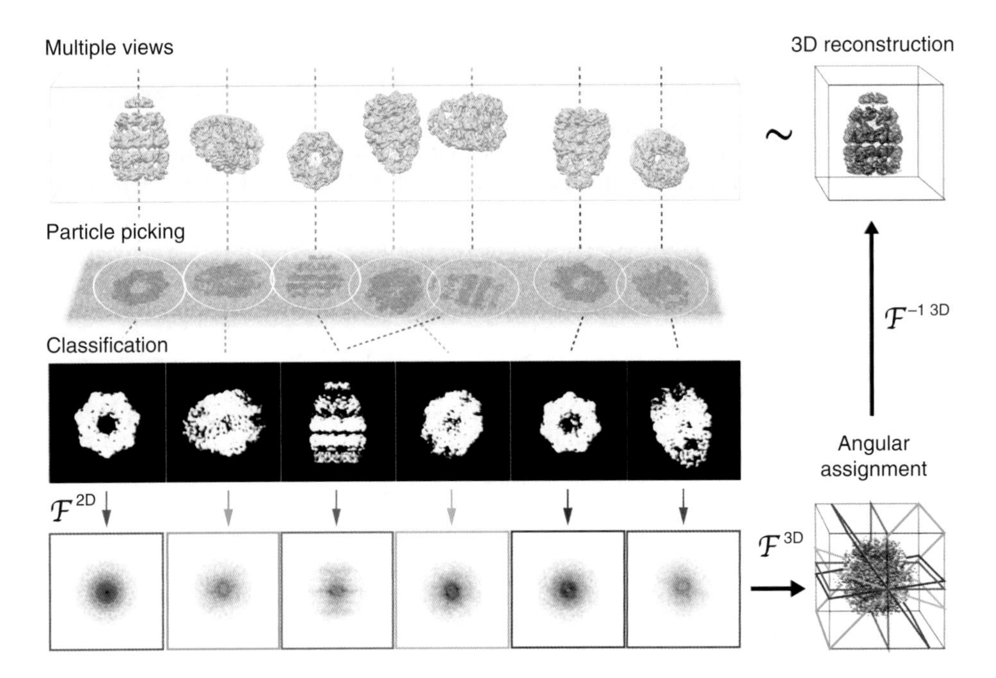

Figure 18.7 Central section theorem and the basic strategy for single particle 3D reconstruction in Fourier space. Single particle projection images contain multiple views of the particle (dashed lines). These views are isolated (particle picking), classified, aligned and averaged. For illustration of the concept, seven simulated particle views are picked (circled) and classified into six simulated classes, as two of the views are related by in-plane rotation and belong to the same class (indicated with blue lines). These so-called class averages have much higher SNR than individual particle views picked from micrographs. The central section theorem states that a 2D Fourier transform of a projection image of an object has the same amplitudes and phases as the central section of the 3D Fourier transform of that object at the plane perpendicular to the projection angle. Multiple projections are assigned spatial angles (Euler angles) in Fourier space (2D Fourier transform images shown with inverted contrast) and populate the 3D Fourier transform (as colour coded) of the object of interest (the example object here is the GroEL/ES protein, corresponding with Figure 18.1). The remaining volume is interpolated from the recorded central slices on to a regular coordinate system (usually Cartesian) and the inverse 3D Fourier transform recovers the structure of the object of interest; ~ denotes similarity to the true object.

Finally, it should be noted that the transition from collecting cryo-TEM images to movies has dramatically increased the amount of data produced in a cryo-TEM experiment and thus the concomitant processing power and storage needed to deal with them. Many of the memory-intense processing operations now take advantage of GPU acceleration [41].

18.3.3.2 Structure Determination Strategies for SPA

Structure determination by cryo-TEM is based on the central section theorem [1], which states that the Fourier transform of a 2D projection from a 3D specimen corresponds to the central section through the 3D Fourier transform of the 3D object. Therefore, the 3D Fourier transform of the object can be built by combining multiple 2D Fourier transforms from all possible angular projections (Figure 18.7). Ultimately, the 3D structure in real space can be calculated using inverse Fourier transformation. Inevitably, this involves a certain level of interpolation, since the angular space is never completely filled with data. Although not all SPA algorithms work in Fourier space, these structural principles still apply.

SPA structure determination therefore requires that the many thousands (sometimes millions) of individual noisy 2D projections of the sample are oriented with respect to each other and combined to yield the 3D structure. In general, the larger the dataset, the higher the expected resolution in the final reconstruction. Many excellent and detailed reviews have been written, both about theoretical considerations at each processing step

[9, 42, 43] and about the particular features associated with individual software packages [44–50]. In addition, as the properties of cryo-TEM images and movies are better understood, it has become possible to automate more steps in the processing pipeline. Software is also increasingly user-friendly, with GUIs being common. Here we aim to summarise qualitatively the key steps involved.

A first step in any SPA processing scheme involves picking individual particles from the micrographs and placing them in a digital 2D box of defined size on which further processing steps will be performed. For the early phases of any new project, manual particle picking is recommended to become familiar with the data. However, automated particle picking is the only realistic option for the very large datasets usually collected for high resolution structural determination. Picking algorithms typically require a target template – often a set of low resolution projections – and usually incorporate some false-positives ('junk') into the data, which can be removed later. After initial structures are calculated from a given dataset, they can be used to generate new projections to reinterrogate the micrographs, potentially extracting more useful particles. As would be expected, particle picking becomes increasingly challenging for smaller and/or more heterogeneous samples.

After this, computational alignment of all the 2D particle views is performed, followed by 2D classification – sometimes relatively crude – of these views using statistical methods, and calculation of 2D averages of similarly appearing particle projections. Junk particles can be easily recognised and removed at this stage. The averaging provides 2D views from the dataset with much improved SNR. This first processing stage provides vital information about the range of views present; it is often the case that molecular complexes assume a more limited range of orientations in the cryo-TEM ice layer, adopting so-called preferred orientations. Further sample optimization or collection of even more data is typically used to deal with this problem. Initial indications about sample heterogeneity – either contamination by other molecules or conformational heterogeneity – may also be evident at this stage. Increasingly, with high quality and large datasets, evidence of secondary structural information is often seen in the 2D classes, providing confidence to proceed. If not, reviewing sample quality or imaging conditions may be required.

The next critical step in the processing pipeline is the determination of the angular relationship between the 2D views to generate the 3D structure. Until recently, most reconstruction algorithms required an initial 3D model of the target structure to provide simulated reference 2D projections against which the 2D data were matched. Such a model needs to be filtered to include only very low-resolution features (a lowpass filter at ~40 Å). This filtering step reduces the very real possibility of the noisy cryo-TEM images aligning to any reference and reproducing its structure, regardless of whether there are genuine similarities between the model and the sample (so-called Einstein-from-noise effect [51]). In the case of samples where such suitable prior structural knowledge exists, this projection matching approach works well and the resulting 3D structures can be validated with the emergence of novel features consistent with the input data. However, the absence of a suitable model for a completely new molecule or complex is a 'catch-22' situation. In these cases, low resolution initial models may be generated using cryo-ET, but often a featureless 3D blob of the appropriate dimensions also works well. The major challenge for very small macromolecular complexes (<100 kD) is the ability to accurately distinguish different 2D views.

Recent innovations in image processing implement the use of maximum likelihood algorithms [44, 49, 52, 53] and have transformed computational approaches to cryo-EM structure determination. Furthermore, stochastic gradient descent allows de novo 3D structure determination, removing the need for an initial reference model and the risk of model bias [54]. Together with the implementation of user-friendly interfaces and GPUs for calculations, 3D reconstructions can now be determined in a matter of hours, rather than over the course of weeks, as was the case previously. 3D classification is now also readily performed, such that SPA experiments typically result in the determination of a set of 3D structures that can capture functionally relevant conformational heterogeneity. Improved efficiency of processing also allows many more processing parameters to be tested to identify the optimal set for a given sample. Expert image processing labs will typically develop processing pipelines that use the best aspects of multiple software packages tuned according to the particular sample.

18.3.3.3 Structure Determination Strategies for Cryo-ET and Use of Subtomogram Averaging

The central challenge in SPA – determining the relationship between different particle views – is not present for cryo-ET data, since the angular relationship between views is defined by the tilt series collection parameters [34]. Nevertheless, imprecisions during sample tilting mean that individual tilt views do not perfectly align with each other. Since the quality of the 3D map depends on accuracy of the projection alignment, this must be optimised prior to 3D reconstruction [55]. Features from within the images themselves may be used to facilitate alignment; alternatively, high contrast gold beads can be included with the sample during vitrification, serving as fiducial markers for alignment, and can be particularly helpful at high tilts. The CTF must also be corrected in each of the tilt images, bearing in mind that – in contrast to SPA samples – the CTF varies across each image due to its tilted configuration and thickness.

A range of computational approaches are implemented for cryo-ET reconstruction, including the popular weighted back-projection approach, in which information from each 2D tilt image is effectively computationally smeared ('projected back') into 3D (Figure 18.6b). Appropriate weighting of different spatial frequencies must be applied to avoid dominance of low resolution features in the final 3D structure. The particular choice of reconstruction algorithm depends on the ultimate goal of the experiment – i.e. extraction of biological information direct from the reconstructed tomogram or further processing of the 3D volume. The final resolutions of cryo-ET reconstructions are typically limited to ~40 Å due to the radiation damage incurred during data collection. Furthermore, a missing wedge of data is intrinsic to cryo-ET structures, leading to distortion of features that must be interpreted with caution.

However, if the 3D tomogram contains multiple copies of the same substructure, these limitations can be largely overcome by averaging subregions, in a procedure known as subtomogram averaging. Defined subregions extracted from the tomographic volume in various orientations in 3D, aligned on common Fourier components to avoid the alignment bias from the missing wedge, may be classified and are then averaged (Figure 18.6c). This effectively removes the missing wedge distortion from the averaged structure, improves the structural SNR and thus its final resolution, sometimes by a substantial amount – see Case Study 2.

18.3.3.4 Resolution Determination and Structure Interpretation

In optical terms, resolution can be thought of as the smallest reliable detail in an image or object that can be distinguished and reflects the SNR in the data, itself a function of dataset size. As mentioned throughout this chapter, there are many factors – relating to both the microscope and the sample – that can limit resolution in cryo-TEM structure determination experiments. In practice, it is not trivial to calculate the resolution of a given reconstruction, and while there are accepted standards in the field, they have limitations, continue to spark debates in the community and should be used judiciously [9].

The most widely used measure of resolution is the Fourier shell correlation (FSC). In the FSC calculation, data contributing to a given structure are randomly split in half and so-called half-maps are computed. The SNR of Fourier components of each half-map are then compared (correlated) across spatial frequencies. The spatial frequency at which the FSC drops below a particular value (more conservatively = 0.5; more commonly = 0.143) is taken as the resolution of the full reconstruction. As is clear from this qualitative description, strictly, the FSC is a measure of data correlation (self-consistency) rather than resolution per se. The ability to see particular features in structures – visualisation of protein α-helices at subnanometre resolution, separation of strands in β-sheets at better than 4 Å resolution, together with densities corresponding to bulky amino acid side chains – is an essential accompaniment to any resolution evaluation. To visualise these protein secondary structure features, the drop of information at higher spatial frequencies due to the envelope function of the microscope must also be corrected for – so-called density sharpening or temperature factor scaling – and is often incorporated into 3D reconstruction algorithms. Numerical resolution measures also cannot capture the variations of resolution that exist in nearly all cryo-TEM-determined structures. These arise from a variety of factors including sample flexibility, heterogeneity and incomplete population of 3D Fourier space due to preferred orientations of SPA samples or the effect of the missing wedge in cryo-ET structures.

After the approximate resolution of a cryo-TEM reconstruction has been established, the reconstruction needs to be interpreted for biological insight. For molecular structure determination experiments, this nearly always involves calculation of an atomic or pseudo-atomic model. The most appropriate method for calculating such a model will be determined by the resolution of the structure. For atomic or near-atomic resolution reconstructions, computational approaches for model building into density – often in tandem with manual manipulation of models – are commonly used, together with validation derived from X-ray crystallography [45]. In 'medium' resolution (5–20 Å) structures, interpretation relies on the availability of existing structural models or predictability of secondary structural regions that can be fitted or docked inside the cryo-TEM derived molecular envelope with variable degrees of precision [46].

On the other hand, although cryo-ET reconstructions of cells are determined at low resolution, they are more complex because the entire contents of the cell are captured [47]. Interpretation of such data is usually achieved via segmentation, in which particular cellular components are highlighted within the cellular volume and their features characterised further. This was previously achieved by hand – a very laborious task – but increasingly, automated segmentation is used, in which, for example, machine-learning algorithms can be trained to identify particular cellular components [48].

18.4 Cryo-TEM Case Studies

18.4.1 Case Study 1: Single Particle Analysis of Microtubule Stabilisation by a Neuronal Microtubule-Associated Protein Doublecortin

Microtubules (MTs) are dynamic polymers of α/β-tubulin dimers (110 kDa) that play a central role in cell division, intracellular transport and cell migration. Individual tubulin dimers can be studied with different structural biology techniques, but MTs are perfectly suited to cryo-TEM structure determination, due to their size (~300 Å diameter and ~μm length).

Tubulin assembles longitudinally in a head-to-tail manner, forming protofilaments (PFs). The cylindrical structure of an MT emerges through lateral association of PFs, preferentially forming homotypic lateral contacts, i.e. α- and β-tubulin lie next to each other. *In vitro*, MTs polymerise with a range of PF numbers, but in cells most MTs have a 13-PF architecture. This architecture necessitates that at one site called the seam, α- and β-tubulin form heterotypic lateral contacts, breaking the otherwise helical symmetry of an MT cylinder. Thus 13-PF MTs are pseudo-helical.

Many proteins in cells bind to MTs, regulating their behaviour and extending their repertoire of functions. One such protein is doublecortin (DCX), which nucleates and stabilises MTs [56]. DCX comprises two ubiquitin-like domains, the N-terminal doublecortin and the C-terminal doublecortin domains (NDC and CDC, respectively) connected by a 42-residue linker. It is specifically expressed in immature neurons, where it is indispensable for neuronal migration [57, 58]. DCX mutations cause defective brain development (lissencephaly or smooth brain) associated with severe intellectual disability and epilepsy.

One of us (CM) has been researching a doublecortin-microtubule (DCX-MT) interaction using cryo-TEM since 2004 [56], initially using data collection on film with a 120 kV FEG microscope and Fourier–Bessel reconstruction methods on the subpopulation of *in vitro* polymerised MTs with strict helical symmetry (PHOELIX [59]). More recently, SPA approaches to MT reconstruction have been introduced [60, 61] that can be applied to pseudo-helical 13-PF MTs. In this approach, the long MT tube is treated as a series of connected single particles. The alignment parameters of the adjacent MT segments are linked by the overall MT architecture, but can be refined separately, allowing distortions in the MT lattice to be corrected. The presence of the seam can also be taken into account. Implementation of SPA approaches led to calculation of structures of 13-PF MTs bound by DCX in the Moores lab, the first sets using 200 kV FEG microscope and film [62, 63]. Most recently, we have used 300 kV microscope and a DED for data collection (Manka and Moores [64]), again using the SPA method for structure determination. Comparison of the DCX-MT structures over time provides an insightful overview of the progress in cryo-TEM technology (Figure 18.8) and how the increasing resolution of a biological structure aids understanding of the biological system in question.

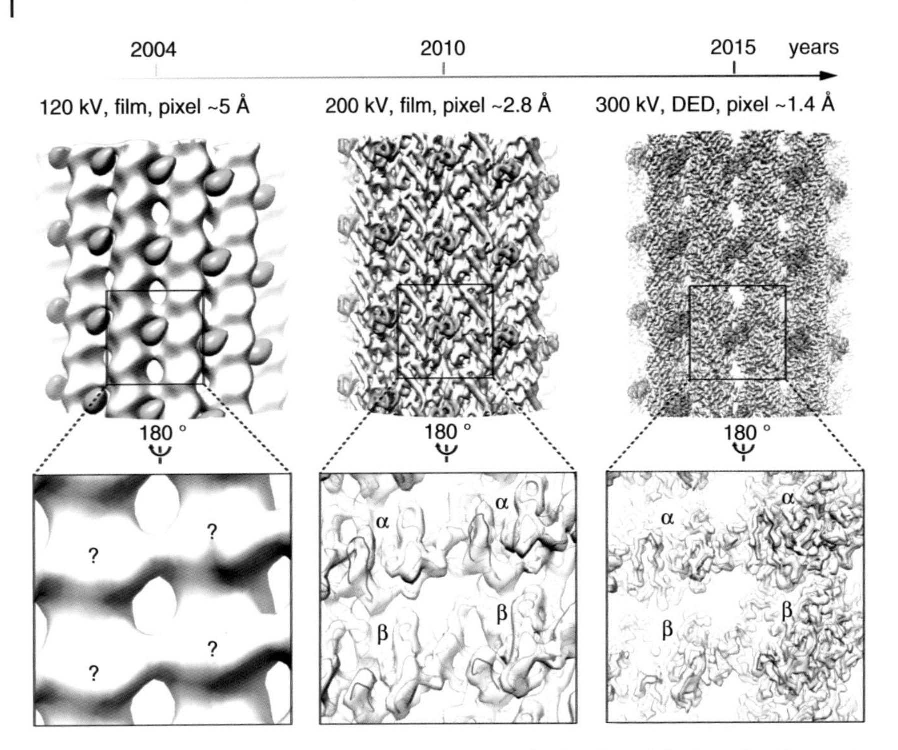

Figure 18.8 Timeline of cryo-TEM 3D reconstructions of DCX-MT and the benefit of increasing resolution. Top row shows side views of MT cylinders (white) decorated with DCX (teal). At the bottom are close-up views of the framed regions of the MT lumen; DCX is not visible in the fenestrations due to depth-cueing. Loops coloured with red (bottom middle and right panels) are the differentiating characteristic between α- and β-tubulin (labelled). Resolution of the structure from 2004 (~20 Å) does not allow differentiation between tubulin monomers (question marks), whereas the one from 2010 (~8 Å) has secondary structure resolution (tubulin model fitted as ribbon), enabling clear localization/identification of tubulin monomers. The most recent structure (~3.8 Å) shows resolved side chain densities (tubulin model fitted as backbone trace) enabling precise examination of intermolecular interfaces. The 2010 and 2015 maps are deposited in the EMDB with the following accession codes: EMD-2095, EMD-3964, respectively.

The DCX-MT structure from 2004 revealed that DCX binds to MTs between PFs, providing the first structural insight into the mechanism of MT nucleation and stabilisation by DCX. From this structure, it became clear that DCX stabilises lateral association of PFs. The structure from 2010 reached subnanometre resolution, providing separation of protein secondary structures. This enabled differentiation between similar α- and β-tubulin subunits based on a difference in the length of one of their loops (Figure 18.1, coloured red). Another facet of MT stabilisation by DCX was then revealed: that it binds the MT lattice in the vertex of four tubulin dimers, thus stabilising not only lateral but also longitudinal lattice contacts. The most recent near-atomic resolution DCX-MT reconstruction allowed direct atomic model refinement in the density. With the resolution reaching ~3.4 Å in the MT lattice core (Figure 18.8), it is now possible to study molecular interfaces in the complex with high precision. Of note, the nucleotides bound to tubulin are also readily visualised in these reconstructions and shed light on GTPase-linked mechanisms of MT dynamic instability [64]. Below we summarise sample preparation, data collection conditions and image processing for the highest-resolution DCX-MT structure.

18.4.1.1 DCX-MT Sample Preparation and Cryo-TEM Data Collection

Human DCX is recombinantly expressed in *Escherichia coli* and incubated with bovine brain tubulin purchased from Cytoskeleton Inc. (Colorado, USA) for 30 minutes at 37 °C to nucleate MTs with the 13-PF architecture promoted by DCX. The polymerised, DCX-bound MTs are applied to a glow-discharged holey

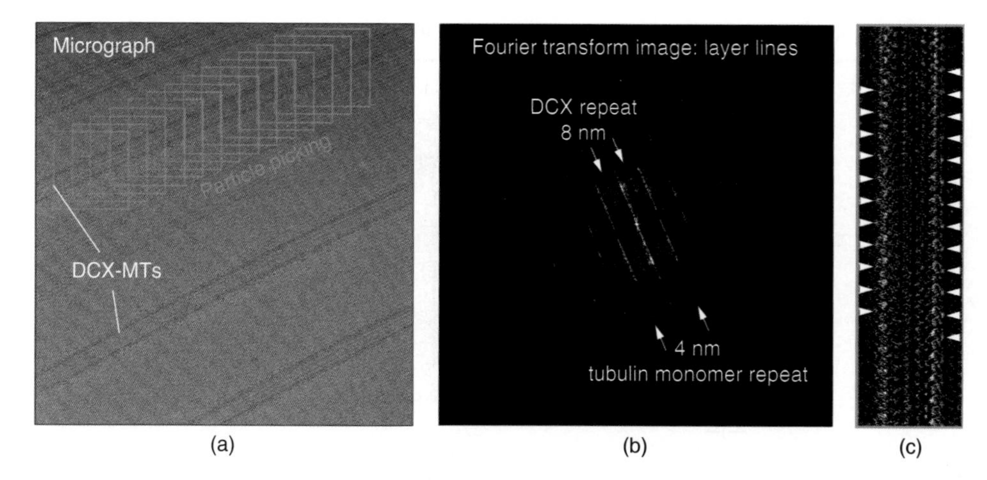

Figure 18.9 DCX-MT particle picking (boxing) and the micrograph quality assessment. (a) Example micrograph of DCX-MTs and the illustration of boxing. (b) Fourier transform of the DCX-MT micrograph shown in (a), with characteristic layer lines; DCX decoration produces 1/8 nm in space layer lines in Fourier space, corresponding to binding every 8 nm in real space. (c) Diagnostic segment average of the DCX-MT boxed in (a), showing MT decoration by DCX every 8 nm (arrowheads).

carbon grid and then transferred to Vitrobot (FEI/Thermo Fisher Scientific) for automated blotting and freezing in liquid ethane. Micrographs are acquired on a 300 kV Polara microscope (FEI) combined with a K2 Summit DED camera (Gatan) operating in counting mode with a dose of 23.5 e-/Å^2 and using an energy filter with a 20 eV slit.

18.4.1.2 Image Processing and 3D Reconstruction

Movie frames were aligned with MotionCor2 [65] before MT segments were cut along MTs with 652 × 652 pixel boxes (Figure 18.9a) using an EMAN software package [66]. The boxes spanned ~11 tubulin dimers and overlapped by ~8 dimers.

These segments were subsequently treated as single-particle input to *Chuff* [60, 62], a custom-designed multiscript processing pipeline using the software Spider [67] and Frealign [68]. The initial seam finding alignment was done in Spider by projection matching to a model 13-PF DCX-MT reference filtered to 30 Å. The CTF parameters were estimated with CTFFIND3 [69] and the CTF correction was performed during local refinement within Frealign, producing isotropic 3D reconstructions. Pseudo-helical symmetry was applied and independently processed half-maps were combined in Relion 1.4 [52] and subjected to its standard post-processing routine, involving: (i) estimation of map resolution based on FSC between the two half-maps, (ii) computation of the average sharpening B-factor using the EMBfactor program [70], and (iii) map sharpening using the computed B-factor value. The resolution of the final map was estimated to be 3.8 Å using 0.143 FSC cut-off criterion. Tubulin atomic models were refined in the cryo-EM density. Molecular visualisations were prepared using UCSF Chimaera [71].

18.4.2 Case Study 2: Insight into the Human Immunodeficiency Virus Type 1 (HIV-1) Assembly Using Subtomogram Averaging

Despite ~30 years of HIV research and anti-retroviral drug development, the acquired immune deficiency syndrome (AIDS) pandemic is still ongoing, with ~37 million people worldwide infected with the still incurable HIV [72]. The virus primarily infects CD4+ T cells in the lymphoid tissues, but also harms bone marrow and thymus, preventing T cell regeneration. Severe depletion of the CD4+ T cells leads to excess pathogen load and chronic inflammation.

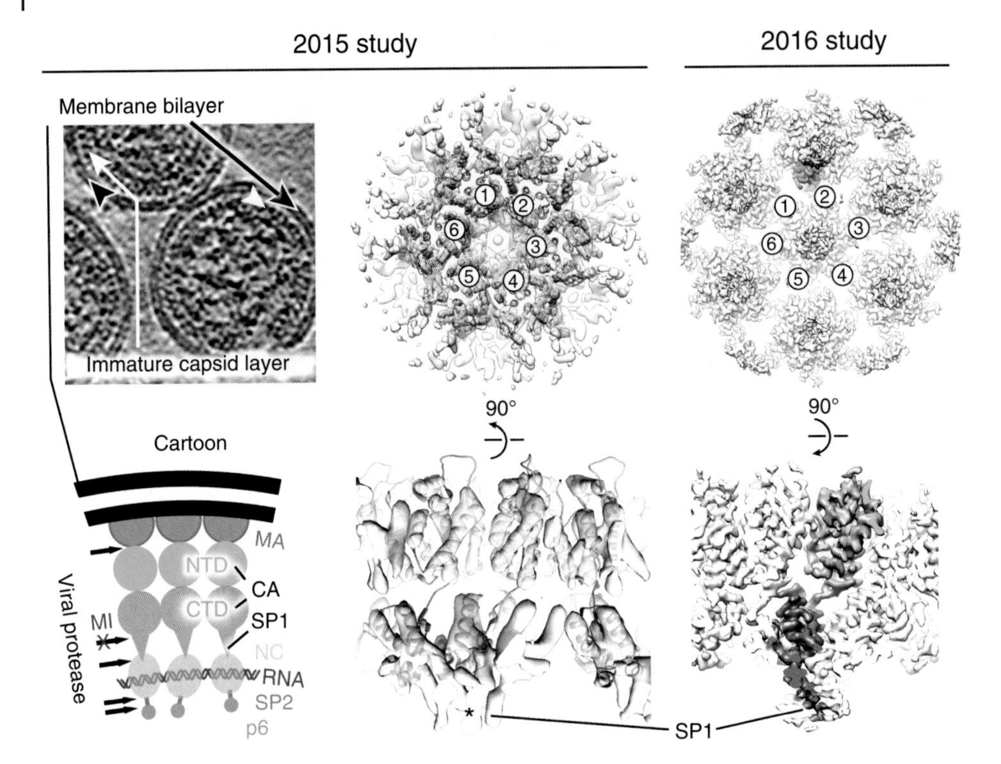

Figure 18.10 Subtomogram average reconstructions of the CA-SP1 lattice of immature HIV-1. Top left, computational slice through a Gaussian-filtered tomogram containing immature HIV-1 particle treated with a maturation inhibitor (MI): amprenavir (APV). (Source: adapted with permission from Springer Nature: Schur et al., 2015 [73]). Bottom left, schematic of the Gag protein domain structure at the plasma membrane with five cleavage sites of viral protease indicated. MA, N-terminal matrix domain; CA-NTD, capsid N-terminal domain; CA-CTD, capsid C-terminal domain; NC, nucleocapsid RNA-binding domain; SP1 and SP2, spacer peptides, p6, C-terminal domain. Middle, CA-SP1 map at 8.8 Å resolution (EMD-2706) viewed from outside the virus (top panel) and from a side (bottom panel). The map is fitted with an atomic model (ribbons) of a single hexameric unit (numbered) surrounded by monomers from the adjacent hexameric units (pdb: 4USN). The unfilled SP1 density is marked with an asterisk. Right, CA-SP1 map at 3.9 Å resolution viewed from inside the virus (top panel) and from a side (bottom panel), with a single CA-SP1 monomer highlighted in colour. The 2015 structure [73] is based on APV-treated viruses and the 2016 structure [74] is based on BVM-treated ΔMACANCSP2 virus-like particles (VLPs) assembled *in vitro*.

The seven stages of the HIV lifecycle can be divided to an early phase – (i) binding to the surface of the host cell, (ii) fusion, (iii) reverse transcription of viral RNA, (iv) integration of viral DNA into host cell genome – and late phase – (v) virus gene expression, (vi) assembly of the main capsid protein Gag (55 kDa) and other proteins at the plasma membrane, and (vii) release of the immature virus. Maturation into the infectious virion is initiated by Gag cleavage at five sites (Figure 18.10, cartoon) by the virus' own protease, triggering dramatic rearrangement of the capsid core structure. This can be prevented by maturation inhibitor (MI) drugs, such as amprenavir (APV) or bevirimat (BVM). Despite therapy, the virus DNA persists indefinitely integrated with the DNA of the long-lived memory CD4+ T cells.

The Gag polyprotein has been a major focus of interest as a mediator of virus maturation. It assembles into a spherically shaped poly-hexameric array at the plasma membrane of the infected cell and bends the membrane, inducing budding. Gag consists of 5 domains: N-terminal matrix domain (MA), followed by two capsid (CA) domains (NTD and CTD), nucleocapsid (NC) domain and the C-terminal p6 domain. The CA and NC domains are connected by spacer-peptide 1 (SP1) and spacer-peptide 2 (SP2) connects the NC domain with the p6 domain (Figure 18.10, cartoon).

The Gag lattice is incomplete (randomly truncated sphere) and exhibits irregular defects; thus, the immature virus particles are heterogeneous in size and morphology, making cryo-ET and subtomogram averaging the ideal method for structural investigation. Using this approach, Briggs and colleagues reported breakthrough 3D reconstructions of the CA-SP1 region within the immature HIV-1 capsid, first in 2015 at subnanometre resolution [73] and then in 2016 – using optimised cryo-ET protocol (see below) – at near-atomic resolution [74].

The first study revealed the positions of the CA domains by fitting all of their alpha-helices to the 8.8 Å resolution density, thus providing insight into the tertiary and quaternary structures of the assembly (Figure 18.10, middle panels). Crucially, the arrangement of capsid N-terminal domain (CA-NTD) was completely different than that predicted from the *in vitro* assembled homologues [75]. This highlights the importance and the power of the in situ subtomogram averaging. However, the highest resolution (3.9 Å) in the 2016 study was achieved using ΔMACANCSP2 virus-like particles (VLPs, without surrounding cell membrane), assembled *in vitro* in the arrangement identical to that in the isolated viruses.

The MA and NC domains did not appear in the averages in either study, as expected from their flexible linkage downstream of the SP1, causing their spatially disordered arrangement relative to the CA-SP1 core. In the 2015 study, the SP1 region seemed to form a six-rod-like bundle around a hollow centre, but no high-resolution nuclear magnetic resonance (NMR) or X-ray structures of that region were known, so it remained unmodelled (Figure 18.1, asterisk). The 2016 study [74] presented the CA-SP1 capsid structure at 3.9 Å resolution, which enabled ab initio modelling of that vital region, including a critical CA-SP1 cleavage site (Figure 18.10, right-hand panels).

Below we focus on the improvements in the cryo-ET and image processing protocol employed in the 2016 study [74] that have led to the resolution boost.

18.4.2.1 Virus Sample Preparation and cryo-ET Data Collection

Production of the APV-treated viruses is described in the 2015 study of Schur et al. [73] and the BVM-treated ΔMACANCSP2 VLPs were generated as described in the 2016 study of Schur et al. [74]. Virus solution containing 10 nm colloidal gold (fiducials) was deposited on a glow-discharged grid and plunge-frozen in liquid ethane using a Vitrobot (FEI/Thermo Fisher Scientific).

Tilt series for both studies were collected on an FEI Titan Krios microscope. In the 2015 study, CCD camera data collection was used while the 2016 used a DED. Crucially, in the 2015 study, the tilt ranged from −45° to +60° in 3° steps, collecting first from 0° to −45° and then from 3° to 60°; whereas in the 2016 study the range was from 0° to 60° and −60° in 3° steps, in a newly developed dose-symmetric tilt scheme (0, +3, −3, −6, +6, +9, −9, −12, etc.) [76]. This methodology ensures that the most information-rich data at low tilts are collected when the least radiation damage has occurred, which can be dose-weighted accordingly.

18.4.2.2 Resolution Improvement Factors

Details of image processing and subtomogram averaging can be found in the Methods sections of the original references. Here, we summarise the main methodological improvements that contributed to the increase in resolution from ~9 Å in the 2015 study to ~4 Å in the 2016 study. The optimised dose-symmetric tilt scheme in the 2016 study allowed a higher electron dose (subject to subsequent filtering) and improved high-resolution information transfer [76]. In the 2015 study [73], the nominal magnification was 42 000, giving a calibrated pixel size of 2.03 Å on the CCD. In the 2016 study [74], this was increased to 105 000, giving a calibrated pixel size of 1.35 Å on the DED – no longer limiting near-atomic resolution. Use of the DED also allowed compensation for beam-induced motion. Thus, image quality in the 2016 study [74] was far superior to that in 2015 [73], with the additional benefit of a larger field of view allowing more efficient collection of a larger dataset. This improved image quality also allowed per-tilt defocus-determination and more precise CTF correction. Lastly, SNR was further improved in reconstructed tomograms with exposure filtering [77].

18.5 Concluding Remarks

Cryo-TEM experiments provide unique insights into the machinery of the cell and enable the study of samples with a very wide range of size and complexity. Cryo-TEM is a rapidly evolving field and incorporation of approaches and technologies from other fields will continue to accelerate innovation in sample handling, hardware and software. The calculation of structures at increasingly high resolution has been inspiring, but it is, of course, important to remember that the usefulness of any 3D structure is determined by the biological insight it provides. A major strength of cryo-TEM is its complementarity with many other techniques. This includes the continuing importance of high resolution structures determined by X-ray crystallography and NMR that support understanding of often heterogeneous and dynamic molecular complexes. At the opposite end of the resolution scale, the ability to correlate light microscopy data with higher resolution cryo-ET views of cells allows deeper and more precise understanding about vital cellular processes. With cryo-TEM now a central technique in structural and cell biology, these are truly exciting times for the field.

Acknowledgements

S.W.M. and C.A.M. are supported by a grant from the Medical Research Council, UK (MR/R000352/1). We thank Giulia Zanetti, Tom Foran (ISMB Birkbeck, London) and Peter Rosenthal (The Francis Crick Institute) for providing invaluable comments on drafts of this chapter.

References

1 De Rosier, D.J. and Klug, A. (1968). Reconstruction of three dimensional structures from electron micrographs. *Nature* 217 (5124): 130–134.
2 Chen, D.-H., Madan, D., Weaver, J. et al. (2013). Visualizing GroEL/ES in the act of encapsulating a folding protein. *Cell* 153 (6): 1354–1365.
3 Slayter, E.M. and Slayter, H.S. (1992). *Light and Electron Microscopy*. New York: Cambridge University Press, Cambridge [England].
4 Hanszen, K.J. (1971). The optical transfer theory of the electron microscope; fundamental principles and applications. In: *Advances in Optical Electron Microscope*, vol. IV (ed. R. Barer and V.E. Cosslett), 1–84. New York and London: Academic Press.
5 Henderson, R. (1995). The potential and limitations of neutrons, electrons and X-rays for atomic resolution microscopy of unstained biological molecules. *Q. Rev. Biophys.* 28 (2): 171–193.
6 Toyoshima, C. and Unwin, N. (1988). Contrast transfer for frozen-hydrated specimens: determination from pairs of defocused images. *Ultramicroscopy* 25 (4): 279–291.
7 Glaeser, R.M. (2013). Invited review article: methods for imaging weak-phase objects in electron microscopy. *Rev. Sci. Instrum.* 84 (11): 111101.
8 Erickson, H.P. and Klug, A. (1971). Measurement and compensation of defocusing and aberrations by Fourier processing of electron micrographs. *Philos. Trans. R. Soc. London, Ser. B* 261 (837): 105–118.
9 Orlova, E.V. and Saibil, H.R. (2011). Structural analysis of macromolecular assemblies by electron microscopy. *Chem. Rev.* 111 (12): 7710–7748.
10 Danev, R. and Baumeister, W. (2017). Expanding the boundaries of cryo-EM with phase plates. *Curr. Opin. Struct. Biol.* 46: 87–94.
11 Danev, R. and Baumeister, W. (2016). Cryo-EM single particle analysis with the Volta phase plate. *eLife* 5: e13046.
12 Passmore, L.A. and Russo, C.J. (2016). Specimen preparation for high-resolution Cryo-EM. *Methods Enzymol.* 579: 51–86.

13 Dubochet, J., Adrian, M., Chang, J.J. et al. (1988). Cryo-electron microscopy of vitrified specimens. *Q. Rev. Biophys.* 21 (2): 129–228.

14 Ohi, M., Li, Y., Cheng, Y., and Walz, T. (2004). Negative staining and image classification – powerful tools in modern electron microscopy. *Biol. Proced. Online* 6: 23–34.

15 Tivol, W.F., Briegel, A., and Jensen, G.J. (2008). An improved cryogen for plunge freezing. *Microsc. Microanal.* 14 (5): 375–379.

16 Thompson, R.F., Walker, M., Siebert, C.A. et al. (2016). An introduction to sample preparation and imaging by cryo-electron microscopy for structural biology. *Methods (San Diego, California)* 100: 3–15.

17 Vinothkumar, K.R. and Henderson, R. (2016). Single particle electron cryomicroscopy: trends, issues and future perspective. *Q. Rev. Biophys.* 49: e13.

18 Rigort, A. and Plitzko, J.M. (2015). Cryo-focused-ion-beam applications in structural biology. *Arch. Biochem. Biophys.* 581: 122–130.

19 Arnold, J., Mahamid, J., Lucic, V. et al. (2016). Site-specific cryo-focused ion beam sample preparation guided by 3D correlative microscopy. *Biophys. J.* 110 (4): 860–869.

20 Kanno, H., Speedy, R.J., and Angell, C.A. (1975). Supercooling of water to −92 °C under pressure. *Science* 189 (4206): 880–881.

21 Studer, D., Humbel, B.M., and Chiquet, M. (2008). Electron microscopy of high pressure frozen samples: bridging the gap between cellular ultrastructure and atomic resolution. *Histochem. Cell Biol.* 130 (5): 877–889.

22 Al-Amoudi, A., Chang, J.-J., Leforestier, A. et al. (2004). Cryo-electron microscopy of vitreous sections. *EMBO J.* 23 (18): 3583–3588.

23 Mahamid, J., Schampers, R., Persoon, H. et al. (2015). A focused ion beam milling and lift-out approach for site-specific preparation of frozen-hydrated lamellas from multicellular organisms. *J. Struct. Biol.* 192 (2): 262–269.

24 Grassucci, R.A., Taylor, D.J., and Frank, J. (2007). Preparation of macromolecular complexes for cryo-electron microscopy. *Nat. Protoc.* 2 (12): 3239–3246.

25 Faruqi, A.R. and Henderson, R. (2007). Electronic detectors for electron microscopy. *Curr. Opin. Struct. Biol.* 17 (5): 549–555.

26 Kühlbrandt, W. (2014). Biochemistry. The resolution revolution. *Science* 343 (6178): 1443–1444.

27 Milazzo, A.-C., Moldovan, G., Lanman, J. et al. (2010). Characterization of a direct detection device imaging camera for transmission electron microscopy. *Ultramicroscopy* 110 (7): 744–747.

28 McMullan, G., Faruqi, A.R., and Henderson, R. (2016). Direct electron detectors. *Methods Enzymol.* 579: 1–17.

29 Brilot, A.F., Chen, J.Z., Cheng, A. et al. (2012). Beam-induced motion of vitrified specimen on holey carbon film. *J. Struct. Biol.* 177 (3): 630–637.

30 Russo, C.J. and Passmore, L.A. (2014). Electron microscopy: ultrastable gold substrates for electron cryomicroscopy. *Science* 346 (6215): 1377–1380.

31 Cheng, Y. (2015). Single-particle cryo-EM at crystallographic resolution. *Cell* 161 (3): 450–457.

32 Subramaniam, S., Earl, L.A., Falconieri, V. et al. (2016). Resolution advances in cryo-EM enable application to drug discovery. *Curr. Opin. Struct. Biol.* 41: 194–202.

33 Cheng, A., Tan, Y.Z., Dandey, V.P. et al. (2016). Strategies for automated cryoEM data collection using direct detectors. *Methods Enzymol.* 579: 87–102.

34 Ben-Harush, K., Maimon, T., Patla, I. et al. (2010). Visualizing cellular processes at the molecular level by cryo-electron tomography. *J. Cell Sci.* 123 (Pt 1): 7–12.

35 Henderson, R., Baldwin, J.M., Ceska, T.A. et al. (1990). Model for the structure of bacteriorhodopsin based on high-resolution electron cryo-microscopy. *J. Mol. Biol.* 213 (4): 899–929.

36 Stahlberg, H., Biyani, N., and Engel, A. (2015). 3D reconstruction of two-dimensional crystals. *Arch. Biochem. Biophys.* 581: 68–77.

37 Abeyrathne, P.D., Chami, M., Pantelic, R.S. et al. (2010). Preparation of 2D crystals of membrane proteins for high-resolution electron crystallography data collection. *Methods Enzymol.* 481: 25–43.
38 Hite, R.K., Schenk, A.D., Li, Z. et al. (2010). Collecting electron crystallographic data of two-dimensional protein crystals. *Methods Enzymol.* 481: 251–282.
39 Rodriguez, J.A. and Gonen, T. (2016). High-resolution macromolecular structure determination by MicroED, a cryo-EM method. *Methods Enzymol.* 579: 369–392.
40 Rodriguez, J.A., Eisenberg, D.S., and Gonen, T. (2017). Taking the measure of MicroED. *Curr. Opin. Struct. Biol.* 46: 79–86.
41 Kimanius, D., Forsberg, B.O., Scheres, S.H., and Lindahl, E. (2016). Accelerated cryo-EM structure determination with parallelisation using GPUs in RELION-2. *eLife* 5: e18722.
42 Frank, J. (2006). *Three-Dimensional Electron Microscopy of Macromolecular Assemblies: Visualization of Biological Molecules in Their Native State.* New York: Oxford University Press, Oxford.
43 Cheng, Y., Grigorieff, N., Penczek, P.A., and Walz, T. (2015). A primer to single-particle cryo-electron microscopy. *Cell* 161 (3): 438–449.
44 Scheres, S.H.W. (2016). Processing of structurally heterogeneous cryo-EM data in RELION. *Methods Enzymol.* 579: 125–157.
45 Grigorieff, N. (2016). Frealign: an exploratory tool for single-particle cryo-EM. *Methods Enzymol.* 579: 191–226.
46 Ludtke, S.J. (2016). Single-particle refinement and variability analysis in EMAN2.1. *Methods Enzymol.* 579: 159–189.
47 Grant, T., Rohou, A., and Grigorieff, N. (2018). cisTEM, user-friendly software for single-particle image processing. *eLife* 7: e35383.
48 van Heel, M., Harauz, G., Orlova, E.V. et al. (1996). A new generation of the IMAGIC image processing system. *J. Struct. Biol.* 116 (1): 17–24.
49 de la Rosa-Trevín, J.M., Otón, J., Marabini, R. et al. (2013). Xmipp 3.0: an improved software suite for image processing in electron microscopy. *J. Struct. Biol.* 184 (2): 321–328.
50 Moriya, T., Saur, M., Stabrin, M. et al. (2017). High-resolution single particle analysis from electron cryo-microscopy images using SPHIRE. *J. Vis. Exp.* 123.
51 Henderson, R. (2013). Avoiding the pitfalls of single particle cryo-electron microscopy: Einstein from noise. *Proc. Natl. Acad. Sci.* 110 (45): 18037–18041.
52 Scheres, S.H.W. (2012). RELION: implementation of a Bayesian approach to cryo-EM structure determination. *J. Struct. Biol.* 180 (3): 519–530.
53 Sigworth, F.J., Doerschuk, P.C., Carazo, J.-M., and Scheres, S.H.W. (2010). An introduction to maximum-likelihood methods in cryo-EM. *Methods Enzymol.* 482: 263–294.
54 Punjani, A., Rubinstein, J.L., Fleet, D.J., and Brubaker, M.A. (2017). cryoSPARC: algorithms for rapid unsupervised cryo-EM structure determination. *Nat. Methods* 14 (3): 290–296.
55 Wan, W. and Briggs, J.a.G. (2016). Cryo-electron tomography and subtomogram averaging. *Methods Enzymol.* 579: 329–367.
56 Moores, C.A., Perderiset, M., Francis, F. et al. (2004). Mechanism of microtubule stabilization by doublecortin. *Mol. Cell.* 14 (6): 833–839.
57 des Portes, V., Pinard, J.M., Billuart, P. et al. (1998). A novel CNS gene required for neuronal migration and involved in X-linked subcortical laminar heterotopia and lissencephaly syndrome. *Cell* 92 (1): 51–61.
58 Gleeson, J.G., Allen, K.M., Fox, J.W. et al. (1998). Doublecortin, a brain-specific gene mutated in human X-linked lissencephaly and double cortex syndrome, encodes a putative signaling protein. *Cell* 92 (1): 63–72.
59 Carragher, B., Whittaker, M., and Milligan, R.A. (1996). Helical processing using PHOELIX. *J. Struct. Biol.* 116 (1): 107–112.
60 Sindelar, C.V. and Downing, K.H. (2007). The beginning of Kinesin's force-generating cycle visualized at 9 Å resolution. *J. Cell Biol.* 177 (3): 377–385.

61 Zhang, R. and Nogales, E. (2015). A new protocol to accurately determine microtubule lattice seam location. *J. Struct. Biol.* 192 (2): 245–254.
62 Fourniol, F.J., Sindelar, C.V., Amigues, B. et al. (2010). Template-free 13-protofilament microtubule-MAP assembly visualized at 8 Å resolution. *J. Cell Biol.* 191 (3): 463–470.
63 Liu, J.S., Schubert, C.R., Fu, X. et al. (2012). Molecular basis for specific regulation of neuronal kinesin-3 motors by doublecortin family proteins. *Mol. Cell.* 47 (5): 707–721.
64 Manka, S.W. and Moores, C.A. (2018). The role of tubulin–tubulin lattice contacts in the mechanism of microtubule dynamic instability. *Nat. Struct. Mol. Biol.* 25 (7): 607–615.
65 Zheng, S.Q., Palovcak, E., Armache, J.-P. et al. (2017). MotionCor2: anisotropic correction of beam-induced motion for improved cryo-electron microscopy. *Nat. Methods* 14 (4): 331–332.
66 Ludtke, S.J., Baldwin, P.R., and Chiu, W. (1999). EMAN: semiautomated software for high-resolution single-particle reconstructions. *J. Struct. Biol.* 128 (1): 82–97.
67 Frank, J., Radermacher, M., Penczek, P. et al. (1996). SPIDER and WEB: processing and visualization of images in 3D electron microscopy and related fields. *J. Struct. Biol.* 116 (1): 190–199.
68 Grigorieff, N. (2007). FREALIGN: high-resolution refinement of single particle structures. *J. Struct. Biol.* 157 (1): 117–125.
69 Mindell, J.A. and Grigorieff, N. (2003). Accurate determination of local defocus and specimen tilt in electron microscopy. *J. Struct. Biol.* 142 (3): 334–347.
70 Rosenthal, P.B. and Henderson, R. (2003). Optimal determination of particle orientation, absolute hand, and contrast loss in single-particle electron cryomicroscopy. *J. Mol. Biol.* 333 (4): 721–745.
71 Pettersen, E.F., Goddard, T.D., Huang, C.C. et al. (2004). UCSF chimera – a visualization system for exploratory research and analysis. *J. Comput. Chem.* 25 (13): 1605–1612.
72 Deeks, S.G., Overbaugh, J., Phillips, A., and Buchbinder, S. (2015). HIV infection. *Nat. Rev. Dis. Primer* 1: 15035.
73 Schur, F.K.M., Hagen, W.J.H., Rumlová, M. et al. (2015). Structure of the immature HIV-1 capsid in intact virus particles at 8.8 Å resolution. *Nature* 517 (7535): 505–508.
74 Schur, F.K.M., Obr, M., Hagen, W.J.H. et al. (2016). An atomic model of HIV-1 capsid-SP1 reveals structures regulating assembly and maturation. *Science* 353 (6298): 506–508.
75 Bharat, T.A.M., Davey, N.E., Ulbrich, P. et al. (2012). Structure of the immature retroviral capsid at 8 Å resolution by cryo-electron microscopy. *Nature* 487 (7407): 385–389.
76 Hagen, W.J.H., Wan, W., and Briggs, J.A.G. (2017). Implementation of a cryo-electron tomography tilt-scheme optimized for high resolution subtomogram averaging. *J. Struct. Biol.* 197 (2): 191–198.
77 Grant, T. and Grigorieff, N. (2015). Measuring the optimal exposure for single particle cryo-EM using a 2.6 Å reconstruction of rotavirus VP6. *eLife* 4: e06980.

Website Resources

https://www.ibiology.org/techniques/transmission-electron-microscopy
https://www.ibiology.org/biophysics/single-particle-cryo-em/#part-1
http://cryo-em-course.caltech.edu/videos
https://www.jove.com/video/52311/do-s-don-ts-cryo-electron-microscopy-primer-on-sample-preparation
https://www.jove.com/video/57199/variations-on-negative-stain-electron-microscopy-methods-tools-for

19

Computer Modelling and Molecular Dynamics Simulation of Biomolecules

Maria Reif and Martin Zacharias

Physics Department T38, Technical University of Munich, James-Franck-Str. 1, 85748 Garching, Germany

19.1 Significance

Computational modelling and simulations have become standard tools in (bio)chemistry. It is important not to use these tools as 'black box' approaches but to understand the underlying physical and algorithmic principles. It requires also careful preparation of the simulation or modelling input by the user and careful analysis of the results. Critical analysis is essential because the computational study of a system always relies on a model representation of the real system and thus comes with approximations that may impact the results [1–3]. A user can, however, only critically analyse the results of a computer modelling or simulation procedure if he/she has sufficient knowledge concerning the limitations of the underlying model. The present chapter will provide a basis for understanding the principles of computer modelling and molecular dynamics (MD) simulation of biomolecules. Further reading suggestions are listed at the end of the chapter.

In general, computational modelling and simulation is widely applied in many areas of science, not only (bio)chemistry. It allows studying systems or processes that are elusive to experimental study, e.g. modelling the crash of a car or aeroplane. In (bio)chemistry, the most important advantage of computational modelling and simulation is that insight complementary to experiment may be provided. For instance, with MD simulations one can 'zoom in' to the dynamic behaviour of biomolecules at atomistic resolution and down to the picosecond-time level and thus gain information that experiments cannot deliver to date. This additional information can be used to suggest novel experiments, e.g. mutations in an enzyme that improve substrate binding or modification of a drug for tighter binding to the receptor.

In 1957, the first MD simulation was performed [4]. The simulated system was very simple, merely consisting of two-dimensional hard disks. In 1964, the first liquid was simulated, argon, for a period of 6 ps [5]. The first molecular liquid, water, was simulated in 1971 for a period of 2 ps [6]. Algorithmic innovations and the development of models representing biomolecules finally allowed the first simulation of a protein in vacuum in 1977 [7] and in water in 1982 for 25 ps [8].

From year to year, researchers are able to study larger (bio)chemical systems for longer simulated time periods, i.e. the time effort required by the computer programs to perform a certain study continues to decrease. This is to a great extent due to hardware advances. Manufacturers keep devising more powerful computer and graphics processing units. The continuous improvements in hardware were, until recently, captured remarkably well by Moore's law [9], stating that the number of transistors in an integrated circuit would double every two years (due to continuous shrinking of transistor size) and hence computer power would also double every two years. This development drove the digital revolution both in science and any other area of life but may have approached an end. Once the atomistic scale has been reached, transistor size cannot be decreased any further along the lines of currently employed technologies. Innovations such as quantum computing [10] or neuromorphic computing [11] are expected to keep advancing the power and influence of computers in science,

Biomolecular and Bioanalytical Techniques: Theory, Methodology and Applications, First Edition. Edited by Vasudevan Ramesh.

technology and daily life. Especially concerning applications in (bio)chemistry, these new technologies appear to offer exciting opportunities [12, 13]. The present chapter, however, deals with the state-of-the-art of molecular modelling and classical MD simulation in (bio)chemistry and the theoretical foundations to provide the reader with an overview and to enable him/her to take further steps in this exciting and rapidly evolving area.

19.2 Theory and Principles

19.2.1 General Considerations

When constructing a computational model of a biomolecular system, four different aspects have to be considered [14]. First, the degrees of freedom (dof) of the system have to be defined. They depend on the spatial resolution at which the system is to be modelled and influence the timescale of motions occurring in the system, as well as, from a practical point of view, the simulation timescale and method. For example, if the chosen dof are atoms including their constituent nuclei and electrons, the fastest motions in the system occur on a timescale of femto- to picoseconds and the simulation method of quantum dynamics [15] may be applied to propagate the dynamics of the system. Given this choice of dof and simulation method, motions that occur on a much larger timescale may be inaccessible due to restrictions in compute power. If, on the other hand, a very coarse-grained depiction of the system is chosen, e.g. the dof are captured by a continuum of velocities, motions in the system occur on a supramolecular to macroscopic timescale and the simulation method of fluid dynamics [16] may be applied to propagate the dynamics of the system. Intermediate between these two regimes is the definition of dof as atoms, small groups of atoms or monomers. Such a description is commonly chosen for bimolecular modelling. Motions in the system then occur on a molecular (picosecond to nanosecond) timescale, the properties of the system can be described with statistical mechanics [17] and its dynamics can be propagated via classical mechanics [18, 19]. In case one is not interested in the dynamics, but in the structural features of a biomolecule, one may model it on the level of its building blocks (monomers, i.e. amino acids or nucleic acid bases) and construct the structure via statistical methods based on database examination (comparative modelling [20, 21]) or 'from scratch' (ab initio modelling [22]).

Second, given a particular set of dof in the system, one has to define how these are interacting, i.e. an energy function has to be defined that returns the potential energy of the system as a function of the coordinates of the dof. The energy function usually used in biomolecular modelling is described in Section 19.3.1.

Third, one has to determine the spatial environment of the model (Figure 19.1). A biomolecule can, for instance, be modelled in vacuum. In most cases this is unrealistic because biomolecules are typically embedded in an aqueous or membrane-like environment and do not preserve their native structure and behaviour in vacuum. Considering that in vacuum there is no dielectric screening of charges, salt bridges and hydrogen bonds are much stronger than in aqueous solution and hence the structure of a biomolecule can deviate significantly from the solvated structure. A biomolecule is more realistically modelled in solution. The solution phase is typically represented by a dielectric continuum of infinite extent (implicit solvation; Section Implicit Solvation) or as a multitude of individual solvent molecules, each modelled in terms of individual atoms or groups of atoms (explicit solvation). The explicit-solvent environment may be represented as a system of finite extent (e.g. a droplet) or of infinite extent, as a periodically replicated array of computational unit boxes, referred to as periodic boundary conditions (PBC). A microscopically small droplet (containing, e.g., 10^5 solvent molecules) gives a poor representation of the true environment of biomolecules in solution because the biomolecule is affected by the presence of the close-by droplet surface. Therefore, biomolecules are typically modelled in a solvent that is represented implicitly or explicitly under PBC.

Lastly, one has to decide how to generate coordinates (configurations) of the model. If one is interested in a static structure, one may obtain coordinates of a biomolecule from a database (e.g. protein databank [23]) or via comparative [20, 21] or ab initio [22] modelling. If one is interested in all configurations of a molecule that are possible at a given temperature, one can sample the entirety of possible configurations with

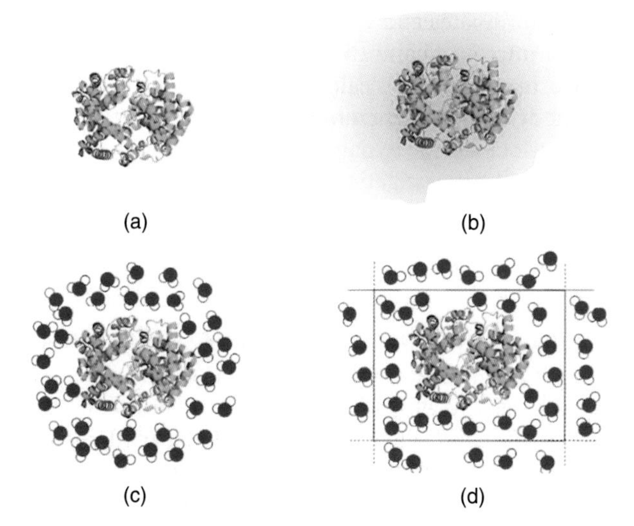

Figure 19.1 Spatial environment of biomolecular models, illustrated for a protein depicted in cartoon representation. Biomolecules may be modelled (a) in vacuum, (b) using an implicit solvent of infinite extent, (c) using explicit solvent molecules in a droplet of finite size, or (d) using explicit solvent molecules under PBC.

a Monte Carlo algorithm [24–27]. This algorithm samples the configurational space of a molecule, such that each configuration occurs with a probability equal to its Boltzmann weight,

$$p = \frac{e^{-\beta U(\mathbf{r}^N)}}{\int d\mathbf{r}^N e^{-\beta U(\mathbf{r}^N)}}, \tag{19.1}$$

where $\beta = (k_B T)^{-1}$, k_B being Boltzmann's constant and T the absolute temperature, $U(\mathbf{r}^N)$ the potential energy of the system as a function of the coordinates of all N particles $\mathbf{r}^N$ and the term in the denominator (configurational integral) is a normalisation constant. A Monte Carlo algorithm to generate biomolecular configurations proceeds as follows. One starts with a molecule with coordinates $\mathbf{r}^{N,\text{old}}$. A random new configuration is explored, e.g. a randomly chosen dihedral angle is rotated by a certain amount. Thus, new coordinates $\mathbf{r}^{N,\text{new}}$ are obtained and one computes the associated energy change $\Delta U = U(\mathbf{r}^{N,\text{new}}) - U(\mathbf{r}^{N,\text{old}})$. The new configuration is accepted with probability

$$p_{acc} = \min\left[1, e^{-\beta \Delta U}\right], \tag{19.2}$$

which means that it is immediately accepted if $\Delta U < 0$ and accepted only with a certain probability, namely p_{acc}, if $\Delta U > 0$. This acceptance criterion is called the Metropolis algorithm [27] and it can be shown that it generates configurations according to their Boltzmann probabilities. The entirety of such configurations is called a Boltzmann-weighted ensemble. Concerning biomolecules, the Monte Carlo/Metropolis algorithm is often used in ab initio protein structure prediction methods [22] because it allows one to efficiently probe possible configurations. A disadvantage of the algorithm is that the obtained configurations are not related to each other via a dynamic sequence, i.e. one does not obtain a movie-like depiction of the time behaviour of the molecule. MD simulation [25, 28], in contrast to mere configurational sampling such as performed by a Monte Carlo algorithm, generates configurations in a time-dependent manner. The dynamics of the system is propagated via an equation of motion (eom) that describes how the new coordinates $\mathbf{r}^N(t + \Delta t)$ depend on the old coordinates a small time increment Δt earlier, $\mathbf{r}^N(t)$, where t denotes time. In practice, this propagation is done with a numerical integrator (Section Integration of the Equation of Motion). The generated ensemble of configurations is Boltzmann-weighted, just as in Monte Carlo sampling. Thus, Monte Carlo sampling performs 'a random walk' through configuration space whereas MD simulation delivers a trajectory in phase space.

The phase space of the system is the entirety of possible coordinates and velocities of its dof. In statistical mechanics, one set of coordinates and velocities is called a microstate and the ensemble of all microstates compatible with given thermodynamic conditions (pressure, temperature) is called a macrostate. The properties of all these configurations may be used to extract thermodynamic information of the system via statistical mechanics (Section 19.2.2). By virtue of involving particle velocities, i.e. dynamic information, MD simulation, in contrast to Monte Carlo sampling, does not only allow the investigation of thermodynamic, but also of dynamic system properties.

19.2.2 Statistical Mechanics: Partition Function and Ensemble Average

Statistical mechanics [17] is the key to the interpretation of the results of an MD simulation [29]. It establishes a relation between the configurations (microstates) sampled in the simulation and macroscopic thermodynamic properties under given thermodynamic conditions. The latter refer to which state variables are kept constant and characterise the ensemble of microstates. For instance, in the canonical ensemble (NVT), microstates are compatible with a constant number of particles (N), volume (V) and temperature (T). The alternate variables, chemical potential, pressure and energy fluctuate. Other ensembles are the microcanonical ensemble (NVE), which pertains to constant number of particles, volume and energy. Here, chemical potential, pressure and temperature fluctuate. In the grandcanonical ensemble (μVT), the chemical potential, volume and temperature are constant and the number of particles, pressure and energy fluctuate. Of most practical relevance is the isothermal-isobaric ensemble (NPT) featuring constant number of particles, pressure and temperature and fluctuating chemical potential, volume and energy.

For simplicity, the following discussion refers to the canonical ensemble. The Boltzmann probabilities in this ensemble are

$$p(\mathbf{r}^N, \mathbf{p}^N) = Q^{-1} e^{-\beta H(\mathbf{r}^N, \mathbf{p}^N)}, \tag{19.3}$$

where h is Planck's constant, $\mathbf{p}^N$ denotes the momenta associated with all dof, $H(\mathbf{r}^N, \mathbf{p}^N)$ is the Hamiltonian of the system (the sum of kinetic and potential energy) and Q is the partition function for the canonical ensemble:

$$Q = (h^{3N} N!)^{-1} \iint \mathrm{d}\mathbf{p}^N \mathrm{d}\mathbf{r}^N e^{-\beta H(\mathbf{r}^N, \mathbf{p}^N)}. \tag{19.4}$$

It involves an integral over all possible positions and momenta, i.e. over the whole phase space and hence cannot be calculated for large systems such as a biomolecule. However, ratios of partition functions are readily accessible via MD simulation (Section Free-Energy Calculations).

The partition function is an important quantity in statistical mechanics because it allows the calculation of macroscopic observables A as ensemble averages $\langle A \rangle$ of the property $A(\mathbf{r}^N, \mathbf{p}^N)$ over the entire phase space,

$$\langle A \rangle = \frac{\iint \mathrm{d}\mathbf{p}^N \mathrm{d}\mathbf{r}^N A(\mathbf{r}^N, \mathbf{p}^N) e^{-\beta H(\mathbf{r}^N, \mathbf{p}^N)}}{\iint \mathrm{d}\mathbf{p}^N \mathrm{d}\mathbf{r}^N e^{-\beta H(\mathbf{r}^N, \mathbf{p}^N)}}. \tag{19.5}$$

This equation expresses the link statistical mechanics provides between the properties of microstates ($A(\mathbf{r}^N, \mathbf{p}^N)$) and the corresponding macroscopic, experimentally measurable, observable. In an MD simulation, however, time averages $\overline{A}$ are calculated,

$$\overline{A} = \frac{1}{N_f} \sum_{i=1}^{N_f} A(t_i) \tag{19.6}$$

where N_f is the number of simulation frames and t_i the time of the ith frame. This is a simple arithmetic average over the properties $A(t_i)$ at each simulation frame, but is valid because a thermostatted (Section Thermo- and Barostatting) MD simulation already samples configurations ($\mathbf{r}^N, \mathbf{p}^N$) with the correct Boltzmann probabilities for the NVT ensemble. The ergodic hypothesis [30, 31] states that in the limit of infinite sampling, i.e. an infinite number of simulation frames N_f, the ensemble average (Eq. (19.5)) is equivalent to a time average (Eq. (19.6) with $N_f \rightarrow \infty$).

19.2.3 Classical Mechanics: Equations of Motion

MD simulations deliver a 'movie' of the simulated system by tracing the time evolution of the system according to an equation of motion (eom). An eom is a relation between particle positions, velocities and accelerations. Newton's eom, which is valid in Cartesian coordinates, is [18, 19]

$$\mathbf{f}^N(\mathbf{r}^N) = \frac{\mathrm{d}\mathbf{p}^N}{\mathrm{d}t} = -\nabla U(\mathbf{r}^N), \tag{19.7}$$

where $\mathbf{f}^N$ is a vector containing the forces on all N particles. In Eq. (19.7), it was assumed that the potential energy is not explicitly time-dependent, i.e. the system is conservative, and that there are no velocity-dependent forces, i.e. there is no friction. Equation (19.7) is a second-order differential equation, which becomes apparent when writing the momentum–time derivative in terms of positions. To ease numerical integration in practice, usually a formalism involving two first-order partial differential equations is used,

$$m_i \frac{\mathrm{d}\mathbf{v}_i}{\mathrm{d}t} = -\frac{\mathrm{d}U(\mathbf{r}^N)}{\mathrm{d}\mathbf{r}_i} \quad \text{and} \quad \frac{\mathrm{d}\mathbf{r}_i}{\mathrm{d}t} = \mathbf{v}_i. \tag{19.8}$$

Here, subscripts i indicate properties referring to particle i. Integration of the eom, together with initial conditions for particle positions and momenta, yields a deterministic evolution of the particle positions and momenta over time. The corresponding series of sampled particle positions and momenta is called a trajectory. Throughout this trajectory, the total energy $H(\mathbf{r}^N, \mathbf{p}^N) = K(\mathbf{p}^N) + U(\mathbf{r}^N)$ is conserved. The kinetic energy is a function of momenta only,

$$K(\mathbf{p}^N) = \sum_{i=1}^{N} \frac{\mathbf{p}_i^2}{2m_i} = \sum_{i=1}^{N} \frac{m_i \mathbf{v}_i^2}{2}, \tag{19.9}$$

where m_i and $\mathbf{v}_i$ are mass and velocity, respectively, of particle i, and the potential energy only depends on particle positions.

Newton's eom does not contain friction. If such non-conservative forces are to be included in the time evolution of the system, e.g. when certain dof are not represented explicitly but implicitly via the effect of their mean fluctuations, one can use Langevin's eom [32, 33],

$$\mathbf{f}(\mathbf{r}_i) = -\frac{\mathrm{d}U(\mathbf{r}^N)}{\mathrm{d}\mathbf{r}_i} + \mathbf{f}_i^R - m_i \gamma \mathbf{v}_i. \tag{19.10}$$

The force $\mathbf{f}(\mathbf{r}_i)$ on particle i thus not only involves a conservative component but also a random and a dissipative one, $\mathbf{f}_i^R$ and $-m_i\gamma\mathbf{v}_i$, respectively, the magnitude of the latter being proportional to a friction coefficient γ. The random force is delta-correlated, has zero mean and its variance is given by

$$\langle f_{i\alpha}^R(t) f_{j\beta}^R(t') \rangle = 2 m_i \gamma_i k_B T \delta_{ij} \delta(t - t'), \tag{19.11}$$

where indices i and j denote particles, α and β denote Cartesian coordinate components, t and t' denote time points, δ_{ij} is the Kronecker-delta and $\delta(x)$ the Dirac-delta function. The magnitude of the variance implies that the two non-conservative forces are kept in balance by the fluctuation-dissipation theorem [33].

Newton's eom is a special case of Lagrange's eom [18, 34], when Cartesian coordinates are used. Lagrange's eom is particularly suited when holonomic (time-independent) constraints on certain coordinates are to be considered in the time evolution of the system or when additional virtual coordinates (dof) are added to the system to generate an extended system. It is formulated in terms of generalised coordinates $\mathbf{q}^N$ and conjugate momenta $\dot{\mathbf{q}}^N$,

$$\frac{\mathrm{d}}{\mathrm{d}t}\left(\frac{\partial L(\mathbf{q}^N,\dot{\mathbf{q}}^N)}{\partial \dot{\mathbf{q}}^N}\right) = \frac{\partial L(\mathbf{q}^N,\dot{\mathbf{q}}^N)}{\partial \mathbf{q}^N}, \tag{19.12}$$

where the Lagrangian $L(\mathbf{q}^N, \dot{\mathbf{q}}^N)$ of the system is defined as the difference of kinetic and potential energy, the former of which now explicitly depends on the generalised coordinates,

$$L(\mathbf{q}^N,\dot{\mathbf{q}}^N) = K(\mathbf{q}^N,\dot{\mathbf{q}}^N) - U(\mathbf{q}^N). \tag{19.13}$$

19.3 Methodology

19.3.1 Molecular Mechanics: Biomolecular Force Fields

Biomolecules are commonly modelled at atomistic resolution using a classical energy function (force field) to describe the interaction between the individual atoms [35–38]. Sometimes, CH, CH_2 or CH_3 groups are modelled as single particles called united atoms. This reduces computational cost because fewer particles are involved and consequently fewer interactions have to be computed. Such a process is called 'coarse-graining'. Many 'coarse-grained' force fields [39–42] exist where the elementary particles are even larger, e.g. entire amino acids, or (a number of) solvent molecules.

A classical force-field description of the system implies several approximations. First, the Hamiltonian is considered time independent and the system is considered to be in the electronic ground state. Second, the Born–Oppenheimer approximation is used, which means that electrons relax instantaneously to the positions of the nuclei and in turn nuclear positions appear stationary to the electrons. Finally, at the core of the classical description is the assumption that the temperature is sufficiently high and that the modelled elementary particles are sufficiently heavy for their motion to be described by Newton's eom. Proton motion and high frequency bond-vibrational motions are not adequately described by a classical force field.

Molecular mechanics describes the potential energy of a molecule with a classical interaction function that is a sum of individual covalent and non-covalent energy terms. Covalent terms describe bond stretching, angle bending, torsional rotations around bonds (dihedral angles) and out-of-plane geometrical restraints. Non-covalent terms describe non-polar and electrostatic interactions. The entirety of the functional form of the individual terms and the involved empirical parameters is called a force field [35–38] (Figure 19.2). Besides the above assumptions, the concept of a molecular-mechanics force field also implies that the total potential

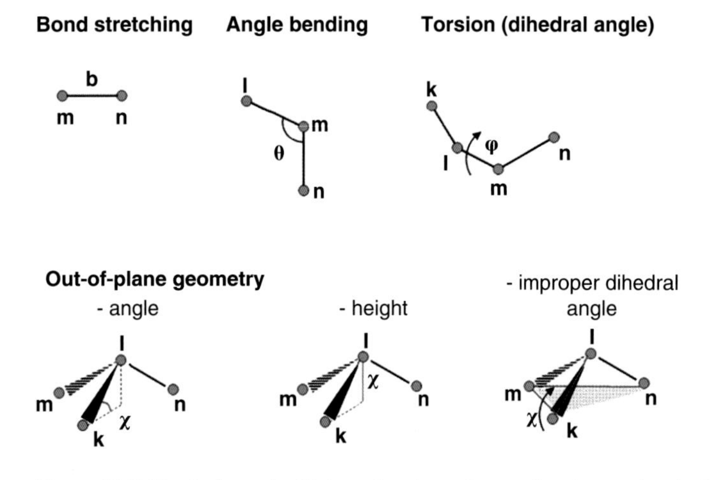

Figure 19.2 Physical covalent interaction terms in a molecular-mechanics force field.

energy of the system is captured by a sum of the individual terms. In the following, the physical covalent and non-covalent interaction terms characterising an atomistically modelled molecule are explained. It is also possible to include non-physical, artificial terms in a force field, e.g. to generate forces pushing apart or pulling together two atoms to achieve a user-defined interatomic separation (Section Experiment-Based Restraints).

Consider two atoms m, n with positions $\mathbf{r}_m$, $\mathbf{r}_n$ connected by a covalent bond. The energy associated with stretching or compressing the bond with respect to the reference bond length b_0 such that the actual bond length is $b(\mathbf{r}_m, \mathbf{r}_n)$ is usually modelled as

$$V^{bond}(\mathbf{r}_m, \mathbf{r}_n) = \frac{1}{2}k^b[b(\mathbf{r}_m, \mathbf{r}_n) - b_0]^2, \tag{19.14}$$

where k^b is a force constant. This is a harmonic potential, the first non-zero term in a Taylor expansion of the energy around the minimum-energy value b_0. The first-order term disappears because the force vanishes when the bond length is exactly b_0. In principle, a harmonic functional form is sufficient to describe the vibrational behaviour of covalent bonds at equilibrium. Covalent bonds are relatively hard dof and only small deviations from the minimum-energy value are possible at room temperature, e.g. around 0.006 nm for a bond between two sp^3-hybridised carbon atoms. One may still keep more terms beyond second-order in the above Taylor expansion. This may be advantageous because anharmonicities are accounted for and hence the real vibrational geometries of a bond may be better reproduced, but the energy calculation is somewhat more computationally intensive and the parameterization is much more cumbersome. Note that Eq. (19.14) cannot describe bond breaking because infinite bond lengths are associated with an infinite energy. Bond breaking events can be modelled classically with other potentials [43] or quantum-mechanically, e.g. with the quantum-mechanical/molecular mechanical (QM/MM) method [44–46], where the reactive part of the system is treated with quantum mechanics and the rest with classical mechanics.

Now, consider three atoms l, m, n with positions $\mathbf{r}_l$, $\mathbf{r}_m$, $\mathbf{r}_n$ connected by two covalent bonds, i.e. forming an angle. The energy associated with bending the angle with respect to the reference angle θ_0 such that the actual angle is $\theta(\mathbf{r}_l, \mathbf{r}_m, \mathbf{r}_n)$, is usually modelled as

$$V^{angle}(\mathbf{r}_l, \mathbf{r}_m, \mathbf{r}_n) = \frac{1}{2}k^a[\theta(\mathbf{r}_l, \mathbf{r}_m, \mathbf{r}_n) - \theta_0]^2, \tag{19.15}$$

i.e. with a harmonic functional form and force constant k^a. Akin to bond lengths, angles are relatively hard dof, but still somewhat softer than bonds. For example, the angle between three sp^3-hybridised carbon atoms can be bent up to around $\pm 10°$ at room temperature.

Now, consider four atoms k, l, m, n with positions $\mathbf{r}_k$, $\mathbf{r}_l$, $\mathbf{r}_m$, $\mathbf{r}_n$ connected by three covalent bonds, i.e. forming two intersecting planes (defined by the first three and last three atoms, respectively). The angle enclosed by the two planes is called a dihedral angle. It may also be envisioned as the torsional rotation of the outermost atoms around the central bond. The energy associated with a torsion is usually modelled as a cosine series,

$$V^{torsion}(\mathbf{r}_k, \mathbf{r}_l, \mathbf{r}_m, \mathbf{r}_n) = \sum_{j=1}^{N_j} k_j^t[1 + \cos[j\varphi(\mathbf{r}_k, \mathbf{r}_l, \mathbf{r}_m, \mathbf{r}_n) - \delta]], \tag{19.16}$$

where j is the multiplicity (number of minima over a rotation of 360°), δ is the phase shift of the cosine terms and the force constants $k_j{}^t$, defined for each term j, determine the barrier height of the rotational potential. For instance, for ethane, $H{-}CH_2{-}CH_2{-}H$, a torsion around the carbon–carbon bond is possible. The multiplicity is three, because there are three possible minima in the staggered geometry. To realistically model the situation in ethane, only the $j = 3$ term is non-zero, i.e. the force constants pertaining to $j = 1$ and 2 vanish and the phase shift δ also vanishes. In contrast to bonds and angles, torsions are soft dof. In general, the whole range of dihedral angles from -180 to 180° can be sampled, except for special situations like double or peptide bonds.

Special geometric situations in a molecule may require the use of so-called out-of-plane geometrical terms. These terms always involve four atoms and enforce planar, tetrahedral or chiral geometries. Consider, for example, the molecule acetone where atoms k, l with positions $\mathbf{r}_k$, $\mathbf{r}_l$, respectively, represent the carbonyl carbon and oxygen atoms, respectively, and atoms m, n with positions $\mathbf{r}_m$, $\mathbf{r}_n$, respectively, represent the two attached methyl carbon atoms. These four atoms are supposed to lie within a plane which can be achieved with a harmonic potential around a reference value χ_0 for the dof χ describing the out-of-plane geometry,

$$V^{oop}(\mathbf{r}_k, \mathbf{r}_l, \mathbf{r}_m, \mathbf{r}_n) = \frac{1}{2}k^{oop}[\chi(\mathbf{r}_k, \mathbf{r}_l, \mathbf{r}_m, \mathbf{r}_n) - \chi_0]^2, \tag{19.17}$$

where k^{oop} is a force constant. The actual value $\chi(\mathbf{r}_k, \mathbf{r}_l, \mathbf{r}_m, \mathbf{r}_n)$ for the out-of-plane dof can be defined by many options. In the acetone example, one option is to use the angle formed by the carbonyl carbon atom, the oxygen atom and the projection of the carbonyl carbon atom in the plane formed by the three other atoms. Other options are to use the height of the carbonyl carbon atom above the latter plane, or the dihedral angle formed by the carbonyl carbon atom, the oxygen atom and the two other carbon atoms. Since this is not a genuine rotational torsion, such dihedral angles are denoted improper dihedral angles.

In summary, the contribution of covalent interactions to the potential energy of a molecule is

$$\begin{aligned} U_c(\mathbf{r}^N) = &\sum_{bonds\,i} U_{bond}(b_i; b_{i,0}, k_i^b) + \sum_{angles\,i} U_{angle}(\theta_i; \theta_{i,0}, k_i^a) \\ &+ \sum_{torsions\,i} U_{torsion}(\varphi_i; \{k_{i,j}^b\}, N_{i,j}, \delta_i) + \sum_{oop\,i} U_{oop}(\chi_i; \chi_{i,0}, k_i^{oop}), \end{aligned} \tag{19.18}$$

where sums over all bonds, angles, torsions and out-of-plane geometries in the system are performed. The parameters required in the covalent force-field terms except for torsions are usually obtained from X-ray crystallography (reference values) and IR spectroscopy (force constants). Parameters for the description of torsions are usually obtained from QM calculations.

The non-polar contribution to non-covalent interactions are van der Waals interactions [47]. They are composed of two contributions, an attractive and a repulsive one. The physical basis of the former are London dispersion forces [48]. These forces arise from fluctuations in the electron cloud of an atom and the corresponding instantaneous dipoles that can induce opposing dipoles in the electron clouds of atoms in the close neighbourhood. They are weak and of a short-range nature. Since the electric field due to the inducing dipole varies as $1/r^3$ and the dipole–dipole interaction varies as $1/r^3$, where r is the interdipole distance, the attractive interaction decays as $1/r^6$. The second, repulsive component accounts for the Pauli exclusion principle [49, 50]. It has, in principle, an exponential distance dependence, but for computational convenience it is usually modelled with a power law. In the Lennard–Jones potential [51], an inverse-twelve power is used because the twelfth power of the interatomic distance can be easily calculated from the sixth power occurring in the attractive term,

$$U_{LJ}(\mathbf{r}_m, \mathbf{r}_n) = \frac{C_{12,mn}}{r_{mn}^{12}} - \frac{C_{6,mn}}{r_{mn}^6} = 4\varepsilon_{mn}\left(\frac{\sigma_{mn}^{12}}{r_{mn}^{12}} - \frac{\sigma_{mn}^6}{r_{mn}^6}\right), \tag{19.19}$$

where $C_{12,mn}$ and $C_{6,mn}$ are constants specific to the interaction between the considered atom pair m, n and r_{mn} is the minimum-image between the two atoms. The minimum-image distance, under PBC, is the minimum distance between two atoms considering the presence of periodic system copies. Sometimes, the Lennard–Jones potential is written in terms of the constants ε_{mn} and σ_{mn}, which describe the well depth and zero of the potential, respectively (Eq. (19.19) and Figure 19.3). To ease parameterisation effort of the Lennard–Jones potential, usually the parameters are specific to certain atom types, for example sp^3-hybridised carbon atoms, carbonyl oxygen atoms, hydroxyl oxygen atoms, etc. In addition, combination rules [52] are used to deduce parameters for heteroatomic interactions based on homoatomic interaction parameters, i.e.

$$\varepsilon_{mn} = f(\varepsilon_{mm}, \varepsilon_{nn}), \quad \sigma_{mn} = f(\sigma_{mm}, \sigma_{nn}), \quad C_{12,mn} = f(C_{12,mm}, C_{12,nn}), \quad C_{6,mn} = f(C_{6,mm}, C_{6,nn}), \tag{19.20}$$

where the function f is, for example, a geometric mean [53]. For biomolecular force fields, the parameterisation is usually based on optimisation with respect to thermodynamic condensed-phase properties such as heats of vaporisation, densities and solvation free energies.

Figure 19.3 The Lennard–Jones potential. Well depth ε_{mn} and zero σ_{mn} are highlighted.

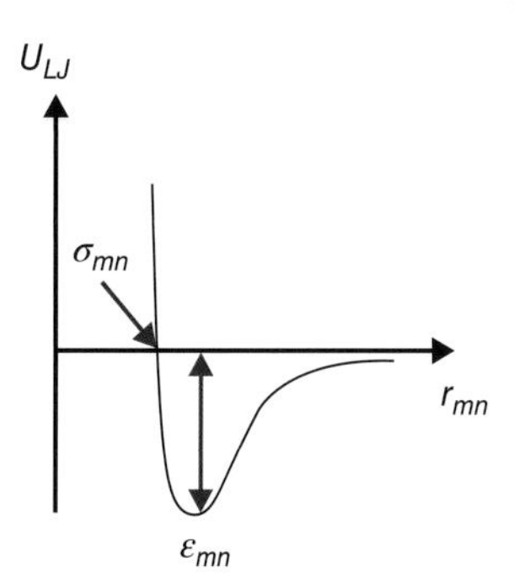

Electrostatic interactions between atoms m, n with partial charges q_m and q_n, respectively, are in principle given by Coulomb's law [54, 55],

$$U_{ele}(\mathbf{r}_m, \mathbf{r}_n) = \frac{q_m q_n}{4\pi\varepsilon_0 r_{mn}}, \tag{19.21}$$

where r_{mn} is again the minimum-image between the two atoms and ε_0 the vacuum permittivity. Usually, the partial charges are parameterised for small groups; e.g. a hydroxyl group in a serine has the same partial charges as the hydroxyl group in a threonine. They are either obtained from QM calculations or optimization with respect to thermodynamic condensed-phase properties such as heats of vaporisation, densities and solvation free energies. In practice, Coulomb's law is not used for simulations with explicit solvent under PBC because it is computationally intensive and gives rise to artefacts when used with a plain nearest-image convention [56]. Other approaches commonly used instead are explained in Section Lattice-Summed Electrostatic Interactions. Note also that the above formalism with fixed partial charges does not account for the effect a change in environment might have on the electron cloud of an atom. Such effects are better reproduced by so-called polarisable force fields [57–61].

In summary, the contribution of non-covalent interactions to the potential energy of a molecule is

$$U_{nc}(\mathbf{r}^N) = \frac{1}{2}\sum_{i=1}^{N}\sum_{\substack{j=1,\\ j\neq i}}^{N} \frac{q_i q_j}{4\pi\varepsilon_0 r_{ij}} + \frac{1}{2}\sum_{i=1}^{N}\sum_{\substack{j=1,\\ j\neq i}}^{N} 4\varepsilon_{ij}\left(\frac{\sigma_{ij}^{12}}{r_{ij}^{12}} - \frac{\sigma_{ij}^{6}}{r_{ij}^{6}}\right), \tag{19.22}$$

where sums over all atom pairs are performed. Note the factors (1/2) to eliminate double-counting of atom pairs.

There is a plethora of biomolecular force fields [35–38]. Currently, the most common ones pertain to the AMBER (e.g. versions ff03 [62], ff14SB [63]), CHARMM (e.g. versions c27 [64], c36 [65]) and GROMOS (e.g. versions 54A7 [66], 54A8 [67]) families. Biomolecular force-field development is in general a complex task because a balance has to be found between computational simplicity and as accurate as possible a description of the energetic and entropic effects governing the behaviour of the molecules [14]. Since free energies of interest are generally not above the order of 10 $k_B T$, and since they result from a summation over around 10^7–10^9 atom pairs, it follows that the individual force-field terms have to be highly accurate to deliver simulation results of chemical accuracy (error $\leq$ 1 kJ/mol).

19.3.2 Molecular Dynamics Simulation of Biomolecules

19.3.2.1 Common Algorithms

Integration of the Equation of Motion Typically, in a biomolecular MD simulation employing a classical molecular-mechanics force field as outlined above, Newton's eom (Eq. (19.8)) is used to describe the dynamics of the particles. The two first-order differential equations (Eq. (19.9)) are integrated numerically.

Since Newton's eom is time-reversible, a numerical integration algorithm should also be time-reversible. The algorithm should also yield as accurate an integration as possible, which is not granted by default because the dynamics is propagated via finite time steps Δt and hence is approximate. The most commonly used integrators are the (Velocity-)Verlet [68, 69] and leapfrog integration algorithms [70]. Here, only the Velocity-Verlet algorithm is explained. It involves a Taylor expansion of positions at time $t + \Delta t$ based on the positions at the previous time step,

$$\mathbf{r}_i(t + \Delta t) = \mathbf{r}_i(t) + \Delta t \mathbf{v}_i(t) + (1/2)(\Delta t)^2 \mathbf{a}_i(t) \tag{19.23}$$

truncated after the second-order term, i.e. there is a truncation error of third order. Here, $\mathbf{v}_i$ and $\mathbf{a}_i$ indicate velocities and accelerations of particle i, respectively. The velocities are not propagated as accurately as positions, namely only to first order and using a mid-point approximation for the acceleration,

$$\mathbf{v}_i(t + \Delta t) = \mathbf{v}_i(t) + (1/2)\Delta t[\mathbf{a}_i(t) + \mathbf{a}_i(t + \Delta t)]. \tag{19.24}$$

Despite the third-order truncation error in positions, the Velocity-Verlet algorithm is a second-order integrator because the global error in positions, i.e. the cumulative error over many iterations, can be shown to be of order two [71]. An important beneficial property of this integrator is that it is symplectic. Symplectic integrators do not lead to systematic energy drifts over many iterations because they reproduce the exact dynamics for a slightly different ('shadow') Hamiltonian [72]. An overview of integration algorithms for MD simulations is given in reference [73].

Thermo- and Barostatting By default, Newton's eom (Eq. (19.8)) delivers an ensemble at constant energy and volume. Typically, when simulating the behaviour of a biomolecule, one seeks conditions close to the experimental situation, i.e. constant temperature and constant pressure. This can be done by application of thermo- and barostatting algorithms. One can distinguish weak-coupling, stochastic and extended-system algorithms. Here, only one thermostatting algorithm of each kind and a weak-coupling barostatting algorithm are presented, which adequately covers the most commonly employed approaches.

The definition of temperature in MD simulation is based on the equipartition theorem [17], which states that each dof is associated with a kinetic energy of half the thermal energy. Considering that there are N_{df} dof in the system, the average kinetic energy K is

$$K = N_{df} \frac{k_B T}{2}. \tag{19.25}$$

Applying this equation at each instantaneous time step in an MD simulation in combination with the instantaneous kinetic energy expressed as a function of particle masses and velocities (Eq. (19.9)), the instantaneous temperature can be calculated.

The weak-coupling thermostat [74] (also called the Berendsen thermostat) relies on the scaling of particle velocities $\mathbf{v}_i(t)$ with a time-dependent scaling factor $\mu_T(t)$, to deliver scaled velocities $\mathbf{v}_{i,scl}(t)$ satisfying a kinetic energy corresponding to the target temperature T_0,

$$\mathbf{v}_{i,scl}(t) = \mu_T(t)\mathbf{v}_i(t). \tag{19.26}$$

It is defined that the temperature should approach the target temperature exponentially according to a relaxation constant τ_T,

$$\frac{\mathrm{d}T(t)}{\mathrm{d}t} = \tau_T^{-1}(T_o - T(t)). \tag{19.27}$$

With Eqs. (19.9) and (19.26), the temperature corresponding to the scaled velocities becomes

$$T_{scl}(t) = \mu_T^2(t)T(t) \approx T(t - \Delta t) + \frac{\Delta t}{\tau_T}[T_o - T(t)] \tag{19.28}$$

and the velocity scaling factor is

$$\mu_T(t) \approx \sqrt{1 + \frac{\Delta t}{\tau_T}\left(\frac{T_o}{T(t)} - 1\right)}, \tag{19.29}$$

where it was assumed that $\Delta t \to 0$. This algorithm is simple to apply but has the flaw of not producing an exact canonical distribution of states [75]; e.g. energy fluctuations are incorrect.

An example for a stochastic thermostat is to use Langevin's eom (Eq. (19.10)). Constant temperature is here achieved by stochastic collisions implied by the random force and the counteracting effect of friction, the exact balance being ascertained by the fluctuation-dissipation theorem (Eq. (19.11)). Stochastic thermostats are naturally employed in implicit-solvent simulations (Section Implicit Solvation). Due to the presence of random forces, they are not suited for the investigation of dynamic properties.

The extended-system thermostat (also called the Nosé–Hoover thermostat) is based on the addition of a virtual 'heat bath'-like dof s (with an associated mass Q) to the system [76, 77]. The entire system containing both physical, real particles and the virtual dof is the extended system. Its Lagrangian reads [76, 77]

$$L = \sum_{i=1}^{N} \frac{m_i}{2} s^2 \mathbf{v}_i^2 - U(\mathbf{r}^N) + \frac{Q}{2}\dot{s}^2 - k_B T g \ln s. \tag{19.30}$$

The first two terms form the physical Lagrangian corresponding to the real particles in the system. The third and fourth terms are the kinetic energy and minus the potential energy of s, respectively. The key of the method lies in the fact that the form of the potential energy and the parameter g are chosen such that a canonical ensemble for the physical subsystem is generated while a microcanonical ensemble for the extended system is retained. It was shown [76, 77] that this is achieved with the choice $g = 3N + 1$, where N is the number of particles in the physical system. During the extended-system simulation, the virtual dof s is propagated along with the physical particles. According to the fluctuation of s, heat flows in or out of the physical subsystem such that a constant temperature is maintained. In practice, the Nosé–Hoover thermostat was shown to have deficiencies for small and stiff systems or far from equilibrium (i.e. the target temperature). In the former situation, it is better to use the so-called Nosé–Hoover chain algorithm [78], which involves successive thermostatting of virtual dof by multiple, rather than only one, virtual variables. An overview of thermostatting algorithms for MD simulations is given in reference [79].

The definition of pressure in MD simulation is based on the total virial W_{tot},

$$W_{tot} = W + \frac{3}{2}PV, \tag{19.31}$$

which consists of the internal virial W and an energy contribution due to system confinement (external virial). The virial theorem [80] states equality of kinetic energy K and W_{tot}; hence

$$P = \frac{2(K - W)}{3V}. \tag{19.32}$$

Applying this equation at each instantaneous time step in an MD simulation in combination with the instantaneous kinetic energy expressed as a function of particle masses and velocities (Eq. (19.9)), the instantaneous volume and the instantaneous internal virial,

$$W(t) = -\frac{1}{2}\sum_i \mathbf{r}_i(t)\cdot\mathbf{f}_i(t), \tag{19.33}$$

a sum of dot products of particle positions and forces acting on the particles, the instantaneous pressure can be calculated. In practice, under PBC, Eq. (19.33) is not used, but $W(t)$ is reformulated in terms of interparticle distances and pairwise forces [29].

The weak-coupling barostat, in analogy to the weak-coupling thermostat, relies on the scaling of particle positions and system dimensions (usually, box-edge lengths L) with a time-dependent scaling factor $\mu_P(t)$ to deliver scaled quantities,

$$\mathbf{r}_{i,scl}(t) = \mu_P(t)\mathbf{r}_i(t) \quad \text{and} \quad L_{scl}(t) = \mu_P(t)L(t) \tag{19.34}$$

satisfying the target pressure. It is defined that the pressure should approach the target pressure exponentially according to a relaxation constant τ_P,

$$\frac{\mathrm{d}P(t)}{\mathrm{d}t} = \tau_P^{-1}(P_o - P(t)). \tag{19.35}$$

With Eq. (19.9) and the definition of isothermal compressibility κ_T, the pressure corresponding to scaled quantities becomes

$$P_{scl}(t) = P(t) - V^{-1}\kappa_T^{-1}\Delta V = P(t) - \kappa_T^{-1}[\mu_P^3(t) - 1] \approx P(t - \Delta t) + \frac{\Delta t}{\tau_P}[P_o - P(t)] \tag{19.36}$$

and the scaling factor is

$$\mu_P(t) \approx \left[1 + \frac{\kappa_T \Delta t}{\tau_P}(P(t) - P_o)\right]^{1/3} \tag{19.37}$$

where it was assumed that $\Delta t \to 0$. This algorithm is simple to apply but, similar to the weak-coupling thermostat, has the flaw of not producing a correct isobaric ensemble; e.g. volume fluctuations are incorrect. Other barostatting algorithms are available, e.g. the Parrinello–Rahman barostat [81].

Bond-Length Constraints Bond-length vibrations in molecules occur with frequencies larger than about $1000\,\mathrm{cm}^{-1}$. This is much larger than k_BT/h at room temperature. Hence, bond vibrations are not excited at room temperature and are not adequately described by a classical harmonic oscillator (Eq. (19.14)). A remedy is to freeze the bond-length vibrations, i.e. to model bond lengths as rigid rods with lengths b_0, an approach referred to as bond-length constraining. Constraining a dof amounts to eliminating this dof because there is no potential or kinetic energy associated with it. Bond-length constraints have the advantage of increased computational efficiency. Since the maximal possible time step (to still have an accurate integration of the eom) in the numerical integrator is about a tenth of the smallest vibrational period and since bond-length vibrations are typically the fastest vibrational motions in the system, constraining bond lengths allows the time step to be increased (usually from 0.5 to 2 fs).

A common algorithm to constrain bond lengths is the SHAKE algorithm [82]. Consider a molecule with N_c bond-length constraints σ_k for bonds between atoms k_1 and k_2 with positions $\mathbf{r}_{k1}, \mathbf{r}_{k2}$, respectively,

$$\sigma_k(\mathbf{r}_{k1}, \mathbf{r}_{k2}) = r_{k1k2}^2 - b_{0,k1k2}^2 = 0 \tag{19.38}$$

These constraints are accounted for via N_c Lagrange multipliers λ_k in the eom (Eq. (19.8)),

$$m_i\mathbf{a}_i = -\frac{\partial U}{\partial \mathbf{r}_i} - \sum_k \lambda_k \frac{\partial \sigma_k}{\partial \mathbf{r}_i} \tag{19.39}$$

The Lagrange multipliers need to be known to determine the constraint forces required to enforce fixed bond lengths. Based on Eqs. (19.38) and (19.39), a matrix equation $\mathbf{A}\boldsymbol{\lambda} = \mathbf{c}$ can be derived, where the λ vector contains all the Lagrange multipliers and terms quadratic in the Lagrange multipliers were discarded [83]. Inversion of $\mathbf{A}$ solves for the λ vector, but this is not commonly done because matrix inversion is computationally expensive and the solution is not exact because quadratic terms were discarded. Only for very small molecules such as water (involving only three constraints), is an iterative procedure of matrix inversions used to solve for the Lagrange multipliers [83]. For larger molecules, instead of trying to simultaneously fulfil all constraints, one assumes decoupled constraints, i.e. that two atoms involved in constraint σ_k are not involved in any other constraint $\sigma_{k'}$. This means that one considers constraints between atoms k_1 and k_2 at a time and solves N_c decoupled equations for the corresponding $\lambda_k(t)$ values. The resetting of particle positions according to the resulting constraint forces (again obtained from neglecting quadratic λ_k terms) leads to constrained positions (here, in the context of a Verlet integrator and with $\mathbf{r}_{k1k2} = \mathbf{r}_{k1} - \mathbf{r}_{k2}$):

$$\mathbf{r}^c_{k_1}(t + \Delta t) = \mathbf{r}^{uc}_{k_1}(t + \Delta t) - (\Delta t)^2\, m^{-1}_{k_1}\, \lambda_k(t)\mathbf{r}_{k_1k_2}(t),$$
$$\mathbf{r}^c_{k_2}(t + \Delta t) = \mathbf{r}^{uc}_{k_2}(t + \Delta t) + (\Delta t)^2\, m^{-1}_{k_2}\, \lambda_k(t)\mathbf{r}_{k_1k_2}(t), \tag{19.40}$$

where superscripts 'c' or 'uc' on atom position vectors denote constrained and unconstrained positions, respectively. Thus, atoms k_1 and k_2 are reset along the direction of the old bond vector, in opposite directions, and proportional to the inverse mass. Because of the decoupling of the constraints, the procedure has to be repeated until all constraints are satisfied. In the case of very large forces, which cause large free-flight steps of certain atoms, it might not be possible to fulfil bond-length constraints via shifting along the old bond-vector direction. In such a situation, the SHAKE algorithm fails.

It was stated above that angles in biomolecules are only slightly softer than bonds. Nevertheless, angles are never constrained in MD simulations because angle constraints were seen to cause artefacts in dynamics, e.g. fewer dihedral angle transitions [84] because the angles in a dihedral angle have to slightly open up for torsional rotation to occur in a normal fashion. In addition, mass-metric tensor effects associated with angle constraints are not negligible [85]. This means that removal of the kinetic and potential energy associated with an angle dof significantly perturbs the probabilities of configurations in the constrained system.

Lattice-Summed Electrostatic Interactions There are numerous ways to calculate electrostatic interactions in a simulation. The reason is that the plain Coulomb potential given by Eq. (19.21) is not applicable in an MD simulation with PBC. It would be required to introduce a nearest-image convention (i.e. each particle only interacts with the nearest neighbour of each other particle), but this leads to artefacts and is computationally expensive [56]. A common remedy is to truncate electrostatic interactions, possibly in combination with shifting or switching schemes, which make the electrostatic interaction between two particles smoothly vanish at the truncation distance [86, 87]. A physically motivated shifting scheme is to include a reaction-field correction that models the environment outside the cut-off sphere of each particle as a dielectric continuum [88]. A commonly used electrostatic interaction scheme not involving a cut-off truncation and constituting a correct calculation of Coulombic electrostatic interactions, albeit under PBC, is lattice summation, where one considers the Coulomb energy of an infinitely periodically replicated cubic unit cell of edge L containing N point charges q_i located at positions $\mathbf{r}_i$,

$$U_{cb} = \frac{1}{2}\frac{1}{4\pi\varepsilon_o}\sum_{\mathbf{n}\in\mathbb{Z}}{}^{\prime}\sum_{i=1}^{N}\sum_{j=1}^{N}\frac{q_iq_j}{r_{ij,n}}, \tag{19.41}$$

where $\mathbf{n} = (n_1, n_2, n_3)$, n_i being integers, are lattice vectors,

$$r_{ij,n} = |\mathbf{r}_{j,n} - \mathbf{r}_i| = |\mathbf{r}_j + \mathbf{n}L - \mathbf{r}_i| \tag{19.42}$$

and the primed sum implies that $i = j$ terms are omitted when $\mathbf{n} = (0, 0, 0)$. The problem is that, in general, the $1/r$-sum over unit cells in Eq. (19.41) is not absolutely convergent. Ewald introduced a splitting function $f(r)$ for the $1/r$ potential [89],

$$1/r = f(r)/r + (1 - f(r))/r, \tag{19.43}$$

which allows to solve for the electrostatic potential (Figure 19.4). Therefore, the original method to calculate electrostatic interactions via a lattice sum is also called Ewald summation.

Expressing the electrostatic potential $\Phi(\mathbf{r})$ in the above periodic system as a solution of the Poisson equation and using Ewald's splitting-function approach,

$$\nabla^2\Phi(\mathbf{r}) = -\frac{1}{\varepsilon_o}\sum_{i=1}^{N}[\rho_i(\mathbf{r}) - \sigma_i(\mathbf{r}) + \sigma_i(\mathbf{r})], \tag{19.44}$$

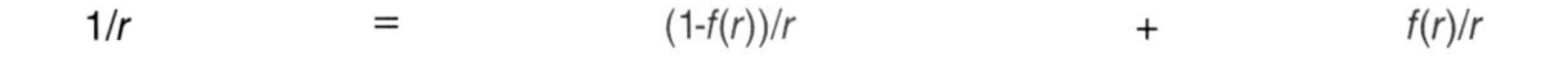

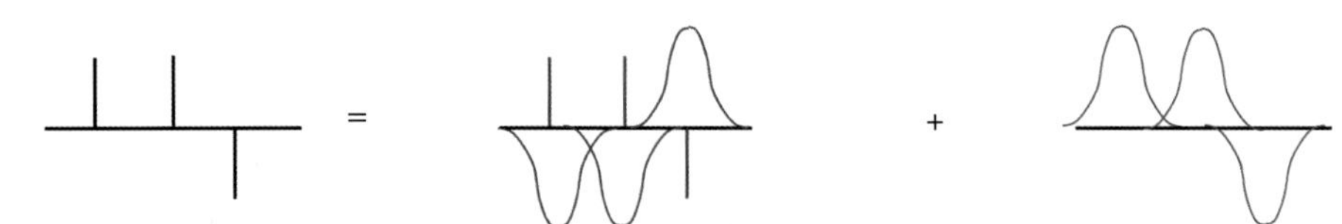

Original charge distribution

- Delta-functions
- Potential: long-range and with singularities

Screened charge distribution

- Screened delta-functions
- Potential: short-range and with singularities

Cancelling charge distribution

- Smooth functions
- Potential: without singularities

Figure 19.4 Illustration of the splitting of the charge density in the Ewald method. A periodic screening charge density is introduced to cause a rapid decay of the 1/r potential. The resulting electrostatic potential is evaluated via summation in real space. The cancelling contribution is evaluated via a Fourier sum in reciprocal space.

where a periodic 'screening' charge density

$$\sigma_i(\mathbf{r}) = \sum_{\mathbf{n}\in\mathbb{Z}} q_i \psi(\mathbf{r} - (\mathbf{r}_i + \mathbf{n}L)) \tag{19.45}$$

depending on so-called charge-shaping functions $\psi(\mathbf{r})$ was used to split the original periodic charge density

$$\rho_i(\mathbf{r}) = \sum_{\mathbf{n}\in\mathbb{Z}} q_i \delta(\mathbf{r} - (\mathbf{r}_i + \mathbf{n}L)), \tag{19.46}$$

where $\delta(\mathbf{x})$ is the three-dimensional Dirac-delta function. Since the Laplace operator is linear, the total electrostatic potential in Eq. (19.44) can be split as well, i.e. $\Phi = \Phi_R + \Phi_K$, with

$$\nabla^2\Phi_R(\mathbf{r}) = -\frac{1}{\varepsilon_o}\sum_{i=1}^{N}[\rho_i(\mathbf{r}) - \sigma_i(\mathbf{r})] \quad \text{and} \quad \nabla^2\Phi_K(\mathbf{r}) = -\frac{1}{\varepsilon_o}\sum_{i=1}^{N}\sigma_i(\mathbf{r}). \tag{19.47}$$

As will be seen below, the potential Φ_R is conveniently evaluated via summation in real space. The potential Φ_K is due to a periodic charge distribution and is conveniently evaluated in reciprocal space (Figure 19.4).

In the original Ewald method, the charge-shaping function is a Gaussian $\gamma(r)$ of width $(2\alpha)^{-1/2}$,

$$\sigma_i(\mathbf{r}) = \sum_{\mathbf{n}\in\mathbb{Z}} q_i \alpha^{3/2}\pi^{-3/2}\exp[-\alpha(\mathbf{r} - (\mathbf{r}_i + \mathbf{n}L))^2], \tag{19.48}$$

resulting in [25]

$$\Phi_R(\mathbf{r}) = \frac{1}{4\pi\varepsilon_o}\sum_{\mathbf{n}\in\mathbb{Z}}\sum_{i=1}^{N} q_i\left(\frac{\mathrm{erfc}(\alpha^{1/2}|\mathbf{r} - (\mathbf{r}_i + \mathbf{n}L)|)}{|\mathbf{r} - (\mathbf{r}_i + \mathbf{n}L)|}\right), \tag{19.49}$$

where the definition of the complementary error function (erfc(x)) was used. Since the latter decays exponentially, the sum over $\mathbf{n}$ may be restricted to $\mathbf{n} = (0, 0, 0)$ and truncation is possible at a certain cut-off distance (called real-space cut-off). The energy associated with Φ_R is obtained from multiplying the charges with Φ_R at the charge locations,

$$U_R = \sum_{i=1}^{N} q_i\Phi_R(\mathbf{r}_i) = \frac{1}{8\pi\varepsilon_o}\sum_{\mathbf{n}\in\mathbb{Z}}{}'\sum_{i=1}^{N}\sum_{j=1}^{N} q_i q_j\left(\frac{\mathrm{erfc}(\alpha^{1/2}|\mathbf{r}_i - (\mathbf{r}_j + \mathbf{n}L)|)}{|\mathbf{r}_i - (\mathbf{r}_j + \mathbf{n}L)|}\right). \tag{19.50}$$

The reciprocal-space potential is long-range, smooth and periodic in **r**. It can be expressed as a Fourier series with Fourier coefficients $\hat{\Phi}_K$,

$$\Phi_K(\mathbf{r}) = V^{-1} \sum_{\mathbf{n}\in\mathbb{Z}} \hat{\Phi}_K(\mathbf{k}) \exp(i\mathbf{kr}), \tag{19.51}$$

where **k** are reciprocal-space vectors, $\mathbf{k} = 2\pi\mathbf{n}/L$. Plugging the reciprocal-space potential (Eq. (19.51)) and the screening-charge distribution (Eq. (19.45)) into the Poisson equation (Eq. (19.47)) leads, for the above Gaussian, to [25]

$$\Phi_K(\mathbf{r}) = \frac{1}{V\varepsilon_o} \sum_{\mathbf{n}\in\mathbb{Z},\mathbf{n}\neq\mathbf{0}} k^{-2} \exp[-k^2/(4\alpha)] \sum_{i=1}^{N} q_i \exp[i\mathbf{k}(\mathbf{r}-\mathbf{r}_i)], \tag{19.52}$$

from which one sees that **k** must not be zero (see below). The energy associated with Φ_K is obtained from multiplying the charges with Φ_K at the charge locations,

$$U_K = \sum_{i=1}^{N} q_i \Phi_K(\mathbf{r}_i) = \frac{1}{2\varepsilon_o V} \sum_{\mathbf{n}\in\mathbb{Z},\mathbf{n}\neq\mathbf{0}} k^{-2} \exp[-k^2/(4\alpha)] \sum_{i=1}^{N}\sum_{j=1}^{N} q_i q_j \exp[i\mathbf{k}(\mathbf{r}_i-\mathbf{r}_j)]. \tag{19.53}$$

The problem with Eq. (19.53) is that the double sum contains the interaction of a continuous Gaussian charge distribution of charge q_i with a point charge q_i at its centre. This overcounts the electrostatic energy. To remove the spurious contribution, one considers the electrostatic potential $\varphi(r)$ due to a spherically symmetric Gaussian charge cloud and the resulting value $\varphi(r=0)$ at its centre [25], giving the self-energy

$$U_{slf} = \frac{1}{2} \sum_{i=1}^{N} q_i \varphi(r=0) = \frac{1}{4\pi\varepsilon_o} \sqrt{\alpha/\pi} \sum_{i=1}^{N} q_i^2. \tag{19.54}$$

Furthermore, the $\mathbf{k} = \mathbf{0}$ term in the reciprocal-space sum (Eq. (19.52)) was omitted. This leads to the average of the reciprocal-space potential being zero. However, the average of the real-space potential is non-vanishing. One hence adds a constant to fulfil the boundary condition of vanishing total average electrostatic potential, $<\Phi(\mathbf{r})> = 0$ (referred to as the Ewald convention). To fulfil this convention, one evaluates the real-space potential average over the unit cell,

$$\langle\Phi(\mathbf{r})\rangle = \frac{1}{4\pi\varepsilon_o}\frac{1}{V}\int_V \mathrm{d}\mathbf{r}\ \Phi_R(\mathbf{r}) = \frac{1}{4\varepsilon_o}\left(\sum_{i=1}^{N} q_i\right)\frac{1}{V\alpha}. \tag{19.55}$$

The interaction of all charges with this potential is removed in the total energy. Note that the average electrostatic potential in Eq. (19.55) is non-zero only for a non-neutral box. Therefore, correcting for it via the Ewald convention may also be interpreted as introducing a homogeneous neutralising background charge that sets the average potential to zero.

The total lattice-sum energy is the sum of the above four contributions,

$$U_{ele} = U_R + U_K - \frac{1}{4\pi\varepsilon_o}\sqrt{\alpha/\pi}\sum_{i=1}^{N} q_i^2 - \frac{1}{8\varepsilon_o}\frac{1}{V\alpha}\left(\sum_{i=1}^{N} q_i\right)^2. \tag{19.56}$$

In practice, in Ewald summation, both real- and reciprocal-space sum are truncated. The parameter α is chosen such that contributions to the real-space sum are negligible for terms separated by a distance larger than the real-space cut-off. Note that a large α entails a narrow Gaussian and a short-range real-space sum. Then the computational cost for the real-space calculation is $O(N)$, but the reciprocal-space sum needs more terms to converge. In general, the reciprocal-space sum scales as $O(N^2)$, which is very inefficient for large systems. It has been shown that an optimal work balance between the two sums can be reached by cleverly adjusting α and the real-space cut-off, leading effectively to an $O(N^{3/2})$ scaling of Ewald summation.

Nowadays, in practice, numerical implementation of lattice summation is used. The so-called particle–mesh approaches exploit the fact that the Poisson equation can be solved efficiently on a grid using fast Fourier transforms (FFT). FFT scale as $N \log(N)$, N being the number of grid points in each dimension. Common particle–mesh approaches are particle-mesh Ewald (PME) [90] or particle–particle particle–mesh Ewald (P3M) [91]. They rely on the same charge-splitting idea as Ewald summation, but instead of directly calculating the Fourier transformation of the charge distribution, one maps the charges on a regular grid. FFT delivers the electrostatic potential on the grid points, which is transformed back to real-space and interpolated to give the electrostatic potential at the actual charge positions.

Lattice summation is a widely used method to calculate electrostatic interactions. One has to keep in mind, however, that the resulting electrostatic interactions are periodic, which may give rise to so-called artificial-periodicity artefacts [92–94]. These are especially pronounced for small box-edge lengths and may significantly perturb the properties of the simulated system. On the other hand, the above-mentioned alternative approach of truncation of electrostatic interactions may also introduce problems because electrostatic interactions are of a very long-range nature. In contrast to the rapidly decaying van der Waals interactions (Section 19.3.1), they decay only with the inverse of the interparticle separation. Therefore, any electrostatic interaction calculation scheme used in MD simulation to date is effective rather than correct and the involved approximations and possibly incurring artefacts must be well understood [95].

19.3.2.2 Special Topics

Free-Energy Calculations Free-energy calculation is an intricate task, because free energies are not simple ensemble averages (Eq. (19.5)) but depend explicitly on the partition function, i.e. the Helmholtz free energy is the thermodynamic potential in the NVT ensemble,

$$A = -k_B T \ln Q(N, V, T) \tag{19.57}$$

and, similarly, the Gibbs free energy G is the thermodynamic potential in the NPT ensemble. Hence, absolute free energies cannot be calculated for realistic biomolecular systems in an explicit solvent. However, ratios of partition functions, i.e. free-energy differences, are readily accessible. For simplicity, the following discussion only refers to the Helmholtz free energy.

There are numerous methods to calculate free energies from MD simulation. Reviews are, for example, provided by references [96–100]. Here, only three popular methods in the context of biomolecular simulation will be presented, thermodynamic integration (TI), free-energy perturbation (FEP) and potential of mean force (PMF) calculations.

Suppose one wants to calculate the free-energy difference between two states 0 and 1, with Hamiltonians H_0 and H_1, respectively. TI relies on a Hamiltonian parameterised as a function of a coupling parameter $0 \leq \lambda \leq 1$ expressing the change between the two Hamiltonians, such that $\lambda = 0$ corresponds to state 0 and $\lambda = 1$ to state 1,

$$H(\lambda; n) = (1 - \lambda)^n H_0 + \lambda^n H_1, \tag{19.58}$$

where n is an integer, $n \geq 1$. This approach was introduced by Kirkwood [101]. It relies on the fact that free energy is a state function and the free-energy difference between two states 0 and 1 corresponds to the reversible work done along any user-defined pathway between the two states. With the Hamiltonian of Eq. (19.58), the free energy can be written as a function of λ, $A(\lambda) = -k_B T \ln(Q(\lambda))$, and it follows that

$$\frac{\partial A}{\partial \lambda} = -k_B T \frac{1}{Q} \frac{\partial Q}{\partial \lambda} = -k_B T \frac{1}{Q} \int \int d\mathbf{p}^N d\mathbf{r}^N \left(-\frac{1}{k_B T} \right) \frac{\partial H(\lambda)}{\partial \lambda} e^{-\frac{H(\mathbf{r}^N, \mathbf{p}^N, \lambda)}{k_B T}} = \left\langle \frac{\partial H}{\partial \lambda} \right\rangle, \tag{19.59}$$

where the definition of statistical-mechanical ensemble averages (Eq. (19.5)) was used in the last step. Equation (19.59) implies that the free-energy derivative at a given value of λ can be calculated as the ensemble

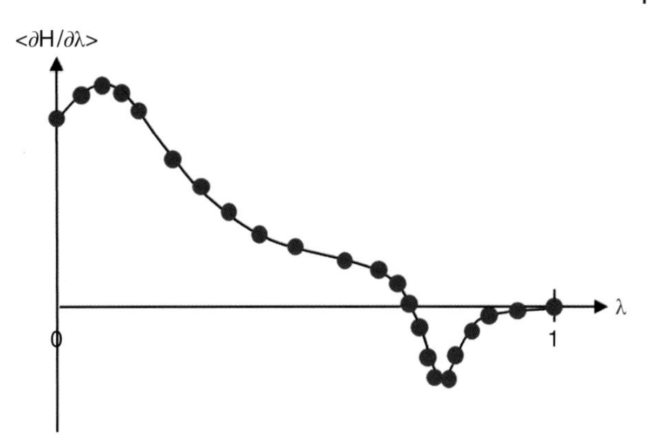

Figure 19.5 Illustration of a typical TI curve (integrand in Eq. (19.61)).

average of the derivative of the Hamiltonian with respect to the coupling parameter λ, sampled at the given value of λ:

$$\left.\frac{\partial A}{\partial \lambda}\right|_{\lambda} = \left\langle \frac{\partial H}{\partial \lambda} \right\rangle_{\lambda}. \tag{19.60}$$

Thus, a free-energy change can be calculated by integrating over λ:

$$\Delta A_{01} = A(\lambda = 1) - A(\lambda = 0) = \int_{\lambda=0}^{\lambda=1} \frac{\partial A}{\partial \lambda} \mathrm{d}\lambda = \int_{\lambda=0}^{\lambda=1} \left\langle \frac{\partial H}{\partial \lambda} \right\rangle_{\lambda} \mathrm{d}\lambda. \tag{19.61}$$

In practice, a force field as presented in Section 19.3.1 delivers analytical expressions for $\partial H/\partial \lambda$ and the integration in Eq. (19.61) is done numerically. An example for the integrand is shown in Figure 19.5. The precision of the TI approach depends on how smooth the TI curve is. In general, the precision of the numerical integral can be increased by increasing the number of λ points (i.e. simulations with different Hamiltonians pertaining to different values of λ) used for the interpolation between the two states, especially in regions of high curvature.

In FEP, one considers two states 0 and 1 differing by a small modification in the Hamiltonian ΔH,

$$H_1 = H_0 + \Delta H. \tag{19.62}$$

The partition functions of the two states, Q_1 and Q_0, are related by

$$Q_1 = \int\int d\mathbf{r}^N d\mathbf{p}^N e^{-\beta(H_0+\Delta H)} = \frac{Q_0}{Q_0}\int\int d\mathbf{r}^N d\mathbf{p}^N e^{-\beta(H_0+\Delta H)} = Q_0 \frac{\int\int d\mathbf{r}^N d\mathbf{p}^N e^{-\beta\Delta H} e^{-\beta H_0}}{Q_0} = Q_0 \langle e^{-\beta \Delta H}\rangle_0 \tag{19.63}$$

where the second step involved multiplication with Q_0 in the numerator and denominator and the notation $<\cdots>_0$ indicates an ensemble average over configurations sampled according to the Hamiltonian H_0. Hence, the free-energy difference between the two states is

$$\Delta A_{01} = A_1 - A_0 = -k_B T \ln \frac{Q_1}{Q_0} = -k_B T \ln \langle \exp[-\beta(H_1 - H_0)] \rangle_0. \tag{19.64}$$

This is an exact perturbation formula and was derived by Zwanzig [102]. It allows the calculation of a free-energy difference from a single simulation. The method is therefore called single-step perturbation. The ensemble average in Eq. (19.64) is, however, only reasonably accurate when ΔH is sufficiently small so that the exponential is sufficiently large, allowing the accumulation of significant contributions to the average. In other words, the configurations sampled with the Hamiltonian H_0 have to be representative of state 1 (i.e. have high Boltzmann factors also when evaluated with the Hamiltonian H_1), which implies that there has to be non-negligible phase-space overlap between the two states (Figure 19.6a). If there is insufficient phase-space

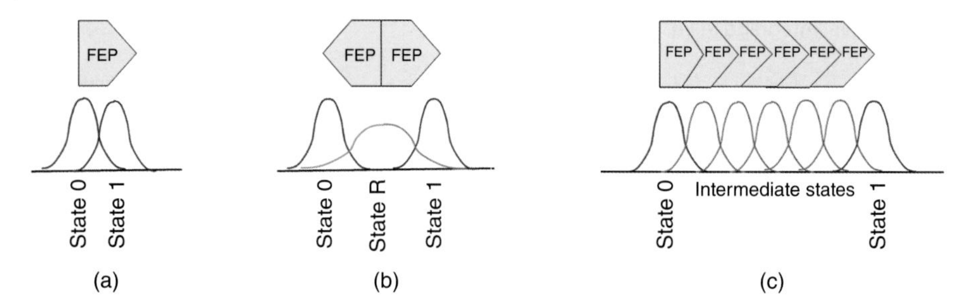

Figure 19.6 Illustration of phase-space overlap issues in FEP. (a) Single-step perturbation. (b) Single-step perturbation involving a reference state. (c) Multistep perturbation.

overlap between states 0 and 1, an artificial state R may be constructed whose phase space overlaps with those of both state 0 and 1 (Figure 19.6b). The perturbation formula (Eq. (19.64)) can then be applied twice to the configurations sampled with Hamiltonian H_R of state R to give the desired free-energy difference as $\Delta A_{01} = \Delta A_{R1} - \Delta A_{R0}$, where

$$\Delta A_{R0} = -k_B T \ln \langle \exp[-\beta(H_0 - H_R)] \rangle_R \text{ and } \quad \Delta A_{R1} = -k_B T \ln \langle \exp[-\beta(H_1 - H_R)] \rangle_R. \tag{19.65}$$

For example, if the free-energy difference of two stereoisomers is sought, the reference state can be an artificial molecule exempt of improper dihedral angle potential at the stereocenter, i.e. it can flip between the two stereoisomers. After the simulation of this molecule, the energy of the sampled configurations is then re-evaluated with force fields including the improper dihedral angle potentials for both stereoisomers, from which the free-energy differences in Eq. (19.65) can be calculated [103].

The two latter approaches (single-step perturbation and free-energy perturbation based on a reference state) are computationally efficient because only a single MD simulation is involved. If it is not possible to find an adequate state R to bridge the phase spaces of state 0 and 1, the perturbation formula may be applied multiple times (Figure 19.6c). This method is called multistep perturbation and relies on the construction of $N - 2$ intermediate states between states 0 and 1 such that the phase-space overlap for two consecutive states i and $i+1$ is sufficiently large to allow accurate evaluation of $\Delta A_{i,i+1}$:

$$\Delta A_{01} = \sum_{i=1}^{N-1} \Delta A_{i,i+1} = -\beta^{-1} \sum_{i=1}^{N-1} \ln \langle \exp[-\beta(H_{i+1} - H_i)] \rangle_i. \tag{19.66}$$

Again, the intermediate states can be artificial. Often, the Hamiltonian is parameterised by a coupling parameter $0 \leq \lambda \leq 1$, similar to TI. For instance,

$$H(\lambda_i) = \lambda_i H_1 + (1 - \lambda_i) H_0 = H_0 + \lambda_i \Delta H. \tag{19.67}$$

The change in the Hamiltonian between two consecutive states i and $i+1$ is $(\lambda_{i+1} - \lambda_i)\ \Delta H$, so

$$\Delta A_{01} = -\beta^{-1} \sum_{i=1}^{N-1} \ln \langle \exp[-\beta(\lambda_{i+1} - \lambda_i)\Delta H] \rangle_{\lambda_i}. \tag{19.68}$$

In general, the perturbation formula (Eq. (19.64)) is widely applied not only to calculate free energies, but also to reevaluate any property (e.g. the number of hydrogen bonds between two residues) for a different Hamiltonian than the one that was used for sampling configurations or to eliminate biasing potential-energy contributions that were employed during a simulation [103]. This process is referred to as 'reweighting' or 'unbiasing'.

TI and FEP are sometimes called 'alchemical' free-energy calculation methods because they allow changes to be performed between molecules that are not feasible in reality. For example, a serine in a protein can be mutated into an alanine along a path involving artificial side chains that are a mixture of an alanine and a

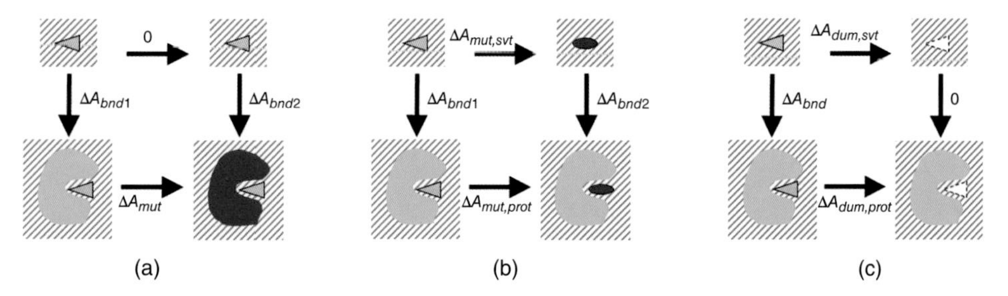

Figure 19.7 Thermodynamic cycles to calculate relative binding free energies (a) for a substrate to two mutants of an enzyme and (b) for two drugs to a receptor, and (c) a thermodynamic cycle to calculate an absolute ligand binding free energy via double decoupling.

serine by presenting a vanishing hydroxyl group and a methylene group slowly turning into a methyl group. Technically, this means that the charges and Lennard–Jones parameters of the hydroxyl group are slowly fading out and that the Lennard–Jones parameters of the methylene group are slowly turning into those of a methyl group. Similarly, covalent parameters may change. Note that the number of particles in an MD simulation has to stay constant. Therefore, vanishing particles in an alchemical TI or FEP calculation does not genuinely eliminate particles but always means removal of corresponding non-bonded parameters. The resulting atoms exempt of charges and Lennard–Jones parameters are called dummy atoms.

In the context of biomolecular simulation, the fact that TI and FEP allow alchemical changes, together with the fact that the free energy is a state function (meaning that around a thermodynamic cycle, the free-energy change is zero), opens the way for computationally efficient approaches to asses relative ligand binding free energies, e.g. the free-energy difference of a substrate binding to two different mutants of an enzyme (Figure 19.7a) or the free-energy difference of two different drugs binding to a receptor (Figure 19.7b), in a thermodynamic cycle. These free-energy changes are the differences of two absolute binding free energies (ΔA_{bnd1} and ΔA_{bnd2}; vertical arrows in Figure 19.7a and b), which can in principle be calculated with a PMF calculation (see below). However, they can also be obtained as differences in alchemical free energies (horizontal arrows in Figure 19.7a and b). In the case depicted in Figure 19.7a, there is only one alchemical free-energy difference, ΔA_{mut}, of mutating the protein when bound to the ligand, so

$$\Delta A_{bnd1} - \Delta A_{bnd2} = -\Delta A_{mut}. \tag{19.69}$$

In the case depicted in Figure 19.7b, the horizontal arrows of the thermodynamic cycle are associated with the free-energy differences $\Delta A_{mut,svt}$ and $\Delta A_{mut,prot}$ of mutating the first ligand into the second ligand, once in solvent and once in the bound state, respectively, so

$$\Delta A_{bnd1} - \Delta A_{bnd2} = \Delta A_{mut,svt} - \Delta A_{mut,prot}. \tag{19.70}$$

Absolute binding free energies can be obtained via TI based on a thermodynamic cycle involving so-called double decoupling (Figure 19.7c), which means that the ligand is mutated into a non-interacting dummy molecule (a molecule corresponding to the real ligand, but with charges and Lennard–Jones parameters set to zero) both in solvent (free-energy change $\Delta A_{dum,svt}$) and in the bound state (free-energy change $\Delta A_{dum,prot}$). The absolute binding free energy ΔA_{bnd} is

$$\Delta A_{bnd} = \Delta A_{dum,svt} - \Delta A_{dum,prot}, \tag{19.71}$$

where it was used that the binding free energy of a dummy molecule vanishes. In practice, mutations to dummy particles involve a complication because the singularity in the Lennard–Jones potential (Eq. (19.19)) gives rise to a divergence of the TI integrand for λ-values approaching the dummy particle ($\lambda = 1$, in the case described here). This effect can be remedied with so-called soft-core potentials [104, 105].

Absolute ligand binding free energies can also be calculated from pathway methods such as PMF-based free-energy calculations. A PMF is the free energy of a system as a function of a reaction coordinate $R(\mathbf{r}^N)$, in general a function of the coordinates of user-defined particles in the system. These coordinates may be internal or Cartesian, the former possibly implying contributions of the Jacobian of the coordinate transformation in the final PMF [106]. To express the free energy as a function of the reaction coordinate, one has to consider the probability that a particular value R' of the reaction coordinate is adopted, the so-called reaction-coordinate probability

$$P_R(R') = \frac{\int\int d\mathbf{p}^N d\mathbf{r}^N \exp[-\beta H(\mathbf{p}^N, \mathbf{r}^N)]\delta(R' - R(\mathbf{r}^N))}{\int\int d\mathbf{p}^N d\mathbf{r}^N \exp[-\beta H(\mathbf{p}^N, \mathbf{r}^N)]} = \frac{Q_R(R')}{Q}. \tag{19.72}$$

where the Dirac-delta function $\delta(x)$ in the numerator selects only those configurations satisfying value R' of the reaction coordinate and the denominator serves as normalisation. This fraction can be written in terms of corresponding partition functions $Q_R(R')$ and Q. The free energy associated with value R' of the reaction coordinate is

$$A_R(R') = -k_B T \ln Q_R(R') = -k_B T \ln P_R(R') - k_B T \ln Q. \tag{19.73}$$

Taking the negative gradient on both sides of Eq. (19.73) with respect to the reaction-coordinate dof (e.g. an interparticle distance), one sees that the negative gradient of the PMF corresponds to an average force of all non-reaction coordinate dof on the reaction coordinate [17].

In general, there are 12 possible ways to calculate a PMF [106], arising from whether (i) the simulation involves no bias, a restraint bias (also called an umbrella bias; see below) or a constraint bias (e.g. constraining a distance between two particles with the SHAKE algorithm; Section Bond-Length Constraints), (ii) the PMF derives from averaging of a force or from the observed probability of states and (iii) whether the reaction coordinate is defined in terms of internal or Cartesian coordinates. In practice, the most frequently chosen approach is to use an umbrella bias on an internal coordinate, along with construction of the PMF from the sampled probability of states. Applying a biasing potential leads to more efficient sampling than in unbiased sampling. For instance, if a PMF along the distance between two molecules such as a ligand and a receptor is sought, an unbiased simulation approach requires simulation of the two molecules for a sufficiently long time until all possible ligand–receptor distances are sampled with their true phase-space probabilities. This is inefficient because if, for example, the binding pocket is buried, it might take extremely long for the ligand to reversibly diffuse in and out of the pocket a sufficient number of times to yield a statistically correct result. Therefore, a biasing potential of the form

$$U^{res}(R(\mathbf{r}^N); R_0) = \frac{1}{2}k^{res}(R(\mathbf{r}^N) - R_0)^2 \tag{19.74}$$

may be used to enforce certain values of the reaction coordinate, here a distance $R(\mathbf{r}^N)$. This is a special potential-energy term that is added to the physical force field and it penalises distances $\mathrm{R}(\mathbf{r}^N)$ deviating from a target distance R_0 according to a harmonic potential with a force constant k^{res}. The corresponding restrained reaction-coordinate probability is

$$P_R^{res}(R') = \frac{\int\int d\mathbf{p}^N d\mathbf{r}^N \exp[-\beta(H(\mathbf{p}^N, \mathbf{r}^N) + U^{res}(R(\mathbf{r}^N); R_0))]\delta(R' - R(\mathbf{r}^N))}{\int\int d\mathbf{p}^N d\mathbf{r}^N \exp[-\beta(H(\mathbf{p}^N, \mathbf{r}^N) + U^{res}(R(\mathbf{r}^N); R_0))]} \tag{19.75}$$

and it can be shown [107] that

$$P_R(R') = P_R^{res}(R')(\langle \exp\left[\beta U^{res}\left(R(\mathbf{r}^N); R_0\right)\right]\rangle_{res})^{-1} \exp[\beta U^{res}(R'; R_0)]. \tag{19.76}$$

Using the relation between the PMF and the reaction-coordinate probability (Eq. (19.73)), the unrestrained PMF is

$$A_R(R') = -k_B T \ln P_R^{res}(R') + k_B T \ln \langle \exp\left[\beta U^{res}\left(R(\mathbf{r}^N); R_0\right)\right]\rangle_{res} - U^{res}(R'; R_0) + \text{constant}. \tag{19.77}$$

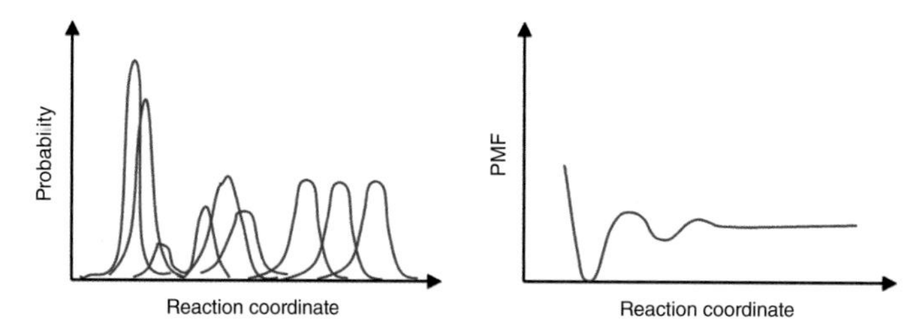

Figure 19.8 Illustration of reaction-coordinate histograms sampled in a PMF calculation (left panel) and the corresponding free-energy profile (right panel).

This approach is widely used and involves multiple simulations with the umbrella restraint bias potential centred at various positions R_0 of the reaction coordinate (often referred to as 'windows'). The probability distributions obtained from the different simulations should overlap to allow reconstruction of the PMF, because in each window, the PMF is only defined within a constant (Eq. (19.77)) and the final PMF is a concatenation of the free energies pertaining to the different windows. The assembly of the total PMF from the raw, biased probability distributions may, for example, be done with the weighted histogram analysis method [108]. Note that, for a given window spacing, smaller force constants k^{res} imply a larger histogram overlap, because the probability distributions are broader, but the disadvantage is that longer simulation times are required to converge these broad distributions. An example for the biased probability distributions and the resulting PMF is displayed in Figure 19.8. Since the PMF is only defined within a constant it can be anchored arbitrarily, for instance such that the bound state corresponds to zero free energy. A binding free energy can then be obtained from such a PMF by taking the negative of the exponential average over the unbound state, as, for example, described in reference [109].

PMF calculations, although possibly not trivial because the choice of reaction coordinate might not be unambiguous and an unfortunate choice may lead to poor sampling and inaccurate results, offer the advantage of beyond-thermodynamic insight because they may reveal barriers along the reaction coordinate and thus convey kinetic information. Because of this, they are not only used for the study of binding processes, but, for example, also for the study of conformational changes. Similar to the TI or multistep FEP method, PMF calculations involve multiple simulations that can be run in parallel. Since free energies pertaining to biomolecular systems are in general not easy to converge (see below), such simulations are often conducted in a Hamiltonian replica exchange framework [110], which enhances configurational sampling (Section Enhanced Sampling).

Enhanced Sampling The power of MD simulation compared to other modelling methods lies in producing a Boltzmann-weighted ensemble of molecular configurations. This is important because one is generally not interested in a biomolecule at zero Kelvin, but under ambient conditions where an ensemble of structures, rather than a single static structure, is representative of the molecule. The configurational phase space of a biomolecule is generally characterised by a potential-energy surface that is rugged due to the presence of a vast number (around 10^4–10^6) of dof differing in their motional behaviour (harmonic, anharmonic, diffusive) and their length- and timescales (0.1–10 nanometres, femtoseconds to milliseconds). Considering the exponential weighting in the Boltzmann factor of the configurations (Eq. (19.1)), it follows that mostly low energy regions of the phase space will contribute significantly to the ensemble. However, in the case of a very high number of high energy configurations, a significant contribution to low free-energy regions of the phase space can come from these high energy configurations because their vast number implies a low entropy. In summary, sampling problems in MD simulation may arise from (i) the ruggedness of the potential-energy surface and hence the difficulty of exhaustive sampling of low energy regions, especially when the barriers separating minima are larger than the thermal energy, and (ii) the difficulty of exhaustive sampling of possibly present high energy/low entropy states [14].

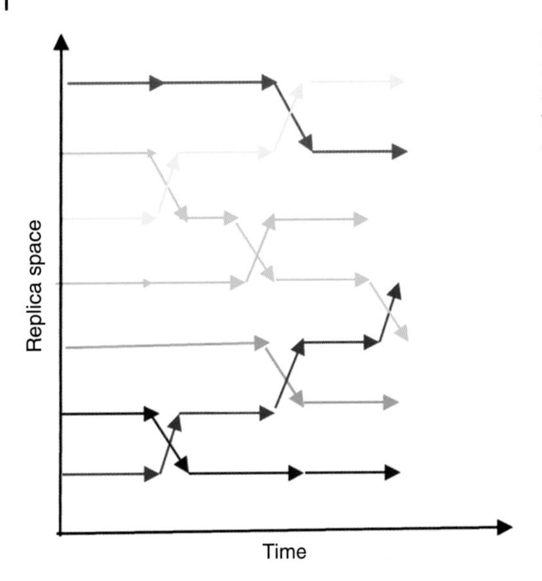

Figure 19.9 Illustration of the exchange of configurations in a replica-exchange simulation. Each line indicates one replica simulation under the control of a separate Hamiltonian (HREMD) or temperature (TREMD). The crossing of arrows indicates a possible exchange of configurations.

A plethora of so-called enhanced sampling methods has been developed to ensure the adequate sampling of the important regions of phase space [111–114]. In biomolecular simulations, often Hamiltonian replica exchange MD (HREMD) simulations [110] are used. One performs not one, but multiple, simulations differing in their Hamiltonian, e.g. different λ-values in a TI-based free-energy calculation or different anchors R_0 of umbrella biasing potentials in a PMF-based free-energy calculation (Figure 19.9). The different simulations are called replicas and any one replica can profit from the sampling achieved at any other replica because, at distinct time intervals, exchanges between neighbouring replicas are attempted according to a Metropolis criterion (Section 19.2.2). The exchange probability w for configurations $\mathbf{x}_i^N$ and $\mathbf{x}_j^N$ (involving positions $\mathbf{r}_i^N$, $\mathbf{r}_j^N$ and corresponding momenta) between replicas i and j is

$$w(\mathbf{x}_i^N \to \mathbf{x}_j^N) = 1 \quad \text{for } \Delta U \leq 0 \quad \text{and} \quad w(\mathbf{x}_i^N \to \mathbf{x}_j^N) = \exp(-\beta \Delta U) \quad \text{for } \Delta U > 0, \tag{19.78}$$

where the energy difference ΔU is

$$\Delta U = U_i(\mathbf{r}_j^N) + U_j(\mathbf{r}_i^N) - U_i(\mathbf{r}_i^N) - U_j(\mathbf{r}_j^N) \tag{19.79}$$

and the potential energy U_i corresponds to the Hamiltonian of replica i. Such a simulation approach is only efficient if the exchange of configurations is global, i.e. occurs throughout all replicas such that, for example, the first replica can ultimately also access configurations pertaining to the last replica. In other words, 'diffusion' of the replicas through Hamiltonian space has to be exhaustive.

An alternative replica-exchange simulation scheme is to make the replicas not differing in Hamiltonian, but temperature (temperature replica exchange MD; TREMD) [115]. Typically, the first replica corresponds to the system at room temperature T_0 and successive replica simulations are performed at continuously increasing temperatures $T_i = T_0 + n\Delta T$, where n is an integer and ΔT the temperature separation between successive replicas. The idea is that the room-temperature simulation can also access high energy regions of phase space or other low energy regions that are separated by high energy barriers through the mixing-in of configurations only accessible in the high-temperature replicas. However, in practice, TREMD simulations using an explicit-solvent representation have been found to be inefficient because the exchange criterion requires low potential-energy differences between neighbouring replicas, which in turn requires close temperature spacing. The latter is because the potential-energy differences tend to be large because they are dominated by the large amount of solvent dof [116]. A large number of replicas, however, increases computational effort because long simulation times are needed to allow thorough mixing of replicas throughout the whole temperature space. Second, it has been found that exchanges in TREMD are mostly due to fluctuations in the solvent

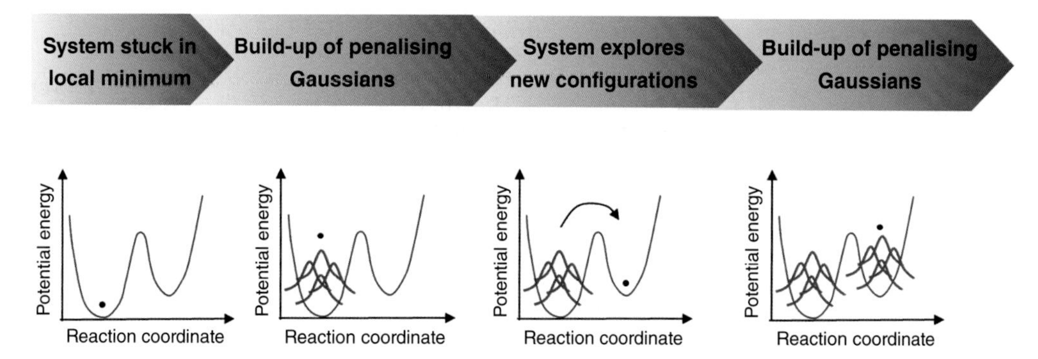

Figure 19.10 Illustration of the build-up of Gaussian penalties in a local-elevation simulation to cross energy barriers and explore new regions of phase space.

[117] and hence do not bring much benefit concerning conformational sampling in the biomolecule. Due to these deficiencies, TREMD is usually only advantageous in the context of implicit-solvent simulations.

Besides the above multireplica approaches, another strategy to enhance sampling is to eliminate the ruggedness of the potential energy surface by filling up the local minima with artificial, positive potential-energy contributions. In the local-elevation (LE) method [118], this is done by progressive addition of biasing potential-energy terms in frequently visited low energy regions (Figure 19.10). Thus, reaccessing those areas is penalised. There are numerous enhanced-sampling methods relying on essentially the same approach, e.g. conformational flooding [119] or metadynamics [120]. In detail, in the LE method, a subset of the dof of the system is chosen whose sampling has to be enhanced, e.g. a side chain dihedral angle. The conformational space of the dof is then discretised and during the simulation one keeps track of when a specific bin is visited and adds a penalising potential-energy term to that bin, e.g. a Gaussian. Clearly, this amounts to a time-dependent Hamiltonian, which is why no equilibrium-thermodynamic information can be gained from a plain LE simulation. However, the resulting smoothened potential energy surface can be used in a second simulation for enhanced configurational sampling, the efficiency being limited only by how fast the 'diffusion' over it occurs. In practice, this is done by using the final built-up penalising potential as an umbrella biasing potential [121]. Note, finally, that the negative of the final built-up penalising potential was shown to approximately correspond to the free-energy surface of the system [122].

Experiment-Based Restraints The biasing potential-energy terms mentioned above in the context of umbrella sampling in PMF calculations or in the LE method are non-physical potential-energy terms that are added to the physical force-field terms to serve a special user-defined purpose. They are therefore also called special force-field terms. There are numerous other special force-field terms, mostly used to restrain a certain dof to a desired target value. This is commonly done via a harmonic potential, e.g. for property A,

$$V(A) = \frac{1}{2}k(A - A_o)^2, \tag{19.80}$$

allowing small deviations from the target value A_0 according to a force constant k. Restraints, as opposed to constraints (as encountered in Section Bond-Length Constraints in the context of bond-length constraints) are thus associated with a non-vanishing kinetic and potential energy of the dof.

Special restraining potential-energy terms are frequently used to generate configurational ensembles compatible with given experimental data, for instance Nuclear Overhauser Effect (NOE)-based distance thresholds [123], ^{3}J-coupling constants [124] or order parameters [125] from nuclear magnetic resonance (NMR) experiments. Common to this data is that they are primary experimental outcomes, i.e. do not involve assumptions or approximations. Also, in these cases, rather than restraining instantaneous values A at each simulation time step, time averages $\overline{A}$ over the elapsed simulation time are restrained,

$$V(\overline{A}) = \frac{1}{2}k(\overline{A} - A_o)^2 \tag{19.81}$$

to account for the fact that the experimental results are also an average over time. In MD simulation, the time average $\overline{A}$ in Eq. (19.81) is computed in a discretised fashion along with exponential dampening of long-time history to keep the restraint responsive to the currently sampled configurations [123–125].

It appears straightforward to construct special potential-energy terms for given properties to establish a connection between ensembles sampled in MD simulations and observed in experiment, but there are several intricacies associated with it [126], e.g. there has to be sufficient sampling in the MD simulation such that all possible solutions fulfilling the experimental restraints are explored.

Implicit Solvation Implicit solvation models [127–131] are an important tool in MD simulations to reduce computational cost (by about one order of magnitude) through elimination of the myriad of solvent dof and instead modelling the solvent as a dielectric continuum with a certain permittivity ε. Thus, longer simulation times and in principle more exhaustive sampling is possible. In addition, due to their efficiency, these tools are useful to simulate a huge number of systems, which is getting more and more important in a time where molecular-level insight is increasingly used to perform research at the interface of structural biology and omics disciplines. Besides their computational efficiency, an advantage of implicit-solvation models is that long-range electrostatic interactions are represented realistically according to Coulomb's law. However, due to the neglect of explicit solvent structure, the results from MD simulations in implicit solvent may not be accurate and only be meaningful for qualitative investigations. In particular, problems are the lack of solute–solvent hydrogen bonding, of dielectric saturation and electrostriction effects around charged solute components, as well as possibly a spurious speeding-up of dynamics.

The most important eom for implicit-solvent simulations is Langevin's eom (Eq. (19.10)). The stochastic force represents random collisions of the solute with solvent molecules, the friction term represents a drag force (the magnitude being determined by the friction coefficient γ, which should be chosen as a function of solvent viscosity) and the physical force represents a mean force not only deriving from the molecular-mechanics force-field interactions within the biomolecule but also including mean solvent contributions. This is done with a potential energy

$$U(\mathbf{r}^N) = U_{mm}(\mathbf{r}^N) + \Delta G_{slv,pol}(\mathbf{r}^N) + \Delta G_{slv,np}(\mathbf{r}^N), \tag{19.82}$$

where the first contribution corresponds to covalent molecular-mechanics force-field terms (Section 19.3.1), together with intrasolute Lennard–Jones interactions and intrasolute Coulombic electrostatic interactions. The latter usually involve screening by the dielectric constant inside the solute ε_P,

$$U_{ele}(\mathbf{r}_m, \mathbf{r}_n) = \frac{q_m q_n}{4\pi\varepsilon_0 \varepsilon_P r_{mn}}. \tag{19.83}$$

The precise value of ε_P is ambiguous and typically chosen within a range of 4–10 [132]. The two last terms in Eq. (19.82) describe the electrostatic and non-polar contribution to the solute solvation free energy in the implicit solvent. These are free energies, i.e. include entropic components, which is why the force deriving from the potential-energy function in Eq. (19.82) is a mean force, accounting for the mean effect of the solvent. A simple way to describe the non-polar solvation free energy is

$$\Delta G_{slv,np}(\mathbf{r}^N) = \sum_{i=1}^{N} a_i p_i, \tag{19.84}$$

where a_i is the solvent-accessible area of atom i and p_i is an atom-specific parameter capturing the non-polar solvation free energy contribution per unit solvent accessible surface area of this atom [133]. The electrostatic solvation free-energy contribution is typically obtained either from the Poisson–Boltzmann (PB) equation [130] or from Generalised Born (GB) models [127]. In the former,

$$\Delta G_{slv,pol}(\mathbf{r}^N) = \frac{1}{2}\sum_{i=1}^{N} q_i(\phi_{slv}(\mathbf{r}_i) - \phi_{vac}(\mathbf{r}_i)), \tag{19.85}$$

where q_i is the partial charge of atom i and $\phi_{slv}(\mathbf{r}_i)$ and $\phi_{vac}(\mathbf{r}_i)$ are the electrostatic potentials at the location of atom i when the solute is solvated and in vacuum, respectively. These electrostatic potentials (in the following denoted $\phi(\mathbf{r})$) are obtained from solving the PB equation

$$\nabla(\varepsilon(\mathbf{r})\nabla\phi(\mathbf{r})) = -\rho(\mathbf{r}) - \sum_i c_i q_i \ \lambda_i(\mathbf{r}) \exp\left[-\frac{q_i\phi(\mathbf{r})}{k_B T}\right], \tag{19.86}$$

where $\rho(\mathbf{r})$ is the solute charge distribution, $\varepsilon(\mathbf{r})$ the position-dependent dielectric permittivity, the sum runs over implicitly modelled ion species i of charge q_i and bulk concentration c_i and $\lambda_i(\mathbf{r})$ is a parameter to describe ion accessibility. The equation is linearised by expanding the exponential term, which is valid for $q_i\phi(\mathbf{r}) \ll k_B\mathrm{T}$, and keeping only the first-order term. Hence,

$$\nabla(\varepsilon(\mathbf{r})\nabla\phi(\mathbf{r})) = -\rho(\mathbf{r}) + \sum_i \frac{c_i q_i^2 \phi(\mathbf{r})\lambda_i(\mathbf{r})}{k_B T} \tag{19.87}$$

The PB equation is usually solved on a grid via finite differences. To this end, the charge distribution, dielectric constant and ion accessibility are mapped on to the grid points and one solves for the electrostatic potential on the grid points along with interpolation of the electrostatic potential to the positions of the solute charges. Two calculations are performed, one in solvent (employing the solvent dielectric permittivity) and one in vacuum (employing a dielectric permittivity of one), finally allowing evaluation of Eq. (19.87).

A computationally more inexpensive method is to use a GB model,

$$\Delta G_{slv,pol} = -\frac{1}{8\pi\varepsilon_0}\left(1 - \frac{1}{\varepsilon}\right)\sum_{i=1}^{N}\sum_{j=1}^{N}\frac{q_i q_j}{f_{GB}(r_{ij})}, \tag{19.88}$$

where

$$f_{GB}(r_{ij}) = \left[r_{ij}^2 + R_i R_j \exp\left(-\frac{r_{ij}^2}{4R_i R_j}\right)\right]^{1/2} \tag{19.89}$$

is a function to account for the influence of the finite separation of the charges on their solvation free energies and R_i is the Generalised-Born radius of atom i, an empirical parameter. Various parameterizations of GB models are available [134].

19.4 Applications

19.4.1 Comparative Protein Structure Modelling

Many proteins in a cell adopt well-defined three-dimensional (3D) structures that determine their biological function. Hence, knowledge of the protein 3D structure is of major biological importance. Experimental protein structure determination, e.g. by X-ray crystallography, requires purification of large amounts of proteins and the ability to crystallise the proteins or protein complexes, which in many cases may be difficult or even impossible. In addition, NMR and cryo-electron microscopy may be used for protein structure determination, but also have certain limitations. Computational modelling and MD simulation can be used to predict protein structure by directly mimicking the folding process. However, MD simulations at atomistic resolution and including surrounding solvent explicitly are still limited to the regime of microseconds, whereas the folding

Sequence-Alignment:

```
Seq.1: GCPRCGQAVYAAEKVIGAGKSWHKSCFRCAKCGKSLESTTL
        ||*|   || |* |   || ||  | *| |||| | | *
Seq.2: KCPKCDKEVYFADRVTSLGKDWH--CLKCEKCGKTLTSGSH
```

Figure 19.11 Illustration of sequence similarity (left panel) and structural similarity (right panel) of two proteins. The sequence alignment identifies identical residues in two protein sequences (indicated as vertical sticks). The structural similarity of proteins is given by best superposition of two protein structures (indicated as red and blue protein Cα atom trace) and measuring the residual deviation of corresponding atoms in the form of a root-mean-square deviation (RMSD).

process of many proteins is in the millisecond to second regime. Nevertheless, with modern force fields and specialised computer hardware [135] it has become possible to directly follow the folding process at atomistic resolution for a number of fast folding proteins [136]. Due to the large compute resources and still time limitations of such simulations this approach is not useful as a general tool for protein structure prediction.

However, comparison of known experimental protein structures indicates reoccurring types of 3D conformations [137]. In particular, proteins with similar sequences are likely to adopt a similar folded structure. This observation forms the basis of molecular modelling approaches known as 'comparative protein modelling', 'protein homology building' or 'template-based 3D modelling' [20, 21, 138]. The similarity of two protein sequences is defined through a sequence alignment, i.e. an algorithm aligning the sequences in two rows such that the number of identical residues in each column is maximised (Figure 19.11). The ratio of the number of identical amino acids after optimal alignment relative to the total number of aligned amino acids defines the sequence identity.

The structural similarity of two proteins is typically measured as a root-mean-square deviation (RMSD) of corresponding atoms after best-possible spatial superposition. As an example, the backbone structure of two proteins with a sequence identity of ~55% and an RMSD of backbone atoms of ~1 Å is shown in Figure 19.11. It is evident that in this particular example the two proteins adopt very similar 3D structures. As a rule of thumb, when comparing natural sequences, analysis of many cases indicates that if the sequence identity of two protein domains is >50% it is highly likely that both domains adopt the same fold with typically a backbone RMSD <1.5 Å. In the case of a sequence identity in the range 30–50% it is still very likely that the same type of fold is formed with RMSD of ~1.5–2.5 Å. If the sequence identity drops to <30% only sometimes, the proteins adopt the same fold with RMSD >2.5 Å, but also often no or only little structural similarity is observed.

The expected close structural similarity for proteins of close sequence similarity can be used to obtain structural models via restrained MD simulations and energy minimization based on a known structure for one of the two proteins (termed the template protein). The observed spatial arrangement of homologous residues in the template protein is encoded in additional force-field terms during a simulation of the homologous target protein. Typically, this can be included by using or slightly modifying force-field terms already available in the original force field described in Section 19.3.1. For example, contacts between residues can be included as additional bonds (distance restraints) using the typical form of a harmonic potential, but instead of defining a single minimum-energy distance one defines a small range of allowed distances between pairs of homologous atoms beyond which a quadratic energy penalty is added to the force-field description of the target protein. With a sufficient number of such restraints, rapid folding of the target protein to a possible structure in close agreement with the known template structure can be obtained [139]. The standard terms in a molecular-mechanics force field prevent that sterically incorrect models (e.g. with atom overlaps or incorrect bonded geometry) are formed (Section 19.3.1). A modelling process including such spatial restraints is illustrated in Figure 19.12. The Modeller program [138] is one of the most common approaches for modelling proteins based on sequence similarity and satisfaction of spatial restraints.

Comparative protein modelling is widely used in molecular and structural biology. In case of close sequence similarity (e.g. ~50%), accurate target structures with backbone RMSD ~1–2 Å from the real structure can be

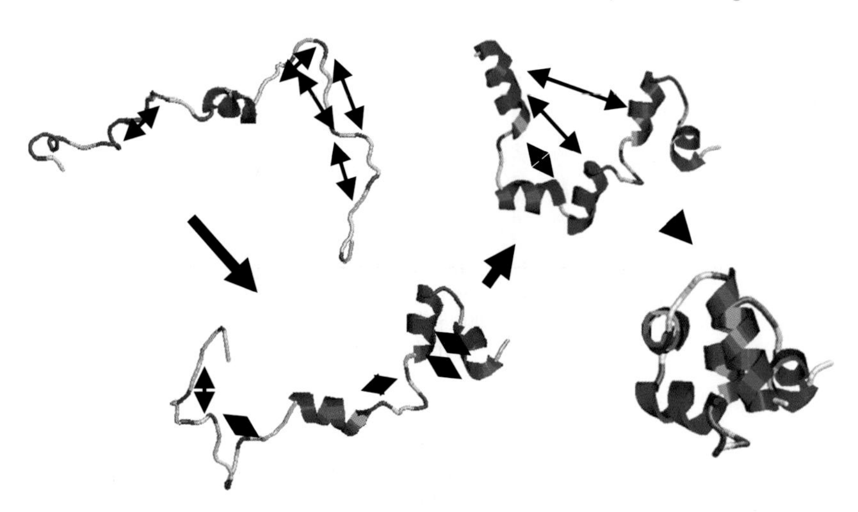

Figure 19.12 Illustration of comparative modelling by satisfaction of spatial restraints. Spatial arrangements of residues observed in the template protein are included in an MD simulation of the target protein as distance or angular restraints that are added to the physical force field of the protein (illustrated as double arrows for residues that are close in space in the template structure). The final comparative model (cartoon model on the right) satisfies all contacts and other spatial restraints extracted from the homologous known structure.

achieved. Since for a given protein fold with a known structure typically many homologous sequences are available, structural information on these homologues can be readily obtained.

19.4.2 Alchemical Free-Energy Simulations to Identify Key Residues of Protein–Protein Interaction

As already outlined in Section Free-Energy Calculations, thermodynamic integration or free energy perturbation approaches allow one to extract free-energy changes associated with processes not feasible in experiment, e.g. alchemical transformation of one chemical group into another. This option can, for example, be used to calculate the influence of mutations on the binding free energy of two interacting proteins and to identify key residues for the interaction. An example is the analysis of the role of hot-spot residues at the interface of colicin E9 and Immunity protein 9 (Im9) [140]. The activity of the bacterial colicin E9 enzyme is controlled by the specific and high-affinity binding of immunity protein Im9. This system is a well-studied model for investigating protein–protein interactions [141]. The E9/Im9 interface structure is illustrated in Figure 19.13.

To identify key residues for the protein interactions and to study the mechanism of high-affinity binding, several Im9 interface residues were alchemically transformed to alanine residues during TI calculations. The simulations followed closely the methodology described in Section Free-Energy Calculations. To calculate the effect of such alchemical mutations of side chains to alanine on the binding affinity it is necessary to perform the mutations both in the complex and the isolated partner protein (for obtaining a full thermodynamic cycle of the transformation). The effect of the side chain mutation on binding can then be calculated as the difference in free energy for the transformation in the complex versus in the isolated unbound partner. Good agreement of the calculated free-energy differences of binding and available experimental data was obtained (Figure 19.13). The simulation study also included analysis of the role of solvent and conformational flexibility of the partner proteins by comparing results in the absence or presence of solvent and with or without positional restraints on the protein partners. Restriction of the conformational flexibility of the proteins resulted in significant changes of the calculated free energies but of similar magnitude for the isolated Im9 and for the complex, and therefore in only modest changes of the relative binding free-energy differences. The largest overall binding free-energy change was obtained for the two Tyr-Ala mutations. However, the physical origin appeared to be different, with solvation changes contributing significantly to the Tyr55Ala mutation and a loss of direct protein–protein interactions dominating the free-energy change due to the Tyr54Ala mutation

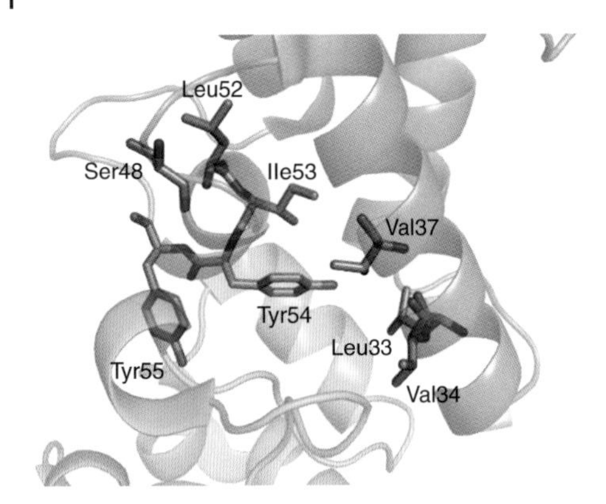

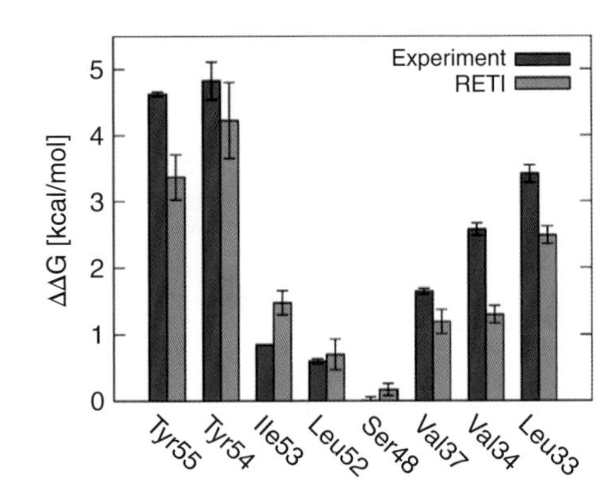

Figure 19.13 (Left panel) E9 (green cartoon)-Im9 (grey cartoon) interface structure with important interface residues indicated as stick models (PDB ID 1EMV). (Right panel) Comparison of calculated and experimental binding free energy upon mutating the residues on the *x*-axis to Ala. The term RETI indicates a combination of thermodynamic integration (TI) and the replica-exchange (RE) method that enhances sampling along the alchemical variable. Source: The figure has been adapted from Figure 1 and 2 of reference [140].

[140]. Hence, in addition to identifying the contribution of interface residues for protein–protein binding, the simulation studies also contribute to a better understanding of the molecular details of the interactions and of the physical origins of the contribution of each residue to protein–protein binding affinity.

19.5 Concluding Remarks

Starting in the 1960s and 1970s, applications of MD simulation and molecular modelling have grown dramatically. In particular, in the last 10 years the introduction of new powerful hardware as well as methodological advances have widened the applicability of MD simulation enormously to investigate many biophysical processes and effects on timescales up to microseconds. Simulation and modelling techniques are involved at many stages of biomolecular structure determination to translate X-ray scattering, NMR or cryo-electron microscopy data into biomolecular structures. It is also possible to apply simulation techniques to follow biomolecular structure formation and association processes.

As outlined above, free-energy simulation methods play an important role to predict or to interpret the effect of chemical modifications on protein–protein and protein–ligand binding. In the field of drug design, molecular simulations and docking methods can help to predict the structure of drug–protein complexes and to estimate the ligand–protein binding affinity. However, it is important to keep in mind that the simulation results are limited by the maximum timescale of the simulation approach and in addition by the accuracy of the underlying force field. The force-field description is an approximation of the true molecular interactions and requires frequent testing by comparison with experimental data and further improvement.

The intention of the present contribution was to give an introduction into the physical and methodological basis for molecular simulations based on a classical molecular-mechanics force field. The description of applications was intended to present examples but cannot cover the entire range of tasks that can be tackled using molecular simulation techniques. In particular, a technique that is increasingly used but not covered in the present contribution is multiscale simulation [46, 142, 143], where different parts of the simulated system are described at different levels of resolution, e.g. combining a (sub)atomistic representation of a solute and its first two solvation shells with a coarse-grained representation of the bulk solvent. A special case of multiresolution simulation is the QM/MM method [44–46], where parts of the system (e.g. an enzyme active site) are modelled quantum-mechanically and the rest of the system is modelled using molecular mechanics (possibly at different levels of resolution). The interested reader is referred to the references for further information.

References

1 Frenkel, D. (2013). Simulations: the dark side. *Eur. Phys. J. Plus* 128: 10/1–10/21.
2 Wong-ekkabut, J. and Karttunen, M. (2016). The good, the bad and the user in soft matter simulations. *Biochim. Biophys. Acta Biomembr.* 1858: 2529–2538.
3 van Gunsteren, W.F., Daura, X., Hansen, N. et al. (2017). Validation of molecular simulation: an overview of issues. *Angew. Chem. Int. Ed.* 57: 884–902.
4 Alder, B.J. and Wainwright, T.E. (1957). Phase transition for a hard sphere system. *J. Chem. Phys.* 27: 1208–1209.
5 Rahman, A. (1964). Correlations in the motion of atoms in liquid argon. *Phys. Rev.* 136: A405–A411.
6 Rahman, A. and Stillinger, F.H. (1971). Molecular dynamics study of liquid water. *J. Chem. Phys.* 55: 3336–3359.
7 McCammon, J.A., Gelin, B.R., and Karplus, M. (1977). Dynamics of folded proteins. *Nature* 267: 585–590.
8 van Gunsteren, W.F. and Karplus, M. (1982). Protein dynamics in solution and in a crystalline environment: a molecular dynamics study. *Biochemistry* 21: 2259–2274.
9 Moore, G.E. (1965). Cramming more components onto integrated circuits. *Electronics* 38: 114–117.
10 Rieffel, E.G. and Polak, W.H. (2014). *Quantum Computing: A Gentle Introduction.* Cambridge, USA: The MIT Press.
11 Furber, S. (2016). Large-scale neuromorphic computing systems. *J. Neural Eng.* 13: 051001/1–051001/14.
12 Lanyon, B.P., Whitfield, J.D., Gillett, G.G. et al. (2010). Towards quantum chemistry on a quantum computer. *Nature Chem.* 2: 106–111.
13 Kassal, I., Whitfield, J.D., Perdomo-Ortiz, A. et al. (2011). Simulating chemistry using quantum computers. *Annu. Rev. Phys. Chem.* 62: 185–207.
14 van Gunsteren, W.F., Bakowies, D., Baron, R. et al. (2006). Biomolecular modeling: goals, problems, perspectives. *Angew. Chem. Int. Ed. Engl.* 45 (25): 4064–4092.
15 Micha, D.A. and Burghardt, I. (2007). *Quantum Dynamics of Complex Molecular Systems.* Heidelberg, Germany: Springer.
16 Wesseling, P. (2000). *Principles of Computational Fluid Dynamics*, 1e. Berlin, Germany: Springer.
17 McQuarrie, D.A. (2000). *Statistical Mechanics.* Sausalito, CA, USA: University Science Books.
18 Goldstein, H. (1980). *Classical Mechanics*, 2e. Reading Massachusetts, USA: Addison Wesley.
19 Newton, I. (1999). *The Principia: Mathematical Principles of Natural Philosophy.* (A new translation by I. B. Cohen and A. Whitman). Berkeley, USA: University of California Press.
20 Lushington, G.H. (2015). Comparative modeling of proteins. *Methods Mol. Biol.* 1215: 309–330.
21 Fiser, A. (2010). Template-based protein structure modeling. *Methods Mol. Biol.* 673: 73–94.
22 Lee, J., Wu, S., and Zhang, Y. (2009). Ab initio protein structure prediction. In: *From Protein Structure to Function with Bioinformatics* (ed. D.J. Rigden), 3–25. Dordrecht, The Netherlands: Springer.
23 The protein data bank. www.rcsb.org.
24 Frenkel, D. (1993). Monte Carlo simulations: a primer. In: *Computer Simulation of Biomolecular Systems, Theoretical and Experimental Applications* (ed. W.F. van Gunsteren, P.K. Weiner and A.J. Wilkinson), 37–66. The Netherlands: ESCOM Science Publishers, B.V., Leiden.
25 Frenkel, D. and Smit, B. (2002). *Understanding Molecular Simulation*, 2e. San Diego: Academic Press, USA.
26 Metropolis, N. and Ulam, S. (1949). The Monte Carlo method. *J. Am. Stat. Assoc.* 44: 335–341.
27 Metropolis, N., Rosenbluth, A.W., Rosenbluth, M.N. et al. (1953). Equation of state calculations by fast computing machines. *J. Chem. Phys.* 21: 1087–1092.
28 van Gunsteren, W.F. (1993). Molecular dynamics and stochastic dynamics simulation: a primer. In: *Computer Simulation of Biomolecular Systems, Theoretical and Experimental Applications* (ed. W.F. van

Gunsteren, P.K. Weiner and A.J. Wilkinson), 3–36. The Netherlands: ESCOM Science Publishers, B.V., Leiden.
29 Allen, M.P. and Tildesley, D.J. (1987). *Computer Simulation of Liquids*. New York, USA: Oxford University Press.
30 Boltzmann, L. (1871). Einige allgemeine Sätze über das Wärmegleichgewicht. *Wien. Ber.* 63: 679–711.
31 Gallavotti, G. (2016). Ergodicity: a historical perspective. Equilibrium and nonequilibrium. *Eur. Phys. J. H* 41: 181–259.
32 Huang, K. (2010). *Introduction to Statistical Physics*, 2e. Boca Raton: Chapman & Hall/CRC Taylor & Francis Group, USA.
33 Langevin, P. (1908). Sur la théorie du mouvement brownien. *C. R. Avad. Sci. Paris* 146: 530–533.
34 Lagrange, J.L. (1788). *Mécanique Analytique*. Pairs, France: Gauthier-Villars.
35 MacKerell, D.A. Jr., (2004). Empirical force fields for biological macromolecules: Overview and issues. *J. Comput. Chem.* 25: 1584–1604.
36 Monticelli, L. and Tieleman, D.P. (2013). Force fields for classical molecular dynamics. *Methods Mol. Biol.* 924: 197–213.
37 Lopes, P.E.M., Guvench, O., and MacKerell, A.D. Jr., (2015). Current status of protein force fields for molecular dynamics simulations. *Methods Mol. Biol.* 1215: 47–71.
38 Hünenberger, P.H. and van Gunsteren, W.F. (1997). Empirical classical interaction functions for molecular simulations. In: *Computer Simulation of Biomolecular Systems, Theoretical and Experimental Applications* (ed. W.F. van Gunsteren, P.K. Weiner and A.J. Wilkinson), 3–82. Dordrecht, The Netherlands: Kluwer/Escom Science Publishers.
39 Kmiecik, S., Gront, D., Kolinski, M. et al. (2016). Coarse-grained protein models and their applications. *Chem. Rev.* 116: 7898–7936.
40 Saunders, M.G. and Voth, G. (2013). Coarse-graining methods for computational biology. *Annu. Rev. Biophys.* 42: 73–93.
41 Voth, G.A. (2008). *Coarse-Graining of Condensed Phase and Biomolecular Systems*. New York, USA: Taylor & Francis, Inc.
42 Ingólfsson, H.I., Lopez, C.A., Uusitalo, J.J. et al. (2013). The power of coarse graining in biomolecular simulations. *WIREs Comput. Mol. Sci.* 4: 225–248.
43 Morse, P.M. (1929). Diatomic molecules according to the wave mechanics. II. Vibrational levels. *Phys. Rev.* 34: 57–64.
44 Ryde, U. (2016). QM/MM calculations on proteins. *Methods Enzymol.* 577: 119–158.
45 Senn, H.M. and Thiel, W. (2009). QM/MM methods for biomolecular systems. *Angew. Chem. Int. Ed. Engl.* 48: 1198–1229.
46 Pezeshki, S. and Lin, H. (2015). Recent developments in QM/MM methods towards open-boundary multi-scale simulations. *Mol. Simul.* 41: 168–189.
47 Stone, A.J. (2000). *The Theory of Intermolecular Forces*. Oxford, UK: Oxford University Press.
48 London, F. (1937). The general theory of molecular forces. *Trans. Faraday Soc.* 33: 8–26.
49 Griffiths, D.J. (2017). *Introduction to Quantum Mechanics*, 2e. Cambridge, UK: University Press.
50 Pauli, W. (1925). Über den Zusammenhang des Abschlusses der Elektronengruppen im Atom mit der Komplexstruktur der Spektren. *Z. Physik* 31: 765–783.
51 Lennard-Jones, J.E. (1937). The equation of state of gases and critical phenomena. *Physica* 4: 941–956.
52 Halgren, T.A. (1992). Representation of van der Waals (vdW) interactions in molecular mechanics force-fields: potential form, combination rules, and vdW parameters. *J. Am. Chem. Soc.* 114: 7827–7843.
53 Good, R.J. and Hope, C.J. (1970). New combining rule for intermolecular distances in intermolecular potential functions. *J. Chem. Phys.* 53: 540–543.
54 Jackson, J.D. (1999). *Classical Electrodynamics*, 3e. New York, USA: Wiley.

55 Coulomb, C.A. (1788). Sur l'électricité et le magnétisme, premier mémoire, construction et usage d'une balance électrique, fondée sur la propriété qu'ont les fils de métal, d'avoir une force de réaction de torsion proportionnelle à l'angle de torsion. *Mém. Acad. Sci.* 1785: 569–577.
56 Adams, D.J. (1979). Computer simulation of ionic systems: the distorting effects of the boundary conditions. *Chem. Phys. Lett.* 62: 329–332.
57 Rick, S.W. and Stuart, S.J. (2002). Potentials and algorithms for incorporating polarizability in computer simulations. *Rev. Comp. Chem.* 18: 89–146.
58 Yu, H. and van Gunsteren, W.F. (2005). Accounting for polarization in molecular simulation. *Comput. Phys. Commun.* 172: 69–85.
59 Warshel, A., Kato, M., and Pisliakov, A.V. (2007). Polarizable force fields: History, test cases, and prospects. *J. Chem. Theory Comput.* 3: 2034–2045.
60 Lopes, P.E., Roux, B., and MacKerell, A.D. Jr., (2009). Molecular modeling and dynamics studies with explicit inclusion of electronic polarizability: theory and applications. *Theor. Chem. Accounts* 124: 11–28.
61 Baker, C.M. (2015). Polarizable force fields for molecular dynamics simulations of biomolecules. *WIREs Comput. Mol. Sci.* 5: 241–254.
62 Duan, Y., Wu, C., Chowdhury, S. et al. (2003). A point-charge force field for molecular mechanics simulations of proteins based on condensed-phase quantum mechanical calculations. *J. Comput. Chem.* 24: 1999–2012.
63 Maler, J.A., Martinez, C., Kasavajhala, K. et al. (2015). ff14SB: improving the accuracy of protein side chain and backbone parameters from ff99SB. *J. Chem. Theory Comput.* 11: 3696–3713.
64 MacKerell, A.D., Feig, M., and Brooks, C.L. (2004). Extending the treatment of backbone energetics in protein force fields: limitations of gas-phase quantum mechanics in reproducing protein conformational distributions in molecular dynamics simulations. *J. Comput. Chem.* 25: 1400–1415.
65 Best, R.B., Zhu, X., Shim, J. et al. (2012). Optimization of the additive CHARMM all-atom protein force field targeting improved sampling of the backbone phi,psi and side-chain chi1 and chi2 dihedral angles. *J. Chem. Theory Comput.* 8: 3257–3273.
66 Schmid, N., Eichenberger, A.P., Choutko, A. et al. (2011). Definition and testing of the GROMOS force-field versions 54A7 and 54B7. *Eur. Biophys. J.* 40: 843–856.
67 Reif, M.M., Hünenberger, P.H., and Oostenbrink, C. (2012). New interaction parameters for charged amino acid side chains in the GROMOS force field. *J. Chem. Theory Comput.* 8: 3705–3723.
68 Verlet, L. (1967). Computer 'experiments' on classical fluids. I. Thermodynamical properties of Lennard-Jones molecules. *Phys. Rev.* 159: 98–103.
69 Swope, W.C., Andersen, H.C., Berens, P.H., and Wilson, K.R. (1982). A computer simulation method for the calculation of equilibrium constants for the formation of physical clusters of molecules: Application to small water clusters. *J. Chem. Phys.* 76: 637–649.
70 Hockney, R.W. (1970). The potential calculation and some applications. *Methods Comput. Phys.* 9: 135–211.
71 Mazur, A.K. (1997). Common molecular dynamics algorithms revisited: accuracy and optimal time steps of Störmer-Leapfrog integrators. *J. Comput. Phys.* 136: 354–365.
72 Toxvaerd, S. (1994). Hamiltonians for discrete dynamics. *Phys. Rev. E* 50: 2271–2274.
73 Bou-Rabee, N. (2014). Time integrators for molecular dynamics. *Entropy* 16: 138–162.
74 Berendsen, H.J.C., Postma, J.P.M., van Gunsteren, W.F. et al. (1984). Molecular dynamics with coupling to an external bath. *J. Chem. Phys.* 81: 3684–3690.
75 Morishita, T. (2000). Fluctuation formula in molecular-dynamics simulations with the weak coupling heat bath. *J. Chem. Phys.* 113: 2976–2982.
76 Hoover, W.G. (1985). Canonical dynamics: equilibrium phase-space distributions. *Phys. Rev. A* 31: 1695–1697.
77 Nosé, S. (1984). A molecular dynamics method for simulations in the canonical ensemble. *Mol. Phys.* 52: 255–268.

78 Martyna, G.J., Klein, M.L., and Tuckerman, M. (1992). Nosé-Hoover chains: the canonical ensemble via continuous dynamics. *J. Chem. Phys.* 97: 2635–2643.

79 Hünenberger, P.H. (2005). Thermostat algorithms for molecular dynamics simulations. *Adv. Polym. Sci.* 173: 105–149.

80 Marc, G. and McMillan, W.G. (1985). The virial theorem. *Adv. Chem. Phys.* 58: 209–361.

81 Parrinello, M. and Rahman, A. (1980). Crystal structure and pair potentials: a molecular-dynamics study. *Phys. Rev. Lett.* 45: 1196–1199.

82 Ryckaert, J.P., Ciccotti, G., and Berendsen, H.J.C. (1977). Numerical integration of the Cartesian equations of motion of a system with constraints: molecular dynamics of n-alkanes. *J. Comput. Phys.* 23: 327–341.

83 Kräutler, V., van Gunsteren, W.F., and Hünenberger, P.H. (2001). A fast SHAKE algorithm to solve distance constraint equations for small molecules in molecular dynamics simulations. *J. Comput. Chem.* 22: 501–508.

84 van Gunsteren, W.F. and Karplus, M. (1982). Effect of constraints on the dynamics of macromolecules. *Macromolecules* 15: 1528–1544.

85 van Gunsteren, W.F. (1980). Constrained dynamics of flexible molecules. *Mol. Phys.* 40: 1015–1019.

86 Steinbach, P.J. and Brooks, B.R. (1994). New spherical-cutoff methods for long-range forces in macromolecular simulation. *J. Comput. Chem.* 15: 667–683.

87 Linse, P. and Andersen, H.C. (1986). Truncation of Coulombic interactions in computer simulations of liquids. *J. Chem. Phys.* 85: 3027–3041.

88 Barker, J.A. and Watts, R.O. (1973). Monte Carlo studies of the dielectric properties of water-like models. *Mol. Phys.* 26: 789–792.

89 Ewald, P.P. (1921). Die Berechnung optischer und elektrostatischer Gitterpotentiale. *Ann. Phys.* 369: 253–287.

90 Darden, T., York, D., and Pedersen, L. (1993). Particle mesh Ewald: An Nlog(N) method for Ewald sums in large systems. *J. Chem. Phys.* 98: 10089–10092.

91 Hockney, R.W. and Eastwood, J.W. (1988). *Computer Simulation Using Particles*, 2e. Bristol, UK: Institute of Physics Publishing.

92 Weber, W., Hünenberger, P.H., and McCammon, J.A. (2000). Molecular dynamics simulations of a polyalanine octapeptide under Ewald boundary conditions: influence of artificial periodicity on peptide conformation. *J. Phys. Chem. B* 104: 3668–3675.

93 Hünenberger, P.H. and McCammon, J.A. (1999). Effect of artificial periodicity in simulations of biomolecules under Ewald boundary conditions: a continuum electrostatics study. *Biophys. Chem.* 78: 69–88.

94 Reif, M.M., Kräutler, V., Kastenholz, M.A. et al. (2009). Explicit-solvent molecular dynamics simulations of a reversibly-folding beta-heptapeptide in methanol: influence of the treatment of long-range electrostatic interactions. *J. Phys. Chem. B* 113: 3112–3128.

95 Kastenholz, M.A. and Hünenberger, P.H. (2006). Computation of methodology-independent ionic solvation free energies from molecular simulations: II. The hydration free energy of the sodium cation. *J. Chem. Phys.* 124: 224501/1–224501/20.

96 Christ, C.D., Mark, A.E., and van Gunsteren, W.F. (2010). Basic ingredients of free energy calculations: a review. *J. Comput. Chem.* 31: 1569–1582.

97 Hansen, N. and van Gunsteren, W.F. (2014). Practical aspects of free-energy calculations: a review. *J. Chem. Theory Comput.* 10: 2632–2647.

98 Mobley, D.L. and Gilson, M.K. (2017). Predicting binding free energies: Frontiers and benchmarks. *Annu. Rev. Biophys.* 46: 531–558.

99 Chipot, C. and Pohorille, A. (2007). *Free Energy Calculations. Theory and Applications in Chemistry and Biology. Springer Series in Chemical Physics*. Springer Verlag.

100 Chipot, C. (2014). Frontiers in free-energy calculations of biological systems. *WIREs Comput. Mol. Sci.* 4: 71–89.
101 Kirkwood, J.G. (1935). Statistical mechanics of fluid mixtures. *J. Chem. Phys.* 3: 300–313.
102 Zwanzig, R.W. (1954). High-temperature equation of state by a perturbation method. I. Nonpolar gases. *J. Chem. Phys.* 22: 1420–1426.
103 Oostenbrink, C. (2012). Free energy calculations from one-step perturbations. In: *Computational Drug Discovery and Design* (ed. R. Baron), 487–499. New York, USA: Humana Press (Springer).
104 Beutler, T.C., Mark, A.E., van Schaik, R. et al. (1994). Avoiding singularities and numerical instabilities in free energy calculations based on molecular simulations. *Chem. Phys. Lett.* 222: 529–539.
105 Zacharias, M., Straatsma, T.P., and McCammon, J.A. (1994). Separation-shifted scaling, a new scaling method for Lennard-Jones interactions in thermodynamic integration. *J. Chem. Phys.* 100: 9025–9031.
106 Trzesniak, D., Kunz, A.P.E., and van Gunsteren, W.F. (2007). A comparison of methods to compute the potential of mean force. *Chem. Phys. Chem.* 8: 162–169.
107 van Gunsteren, W.F., Beutler, T.C., Fraternali, F. et al. (1993). Computation of free energy in practice: choice of approximations and accuracy limiting factors. In: *Computer Simulation of Biomolecular Systems, Theoretical and Experimental Applications*, vol. 2, 315–367. The Netherlands: ESCOM Science Publishers, B.V., Leiden.
108 Kumar, S., Bouzida, D., Swendsen, R.H. et al. (1992). The weighted histogram analysis method for free-energy calculations on biomolecules. I. The Method. *J. Comput. Chem.* 13: 1011–1021.
109 Doudou, S., Burton, N.A., and Henchman, R.H. (2009). Standard free energy of binding from a one-dimensional potential of mean force. *J. Chem. Theory Comput.* 5: 909–918.
110 Sugita, Y., Kitao, A., and Okamoto, Y. (2000). Multidimensional replica-exchange method for free-energy calculations. *J. Chem. Phys.* 113: 6042–6051.
111 Yang, L., Liu, C.W., Shao, Q. et al. (2015). From thermodynamics to kinetics: enhanced sampling of rare events. *Acc. Chem. Res.* 48: 947–955.
112 Bernardi, R.C., Melo, M.C.R., and Schulten, K. (2015). Enhanced sampling techniques in molecular dynamics simulations of biological systems. *Biochim. Biophys. Acta* 1850: 872–877.
113 Miao, Y. and McCammon, J.A. (2016). Unconstrained enhanced sampling for free energy calculations of biomolecules: a review. *Mol. Simul.* 42: 1046–1055.
114 Luitz, M., Bomblies, R., Ostermeir, K., and Zacharias, M. (2015). Exploring biomolecular dynamics and interactions using advanced sampling methods. *J. Phys. Condens. Matter* 27: 323101/1–323101/13.
115 Sugita, Y. and Okamoto, Y. (1999). Replica-exchange molecular dynamics method for protein folding. *Chem. Phys. Lett.* 134: 141–151.
116 Fukunishi, H., Watanabe, O., and Takada, S. (2002). On the Hamiltonian replica exchange method for efficient sampling of biomolecular systems: application to protein structure prediction. *J. Chem. Phys.* 116: 9058–9067.
117 Periole, X. and Mark, A.E. (2007). Convergence and sampling efficiency in replica exchange simulations of peptide folding in explicit solvent. *J. Chem. Phys.* 126: 014903/1-014903/11.
118 Huber, T., Torda, A.E., and van Gunsteren, W.F. (1994). Local elevation: a method for improving the searching properties of molecular dynamics simulation. *J. Comput. Aided Mol. Des.* 8: 695–708.
119 Grubmüller, H. (1995). Predicting slow structural transitions in macromolecular systems: conformational flooding. *Phys. Rev. E* 52: 2893–2906.
120 Laio, A. and Parrinello, M. (2002). Escaping free-energy minima. *Proc. Natl Acad. Sci. USA* 99: 12562–12566.
121 Hansen, H.S. and Hünenberger, P.H. (2010). Using the local elevation method to construct optimized umbrella sampling potentials: calculation of the relative free energies and interconversion barriers of glucopyranose ring conformers in water. *J. Comput. Chem.* 31: 1–23.
122 Engkvist, O. and Karlström, G. (1996). A method to calculate the probability distribution for systems with large energy barriers. *Chem. Phys.* 213: 63–76.

123 Torda, A.E., Scheek, R.M., and van Gunsteren, W.F. (1990). Time-averaged nuclear Overhauser effect distance restraints applied to tendamistat. *J. Mol. Biol.* 214: 223–235.

124 Torda, A.E., Brunne, R.M., Huber, T. et al. (1993). Structure refinement using time-averaged J-coupling constant restraints. *J. Biomol. NMR* 3: 55–66.

125 Hansen, N., Heller, F., Schmid, N., and van Gunsteren, W.F. (2014). Time-averaged order parameter restraints in molecular dynamics simulations. *J. Biomol. NMR* 60: 169–187.

126 Van Gunsteren, W.F., Allison, J.R., Daura, X. et al. (2016). Deriving structural information from experimentally measured data on biomolecules. *Angew. Chem. Int. Ed.* 55: 15990–16010.

127 Bashford, D. and Case, D.A. (2000). Generalized Born models of macromolecular solvation effects. *Annu. Rev. Phys. Chem.* 51: 129–152.

128 Kleinjung, J. and Fraternali, F. (2014). Design and application of implicit solvent models in biomolecular simulations. *Curr. Opin. Struct. Biol.* 25: 126–134.

129 Chen, J., Brooks, C.L., and Khandogin, J. (2008). Recent advances in implicit solvent-based methods for biomolecular simulations. *Curr. Opin. Struct. Biol.* 18: 140–148.

130 Baker, N.A. (2004). Poisson-Boltzmann methods for biomolecular electrostatics. *Methods Enzymol.* 383: 94–118.

131 Baker, N.A. (2005). Improving implicit solvent simulations: a Poisson-centric view. *Curr. Opin. Struct. Biol.* 15: 137–143.

132 Mellor, B.L., Cortes, E.C., Busath, D.D., and Mazzeo, B.A. (2011). Method for estimating the internal permittivity of proteins using dielectric spectroscopy. *J. Phys. Chem. B* 115: 2205–2213.

133 Chen, J. and Brooks, C.L. 3rd, (2008). Implicit modeling of nonpolar solvation for simulating protein folding and conformational transitions. *Phys. Chem. Chem. Phys.* 10 (4): 471–481.

134 Nguyen, H., Roe, D.R., and Simmerling, C. (2013). Improved generalized Born solvent model parameters for protein simulations. *J. Chem. Theory Comput.* 9: 2020–2034.

135 Shaw, D.E., Dror, R.O., Salmon, J.K. et al. (2009). Millisecond-scale molecular dynamics simulations on Anton. In: *SC '09: Proceedings of the Conference on High Performance Computing Networking, Storage and Analysis*, 1–11. New York, NY, USA: ACM.

136 Shaw, D.E., Maragakis, P., Lindorff-Larsen, K. et al. (2010). Atomic-level characterization of the structural dynamics of proteins. *Science* 330: 341–346.

137 Schaeffer, R.D. and Daggett, V. (2011). Protein folds and protein folding. *Protein Eng. Des. Sel.* 24: 11–19.

138 Webb, B. and Sali, A. (2014). Comparative protein structure modeling using MODELLER. *Curr. Protoc. Bioinformatics* 47: 5.6.1–5.6.32.

139 Smith-Brown, M.J., Kominos, D., and Levy, R.M. (1993). Global folding of proteins using a limited number of distance constraints. *Protein Eng.* 6 (6): 605–614.

140 Luitz, M. and Zacharias, M. (2013). Role of tyrosine hot-spot residues at the interface of colicin E9 and immunity protein 9: a comparative free energy simulation study. *Proteins: Struct. Funct. Bioinf.* 81: 461–468.

141 Meenan, N.A., Sharma, A., Fleishman, S.J. et al. (2010). The structural and energetic basis for high selectivity in a high-affinity protein-protein interaction. *Proc. Natl Acad. Sci. USA.* 107 (22): 10080–10085.

142 Shen, L. and Yang, W. (2016). Quantum mechanics/molecular mechanics method combined with hybrid all-atom and coarse-grained model: Theory and application on redox potential calculations. *J. Chem. Theory Comput.* 12: 2017–2027.

143 Meier, K., Choutko, A., Dolenc, J. et al. (2013). Multi-resolution simulation of biomolecular systems: a review of methodological issues. *Angew. Chem. Int. Ed. Engl.* 52: 2820–2834.

Further Reading

Allen, M.P. and Tildesley, D.J. (1989). *Computer Simulation of Liquids*. Reprint edition. Oxford Science Publications.
Frenkel, D. and Smit, B. (2001). *Understanding Molecular Simulation*, 2e. Academic Press.
Hinchliffe, A. (2008). *Molecular Modelling for Beginners*, 2e. Wiley.
Leach, A.R. (2001). *Molecular Modelling. Principles and Applications*, 2e. Pearson.
Rapaport, D.C. (2004). *The Art of Molecular Dynamics Simulation*, 2e. Cambridge University Press.
Schlick, T. (2010). *Molecular Modeling and Simulation: An Interdisciplinary Guide*, 2e. Springer.
Tuckerman, M.E. (2010). *Statistical Mechanics: Theory and Molecular Simulation*, 1e. Oxford University Press.

Index

Biomolecular and Bioanalytical Techniques: Theory, Methodology and Applications, First Edition. Edited by Vasudevan Ramesh.

h

i

k

l

m

q

r